Lecture Notes in Computer Sci

T0238669

Commenced Publication in 1973
Founding and Former Series Editors:
Gerhard Goos, Juris Hartmanis, and Jan van Leeuwen

Editorial Board

Hoon Hong   Chee Yap (Eds.)

Mathematical Software – ICMS 2014

4th International Congress
Seoul, South Korea, August 5-9, 2014
Proceedings

 Springer

Volume Editors

Hoon Hong
North Carolina State University
Department of Mathematics
Raleigh, NC, USA
E-mail: hong@ncsu.edu

Chee Yap
New York University
Courant Institute
Department of Computer Science
New York, NY, USA
E-mail: yap@cs.nyu.edu

ISSN 0302-9743   e-ISSN 1611-3349
ISBN 978-3-662-44198-5   e-ISBN 978-3-662-44199-2
DOI 10.1007/978-3-662-44199-2
Springer Heidelberg New York Dordrecht London

Library of Congress Control Number: 2014943438

LNCS Sublibrary: SL 1 – Theoretical Computer Science and General Issues

Typesetting: Camera-ready by author, data conversion by Scientific Publishing Services, Chennai, India

Printed on acid-free paper

Springer is part of Springer Science+Business Media (www.springer.com)

Hoon Hong   Chee Yap (Eds.)

Mathematical Software – ICMS 2014

4th International Congress
Seoul, South Korea, August 5-9, 2014
Proceedings

 Springer

Volume Editors

Hoon Hong
North Carolina State University
Department of Mathematics
Raleigh, NC, USA
E-mail: hong@ncsu.edu

Chee Yap
New York University
Courant Institute
Department of Computer Science
New York, NY, USA
E-mail: yap@cs.nyu.edu

ISSN 0302-9743   e-ISSN 1611-3349
ISBN 978-3-662-44198-5   e-ISBN 978-3-662-44199-2
DOI 10.1007/978-3-662-44199-2
Springer Heidelberg New York Dordrecht London

Library of Congress Control Number: 2014943438

LNCS Sublibrary: SL 1 – Theoretical Computer Science and General Issues

Typesetting: Camera-ready by author, data conversion by Scientific Publishing Services, Chennai, India

Printed on acid-free paper

Springer is part of Springer Science+Business Media (www.springer.com)

Preface

The 4th International Congress on Mathematical Software (ICMS 2014) was held during August 5–9, 2014, at Hanyang University in Seoul, Korea. It continued the tradition of being held every four years as a satellite conference of the International Congress of Mathematicians, which was also held in Seoul. There were five invited plenary talks and 125 contributed talks. From the abstracts of these talks, 106 were submitted and accepted as extended abstracts for the present proceedings.

Mathematics has many interrelated branches. It is commonly observed that despite this diversity, there is deep unity in mathematical thinking. One emerging thread across all branches of mathematics is the notion of effectivity or computation. Mathematical theories often predict the existence of objects with certain properties and it might be important to find the objects. Finding such objects calls for a finite procedure or algorithm, which we implement in software. Conversely, to formulate conjectures or new mathematical theories, we may want to explore the space of such objects. Searching the space also requires software. Mathematics has increasing overlap with disciplines such as computer science, the emerging area of computational sciences and engineering and various application areas. Again the idea of computation is a key factor in this convergence.

Thus, the computational phenomenon brings mathematics into direct contact with technology, resulting in the creation of certain residue or artifacts that we call mathematical software. Bruno Buchberger in his invited paper here goes further, with the bold assertion: *Mathematics is essentially software.* We in the International Conference of Mathematical Software believe that the appearance of mathematical software is one of the most important modern developments in mathematics, and this phenomenon should be studied as a coherent whole. Our vision for ICMS is to serve as the major forum for mathematicians, scientists, programmers, and developers who are interested in software.

Software is not static: Anyone who uses software knows that its typical "half-life" is frustratingly short: *But it compiled properly just last year!* There is constant renewal, development, and disruptive changes. It is partly caused by new mathematical advances, but often the pressure is from technological changes, e.g., the appearance of graphics processing units (GPUs). How do we produce software in such an environment? Software requires algorithms, but to realize algorithms we need organizational principles and tools. There are issues of numerical robustness, scalability, usability, maintainability, best practice, and efficiency of software. These are standard topics in computer science, but the nature of mathematical software also presents unique issues.

What are these unique issues? We cite one from the invited abstract of Wolfram Decker below: "… *the implementation of an advanced and more abstract computational machinery often depends on a long chain of more specialized algorithms and efficient data structures at various levels.*" In Decker's work on computer algebra systems, such a chain might involve concepts and algorithms from commutative algebra, algebraic geometry, arithmetic algebraic geometry, and singularity theory. Any working mathematician will instantly recognize this as another example of the said interrelatedness of mathematics. It is true that any complex real-world application will exhibit such interconnectedness. But in mathematics, the interconnection is more precise, even axiomatizable.

Such issues are reflected in the extended abstracts collected in this volume. They cover wide-ranging mathematical areas and software issues. They are organized according to various sessions at ICMS 2014. Furthermore, the range of mathematical software represented at ICMS 2014 bodes well for our vision. It shows strong demand for mathematical software. Such software may be classified as either large monolithic and comprehensive systems or boutique software with more specialized targets. A more recent third class is suggested by Decker's example above: aggregative systems that aim to provide a common framework for two or more self-contained systems. We predict that all these varieties of software will continue to multiply and diversify.

To ensure that ICMS continues to play a positive role in future of this field, for the first time, the organizers adopted a set of bylaws to govern ICMS 2014 and beyond. This is a minimal set of rules to standardize the organization and whose interpretation is guided by past ICMS practice. They were tentatively adopted by the Advisory Board and will be presented for ratification at the first business meeting of ICMS during the conference.

We are thankful to all the individuals whose effort and support make ICMS 2014 possible: the plenary speakers, all the contributors of abstracts and extended abstracts, the special session organizers, and the LNCS team at Springer under the leadership of Alfred Hofmann. We have benefited from the experience of the ICMS Advisory Board chaired by Professors Nobuki Takayama and Andres Iglesias. Last but not least, we acknowledge the work of the local chair, Professor Deok-Soo Kim, and his committee at Hanyang University. Our plenary speakers were funded by a generous grant from the National Institute for Mathematical Sciences of Korea.

June 2014

Hoon Hong
Chee Yap

Bylaws of ICMS

The motivation for these bylaws is to guide the future directions and governance of the International Congress of Mathematical Software (ICMS). Ultimately, we hope to build a community of researchers and practitioners centered around the aims of the first three ICMS, namely, "mathematical software" viewed as a scientific activity. Such a community is closely allied with areas such as algorithms and complexity, software engineering, computational sciences, and of course all of constructive mathematics including computer algebra and numerical computation. But mathematical software has unique (evolving) characteristics that ICMS aims to foster and support. To build such a community, we need continuity and some rules governing the central activity of our research community, namely, the ICMS conference. The following proposal is based on, and is consistent with, the historical patterns observed in the first three ICMS events (2002, 2006, 2010). The proposal is deliberately minimal and under-specified. Therefore the interpretations should be guided by historical patterns.

Bylaw 1: Composition of Organizers

Each ICMS conference shall have the following organizational positions:

1. Advisory Board
2. General Chair
3. Program Chair
4. Local Chair
5. Secretary
6. Program Committee
7. Local Committee

Chair could also mean Co-chairs.

Bylaw 2: Appointments

1. The Advisory Board shall consist of the General Chair, Program Chair, and Local Chair of the previous two conferences, and any other members that they shall appoint. The General Chair of the last-but-one ICMS shall serve as the chair for the current Advisory Board. All appointments to the Advisory Board last for two ICMS conferences.
2. The Advisory Board appoints the Secretary.
3. The Advisory Board appoints the General Chair for the next conference.
4. The General Chair, in consultation with the Advisory Board, appoints the Program Chair and Local Chair.

5. The Program Chair, in consultation with the General Chair and the Advisory Board, appoints the Program Committee members.
6. The Local Chair, in consultation with the General Chair, appoints the Local Committee members.

Bylaw 3: Duties

1. The Advisory Board Chair will hold an ICMS business meeting during each conference.
2. The Secretary shall maintain a permanent ICMS website for past activities, and also a list of names and emails of attendees of past ICMS conferences.

Bylaw 4: Amendments

1. The bylaws can be amended by ballot, either at the ICMS business meeting or by email.
2. Persons who have registered for at least one of the three preceding ICMS conferences are eligible to vote.

Appendix: Remarks on the Bylaws

1. The bylaw is self-described as minimal and under-specified; both are viewed as positive qualities. This appendix will comment on the bylaws using their historical (non-binding) interpretations. It will also motivate the exclusion of certain items in the bylaws.
2. Historically, ICMS was organized as a satellite of ICM. Like ICM, ICMS is held every 4 years. But even in our short history, there was a break in this pattern in 2010. Looking forward, there are good arguments to have biennial meetings (e.g., this is better for community building).
3. We do not specify the format of ICMS, believing it to be the prerogative of the General Chair and the Program Chair to shape it to best serve the community. Historically, the program has centered around plenary speakers and special sessions organized by experts in the area of interest.
4. The term "software" is a unique characteristic of ICMS that distinguishes it from the allied areas mentioned in the bylaw. We are not only interested in "paper algorithms" but in their implementation and in their software environment. We want to foster software development as a scientific activity, to promote the publication of software-like paper publications, and to establish standards for such activities. Past ICMSs have had an important component of software tutorial and demonstrations and distribution of free software (e.g., Knoppix CD).

5. The ICMS positions listed in the bylaw do not exclude additional positions: the positions of a treasurer and a "documentation chair" for software have been suggested. But we refrain from mandating such positions in the bylaw.
6. Term limit for appointment to ICMS posts is generally a good thing. Again, we do not encode this into the bylaw, as we recognize many good reasons to make exceptions. For example, a competent "document chair" should probably be given a life appointment.
7. The idea of a permanent repository for ICMS is assumed in the bylaw. Nobuki Takayama has a website that might be considered as the starting point. The General Chair and Program Chair should each deposit a report for the activities of their particular ICMS in this repository.

Organization

Executive Committee

General Chair

Chee Yap   New York University, USA

Program Chair

Hoon Hong   North Carolina State University, USA

Local Chair

Deok-Soo Kim   Hanyang University, South Korea

Program Committee

Hirokazu Anai   Kyushu University, Japan
Nikolaj Bjorner   Microsoft Research
Jonathan Borwein   University of Newcastle, Australia
Bruno Buchberger   RISC Johannes Kepler University, Austria
Changbo Chen   Chinese Academy of Sciences
Xiaoyu Chen   Beihang University, China
Jin-San Cheng   Chinese Academy of Sciences
Heiko Dietrich   Monash University, Australia
Arie Gurfinkel   Carnegie Mellon Software Engineering
                                 Institute, USA
Jonathan Hauenstein   North Carolina State University, USA
Hoon Hong (Chair)   North Carolina State University, USA
Jeff Hooper   Acadia University, Canada
Alexander Hulpke   Colorado State University, USA
Andreas Iglesias   University of Cantabria, Spain
Hidenao Iwane   Fujitsu Laboratories
Michael Kerber   Max Planck Institute for Informatics, Germany
Jon-Lark Kim   Sogang University, South Korea
Alexander Maletzky   RISC Johannes Kepler University, Austria
David Monniaux   Verimag, France
Antonio Montes   Universitat Politècnica de Catalunya, Spain
Marc Moreno Maza   University of Western Ontario, Canada
Marco Pollanen   Trent University, Canada
Yosuke Sato   Tokyo University of Science, Japan
Vikram Sharma   The Institute of Mathematical Sciences, India
Andrew Sommese   University of Notre Dame, USA
Setsuo Takato   Toho University, Japan

James Wan   Singapore University of Technology and Design
Dingkang Wang   Chinese Academy of Sciences
Dongming Wang   CNRS, France and Beihang University, China
Wolfgang Windsteiger   RISC Johannes Kepler University, Austria

Topical Session Organizers

1. Mathematical Theory Exploration
   (Bruno Buchberger and Wolfgang Windsteiger)

2. Computational Group Theory
   (Heiko Dietrich and Alexander Hulpke)

3. Coding Theory
   (Jon-Lark Kim)

4. Computational Topology
   (Michael Kerber)

5. Numerical Algebraic Geometry
   (Jonathan Hauenstein and Andrew Sommese)

6. Geometry
   (Xiaoyu Chen and Dongming Wang)

7. Curves and Surfaces
   (Jin-San Cheng and Vikram Sharma)

8. Quantified Reasoning
   (Nikolaj Bjorner, Arie Gurfinkel, and David Monniaux)

9. Special Functions and Concrete Mathematics
   (Jonathan M. Borwein and James G. Wan)

10. Groebner Bases
    (Bruno Buchberger and Alexander Maletzky)

11. Triangular Decompositions of Polynomial Systems
    (Changbo Chen and Marc Moreno Maza)

12. Parametric Polynomial Systems
    (Hirokazu Anai, Hidenao Iwane, Antonio Montes, Yosuke Sato, and Dingkang Wang)

13. Mathematical Web/Mobile Interfaces, Editing and Scientific Visualization
    (Marco Pollanen, Andres Iglesias, Setsuo Takato, and Jeff Hooper)

14. General
    (Chee Yap and Hoon Hong)

Local Committee

Kyung-Joon Cha	Hanyang University
Joo Sup Chang	Hanyang University
Deok-Soo Kim (Chair)	Hanyang University
Donguk Kim	Gangneung-Wonju National University
Jae-Kwan Kim	Voronoi Diagram Research Center
Taewan Kim	Seoul National University
Yonggu Kim	Chonnam University
Joonyoung Park	Dongguk University
Joonghyun Ryu	Voronoi Diagram Research Center
Hayong Shin	Korea Advanced Institute of Science and Technology

Advisory Board

Nobuki Takayama (Co-chair)	Kobe University, Japan
Andres Iglesias (Co-chair)	University of Cantabria, Spain
Xiao-Shan Gao	Chinese Academy of Sciences, PR China
Komei Fukuda	ETH Zurich, Switzerland
Joris Van der Hoeven	École Polytechnique Paris, France
Michael Joswig	Technical University Berlin, Germany
Jaime Gutierrez	University of Cantabria, Spain
Masayuki Noro	Rikkyo University, Japan

Sponsoring Institutions

National Institute for Mathematical Sciences (NIMS)
Hanyang University
Hanyang Computational Science Center
Voronoi Diagram Research Center
Society of CAD/CAM Engineers

Abstracts of the Invited Talks

Invited Plenary Speakers and Talks

Invited Plenary Speakers[1]

Jonathan Borwein	University of Newcastle
Bruno Buchberger	RISC Johannes Kepler University
Wolfram Decker	Technische Universität Kaiserslautern
Andrew Sommese	University of Notre Dame
Kokichi Sugihara	Meiji University
Lloyd N. Trefethen	University of Oxford

Abstracts of Invited Plenary Talks

1. *Computer Discovery and Visual Theorems in Mathematics*
   Jonathan Borwein (University of Newcastle)

   Long before current graphic, visualization and geometric tools were available, John E. Littlewood, 1885-1977, wrote in his delightful *Miscellany*:

   > *A heavy warning used to be given [by lecturers] that pictures are not rigorous; this has never had its bluff called and has permanently frightened its victims into playing for safety. Some pictures, of course, are not rigorous, but I should say most are (and I use them whenever possible myself).*

   Over the past five years, the role of visual computing in my own research has expanded dramatically. In part this was made possible by the increasing speed and storage capabilities—and the growing ease of programming—of modern multi-core computing environments. But, at least as much, it has been driven by my group's paying more active attention to the possibilities for graphing, animating or simulating most mathematical research activities.

   I shall describe diverse work from my group in transcendental number theory (normality of real numbers), in dynamic geometry (iterative reflection methods), probability (behaviour of short random walks), and matrix completion problems (especially, applied to protein confirmation). While all of this involved significant numerical-symbolic computation, I shall focus on the visual components.

---

[1] Wolfram was unable to present his talk after an unfortunate accident.

2. *Soft Math*
   Math Soft
   Bruno Buchberger (RISC Johannes Kepler University)

In this talk we argue that mathematics is essentially software. In fact, from the beginning of mathematics, it was the goal of mathematics to automate problem solving. By systematic and deep thinking, for problems whose solution was difficult in each individual instance, systematic procedures were found that allow to solve each instance without further thinking. In each round of automation in mathematics, the deep thinking on spectra of problem instances is reflected by deep theorems with deep proofs.

In 20th century, the systematic procedures for spectra of problems became physically tangible as algorithms / software for the universal computer (which itself essentially is a mathematical invention). In 21st century, the rounds of automation in mathematics reach higher and higher levels and move more and more to the meta-level of mathematics, i.e. to the automation of mathematical thinking itself.

In this talk, we illustrate the evolution of mathematics towards higher and higher levels of automation of its own problem solving and thinking process by a couple of examples of increasing sophistication starting from calculation with Roman numbers up to the automatic invention of algorithms like the speaker's algorithm for computing Gröbner bases.

As a practical experience of Gödel's Incompleteness Theorem, there is no upper bound to the sophistication of higher and higher rounds in mathematical automation. Thus, mathematicians live in the best of all worlds: They can embark on more and more challenging problems that need an algorithmic solution knowing that, after an algorithmic solution has been achieved, there will always be room for more sophistication and human mathematical invention. Thus, in a sense, mathematicians will never become jobless.

By the intellectual power in the automation of mathematical problem solving, visible as "software", mathematics is also in the center of the spiral of automation in all science, technology and economy and is the silent driving force behind the spiral of innovation.

Unfortunately, the intellectual attractiveness and practical relevance of mathematics is hard to explain to outsiders. Or, more precisely, mathematicians often do a very bad job for explaining the fundamental role of mathematics clearly enough to outsiders, who in fact are the insiders of modern society. However, it is very important for the further development of mathematics that the role of mathematics for modern science, technology, economy, and welfare is made public. We will also present a few ideas about this political aspect of being a mathematician in today's society.

3. *Challenges in the Development of Open Source Computer Algebra Systems*
   Wolfram Decker (Technische Universität Kaiserslautern)

Computer algebra is facing new challenges as mathematicians are inventing new and more abstract tools to answer difficult problems and connect apparently remote fields of mathematics. On the mathematical side, while we wish to provide cutting-edge techniques for application areas such as commutative algebra, algebraic geometry, arithmetic algebraic geometry, singularity theory, and many more, the implementation of an advanced and more abstract computational machinery often depends on a long chain of more specialized algorithms and efficient data structures at various levels. On the software development side, for cross-border approaches to solving mathematical problems, the efficient interaction of systems specializing in different areas is indispensable; handling complex examples or large classes of examples often requires a considerably enhanced performance. Whereas the interaction of systems is based on a systematic software modularization and the design of mutual interfaces, a new level of computational performance is reached via parallelization, which opens up the full power of multi-core computers, or clusters of computers.

In my talk, I will report on the ongoing collaboration of groups of developers of several well-known open source computer algebra systems: GAP, which pays particular emphasis to group theory, SINGULAR, a system for applications in algebraic geometry and singularity theory, and POLYMAKE, a software for convex geometry. In presenting computational tools relying on this collaboration, and some of the mathematical challenges which lead us to develop such tools, I will in particular highlight the HOMALG project which provides an abstract structure and algorithms for Abelian categories, aiming at concrete applications ranging from linear control theory to commutative algebra and algebraic geometry.

I will also comment on progress in the design of parallel algorithms for basic tasks in commutative algebra and algebraic geometry such as primary decomposition, normalization, finding adjoint curves, or parametrizing rational curves.

4. *Numerical Algebraic Geometry: Theory and Practice*
   Andrew Sommese (University of Notre Dame)

The goal of Numerical Algebraic Geometry is to carry out algebraic geometric calculations in characteristic zero using numerical analysis algorithms. This comes down to numerical algorithms to compute and manipulate solution sets of polynomial systems. Numerical Algebraic Geometry is a natural outgrowth of the continuation methods to compute isolated complex solutions of systems of polynomials with complex coefficients. There are a wide range of applications including solution of chemical systems, kinematics, numerical solution of systems of nonlinear differential equations, and computation of algebraic geometric invariants.

Bertini is open-source C software, developed by Bates (Colorado State U.), Hauenstein (Notre Dame), Sommese (Notre Dame), and Wampler (General Motors R. & D.), to carry out Numerical Algebraic Geometry computations. Bertini will be rewritten to make it a better tool for users. Bertini dates from over a decade ago, and from this experience we have identified several possibilities for significant improvements. One goal is to change some of the data structures and add internal functionality that will give the user the ability to write scripts and interface with other software.

In this talk, I will give an overview of Numerical Algebraic Geometry. I will consider the theoretical algorithms underlying the area in the light of the practical issues that arise when implementing the algorithms in the current and the future Bertini.

5. *Principle of Independence for Robust Geometric Software Learned by the Human Visual Computation*
Kokichi Sugihara (Meiji University)

Straightforward implementation of geometric algorithms usually results in unstable software because numerical errors generate inconsistency and make the software to fail. The human brain, on the other hand, can manage image data robustly although the resolution, precision and computation speed are all poor when compared with the electric computers.

In this talk we try to characterize the sources of the robustness of the human visual computation through optical illusion, and consider to utilize them for designing robust geometric software. In particular we point out that the human brain is persistent and similar persistency can be implemented to make geometric software robust. In other words, we treat only independent set of information and thus avoid inconsistency. This idea includes the topology-oriented approach which we have studied for robust geometric computation.

6. *CHEBFUN as a software project*
Lloyd N. Trefethen (University of Oxford)

Chebfun is a software system for numerical computation with functions. The starting point in 2002 was the idea of overloading Matlab's discrete objects (vectors, matrices,...) to continuous analogues (functions, operators,...). For speed and robustness, however, everything remains numerical, based on piecewise Chebyshev expansions. Thus another way to view Chebfun is as an extension of the rounding-to-16-digits idea of floating point arithmetic from numbers to functions.

From this starting point, Chebfun has moved in many directions, including linear and nonlinear ODEs, time-dependent PDEs, edge-detection, computation with functions defined on rectangles in 2D, quadrature and orthogonal polynomials, Legendre-Chebyshev conversions, rootfinding in 1D and 2D, and rational approximation. Besides its convenience for practical desktop computing, it is also an excellent tool for exploring many topics of approximation theory.

Though this talk will touch on these mathematical topics, the organizing principle will be software. How does a project like this evolve over a decade from one programmer and his PhD supervisor to a dozen developers linked by GitHub with users around the world? How do you keep the project under control and make it strong for the future even when most of the work is done by students and postdocs passing through transiently and busy building their research careers? These challenges are new to me in the past decade, and I have as many questions as answers.

Table of Contents

Coding Theory

Computational Topology

Numerical Algebraic Geometry

Geometry

Curves and Surfaces

Quantified Reasoning

Triangular Decompositions of Polynomial Systems

Parametric Polynomial Systems

Mathematical Web/Mobile Interfaces and Visualization

General Session

Experimental Computation
and Visual Theorems

Jonathan M. Borwein

CARMA, University of Newcastle, NSW
jon.borwein@gmail.com
www.carma.newcastle.edu.au

Abstract. Long before current graphic, visualisation and geometric tools were available, John E. Littlewood (1885-1977) wrote in his delightful *Miscellany*[1]:

> A heavy warning used to be given [by lecturers] that pictures are not rigorous; this has never had its bluff called and has permanently frightened its victims into playing for safety. Some pictures, of course, are not rigorous, but I should say most are (and I use them whenever possible myself). [p. 53]

Over the past five years, the role of visual computing in my own research has expanded dramatically. In part this was made possible by the increasing speed and storage capabilities—and the growing ease of programming—of modern multi-core computing environments.

But, at least as much, it has been driven by my group's paying more active attention to the possibilities for graphing, animating or simulating most mathematical research activities.

Keywords: visual theorems, experimental mathematics, randomness, normality of numbers, short walks, planar walks, fractals, protein confirmation.

1 Introduction

I first briefly discuss what is meant both by *visual theorems* and by *experimental computation*. I then turn to *dynamic geometry* (iterative reflection methods [1]) and *matrix completion problems*[2] (applied to protein confirmation [3]). (See Case studies I and II.) I end with description of recent work from my group in *probability* (behaviour of short random walks [6,8]) and *transcendental number theory* (normality of real numbers [2]). (See Case studies III.)

---

[1] J.E. Littlewood, *A mathematician's miscellany*, London: Methuen (1953); J. E. Littlewood and Béla Bollobás, ed., *Littlewood's miscellany*, Cambridge University Press, 1986.

[2] See http://www.carma.newcastle.edu.au/jon/Completion.pdf and http://www.carma.newcastle.edu.au/jon/dr-fields11.pptx.

H. Hong and C. Yap (Eds.): ICMS 2014, LNCS 8592, pp. 1–8, 2014.

1.1 Some Early Conclusions: So I Am Sure They Get Made

1. Maths can be done *experimentally*[3] (it is fun) using computer algebra, numerical computation and graphics: SNaG Computations, tables and pictures are experimental data but you can not stop thinking.
2. Making mistakes is fine as long as you learn from them, and keep your eyes open (conquer fear).
3. You can not use what you do not know and what you know you can usually use. Indeed, you do not need to know much before you start research (as we shall see).

2   Visual Theorems and Experimental Mathematics

In a 2012 study *On Proof and Proving* [10] the International Council on Mathematical Instruction wrote:

> The latest developments in computer and video technology have provided a multiplicity of computational and symbolic tools that have rejuvenated mathematics and mathematics education. Two important examples of this revitalization are experimental mathematics and visual theorems.

By a *visual theorem*[4] I mean a picture or animation which gives one confidence that a desired result is true in Gianqunto's sense that it represents "coming to believe it in an independent, reliable, and rational way" (either as discovery or validation) as described in [4]. While we have famous pictorial examples purporting to show all triangle are equilateral, there are equally many or more bogus symbolic proofs that $1 + 1 = 1$. In all cases 'caveat emptor'.

Modern technology properly mastered allows for a much richer set of tools for discovery, validation, and even rigorous proof than our precursors could have ever imagined would come to pass—and it is early days. The same ICMI study [10], quoting [5, p. 1], says enough about the meaning of *experimental mathematics* for our curernet purposes:

> Experimental mathematics is the use of a computer to run computations— sometimes no more than trial-and- error tests—to look for patterns, to identify particular numbers and sequences, to gather evidence in support of specific mathematical assertions that may themselves arise by computational means, including search.
>
> Like contemporary chemists — and before them the alchemists of old—who mix various substances together in a crucible and heat them to a high temperature to see what happens, today's experimental mathematicians put a hopefully potent mix of numbers, formulas, and algorithms into a computer in the hope that something of interest emerges.

---

[3] DHB and JMB, "Exploratory Experimentation in Mathematics" (2011), www.ams.org/notices/201110/rtx111001410p.pdf

[4] See http://vis.carma.newcastle.edu.au/.

3   Case Studies

We turn to three sets of examples:

3.1   Case Study Ia: Iterative Reflections

Let $S \subset R^m$. The (nearest point or metric) *projection* onto S is the (set-valued) mapping, $P_S x := \mathrm{argmin}_{s \in S} \|s - x\|$. The *reflection* with respect to S is then the (set-valued) mapping, $R_S := 2P_S - I$. Iterative projection methods have a long and successful history. The basic model [1,3] finds a point in $A \cap B$ assuming information about the projections on A and B is accessible. The corresponding reflection methods are more recent and appear more potent.

Theorem 1 (Douglas–Rachford (1956–1979)). *Suppose* $A, B \subset R^m$ *are closed and convex. For any* $x_0 \in R^m$ *define*

$$x_{n+1} := T_{A,B} x_n \text{ where } T_{A,B} := \frac{I + R_B R_A}{2}.$$

If $A \cap B \neq \emptyset$, *then* $x_n \to x$ *such that* $P_A x \in A \cap B$. *Else* $\|x_n\| \to +\infty$.

The method also applies to a good model for *phase reconstruction*, namely for B affine and A a boundary 'sphere'. In this case we have some local and many fewer global convergence results; but much empirical evidence— both numeric and geometric (using *Cinderella, Maple* and SAGE).

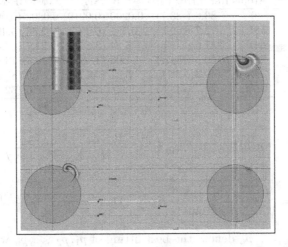

Cinderella applet[5] showing 20000 starting points coloured by distance from y-axis after $0, 7, 14, 21$ steps. Is this a *"generic* visual theorem" showing global convergence off the (chaotic) y-axis? Note the *error*—scattered red points—from using 'only' 14 digit computation.

---

[5] See http://carma.newcastle.edu.au/jon/expansion.html.

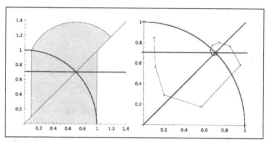

Proven region of convergence in grey showing what we can *prove* (L) is less than what we can *see* (R).

3.2   Case Study Ib: Protein Confirmation

Proteins are large biomolecules comprising multiple amino acid chains.[6] Proteins participate in virtually every cellular process and Protein structure $\rightarrow$ predicts how functions are performed. NMR spectroscopy (Nuclear Overhauser effect[7]) can determine a subset of interatomic distances without damage (under 6Å). This can profitably be viewed as a non-convex *low-rank Euclidean distance matrix completion* problem. We use only interatomic distances below 6Å typically constituting less than 8% of the total nonzero entries of the distance matrix and use our reflection method to extrapolate the rest.

Six Proteins: average (maximum) errors from five replications.

Protein	# Atoms	Rel. Error (dB)	RMSE	Max Error
1PTQ	40	-83.6 (-83.7)	0.0200 (0.0219)	0.0802 (0.0923)
1HOE	581	-72.7 (-69.3)	0.191 (0.257)	2.88 (5.49)
1LFB	641	-47.6 (-45.3)	3.24 (3.53)	21.7 (24.0)
1PHT	988	-60.5 (-58.1)	1.03 (1.18)	12.7 (13.8)
1POA	1067	-49.3 (-48.1)	34.1 (34.3)	81.9 (87.6)
1AX8	1074	-46.7 (-43.5)	9.69 (10.36)	58.6 (62.6)

Here

$$\text{Rel.error(dB)} := 10 \log_{10} \left(\frac{\|P_{C_2} P_{C_1} X_N - P_{C_1} X_N\|^2}{\|P_{C_1} X_N\|^2} \right),$$

$$\text{RMSE} := \sqrt{\frac{\sum_{i=1}^{m} \|\hat{p}_i - p_i^{true}\|_2^2}{\#\text{ofatoms}}}, \qquad \text{Max} := \max_{1 \le i \le m} \|\hat{p}_i - p_i^{true}\|_2.$$

The points $\hat{p}_1, \hat{p}_2, \ldots, \hat{p}_n$ denote the best fitting of $p_1, p_2, \ldots, p_n$ when rotation, translation and reflection is allowed.

The numeric estimates do not well segregate good and poor reconstructions so we ask what the reconstructions *look* like?

---

[6] RuBisCO (responsible for photosynthesis) has 550 amino acids (smallish).

[7] A coupling which occurs through space, rather than chemical bonds.

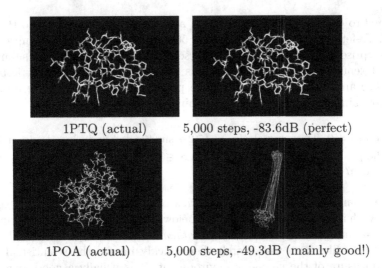

1PTQ (actual)   5,000 steps, -83.6dB (perfect)

1POA (actual)   5,000 steps, -49.3dB (mainly good!)

The picture of 'failure' suggests many strategies for success. What do reconstructions *look* like?[8] There are many projection methods, so it is fair to ask why we use Douglas-Rachford? The two sets of images below show the striking difference in the two methods.

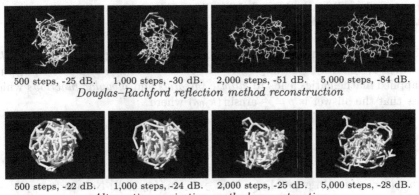

500 steps, -25 dB.   1,000 steps, -30 dB.   2,000 steps, -51 dB.   5,000 steps, -84 dB.
Douglas–Rachford reflection method reconstruction

500 steps, -22 dB.   1,000 steps, -24 dB.   2,000 steps, -25 dB.   5,000 steps, -28 dB.
Alternating projection method reconstruction

Yet the method of alternating projections works very well for optical abberation correction (originally on the Hubble telescope and now on amateur telescopes attached to latops). And we still struggle to understand why and when these methods work on different convex problems?

3.3   Case Study II: Trefethen's 100 Digit Challenge

In the January 2002 issue of *SIAM News*, Nick Trefethen presented ten diverse problems used in teaching *modern* graduate numerical analysis students at Oxford University, the answer to each being a certain real number. Readers were

---

[8] Video of the first 3,000 steps of the 1PTQ reconstruction is at http://carma.newcastle.edu.au/DRmethods/1PTQ.html.

challenged to compute ten digits of each answer, with a $100 prize to the best entrant. Trefethen wrote, "If anyone gets 50 digits in total, I will be impressed." To his surprise, a total of 94 teams, representing 25 different nations, submitted results. Twenty received a full 100 points (10 correct digits for each problem). Bailey, Fee and I quit at 85 digits! The problems and solutions are dissected most entertainingly in [9]. We shall examine the two final problems.

Problem #9. The integral $I(a) = \int_0^2 [2 + \sin(10\alpha)]x^\alpha \sin\left(\frac{\alpha}{2-x}\right) dx$ depends on the parameter α. What is the value $\alpha \in [0, 5]$ at which $I(\alpha)$ achieves its maximum?

The maximum α is expressible in terms of a *Meijer-G function*—a special function with a solid history that we use below. While knowledge of this function was not common among contestants, *Mathematica* and *Maple* both will figure this out; help files or a web search then quickly inform the scientist. This is another measure of the changing environment. It is usually a good idea—and not at all immoral—to *data-mine*.

Problem #10. A particle at the center of a 10×1 rectangle undergoes Brownian motion (i.e., 2-D random walk with infinitesimal step lengths) till it hits the boundary. What is the probability that it hits at one of the ends rather than at one of the sides?

Bornemann starts his remarkable solution by exploring *Monte-Carlo methods*, which are shown to be impracticable. A tour through many areas of pure and applied mathematics leads to *elliptic integrals* and *modular functions* which *proves* that the answer is $p = \frac{2}{\pi} \arcsin(k_{100})$ where

$$k_{100} := \left(\left(3 - 2\sqrt{2}\right)\left(2 + \sqrt{5}\right)\left(-3 + \sqrt{10}\right)\left(-\sqrt{2} + \sqrt[4]{5}\right)^2\right)^2,$$

is a *singular value*. [In general $p(a, b) = \frac{2}{\pi} \arcsin(k_{(a/b)^2})$.] No one (except harmonic analysts perhaps) anticipated a closed form—let alone one like this. This analysis can be extended to some other shapes, and the computation has been performed by Nathan Cilsby for self-avoiding walks.

3.4   Case Study IIIa: Short Walks

The final set of studies expressedly involve random walks. Our group, motivate initially by multi-dimensional quadrature techniques for higher precision than Monte Carlo can provide looked at the moments and densities of n-step walks of unit size with uniform random angles [6,8]. Intensive numeric-symbolic and graphic computing lead to some striking new results for a century old problem. Here we mention only two. Here p_n is the radial density of the n-step walk $(p_n(x) \sim \frac{2x}{n} e^{-x^2/n})$.

The densities p_3 (L) and p_4 (R) and simulations.

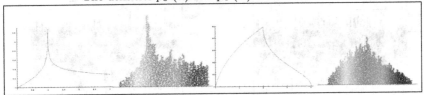

We first discovered $\sigma(x) := \frac{3-x}{1+x}$ is an *involution* on $[0,3]$ ($[0,1] \mapsto [1,3]$):

$$p_3(x) = \frac{4x}{(3-x)(x+1)} p_3(\sigma(x)). \tag{1}$$

So $\frac{3}{4}p_3'(0) = p_3(3) = \frac{\sqrt{3}}{2\pi}$, $p(1) = \infty$. We then found and proved that:

$$p_3(\alpha) = \frac{2\sqrt{3}\alpha}{\pi(3+\alpha^2)} \, {}_2F_1\left(\frac{1}{3}, \frac{2}{3}, 1 \middle| \frac{\alpha^2(9-\alpha^2)^2}{(3+\alpha^2)^3}\right) = \frac{2\sqrt{3}}{\pi} \frac{\alpha}{\mathrm{AG}_3(3+\alpha^2, 3(1-\alpha^2)^{2/3})} \tag{2}$$

where AG_3 is the *cubically convergent* mean iteration (1991): $\mathrm{AG}_3(a,b) :=$ $\lim_n a_n = \lim_n b_n$ with $a_{n+1} = \frac{a_n + 2b_n}{3}$ and $b_{n+1} = \sqrt[3]{b_n \cdot \frac{a_n^2 + a_n b_n + b_n^2}{3}}$, starting with $a_0 = a, b_0 = b$. More surprisingly we ultimately get a modular closed form:

$$p_4(\alpha) = \frac{2}{\pi^2} \frac{\sqrt{16-\alpha^2}}{\alpha} \operatorname{Re} {}_3F_2\left(\frac{1}{2}, \frac{1}{2}, \frac{1}{2}, \frac{5}{6}, \frac{7}{6} \middle| \frac{(16-\alpha^2)^3}{108\,\alpha^4}\right). \tag{3}$$

Crucially, for $\operatorname{Re} s > -2$ and s not an odd integer the corresponding *moment functions* [6], W_3, W_4 have Meijer-G representations

$$W_3(s) = \frac{\Gamma(1+\frac{s}{2})}{\sqrt{\pi}\,\Gamma(-\frac{s}{2})} G_{33}^{21}\left(\begin{matrix} 1,1,1 \\ \frac{1}{2}, -\frac{s}{2}, -\frac{s}{2} \end{matrix}\middle| \frac{1}{4}\right), \quad W_4(s) = \frac{2^s\,\Gamma(1+\frac{s}{2})}{\pi\,\Gamma(-\frac{s}{2})} G_{44}^{22}\left(\begin{matrix} 1, \frac{1-s}{2}, 1, 1 \\ \frac{1}{2}, -\frac{s}{2}, -\frac{s}{2}, -\frac{s}{2} \end{matrix}\middle| 1\right).$$

3.5 Case Study IIIb: Number Walks

Our final studies concern representing base-b representations of real numbers as planar walks. For simplicity we consider only binary or hex numbers and use two bits for each direction: 0 = right, 1=up, 2=left, and 3=down [2]. This allows us to compare the statistics of walks on any real number to those for pseudo-random walks[9] of the same length. For now we illustrate only the comparison between the number of points visited by 10,000 million-step pseudo-random walks and for 10 trillion bits of π chopped up into 10,000 walks.

---

[9] Python uses the *Mersenne Twister* as the core generator. It has a period of $2^{19937} - 1 \approx 10^{6002}$.

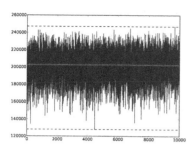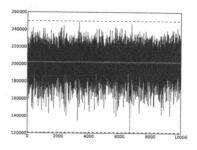

Number of points visited by 10,000 million-step base-4 random walks (L) and π (R)

3.6 Case Study IIIc: Normality of Stoneham Numbers

A real constant α is *b-normal* if, given integer $b \geq 2$, every m-long string of digits appears in the base-b expansion of α with precisely the expected limiting frequency $1/b^m$. Borel showed that almost all irrational real numbers are b-normal in any base but no really explicit numbers (e.g., $e, \pi\sqrt{2}$) have been proven normal. In our final study we shall detail the discovery of the next theorem.

The *Stoneham numbers* are defined by $\alpha_{b,c} = \sum_{n=1}^{\infty} \frac{1}{c^n b^{c^n}}$.

Theorem 2 (Normality of Stoneham constants). *For coprime pairs $b \geq 2, c \geq 2$, the constant $\alpha_{b,c}$ is b-normal, while if $c < b^{c-1}$, $\alpha_{b,c}$ is bc-nonnormal.*

Since $3 < 2^{3-1} = 4, \alpha_{2,3}$ is 2-normal but 6-nonnormal ! This yields the first concrete transcendental to be shown normal in one base yet abnormal in another.

References

1. Aragon, F., Borwein, J.M.: Global convergence of a non-convex Douglas-Rachford iteration. J. Global Optim. 57(3), 753–769 (2013)
2. Aragon, F., Bailey, D.H., Borwein, J.M., Borwein, P.B.: Walking on real numbers. Mathematical Intelligencer 35(1), 42–60 (2013)
3. Aragon, F., Borwein, J.M., Tam, M.: Douglas-Rachford feasibility methods for matrix completion problems. ANZIAM Journal (accepted March 2014)
4. Bailey, D.H., Borwein, J.M.: Exploratory Experimentation and Computation. Notices of the AMS 58(10), 1410–1419 (2011)
5. Borwein, J., Devlin, K.: The Computer as Crucible: an Introduction to Experimental Mathematics. AK Peters (2008)
6. Borwein, J.M., Straub, A.: Mahler measures, short walks and logsine integrals. Theoretical Computer Science 479(1), 4–21 (2013)
7. Borwein, J.M., Skerritt, M., Maitland, C.: Computation of a lower bound to Giuga's primality conjecture. Integers 13 (2013), Online September 2013 at #A67, http://www.westga.edu/~integers/cgi-bin/get.cgi
8. Borwein, J.M., Straub, A., Wan, J., Zudilin, W. (with an Appendix by Don Zagier): Densities of short uniform random walks. Can. J. Math. 64(5), 961–990 (2012)
9. Bornemann, F., Laurie, D., Wagon, S., Waldvogel, J.: The SIAM 100-Digit Challenge: A Study In High-accuracy Numerical Computing. SIAM, Philadelphia (2004)
10. Hanna, G., de Villiers, M. (eds.) ICMI, Proof and Proving in Mathematics Education, The 19th ICMI Study. New ICMI Study Series, vol. 15. Springer (2012)

Soft Math
Math Soft

Bruno Buchberger

RISC, Johannes Kepler University, Linz, Austria
bruno.buchberger@risc.jku.at
http://www.risc.jku.at/home/buchberg

In this talk we argue that mathematics is essentially software. In fact, from the beginning of mathematics, it was the goal of mathematics to automate problem solving. By systematic and deep thinking, for problems whose solution was difficult in each individual instance, systematic procedures were found that allow to solve each instance without further thinking. In each round of automation in mathematics, the deep thinking on spectra of problem instances is reflected by deep theorems with deep proofs.

In 20th century, the systematic procedures for spectra of problems became physically tangible as algorithms / software for the universal computer (which itself essentially is a mathematical invention). In 21st century, the rounds of automation in mathematics reach higher and higher levels and move more and more to the meta-level of mathematics, i.e. to the automation of mathematical thinking itself.

In this talk, we illustrate the evolution of mathematics towards higher and higher levels of automation of its own problem solving and thinking process by a couple of examples of increasing sophistication starting from calculation with Roman numbers up to the automatic invention of algorithms like the speaker's algorithm for computing Gröbner bases.

As a practical experience of Gödel's Incompleteness Theorem, there is no upper bound to the sophistication of higher and higher rounds in mathematical automation. Thus, mathematicians live in the best of all worlds: They can embark on more and more challenging problems that need an algorithmic solution knowing that, after an algorithmic solution has been achieved, there will always be room for more sophistication and human mathematical invention. Thus, in a sense, mathematicians will never become jobless.

By the intellectual power in the automation of mathematical problem solving, visible as "software", mathematics is also in the center of the spiral of automation in all science, technology and economy and is the silent driving force behind the spiral of innovation.

Unfortunately, the intellectual attractiveness and practical relevance of mathematics is hard to explain to outsiders. Or, more precisely, mathematicians often do a very bad job for explaining the fundamental role of mathematics clearly enough to outsiders, who in fact are the insiders of modern society. However, it is very important for the further development of mathematics that the role of mathematics for modern science, technology, economy, and welfare is made

H. Hong and C. Yap (Eds.): ICMS 2014, LNCS 8592, pp. 9–15, 2014.

public. We will also present a few ideas about this political aspect of being a mathematician in today's society. Some of these ideas are:

- The *university* is the place from where the deep understanding of the essence and role of mathematics must emanate and penetrate all branches of science, technology, economy, and society. Also, university teaching of mathematics must be the solid ground on which the appropriate ways of teaching mathematics at all levels and in all branches of the educational systems can be based.
- For being able to play this role, *university professors* must feel a high responsibility for mastering all the different aspects of mathematics (modeling of real world, invention of abstract mathematical knowledge, algorithmic problem solving, formal reasoning, interpretation of mathematical results in real world) in whatever field of mathematics they are working. In a provocative slogan, I request that every math professor at university level should be a master "both in proving as well as in programming". A math professor at university level must also be a master of language who is able to "speak mathematics" in whatever language our addressees from science, technology, economy, medicine, politics, etc. speak.
- Independent of specific contents, the essence of mathematics is the "*art of explaining*", i.e. the art of making complicated things (Latin "ex": sticking out like a high mountain) simple ("plain", "flat"). The art of explaining is the essence of both research and teaching. Explanation is only possible by deep thinking. Complicated things do not become "plain" by themselves but only by looking at them from many different angles, trying out various different ways of abstraction and simplification etc. It is the "*miracle of mathematics*" that by thinking hard and deeply once (in finite time) about the general version of a problem one may find a solution (method, algorithm) that allows us not to think any more in the infinitely many concrete instances of the problem.
- It is exactly this explanatory power of mathematics – the power of reducing something complex to something simpler – that constitutes *the universal value of mathematics*. University mathematics has a high responsibility to explain and demonstrate this universal value to all branches of science and technology but, more importantly, also to all institutional instances of society. As an echo, mathematics will receive the attention and appreciation of society it deserves.
- In fact, through my experience of working as the head of Softwarepark Hagenberg and, hence, based on hundreds of concrete interactions I had with (big and small) companies, research institutes, public institutions, investors, banks, educational institutions, political institutions and administration, I am coming more and more to the conclusion that the biggest waste of money and time in today's society is caused by *the lack of being able to "ex-plain"*: lawyers who can not really explain the essence of complicated formulations to their customers; customers of software companies who cannot really specify clearly what in the end the new software should do; software experts

who do not know how to force a customer to reveal what his problem really is; members of governance boards who do not know what the numbers in a balance sheet really mean and who do not dare to ask; doctors who do not really listen to the patient and patients who cannot really express what their problem is; civil servants who only apply rules instead of trying to solve the problem of the citizen; etc.

Since our systems grow in size, the lack of explanation skill has more and more drastic effects. We only get shocked by big scandals like the finance crisis or the NSA disaster but millions of small, hidden scandals that waste or destroy time, money, motivation, welfare, ... happen in every moment. I think that only a small part of the negative effects are generated by criminal energy, the biggest part is just generated by not being able to explain things clearly with an attempt at expressing what we want to express in the language of the person to whom we talk. I think that, more than anything else, providing mathematics as the "art of explaining" could and should be our greatest gift to the world.

- The *actual situation of math research and teaching at many universities* is very different from the ideals sketched above: Many mathematicians unlearn to speak with "normal" people and become masters in making simple things complicated and erect a language wall around themselves that let outsiders soon give up. More shockingly, children and students get the impression that mathematics is a closed esoteric world with little relevance for real life and is something very special rather a universal and actually very practical skill of how to handle complex situations by systematic "ex-planation".

- The *"spiral of marginalization of mathematics"* is disastrous, both for society and mathematics: If university math conveys an unclear and narrow picture about the essence and role of mathematics to university math students, the math graduates will convey an even more foggy and more limited picture to the environment in which they will work (in particular also to the children and students they teach at elementary and high school level). Children and students grow up with a distorted and limited, even threatening, experience about mathematics and universities will have to accept that the level of mathematical capabilities of incoming students is less and less sufficient for what is needed for the next round of scientific and technological progress. So, the negative spiral is closed and leads to a disaster. (In the same way, companies with employees who have a distorted and limited picture of mathematics will not find their way to using mathematics if they face challenging problems of technological, industrial or organizational innovation.)

- This *downward spiral must be interrupted.* Of course, one could identify anyone at any level of this spiral as being responsible for improving the situation. However, since I do not like this "circular shift of responsibility" which is prevalent in today's society (in many different areas as, for example, financial crisis, environmental problems, bureaucratic overflow etc.), I would like to identify university mathematics (concretely, *university professors*) as the crucial group of people who must manage to turn the downward spiral into an upward spiral: Mathematics professors at the universities must develop in themselves the comprehensive picture and methodological competence I tried to sketch above in order to establish mathematics as the central thinking and problem solving technology of current and future science, technology, economy and society. At the same time, university professors have to make sure that their students (in particular the PhD students who will shape the next generation of mathematics), despite the enormous intellectual pressure they face when trying to become professional in the technicalities of mathematics, are motivated and attracted to take their time for developing a comprehensive view of the various different layers and aspects of mathematics and for acquiring and cultivating the universally applicable thinking and problem solving skill, including the "the art of explaining".

– In a comprehensive view of mathematics, in a modern setting, *software* plays a particularly important role. Software is the endpoint in the materialization of mathematical theories: From mathematical knowledge (proven insight; theorems) through mathematical algorithms (whose proven correctness is based on theorems) to their implementation as programs executable on machines. Thus, software realizes the fundamental aspiration of mathematics to solve problems (in their general specification) by deep thinking in general terms (but finite amount of time) until a general method can be established whose execution for the potentially infinitely many instances does not need thinking any more, i.e. can be given to "non-thinkers", i.e. ultimately to computers. If mathematics leaves out the aspect of software, it deprives itself of an essential aspect. One may even say, it deprives itself of *the* essential aspects, if one adheres to a view of mathematics ("demand driven view") where one starts from a problem and goes to the world of theorems for obtaining knowledge that will allow to solve the problem. If one adheres to a view of mathematics where one starts from systematic investigation of what is true in a certain theory ("curiosity driven view") and then asks whether the new knowledge has applications for problem solving, then the aspect of software is essential but secondary. In my comprehensive view, the demand driven and the curiosity driven aspect are equally important, they interact in a spiral, none of them is first or second, higher or lower, less or more important but, rather, they constitute two sides of the same medal and are both indispensable for making the object a medal.

– Today, however, mathematics and software are even more intimately connected by being the meta-theory for each other. The intellectual process of developing software (even for applications that do not need any non-trivial mathematics) is a formal process whose quality heavily depends on systematic and sound design procedures. This process, more and more, is understood as a technological process much like the production of industrial goods that can be and should be handled by mathematical methods. Essential aspects of this process like model based specification, verification, complexity analysis, component based design and implementation, data structures, generic programming need both the mathematical expertise gained from the application of mathematics to processes in science and technology as well as the deep insights gained in 20th century mathematical logic. Here, I call this new area (formal theory and technology of designing and implementing software) *"software science"*.

In principle, what I said here about the role of mathematics for the process of software design and development is equally true for the process of designing new hardware. Only that, of course, more and more sophisticated insights about the physical reality go into hardware research and design and form the challenging limitations and challenging new opportunities for realizing the principle of "universal, program based computing", which is a fundamentally mathematical concept. In fact, software and hardware design grow more and more together and the mathematical challenge for both aspects is basically identical.

– Conversely, the best tools of current software technology are now available, and heavily used, for building up impressive *mathematical software systems* (like *Mathematica*, Maple etc.) that comprise basically all the algorithmic mathematical methods that have been worked out over the past centuries and decades – many of them based on deep and new mathematical theories – in comprehensive, nicely structured, extensible and programmable algorithm libraries. Some of these systems specialize to particular areas of mathematics (like CoCoA for polynomial ideal theory) and try to be comprehensive for this area.

– Strategically, I think that mathematics is well advised to consider both software science as well as the professional implementation of mathematical software as an integral part of mathematics. I hope I was able to make clear that this is essential from a philosophical and theoretical point of view in order to preserve and expand mathematics as the comprehensive, sound and strong "thinking technology" of mankind. However, it is also essential from a political point of view. *If mathematics gives up software* (or, even worse, does not even embark on software both as a science and as a tool) mathematics will be marginalized in current science, technology, economy, and society in general. And others will take over (or already have taken over) who will do the job – with less rigor and less potential for the "arts of ex-plaining" but more feeling for how to earn money and how to be indispensable and influential. This will also have the consequence that only very few youngsters will find their way to studying mathematics and those who will still get there will find themselves in a difficult position in society and, also, will encounter a mathematics deprived of essential ingredients so that another downward spiral is generated.

– Mathematics, in its comprehensive view and not only as a collection of results but more essentially as a particular thinking culture, is an essential part of the *heritage of mankind*. If we do not manage to attract brilliant and enthusiastic young people to embark on the adventure of mathematics as a research topic, global society will grow into a situation where the sophisticated results of mathematics – packed in software ("apps", ...) – will be available to everyone everywhere at any time but nobody will be able to understand the mathematical principles and results behind these external tools – both in their potential but also in their limitations. (It is alarming that in media, politics and even schools – and even universities people are still speaking about the "the computer can do or cannot do or never will be able to do ... this or that" whereas "human can do or cannot do this or that ...". Of course, sometimes, this is just a sloppy way of speaking but most of the times this expresses a fundamental misunderstanding about the intelligence behind computers, which is human mathematical intelligence as it always was.)

Similarly, if essential mathematical skills (in the comprehensive sense sketched above) are not any more available in the masses as a reliable part of general education, neither technological and economic innovation, nor growth in common welfare, nor an increase in democratic awareness, democratic culture and societal evolution will be possible.

– A comprehensive view of mathematics that also sees the object level of mathematics permanently together with the meta-level (the level in which we try to automate and support the thinking process of doing mathematics in a certain area) will change the way of how, in the future, we will organize the quality control and archiving process of mathematics, which currently is the noble purpose of *mathematical journals*. It is near at hand that, practically, by the advent of the web, the process of archiving, documenting, retrieving mathematical knowledge will soon change and printed versions of mathematical journals will play less and less role or will just be a by-product of the archiving process. In contrast, quality control by the *anonymous peer reviewing* process, in my view, will become even more important in the computer age. This process is a crucial intellectual invention, which is only a couple of centuries old and, in fact, was the crucial reason why science and technology evolved in such a breath-taking speed with such innovative power. Even if much of new research is just "published" by uploading a paper onto the web, it will always make a difference whether something is exposed to anonymous criticism or not (well knowing and emphasizing that there is nothing like "an absolute instance for determining truth"). Hence, the question is how this process of anonymous peer reviewing will change and can be made more efficient, reliable, transparent, flexible, structured.

I am optimistic and (in the frame of my *Theorema* Project) I am actively working on how the enormous advances which were made in automated reasoning could be used for supporting the anonymous refereeing procedure of new mathematical results. In fact, I think that the most essential criteria for the scientific value of new results (like importance, originality, correctness, completeness) can well be decided (at least partly) by formal methods. Also, using a formal approach, (mathematical) journals will turn more and more from passive knowledge bases to globally accessible interactive knowledge purifiers, expanders and generators.

– Finally, let me briefly touch the question of *mathematical software in education*. Still, a kind of battle is going on between the "purists" who believe that the use of mathematical software (the big brother of the "pocket calculators") will spoil the mathematical understanding and the "populists" who believe that many parts of mathematical education are obsolete because mathematical software is available for solving problems by a mouse click. I think that both these views are fundamentally wrong and based on a limited understanding of mathematics. Mathematics is a process that goes from a first vague understanding of a problem via deep thinking to a clear understanding of the problem hand in hand with the development and proof of mathematical knowledge that allows to solve the problem in a systematic way (in the ideal case by an algorithm). This process iterates through higher and higher levels of mathematical knowledge and mathematical problem solving.

Good math education repeats this invention process with the student. It should give the student the chance to understand the invention in the first phase (which I call the "white-box phase"). In this phase, it would be a silly short-cut if, instead of developing understanding, one would show the student just which button has to pressed (or which function has to be called) using existing software. In the second phase (which I call the "black-box phase"), when the method is understood, the method should be programmed (in the ideal case when students are at an age in which they can write programs themselves) and one should give the student the joyful experience that, from now on, the problem discussed does not need any more hard work but rather, in each instance, can now be solved by calling the algorithm. I formulated this "*White Box / Black Box Principle*" already in 1989. However, it seems that, unfortunately, the battle between purists and populists is still going on (in particular, in the math didactics community).

Flyspecking Flyspeck

Mark Adams

Proof Technologies Ltd, UK

Abstract. The formalisation of mathematics by use of theorem provers has reached the stage where previously questioned mathematical proofs have been formalised. However, sceptics will argue that lingering doubts remain about the efficacy of these formalisations. In this paper we motivate and describe a capability for addressing such concerns. We concentrate on the nearly-complete Flyspeck Project, which uses the HOL Light system to formalise the Kepler Conjecture proof. We first explain why a sceptic might doubt the formalisation. We go on to explain how the formal proof can be ported to the highly-trustworthy HOL Zero system and then independently audited, thus resolving any doubts.

1 The Flyspeck Project

Tom Hales' proof of the Kepler Conjecture consists of 300 pages of mathematical text, and uses the results of executing three bespoke computer programs consisting of tens of thousands of lines of computer source code. When submitted for publication, the referees held back from giving a full endorsement, complaining that the proof was too complex to check in its entirety. Hales' response was to instigate the Flyspeck Project [1] to settle the matter once and for all, using the HOL Light [2] theorem prover for the HOL logic [3] to formalise his proof.

The project is now nearing completion, with the formalisation of the results of one of the three computer programs being the only incomplete aspect. As predicted by Hales, the project has consumed around 20 man years of effort, the bulk of which has been concerned with formalising the mathematical text. This aspect of the project was carried out by an international team of mathematicians, with about 15 contributors.

For the text formalisation, the text was broken down into around 700 distinct lemmas, each with a bounty attached that was awarded on completion of the lemma's formal proof. Formal proofs were submitted as HOL Light proof script files (written in ML source code), and rerun by Hales before being incorporated into the project repository. Within these proof script files, almost anything was acceptable, so long as no new axioms[1] were added and the desired lemma result was assigned as the value of a pre-arranged ML identifier. Now complete, the text formalisation consists of around 200 proof scripts with a total of around 450,000 non-comment/blank lines of ML, resulting in around half a billion primitive inferences when processed through HOL Light.

---

[1] In this paper, by *axiom* we mean an extension to a theory using a general extension command that does not enforce conservative extension.

H. Hong and C. Yap (Eds.): ICMS 2014, LNCS 8592, pp. 16–20, 2014.

Once the results of the third computer program have been formalised, the statement of the Kepler Conjecture can be proved as a final theorem within HOL Light, by bringing together the four components of the project. Assuming that HOL Light itself is sound, if processing all the proof scripts through HOL Light results in the final theorem being proved and no new axioms, then the Kepler Conjecture has been formally proved. Or so the optimists would have us believe.

2   The Sceptics' Concerns

Sceptics would disagree, and list many concerns that need to be addressed.[2] Note that these concerns apply equally to any mathematics formalisation project.

Firstly, has a final theorem actually been proved in the theorem prover? Perhaps the proof scripts simply fail to produce a final theorem when processed altogether in one session. This is easy to address, but still must be done. The scripts must been rerun by someone who is independent of the formalisation project, to check that everything successfully processes. We call the person performing this role a *proof auditor*, or *auditor* for short.

Secondly, how do we know that the statement of the final theorem means what it is purported to mean? Maybe there is a subtle problem in the statement, or in the definition of one of the constants used in it, or in the definition of one of the constants used in one of the definitions. The final theorem and its tree of dependent definitions all need to be reviewed in minute detail by the auditor. This task is far from trivial in a large proof formalisation, where there is typically substantial supporting theory referred to in the statement of the final theorem (although in Flyspeck this is deliberately minimized to reduce the risk), and where the exact equivalence of the final theorem with the original mathematical result is not obvious. Thus the auditor must be an expert in the field of mathematics formalisation.

Thirdly, have any axioms been added that make the theorem prover's theory inconsistent? This question should be easy to resolve in projects such as Flyspeck where no axioms are supposed to be added.

Fourthly, maybe the settings for displaying concrete syntax (e.g. the fixity settings for the constants used in the statement of the theorem) have been configured in some way that happens make the statement get misinterpreted to mean something different. The auditor needs to be aware of all the settings that can alter the display of theorems, and how these settings have been configured.

Fifthly, can we really trust the theorem prover to correctly record and display all this information that is required for the review? Are there flaws in the implementation of the theorem prover that mean statements can get confusingly displayed (e.g. as in Pollack-inconsistency [5]), or that definitions or axioms don't get correctly recorded? The auditor needs to somehow address such concerns.

Sixthly, can we really trust that the theorem prover is sound? Is it possible that there is a subtle programming error in the implementation of one of its inference rules, or in the setting up of its theory?

---

[2] For an alternative discussion of these issues, see [4].

Seventhly, can we really trust that the theorem prover implementation guarantees that it is sufficient to consider just the soundness of the system and its state after having processed a proof script to ascertain whether a theorem has been proved. Is it possible for a proof script to make the theorem prover unsound, thus requiring the auditor to consider the proof script in their review?

Finally, just in case anyone is thinking of the improbability of the above concerns happening purely by accident, can we really trust that someone involved in the project has not maliciously exploited a vulnerability for their own ends? If a contributor knew about a back door to creating theorems, perhaps they would be tempted to exploit this to get their bounty payment more easily. Or perhaps the manager might be tempted to exploit a flaw to get the project completed on time or on budget. In their review, the auditor must assume malicious intent, rather than use arguments about the improbability of innocent error.

3   Problems with HOL Light

HOL Light is one of the simplest and widely-studied theorem provers. It implements the HOL logic, one of the simplest, widely-understood and uncontroversial formal logics. Furthermore, it has an LCF-style kernel [6], whereby the type system of its implementation language is used to enforce that all proofs ultimately execute purely in terms of the kernel's 10 primitive inference rules. Finally, its kernel has itself been formally verified correct. As theorem provers go, it is one of the highest regarded for trustworthiness.

However, concentrating on the sceptics' concerns about the theorem prover being used (the fifth, sixth and seventh from the previous section), HOL Light does not fare well.[3] We know of no problems with respect to the sixth concern, but for the other two there are various. Note that other well-known theorem provers have their own problems and overall fare no better than HOL Light. We concentrate on HOL Light's weaknesses in this paper because it is the system used in Flyspeck.

Firstly, it does not quite record all definitions: type constants defined by directly using the kernel interface are not recorded. Thus any type constants defined in proof scripts cannot be trusted to have any given definition.

Secondly, there are various flaws in the way it displays HOL concrete syntax that make it Pollack-inconsistent. One such flaw is that type annotation is never used in displayed expressions. This can cause various kinds of confusion, for example an expression may contain two variables with the same name but different type that will appear to be the same variable, or a theorem may appear to be universally true for variables of any type when it has actually only been proved for variables of a specific type (e.g. a type with just one element). Another flaw is that overloaded names are not distinguished, so a variable with the same name as a constant will appear to be that constant, or an expression containing a variable with the same name as a reserved word might appear to be some expression

---

[3] Our observations apply to all recent versions of HOL Light, including the most recent at the time of writing, SVN revsion 193.

with a completely different syntactic form. Note that circumventing these issues by displaying primitive syntax carries its own risks, because expressions then become much more difficult to read, and thus easier to misinterpret.

Thirdly, HOL Light's LCF-style kernel is not completely watertight: it is possible to process a proof script that will result in unsound deduction. These vulnerabilities stem from HOL Light not addressing aspects of its implementation language, a dialect of ML called OCaml, which is also the language its proof scripts are written in. One vulnerability is that HOL Light does not protect against OCaml's mutable strings, and so the name of a HOL constant can be altered by the user simply by altering the string storing the name (see Figure 1). Another is that OCaml has an (undocumented) function called `Obj.magic` that can be used to subvert the OCaml type system and thus bypass the kernel.

```
let t = fst (dest_const (concl TRUTH));;
t.[0] <- 'F';;
let FALSE = EQ_MP (REFL 'F') TRUTH;;
t.[0] <- 'T';;
```

Fig. 1. Exploiting OCaml string mutability to prove false in HOL Light

4   A Proof Auditing Capability

We now describe how components from the Common HOL Project [7], for assisting portability between HOL theorem provers, can be employed to support the process of auditing large formal proofs performed in HOL theorem provers. Common HOL is based around a set of basic theory and inference rules that is common to all HOL systems.

One component of Common HOL is the HOL Zero theorem prover. Unlike the other HOL systems, this is not designed for developing formal proofs, but as a HOL proof checker, i.e. for checking formal proofs developed on other HOL systems. It has been carefully designed to excel at trustworthiness, with a simple and well-documented implementation, an LCF-style kernel and no known soundness-related flaws. It has a parser and pretty printer for concrete syntax, but unlike any other HOL system it is Pollack-consistent. Even though it is implemented in OCaml, it addresses the associated vulnerabilities, for example it protects against mutable strings by making copies at suitable points. There is even a bounty of $100 for discovering soundness-related flaws, and a list of exposed flaws is published on the website (the most recent flaw was in 2011).

Another component is a proof porting capability. This records a proof on one HOL system during the execution of its proof script, recording it in terms of Common HOL theory and inference rules. This recorded proof can then be exported as a proof object file, and then imported into another HOL system, where it can be replayed to recreate the original proof. Common HOL proof exporters and importers currently exist for HOL Light and HOL Zero, and so it is possible to port proofs between these two systems. This capability can port massive formal proofs, involving hundreds of millions of basic steps, with ease. The entire Flyspeck text formalisation is ported using only modest hardware and in less than twice the time it takes to process in HOL Light.

By combining these components, it is possible create an effective proof auditing capability. Proofs carried out on one HOL theorem prover can be quickly ported to HOL Zero. Because we can assume HOL Zero is sound, replaying the ported proof on HOL Zero establishes that the proof does not exploit unsoundness in the original system. Furthermore, because we can trust that HOL Zero records all axioms and definitions correctly and displays them unambiguously, the auditor can concentrate on reviewing the content of these, displayed in human-readable concrete syntax, rather than worry about flaws in the theorem prover. And because the input is a proof object file, rather than an ML proof script, the auditor need not worry about whether arbitrary ML could play havoc with the LCF-style kernel.

5   Conclusion

The Common HOL Project provides the necessary components for an effective proof auditing capability for proofs based on HOL. Proofs carried out in one HOL system can be quickly ported to HOL Zero, which offers a suitably trustworthy environment for performing an audit of the proof. We recommend that mathematics formalisation projects such as the Flyspeck Project are audited using this capability, to resolve any lingering doubts sceptics might have about their efficacy.

References

1. Hales, T.: Introduction to the Flyspeck Project. In: Mathematics, Algorithms, Proofs. Dagstuhl Seminar Proceedings, vol. 05021. Internationales Begegnungs- und Forschungszentrum für Informatik (2006)
2. Harrison, J.: HOL Light: An Overview. In: Berghofer, S., Nipkow, T., Urban, C., Wenzel, M. (eds.) TPHOLs 2009. LNCS, vol. 5674, pp. 60–66. Springer, Heidelberg (2009)
3. Gordon, M.: An Introduction to the HOL System. In: Proceedings of the 1991 International Workshop on the HOL Theorem Proving System and its Applications. IEEE Computer Society Press (1992)
4. Pollack, R.: How to Believe a Machine-Checked Proof. In: Twenty Five Years of Constructive Type Theory. Oxford University Press (1998)
5. Wiedijk, F.: Pollack-Inconsistency. Electronic Notes in Theoretical Computer Science, vol. 285. Elsevier Science (2012)
6. Gordon, M., Milner, R., Wadsworth, C.P.: Edinburgh LCF. LNCS, vol. 78. Springer, Heidelberg (1979)
7. Proof Technologies Ltd. website, http://www.proof-technologies.com/

Symbolic Computing Package for Mathematica for Versatile Manipulation of Mathematical Expressions

Youngjoo Chung

Gwangju Institute of Science and Technology, Korea
ychung@gist.ac.kr
http://symbcomp.gist.ac.kr

Abstract. Symbolic Computing package is an add-on package that facilitates symbolic computation in Mathematica. It enables display and interpretation of derivatives, integrals, sums, products, vector operators, brackets, and various forms of subscripts and superscripts using the traditional mathematical notation based on the low-level box language and contains over 700 functions for notation, algebraic manipulation and evaluation of mathematical expressions. The package function categories include: basic algebra, complex variables, differential calculus, elementary functions, equation solving, equations, formula manipulation, Fourier analysis, function analysis, integral calculus, operator analysis, polynomials and series, products, sums, trigonometric functions, vectors and matrices. The package has its own interpreter language, complete on-line documentation and two palettes for entering mathematical expressions and execution control of functions. This provides a powerful platform for streamlined manipulation of all or parts of an expression and will significantly enhance the capabilities of the kernel and user-defined functions. Development of the package and its applications to various topics of mathematics and related disciplines will be presented.

Keywords: Symbolic Computing, Formula Manipulation, Mathematica.

1 Introduction

Symbolic computing [1], in contrast to numerical computing, is computation with variables and constants according to the rules of algebra for manipulation and evaluation of mathematical expressions. This will lead to dramatic improvement of analytical calculation and can be applied to research and education of mathematics, physics and various other science and engineering disciplines. Other advantages include minimization of human errors and improvement of accuracy during calculation by using computer software that incorporates known algorithms and mathematical identities.

Symbolic Computing package [2] is an add-on package that facilitates symbolic computation in Mathematica [3]. With over 700 functions that enable traditional

H. Hong and C. Yap (Eds.): ICMS 2014, LNCS 8592, pp. 21–25, 2014.

mathematical notation, algebraic manipulation and evaluation of various mathematical expressions, the package allows the users to focus on the principles instead of time-consuming and error-prone calculations and provides good readability and minimization of human errors during calculations. Expressions that closely resemble the traditional mathematical style, e.g., subscripts, superscripts and vector notations, can be used and this will replace a lot of hand calculations. Using this approach, materials and references including derivation of the mathematical formulas can be contained in a single document.

This paper will describe the main components and key features of the package, examples of the usage of the functions for formula manipulation, the interpreter language implemented in the SCMAF function, and some of the technical details underlying the design of the package.

2   Main Components and Key Features

The development objective of the Symbolic Computing package is to design and implement a symbolic computing system based on Mathematica that can freely manipulate various mathematical expressions using traditional notations and deferred on-demand evaluation. The package can be used for algebraic manipulation of formulas using symbolic computing and it provides seamless integration with the computing environment of Mathematica.

There are over 700 functions in the package and all functions have the prefix SC, which stands for "Symbolic Computing," e.g., SCDerivExpand and SCEvalInt. The package function categories include: basic algebra, complex variables, differential calculus, elementary functions, equation solving, equations, formula manipulation, Fourier analysis, function analysis, integral calculus, operator analysis, polynomials and series, products, sums, trigonometric functions, vectors and matrices.

The package also has its own interpreter language implemented in the SCMAF function, which is an abbreviation of SCMapApplyFunctions. It provides a powerful platform for streamlined manipulation of all or parts of an expression. The kernel and user-defined functions are called in sequence for step-by-step controlled manipulation of expressions and the features provided by SCMAF significantly enhances the capabilities of the functions. User-defined functions can still be used separately independent of SCMAF, in which case the features provided by the options of SCMAF cannot be used. The basic syntax of SCMAF is shown below.

SCMAF[$expr$,
$\quad f_1, \{x_{11}, x_{12}, ...\}, lopts_1,$
$\quad f_2, \{x_{21}, x_{22}, ...\}, lopts_2,$
$\quad f_3, x_{31}, lopts_3,$
$\quad f_4, \{\{x_{411}, x_{412}, ...\}, sopts_{41}, \{x_{421}, x_{422}, ...\}, sopts_{42}, ...\}, lopts_4,$
$\quad ...,$
$\quad gopts],$

where f_1, f_2, ... are functions to operate on the expressions in sequence, $\{x_{11}, x_{12}, ...\}$, $\{x_{21}, x_{22}, ...\}$, ... are the argument lists, and $lopts_i$, $sopts_{ij}$ are local options applied to the function f_i and the preceding argument list. $gopts$ is the global options for the SCMAF function. By default, the first arguments x_{11}, x_{21}, ... are replaced according to $x_{11} \rightarrow f_1[x_{11}, x_{12}, ...]$, $x_{21} \rightarrow f_2[x_{21}, x_{22}, ...]$, etc., and the MainArg local option can be used to change the main argument.

The package has on-line documentation of all functions accessible through the Mathematica help browser and two palettes for entering mathematical formulas and navigating through the functions in SCMAF.

3   Formula Manipulation

A large portion of the functions in the package is for manipulation of mathematical expressions. For this purpose, expressions like integrals, products and sums entered in 2-D format are not evaluated so that further operations can be done. Even though this can be sometimes achieved, e.g. for integrals, by providing explicit specification of the function arguments when the integral cannot be evaluated, they are often omitted in practice, and in such cases, evaluation should be deferred. Merging derivatives and integrals using the package functions SCMergeDerivs and SCMergeInts, respectively, is shown in Fig. 1, where f and g are implicitly functions. Inside the shaded box is the user input and the output is shown below. Note that evaluation of the expressions is not done so that the merging operations can be performed.

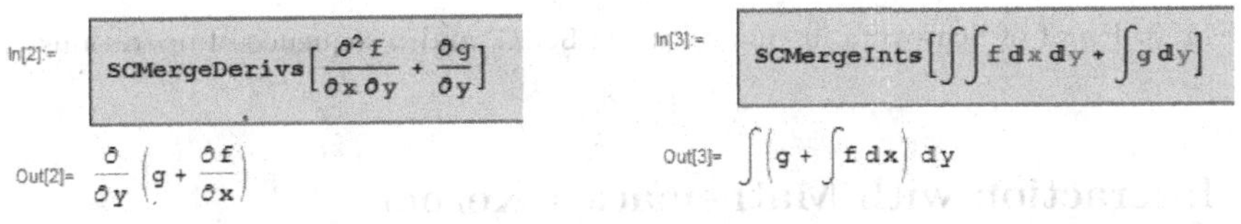

In[2]:= $\mathtt{SCMergeDerivs}\left[\dfrac{\partial^2 f}{\partial x\,\partial y} + \dfrac{\partial g}{\partial y}\right]$

Out[2]= $\dfrac{\partial}{\partial y}\left(g + \dfrac{\partial f}{\partial x}\right)$

In[3]:= $\mathtt{SCMergeInts}\left[\displaystyle\int\!\!\int f\,\mathrm{d}x\,\mathrm{d}y + \int g\,\mathrm{d}y\right]$

Out[3]= $\displaystyle\int\left(g + \int f\,\mathrm{d}x\right)\mathrm{d}y$

Fig. 1. Merging derivatives and integrals

The package functions can have options according to the standard Mathematica syntax. As an example, Fig. 2 shows transformation of a second-order partial derivative using the variable transformation $(x, t) \rightarrow (\xi, \eta)$. The Apply option specifies the function to apply to the result for post-processing.

Figure 3 shows an example of using SCMAF for the proof of Schwarz's inequality in the complex vector space. It starts with the obvious statement $\|\mathbf{a} + \lambda\mathbf{b}\|^2 \geq 0$, expands it in complex vector space and a sequence of operations are done to arrive at the desired result $\|\mathbf{a}^* \cdot \mathbf{b}\| \leq \|\mathbf{a}\|\,\|\mathbf{b}\|$.

In[4]:= `SCTransDeriv[ ∂²φ/∂x² , TransVar → {{x, t}, {ξ, η}}, Apply → Expand]`

Out[4]= $\dfrac{\partial^2 \eta}{\partial x^2}\dfrac{\partial \phi}{\partial \eta} + \dfrac{\partial^2 \xi}{\partial x^2}\dfrac{\partial \phi}{\partial \xi} + \left(\dfrac{\partial \eta}{\partial x}\right)^2 \dfrac{\partial^2 \phi}{\partial \eta^2} + \left(\dfrac{\partial \xi}{\partial x}\right)^2 \dfrac{\partial^2 \phi}{\partial \xi^2} + 2\dfrac{\partial \eta}{\partial x}\dfrac{\partial \xi}{\partial x}\dfrac{\partial^2 \phi}{\partial \xi \partial \eta}$

Fig. 2. Transformation of a partial derivative

In[5]:=

$\|\bar{a} + \lambda\, \bar{b}\|^2 \geq 0$

`SCMAF[%, SCToConjugate, At[1],`

`SCVecExpand, At[1],`

`RA, {At[1], ā*.b̄ = |ā*.b̄| e^{i α}}, Complex → {λ, ā, b̄},`

`RA, {At[1], λ = r e^{-i α}}, Complex → {ā, b̄}, RA → A_*.A_ = ‖A‖², ,`

`SCMergePoly, {At[1], r},`

`RA, {At[1], At[1] = ‖ā‖² - |ā*.b̄|²/‖b̄‖²}, ,`

`SCMultiply, {At[1], ‖b̄‖²},`

`SCIneqSolve, {All, |ā*.b̄|²}, , ,`

`SCEqApply, {All, √π &}]`

Out[5]= $\|\bar{a} + \lambda\, \bar{b}\|^2 \geq 0$

$r\,|\bar{a}^*.\bar{b}| + e^{i α} r\, \bar{b}^*.\bar{a} + \|\bar{a}\|^2 + r^2\,\|\bar{b}\|^2 \geq 0$

$\|\bar{a}\|^2 - \dfrac{|\bar{a}^*.\bar{b}|^2}{\|\bar{b}\|^2} \geq 0$

Out[6]= $|\bar{a}^*.\bar{b}| \leq \|\bar{a}\|\,\|\bar{b}\|$

Fig. 3. Proof of Schwarz's inequality using `SCMAF` and a sequence of operations

4   Interaction with Mathematica Kernel

The Symbolic Computing package has a large collection of functions designed to perform manipulation of various forms of mathematical expressions using the traditional notation. It utilizes the functionalities of the Mathematica kernel as much as possible and integrates with it in a seamless manner by not interfering with its operation.

However, in order to accomplish the package's main objectives of formula manipulation, certain forms of expressions like integrals, products and sums entered in 2-D form are not evaluated until explicitly asked by the user by invoking the corresponding kernel functions, as compared to the normal Mathematica convention. The package also contains various known mathematical algorithms and identities in addition to those already built in the kernel, which, combined with the effective user command interface, significantly facilitates manipulation and transformation of expressions.

5   Implementation Details

The notations are implemented primarily using the low-level box language functions `MakeExpression` for input and `MakeBoxes` for output. Great care needs to be taken when defining the rules for these functions in order not to cause inconsistencies.

The `SCMAF` function, whose example is shown in Fig. 3, provides a mechanism of precisely specifying the parts of the expression to which the functions are applied. Position specifications and patterns can be used as well as any Mathematica expressions. The `Base` and `Target` options can also be used to complement this functionality.

The syntaxes of the package functions were designed in conformity with the conventions of Mathematica. Minimum information is required for the input and others are specified using options with the default values that are most commonly used. In a typical usage shown in Fig. 4, the Laplacian of the vector field function $\mathbf{A} = \hat{r}f(r)\sin\phi$ is evaluated in the spherical coordinate system (r, θ, ϕ). The functions given by the `Apply` option are applied in the reverse order like the kernel function `Composition`.

In[20]:=
$$\texttt{SCEvalVecOp}\left[\left(\nabla^2\,\bar{A}\right).\hat{r} + \frac{1}{r}\,\hat{r}.\nabla\bar{A}.\hat{\phi},\ \bar{A} = \hat{r}\,\texttt{f[r]}\,\texttt{Sin[ϕ]},\ \{r,\,\theta,\,\phi\},\right.$$
$$\left.\texttt{"Spherical"},\ \{\hat{r},\,\hat{\theta},\,\hat{\phi}\},\ \texttt{Apply} \to \{\texttt{Simplify},\ \texttt{Expand}\},\ \texttt{AbbrevFunc} \to \texttt{True}\right]$$

Out[20]=
$$\frac{1}{r^2}\left(r\,\texttt{Sin[ϕ]}\left(2\,\frac{df}{dr} + r\,\frac{d^2f}{dr^2}\right) + f\left(\texttt{Cos[ϕ]}\,\texttt{Csc[θ]} - \left(2 + \texttt{Csc[θ]}^2\right)\texttt{Sin[ϕ]}\right)\right)$$

Fig. 4. Evaluation of the radial component of the Laplacian of the vector **A** using `SCEvalVecOp` and some options

References

1. Buchberger, B.: Symbolic Computation (An Editorial). J. of Symb. Comput. 1, 1–6 (1985)
2. Chung, Y.: Symbolic Computing Package for Mathematica (2012), http://symbcomp.gist.ac.kr
3. Wolfram, S.: Mathematica, Wolfram Research, Inc., http://www.wolfram.com

Representing, Archiving, and Searching the Space of Mathematical Knowledge

Mihnea Iancu, Michael Kohlhase, and Corneliu Prodescu

Jacobs University Bremen, Germany
http://kwarc.info

1 Introduction

There is an interesting duality between the forms and extents of mathematical knowledge that is verbally expressed (published in articles, scribbled on blackboards, or presented in talks/discussions) and the forms that are needed to successfully extend and apply mathematics. To "do mathematics", we need to judge the veracity, extract the relevant structures, and reconcile them with the context of our existing knowledge – recognizing parts as already known and identifying those that are new to us. In this process we may abstract from syntactic differences, and even employ interpretations via non-trivial mappings as long as they are meaning-preserving.

This mathematical practice of viewing an object of class A as one of class B – which we call **framing** – is an essential part of **mathematical literacy** – the skillset that identifies mathematical training. Indeed, framing is at the heart of understanding – seeing the network structure of math knowledge – and applying it. The essence of mathematical literacy is depicted in the figure on the right: trained mathematicians have access to a large, structured space of knowledge – we call it the **Mathematical Knowledge Space** (MKS) – that is induced via framing from a small core of represented knowledge. Unfortunately, mathematical software systems currently show only a very small degree of mathematical literacy. In this paper we present MMT theory graphs as a modular representation paradigm for mathematical knowledge, MathHub.info as an archive system that supports MMT-encoded knowledge, and ♭SEARCH as an example of a math-literate search engine.

2 Representing the Math Knowledge Space in MMT

We will now present the OMDoc/MMT format [5] which focuses on the network structure of mathematical knowledge and makes framing a central representational concern: MMT groups symbols facts into **theories** and represents (potential) framings as *theory morphisms*, which interlink theories into a **theory graph**. Theory morphisms are mappings between theories which map axioms of the source theory to theorems of the target theory. This ensures that all theorems of the source theory induce theorems of the target theory.

H. Hong and C. Yap (Eds.): ICMS 2014, LNCS 8592, pp. 26–30, 2014.
© Springer-Verlag Berlin Heidelberg 2014

To understand the setup, consider the theory graph in Fig. 1. The right side of the graph introduces the elementary algebraic hierarchy building up algebraic structures step by step up to rings; the left side contains a construction of the integers. In this graph, the nodes are *theories*, the solid edges are *imports* and the dashed edges are *views*.

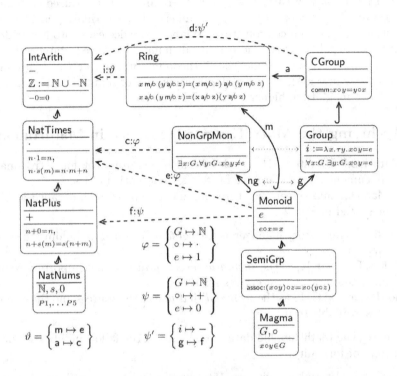

Fig. 1. A MMT Graph for Elementary Algebra

Importantly, every MMT symbol and statement is identified by a canonical, globally unique URI (called its **MMT URI**). Theories and views can be referenced relative to the URI of the containing theory graph, and symbol declarations by the URI of the containing theory, separated by ?. For instance, if U is the URI of the theory graph in Fig. 1 then the theory NatPlus and its symbol + have URIs U?NatPlus and, respectively, U?NatPlus?+.

Theory inheritance is realized by *structures*, which are named imports (and defined using theory morphisms). *Includes* are trivial structures which are unnamed and total. Symbol declarations induced by structures and views can be referenced relative to their name, separated by /. For instance, the addition operation from Ring can be referenced with U?Ring?a/○.

The definition of the theory Ring makes use of two MMT structures: m (for the multiplicative operations) and a (for the additive operations). To complete the ring we only need to add the two distributivity axioms in the inherited operators m/∘ and a/∘. Furthermore, a theory morphism, f, is used to represent that natural numbers with addition (NatPlus) form a monoid (Monoid).

It is a special feature of MMT that assignments can also map morphisms into the source theory to morphisms into the target theory. We use this to specify the morphism c modularly (in particular, this allows to re-use the proofs from e and c). Note that already in this small graph, there are a lot of induced statements. For instance, the associativity axiom is inherited seven times (via inclusions; twice into Ring) and induced four times (via views; twice each into NatArith and IntArith). All in all, we have more than an hundred induced statements from the axioms alone. If we assume just 5 theorems proven per theory (a rather conservative estimation), then we obtain a number of induced statements that is an order of magnitude higher.

3   Archiving the Math Knowledge Space in **MathHub.info**

The MathHub.info system [1] is a development environment for active mathematical documents and an archive for flexiformal mathematics.

The MathHub.info system has three main components (the detailed architecture is presented in Fig. 2):

- the GitLab repository manager as the versioned *data store* holding the source documents
- the MMT system [4] as the *semantic service provider* that imports the source documents and provides services for them
- and the Drupal CMS as the *frontend* that makes the sources and the semantic services available to users.

Currently, the MathHub.info data store contains the following libraries of various degrees of formality:

- the SMGloM termbase with ca. 1500 small sTeX files containing definitions of mathematical terminology and notation definitions.
- ca. 6500 files with sTeX-encoded teaching materials (slides, course notes, problems, and solutions) in Computer Science,
- the LATIN logic atlas with ca. 1000 meta-theories and logic morphisms,
- the Mizar Mathematical Library of ca. 1000 articles with ca. 50.000 theorems, definitions, and proofs, and
- a part of the HOL Light Library with 22 theories and over 2800 declarations.

We have MMT importers for all MathHub.info libraries and, therefore, MMT services become available for them. Current services including HTML presentation, querying, type checking and change management.

On the frontend side, Drupal natively supplies uniform theming, user management, discussion forums, etc. We extend it with dedicated modules to connect with the source documents in GitLab (for editing) as well as the imports in MMT (for MMT services, e.g. HTML presentation). Moreover, the JavaScript library JOBAD makes the documents active by interfacing with MMT services to enable complex in-browser interactions.

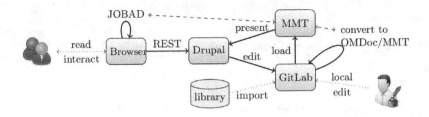

Fig. 2. The MathHub.info Architecture

4   Searching for Induced Statements

To search for induced statements, we use our MATHWEBSEARCH system [3], which indexes formula-URL pairs and provides a web interface querying the formula index via unification. This can be used for

Instance Search e.g. to find all instance of associativity we can issue the query $\forall x, y, z : \boxed{S}.(x\,\boxed{\text{op}}\,y)\,\boxed{\text{op}}\,z = x\,\boxed{\text{op}}\,(y\,\boxed{\text{op}}\,z)$, where the $\boxed{\text{-}}$ are query variables that can be instantiated in the query. In the library from Fig. 1 we would find the commutativity axiom SemiGrp/assoc, its directly inherited versions in Monoid, Ring and in particular the version u?IntArith?c/g/assoc.

Applicable Theorem Search where universal variables in the index can be instantiated as well; this was introduced for a non-modular formal libraries in [2]. Here we could search for $3 + 4 = \boxed{R}$ and find the induced statement u?IntArith?c/comm with the substitution $R \mapsto 4 + 3$, which allows the user to instantiate the query and obtain the equation $3 + 4 = 4 + 3$ together with the justification u?IntArith?c/comm that can directly be used in a proof.

Realizing ♭SEARCH on top of MATHWEBSEARCH has two parts:

- The search engine proper is very simple: instead of harvesting formulae directly from a formal digital library, we flatten the library first, and then harvest formulae. Flattening is the process of explicating all induced statements in an OMDoc/MMT theory graph, a central service of the MMT system, and defining feature of the ♭SEARCH system. Note that the MMT URIs of statements do not change during flattening, so they can directly be utilized as search hits in ♭SEARCH.
- For the presentation of search hits, we cannot simply rely on the MMT system to dereference the MMT URIs (which would indeed compute the induced statements), but we have to use the structure of the OMDoc/MMT theory graph to explain the path between the search hits and the represented knowledge. Luckily the MMT URIs contain enough information to compute this. Fig. 3 shows a ♭SEARCH result in action: ♭SEARCH found the induced statement of associativity of $+$ on $\mathbb{Z}$ and uses the combinations of morphisms m and i from Fig. 1 to justify the hit.

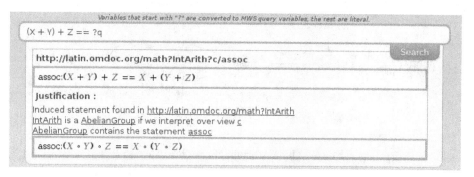

Fig. 3. The ♭SEARCH Web Interface for MathHub.info

5   Conclusion and Future Work

We have presented a unified framework for representing the inherent network structure of mathematical knowledge – OMDoc/MMT –, for enabling mathematically literate services – MathHub.info – and substantiated this with a model service – ♭SEARCH. The OMDoc/MMT language has been validated in large-scale representation and translation experiments, the systems are in a late prototype state; ♭SEARCH is fully integrated into MathHub.info and can be used on the MathHub.info content directly (though results depend on the modular structure). We expect to open MathHub.info for general use in this year, when the system has stabilized.

References

1. Iancu, M., Jucovschi, C., Kohlhase, M., Wiesing, T.: System description: Math-Hub.info. In: Watt, S., Davenport, J., Sexton, A., Sojka, P., Urban, J. (eds.) CICM 2014. LNCS, vol. 8543, pp. 431–434. Springer, Heidelberg (2014)
2. Iancu, M., Kohlhase, M., Rabe, F., Urban, J.: The Mizar mathematical library in OMDoc: Translation and applications. Journal of Automated Reasoning 50(2), 191–202 (2013)
3. MathWebSearch searching mathematics on the web
4. Rabe, F.: MMT – a module system for mathematical theories
5. Rabe, F., Kohlhase, M.: Information & Computation (230), 1–54 (2013)

Early Examples of Software in Mathematical Knowledge Management

Patrick Ion[1,2]

[1] Mathematical Reviews [AMS], USA
ion@ams.org
[2] University of Michigan, USA
pion@umich.edu
http://www-personal.umich.edu/~pion/

Abstract. There are new roles for software in mathematical knowledge management (MKM). Three simple initial examples of MKM roles will be considered here. The first is software applied to the Mathematical Subject Classification (MSC). The second example is MathML (Mathematics Markup Language), a standard from the W3C, now in its third edition, and hoping to become an ISO standard. The third example of software in the service of mathematical knowledge is the use of programs to analyze the nature of our subject as represented by its literature seen as a network. How these tools have already been deployed makes clear that mathematical knowledge management, even in its primitive present form, can aid further development of mathematics. These examples above are just starting points.

Keywords: mathematical knowledge management, mathematical subject classification, MSC, mathematical markup language, MathML, SKOS, network analysis.

1 Introduction

Modern computers are often employed to do the calculations needed for mathematics, whether numerical or symbolic. It can be claimed that's why they were invented. There are also roles for software in mathematical knowledge management (MKM). An obvious one developing new tools to help access the mathematical literature, which is the main way mathematical knowledge has been expressed and archived until recently. Three simple initial examples of MKM roles will be considered here. The first is software applied to the Mathematical Subject Classification (MSC) which is a traditional way of organizing literature holdings.

2 Mathematics Subject Classification

A simple hierarchy of areas and sub-areas of mathematics is the basic structure of most traditional classifications of mathematics (or any other field) and, in particular of the Mathematics Subject Classification (MSC) developed in the 1960s

H. Hong and C. Yap (Eds.): ICMS 2014, LNCS 8592, pp. 31–35, 2014.

and now jointly maintained by Mathematical Reviews (MR) and Zentralblatt für Mathematik (ZB). Both these secondary knowledge services are probably now better known for their online databases MathSciNet [MathSciNet:website] and zbMATH [zbMATH:website].

In its present form the MSC skeleton is a rather flat three-level tree: 1 root, 63 top-level areas, 528 secondary areas, and 5606 leaf nodes. This gives an idea of its size. Over the years the MSC has been revised several times since the fields of mathematics vary in their importance and new views of mathematics and new concepts arise. Also mathematics can be said to have lost areas once regarded as within its purview, such as many of its classical applications, and especially statistics (applied statistics), and most recently computer science or informatics.

The most recent revision of the MSC resulted in the current MSC2010 which has its own web site, [MSC2010:website], used in preparing the revision during 2006–2009. Until next revision, MSC2020 — which I tend think of as the "Full Hindsight Revision", the web site will be the public archive of information about the MSC and deliberations about its development, and a proposed location for some services based on the MSC.

There are a several aspects of the details of the MSC that make for complications in realizing the MSC in software; for instance:

- It was developed from input by the highly varied mathematical community interpreted by, on the order of, a hundred mathematical editors from MR and ZB over decades.
- There are numerous additional relationships going beyond inclusion between areas identified.
- There is tension between the simple form of the MSC tree and faceting expressed in what has been developed by a heterogeneous collection of contributing authors.
- There is reuse of terminology in the node descriptions: linguistic overloading with mathematics.
- There are multilingual problems that arise from the international desire for translations of the MSC (e.g. Chinese, Russian, and Italian so far).
- The master MSC versions had been encoded since 1984, for the dominant purposes of printing them, using TeX typesetting system (which is admittedly a full macro computing language, in principle).
- There is a desire to record the evolution of the MSC over the years.
- It is not trivial to maintain an evolving labeled tree.
- There are mathematical formulas present in the descriptions.

The revision to MSC2010 was taken as a good time to change the authoritative source from the MSC to a form more promising for the Semantic Web. But it was also a time in which suggestions from the mathematical community could be collected through a web site, stored in a MySQL database so that all would be dealt with, and the changes being adopted could all be exposed with MediaWiki to public view.

MSC2010 information is now held in a master file encoded using SKOS (Simple Knowledge Organization System) [SKOS:website] which is a World Wide Web

Consortium standard [W3C:website]. Conversion to SKOS was done using Perl and Python scripts. But this eventually involved some small customizations of the vocabulary, which was envisaged by SKOS. Our needs went beyond the paradigms seen from the use by the US Library of Congress in converting their LC Subject Headings to SKOS (they have over 250,000 of those).

Of course, it is probably MR and ZB, and the traditional publishing world who make most use of the MSC in the course of their daily workflows. However, it does play a role in traditional searches for mathematical material, and can be used to make phrase-based searching more nuanced. To encourage the creation of more tools using the MSC the authoritative information is offered publicly in many forms: [MSC2010:SKOS], RDF/XML, Turtle, N-Triples, TriX, and JSON, as well as on-screen display versions in English, Chinese, Russian, Italian and the [MSC2010:MediaWiki] and the [MSC2010:TiddlyWiki]. There the beginnings of SPARQL access with prototype examples. In addition there are some classic text, TeX and PDF forms as well as a KWIC index on the site.

3   Mathematics Markup Language — MathML

One special aspect of the MSC was the inclusion of some mathematical formulas. Mathematical expressions are nowadays properly encoded for the web in MathML, our second example. MathML [MathML3:spec] is also a standard from the W3C, now in its third edition, and hoping to become an ISO standard soon.

MathML is a markup language for mathematical expressions. It was originally developed, starting in 1998, as an XML [XML:spec] vocabulary to support mathematical publication in the modern information world and was apparently oriented toward XHTML for the rest of the documents where formulas were to be found. As such MathML specifies a class of labeled rooted planar trees, but the details are significant. The changing web standards landscape and the rise of HTML5, an extensive rework of HTML, have shown that MathML can work in the new context with surprisingly little adjustment. The purpose of MathML is to capture both presentational aspects and some of the semantics of mathematics, so MathML is in the tradition of the efforts at pasigraphy reported at the first ICM in 1897 [Schröder:1897][Peano:1894], and also harks back to Leibniz's *calculus ratiocinator*. In its newest version 3.0 (Second edition) of 10 April 2014 MathML plays very well with HTML5. In turn HTML5 recognizes the `math` element from the MathML namespace and specifies that the semantics of markup within that element shall be defined by the MathML specification (and other applicable specifications) (Section 4.7.14 in the HTML 5.0 specification, currently a Candidate Recommendation of the W3C as of this writing [HTML5:spec]).

The adoption of a new standard does take years—on average about a decade and a half according to Andrew Odlyzko. MathML is thus not doing so badly. The example of TeX which may be thought to have taken the mathematical community by storm can be said to have taken over a decade to really catch on from its first edition in SAIL in 1978, or from the first complete rewrite in Web, a literate programming extension of Pascal. TeX did have the advantage of being a complete package for document composition not a specification attempting to be part of a larger context of specifications and technology still under very active development. It was also the work of a single genius, Donald Knuth, and not of a changing committee.

However, MathML is being utilized, not just by MR and ZB, who have been relatively early adopters as they were for TEX but by publishers with XML work-flows, and by those who have to have the assurance of using a publicly adopted standard, not a proprietary one, which is coherent with the Web. While rendering support within browsers remains a problem area despite years of lobbying, it is improving—mathematics, and indeed technical documentation, is just a much less lucrative business than advertising and entertainment. In the meantime a technology originally intended as a stopgap, namely MathJax [MathJax:website], has brilliantly provided, through JavaScript, rendering support of a surprisingly high quality and reliability in almost all browsers. Perhaps its initial success can be ascribed to its being a single person's conception and work: Davide Cervone, then advised by his friend Robert Miner, who contributed a great deal to the development of MathML and regrettably died young.

4   Networks of Mathematics

The third example of software in the service of mathematical knowledge is the use of programs to analyze the nature of our subject as represented by its literature. Possibly the oldest consideration of this sort is the Erdős number, which comes from the co-authorship graph of mathematical papers. Later and more thorough analyses have been done of other networks representing mathematics' publications, whether in terms of co-authorship or co-citation, or in relation to subject areas (using the MSC), e.g. [Brunson:2013]. Further studies have begun [Dubois:2013] [Borjas:2012], leading to such modern topics as persistent homology [Bampasidou:2014] and A-theory [Babson:2006][Atkin:1974]. Machine processing of the corpus of mathematics as a natural language has also started. Analysis of the use of formulas depends on a standard notation such as MathML.

5   Conclusion

Finally let it be pointed out that the MSC and MathML are already extensively used in such places as [Wikipedia], [PlanetMath], and the [EuDML] as well as essentially in the publishing world, MathSciNet and zbMATH. The easing of access to recorded mathematical knowledge offered a possible World Heritage DML, and even use of MSC and MathML in [swMATH], make clear that mathematical knowledge management, even in its primitive present form, can aid further development of mathematics. The examples above are just starting points.

References

[Atkin:1974] Atkin, R.H.: An algebra for patterns on a complex, I. Internat. J. Man-Machine Stud. 6, 285–307 (1974); II. Internat. J. Man-Machine Stud. 8, 483–448 (1976)

[Babson:2006] Babson, E., Barcelo, H., de Longueville, M., Laubenbacher, R.: Homotopy theory of graphs. J. Algebr. Comb. 24, 31–44 (2006), doi:10.1007/s10801-006-9100-0

[Bampasidou:2014] Bampasidou, M., Gentimis, T.: Modeling collaborations with persistent homology. Preprint arXiv:1403.5346v1

[Borjas:2012] Borjas, G., Doran, K.: The Collapse of the Soviet Union and the Productivity of American Mathematicians. Quarterly Journal of Economics 127(3), 1143–1203 (2012)

[Brunson:2013] J.C. Brunson, S. Fassino, A. McInnes, M. Narayan, B. Richardson, C. Franck, P. Ion and R. Laubenbacher: Laubenbacher: Scientometrics (2013), doi: 10.1007/s11192-013-1209-z

[Dubois:2013] Dubois, P., Rochet, J.-C., Schlenker, J.-M.: Productivity and Mobility in Academic Research: Evidence from Mathematicians. IDEI Working Paper, n. 606 (October 2010), (revised March 2013)

[HTML5:spec] Berjon, R., Faulkner, S., Leithead, T., Navara, E.D., O'Connor, E., Pfeiffer, S., Hickson, I.: HTML 5, A vocabulary and associated APIs for HTML and XHTML. W3C Candidate Recommendation (February 04, 2014), http://www.w3.org/TR/html5/

[MathML3:spec] Carlisle, D., Ion, P., Miner, R.: Mathematical Markup Language (MathML) Version 3.0, 2nd edn. W3C Recommendation (April 10, 2014), http://www.w3.org/TR/2014/REC-MathML3-20140410/

[Peano:1894] Peano, G.: Formulaire de mathématiques. t. I-V. Turin, Bocca frères, Ch. Clausen, (1858-1932) 1894–1908

[Schröder:1897] Schröder, E.: Über Pasigraphie, ihren gegenwärtigen Stand und die pasigraphische Bewegung in Italien. 147–162 of Verhandlungen des ersten Internationalen Mathematiker-Kongresses in Zürich vom 9. bis 11. August (1897)

[XML:spec] Bray, T., Paoli, J., Sperberg-McQueen, C.M., Maler, E., Yergeau, F. (eds.): Extensible Markup Language (XML) 1.0 , 5th edn. W3C Recommendation, (November 26, 2008), http://www.w3.org/TR/xml/

[Web References] valid 15 May 2014 or thereafter

[EuDML] EuDML - The European Digital Mathematics Library, https://eudml.org/

[MathJax:website] MathJax Home Page, http://www.mathjax.org

[MathSciNet:website] Database, http://www.ams.org/mathscinet/

[MSC2010:website] Mathematics Subject Classification, http://msc2010.org

[MSC2010:MediaWiki] http://msc2010.org/mscwiki/

[MSC2010:SKOS] http://msc2010.org/resources/MSC/2010/info/

[MSC2010:TiddlyWiki] http://msc2010.org/MSC-2010-server.html

[PlanetMath] PlanetMath: math for the people, http://planetmath.org/

[SKOS:website] http://www.w3.org/2004/02/skos/

[swMATH] An information system for mathematical software, http://swmath.org/

[W3C:website] World Wide Web Consortium, http://www.w3.org/

[Wikipedia] Wikipedia, http://wikipedia.org/

[zbMATH:website] Database, http://www.ams.org/mathscinet/

Discourse-Level Parallel Markup and Meaning Adoption in Flexiformal Theory Graphs

Michael Kohlhase and Mihnea Iancu

Jacobs University Bremen, Germany
{m.kohlhase,m.iancu}@jacobs-university.de
http://kwarc.info

1 Introduction

Representation formats based on theory graphs have been successful in formalized mathematics as they provide valuable logic-compatible modularity and foster reuse. Theories – sets of symbols and axioms – serve as modules and theory morphisms – truth-preserving mappings from the (language of the) source theory to the target theory – formalize inheritance and applicability of theorems. The MMT [4] system re-developed the formal part of the OMDoc theory graph into a foundation-independent meta-system for formal mathematics and implemented it in the MMT API.

But full formalization of mathematics is tedious in the best of situations, often prohibitively costly. Moreover, it forces commitment to irrelevant foundational choices. As it is also unnecessary for many applications, we are currently extending the MMT format to allow content of flexible formality (which we call **flexiformal content**) in an effort to regain the original OMDoc coverage for OMDoc2.

In flexiformal representation formats, the basic inventory of theory graph notions from MMT is insufficient due to the presence of natural language and presentation markup in formulae: these are – in the absence of AI techniques – opaque to formal methods. As a consequence, we need other means of assigning meaning to them.

In this paper, we study two interrelated mechanisms for that:

- *extending parallel markup* (fine-grained cross-referencing between presentation and content markup) to the discourse level and
- *meaning adoption via postulated views.*

The first gives meaning to informal statements (definitions, theorems, proofs) by linking them with formal counterparts. The second, by *adopting* the semantics from another (more) formal theory.

2 Parallel Markup

The idea of parallel markup has been pioneered in the MathML format [1], which has two sub-languages: presentation MathML for the layout of mathematical

H. Hong and C. Yap (Eds.): ICMS 2014, LNCS 8592, pp. 36–40, 2014.

formulae and content MathML for the specification of the functional structure
(the "operator tree"). For parallel markup of a formula, MathML combines the
presentation and content trees in a single XML tree and marks up corresponding
subtrees by cross-references. Parallel markup supports two workflows:

- *formalization*: the annotation of presentation formulae with (multiple) for-
  malizations and
- *presentation*: the annotation of content formulae with (multiple) presentations.

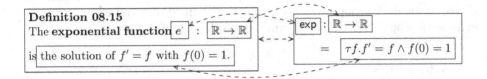

Fig. 1. Parallel Markup for a Definition

The duality between presentation and content captured by MathML is not
restricted to the formulae level. At the statement level, inside mathematical
documents, it manifests as the duality between narration and content. The narrative structure of mathematical text is often different from the structure of it's
formalization. Definitions may refer to concepts not yet introduced, proofs may
omit or reorder reasoning steps. Moreover, narrative mathematical texts often
opt for conciseness in detriment of rigor and rely to the intuitions of the reader
to infer the meaning and resolve ambiguities. The same motivates the choice of
notations at the formula level.

Therefore, co-representing both aspects of mathematics is a fundamental challenge not only at the formula level but also at the statement level. Consequently,
in OMDoc2 we extend parallel markup to all levels. A fundamental difference is
that, at the formula level, the human-oriented representation is given by notations, while at the statement level it is given by the mathematical vernacular. It
is also much more flexible in expression.

Here, we see the presentation of formulae to correspond to the narrative
(human-oriented) representation at the general level. Then parallel markup in
OMDoc2 amounts to co-representing the semantic and narrative aspects of mathematics in one format. From the OMDoc2 perspective, mathematical documents
where both aspects are adequately marked up are the ideal flexiformal structures.
Therefore, while supporting the same formalization and presentation workflows
as MathML, we use *flexiformalization* to refer to the process of adding semantic
or narrative information to a document.

We extend OMDoc2 with parallel markup by supplementing the narrative-
oriented notions of OMDoc (e.g. assertions, theorems, definitions, proofs, etc)
with the content-oriented notations of MMT (e.g. constants, assignments, structures). This establishes the appropriate containers needed for parallel markup.

We use the pre-existing OMDoc metadata infrastructure [3] to mark up the cross-references that make up the parallelism relation, since this is more flexible and can be harvested by standard metadata harvesters.

We see an instance of statement-level parallel markup in Figure 1. The narrative structure (a paragraph with a numbered classification) is given on the left; one phrase (the definiendum) is marked by emphasis, other parts – the notation and the definiens are left implicit. The OMDoc narrative elements provide markup for these – here indicated by boxes – and metadata for cross-references – here dashed arrows. On the right side we see the MMT formalization as a typed constant declaration with a definiens – τf. is the description operator that defines the constant exp as "that function f, such that …". The three parts of the constant declaration are given syntactically in MMT, and can – with the extensions proposed here – be integrated into the parallel markup.

3   Adoptions

Adoptions introduce a new kind of theory morphism that differs from the two primary MMT ones in its dynamics. Currently MMT has

structures which contribute to the specification of a target theory by importing the symbols and axioms of the source theory (modulo a mapping).

views which establish a meaning-preserving mapping between two pre-existing theories by satisfying proof obligations (proving the translated axioms of the source) in the target theory. Semantically they show that the target is a specialization or implementation of the source.

For meaning adoption we have a situation that is somewhere in-between. For instance, we have the situation of a recap in the introduction of a paper. This briefly introduces the concepts and properties necessary to make the paper self-contained without giving a full development. Instead, their meaning is established by adopting the referenced development – which we assume to be in the form of a theory (graph) for this discussion. The reader can remember or accept the content of the recap or read up on the referenced source. In a theory graph-based setting we want to understand the relation between a recap and its source as a constitutive relation; we call it an **adoption**. An adoption behaves dynamically like a structure in that it adds to the specification of its target – like a structure it does not have/need proof obligations, but logically acts like a view from the target to the source, in that it makes the recap a specialization of the full development.

Consider the situation where the symbols of a theory S are imported in T via a partial inclusion i, and their meaning is specified via a *postulated* view $v : T \to S$. Then S and T form an *adoption* and there is an *adoption morphism* i/v from S to T.

4   Scenario: Grounding Course Materials by Adoption

We will now look at a typical situation where parallel markup and adoption happen and work out the details of postulated views and the influence on property and symbol inheritance.

Take for instance a course which introduces (naive) set theory informally, but grounds itself in a formal, modular definition of axiomatic set theory. Then we have the situation in Figure 2. On the right hand side, we have a careful introduction in the form of a modular theory graph starting at a theory ZFset that introduces membership relation and the axioms of existence, extensionality, and separation and defines the set constructor $\{\cdot|\cdot\}$ from these axioms. On the left we have a theory SET that adopts the symbols $\in$ and $\{\cdot|\cdot\}$ via a partial inclusion a_1 from ZFset to SET but "defines" them by alluding to the intuitions of the students. Note that such a partial inclusion always gives rise to a view in the opposite direction, here the view v_1 from SET to ZFset. We cannot discharge the proof obligations in v_1, since the definition of the set constructor $\{\cdot|\cdot\}$ is opaque – i.e. given as natural language, which is not subject to formal methods. As introduced above we think of v_1 as a "definitional view" that gives meaning to the opaque parts in SET: the proof obligations have to be met in order for the diagram to commute (which is an invariant we want to maintain). Then Set and ZFset form an adoption morphism.

The setting also accommodates parallel markup nicely. The "definition" of $\{\cdot|\cdot\}$, which is similar to the left side of Figure 2 fits into theory SET and the formal counterpart into ZFset. Note that the "top-level" parallel relation between the narrative definition and the constant declaration is directly induced by the adoption morphism a_1/v_1.

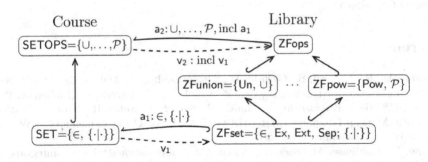

Fig. 2. A course grounded in a modular Library

The informal course materials continue by introducing the set operations ranging from set union to the power set in one go in the theory SETOPS. On the library side, we introduce the set theory axioms one by one, derive the respective operators from them, and at the end collect all the material in the theory ZFops. At this point, we can justify the theory SETOPS via a partial inclusion of the symbols $\cup$, ..., $\mathcal{P}$ from ZFops, which gives rise to another definitional view v_2 to ZFops. As MMT allows inclusions between morphisms we can reuse the assignments from v_1 when defining v_2, as indicated by v_2 : incl v_1.

We observe that the two theory graphs are self-contained: the course materials can be understood without knowing about the library; in particular, the membership relation used in the definition of the union operator in SETOPS is from theory SET. This self-containedness is important for intra-course didactics, but it has the problem that the courses become insular; how are students going to communicate with mathematicians who have learned their maths from other courses? Here is where the views v_i come in. Say the other mathematicians have course theories $\overline{SET}$ and $\overline{SETOPS}$ with views $\overline{v_1}$ and $\overline{v_2}$ into the same library, then the views v_1 and $\overline{v_1}$ induce a partial isomorphisms between SET and $\overline{SET}$ in the sense of [2] (and correspondingly between SETOPS and $\overline{SETOPS}$) that justify communication.

5 Conclusion

Most mathematical knowledge is laid down as mathematical vernacular in the form of rigorous, but informal documents. Theory-graph-based approaches, have been very successful in providing modularity and avoiding redundancy in formal methods, but cannot directly be applied to informal content since that is opaque to formal methods. In particular informal content does not offer the substructures to anchor the meaning-giving relations to. In this paper, we present two interrelated measures to alleviate this problem: 1. statement and formulae-level parallel markup 2. meaning adoption via postulated views, and show their contribution on examples. We are already testing these two in a semantic, multilingual glossary of mathematics and course materials for an introductory course on computer science.

References

1. Ausbrooks, R., Buswell, S., Carlisle, D., Chavchanidze, G., Dalmas, S., Devitt, S., Diaz, A., Dooley, S., Hunter, R., Ion, P., Kohlhase, M., Lazrek, A., Libbrecht, P., Miller, B., Miner, R., Sargent, M., Smith, B., Soiffer, N., Sutor, R., Watt, S.: Mathematical Markup Language (MathML) version 3.0. W3C Recommendation, World Wide Web Consortium, W3C (2010)
2. Rabe, F., Kohlhase, M., Sacerdoti Coen, C.: A foundational view on integration problems. In: Davenport, J.H., Farmer, W.M., Urban, J., Rabe, F. (eds.) Calculemus/MKM 2011. LNCS, vol. 6824, pp. 107–122. Springer, Heidelberg (2011)
3. Lange, C., Kohlhase, M.: A mathematical approach to ontology authoring and documentation. In: Carette, J., Dixon, L., Coen, C.S., Watt, S.M. (eds.) Calculemus/MKM 2009. LNCS (LNAI), vol. 5625, pp. 389–404. Springer, Heidelberg (2009)
4. Rabe, F., Kohlhase, M.: Information & Computation (230), 1–54 (2013)

Complexity Analysis of the Bivariate Buchberger Algorithm in *Theorema*[*]

Alexander Maletzky[1] and Bruno Buchberger[2]

[1] Doctoral College "Computational Mathematics" and RISC,
Johannes Kepler University, Linz, Austria
alexander.maletzky@dk-compmath.jku.at
https://www.dk-compmath.jku.at/people/alexander-maletzky
[2] RISC, Johannes Kepler University, Linz, Austria
bruno.buchberger@risc.jku.at
http://www.risc.jku.at/home/buchberg

Abstract. In this talk we present the formalization and formal verification of the complexity analysis of Buchberger's algorithm in the bivariate case in the computer system *Theorema* as a case study for using the system in mathematical theory exploration.

We describe how Buchberger's original complexity proof for Groebner bases can be carried out within the *Theorema* system. As in the original proof, the whole setting is transferred from rings of bivariate polynomials over fields to the discrete space of pairs of natural numbers by mapping each polynomial to the exponent vector of its leading monomial. The complexity analysis is then carried out in the discrete space, mostly by means of combinatorial methods that require many tedious case distinctions, making this proof a natural candidate for automated theorem proving. However, following our *Theorema* philosophy, we do not expect general theorem provers (like resolution provers) to carry out this task in a natural and efficient way. Rather, we designed and implemented a special prover for such proofs. We show how the *Theorema* philosophy of working in parallel both on the meta level (designing and implementing special provers) and on the object level (design of the notions and theorems) of a theory can lead to a new quality and style of mathematical research.

Keywords: Groebner basis, Buchberger algorithm, mathematical theory exploration, complexity analysis, Theorema.

1 Introduction

The purpose of this talk is to present a major case study in how mathematical theory exploration can be carried out in the *Theorema* system: *Theorema* [16,10] is a system which was initiated by Bruno Buchberger and developed in

---

[*] This research was funded by the Austrian Science Fund (FWF): grant no. W1214-N15, project DK1.

H. Hong and C. Yap (Eds.): ICMS 2014, LNCS 8592, pp. 41–48, 2014.

his *Theorema* group at RISC since the the mid-nineties. It uses the computer algebra system *Mathematica* [13] as software frame. Its user interface is currently re-designed and -implemented (*Theorema* Version 2.0). The case study that is presented here explores the complexity of Buchberger's algorithm [1,2,7] for computing Groebner bases of polynomial ideals over fields in the bivariate case.

It is important to note already at this point that the underlying theory (i.e. the complexity analysis) is not "new" in the sense that it was developed only recently with the help of the *Theorema* system, but in fact it was already developed more than 30 years ago by Buchberger in [3,4,5]. This, however, allows one to observe one of the essential strategies of *Theorema*: It is easily possible to take an existing theory produced step by step in ordinary mathematical notation, and convert it into a completely formal version in almost exactly the same (natural mathematical) notation in *Theorema* with hardly any effort. Significant portions of the proofs can then be generated automatically by using existing *Theorema* provers and designing a few others (which might be used later again in similar but different theories).

The focus of this talk is not on *Theorema* itself – how it is implemented, how it works, etc. – but mainly on how it can be *used* in mathematical theory exploration, i.e. in the everyday-life of "working mathematicians".

2 Theoretic Background

The case study in this paper is concerned with the analysis of the complexity of Buchberger's algorithm [1,2,7] in the bivariate case. Buchberger's algorithm computes so-called *Groebner bases* of polynomial sets over fields. A number of fundamental problems for polynomial ideals can be solved once a Groebner basis for the ideal is known.

Hence, deriving bounds on the complexity of this algorithm has been of interest since the introduction of Groebner bases: Even the very first presentation of the algorithm in Buchberger's 1965 PhD thesis already contained a rough analysis. Later [3,4,5] Buchberger concentrated especially on the bivariate case and managed to derive tight bounds on the degrees of the polynomials in the Groebner basis both in the case of using graded admissible term orderings and pure lexical orderings. These degree-bounds are expressed in terms of the degrees of the polynomials in the input basis.

For the sake of completeness it has to be mentioned that it is well-known already for a long time that Buchberger's algorithm has double exponential time- and space complexity in the number of indeterminates [14], and that the degrees of the polynomials in the Groebner basis resulting from an application of Buchberger's algorithm are polynomial in the maximum degree of the polynomials in the input set, if the number of indeterminates is fixed [12,15].

The complexity analysis in [3] and [5] proceeds in the following way: First of all, the whole problem setting is transferred from $K[x, y]$, the ring of bivariate polynomials over the field K, to the discrete space $\mathbb{N}^2$ by mapping each non-zero polynomial to the exponent vector of its leading monomial w.r.t. some graded admissible ordering. The rest of the elaboration is combinatorial, mainly distinguishing between all possible cases that might occur during the algorithm. None of these cases requires deep mathematical thinking so that the exploration lends itself to automated theorem proving (It should be noted, however, that the set-up and flow of the proof - which is basically the invention of a suitable degree invariant in the main loop of the algorithm - is non-trivial).

We followed the ideas of [3,5] in our formalization, with some slight deviations:

1. The domain of the exponents is not restricted to $\mathbb{N}$, but to so-called *totally-ordered Abelian monoids D*,
2. As much as possible, the number of indeterminates n is not restricted to two since some results also hold for general n
3. In the bivariate case, a different partition of the "exponent space" D^2 is chosen which is different from the one in [3]; In fact, it is not a partition, but only a cover.

The first two deviations were made for the purpose of making everything as general as possible. A totally-ordered Abelian monoid is a commutative semigroup with unit, where in addition

- The monoid operation possesses the so-called *cancellation property*, meaning that $x + z = y + z$ is always equivalent to $x = y$.
- A total order relation $\leq$ is defined, which also has the cancellation property in the sense that $x + z \leq y + z$ is always equivalent to $x \leq y$.

It is quite easy to see that $\mathbb{N}$ is such a totally-ordered Abelian monoid, as are $\mathbb{Z}$, $\mathbb{Q}$, $\mathbb{R}$ and even $\mathbb{C}$ with a lexicographic ordering.

The third deviation is a simplification: It turns out that the proof of the main theorem in [3] can be simplified a bit, and that a big part of the proof of the main theorem in [5] becomes superfluous, if our new partition (or cover) is used[1].

3 Formalizing the Theory in *Theorema*

Formalizing a mathematical theory in *Theorema* does not require any knowledge that goes beyond the mathematical knowledge and mathematical thinking culture of a "working mathematician". In particular, no specific programming language needs to be known and no special syntax has to be learned. Rather, *Theorema* syntax is just a "cultivated" version of ordinary ("two-dimensional") mathematical syntax. However, it is "formal" in the sense that it can be processed by algorithmic inference techniques.

As an example, consider the aforementioned criterion that detects unnecessary steps in Buchberger's algorithm (the *chain criterion*):

---

[1] For the readers familiar with the proof strategy in [3,5]: There, the focus is very much on *contours* of sets of points in $\mathbb{N}^2$, and quite some effort is needed to reduce the general case to the case of contours. This is not needed at all.

Definition 1. *For all x, y and A:*

$$\mathrm{CHAINCRIT}(x, y, A) :\Leftrightarrow \neg \underset{1 \le j \le |A|}{\exists} \left(\bigwedge \left\{ \begin{array}{l} A_j | z \\ \deg(\mathrm{lcm}(x, A_j)) < \deg(z) \\ \deg(\mathrm{lcm}(A_j, y)) < \deg(z) \end{array} \right. \right)$$

where z denotes $\mathrm{lcm}(x, y)$.

This textbook-style definition already comes very close to the *Theorema* syntax: There, one also has quantifiers, abbreviations, subscripts, and many other syntactic constructs available. Hence, for reading and writing *Theorema* definitions and theorems, one does *not* have to get acquainted to a new, unnatural notation first.

3.1 Details of the Formalization

Formalizing a theory in *Theorema* is not straightforward in the sense that many decisions have to be made regarding *how* the theory should be formalized and which goals one wants to achieve. The need for making decisions is not a deficiency of the *Theorema* formal approach to mathematics but, rather, a system like *Theorema* *should* allow to set up a theory in many different "views" and styles according to the tastes and exploration goals of the person working with the system.

One decision we had to make, for instance, was about using *functors* [6,17,8] for building up towers of domains in a structured way; In particular, as already indicated above, we did (and do) not want to restrict ourselves to the case of pairs of exponents over $\mathbb{N}$. Hence, the first idea at hand is to use a functor that maps domains D and natural numbers n to the domain of exponent vectors of length n over domain D and defines all the necessary operations on them (like CHAINCRIT). However, later it turned out that in each part of the theory always *one* particular domain D and dimension n are fixed anyway, meaning that even in proofs one does not have to fall back to other choices of D or n. Thus, a functor is not needed, and so we dropped it and introduced "global constants" for D and n instead.

3.2 Computations

If one wants to actually carry out computations involving notions such as CHAINCRIT in *Theorema*, there is no need to do anything further than entering the definition into the system, in a form which is very close to usual textbook notation (c. f. definition 1). As soon as this is done, one can immediately compute with the notion, which is because the equational part of higher-order predicate logic (the rewrite mechanism that successively replaces equals by equals (in a directed way) until no more replacements are possible) can be considered as the interpreter of a universal programming language. In other words, part of the (*Theorema* version of) predicate logic *is* a programming language.

For instance, if one wants to check whether the chain criterion holds for exponent vectors $\langle 10, 0 \rangle$ and $\langle 0, 12 \rangle$ and tuple $\langle \langle 10, 0 \rangle, \langle 11, 10 \rangle, \langle 0, 12 \rangle \rangle$ of exponent vectors, one basically just has to type in

$$\texttt{chainCrit}[\langle 10, 0 \rangle, \langle 0, 12 \rangle, \langle \langle 10, 0 \rangle, \langle 11, 10 \rangle, \langle 0, 12 \rangle \rangle]$$

and hit shift+enter - Voilà! The result will be `True`, meaning that the chain criterion indeed holds.

Note that *Theorema* provides built-in support for tuples: Tuples are simply represented as sequences of expressions enclosed in angle brackets. Either the individual elements are given explicitly, or a quantifier may be used to construct the elements of the tuple. For the sake of convenience we decided to represent exponent vectors as tuples, too.

4   Designing a New Prover in *Theorema*

One of the main ideas behind *Theorema* is the the philosophy that automated reasoning can practically only be carried out if an entire hierarchy of special provers is at the disposal of the user, each designed for proving theorems in a certain theory. This is in contrast to having only one single proving technique (e. g. resolution) available, which, theoretically, would be sufficient but does not generate short and structured proofs. The key strategy for this approach is "proving *by* intermediate principles", introduced in [9].

Therefore, we also decided to create a new prover for our own purpose, which should be capable of proving theorems in the present theory of complexity analysis. This prover is, in particular, able to handle tuples, total order relations, associative-commutative operations, and functions related to minimum and maximum in a way which is both correct and concise.

Creating a new prover in *Theorema* is a bit more involved than formalizing a theory: Since it operates on objects of the *object* level (formulas), the prover itself is an object of the *meta* level. We chose *Mathematica* as the meta-language for *Theorema*, which means that new provers have to be implemented directly in *Mathematica*. Thus, users who want to add new provers to *Theorema*, must know how to program in *Mathematica*. If one knows (basic) *Mathematica*, writing a prover is again easy: It only consists of two parts: The first part is designing a collection of inference rules, each transforming one proof situation (given by a list of formulas constituting the current *knowledge* and a single formula constituting the current proof *goal*) into new proof situations in a style which very much resembles sequent calculus proving. The second part consists of finding a good strategy that guides the proof search, i. e. decides in which order the rules are tried, whether all applicable rules or only the first one are applied, etc. The two parts are independent of each other in the sense that one can combine the inference rules and strategies in any way; In particular, when creating a new prover it is possible to only specify the inference rules but use an already existing strategy, or the other way round.

Here, a subtle problem of automated proving has to be mentioned: Before one can really *trust* the output produced by an automated prover, one first has to *verify* the prover itself, i.e. prove it correct. Otherwise, there will always be a logical gap in the computer-supported treatment of formalized mathematics. Now, since provers in *Theorema* have to be implemented in *Mathematica* (on the meta level) and can thus not be the subject of computations of whatever kind on the object level in *Theorema*, this implies that *Theorema* cannot be used for verifying its own provers. Although some research has already been and is still being conducted to overcome this issue in *Theorema* (c.f. [11]), at the present stage one still has to live with it. And, of course, mathematical proving without a proving system has to "trust" that the human prover is correct in each and any individual proof step.

5   Verifying the Theory in *Theorema*

As soon as both the formalization has been done and a suitable prover has either been implemented or chosen from a list of already existing ones, one can immediately start proving. For this, one just has to set up the proof task, i.e. select the formula one wants to prove (the proof goal), the formulas one wants to use (the knowledge base), and some other options depending on the prover and proof strategy selected (e.g. which of the inference rules one really wants to make use of, parameters concerning search time and -depth, or the degree of user interaction). Finally, one clicks a button and waits until a result is obtained - unless some of the inference rules require user interaction, such as finding witnesses for existentially quantified proof goals or selecting the proof branch that looks most promising.

Indeed, the prover we created in the frame of our complexity analysis relies on such user interaction to some extent: Apart from the usual tasks that might come to one's mind and that have been pointed out above, like instantiating quantifiers in a clever way, we also allow the user to

- select an implication in the knowledge base, first prove its premise in a sub-proof, and then continue with the original goal, having the consequence of the implication among the assumptions, and to
- "exchange" the current goal and a formula in the knowledge base by putting their negations "on the other side", i.e. from goal to knowledge and from knowledge to goal, respectively.

Both of these strategies proved to be quite convenient on several occasions.

Another feature of our prover is that it heavily makes use of (conditional) rewriting of terms and formulas by rewrite rules originating from formulas in the knowledge base. For instance, if

$$\underset{x,y,A}{\forall} \text{CHAINCRIT}(x, y, A \curlywedge x) \Leftrightarrow \text{CHAINCRIT}(x, y, A)$$

is known[2], then every (sub-)formula of the form

$$\text{CHAIN}\textsc{Crit}(x, y, A \curvearrowleft x)$$

in the proof situation (with x, y and A arbitrary terms) can actually be replaced by

$$\text{CHAIN}\textsc{Crit}(x, y, A)$$

All this is not something we invented only for our own purpose, but rather it is once again a fundamental concept in the philosophy of *Theorema*.

6 Conclusion

We want to demonstrate that the *Theorema* system, whose user-interface is currently redesigned, provides a good approach for doing computer-supported mathematics in a formal and formally verified way. Not only does computer-supported theory exploration have all the advantages it is expected to have (automatically finding proofs, performing computations, having well-structured theories), but it also helps in improving the mathematical contents: We could easily generalize the domain of exponents from $\mathbb{N}$ to totally-ordered Abelian monoids, and we also realized that big parts of Buchberger's original elaboration are in fact superfluous, meaning that some of the proofs could drastically be shortened. From a methodological point of view, this is interesting because - as expected in our philosophy - formal treatment of mathematics will often lead also to improvements and purification of the mathematical ideas themselves.

References

1. Buchberger, B.: Ein Algorithmus zum Auffinden der Basiselemente des Restklassenringes nach einem nulldimensionalen Polynomideal (An Algorithm for Finding the Basis Elements in the Residue Class Ring Modulo a Zero Dimensional Polynomial Ideal). PhD thesis, Mathematical Institute, University of Innsbruck, Austria 1965: English translation in J. of Symbolic Computation, Special Issue on Logic, Mathematics, and Computer Science: Interactions 41(3-4), 475–511 (2006)
2. Buchberger, B.: A Criterion for Detecting Unnecessary Reductions in the Construction of Groebner Bases. In: Ng, K.W. (ed.) EUROSAM 1979 and ISSAC 1979. LNCS, vol. 72, pp. 3–21. Springer, Heidelberg (1979)
3. Buchberger, B., Winkler, F.: Miscellaneous Results on the Construction of Groebner-Bases for Polynomial Ideals. Technical Report 137, Johannes Kepler University Linz, Technisch-Naturwissenschaftliche Fakultaet, Insitut fuer Mathematik (June 1979)
4. Buchberger, B.: A Note on the Complexity of Constructing Groebner-Bases. In: van Hulzen, J.A. (ed.) ISSAC 1983 and EUROCAL 1983. LNCS, vol. 162, pp. 137–145. Springer, Heidelberg (1983)

---

[2] "$A \curvearrowleft x$" denotes appending object x to tuple A.

5. Buchberger, B.: Miscellaneous Results on Groebner-Bases for Polynomial Ideals II. Technical Report 83-1, Department of Computer and Information Sciences, University of Delaware (1983)
6. Buchberger, B.: Mathematica as a Rewrite Language. In: Ida, T., Ohori, A., Takeichi, M. (eds.) Functional and Logic Programming (Proceedings of the 2nd Fuji International Workshop on Functional and Logic Programming, Shonan Village Center, November 1-4), pp. 1–4. World Scientific, Singapore (1996)
7. Buchberger, B.: Introduction to Groebner Bases. London Mathematical Society Lecture Notes Series, vol. 251. Cambridge University Press (April 1998)
8. Buchberger, B.: Groebner Rings in Theorema: A Case Study in Functors and Categories. Technical Report 2003-49, Johannes Kepler University Linz, Spezialforschungsbereich F013 (November 2003)
9. Buchberger, B.: Proving by First and Intermediate Principles, November 1-2, Invited Talk at Workshop on Types for "Mathematics/Libraries of Formal Mathematics", University of Nijmegen, The Netherlands (2004)
10. Buchberger, B., Crăciun, A., Jebelean, T., Kovács, L., Kutsia, T., Nakagawa, K., Piroi, F., Popov, N., Robu, J., Rosenkranz, M., Windsteiger, W.: Theorema: Towards Computer-Aided Mathematical Theory Exploration. Journal of Applied Logic 4(4), 470–504 (2006)
11. Giese, M., Buchberger, B.: Towards Practical Reflection for Formal Mathematics. RISC Report Series 07-05, Research Institute for Symbolic Computation (RISC), University of Linz, Schloss Hagenberg, 4232 Hagenberg, Austria (2007)
12. Giusti, M.: Some effectivity problems in polynomial ideal theory. In: Fitch, J. (ed.) EUROSAM 1984 and ISSAC 1984. LNCS, vol. 174, pp. 159–171. Springer, Heidelberg (1984)
13. Wolfram Mathematica, http://www.wolfram.com/mathematica/
14. Mayr, E.W., Meyer, A.R.: The complexity of the word problems for commutative semigroups and polynomial ideals. Advances in Mathematics 46(3), 305–329 (1982)
15. Michael Moeller, H., Mora, F.: Upper and lower bounds for the degree of Groebner bases. In: Fitch, J. (ed.) EUROSAM 1984 and ISSAC 1984. LNCS, vol. 174, pp. 172–183. Springer, Heidelberg (1984)
16. The Theorema system, http://www.risc.jku.at/research/theorema/description/
17. Windsteiger, W.: Building Up Hierarchical Mathematical Domains Using Functors in THEOREMA. In: Armando, A., Jebelean, T. (eds.) Electronic Notes in Theoretical Computer Science. ENTCS, vol. 23, pp. 401–419. Elsevier (1999)

Theorema 2.0: A System for Mathematical Theory Exploration

Wolfgang Windsteiger

RISC / JKU Linz, Austria
Wolfgang.Windsteiger@risc.jku.at
http://www.risc.jku.at/home/wwindste

Abstract. Theorema 2.0 stands for a re-design including a complete re-implementation of the Theorema system, which was originally designed, developed, and implemented by Bruno Buchberger and his Theorema group at RISC. In this talk, we want to present the current status of the new implementation, in particular the new user interface of the system.

Keywords: Theorema, mathematical assistant system, automated theorem proving, theory exploration, user interfaces, GPL.

1 Introduction

Theorema 2.0 is—like its predecessor versions—based on Mathematica, which means that it is implemented in the Mathematica programming language and that it uses the Mathematica notebook front end as its user interface. Unlike the command-oriented interaction pattern typically propagated in Mathematica applications, Theorema 2.0 is heavily based on the graphical user interface capabilities supported in recent versions of Mathematica. As a result, the user needs the keyboard only for typing the mathematics (definitions, theorems, explanatory text) into the system, all actions to be performed are guided by the graphical user interface. This approach fosters the convergence of *writing formal mathematics* towards *writing normal mathematics*, because the overhead when *writing a Theorema document* compared to *writing a standard mathematical document* shrinks to almost zero. Moreover, the learning curve for using a mathematical assistant system is considerably flattened and the system will be more attractive, in particular for beginners.

A first version of Theorema 2.0 has already been presented in [Win12], where an emphasis was put on the new graphical user interface. In this presentation, we report on improvements and further extensions, but also on some new developments in the system that are not connected directly to the user interface.

Theorema 2.0 runs on all platforms, on which Mathematica is available. Mathematica is needed to run the system, but the Theorema system itself is open source licensed under GPL and is available at GitHub.

H. Hong and C. Yap (Eds.): ICMS 2014, LNCS 8592, pp. 49–52, 2014.

2   How to Use the Theorema System

When using (mathematical) software it is important for the user to exactly understand, for which intended purpose the software has been developed. Of course, there are examples of "legitimate fruitful abuse" (e.g. using a spreadsheet program to illustrate iterative algorithms when teaching mathematics) but in general the user is better off when she uses the software in line with the developers' intentions.

Much of mathematical software falls into the category of *algorithm libraries*, i.e. collections of algorithms for certain more or less well described application areas, like linear algebra, polynomial equations, geometry, differential equations, first order theorem proving, and the like. For each of the algorithms there is an input-output-specification and the systems differ in the range of problems that can be solved, the computational efficiency, or the input/output format. For the Theorema system, the situation is a bit more complex since Theorema tries to be a *mathematical assistant system* that supports the mathematician during all her mathematical activities, from first scratch work on some topic, through giving definitions of mathematical notions, formulating conjectures, proving theorems, formulating algorithms, executing algorithms on concrete input data, organizing the knowledge in order to reuse it in the future, composing lecture notes until finally writing a proper mathematical publication.

Although computer-support for automated or interactive theorem proving is in our main focus, the acceptance of a mathematical assistant system does not depend solely on the power of the prover. The huge variety of different working styles and habits is a major challenge for the user interface. Fig. 1 shows the new interface of Theorema 2.0, which consists of one or more *Theorema notebook documents* (left) and the *Theorema commander* (right).

In addition to standard Mathematica notebook features, a Theorema notebook supports *Theorema environments*, which contain blocks of formal mathematics such as definitions or theorems. The name of the environment ("Facts" in Fig. 1) can be freely chosen, it serves only structuring purposes and carries no semantics. Inside an environment, formal mathematics is written in cells of a particular style defined by the Theorema system, in fact by the *Theorema stylesheet* that is required to be used for Theorema notebooks. Formal mathematics is written in a very rich version of the language of predicate logic in common two-dimensional notation and must be executed (like Mathematica input in a standard Mathematica notebook) in order to become known to Theorema within the current session. It is important to note that the stylesheet does not only define the optical appearance of formal mathematics cells but also their functionality. We use the possibility to define actions to be executed before and after the cell content is processed and only so it is guaranteed that Theorema input is processed correctly. An important consequence of this setting is that Theorema does not interfere Mathematica in any way, the whole functionality of Mathematica can be accessed in standard Mathematica input cells. Every formal math cell carries a label through which the formula can be referenced (e.g. in a proof). In order to accommodate common practice, formal mathematics can be intermixed with plain informal text as shown in Fig. 1 also.

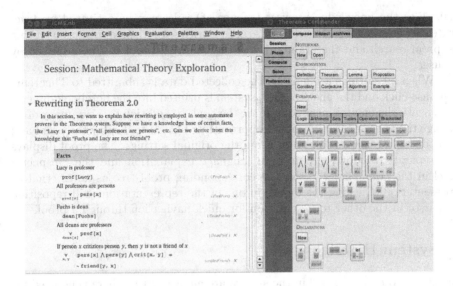

Fig. 1. Theorema 2.0 user interface

The Theorema commander is responsible to guide the user through all sorts of *activities* (to be selected in the left column in the commander) to be performed on the formal mathematics written in the notebooks. For every activity chosen, the commander opens a wizard that guides the user through the concrete *actions* in that activity. The "Session"-activity, for instance, has the actions "compose", "inspect", and "archives" helping to compose notebook content, inspect the formulas available in the current session, and setting up of knowledge archives, respectively.

Example 1 (How to prove a theorem). In order to prove a theorem, the system needs the theorem (= the proof goal), the knowledge base available, and the proving method to be applied (since Theorema is a multi-method system). The prove-activity with its actions "goal", "knowledge", "built-in", "prover", "submit", and "inspect" guides the user through this process.

1. The goal is specified by simply selecting the cell containing the theorem in the notebook.

2. The knowledge base consists of a) user-defined knowledge contained in some environments possibly spread over several notebooks and b) Theorema built-in knowledge that can be added to the knowledge base, e.g. on built-in arithmetic operations. For composing user-defined knowledge, Theorema 2.0 provides the *knowledge browser*, which contains for each Theorema notebook available in the current session a structured outline, in which all (groups of) formulas that should go into the knowledge base can be checked by mouse-click. A similar mechanism is used to select built-ins.

3. A prover in Theorema 2.0 is a collection of inference rules. Rules are grouped into categories (e.g. quantifier rules) that are displayed in the *rule browser*. In

analogy to the knowledge browser (groups of) rules can be activated or deactivated by mouse-click. In addition, rule priorities for their application during the proof search and the granularity of the resulting proof can be adjusted by the user.

4. When all settings are finished the collected data is submitted to Theorema by mouse-click and the answer of the system is printed into the notebook directly underneath the environment containing the goal. In addition to a summary of all settings the answer contains most importantly a button to display the proof and a button to regenerate the proof using the original settings. The proof displays in a separate window with natural language explanation and the "inspect"-panel in the commander shows the corresponding proof tree as an alternative representation. Clicking the mouse in one of the representations will reposition the cursor in the other representation for quick navigation through a proof.

3   System Highlights

All actions to be performed in the system are mouse-driven, there is no need for the user to call complicated functions with lots of parameters in order to initiate some action. The interaction pattern should be more like using a web shop in the internet.

The proof methods are highly configurable through the Theorema commander. It should be easy for the user to adjust the behavior of the system as appropriate for a concrete problem.

Computation is an integral component in the provers. Every formula is silently simplified by computation as soon as it enters a proof. The computational knowledge applied is subject to user configuration, no user is forced to use Theorema built-in knowledge in a proof.

Formula input is supported both through palettes with mouse-click and through keyboard shortcuts. Structural input of formulas following their tree structure is supported, "invisible parentheses" guarantee correct grouping without any need for operator precedences.

Theorema notebooks can be setup to contain *theories*, which can be exported and stored in a format to be later imported and re-used in other notebooks or theories. The namespaces are separated such that naming collisions between theories are avoided.

Referance

[Win12] Windsteiger, W.: Theorema 2.0: A Graphical User Interface for a Mathematical Assistant System. In: Kaliszyk, C., Lth, C. (eds.) Proceedings of the10th International Workshop on User Interfaces for Theorem, Provers, Bremen, Germany, July 11. Electronic Proceedings in Theoretical Computer Science, vol. 118, pp. 72–82. Open Publishing Association (2012)

New Approaches in Black Box Group Theory

Alexandre Borovik[1] and Şükrü Yalçınkaya[2]

[1] University of Manchester, United Kingdom
alexandre@borovik.net
[2] İstanbul University, Turkey
sukru.yalcinkaya@istanbul.edu.tr

Abstract. We introduce a new approach in black box group theory which deals with black box group problems in the category of black boxes and their morphisms. This enables us to enrich black box groups by actions of outer automorphisms such as Frobenius maps or graph automorphisms of simple groups of Lie type. As an application of this new technique, we present a number of new results, including a solution of an old problem about constructing unipotent elements in groups of Lie type of odd characteristic.

Keywords: Black box groups, projective geometry, classical groups, unipotent elements.

1 Black Box Groups

A black box group $\mathbf{X}$ is a black box (or an oracle, or an algorithm) operating on $0 - 1$ strings of uniform length which encrypt elements of some finite group G. The procedures performed by a black box are specified as follows.

BB1 $\mathbf{X}$ produces strings of fixed length $l(\mathbf{X})$ encrypting random (almost) uniformly distributed elements from G; this is done in probabilistic time polynomial in $l(\mathbf{X})$.

BB2 $\mathbf{X}$ computes, in probabilistic time polynomial in $l(\mathbf{X})$, a string encrypting the product of two group elements given by strings or a string encrypting the inverse of an element given by a string.

BB3 $\mathbf{X}$ decides, in probabilistic time polynomial in $l(\mathbf{X})$, whether two strings encrypt the same element in G—therefore identification of strings is a canonical projection

$$\mathbf{X} \xrightarrow{\ \pi\ } G.$$

In this situation we say that $\mathbf{X}$ encrypts the group G.

A natural question here is to determine the isomorphism type of a black box group $\mathbf{X}$ or, if it is known, find an isomorphism between $\mathbf{X}$ and its natural copy. To that end, we need additional assumptions about $\mathbf{X}$, which we are keeping to a minimum by adopting an additional axiom.

H. Hong and C. Yap (Eds.): ICMS 2014, LNCS 8592, pp. 53–58, 2014.
© Springer-Verlag Berlin Heidelberg 2014

BB4 We are given a *global exponent* of $\mathbf{X}$, that is, a natural number E such that $\pi(x)^E = 1$ for all strings $x \in \mathbf{X}$ while computation of x^E is computationally feasible (say, $\log E$ is polynomially bounded in terms of $\log|G|$).

Note that axioms **BB1–BB4** hold, for example, in matrix groups over finite fields where we can take for E the exponent of the ambient $\mathrm{GL}_n(q)$.

In this paper, we assume **BB1–BB4** and are concerned with *structure recovery* of black box groups $\mathbf{X}$ encrypting an explicitly given group G of Lie type over $\mathbb{F}_q$, that is, with constructing, in probabilistic polynomial time in $\log|G|$,

– a black box field $\mathbf{K}$ encrypting $\mathbb{F}_q$, and
– a morphism $\Psi : G(\mathbf{K}) \to \mathbf{X}$.

Unlike the constructive recognition algorithms of black box groups [7–11], we shall note here that we are not using a discrete logarithm oracle or an $\mathrm{SL}_2(q)$-oracle, see [4] for a detailed discussion of the hierarchy of black box group problems.

2   Morphisms and Automorphisms

Let $\mathbf{X}$ and $\mathbf{Y}$ be two black box groups encrypting the groups G and H, respectively. We say that a map ζ, which assigns strings from $\mathbf{X}$ to $\mathbf{Y}$, is a *morphism* of black box groups if

– the map ζ is computable in probabilistic time polynomial in $l(\mathbf{X})$ and $l(\mathbf{Y})$; and
– there is an abstract homomorphism $\phi : G \to H$ such that the following diagram is commutative:

$$
\begin{array}{ccc}
\mathbf{X} & \xrightarrow{\ \zeta\ } & \mathbf{Y} \\
\pi_{\mathbf{X}} \downarrow & & \downarrow \pi_{\mathbf{Y}} \\
G & \xrightarrow{\ \phi\ } & H
\end{array}
$$

where $\pi_{\mathbf{X}}$ and $\pi_{\mathbf{Y}}$ are the canonical projections of $\mathbf{X}$ and $\mathbf{Y}$ onto G and H, respectively.

In this case we say that a morphism ζ encrypts the homomorphism ϕ. Observe that replacing a given generating set of a black box group $\mathbf{X}$ by a more suitable one means that we construct a new black box $\mathbf{Y}$ and work with the corresponding morphism $\mathbf{Y} \to \mathbf{X}$.

The first result based on this new philosophy is "amalgamation of local automorphisms":

Theorem 1. [4, Theorem 5.1] *Let $\mathbf{X}$ be a black box group encrypting a group G. Assume that G contains subgroups $G_1, \ldots, G_l$ invariant under an automorphism $\alpha \in \mathrm{Aut}\, G$ and that these subgroups are encrypted in $\mathbf{X}$ as black boxes $\mathbf{X}_i$, $i = 1, \ldots, l$, supplied with morphisms*

$$
\phi_i : \mathbf{X}_i \longrightarrow \mathbf{X}_i
$$

which encrypt restrictions $\alpha|_{G_i}$ of α on G_i. Assume also that $\langle G_i, i = 1, \ldots, l \rangle = G$. Then we can construct, in time polynomial in $l(\mathbf{X})$, a morphism $\phi : \mathbf{X} \longrightarrow \mathbf{X}$ which encrypts α.

This theorem can be applied, for example, to groups of Lie type and systems of root SL_2-subgroups corresponding to the nodes in the associated Dynkin diagrams. That way, we construct the following automorphisms of groups of Lie type.

(1) Frobenius maps on groups of Lie type of odd characteristic [4];
(2) Graph automorphisms of $SL_n(q)$, $D_n(q)$ (including the triality of $D_4(q)$), $F_4(q)$, and $E_6(q)$ (for odd q) [6].

Interestingly, construction of graph automorphisms in black box groups of Lie type of odd characteristic does not use information about the underlying field. Further manipulation with morphisms between root $SL_2(q)$-subgroups yields, for example, the following (field-independent) black box embeddings constructed in time polynomial in $\log q$ and n:

- $SU_n(q) \hookrightarrow SL_n(q^2)$;
- $G_2(q) \hookrightarrow SO_7(q) \hookrightarrow SO_8^+(q) \hookrightarrow SL_8(q)$;
- $^3D_4(q) \hookrightarrow SO_8^+(q) \hookrightarrow SL_8(q)$;

These embeddings are implemented in GAP for various fields but notably we construct the embedding $SU_3(p) \hookrightarrow SL_3(p^2)$ for the 60 digit prime

$$p = 622288097498926496141095869268883999563096063592498055290461.$$

Notice that the size of $SL_3(p^2)$ is bigger than 10^{960}.

Another very important corollary of Theorem 1 is that if the action of an involutive automorphism a of G is known on some a-invariant subgroups of G generating G, then we can transfer the action of a on these subgroups to whole group G. We call this process a *reification* of a. More precisely, we have

Theorem 2. [4, Theorem 7.1] *Let* $\mathbf{X}$ *be a black box group encrypting a finite group* G. *Assume that* G *admits an involutive automorphism* $a \in \operatorname{Aut} G$ *and contains* a-invariant subgroups $H_1, \ldots, H_n$ *where* a *either inverts or centralizes each* H_i.

Assume also that we are given black boxes $\mathbf{Y}_1, \ldots, \mathbf{Y}_n$ *encrypting subgroups* $H_1, \ldots, H_n$. *Then we can construct, in polynomial time,*

- *a black box for the structure* $\{\mathbf{Y}, \alpha\}$, *where* $\mathbf{Y}$ *encrypts* $H = \langle H_1, \ldots, H_n \rangle$ *and* α *encrypts the restriction of* $a|_H$ *of* a *to* H;
- *a black box subgroup* $\mathbf{Z}$ *encrypting* $\Omega_1(Z(C_H(a)))$, *the subgroup generated by involutions from* $Z(C_H(a))$;
- *if, in addition, the automorphism* $a \in G$ *and* $H = G$ *then* α *is induced by one of the involutions in* $\mathbf{Z}$.

An immediate application of Theorem 2 is that we can append a diagonal automorphism d of $PSL_2(q)$ to a black box group $\mathbf{X}$ encrypting $PSL_2(q)$ to obtain a black box group $\mathbf{Y} = \mathbf{X} \rtimes \langle \delta \rangle$ encrypting $PGL_2(q)$, where δ encrypts d, see [5] for details. This construction plays a crucial role in the proof of Theorem 3 below.

In addition, if a is an *inner* involutive automorphism in a group G of small 2-rank, after reification it can be identified with a string in X.

It turns out that construction of an involution in black box groups encrypting $PGL_2(2^k)$ by Kantor and Kassabov [12] is a special case of Theorem 2, see [4] for further discussion. Moreover, the construction of a black box projective plane is based on reification of involutions.

3   PSL₂(q): Structure Recovery and Unipotent Elements

This is our principal result.

Theorem 3. *[5] Let $\mathbf{Y}$ be a black box group encrypting $PSL_2(\mathbb{F}_q)$ for $q = p^k$ of known odd characteristic p. Then we construct, in probabilistic time polynomial in $\log q$,*

- *a black box group $\mathbf{X}$ encrypting $PGL_2(\mathbb{F}_q)$ and an effective embedding*

$$\mathbf{Y} \hookrightarrow \mathbf{X};$$

- *a black box field $\mathbf{K}$ of order q, and*
- *polynomial in $\log q$ time isomorphisms*

$$\begin{array}{c} \mathbf{Y} \\ \downarrow \\ PGL_2(\mathbb{F}_q) \twoheadrightarrow PGL_2(\mathbf{K}) \twoheadrightarrow \mathbf{X} \longrightarrow SO_3(\mathbf{K}) \end{array}$$

where $\mathbb{F}_q$ is the standard explicitly given field of order q.

Construction of unipotent elements in $\mathbf{X}$ is an automatic corollary, but can be actually done at early stages of the proof of this theorem.

Our approach to the proof is recovery, within $\mathbf{X}$, of geometric structures arising from the adjoint representation of the group $PGL_2(q)$ on its Lie algebra $\mathfrak{sl}_2$ seen as an inner product space with respect to its Killing form—this explains appearance of the morphism

$$\mathbf{X} \longrightarrow SO_3(\mathbf{K})$$

in the statement of Theorem 3.

Our proof in [5] starts by exploiting the fact that the set of involutions in $\mathbf{X}$ is the set $\mathfrak{I} = \mathfrak{P} \setminus \mathfrak{Q}$ of regular points in projective plane $\mathfrak{P}$ over $\mathfrak{sl}_2$ with a quadric $\mathfrak{Q}$ (coming from the Killing form), and the points in $\mathfrak{Q}$ are the Borel subgroups in $\mathbf{X}$. There are also two types of lines in $\mathfrak{P}$: regular and parabolic. The regular lines are the polar images of regular points

$$\pi(t) = \{x \in \mathfrak{I} \mid [t,x] = 1 \text{ and } t \neq x\}. \tag{1}$$

The parabolic lines correspond to Borel subgroups B in $\mathbf{X}$ and consist of involutions inverting a maximal unipotent subgroup U of B, together with U itself seen as a point in $\mathfrak{P}$.

It turns out that the set $\mathfrak{I}$ is a finite symmetric space with the conjugation operation $\circ$, for $s, t \in \mathfrak{I}$, $s \circ t = t^s$, forming a finite field analogue of the real hyperbolic (Lobachevsky) plane viewed as a symmetric space. The black box field $\mathbf{K}$ is built by applying the Hilbert's coordinatization on this Lobachevsky plane $\mathfrak{I}$. The analysis of the action of $\mathbf{X}$ on $\mathfrak{I}$ produces the morphism

$$\mathbf{X} \longrightarrow \mathrm{SO}_3(\mathbf{K}).$$

Constructing the black box field by coordinatizing the Lobachevsky plane enables us to construct arbitrary elements in $\mathfrak{P}$ with specified coordinates. In particular, we can construct unipotent elements in $\mathbf{X}$ encrypting $\mathrm{PGL}_2(q)$, which are precisely the points on the quadric $\mathfrak{Q}$.

4   Toolbox in Lobachevsky Plane

It is shown in [5] that the following procedures are performed in time polynomial in $\log q$ inside the Lobachevsky plane constructed in $\mathbf{X}$. So we construct a black box that

(a) produces uniformly distributed points from $\mathfrak{I}$;
(b) checks the equality of points;
(c) checks collinearity of triples of points;
(d) for any two points $s, t \in \mathfrak{I}$, computes the half turn of t around s, which we denote by $s \circ t$;
(e) for any involution $t \in \mathfrak{I}$, produces uniformly distributed regular points in the polar image of t:

$$\varpi(t) = \{ s \in \mathfrak{I} \mid s \circ t = t \text{ and } s \neq t \};$$

(f) for any two distinct points $s, t \in \mathfrak{I}$, produces uniformly distributed regular points on the line $s \vee t$ through s and t;
(g) for a regular line through two distinct points s and t, constructs its pole, which is the involution commuting with both s and t;
(h) for any two distinct lines $\mathbf{k}$ and $\mathbf{l}$, finds its intersection point $\mathbf{k} \wedge \mathbf{l}$ or, if the lines $\mathbf{k}$ and $\mathbf{l}$ do not intersect in $\mathfrak{I}$ and therefore their intersection point z belongs to $\mathfrak{Q}$, computes the unipotent element.
(i) for a point $s \in \mathfrak{I}$, computes the polar projection

$$\xi_s : \mathfrak{I} \setminus \{s\} \longrightarrow \pi(s)$$
$$x \mapsto \pi(x) \wedge (s \vee x);$$

(j) for any two points $s, t \in \mathfrak{I}$ conjugate under the action of $\mathbf{X}$, finds $r \in \mathfrak{I}$ such that $r \circ s = t$;

(k) represents any element of $\mathbf{X}$ as a product of two involutions from $\mathbf{X}$.

As an example, we show how we draw the line passing through two distinct points $s, t \in \mathfrak{I}$ as in item (f). For an involution $x \in \mathbf{X}$ denote by $\mathbf{T}_x$ the maximal torus in $C_{\mathbf{X}}(x)$.

If $z = st$ is a unipotent element then $\langle z^{\mathbf{T}_s} \rangle s$ is a parabolic line. Otherwise observe that it suffices to construct the involution $j := j(s, t)$ which commutes with both s and t. Indeed, the line passing through s and t is the coset $\mathbf{T}_j w$ where w is an involution inverting $\mathbf{T}_j$, see Equation (1). If $z = st$ has even order, then j is the involution in $\langle z \rangle$ which can be constructed by using square and multiply method. However, if z has odd order, then we can not construct j immediately but we know its action on $\mathbf{X}$:

- j centralizes $\langle z \rangle$,
- j inverts every element in the torus $\mathbf{T}_s$.

Hence, $j(s, t)$ can be reified from these two conditions by using Theorem 2.

References

1. Borovik, A.V., Yalçınkaya, Ş.: Construction of Curtis-Phan-Tits system for black box classical groups, Available at arXiv:1008.2823v1
2. Borovik, A.V., Yalçınkaya, Ş.: Steinberg presentations of black box classical groups in small characteristics, Available at arXiv:1302.3059v1
3. Borovik, A.V., Yalçınkaya, Ş.: Fifty shades of black, Available at arXiv:1308.2487
4. Borovik, A.V., Yalçınkaya, Ş.: Black box, white arrow, Available at arXiv:1404.7700
5. Borovik, A.V., Yalçınkaya, Ş.: Revelations and reifications: Adjoint representations of black box groups $PSL_2(q)$ (in preparation)
6. Borovik, A.V., Yalçınkaya, Ş.: Subgroup structure and automorphisms of black box classical groups (in preparation)
7. P. A. Brooksbank, *A constructive recognition algorithm for the matrix group* $\Omega(d, q)$, Groups and Computation III (W. M. Kantor and Á. Seress, eds.), Ohio State Univ. Math. Res. Inst. Publ., vol. 8, de Gruyter, Berlin, 2001, pp. 79–93.
8. Brooksbank, P.A.: Fast constructive recognition of black-box unitary groups. LMS J. Comput. Math. 6, 162–197 (2003)
9. Brooksbank, P.A.: Fast constructive recognition of black box symplectic groups. J. Algebra 320(2), 885–909 (2008)
10. Brooksbank, P.A., Kantor, W.M.: On constructive recognition of a black box $PSL(d, q)$. In: Kantor, W.M., Seress, Á. (eds.) Groups and Computation III, vol. 8, pp. 95–111. Ohio State Univ. Math. Res. Inst. Publ., de Gruyter, Berlin (2001)
11. Brooksbank, P.A., Kantor, W.M.: Fast constructive recognition of black box orthogonal groups. J. Algebra 300(1), 256–288 (2006)
12. Kantor, W.M., Kassabov, M.: Black box groups $PGL(2, 2^e)$, arXiv:1309.3715v2
13. Yalçınkaya, Ş.: Black box groups. Turkish J. Math. 31(suppl.), 171–210 (2007)

A GAP Package for Computing
with Real Semisimple Lie Algebras

Heiko Dietrich[1], Paolo Faccin[2], and Willem A. de Graaf[3]

[1] School of Mathematical Sciences, Monash University, Australia
heiko.dietrich@tu-bs.de
[2] Department of Mathematics, University of Trento, Povo (Trento), Italy
faccin@science.unitn.it
[3] Department of Mathematics, University of Trento, Povo (Trento), Italy
deGraaf@science.unitn.it

Abstract. We report on the functionality and the underlying theory of
the GAP package CoRelG (*Computing with Real Lie Groups*)[1]; it provides functionality to construct real semisimple Lie algebras, to check for
isomorphisms, and to compute Cartan decompositions, Cartan subalgebras, and nilpotent orbits.

Keywords: GAP, real semisimple Lie algebras.

1 Introduction

An n-dimensional Lie algebra over a field $\mathbb{F}$ is an n-dimensional $\mathbb{F}$-vector space
$\mathfrak{g}$, furnished with a bilinear multiplication

$$[-,-]\colon \mathfrak{g} \times \mathfrak{g} \to \mathfrak{g}, \quad (a,b) \mapsto [a,b]$$

which satisfies $[u,u] = 0$ and $[u,[v,w]] + [v,[w,u]] + [w,[u,v]] = 0$ (Jacobi iden-
tity) for all $u,v,w \in \mathfrak{g}$. Studied originally over the complex field $\mathbb{F} = \mathbb{C}$, Lie
theory originated in the 19-th century in the work of the Norwegian mathemati-
cian Sophus Lie. Since then it has developed tremendously and it has become
one of the central areas of 20-th and 21-st century mathematics, finding many
applications in such diverse fields as physics, geometry, and group theory. In the
second half of the 20-th century, the development of the computer provided a
new research tool in Lie theory. Algorithms were developed and implemented on
computer for various tasks related to Lie theory. Initially, this mostly concerned
the combinatorial formulae for investigating representations of Lie groups due to,
for example, Weyl and Freudenthal. The success of this endeavour has led to a
new field of research, called Computational Lie Theory, which is concerned with
the development of algorithms in Lie theory, their implementation on computer,
and their application to theoretical problems. Over the past decades several computer programs in this area have emerged, for example, LiE [3], GAP [8], and

---

[1] This work and the first author were supported by a Marie-Curie Fellowship (grant
no. PIEF-GA-2010-271712) and an ARC-DECRA Fellowship, project DE140100088.

H. Hong and C. Yap (Eds.): ICMS 2014, LNCS 8592, pp. 59–66, 2014.

MAGMA [1]. The last two programs are large computer algebra systems having well developed libraries for Computational Lie Theory.

The main focus in Computational Lie Theory has been on complex semisimple Lie algebras and Lie groups, and their representations. However, an important branch of Lie theory deals with real Lie groups and algebras. These are of paramount importance in physics and differential geometry. Probably due to the difficulty with dealing with the field of real numbers, which leads to various phenomena of non-splitness, there has not been much attention to real Lie groups in Computational Lie Theory. This changed at the beginning of the 21-st century, when a large group in the United States set up a research program to study real Lie groups by computational means. This is known as the Atlas project [2], and its main goal is to study the unitary dual of a real Lie group. An important problem in real Lie theory, not addressed by the Atlas project, is the classification of the orbits of a real Lie group acting on a vector space.

In this paper, we report on our GAP-package CORELG [7], for working with real semisimple Lie algebras given by a multiplication table (which the Atlas software does not do). As described in detail in the book [9], defining a Lie algebra by its structure constants allows for a detailed investigation of its structure. We remark that efficient algorithms for dealing with complex semisimple Lie algebras are already available in the GAP-package SLA [10].

1.1   Notation

The aim of this section is to introduce necessary notation; we refer to any standard book (for example, [12], [13], and [14]) for details and proofs. The *structure constants* of an n-dimensional Lie algebra $\mathfrak{g}$ with basis $\{v_1, \ldots, v_n\}$ are $\{c_{a,b}^{(k)}\}_{1 \leq a,b,k \leq n}$, defined by

$$[v_a, v_b] = \sum_{k=1}^{n} c_{a,b}^{(k)} v_k.$$

The Lie algebra $\mathfrak{g}$ is *semisimple* if it has no nontrivial abelian ideals, or, equivalently, if it is the direct sum of *simple* Lie algebras, that is, nonabelian Lie algebras which have no nontrivial ideals. The *adjoint* of $g \in \mathfrak{g}$ is the map $\mathrm{ad}_\mathfrak{g}(g) \colon \mathfrak{g} \to \mathfrak{g}, h \mapsto [g, h]$. The *Killing form* of $\mathfrak{g}$ is the bilinear map $\kappa_\mathfrak{g} \colon \mathfrak{g} \times \mathfrak{g} \to \mathbb{C}$, $\kappa_\mathfrak{g}(g, h) = \mathrm{trace}(\mathrm{ad}_\mathfrak{g}(g) \circ \mathrm{ad}_\mathfrak{g}(h))$. Each semisimple Lie algebra $\mathfrak{g}$ defined over $\mathbb{C}$ has a *Cartan subalgebra* $\mathfrak{H} \leq \mathfrak{g}$, which is a maximal abelian subalgebra consists of semisimple elements, that is, each $h \in \mathfrak{H}$ has a diagonalisable adjoint. This gives rise to the *root space decomposition*

$$\mathfrak{g} = \mathfrak{H} \oplus \bigoplus\nolimits_{\alpha \in \Phi} \mathfrak{g}_\alpha \quad \text{where} \quad \mathfrak{g}_\alpha = \{g \in \mathfrak{g} \mid \forall h \in \mathfrak{H} \colon [h, g] = \alpha(h)g\};$$

where $\Phi \subseteq \mathfrak{H}^*$ consists of all linear maps $\mathfrak{H} \to \mathbb{C}$ such that $\mathfrak{g}_\alpha \neq \{0\}$. Since $\kappa_\mathfrak{g}$ is non-degenerate, for each $\alpha \in \Phi$ there exists $t_\alpha \in \mathfrak{H}$ with $\alpha(-) = \kappa(t_\alpha, -)$; for $\alpha, \beta \in \Phi$ define $(\alpha, \beta) = \kappa_\mathfrak{g}(t_\alpha, t_\beta) = \alpha(t_\beta)$. Now $V = \mathrm{Span}_\mathbb{R}(\Phi)$ is an Euclidean space with inner product $(-, -)$; this is a *root system*. The theory of abstract root systems shows that there exist a *basis of simple roots* $\Pi = \{\alpha_1, \ldots, \alpha_\ell\} \subseteq \Phi$ (which also is a vector space basis of V), and an associated Weyl group, Cartan matrix, and Dynkin diagram. The Cartan-Killing-Dynkin classification of simple complex Lie algebras states a one-to-one correspondence between the isomorphism types of these Lie algebras and the isomorphism types of Dynkin diagrams. The Dynkin diagrams are classified by their type: there are four infinite families A_n $(n \geq 1)$, B_n $(n \geq 2)$, C_n $(n \geq 3)$, and D_n $(n \geq 4)$, and five exceptional types G_2, F_4, E_6, E_7, and E_8.

If $\mathfrak{g}$ is a simple Lie algebra defined over the real numbers, then either $\mathfrak{g}$ is a simple complex Lie algebra considered as real, or the complexification $\mathfrak{g}^c = \mathfrak{g} \otimes_{\mathbb{R}} \mathbb{C}$ of $\mathfrak{g}$ is a simple complex Lie algebra. In the latter case, $\mathfrak{g}$ is a *real form* of the simple complex Lie algebra $\mathfrak{g}^c$. Each complex simple Lie algebra has, up to isomorphism, only finitely many real forms. A Cartan subalgebra of a real semisimple Lie algebra $\mathfrak{g}$ is a nilpotent self-normalising subalgebra $\mathfrak{h} \leq \mathfrak{g}$; its complexification $\mathfrak{h}^c$ is a Cartan subalgebra of $\mathfrak{g}^c$.

2   Applications

In this section, we describe some of the new functionality provided by our software package CoReLG [7]; we give details on the underlying theory in Section 3. Our algorithms and implementations allow to investigate real (semi)simple Lie algebras computationally: one can compute Cartan decompositions, Cartan subalgebras, nilpotent orbits, and isomorphisms between Lie algebras. The field SqrtField in the example output of our algorithms is the infinite-dimensional number field $\mathbb{Q}(\sqrt{-1}, \sqrt{2}, \sqrt{3}, \sqrt{5}, \sqrt{7} \ldots)$; we give more details in Section 4.

2.1   Construction of Simple Real Lie Algebras

For every type of simple complex Lie algebra (A_n, B_n, C_n, D_n, G_2, F_4, E_6, E_7, E_8) there exist, up to isomorphism, only finitely many real forms. We provide functions RealFormsInformation, IdRealForm, and RealFormById which construct these real simple Lie algebras; the example below considers type A_3.

```
gap> RealFormsInformation("A",3);

  There are 5 simple real forms with complexification A3
    1 is of type su(4), compact form
    2 - 3 are of type su(p,4-p) with 1 <= p <= 2
    4 is of type sl(2,H)
    5 is of type sl(4,R)
  Index '0' returns the realification of A3

gap> L := RealFormById("A",3,4);
<Lie algebra of dimension 15 over SqrtField>
gap> IdRealForm(L);
[ "A", 3, 4 ]
```

2.2   Cartan Subalgebras

We provide a function `CartanSubalgebrasOfRealForm` which constructs, up to conjugacy, all Cartan subalgebras of a real semisimple Lie algebra.

```
gap> L   := RealFormById("F",4,2);;
gap> CSA := CartanSubalgebrasOfRealForm(L);;
gap> Size(CSA);
8
gap> CSA[1];
<Lie algebra of dimension 4 over SqrtField>
```

2.3   Isomorphisms

The function `IsomorphismOfRealSemisimpleLieAlgebras` constructs, if exists, an isomorphism between two given real semisimple Lie algebras. The function `VoganDiagram` outputs the associated Vogan diagram, which determines the isomorphism type of the real form.

```
gap> L   := RealFormById( "F", 4, 2 );;
gap> sc  := StructureConstantsTable(Basis(L));;
gap> K   := LieAlgebraByStructureConstants(SqrtField,sc);;
gap> iso := IsomorphismOfRealSemisimpleLieAlgebras(K,L);
<Lie algebra isomorphism between Lie algebras of dimension 52>
gap> Display(VoganDiagram(L));
F4:   2---(4)=>=3---1
Involution: ()
```

2.4   Nilpotent Orbits

The nilpotent orbits of a real simple Lie algebra $\mathfrak{g}$ are the G-orbits of nilpotent elements in $\mathfrak{g}$, where G is the adjoint group of $\mathfrak{g}$. If $\mathfrak{g}$ has rank at most 8, then the function `NilpotentOrbitsOfRealForm` computes representatives of the nilpotent orbits of $\mathfrak{g}$. These orbits have been precomputed and are stored in a database; they are constructed as orbits in $\mathfrak{g}$ by using the isomorphism functionality described above. The function `RealCayleyTriple` returns a so-called $\mathfrak{sl}_2$-triple defining the orbit; its third component is a nilpotent element representing the orbit.

```
gap> L:=RealFormById("A",3,3);;
gap> orb:=NilpotentOrbitsOfRealForm(L);;
gap> Length(orb);
9
gap> o:=orb[2];
<nilpotent orbit in Lie algebra>
gap> RealCayleyTriple(o);
[ (-1/4)*v.8+(-1/4)*v.14, (1/2)*v.2, (-1/4)*v.8+(1/4)*v.14 ]
```

Our algorithms can answer the following questions: Let $\mathfrak{g}$ be a real semisimple Lie algebra with semisimple subalgebra $\mathfrak{a} \leq \mathfrak{g}$; let $\mathfrak{s}$ be the semisimple part of the centraliser of $\mathfrak{a}$ in $\mathfrak{g}$; what is the structure of $\mathfrak{s}$, that is, its Cartan subalgebras, Cartan decompositions, and its isomorphism type? The following example considers the semisimple part $\mathfrak{c}$ of the centraliser of a subalgebra $\mathfrak{a} \leq \mathfrak{g}$ with $\mathfrak{a}$ and $\mathfrak{g}$ real forms of type A_1 and E_7, respectively.

```
gap> L:=RealFormById("E",7,2);;
gap> ch:=ChevalleyBasis(L);;
gap> A:=Subalgebra(L,[ch[1][1],ch[2][1],ch[3][1]],"basis");;
gap> C:=LieDerivedSubalgebra(LieCentraliser(L,A));;
gap> IdRealForm(C);
[ "D", 6, 5 ]
gap> Length(CartanSubalgebrasOfRealForm(C));
4
gap> Display(VoganDiagram(C));
                    (5)
                   /
D6:   1---2---3---4
                   \
                    6


Involution: ()
```

3   Underlying Theory

We comment on the underlying theory for the tasks described in Section 2, see also [5] and [6].

3.1   Construction of Simple Real Lie Algebras

The classification of the simple real Lie algebras is known, and, up to isomorphism, the real forms of a simple complex Lie algebra $\mathfrak{g}$ can be constructed as follows. The first step is straightforward and requires to construct the so-called *compact real form* $\mathfrak{c}$ of $\mathfrak{g}$. The associated *(compact) real structure* is $\tau\colon \mathfrak{g} \to \mathfrak{g}$, $a + \imath b \mapsto a - \imath b$, where $a, b \in \mathfrak{c}$; here we write $\mathfrak{g} = \mathfrak{c} \oplus \imath\mathfrak{c}$. Let $\theta \in \mathrm{Aut}(\mathfrak{g})$ be an automorphism of order 2, commuting with τ, and denote by $\mathfrak{c}_\pm$ the ± 1-eigenspace of the restriction of θ to $\mathfrak{c}$. Now $\mathfrak{r}_{\tau,\theta} = \mathfrak{c}_+ \oplus \imath\mathfrak{c}_-$ is a real form of $\mathfrak{g}$, and every real form of $\mathfrak{g}$ is isomorphic to $\mathfrak{r}_{\tau,\theta}$ for some θ. Moreover, $\mathfrak{r}_{\tau,\theta} \cong \mathfrak{r}_{\tau,\theta'}$ if and only if θ and θ' are conjugate in $\mathrm{Aut}(\mathfrak{g})$. Involutionary automorphisms of $\mathfrak{g}$ are classified, up to conjugacy, in terms of Vogan diagrams; running over these automorphisms yields all real forms of $\mathfrak{g}$ up to isomorphism.

Note that $\mathfrak{r}_{\tau,\theta} = \mathfrak{k} \oplus \mathfrak{p}$ where $\mathfrak{k}$ and $\mathfrak{p}$ are the 1- and (-1)-eigenspace, respectively, of the restriction of θ to $\mathfrak{r}_{\tau,\theta}$. This decomposition is a *Cartan decomposition* of $\mathfrak{r}_{\tau,\theta}$ with *Cartan involution* θ; it is unique up to conjugacy.

For a given simple complex Lie algebra and fixed Cartan subalgebra and *Chevalley basis*, there exist *canonical choices* for the compact real form and the involutionary automorphisms. The real forms, $\mathfrak{r}_{\tau,\theta}$, obtained from these choices are called real forms in canonical form. (We remark that the structure of the canonical automorphism is encoded in the Vogan diagram of the real form, see `VoganDiagram` above.)

The challenge to construct real forms efficiently is to determine the multiplication table of $\mathfrak{r}_{\tau,\theta}$ by theoretical means, which allows one to write down this table (and thus to define $\mathfrak{r}_{\tau,\theta}$) directly, avoiding all computations. This is a straightforward, but tedious undertaking; it requires to determine the structure constants of a suitable basis of each $\mathfrak{r}_{\tau,\theta}$.

3.2   Cartan Subalgebras

If $\mathfrak{g}$ is a complex simple Lie algebra, then, up to conjugacy under its adjoint group, there is a unique Cartan subalgebra in $\mathfrak{g}$. In contrast, there is no unique Cartan subalgebra in a real simple Lie algebra. However, up to conjugacy, there are only finitely many Cartan subalgebras; they have been classified by Kostant (1955) and Sugiura (1959). For our implementation, we devised a constructive version of Sugiura's classification theorem; it depends on the notion of strongly orthogonal sets of roots.

3.3   Isomorphisms

Let $\mathfrak{g}$ be a simple real Lie algebra. By constructing and analysing its complexification $\mathfrak{g}^c$, we know the type of $\mathfrak{g}^c$. In particular, $\mathfrak{g}$ is isomorphic to some canonical real form $\mathfrak{r} = \mathfrak{r}_{\tau,\theta}$, with $\mathfrak{r}^c \cong \mathfrak{g}^c$, for some involutionary automorphism $\theta \in \mathrm{Aut}(\mathfrak{r}^c)$, commuting with the compact real structure τ of $\mathfrak{r}^c$. Our approach is to construct an isomorphism $\mathfrak{g}^c \to \mathfrak{r}^c$ which is compatible with the associated real structures of $\mathfrak{g}$ and $\mathfrak{r}$; such an isomorphism clearly induces an isomorphism between the real forms $\mathfrak{g}$ and $\mathfrak{r}$. We construct this isomorphism in several steps. First, for each Lie algebra $\mathfrak{g}$ and $\mathfrak{r}$, we construct a so-called maximally compact Cartan subalgebra and a Cartan involution stabilising this Cartan subalgebra. (Our implementations provide this functionality.) With respect to this Cartan subalgebra and chosen basis of simple roots, we construct a so-called Chevalley basis and canonical generating set; this allows us to define an explicit isomorphism $\varphi \colon \mathfrak{g}^c \to \mathfrak{r}^c$. The next step is to modify φ (by means of defining it with respect to a modified canonical generating set) so that φ is compatible with the Cartan involutions of $\mathfrak{g}$ and $\mathfrak{r}$, respectively. We achieve this by acting with the Weyl groups of $\mathfrak{g}^c$ and $\mathfrak{r}^c$ on the respective canonical generating sets. Once such a compatible φ is found, one can easily modify it again so that it is also compatible with the respective real structures. This completes the construction of an isomorphism $\mathfrak{g} \to \mathfrak{r}$.

The above construction assumes we know that $\mathfrak{g} \cong \mathfrak{r}$. In practice, we only work with $\mathfrak{g}$ and, using the above approach, find a suitable canonical generating set so that the Cartan involution acts on it in a *standard way*; in other words, we construct *standard parameters* for $\mathfrak{g}$, such that two real simple Lie algebras are isomorphic if and only if their standard parameters coincide. In this case, it is straightforward to write down an explicit isomorphism. By construction, the canonical forms $\mathfrak{r}_{\tau,\theta}$ have standard parameters.

3.4 Nilpotent Orbits

The nilpotent orbits of a simple complex Lie algebra are determined by the Dynkin-Kostant and Bala-Carter classifications (see [4]): the nonzero nilpotent elements are in one-to-one correspondence to certain semisimple elements (*characteristic elements*), which are in one-to-one correspondence to certain weighted Dynkin diagrams. The situation is more complicated for a simple real Lie algebra $\mathfrak{g}$. By the Kostant-Sekiguchi correspondence, if $\mathfrak{g} = \mathfrak{k} \oplus \mathfrak{p}$ is a Cartan decomposition, then the nilpotent orbits in $\mathfrak{g}$ are in one-to-one correspondence to the nilpotent K-orbits in $\mathfrak{p}^c$, where K is the adjoint group of $\mathfrak{k}^c$. There exist efficient implementations for computing the K-orbits in $\mathfrak{p}^c$ (see [11]). However, the Kostant-Sekiguchi correspondence is non-constructive, and obtaining explicit orbit representatives in $\mathfrak{g}$ is difficult. We used ad hoc computations and Gröbner bases to make this correspondence explicit for Lie algebras of rank at most 8.

4   Technical Problems

Our implementations face three technical (and theoretical) limitations.

Firstly, for a given real semisimple Lie algebra $\mathfrak{g}$, we have to construct a Cartan subalgebra of $\mathfrak{g}^c$ and a corresponding root system. While there exist efficient algorithms to construct Cartan subalgebras, computations of the associated root systems may fail because the algorithm does not succeed in splitting the Cartan subalgebra over a small-degree extension of the base field. The problem of finding Cartan subalgebras which can be split is very difficult.

Secondly, our current approach for making the Kostant-Sekiguchi correspondence explicit requires the use of ad hoc computations using Gröbner bases. Even though we automated these computations systematically, the complexity of Gröbner basis computations limits the scope of this approach. This, and some limitations of the algorithms in [11], are the reason why our databank of nilpotent orbits currently contains only Lie algebras of rank at most 8.

The final limitation is concerned with the base field. In order to define a Lie algebra by a multiplication table over the reals, it usually suffices to take a subfield of the real field as base field. However, many algorithms need a Chevalley basis which is defined over the complex numbers; therefore, we require that the base field also contains the imaginary unit $\imath$. Other procedures, for example, the isomorphism test, requires the computation of square roots. Thus, in practise, the base field of our real Lie algebras is $\mathbb{Q}^{\surd} = \mathbb{Q}(\imath, \sqrt{2}, \sqrt{3}, \sqrt{5}, \ldots)$, the Gaussian rationals with all $\sqrt{p}$, p a prime, adjoined. We have implemented the arithmetic of this field in GAP, and realised it as the field `SqrtField`. We remark that, in theory, a computation with our implementation can fail because we cannot construct a particular square root; our observation is that this happens rather sporadically.

References

1. Bosma, W., Cannon, J., Playoust, C.: The Magma algebra system. I. The user language, Computational algebra and number theory (London, 1993). J. Symbolic Comput. 24, 235–265 (1997)
2. du Cloux, F., van Leeuwen, M.: Software for structure and representations of real reductive groups, v. 0.4.6, http://www.liegroups.org
3. Cohen, A.M., van Leeuwen, M.A.A., Lisser, B.: LiE a Package for Lie Group Computations. CAN, Amsterdam (1992)
4. Collingwood, D.H., McGovern, W.M.: Nilpotent orbits in semisimple Lie algebras. Van Nostrand Reinhold Mathematics Series. Van Nostrand Reinhold Co., New York (1993)
5. Dietrich, H., de Graaf, W.A.: A computational approach to the Kostant-Sekiguchi correspondence. Pacific J. Mathematics 265, 349–379 (2013)
6. Dietrich, H., Faccin, P., de Graaf, W.A.: Computing with real Lie algebras: real forms, Cartan decompositions, and Cartan subalgebras. J. Symbolic Comp. 56, 27–45 (2013)
7. Dietrich, H., Faccin, P., de Graaf, W.A.: CoReLG – computing with real Lie groups. A GAP4 package (2014), http://www.science.unitn.it/~corelg
8. The GAP Group, GAP – Groups, Algorithms, and Programming, Version 4.5.5 (2012), www.gap-system.org
9. de Graaf, W.A.: Lie Algebras: Theory and Algorithms. North-Holland Math. Lib., vol. 56. Elsevier Science (2000)
10. de Graaf, W.A.: SLA – computing with Simple Lie Algebras. A GAP4 package (2012), http://www.science.unitn.it/~egraaf/sla.html
11. de Graaf, W.A.: Computing representatives of nilpotent orbits of θ-groups. J. Symbolic Comput. 46, 438–458 (2011)
12. Humphreys, J.E.: Introduction to Lie algebras and representation theory. Second printing, revised. Graduate Texts in Mathematics, vol. 9. Springer, New York (1978)
13. Knapp, A.W.: Lie groups beyond an introduction, 2nd edn. Progress in Mathematics, vol. 140. Birkhäuser (2002)
14. Onishchik, A.L.: Lectures on Real Semisimple Lie Algebras and Their Representations. ESI Lectures in Mathematics and Physics. European Mathematical Society (EMS), Zürich (2004)

Bacterial Genomics and Computational Group Theory: The BioGAP Package for GAP

Attila Egri-Nagy, Andrew R. Francis, and Volker Gebhardt

Centre for Research in Mathematics
School of Computing, Engineering and Mathematics
University of Western Sydney, Australia
{A.Egri-Nagy,A.Francis,V.Gebhardt}@uws.edu.au

Abstract. Bacterial genomes can be modelled as permutations of conserved regions. These regions are sequences of nucleotides that are identified for a set of bacterial genomes through sequence alignment, and are presumed to be preserved through the underlying process, whether through chance or selection. Once a correspondence is established between genomes and permutations, the problem of determining the evolutionary distance between genomes (in order to construct phylogenetic trees) can be tackled by use of group-theoretical tools. Here we review some of the resulting problems in computational group theory and describe BioGAP, a computer algebra package for genome rearrangement calculations, implemented in GAP.

Keywords: computational group theory, bacterial genomics.

1 Introduction

We aim to introduce biologically inspired computational problems to a group theory and wider mathematics and computer science audience. The biological background (bacterial genomics, phylogenetics) is briefly described, the genome-permutation correspondence established, then more specific models introduced. We identify the types of computational problems and describe implemented solutions for some models in our software package BioGAP[7].

Bacteria

Bacteria are single-cell microorganisms. Despite their small size they are very important: they live everywhere, they take part in nutrient recycling, they are important for the human body and also cause illnesses, and they form a biomass bigger than the combined biomass of all plants and animals. Therefore, understanding their evolutionary development is crucial.

The bacterial genome is circular; a remarkable property with important consequences for algebraic modelling. Bacterial evolution happens at three different levels [11]:

H. Hong and C. Yap (Eds.): ICMS 2014, LNCS 8592, pp. 67–74, 2014.

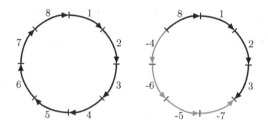

Fig. 1. Reference genome and the signed permutation $[1, 2, 3, -7, -6, -5, -4, 8]$

1. *Local* changes such as *single nucleotide polymorphisms* (SNPs).
2. *Regional* changes include inversion, translocation, deletion and others.
3. *Topological* changes that produce knots and links in the DNA.

Here we are interested in regional changes and this level of description is particularly suitable for abstract algebraic modelling.

The Genome-Permutation Correspondence

Conserved regions are sequences of nucleotides that are identified for a set of bacterial genomes through sequence alignment, and are presumed to be preserved through the underlying process, whether through chance or selection. Therefore, an evolutionary event can be described as a rearrangement of these regions [10]. For more precise models, we also need to track the orientation of the regions.

According to biological observations, the most frequent rearrangement type is *inversion*, cutting out a segment, turning it around and gluing it back. For example, the regions can be numbered so the "reference" genome is represented by the sequence $[1, 2, 3, 4, 5, 6, 7, 8]$ for 8 regions. An inversion of the segment between regions 4 and 7 (inclusive) is then either $[1, 2, 3, 7, 6, 5, 4, 8]$ (unsigned), or $[1, 2, 3, -7, -6, -5, -4, 8]$ (signed), see Fig. 1.

It is unfortunate that the genome rearrangement terminology was developed to some extent independently from the algebraic terminology. For instance, transposition in group theory means swapping neighboring points as a permutation, while in genomics it is the operation of cutting out a segment and gluing it in somewhere else.

Phylogenies and Genomic Distance

One of the key questions in biology is to establish relationships between organisms and species, that is, reconstructing their evolutionary history. A commonly used way to do this is by drawing a phylogenetic tree. With some tree construction methods (e.g. neighbour-joining or UPGMA [12]) this requires calculating the distance between individual genomes, where we assume that the distance is the shortest path. Using the genome-permutation correspondence the distance calculation can be turned into a combinatorial group theory problem, namely calculating the word distance between group elements.

The assumption that the evolutionary history is the distance is just the first approximation. Biological intuition says that the number of these shortest paths can also have a significant role [18,17]. Thinking in terms of random walks, in case if there are many shortest paths we overestimate, if there are only a few then we underestimate the length of the sequence of evolutionary events.

Computational Tasks

Let G be a group with generators $S = \{s_1, \ldots, s_n\}$, so that every element of G can be written as a product of the generators and their inverses. S^* is the free monoid generated by S, which is the set of all *words*, i.e. finite sequences of the elements of S. The empty word is denoted by ε. The group element realized by the word w is denoted by $\overline{w}$, thus $w \in S^*$ and $\overline{w} \in G$.

The *geodesic distance* is defined by $d_S(g_1, g_2) = |u|$, where u is a minimal length word in S^*, called a *geodesic* word, with the property that $g_1 \overline{u} = g_2$, also denoted by $g_1 \xrightarrow{u} g_2$. $\text{Geo}_S(g_1, g_2)$ is the set of all geodesic words from g_1 to g_2. If no confusion arises, then we will use $d(g_1, g_2)$ and $\text{Geo}(g_1, g_2)$. The length of a group element g is defined by its distance from the identity: $\ell(g) = d(1_G, g)$ and we also write $\text{Geo}(g)$ instead of $\text{Geo}(1_G, g)$, where 1_G is the identity of the group. In particular $\ell(1_G) = 0$ and $\text{Geo}(1_G) = \{\varepsilon\}$.

In groups, due to a simple translation principle, it is convenient and enough to study the geodesics starting from the identity only. For group elements $g_1, g_2 \in G = \langle S \rangle$ we write $g_1 \leq_S g_2$ if $\exists w = uv \in S^*$ such that $\overline{w} = g_2, \overline{u} = g_1, w \in \text{Geo}(g_2)$, i.e. there is a geodesic from the identity to g_2 and g_1 is on it. Using terminology from order theory [5] we can call the set of group elements occurring in $\text{Geo}(1_G, g)$ with this partial order an *interval*, and denote it $[1_G, g]$. Figure 2 demonstrates in a simple example that group elements of the same length can have different intervals.

The *Cayley graph* $\Gamma(G, S)$ of G with respect to the generating set S is the directed graph with group elements as nodes and the labeled edges encoding the action of G on itself. Thus $g \xrightarrow{s} gs$ is an edge, $s \in S$. The *diameter* of G is defined by $\text{diam}(G) = \max_{g \in G} \ell(g)$. With the Cayley graph formalism we can rephrase the phylogeny reconstruction problem as follows: a *phylogeny* on $X \subseteq G$ is a minimal connected subgraph of the Cayley graph containing X.

Using these formal definitions we can identify two basic types of group-theoretical computational tasks.

1. **Distance**: calculating the length of a shortest path between two group elements g_1 and g_2.
2. **Intervals**: construct $\text{Geo}_S(g_1, g_2)$, the set of all geodesic paths or estimate its size.

Both problems have trivial algorithmic solutions. The distance can be found by an orbit calculation [14] and the interval is defined by all reduced words

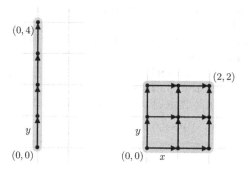

Fig. 2. In $\mathbb{Z}^2$ two group elements with the same length can have intervals of different size. $\big\|[(0,0),(0,4)]\big\| = 5$, $\big\|[(0,0),(2,2)]\big\| = 9$.

realizing a group element in a presentation. However, the number of regions we need to calculate with are beyond the limits of simple brute-force methods. Given the impossibility of searching the entire group when the number of regions is large (for example, genomes with 60 regions can correspond to groups of order $\sim 10^{100}$), it is necessary to exploit the algebraic structure.

The groups involved in this research are well-studied in mathematics. The group of unsigned permutations of the set $\mathbf{n} = \{1, \ldots, n\}$ is the *symmetric* group $\mathcal{S}_n$ and it is a subgroup of the group of signed permutations (the *hyperoctahedral* group, or Coxeter group of type B). However, the generating sets derived from models of the biological processes are different from standard generating sets, rendering many previous mathematical results inapplicable.

Also, the properties of Cayley graphs are studied in geometric group theory, but the focus there is on infinite groups, and the results are often trivial when applied to finite groups (e.g. hyperbolicity).

2   Inversion Systems

Following [8], in general we define an *inversion system* to be a tuple $(G, \mathcal{I})$ where G is a permutation group and $\mathcal{I}$ is a set of inversions such that $\langle \mathcal{I} \rangle = G$, i.e. $\mathcal{I}$ generates G. We need to keep the generating set, therefore algorithms that change generating sets are not directly applicable.

If all inversions are allowed and we ignore orientation, then we have the group of all permutations of $\mathbf{n}$, namely the symmetric group $\mathcal{S}_n$. The generators are all possible inversions of segments on the circle, and the metric is the word length, up to the action of the dihedral group. This is the model considered by [20], and for which they obtained bounds on the distance. A significant number of extensions and improvements have followed this path with great success [15,13,1]. The signed version of this model, in which regions are regarded as having orientation, gives rise to the hyperoctahedral group, and is the most widely studied model in the inversion distance literature.

On the circle, an inversion of one region is equivalent to the inversion of the *complementary* region. One consequence is that one may consider a select region to be fixed, and only consider inversions that do not move it. This enables treatment of the problem as if it were a linear chromosome. This is the basis for many efficient methods currently available, including the use of breakpoints [15] and methods using the breakpoint graph due to [2].

The Two-Inversion Unsigned Model

Biological observations suggest that not all inversions are equally probable [4], therefore the obvious next step is to restrict the size of the set of generating inversions. In the unsigned case, considering only inversions of size two we have transpositions (in the algebraic sense) as generators, for n regions $\mathcal{I} = \{s_1, \ldots, s_n\}$ where $s_n = (n, 1)$. Omitting s_n gives the Coxeter generators for the symmetric group [3], and an inversion system that is well-understood mathematically with an easy length calculation by the bubblesort algorithm [16]. However, from the biological point of view we need the generator s_n, since we deal with circular genomes. It turns out that to some extent the idea of counting the number of crossings to get the word distance can be salvaged if we go to the affine symmetric group – sort of unrolling the circle on an infinite line in both directions (see [8] for details). This example shows that small changes in the generating set can lead to a different situation, where non-trivial group theoretical knowledge and technical arguments need to be applied.

3 Functionality of the BioGAP Package

To attack these problems we have developed a software package called BioGAP [7]. Although the package was specially written to support our research, we have made the code as general as possible.

BioGAP is a package for GAP, which is a system for computational discrete algebra, with particular emphasis on computational group theory. Implementing our algorithms in GAP gives us access to the advanced methods of computational group theory. Relying on well-tested library functions provided by GAP we can ensure correctness and save development time.

There are three main functionalities of BioGAP:

- Calculating with signed permutations;
- Geodesic paths in Cayley graphs; and
- Visualisation.

Signed Permutations

We constructed a new data type for signed permutations. Following the software design principle of reusing, we implemented signed permutations of n points as unsigned permutations of $2n$ points. Signed points i and $-i$ are coded as consecutive integers, so it is easy to fix a point regardless its orientation, and it is possible to do the mapping without knowing the number of points (an alternative would be to represent positive points as $(1,\ldots,n)$ then $(n+1,\ldots,2n)$). Here is the coding and decoding.

$$C(i) = \begin{cases} 2i & \text{if } i > 0 \\ 2|i| - 1 & \text{if } i < 0 \end{cases}$$

$$D(k) = \left\lfloor \frac{k+1}{2} \right\rfloor \cdot (-1)^{k \bmod 2}$$

Here is a short calculation demonstrating how signed permutations are defined by their list of images.

```
gap> p := SignedPermutation([1,-3,-2,4,5]);
(-2,3)(2,-3)
gap> q := SignedPermutation([1,2,-5,-4,-3]);
(-3,5)(3,-5)(-4,4)
gap> p*q;
(-2,3,-5)(2,-3,5)(-4,4)
gap> ImageListOfSignedPerm(last);
[ 1, -3, -5, -4, 2 ]
```

The cyclic notation's redundancy comes from the definition of signed permutation, namely that for a signed permutation $p(-i) = -p(i)$ for all $i \in \mathbf{n}$.

Geodesic Paths

Assuming that we can calculate the length efficiently, there is a straightforward algorithm for constructing the interval. For finding the geodesics, instead of a brute-force search in every direction, we can quickly discard those paths where the sum of the distance from the start and the remaining distance to the destination is more than the length of a shortest path. This allows us both to study small but non-trivial cases of intervals in groups (e.g. Fig. 3) and also to count or estimate the number of geodesics without constructing the interval explicitly. For a group given by a presentation, an alternative approach would be to start with a geodesic word and systematically generate equivalent words by applying the defining relations.

Visualisation

GRAPHVIZ is a widely used graph visualisation package [9], also used by several GAP packages. TIKZ/PGF [19] is a package for producing vector graphics for LATEX. The underlying idea is that a function generates source code in the dot language or in LATEX for the given mathematical object. Then the actual figure can be generated separately to be included in papers, or using the VIZ package [6] immediately displayed on screen from the GAP command line. Figures 1 and 3 were both auto-generated using TIKZ and GRAPHVIZ respectively.

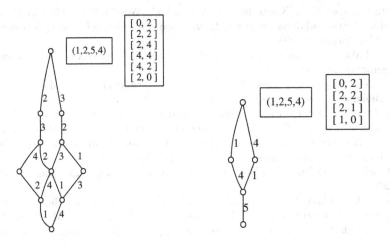

Fig. 3. Intervals $[1,(1,2,5,4)]$ using transposition generating sets $\{s_1, s_2, s_3, s_4\}$ (linear genome) and $\{s_1, s_2, s_3, s_4, s_5\}$ (circular genome). The bottom node is the identity of the group. The edge labels encode the generators. The box contains the number of outgoing and incoming edges at each level of the ranked poset.

Availability

The BioGAP package is still under active development. Its source code is available from `https://bitbucket.org/egri-nagy/biogaphttps://bitbucket.org/egri-nagy/biogap`, and on this site support can also be obtained.

References

1. Bader, D.A., Moret, B.M.E., Yan, M.: A linear-time algorithm for computing inversion distance between signed permutations with an experimental study. Journal of Computational Biology 8(5), 483–491 (2001)
2. Bafna, V., Pevzner, P.A.: Genome rearrangements and sorting by reversals. In: IEEE Proceedings of the 34th Annual Symposium on the Foundations of Computer Science, pp. 148–157 (1993)
3. Björner, A., Brenti, F.: Combinatorics of Coxeter Groups. Graduate Texts in Mathematics. Springer (2005)
4. Darling, A.E., Miklós, I., Ragan, M.A.: Dynamics of genome rearrangement in bacterial populations. PLoS Genetics 4(7) (2008)
5. Davey, B.A., Priestley, H.A.: Introduction to Lattices and Order, 2nd edn. Cambridge University Press (2002)
6. Delgado, M., Egri-Nagy, A., Mitchell, J.D., Pfeiffer, M.: VIZ – GAP package for visualisation (2014), `https://bitbucket.org/james-d-mitchell/viz`
7. Egri-Nagy, A., Francis, A.R.: BioGAP – software package for genome rearrangement calculations (2014), `https://bitbucket.org/egri-nagy/biogap`
8. Egri-Nagy, A., Gebhardt, V., Tanaka, M.M., Francis, A.R.: Group-theoretic models of the inversion process in bacterial genomes. Journal of Mathematical Biology, 23 pages (June 2013) (in press)

9. Ellson, J., Gansner, E.R., Koutsofios, E., North, S.C., Woodhull, G.: Graphviz and dynagraph – static and dynamic graph drawing tools. In: Graph Drawing Software, pp. 127–148. Springer (2003)

10. Fertin, G., Labarre, A., Rusu, I., Tannier, É., Vialette, S.: Combinatorics of genome rearrangements. MIT Press (2009)

11. Francis, A.R.: An algebraic view of bacterial genome evolution. Journal of Mathematical Biology, 26 pages (December 2013) (in press)

12. Gascuel, O. (ed.): Mathematics of Evolution and Phylogeny. OUP, Oxford (2005)

13. Hannenhalli, S., Pevzner, P.A.: Transforming men into mice (polynomial algorithm for genomic distance problem). In: Proceedings of the 36th Annual Symposium on Foundations of Computer Science, pp. 581–592 (1995)

14. Holt, D., Eick, B., O'Brien, E.: Handbook of Computational Group Theory. CRC Press (2005)

15. Kececioglu, J., Sankoff, D.: Exact and approximation algorithms for the inversion distance between two chromosomes. In: Apostolico, A., Crochemore, M., Galil, Z., Manber, U. (eds.) CPM 1993. LNCS, vol. 684, pp. 87–105. Springer, Heidelberg (1993)

16. Knuth, D.E.: The Art of Computer Programming: Sorting and searching. The Art of Computer Programming. Addison-Wesley (1998)

17. Miklós, I., Darling, A.E.: Efficient sampling of parsimonious inversion histories with application to genome rearrangement in yersinia. Genome Biology & Evolution 2009 (2009)

18. Siepel, A.C.: An algorithm to enumerate all sorting reversals. In: Proceedings of the Sixth Annual International Conference on Computational Biology (2002)

19. Tantau, T.: TikZ and PGF manual for version 2.10. University of Lübeck (2010), http://texample.net/

20. Watterson, G.A., Ewens, W.J., Hall, T.E., Morgan, A.: The chromosome inversion problem. Journal of Theoretical Biology 99(1), 1–7 (1982)

SgpDec: Cascade (De)Compositions of Finite Transformation Semigroups and Permutation Groups

Attila Egri-Nagy[1,3], James D. Mitchell[2], and Chrystopher L. Nehaniv[3]

[1] Centre for Research in Mathematics,
School of Computing, Engineering and Mathematics,
University of Western Sydney, Australia
A.Egri-Nagy@uws.edu.au
www.egri-nagy.hu
[2] School of Mathematics and Statistics,
University of St Andrews, United Kingdom
jdm3@st-and.ac.uk
http://www-groups.mcs.st-andrews.ac.uk/~jamesm/
[3] Centre for Computer Science & Informatics Research,
University of Hertfordshire, United Kingdom
C.L.Nehaniv@herts.ac.uk
http://homepages.herts.ac.uk/~comqcln/

Abstract. We describe how the SGPDEC computer algebra package can be used for composing and decomposing permutation groups and transformation semigroups hierarchically by directly constructing substructures of wreath products, the so called cascade products.

Keywords: transformation semigroup, permutation group, wreath product, Krohn-Rhodes Theory.

1 Introduction

Wreath products are widely used theoretical constructions in group and semigroup theory whenever one needs to build a composite structure with hierarchical relations between the building blocks. However, from a computational and engineering perspective they are less useful since wreath products are subject to combinatorial explosions and we are often interested only in substructures of them. Cascade products precisely build these substructures by defining the hierarchical connections explicitly. As input, given a group or a semigroup with unknown internal structure, the goal of cascade decomposition algorithms is to come up with a list of simpler building blocks and put them together in a cascade product, which realizes in some sense the original group or semigroup. Roughly speaking, for permutation groups, cascade product decompositions can be interpreted as putting the inner workings of the Schreier-Sims algorithm (generalized to any subgroup chain) into an external product form, therefore one can build cascade products isomorphic to the group being decomposed. For semigroups,

H. Hong and C. Yap (Eds.): ICMS 2014, LNCS 8592, pp. 75–82, 2014.
© Springer-Verlag Berlin Heidelberg 2014

Krohn-Rhodes decompositions [14] can be computationally represented by cascade products of transformation semigroups.

In this paper we describe how the GAP [9] package SGPDEC [6] implements cascade products and decomposition algorithms and we also give a few simple example computations. This description of the package only focuses on the core functionality of the package.

A *transformation* is a function $f : X \to X$ from a set to itself, and a *transformation semigroup* (X, S) of degree n is a collection S of transformations of X closed under function composition, $|X| = n$. In case S is a group of permutations of X, we call (X, S) a *permutation group*. Using automata theory terminology sometimes we call X the *state set*, often represented as a set of integers $\mathbf{n} = \{0, \ldots, n-1\}$. We write x^s to denote the new state resulting from applying a transformation $s \in S$ to a state $x \in X$.

2   Cascade Product by a Motivating Example

To motivate the definition of the cascade product, we consider how the mod-4 counter, the cyclic permutation group $(\mathbf{4}, \mathbb{Z}_4)$, can be constructed from two mod-2 counters. The direct product $\mathbb{Z}_2 \times \mathbb{Z}_2$ contains no element of order 4. Since $\mathrm{Aut}(\mathbb{Z}_2)$ is trivial there is only one semidirect product of $\mathbb{Z}_2$ and $\mathbb{Z}_2$, which equals their direct product. Their wreath product, $\mathbb{Z}_2 \wr \mathbb{Z}_2 \cong D_4$, the dihedral group of the square can be used to emulate a mod-4 counter, since $\mathbb{Z}_4 \hookrightarrow D_4$. But this construction is not efficient, beyond the required rotations the dihedral group has the flip-symmetry as well, doubling the size of the group. However, we would like to have a product construction that is isomorphic to $(\mathbf{4}, \mathbb{Z}_4)$.

This motivates the definition of *cascade products*: efficient constructions of substructures of wreath products, induced by explicit dependency functions [5]. Essentially, cascade products are transformation semigroups glued together by functions in a hierarchical tree. More precisely, let $((X_1, S_1), \ldots, (X_n, S_n))$ be a fixed list of transformation semigroups, and dependency functions of the form

$$d_i : X_1 \times \ldots \times X_{i-1} \to S_i, \quad \text{for } i \in \{1 \ldots n\}.$$

A *transformation cascade* is then defined to be an n-tuple of dependency functions $(d_1, \ldots, d_n)$, where d_i is a dependency function of level i. If no confusion arises, on the top level we can simply write $d_1 \in S_1$ instead of $d_1(\varnothing) \in S_1$. The cascade action is defined coordinatewise by $x_i^{d_i(x_1, \ldots, x_{i-1})}$, applying the results of the evaluated dependency functions (see Fig. 1), so that the cascade product can be regarded as a special transformation representation on the set $X_1 \times \ldots \times X_n$. The hierarchical structure allows us to conveniently distribute computation among the components (X_i, S_i), and perform abstractions and approximations of the system modelled as a cascade product. Then if W is a set of transformation cascades $(X_1, S_1) \wr_W \cdots \wr_W (X_n, S_n)$ denotes the transformation semigroup $(X_1 \times \cdots X_n, \langle W \rangle)$, where $\langle W \rangle$ is the semigroup of transformation cascades generated by W.

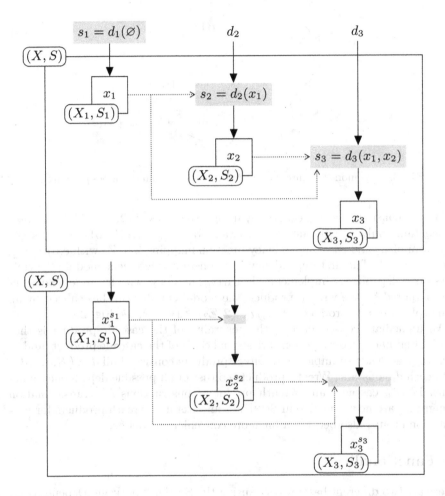

Fig. 1. Action in a cascade product of components $[(X_1, S_1), (X_2, S_2), (X_3, S_3)]$. The current state (x_1, x_2, x_3) (top) is transformed to the new state $(x_1^{s_1}, x_2^{s_2}, x_3^{s_3})$ (bottom) by the transformation cascade (d_1, d_2, d_3). The component actions s_i are calculated by evaluating the dependency functions of (d_1, d_2, d_3) on the states of the components above. The evaluations are highlighted and they happen at the same time. The dependencies, where the state information travels, are denoted by dotted lines.

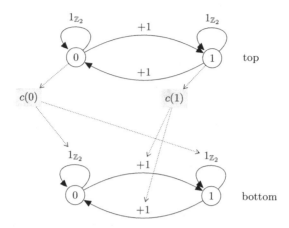

Fig. 2. Two mod-2 counters cascaded together to build a mod-4 counter

We can construct $(\mathbf{4}, \mathbb{Z}_4)$ exactly by using two copies of $(\mathbf{2}, \mathbb{Z}_2)$. The generator set contains only one permutation cascade $W = \{(+1, c)\}$, where $+1$ is the generator of $\mathbb{Z}_2$ and c is a dependency function mapping $\mathbf{2}$ to $\mathbb{Z}_2$ with $c(0) = 1_{\mathbb{Z}_2}$, and $c(1) = +1$. The first dependency is a constant (increment modulo 2) while the second dependency implements the carry. Therefore, with fewer dependencies than required by the wreath product, the mod-4 counter can be realized by an isomorphic cascade product: $(\mathbf{2}, \mathbb{Z}_2) \wr_W (\mathbf{2}, \mathbb{Z}_2) \cong (\mathbf{4}, \mathbb{Z}_4)$, see Fig. 2.

An immediate consequence of the generality of the cascade product is that several well-known constructions are special cases of the cascade product, and as such they are easy to implement. Direct products consist of all $d = (d_1, \ldots, d_n)$ with each d_i constant. Wreath products consist of all possible dependency functions. Direct, cascade, and wreath products constructions for transformation semigroups are now available in SGPDEC, and iterated wreath products for permutation groups also became a bit more convenient to define.

3   Functionality

There are two different basic ways of using the SGPDEC package. Depending on whether the starting point is a complex structure or a set of (simple) building blocks, we can do *decomposition* or *composition*.

3.1   Composition and Construction

The questions we aim to answer by constructing cascade products can be of the following types.

1. What is the (semi)group generated by a given set of transformation cascades?
2. What can be built from a given set of (simple) components?

The usual scenario is that for a list of components we give a set of cascades as a generating set. For instance, the quaternion group $Q = \langle i, j \rangle$ is not a semidirect product, but it embeds into the full cascade product $(\mathbf{2}, \mathbb{Z}_2) \wr (\mathbf{2}, \mathbb{Z}_2) \wr (\mathbf{2}, \mathbb{Z}_2)$, a group with 128 elements. Therefore, it can be built from copies of $\mathbb{Z}_2$. The dependency functions can only have two values, thus to define cascade permutations it is enough to give only those arguments that give $+1$ (the generator of $\mathbb{Z}_2$). A cascade permutation realizing i is defined by the dependency functions (d_1, d_2, d_3) where $d_2(0) = d_2(1) = d_3(0,0) = d_3(1,1) = +1$ and all other arguments map to the identity. Similarly, a cascade realizing j is defined by (d'_1, d'_2, d'_3) where $d'_1(\varnothing) = d'_3(0,0) = d'_3(0,1) = +1$, (see Fig. 3, note that the state values are shifted by 1). One can check that these two order 4 elements generate the 8-element quaternion group Q. Therefore by $W = \{(d_1, d_2, d_3), (d'_1, d'_2, d'_3)\}$ we have

$$(Q, Q) \cong (\mathbf{2}, \mathbb{Z}_2) \wr_W (\mathbf{2}, \mathbb{Z}_2) \wr_W (\mathbf{2}, \mathbb{Z}_2).$$

```
gap> Z2:=CyclicGroup(IsPermGroup,2);
Group([ (1,2) ])
gap> d:=Cascade([Z2,Z2,Z2],[[[1],(1,2)],[[2],(1,2)],
                           [[1,1],(1,2)],[[2,2],(1,2)]]);
<perm cascade with 3 levels with (2, 2, 2) pts, 4 dependencies>
gap> dprime:=Cascade([Z2,Z2,Z2],[[[],(1,2)],[[1,1],(1,2)],[[1,2],(1,2)]]);
<perm cascade with 3 levels with (2, 2, 2) pts, 3 dependencies>
gap> StructureDescription(Group([d,dprime]));
"Q8"
```

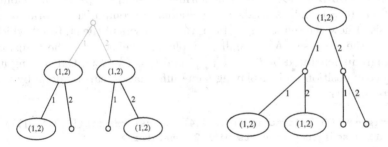

Fig. 3. Generators of a cascade representation of the quaternion group in a tree form. The edge labels are states, while the nodes contain the action. Empty node corresponds to the identity. The gray part of the tree is fixed.

3.2 Decomposition

1. What are the basic building blocks of a given (semi)group?
2. How can we represent it as a cascade product?

A typical scenario is that for a given composite semigroup or group we choose a decomposition algorithm which returns a cascade product.

Frobenius-Lagrange Decomposition. In the case of groups the decomposition uses the idea behind induction in representation theory (see e.g. [1]), so it traces back to Frobenius. Indeed, a special case of them comprises the well-known Krasner-Kaloujnine embeddings [13]. All we need here is just standard group theory, namely the action on cosets, hence the name *Frobenius-Lagrange Decomposition*.

How would someone come up with the generators cascades of the quaternion group in Section 3.1? The easiest solution is to use this group decomposition.

```
gap> Q := QuaternionGroup(IsPermGroup,8);
Group([ (1,5,3,7)(2,8,4,6), (1,2,3,4)(5,6,7,8) ])
gap> CQ := FLCascadeGroup(Q);
<cascade group with 2 generators, 3 levels with (2, 2, 2) pts>
```

The actual implementation takes a subgroup chain as input (chief series by default) and form the components by examining the coset space actions derived from the chain. Therefore, the decomposition method can be considered as a generalized Schreier-Sims algorithm [12].

Coordinatewise calculation in a cascade product can also be thought of as a sequence of refining approximate solutions. For instance, each completed step of an algorithm for solving the Rubik's Cube corresponds to calculating the desired value at a hierarchical level of some cascade product representation and it gives a configuration 'closer' to the solved state.

Holonomy Decomposition. For transformation semigroups the holonomy method [16,17,10,7,11,15,4] is used. The holonomy decomposition works by a close analysis of how the semigroup acts on those subsets of the state set which are images of the state set. As a small example let's define T as the transformation semigroup generated by $t_1 = \left(\begin{smallmatrix} 1 & 2 & 3 & 4 \\ 3 & 2 & 4 & 4 \end{smallmatrix}\right)$ and $t_2 = \left(\begin{smallmatrix} 1 & 2 & 3 & 4 \\ 3 & 3 & 1 & 3 \end{smallmatrix}\right)$. Calculating its holonomy decomposition and displaying some information can be done by the following commands:

```
gap> T:=Semigroup([Transformation([3,2,4,4]),Transformation([3,3,1,3])]);
<transformation semigroup on 4 pts with 2 generators>
gap> HT := HolonomyCascadeSemigroup(T);
<cascade semigroup with 2 generators, 3 levels with (2, 2, 4) pts>
gap> DisplayHolonomyComponents(SkeletonOf(HT));
1: 2
2: 2
3: (2,C2) 2
```

The displayed information tells us that this 13-element semigroup can be realized as the cascade product of four copies of the transformation monoid of constant maps of two points and one instance of $\mathbb{Z}_2$. The components are put together in a 3-level cascade product.

Holonomy decompositions are useful whenever a finite state-transition model of some process needs to be analyzed (e.g. [3]).

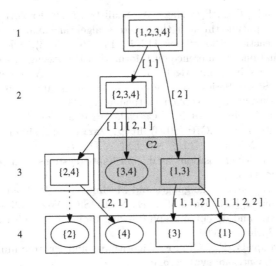

Fig. 4. Tiling picture – the internal details of the holonomy decomposition of the transformation semigroup T generated by t_1 and t_2. The numbers on the left denote the hierarchical levels (level 4 consists of singleton sets and it is needed by the holonomy algorithm but not a component of the cascade product). Outer boxes contain subsets that are mutually reachable from each other under the semigroup action. The arrows indicate how a subset is 'tiled' by its subsets, the arrow labels contain words (sequences of generators) that take a subset to one of its tiles. Dotted arrow means the tile is not an image. Roughly, the holonomy algorithm finds the components by checking the action of the semigroup on a set of tiles.

3.3 Visualization

SgpDec uses GraphViz, a widely used graph drawing package [8], for visualisation purposes. The underlying idea is that a function generates source code in the dot language for the given mathematical object. Then the actual figure can be generated separately to be included in papers, or using the Viz package [2] immediately displayed on screen from the Gap command line. Figure 3 and 4 were both auto-generated using GraphViz.

Acknowledgment. The work reported in this article was funded in part by the EU project BIOMICS, contract number CNECT-318202. This support is gratefully acknowledged.

References

1. Alperin, J.L., Bell, R.B.: Groups and Representations. Springer (1995)
2. Delgado, M., Egri-Nagy, A., Mitchell, J.D., Pfeiffer, M.: VIZ – GAP package for visualisation (2014), https://bitbucket.org/james-d-mitchell/viz

3. Dini, P., Nehaniv, C.L., Egri-Nagy, A., Schilstra, M.J.: Exploring the concept of interaction computing through the discrete algebraic analysis of the belousov-zhabotinsky reaction. Biosystems 112(2), 145–162 (2013), Selected papers from the 9th International Conference on Information Processing in Cells and Tissues

4. Dömösi, P., Nehaniv, C.L.: Algebraic Theory of Finite Automata Networks: An Introduction. SIAM Series on Discrete Mathematics and Applications, vol. 11. Society for Industrial and Applied Mathematics (2005)

5. Egri-Nagy, A., Nehaniv, C.L.: Cascade Product of Permutation Groups. arXiv:1303.0091v3 [math.GR] (2013), http://arxiv.org/abs/1303.0091

6. Egri-Nagy, A., Nehaniv, C.L., Mitchell, J.D.: SGPDEC – software package for hierarchical decompositions and coordinate systems, Version 0.7+ (2013), http://sgpdec.sf.net

7. Eilenberg, S.: Automata, Languages and Machines, vol. B. Academic Press (1976)

8. Ellson, J., Gansner, E.R., Koutsofios, E., North, S.C., Woodhull, G.: Graphviz and dynagraph static and dynamic graph drawing tools. In: Graph Drawing Software, pp. 127–148. Springer (2003)

9. The GAP Group: GAP – Groups, Algorithms, and Programming, Version 4.7.1 (2013), http://www.gap-system.org

10. Ginzburg, A.: Algebraic Theory of Automata. Academic Press (1968)

11. Holcombe, W.M.L.: Algebraic Automata Theory. Cambridge University Press (1982)

12. Holt, D., Eick, B., O'Brien, E.: Handbook of Computational Group Theory. CRC Press (2005)

13. Krasner, M., Kaloujnine, L.: Produit complet des groupes de permutations et problème d' extension de groupes. Acta Scientiarium Mathematicarum (Szeged) 14, 39–66 (1951)

14. Krohn, K., Rhodes, J.: Algebraic Theory of Machines. I. Prime Decomposition Theorem for Finite Semigroups and Machines. Transactions of the American Mathematical Society 116, 450–464 (1965)

15. Wells, C.: A Krohn-Rhodes theorem for categories. Journal of Algebra 64, 37–45 (1980)

16. Paul Zeiger, H.: Cascade synthesis of finite state machines. Information and Control 10, 419–433 (1967), plus erratum

17. Paul Zeiger, H.: Cascade Decomposition Using Covers. In: Arbib, M.A. (ed.) Algebraic Theory of Machines, Languages, and Semigroups, ch. 4, pp. 55–80. Academic Press (1968)

Approximating Generators for Integral Arithmetic Groups

Bettina Eick

TU Braunschweig, Germany
beick@tu-bs.de
www.icm.tu-bs.de/~beick

Abstract. We exhibit an implementation in the computer algebra system GAP of a method to *approximate* generators of an integral arithmetic group.

Keywords: arithmetic groups, finite approximation.

1 Introduction

A rational algebraic group $G(\mathbb{Q})$ is a subgroup of $GL(n, \mathbb{Q})$ defined by a finite set of polynomial equations $f_1, \ldots, f_l \in \mathbb{Q}[x_1, \ldots, x_{n^2}]$. More precisely, if $\overline{g} \in \mathbb{Q}^{n^2}$ denotes the list of the entries of $g \in GL(n, \mathbb{Q})$, then $G(\mathbb{Q}) = \{g \in GL(n, \mathbb{Q}) \mid f_i(\overline{g}) = 0 \text{ for } 1 \leq i \leq l\}$. Further, we write

$$G(\mathbb{Z}) = G(\mathbb{Q}) \cap GL(n, \mathbb{Z}).$$

A subgroup $H \leq GL(n, \mathbb{Q})$ is an arithmetic group in $G(\mathbb{Q})$ if H is commensurable with $G(\mathbb{Z})$; that is, the intersection $H \cap G(\mathbb{Z})$ has finite index in H and in $G(\mathbb{Z})$. The group H is an *integral arithmetic group* in $G(\mathbb{Q})$ if H is arithmetic in $G(\mathbb{Q})$ and satisfies $H \leq GL(n, \mathbb{Z})$. Thus H is an integral arithmetic group if and only if H is a subgroup of finite index in $G(\mathbb{Z})$.

Arithmetic groups have interesting applications. For example, many interesting linear groups such as the special linear groups $SL(n, \mathbb{Z})$ and the orthogonal groups $O(n, \mathbb{Z})$ are arithmetic groups. Further, Baues and Grunewald [1] prove that the outer automorphism group of a polycyclic-by-finite group is an arithmetic group and Grunewald and Segal [5] show that the automorphism group of a finitely generated nilpotent group is closely related to an arithmetic group.

Borel and Harish-Chandra [2] prove that every arithmetic group is finitely generated. However, despite the interest in the topic, it is still difficult to determine explicit generators for an arithmetic group described by its associated polynomials $f_1, \ldots, f_l$ and further information how it arises from $G(\mathbb{Z})$. Grunewald and Segal [4] introduce a (theoretical, but not practical) algorithm to determine a finite set of generators for an arithmetic group. Hence the problem of finding generators for an arithmetic group is algorithmically decidable. De Graaf and Pavan [3] exhibit a practical algorithm to determine generators for a unipotent arithmetic group. But a practical algorithm for the general case is not available.

H. Hong and C. Yap (Eds.): ICMS 2014, LNCS 8592, pp. 83–86, 2014.

2 Aims

The deterministic construction of generators for $G(\mathbb{Z})$ seems out of reach at current. Thus our aim is to *approximate* generators of $G(\mathbb{Z})$ and to implement methods for this purpose as software package for the computer algebra system GAP [6].

The main function of this software package takes as input a set of polynomials $f_1, \ldots, f_l$ associated to a rational algebraic group and a sequence of natural numbers $S = (s_1, \ldots, s_k)$. It determines a finite generating set for a group $G_S(\mathbb{Z})$ so that the following holds:

- $G_S(\mathbb{Z}) \leq G(\mathbb{Z})$ for each sequence S.
- $G_S(\mathbb{Z}) \leq G_{S'}(\mathbb{Z})$ if $S' = (s_1, \ldots, s_k, s_{k+1})$ extends $S = (s_1, \ldots, s_k)$.
- $G_S(\mathbb{Z}) = G(\mathbb{Z})$ for almost all sequences of length 1.

Once generators (or approximate generators) for $G(\mathbb{Z})$ are available, it is possible to determine generators (or approximate generators) for an integral arithmetic subgroup of $G(\mathbb{Z})$ if this subgroup is suitably described. In many applications these subgroups are described as stabiliser in $G(\mathbb{Z})$ of some explicitly described action. In this case on can apply a stabiliser algorithm to determine or approximate the desired integral arithmetic subgroup of $G(\mathbb{Z})$.

3 Approaches

Suppose that a list of polynomials $f_1, \ldots, f_l \in \mathbb{Q}[x_1, \ldots, x_{n^2}]$ is given. We assume throughout that the polynomials define a rational algebraic group; that is, the zeros of $f_1, \ldots, f_l$ in $GL(n, \mathbb{Q})$ form a subgroup of $GL(n, \mathbb{Q})$.

First we introduce some notation. For $m \in \mathbb{N}$ let $GL(n, m) = GL(n, \mathbb{Z}/m\mathbb{Z})$ and denote

$$\hat{GL}(n, m) = \{g \in GL(n, m) \mid det(g) \in \{-1, 1\}\}.$$

Next, let $\rho_m : \mathbb{Z} \to \mathbb{Z}/m\mathbb{Z}$. This natural epimorphism induces a homomorphism

$$\rho_m : GL(n, \mathbb{Z}) \to \hat{GL}(n, m).$$

Let $t \in GL(n, m)$. The matrix $s \in \mathbb{Z}^{n \times n}$ is called the normalised preimage of t, if $\rho_m(s) = t$ and the entries $s_{i,j}$ of s satisfy $-m/2 < s_{i,j} \leq m/2$. Note that this defines the normalised preimage of t uniquely. Also note that $\rho_m(det(s)) = det(t)$ and thus $det(s) \neq 0$. Hence s is an element of $GL(n, \mathbb{Q})$. As s is an integral matrix, it follows that s is an element of $GL(n, \mathbb{Z})$ if and only if $det(s) \in \{-1, 1\}$. We denote

$$\sigma_m : GL(n, m) \to GL(n, \mathbb{Q}) : t \mapsto s.$$

The following is a basic outline of our proposed method. Let $S = (s_1, \ldots, s_k)$ denote a sequence of natural numbers and write $m_i = s_1 \cdots s_i$ for $0 \leq i \leq k$.

(1) Initialise $U = \langle 1 \rangle \leq GL(n, \mathbb{Z})$.
(2) For i in $(1, \ldots, k)$ do
    (a) Determine $V_i = \langle g \in \hat{G}L(n, m_i) \mid f_i(\overline{g}) \equiv 0 \bmod m_i$ for $1 \leq i \leq l \rangle$.
    (b) Let $U_i = \rho_{m_i}(U)$ be the image of U in $\hat{G}L(n, m_i)$.
    (c) Let T_i be a transversal of U_i in V_i.
    (d) For each $t \in T_i$ do
        (i) determine the normalised preimage $s = \sigma_{m_i}(t)$.
        (ii) check whether $s \in G(\mathbb{Z})$ by evaluating $f_1(\overline{s}), \ldots, f_l(\overline{s})$ and $det(s)$.
        (iii) if yes, then replace U by $\langle U, s \rangle$.
(3) Return U.

In the talk we discuss variations of this method as well as open problems around it and possible connections to ideas from algebraic geometry.

An extreme case is the case of a sequence of length 1. In this case, the following straightforward observation can be made.

Theorem 1. *For almost all $m \in \mathbb{N}$ it follows that $G_{(m)}(\mathbb{Z}) = G(\mathbb{Z})$.*

Proof. By construction, $G_S(\mathbb{Z}) \subseteq G(\mathbb{Z})$ for any sequence S. It remains to investigate the converse inclusion. Recall that the group $G(\mathbb{Z})$ is finitely generated by the theorem of Borel and Harish-Chandra [2]. Let $g_1, \ldots, g_r$ denote a set of generators for $G(\mathbb{Z})$. Let m' denote the maximum among the absolute values of the entries of these generators and choose $m > 2m'$. Then $g_i = \sigma_m(\rho_m(g_i))$ for $1 \leq i \leq r$ holds and hence the above algorithm detects each of these generators.

Remark 1. If a bound m on the absolute values of the entries of a finite generating set of $G(\mathbb{Z})$ could be determined from $f_1, \ldots, f_l$, then the above method could be turned into a deterministic method to determine a finite generating set of $G(\mathbb{Z})$.

4   An Example

For an example calculation we consider the group

$$G = \langle g_1, \ldots, g_5 \mid [g_1, g_2] = g_5, [g_3, g_4] = g_5^2, [g_i, g_j] = 1$$
$$\text{for all } i < j \text{ with } (i, j) \neq (1, 2), (3, 4) \rangle.$$

Then G is a torsion-free nilpotent group of class 2 with $G' = Z(G) = \langle g_5 \rangle$. There exists a natural homomorphism

$$\nu : Aut(G) \to Aut(G/G') : \alpha \mapsto \alpha_{G/G'}.$$

The kernel of ν consists of those automorphism inducing the identity on G/G'. This kernel is naturally isomorphic to $Z^1(G/G', G')$ and thus is free abelian of rank 4. The critical part of the determination of $Aut(G)$ is the construction of the image of ν. Let

$$D = \begin{pmatrix} 0 & -1 & 0 & 0 \\ 1 & 0 & 0 & 0 \\ 0 & 0 & 0 & -2 \\ 0 & 0 & 2 & 0 \end{pmatrix}.$$

Then the entries $d_{i,j}$ of D satisfy $[g_j, g_i] = g_5^{d_{i,j}}$. It is not difficult to observe that

$$image(\nu) \cong M(D) = \{g \in GL(4, \mathbb{Z}) \mid gDg^T = D\}.$$

Note that $M(D)$ is an arithmetic group. We apply (a variation of) the algorithm proposed above using the sequence $S = (2, 2, 2)$ and obtain a subgroup U of $G(\mathbb{Z})$ generated by 16 generators. Among these 16 generators are the following two matrices

$$m_1 = \begin{pmatrix} 1 & -1 & 0 & 1 \\ 0 & 1 & 0 & 0 \\ 0 & -2 & -1 & 0 \\ 0 & -2 & -1 & -1 \end{pmatrix} \quad \text{and} \quad m_2 = \begin{pmatrix} -2 & 1 & -1 & -1 \\ -1 & 0 & 0 & 0 \\ -2 & 0 & 0 & 1 \\ 0 & 0 & -1 & -1 \end{pmatrix}.$$

Thus $\langle m_1, m_2 \rangle \leq U \leq M(D)$ by construction. We note that the images of $\langle m_1, m_2 \rangle$ and U in $GL(4, p)$ for the primes $p \in \{2, 3, 5, 7, 11, 13, 17, 19\}$ agree. Based on this, we propose the following conjecture.

Conjecture 1. The matrices m_1 and m_2 generate $M(D)$.

References

1. Baues, O., Grunewald, F.: Automorphism groups of polycyclic-by-finite groups and arithmetic groups. Publ. Math. Inst. Hautes Études Sci. 104, 213–268 (2006)
2. Borel, A., Harish-Chandra: Arithmetic subgroups of algebraic groups. Ann. of Math. (2) 75, 485–535 (1962)
3. de Graaf, W.A., Pavan, A.: Constructing arithmetic subgroups of unipotent groups. J. Algebra 322(11), 3950–3970 (2009)
4. Grunewald, F., Segal, D.: Some general algorithms, I: Arithmetic groups. Ann. Math. 112, 531–583 (1980)
5. Grunewald, F., Segal, D.: Some general algorithms, II: Nilpotent groups. Ann. Math. 112, 585–617 (1980)
6. The GAP Group. GAP – Groups, Algorithms and Programming, Version 4.4 (2005), http://www.gap-system.org

Software for Groups: Theory and Practice

Alexander Hulpke

Department of Mathematics,
Colorado State University,
1874 Campus Delivery,
Fort Collins, CO, 80523, USA
hulpke@math.colostate.edu
www.math.colostate.edu/~hulpke

Abstract. Some problems in algorithmic group theory have good, well-understood, solutions as far as theory is concerned. This often is because of considering problems from other areas as solved, or because of ignoring practical aspects that are rarely if ever spelled out. This article considers some such aspects that have turned out to be difficult in the context of writing group theoretic software.

1 Introduction

Solving group theoretic questions has been one of the first tasks that computers have been put to. Indeed, already in 1945 Alan Turing writes [14, p.3]:

> [...] switch from calculating the energy levels of the neon atom to the enumeration of groups of order 720.

Since then many group theoretic algorithms have been developed. Trying to give even only a summary is beyond the scope of this note, a good starting point for the interested reader is the monograph [8].

Today, computer calculations have become part of the group theorists toolkit. This development has been aided by the availability of systems – namely Cayley [5], successor Magma [2], as well as GAP [7] – that are broad in scope and don't require knowledge of the details of the underlying algorithms.

The reduction of a problem to calculations done on a computer has become mostly uncontroversial, criticisms being not so much on the side of correctness but based on esthetics and the lack of insight in quoting a computer result.

Nevertheless software does not always do in practice, what in theory should be easy (or what has been proven as being easy: The asymptotic behavior described by complexity estimates might consider any practically feasible case as a constant). Sometimes there is a notable chasm between theoretical results and practical software. (Vice versa some problems have good practical solutions for any case that ever occurred in practice, but seem to be theoretically intractable.)

H. Hong and C. Yap (Eds.): ICMS 2014, LNCS 8592, pp. 87–91, 2014.

Many of these issues should be of interest to other areas of mathematical software, be it as a confirmation of issues other software also faces, as possibly interesting problems in other areas of mathematics, or even as application of a solution method that is not yet know in group theory.

2   Deciding on a Method

Often there are different algorithms for achieving the same result, the better performing one being applicable only under particular circumstances. When using these algorithms in a system in the context of a larger calculation it therefore is desirable to have the system select the best method available to perform a particular calculation. The type system [4] of **GAP4** tries to do this by associating bits to known properties, it then selects amongst multiple applicable methods the one that meets most bits as being the most specialized one. In basic situations this works well. However it is easy to construct situations in which a property test is potentially expensive (for example testing for finiteness might not even terminate). Still, in some cases it would be far better to first test the property, than having to use a slower, more general, algorithm.

Q1: How does one decide on the "typical" case when algorithms test for properties that can be expensive, but can also speed up calculations?

Q2: Can one indicate to the system how expensive a certain calculations will be, to let it decide on whether a particular test is worth doing first?

Q3: If not, can/should one alert the user about such situations, without overwhelming the user who knows to be in a case where tests don't help?

A similar question comes up when considering the resources to utilize. A trade-off between speed and memory is often possible, however systems with large amounts of memory often are shared between tasks and a system should not hamper other users by always choosing to sacrifice memory for speed.

Q4: How much memory should the system use if it would speed up calculations?

Q5: How many processors should a parallel calculation use if a calculation can be parallelized?

Note that the prototypical group theoretic calculation of identifying the orbit in which an object lies often has super-linear speedup on a parallel setting if orbits are given by representatives. This is because a test for mapping one object to another can terminate if it succeeds, but needs to exhaust the full space to show nonequivalence. Thus once one mapping test succeeds, all other parallel tests can be terminated. (Q5 thus cannot simply be answered by Amdahl's law.)

3   Integer Matrix Normal Forms

The issue of exact integer values becoming large has been long recognized in computer algebra as requiring care in the design of algorithms. It usually is not relevant when working with finite groups, but as soon as groups become potentially infinite it raises its head.

The prototype of this is the case of determining the quotient G/G' of a finitely presented group G. Writing the relators of G in abelianized form (i.e. $aba = b^2$ becomes a^2b^{-1}) as rows in a matrix A, we determine the Smith Normal Form $S = P \cdot A \cdot Q$ of A. The diagonal entries of S determine the structure of G/G', the entries of Q^{-1} the homomorphism from G to an abelian group with this structure in normal form.

To make such calculations effective we would not only like this calculation to run fast, but even more to keep the entries of Q^{-1} small.

Q6: Calculate a Smith Normal Form while minimizing the transforming matrices (in a suitable norm).

In recent work [6] this same problem came up in a slightly different context: Let $G \leq \mathrm{SL}_n(\mathbb{Z})$ and $\varphi \colon G \to \mathrm{SL}_n(\mathbb{Z}/\mathbb{Z}_m)$ be the natural reduction modulo m. We want to compute pre-images under φ, that is given a matrix A such that $\det(A) \equiv 1 \pmod{m}$, find $B \equiv A \pmod{m}$ with $\det(B) = 1$. We can find B by decomposing $A = P^{-1}SQ^{-1}$, then the nonidentity determinant will be in S and $B = P^{-1}Q^{-1} \equiv A \pmod{m}$ has determinant 1. To keep the coefficients in B small we would like to keep those of P and Q small as well.

4   Polynomial Equations

A principal tool for working with finitely presented groups is to find quotients (or to enlarge existing quotients). By using ring structures to represent the quotient group over, the relations for the group then produce equations over the ring that seem to be amenable to Gröbner Basis methods. Assume that $F = \langle x_1, \ldots, x_n \rangle$ is a free group, $R \lhd F$ is the normal subgroup generated by the relators $r_1(\mathbf{x}), \ldots, r_k(\mathbf{x})$, and $G = F/R$. Then a map $x_i \mapsto y_i$ given on the group generators extends to a homomorphism if and only if $r_j(\mathbf{y}) = 1$ for all j.

We first consider the case considered in [13]. Here the y_i are taken to be in a matrix group as matrices with variable entries. (Assumptions on the module structure are used to reduce the number of variables.) The conditions $r_j(\mathbf{y}) = 1$ and for the determinants being nonzero then give a system of polynomial equations in these variables. As multiple homomorphisms might be possible, there typically are multiple solutions in multiple characteristics, but not all solutions correspond to homomorphisms with different kernels.

Q7: Describe the different group quotients given by such a variety.

A second case is that of so-called *quotient algorithms* [12,11]: We assume knowledge of a homomorphism $\varphi \colon G \to H$ to some (often finite) group H and are looking for a homomorphism $\psi \colon G \to K$ with $\ker \psi < \ker \varphi$ and $N = \ker \varphi/\ker \psi$ a vector space (or $\mathbb{Z}$-module). The image of ψ thus can be considered as an extension of N by H. The relations for G then become equations for N as an H-module. (For example, suppose $aba = 1$ is a relation for G and $a \mapsto g \cdot n$, $b \mapsto hm$ with n, m variable entries in N. Then $1 = aba \mapsto gnhmgn = ghg \cdot n^{hg}m^gn$ with ghg evaluating to a constant in N.) The resulting module presentation describes the largest possible choice for N as a quotient for G.

To describe ψ concretely, it is necessary to determine a basis for N and matrices for the action of G on N. This is essentially a non-commutative Gröbner basis calculation. Again it is of interest to solve this in all characteristics simultaneously

Q8: Given a module presentation, find a basis, as well as matrices describing the action of algebra generators.

5   Theoretical Knowledge

The classification of finite simple groups is the 800lb gorilla in the background of finite group theory. Unsurprisingly, many algorithms reduce to the case of simple groups as base case. In this situation the algorithm then will use *constructive recognition* to construct an isomorphism to the group in its "textbook" form (say A_n represented as even permutations of degree n) and fetch tabulated information about the group that is "taken from the literature". Examples would be the structure of the outer automorphism group, or a list of representatives of maximal subgroups.

Clearly one would like a computer system to have access to such data. (Indeed some results of prior (computer) classifications – such as groups of small order [1], or transitive permutation groups [9] – probably only can be disseminated in form of a computer data base.)

Given the heterogeneity of information, such data bases most likely have to contain not just "numerical" information, but also procedures that construct parameterized objects. Often there are extensive case distinctions depending on parameter classes.

As such data then is used within algorithms to make classification decisions, correctness is of utmost importance. However, there is often a history of theoretical classifications that had to be corrected later. (A prominent example of this is the classification of maximal subgroups of classical groups [10], which just recently was redone [3].) For the software

Q9: How do we ensure/check that a theoretical classification was implemented correctly, in particular if it involves many special cases?

Q10: How do we indicate that a substantial theoretical classification was used in a result? How do we deal with (keeping track, changing code, correcting previously wrong results) subsequent corrections of the classification?

6   Making the Software Usable

Finally, to make software actually usable, there are practical aspects. These include not only maintenance and bug fixes, but already the initial step of getting it to run on a user's machine.

Q11: Provide a user-friendly distribution and an installation process on common platforms that follows standard conventions.

Work towards this goal is crucial for software to stay usable as a research tool. It unfortunately don't fit well in the current model of publications and research grants the community uses to acknowledge and reward work. Finding appropriate ways to sustain such work is a task the mathematical community will need to solve in the future if mathematical software is to blossom.

References

1. Besche, H.U., Eick, B., O'Brien, E.A.: A millennium project: constructing small groups. Internat. J. Algebra Comput. 12(5), 623–644 (2002)
2. Bosma, W., Cannon, J., Playoust, C.: The MAGMA algebra system I: The user language. J. Symbolic Comput. 24(3/4), 235–265 (1997)
3. Bray, J.N., Holt, D.F., Roney-Dougal, C.M.: The maximal subgroups of the low-dimensional finite classical groups. London Mathematical Society Lecture Note Series, vol. 407. Cambridge University Press, Cambridge (2013); With a foreword by Martin Liebeck
4. Breuer, T., Linton, S.: The GAP 4 type system – Organising algebraic algorithms. In: Gloor, O. (ed.) Proceedings of the 1998 International Symposium on Symbolic and Algebraic Computation, pp. 38–45. ACM Press (1998)
5. Cannon, J.J.: An introduction to the group theory language, Cayley. In: Atkinson, M.D. (ed.) Computational Group Theory, pp. 145–183. Academic Press (1984)
6. Detinko, A., Flannery, D., Hulpke, A.: Algorithms for arithmetic groups with the congruence subgroup property (submitted, 2014)
7. The GAP: Group, GAP – Groups, Algorithms, and Programming, Version 4.6.3 (2013), http://www.gap-system.org
8. Holt, D.F., Eick, B., O'Brien, E.A.: Handbook of Computational Group Theory. Chapman & Hall/CRC, Boca Raton (2005)
9. Hulpke, A.: Constructing transitive permutation groups. J. Symbolic Comput. 39(1), 1–30 (2005)
10. Kleidman, P., Liebeck, M.: The subgroup structure of the finite classical groups. London Mathematical Society Lecture Note Series, vol. 129. Cambridge University Press, Cambridge (1990)
11. Lo, E.H.: A polycyclic quotient algorithm. J. Symbolic Comput. 25(1), 61–97 (1998)
12. Niemeyer, A.C.: A finite soluble quotient algorithm. J. Symbolic Comput. 18(6), 541–561 (1994)
13. Plesken, W., Souvignier, B.: Analysing finitely presented groups by constructing representations. J. Symbolic Comput. 24(3-4), 335–349 (1997); Computational algebra and number theory, London (1993)
14. Turing, A.: Proposed electronic calculator, Paper e.882, National Physics Laboratory (1945),
http://www.alanturing.net/turing_archive/archive/p/p01/p01.php

Computation of Genus 0 Belyi Functions

Mark van Hoeij[1] and Raimundas Vidunas[2]

[1] Department of Mathematics, Florida State University, Tallahassee,
Florida 32306, USA
hoeij@math.fsu.edu
http://www.math.fsu.edu/~hoeij/
[2] Lab of Geometric & Algebraic Algorithms,
Department of Informatics & Telecommunications,
National Kapodistrian University of Athens, Panepistimiopolis 15784, Greece
rvidunas@gmail.com
http://users.uoa.gr/~rvidunas

Abstract. A tool package for computing genus 0 Belyi functions is presented, including simplification routines, computation of moduli fields, decompositions, dessins d'enfant. The main algorithm for computing the Belyi functions themselves is based on implied transformations of the hypergeometric differential equation to Fuchsian equations, preferably with few singular points. This gives a fast way to compute the Belyi functions (of degree 60 and beyond) with nearly regular branching patterns.

Keywords: Belyi functions, Heun functions, pull-back transformations.

1 Introduction

Although Belyi functions is a captivating field of research in algebraic geometry, Galois theory and related fields, their computation is still considered hard. Grothendieck [1, pg. 248] doubted that *"there is a uniform method for solving the problem by computer"*. The offered software efficiently computes Belyi functions of genus 0 with nearly regular branching patterns (or dessins d'enfant). Recall that a *Belyi function* of genus 0 is a rational function $\varphi : \mathbb{P}^1 \to \mathbb{P}^1$ that branches only in the 3 fibers $\varphi(x) \in \{0, 1, \infty\}$.

The mentioned near-regularity is defined as follows. Given positive integers k, ℓ, m, a Belyi function $\varphi : \mathbb{P}^1 \to \mathbb{P}^1$ is called (k, ℓ, m)-*regular* if all points above $\varphi = 1$ have the branching order k, all points above $\varphi = 0$ have the branching order ℓ, and all points above $\varphi = \infty$ have the branching order m. Given yet another positive integer n, a Belyi function $\varphi : \mathbb{P}^1_x \to \mathbb{P}^1_z$ is called (k, ℓ, m)-*minus-n-regular* if, with exactly n exceptions, all points above $\varphi = 1$ have the branching order k, all points above $\varphi = 0$ have the branching order ℓ, and all points above $\varphi = \infty$ have the branching order m. An example of a $(2, 3, 9)$-minus-4-regular function is the degree 12 Belyi function[1].

---

[1] One can check that the numerator of $\varphi(x) - 1$ is a square. By the Hurwitz formula or checking $\varphi'(x)$, there is no branching in other fibers. The 4 exceptional points are $x = \infty$, $x = 0$, $x = 9$, $x = 1$.

H. Hong and C. Yap (Eds.): ICMS 2014, LNCS 8592, pp. 92–98, 2014.

$$\varphi(x) = \frac{64x^2(x-3)^9(x-9)}{27(x-1)(8x^3 - 72x^2 - 27x + 27)^3}.$$

The nearly regular Belyi functions transform between Fuchsian equations with a small number (say, ≤ 4) of singularities. Generally, a (k, ℓ, m)-minus-n-regular Belyi function with $1 \notin \{k, \ell, m\}$ pulls-back a hypergeometric equation with the local exponent differences $1/k, 1/\ell, 1/m$ to a Fuchsian equation with n singularities (after a proper normalization). In particular, minus-3-regular functions give pull-back transformations between hypergeometric equations, while minus-4-regular functions give pull-backs to Heun equations[2]. In fact, the main algorithm of our software utilizes the implied pull-back transformations of the hypergeometric differential equation.

The software was developed while classifying [5] hypergeometric-to-Heun transformations in the $1/k + 1/\ell + 1/m < 1$ case. The maximal degree of these transformations is 60. The software computes the degree 60 functions (two cubic Galois orbits) within 2 min. on Maple 15 and 2.66 GHz Mac. The complete list of these minus-4-regular functions consists of 366 Galois orbits. The largest Galois orbit (i.e., the degree of the moduli field) has size 15, for a degree 37 branching pattern; it is computed within 5 min. The largest computation (14 min.) is for a degree 44 branching pattern with Galois orbits of size 3 and 13. Extensive results of these computations are presented in [4], along with the main Maple program in ComputeBelyi.mpl, and several other Maple routines for simplification, decomposition routines, and computation of dessins d'enfant, moduli fields and (if applicable) the obstruction conic for realization fields.

2   Computing Belyi Functions

Most straightforwardly, a (k, ℓ, m)-minus-n-regular Belyi function is determined by the polynomial identity

$$P^\ell U = Q^m V + R^k W \tag{1}$$

as $\varphi(x) = P^\ell U / Q^m V$. Here P, Q, R are monic polynomials in $\mathbb{C}[x]$ whose roots are the points with the prescribed branching orders ℓ, m, k; and U, V, W are polynomials whose roots are the n exceptional points. The polynomials P, Q, R should not have multiple or common roots, and one of U, V, W may be assumed to be monic. The degrees of these polynomials are set by the branching pattern and the assignment[3] of $x = \infty$. Their coefficients are to be determined. The

---

[2] Examples of (k, ℓ, m)-regular functions the well-known Galois coverings $\mathbb{P}^1 \to \mathbb{P}^1$ of degree 12, 24, 60 with the tetrahedral, octahedral or icosahedral monodromies, respectively. We take $k = 2$, $\ell = 3$, $m \in \{3, 4, 5\}$ for that.

[3] It is convenient to assign $x = \infty$ to a *bachelor point*, i.e., a point with a unique branching order for its fiber. Due to Möbius transformations, we may assign $x = 0$ and $x = 1$ to other bachelor points if those exist. It is usually efficient to assign $x = \infty$ even if there are no bachelor points, though then solutions can be obtained over larger number fields than necessary. Rather than assigning $x = 0$, we may assume the shape $x^d + \star x^{d-2} + \ldots$ of one of the polynomials. If $x = 1$ is not assigned, the undetermined coefficients are weighted-homogeneous by the scaling $x \mapsto \alpha x$.

straightforward method just expands (1) and compares the coefficients to x. This is not practical for Belyi functions of degree ≥ 12. One reason is numerous *parasitic* [2] solutions where the three terms in (1) have common roots. Parasitic solutions may even arise in families of positive dimension.

Differentiation helps to compute Belyi functions more efficiently, as is occasionally demonstrated in the literature. In particular, the roots of $\varphi'(x)$ are the branching points above $\varphi = 0$ and $\varphi = 1$ with the multiplicities reduced by 1. We conclude the following shape of the logarithmic derivatives of $\varphi(x)$ and $\varphi(x) - 1$:

$$\frac{\varphi'(x)}{\varphi(x)} = h_1 \frac{R^{k-1} W}{P Q F}, \qquad \frac{\varphi'(x)}{\varphi(x) - 1} = h_2 \frac{P^{\ell-1} U}{Q R F}. \tag{2}$$

Here h_1, h_2 are constants, and F is the product of irreducible factors of $U V W$, each to the power 1. If $x = \infty$ lies above $\varphi = \infty$ then

$$h_1 = h_2 = [\text{ the branching order at } x = \infty], \tag{3}$$

as this is the residue of both logarithmic derivatives at $x = \infty$. On the other hand,

$$\frac{\varphi'(x)}{\varphi(x)} = \ell \frac{P'}{P} + \frac{U'}{U} - m \frac{Q'}{Q} - \frac{V'}{V}, \quad \text{etc.} \tag{4}$$

Pairs of expressions for the logarithmic derivatives of $\varphi(x)$, $\varphi(x) - 1$ lead to a stronger system of algebraic equations, usually of smaller degree and with less parasitic solutions. If $k = 2$, the polynomial R can be eliminated symbolically. This ansatz does not use the location $\varphi = 1$ of the third fiber. Therefore U, V, W can be assumed to be monic as well, and then (1) can be adjusted by constant multiples at the latest stage.

To get an even more restrictive system of algebraic equations, we utilize the near-regularity of $\varphi(x)$. The Gauss hypergeometric equation is

$$\frac{d^2 y(z)}{dz^2} + \left(\frac{C}{z} + \frac{A + B - C + 1}{z - 1} \right) \frac{dy(z)}{dz} + \frac{A B}{z (z - 1)} y(z) = 0. \tag{5}$$

This is a canonical Fuchsian equation with three singularities: $z = 0$, $z = 1$, $z = \infty$. The local exponent differences[4] at them are $1 - C$, $C - A - B$, $A - B$, respectively. We assume these numbers to equal $1/\ell, 1/k, 1/m$, respectively. Let us apply a *pull-back transformation* of the form

$$z \longmapsto \varphi(x), \qquad y(z) \longmapsto Y(x) = \theta(x) \, y(\varphi(x)), \tag{6}$$

where $\varphi(x)$ is a rational function, and $\theta(x)$ is a radical function (an algebraic root of a rational function). A blunt symbolic computation gives this lemma.

---

[4] In particular, the local exponents at $z = 0$ equal 0 and $1 - C$, as visible from the general local solutions $_2F_1(A, B, C; z)$ and $z^{1-C} \, _2F_1(1 + A - C, 1 + B - C, 2 - C; z)$.

Lemma 1. *Let $\varphi(x)$ be a Belyi map determined by (1). The hypergeometric equation (5) with*

$$A = \frac{1}{2}\left(1 - \frac{1}{k} - \frac{1}{\ell} - \frac{1}{m}\right), \quad B = \frac{1}{2}\left(1 - \frac{1}{k} - \frac{1}{\ell} + \frac{1}{m}\right), \quad C = 1 - \frac{1}{\ell}$$

is transformed to the following differential equation under the pull-back transformation $z \mapsto \varphi(x)$, $y(z) \mapsto (Q^m V)^A Y(\varphi(x))$:

$$\frac{d^2 Y(x)}{dx^2} + \left(\frac{F'}{F} - \frac{U'}{\ell U} - \frac{V'}{m V} - \frac{W'}{k W}\right)\frac{Y(x)}{dx} +$$

$$+ A\left[B\left(\frac{h_1 h_2 \, P^{\ell-2} R^{k-2} U W}{Q^2 F^2} - \frac{m^2 Q'^2}{Q^2} - \frac{V'^2}{V^2}\right) + \frac{m Q''}{Q} + \frac{V''}{V} +\right.$$

$$\left. + \left(\frac{1}{k} + \frac{1}{\ell}\right)\frac{m Q' V'}{Q V} + \left(\frac{m Q'}{Q} + \frac{V'}{V}\right)\left(\frac{F'}{F} - \frac{U'}{\ell U} - \frac{V'}{V} - \frac{W'}{k W}\right)\right] Y(x) = 0.$$

On the other hand, transformation (6) multiplies the exponent differences by the branching order of $\varphi(x)$ at each x-point. The points with the respectively prescribed branchings k, ℓ, m have the exponent difference 1, and become non-singular with proper $\theta(x)$. The transformed equation will have n singularities. The location of its singularities is determined by U, V, W. There will be $n - 3$ new variables, the *accessory parameters* of the target equation. The terms to $dY(x)/dx$ are always the same for the target and symbolically computed equations. But comparison of the terms to $Y(x)$ gives new algebraic equations between the undetermined coefficients, unless $A = 0$. The key contribution of this trick is to simplify Q from the denominator of the $Y(x)$ coefficient.

Combination of the logarithmic derivative ansatz and Lemma 1 usually allows straightforward elimination of most accessory parameters and coefficients of P, Q, R. If $x = \infty$ is assigned above $\varphi = \infty$, usually only $2n - 5$ variables are left for hard Gröbner basis computations. Increasing n by 1 basically adds two new variables: the location of the new singularity and an accessory parameter. If $k = 2$, $\ell = 3$ and $m \neq 6$, the polynomials R, P can be eliminated symbolically.

3   Examples

Consider computation of degree 15 Belyi functions with the branching pattern[5] $6\,[2] + 1 + 1 + 1 = 5\,[3] = 2\,[7] + 1$. We assign the bachelor non-branching point to $x = \infty$. The points of the same order are described by monic polynomials P, Q, R, W without multiple roots, of degree 5, 2, 6, 3 respectively. We have $U = V = 1$. Rather than fixing $x = 0$, we normalize $Q = x^2 + c$ by a translation $x \mapsto x + \beta$. Now only scaling Möbius transformations $x \to \alpha x$ are left to act. The logarithmic derivative ansatz gives

$$R = 3P'Q - 7PQ', \qquad P^2 = 2QR'W + QRW' - 7Q'RW.$$

---

[5] We are thus looking for $(2, 3, 7)$-minus-4 regular functions. Following the notation in [5], the regular branching orders are given in square brackets.

Lemma 1 gives

$$\frac{13}{84}\left(\frac{P}{Q^2 W} - \frac{49Q'^2}{Q^2}\right) + \frac{7Q''}{Q} + \frac{7Q'W'}{2QW} = \frac{135\,(x-q)}{28W},$$

where q is an accessory parameter. Symbolic elimination of R, P on Maple gives a non-linear differential equation for $W = x^3 + w_1 x^2 + w_2 x + w_3$, and then an expanded polynomial expression of degree 8 in x. The coefficients are weighted-homogeneous equations for the undetermined c, w_1, w_2, w_3, q. Maple solves them immediately. There are 4 Galois orbits of solutions, 3 of them parasitic[6]. The proper solution is defined over a cubic field K. A common problem with computing Belyi functions is simplification of the moduli field and of the Belyi function itself. We use LLL techniques techniques to find a small field polynomial, $\xi^3 + 2\xi^2 + 6\xi - 8$ in this case. The shortest expression[7] for $\varphi(x)$ we found is

$$\varphi(x) = \frac{1162 + 4282\xi + 1523\xi^2}{27} \frac{(14x + 12 + \xi - \xi^2)^3}{3x + 3 - \frac{1}{2}\xi^2} \times$$

$$\frac{\left(x^4 + x^3 - (1 + 2\xi + \frac{1}{2}\xi^2)x^2 - (5 + \xi + \frac{3}{2}\xi^2)x - 4\right)^3}{\left(5x^2 + (13 + 3\xi - \frac{1}{2}\xi^2)x + 4 + 4\xi + \xi^2\right)^7}. \qquad (7)$$

This Belyi function can be composed with a quartic covering to obtain a degree 60 covering with the branching pattern $30\,[2] = 20\,[3] = 8\,[7] + 1 + 1 + 1 + 1$.

The implied diffferential relations explain appearance of Chebyshev and Jacobi polynomials in certain Belyi functions. As is known, Chebyshev polynomials

$$T_n(x) = \cos(n \arccos x), \qquad U_n(x) = \frac{\sin(n \arccos x)}{\sin x}. \qquad (8)$$

define the Belyi functions

$$\varphi(x) = T_n(1 - 2x)^2, \qquad \varphi(x) - 1 = 4\,x\,(x-1)\,U_{n-1}(1 - 2x)^2 \qquad (9)$$

---

[6] Two parasitic solutions have $c = w_2 = w_3 = 0$ but different q/w_1. The third solution is a degree 9 Belyi function with the branching pattern $4\,[2] + 1 = 3\,[3] = [7] + 1 + 1$, defined over $\mathbb{Q}(\sqrt{-7})$. The moduli field is $\mathbb{Q}$ actually, but the quadratic extension occurs because $x = \infty$ is assigned to a non-bachelor point (from the parasitic perspective). For comparison, the logarithmic derivative ansatz alone gives 9 Galois orbits of parasitic solutions, of total degree 14. A degree 54 example in [6, §2.4] is computed with just 3 parasitic Galois orbits, but the logarithmic derivative ansatz would give hundreds of parasitic solutions.

[7] Our standard routine focuses on finding the shortest expression for W up to Möbius transformations. It finds a small defining polynomial for the minimal degree 9 field (over $\mathbb{Q}$) containing the roots of W. This polynomial necessarily has a cubic factor over K, which is identified with W by a Möbius transformation that matches the 3 roots. Expression (7) was obtained by observing that P factors over K into a linear and a quartic factor P^*. We looked for a small defining polynomial P' for the degree 12 field for the roots of P^*, with the additional condition that the j-invariant of the four roots (of a quartic factor of P' over K) equals the j-invariant for P^*. This polynomial P' was found immediately from the LLL basis of the field computation. A Möbius transformation identifies the short quartic factor of P' with P^*.

with the linear dessins d'enfant like •—○—•—○—•—○—•—○—•. This follows from the logarithmic derivative ansatz already: assuming $\varphi(x)$, $\varphi(x) - 1$ proportional to $F^2, x(x - 1)G^2$ (with F, G of degree n, $n - 1$, respectively) leads to

$$2n\,G = 2\,F', \qquad 2n\,F = (2x - 1)\,G + 2\,x\,(x - 1)\,G', \qquad (10)$$

and then to hypergeometric equations for the Chebyshev polynomials; see [6, §5.1] for details. Jacobi polynomials $J_N^{(\alpha,\beta)}$ with α, β proper half-integers[8] (or rational numbers) appear in Belyi pull-backs to hypergeometric equations with the $\mathbb{Z}/2\mathbb{Z}$ (or a finite cyclic) monodromy. This is shown in [6, §5.3] by relating the Belyi functions to Schwarz maps for the involved hypergeometric equations. This generalizes the observation of Jacobi polynomials in Belyi functions in [3].

4 Software Usage

The main Maple routine is ComputeBelyi() in [4, ComputeBelyi.mpl]. Here is the input specification:

$$\text{ComputeBelyi}(d, [k, \ell, m], [B_k, B_\ell, B_m], x);$$

where d is the degree; B_k, B_ℓ, B_m are lists of branching orders in the 3 fibers. To force the assignment $x = \infty$, list that branching order multiplied by -1. The assigned branchings k, ℓ, m can be skipped in B_k, B_ℓ, B_m, respectively. In each B_k, B_ℓ, B_m, one exceptional branching $< k, \ell, m$ (respectively) can be skipped. Empty B_m (and then empty B_ℓ) can be skipped. B_k can be written as a sequence of numbers rather than a list.

The output is the list $[G_j]$, where each $G_j = [\varphi, \ldots]$ represents a Galois orbit. If the moduli field of G_j is not a realization field, the second member of the list G_j is the triple $[C(u, v), u, v]$, where $C(u, v) = 0$ is the obstruction conic. The remaining members of G_j are the polynomials defining the realization field of φ (and the extra variables in the expression of φ).

Our website [4] offers a set of routines to simplify Belyi functions, compute their moduli fields, obstruction conics, decompositions and dessins d'enfant, Less trivial algorithms are described in [6, §4]. We also developed a script language (executable on Maple) to draw dessins in the Latex or the EPS format.

References

1. Grothendieck, A.: Esquisse d'un Programme. In: Schneps, L., Lochak, P. (eds.) Geometric Galois Actions. London Math. Soc. Lecture Notes, vol. 242, pp. 5–48 (1984); English transl. 243–283. Cambridge University Press (1997)
2. Kreines, E.: On families of geometric parasitic solutions for Belyi systems of genus zero. Fundamentalnaya i Priklandaya Matematika 9(1), 103–111 (2003)

---

[8] Not necessarily $\alpha, \beta > -1$, so that orthogonality of Jacobi polynomials plays only a formal role. Correspondingly, the dessins d'enfant are not wholly on the real line.

3. Magot, N.: Cartes planaires et fonctions de Belyi: Aspects algorithmiques et expérimentaux. PhD thesis, Université Bordeaux I (1997)
4. van Hoeij, M., Vidunas, R.: Online data for [5],
   http://www.math.fsu.edu/~hoeij/Heun
5. van Hoeij, M., Vidunas, R.: Belyi functions for hyperbolic hypergeometric-to-Heun transformations. Available at arxiv:1212.3803 (2012)
6. van Hoeij, M., Vidunas, R.: Algorithms and differential relations for Belyi functions. Available at arxiv:1305.7218 (2013)

On Computation of the First Baues–Wirsching Cohomology of a Freely-Generated Small Category

Yasuhiro Momose[1] and Yasuhide Numata[2]

[1] Shinshu University, Japan
momose@math.shinshu-u.ac.jp
[2] Shinshu University, Japan
nu@math.shinshu-u.ac.jp

Abstract. The Baues–Wirsching cohomology is one of the cohomologies of a small category. Our aim is to describe the first Baues–Wirsching cohomology of the small category generated by a finite quiver freely. We consider the case where the coefficient is a natural system obtained by the composition of a functor and the target functor. We give an algorithm to obtain generators of the vector space of inner derivations. It is known that there exists a surjection from the vector space of derivations of the small category to the first Baues–Wirsching cohomology whose kernel is the vector space of inner derivations.

Keywords: Finite quivers, path algebras, category algebras, inner derivations, Gaussian elimination.

1 Introduction

Baues and Wirsching [1] introduced a cohomology of a small category, which is called nowadays the Baues–Wirsching cohomology. It is known that the Baues–Wirsching cohomology is a generalization of some cohomologies; e.g., the cohomology of a group G with coefficients in a left G-module, the singular cohomology of the classifying space of a small category with coefficients in a field, and so on. Let k be a field and D a natural system on a small category $\mathcal{C}$; that is, a functor from the category of factorizations in $\mathcal{C}$ to the category k-Mod of left k-modules. The n-th Baues–Wirsching cohomology of $\mathcal{C}$ with coefficients in D is denoted by $\mathrm{H}^n_{BW}(\mathcal{C}, D)$. For an equivalence $\phi : \mathcal{C} \to \mathcal{C}'$ of small categories and a natural system D on $\mathcal{C}$, Baues and Wirsching showed that the k-linear map $\tilde{\phi} : \mathrm{H}^n_{BW}(\mathcal{C}, D) \to \mathrm{H}^n_{BW}(\mathcal{C}', \phi^* D)$ induced by ϕ is an isomorphism for $n \in \mathbb{Z}$. The Baues–Wirsching cohomology is an invariant for the equivalence of small categories in this sense.

Assume that $\mathcal{C}$ is freely generated by a finite quiver and that $D = \check{D} \circ t$ is the composition of $\check{D}$ and the target functor t. In this case, it is known that $\mathrm{H}^n_{BW}(\mathcal{C}, D)$ vanishes for $n \geq 2$ and that $\mathrm{H}^0_{BW}(\mathcal{C}, D)$ is isomorphic to the limit $\lim_{\mathcal{C}} \check{D}$. Therefore, we focus on the first cohomology $\mathrm{H}^1_{BW}(\mathcal{C}, D)$. Let $k\mathcal{C}$ be

H. Hong and C. Yap (Eds.): ICMS 2014, LNCS 8592, pp. 99–105, 2014.

the category algebra of $\mathcal{C}$, i.e. the algebra whose basis is formed by the set of morphisms of $\mathcal{C}$ and whose multiplication is the composition of morphisms (if the morphisms are not composable, then the multiplication is zero). Since $\mathcal{C}$ is generated by Q, the category algebra is the path algebra kQ. Define the functor $\pi_{\mathcal{C}}$ from $k\mathcal{C}$-Mod to the category k-Mod$^{\mathcal{C}}$ of functors from $\mathcal{C}$ to k-Mod as follows: $\pi_{\mathcal{C}}$ maps an object M in $k\mathcal{C}$-Mod to the functor which maps $x \in$ ob($\mathcal{C}$) to id$_x \cdot M$ and which maps $u \in$ mor($\mathcal{C}$) to the left multiplicative map of u; and $\pi_{\mathcal{C}}$ maps a morphism f in $k\mathcal{C}$-Mod to the natural transformation $\{f|_{\mathrm{id}_x \cdot M}\}_{x \in \mathrm{ob}(\mathcal{C})}$. Since the set of objects in $\mathcal{C}$ is finite, $\pi_{\mathcal{C}}$ is an equivalence of categories. (See [5].) Our algorithm introduced in this article computes the first cohomology $\mathrm{H}^1_{BW}(\mathcal{C}, \pi_{\mathcal{C}}(N) \circ t)$ for a left $k\mathcal{C}$-module N.

By [1], we can regard $\mathrm{H}^n_{BW}(\mathcal{C}, \pi_{\mathcal{C}}(N) \circ t)$ as the Hochschild–Mitchell cohomology. Since $\mathcal{C}$ is generated by Q and $\pi_{\mathcal{C}}$ is an equivalence of categories, $\mathrm{H}^1_{BW}(\mathcal{C}, \pi_{\mathcal{C}}(N) \circ t)$ is the first Hochschild cohomology of the path algebra kQ with coefficients in N. In [4], Han gave an algorithm to compute Hochschild cohomologies of a monomial algebra A with coefficients in A. Han uses computation of the cyclic cohomology of A in his algorithm. In [3], Green and Solberg gave an algorithm to construct projective resolutions of modules over quotients of path algebras. In [2], Buchweitz, Green and Solberg say that the resolution is used to give a minimal projective resolution of the quotients algebra over the enveloping algebra and they research on the multiplicative structure of the Hochschild cohomology ring. We do not use a cyclic cohomology nor a projective resolution in our algorithm, but we give generators of the vector space of inner derivations in a combinatorial way.

The authors gave a description of the first Baues–Wirsching cohomology in the case where $\mathcal{C}$ is a B_2-free poset [6]. The algorithm in this paper is a generalization of the idea of the special case.

This article is organized as follows: In Section 2.1, we define some notation. In Section 2.2, we give algorithms. In Section 3, we show our main result.

2 Definition

2.1 Definition of the First Baues–Wirsching Cohomology

We define some notation on the first Baues–Wirsching cohomology in this section.

Let P and Q be finite sets, s and t maps from Q to P. We call the set Q equipped with the triple $(P; s, t)$ a *finite quiver*. We call an element of P a *vertex* and call an element of Q an *arrow*. An arrow $f \in Q$ such that $s(f) = a$ and $t(f) = b$ is denoted by $f : a \to b$. We call a sequence $f_1 \cdots f_l$ of arrows a *path of length l* if $s(f_i) = t(f_{i+1})$ for all i. A path $f_1 \cdots f_l$ such that $t(f_1) = s(f_l)$ is called a *cycle*. We say that a quiver Q is *acyclic* if Q has no cycle. Let Q' be a subset of Q and P' a subset of P. We call the set Q' equipped with the triple $(P'; s|_{Q'}, t|_{Q'})$ a *subquiver* of Q if $s(Q')$ and $t(Q')$ are subsets of P'.

Let Q be a finite quiver. The category defined in the following manner is called the *small category freely generated by Q*:

- the set of objects is the set of vertices of Q;
- a morphism from x to y is a path from x to y;
- the identity id_x is the path from x to x of length 0; and
- if $s(f) = t(g)$, then the composition of morphisms f and g is the concatenation of paths f and g.

Let $\mathcal{C}$ be a small category freely generated by Q. The category $\mathcal{F}(\mathcal{C})$ defined in the following manner is called the *category of factorizations in $\mathcal{C}$*:

- the objects are morphisms in $\mathcal{C}$;
- a morphism from α to β is a pair (u, v) of morphisms in $\mathcal{C}$ such that $\beta = u \circ \alpha \circ v$; and
- the composition of (u', v') and (u, v) is defined by $(u', v') \circ (u, v) = (u' \circ u, v \circ v')$.

A covariant functor from $\mathcal{F}(\mathcal{C})$ to k-Mod is called a *natural system* on a small category $\mathcal{C}$. Let D be a natural system on the small category $\mathcal{C}$. For $\alpha \in \mathrm{ob}(\mathcal{F}(\mathcal{C}))$, D_α denotes the k-module corresponding to α. For a pair (u, v) of composable morphisms, we define u_* and v^* by

$$u_* = D(u, \mathrm{id}_{s(v)}) : D_v \to D_{uov},$$
$$v^* = D(\mathrm{id}_{t(u)}, v) : D_u \to D_{uov}.$$

Let $d : \mathrm{mor}(\mathcal{C}) \to \prod_{\varphi \in \mathrm{mor}(\mathcal{C})} D_\varphi$ be a map such that $d(f) \in D_f$ for each $f \in \mathrm{mor}(\mathcal{C})$. We call d a *derivation* from $\mathcal{C}$ to D if $d(f \circ g) = f_*(dg) + g^*(df)$ for each pair (f, g) of composable morphisms. We define $\mathrm{Der}(\mathcal{C}, D)$ to be the k-vector space of derivations from $\mathcal{C}$ to D. We call d an *inner derivation* from $\mathcal{C}$ to D if there exists an element $(n_x)_{x \in \mathrm{ob}(\mathcal{C})} \in \prod_{x \in \mathrm{ob}(\mathcal{C})} D_{\mathrm{id}_x}$ such that $d(f) = f_*(n_{s(f)}) - f^*(n_{t(f)})$ for each $f \in \mathrm{mor}(\mathcal{C})$. We define $\mathrm{Ider}(\mathcal{C}, D)$ to be the k-vector space of inner derivations from $\mathcal{C}$ to D. The first Baues–Wirsching cohomology $\mathrm{H}^1_{BW}(\mathcal{C}, D)$ is the quotient space $\mathrm{Der}(\mathcal{C}, D)/\mathrm{Ider}(\mathcal{C}, D)$.

Remark 1. Let Q be a quiver, $\mathcal{C}$ a small category freely generated by Q, N a $k\mathcal{C}$-module, t the target functor, and $\tilde{D}$ the natural system $\pi_{\mathcal{C}}(N) \circ t$. For a pair (u, v) of composable morphisms, u_* (*resp.* v^*) maps $m \in \tilde{D}_v = \mathrm{id}_{t(v)} \cdot N$ (*resp.* $n \in \tilde{D}_u = \mathrm{id}_{t(u)} \cdot N$) to $u \cdot m \in \tilde{D}_{uov} = \mathrm{id}_{t(u)} \cdot N$ (*resp.* $n \in \tilde{D}_{uov} = \mathrm{id}_{t(u)} \cdot N$).

2.2 Definition of Algorithms

In this section, we give algorithms to obtain generators of $\mathrm{Ider}(\mathcal{C}, D)$.

Let Q be a finite quiver, and P the set of vertices of Q. For subsets Q_1, Q_3 of Q and a subset $\hat{P}$ of P, we define the set $H(\hat{P}; Q, Q_1, Q_3)$ to be

$$\left\{ h \in Q_3 \,\middle|\, \begin{array}{l} t(h) \in \hat{P}. \\ hp \text{ is not a cycle in } Q \text{ for any path } p \text{ in } Q_1. \end{array} \right\}.$$

For subsets Q_1, Q_2 of Q and $h \in H(\hat{P}; Q, Q_1, Q_3)$, we define the set $G(Q_1, Q_2; h)$ to be

$$\left\{ g \in Q_2 \,\middle|\, \begin{array}{l} \text{There exists a cycle in } Q_1 \cup Q_2 \cup \{ h \} \\ \text{which contains } g \text{ and } h. \end{array} \right\}.$$

Definition 1 (Algorithm to construct a decomposition of a quiver).

Input *a finite quiver Q.*
Output $((a_i)_{i=1}^l; (b_i)_{i=1}^m; (f_i)_{i=1}^l; (g_i)_{i=1}^n; (h_i)_{i=1}^r)$.
Procedure

1. *Let P be the set of vertices of Q.*
2. *Let $\check{P} = \emptyset$, $\hat{P} = P$, $Q_1 = \emptyset$, $Q_2 = \emptyset$, $Q_3 = Q$.*
3. *While $H(\hat{P}; Q, Q_1, Q_3) \neq \emptyset$, do the following:*
   (a) *Choose an element $h \in H(\hat{P}; Q, Q_1, Q_3)$.*
   (b) *Let $Q' = ((Q_1 \cup Q_2) \setminus G(Q_1, Q_2; h)) \cup \{\, h \,\}$.*
   (c) *Let $\bar{Q}$ be a maximal acyclic subquiver of Q including Q'.*
   (d) *Let $\check{P} = \{\, a \in P \mid \exists f \in \bar{Q} \text{ such that } t(f) = a. \,\}$.*
   (e) *Let $\hat{P} = P \setminus \check{P}$.*
   (f) *For each $a \in \check{P}$, choose $f_a \in \bar{Q}$ so that $t(f_a) = a$.*
   (g) *Let $Q_1 = \{\, f_a \mid a \in \check{P} \,\}$, $Q_2 = Q' \setminus Q_1$, and $Q_3 = Q \setminus Q'$.*
4. *Let $l = |\check{P}|$. For $i = 1, \ldots, l$, do the following:*
   (a) *Choose a vertex $x \in \check{P}$ such that there exists no arrow in Q_1 whose source is x.*
   (b) *Let $a_i = x$.*
   (c) *For $\alpha \in Q_1$ so that $t(\alpha) = x$, let $f_i = \alpha$.*
   (d) *Let $\check{P} = \check{P} \setminus \{\, x \,\}$, and $Q_1 = Q_1 \setminus \{\, \alpha \,\}$.*
5. *Let $\{\, b_1, \ldots, b_m \,\} = \hat{P}$.*
6. *Let $\{\, g_1, \ldots, g_n \,\} = Q_2$.*
7. *Let $\{\, h_1, \ldots, h_r \,\} = Q_3$.*

Remark 2. In Step 3 in Definition 1, $|H(\hat{P}; Q, Q_1, Q_3)|$ strictly decreases since $|\hat{P}|$ decreases in each step. Hence Step 3 is a finite procedure.

Remark 3. Let $((a_i)_{i=1}^l; (b_i)_{i=1}^m; (f_i)_{i=1}^l; (g_i)_{i=1}^n; (h_i)_{i=1}^r)$ be an output of Definition 1. Let

$$\check{P} = \{\, a_1, \ldots, a_l \,\},$$
$$\hat{P} = \{\, b_1, \ldots, b_m \,\},$$
$$Q_1 = \{\, f_1, \ldots, f_l \,\},$$
$$Q_2 = \{\, g_1, \ldots, g_n \,\}, \text{and}$$
$$Q_3 = \{\, h_1, \ldots, h_r \,\}.$$

The set $\check{P} \coprod \hat{P}$ is a decomposition of P. The set $Q_1 \coprod Q_2 \coprod Q_3$ is also a decomposition of Q. By Step 4 in Definition 1, a_i corresponds to the target of f_i for $i = 1, \ldots, l$. Hence if there exists a path from a_j to a_i or a path from b_j to a_i in Q_1, then the path is unique. Since the quiver $Q_1 \cup Q_2$ is a maximal acyclic subquiver of Q, we can regard $\check{P}$ as a poset. Moreover, if $a_j \leq a_i$ in the poset $\check{P}$, then the inequality $i \leq j$ holds. If Q is a finite acyclic quiver, then Q_3 is the empty set. By Step 3 in Definition 1, for h_i so that $t(h_i) \in \hat{P}$, there exists a path p in Q_1 such that $h_i p$ is a cycle in Q.

Definition 2 (Algorithm to construct generators of the vector space of inner derivations).

Input $((a_i)_{i=1}^l; (b_i)_{i=1}^m; (f_i)_{i=1}^l; (g_i)_{i=1}^n; (h_i)_{i=1}^r)$.
Output (V, W).
Procedure

1. Let $Q_1 = \{ f_1, \ldots, f_l \}$.
2. (We define elements $v_{i,j}$ in the path algebra kQ.) For $j = 1, \ldots, l$, do the following:
   (a) For $i = 1, \ldots, l$, let $v_{i,j} = 0$.
   (b) Let $v_{j,j} = \mathrm{id}_{a_j}$.
   (c) For $i = 1, \ldots, n$, do the following:
      i. Let $v_{l+i,j} = 0$.
      ii. If there exists a path p from a_j to $t(g_i)$ in Q_1, then let $v_{l+i,j} = v_{l+i,j} + p$.
      iii. If there exists a path p from a_j to $s(g_i)$ in Q_1, then let $v_{l+i,j} = v_{l+i,j} - g_i p$.
   (d) For $i = 1, \ldots, r$, do the following:
      i. Let $v_{l+n+i,j} = 0$.
      ii. If there exists a path p from a_j to $t(h_i)$ in Q_1, then let $v_{l+n+i,j} = v_{l+n+i,j} + p$.
      iii. If there exists a path p from a_j to $s(h_i)$ in Q_1, then let $v_{l+n+i,j} = v_{l+n+i,j} - h_i p$.
3. Let $V = (v_{i,j})_{1 \le i \le l+n+r,\ 1 \le j \le l}$.
4. (We define elements $w_{i,j}$ in the path algebra kQ.) For $j = 1, \ldots, m$, do the following:
   (a) For $i = 1, \ldots, l$, let $w_{i,j} = 0$.
   (b) For $i = 1, \ldots, n$, do the following:
      i. Let $w_{l+i,j} = 0$.
      ii. If there exists a path p from b_j to $t(g_i)$ in Q_1, then let $w_{l+i,j} = w_{l+i,j} + p$.
      iii. If there exists a path p from b_j to $s(g_i)$ in Q_1, then let $w_{l+i,j} = w_{l+i,j} - g_i p$.
   (c) For $i = 1, \ldots, r$, do the following:
      i. Let $w_{l+n+i,j} = 0$.
      ii. If there exists a path p from b_j to $t(h_i)$ in Q_1, then let $w_{l+n+i,j} = w_{l+n+i,j} + p$.
      iii. If there exists a path p from b_j to $s(h_i)$ in Q_1, then let $w_{l+n+i,j} = w_{l+n+i,j} - h_i p$.
5. Let $W = (w_{i,j})_{1 \le i \le l+n+r,\ 1 \le j \le m}$.

Remark 4. Let (V, W) be the output of Definition 2 for some input. The matrix $(v_{i,j})_{1 \le i \le l,\ 1 \le j \le l}$ is the identity matrix, i.e., the diagonal matrix whose entries one ($\mathrm{id}_{a_1}, \ldots, \mathrm{id}_{a_l}$). The matrix $(w_{i,j})_{1 \le i \le l,\ 1 \le j \le m}$ is the zero matrix.

3 Our Main Result

We show our main result in this section. Our main result computes the first Baues–Wirsching cohomology via the column echelon matrix obtained by our algorithm.

Let Q be a finite quiver, $\mathcal{C}$ a small category freely generated by Q. Fix a left $k\mathcal{C}$-module N, and consider the natural system $\tilde{D} = \pi_{\mathcal{C}}(N) \circ t$.

Let $T = ((a_i)_{i=1}^l; (b_i)_{i=1}^m; (f_i)_{i=1}^l; (g_i)_{i=1}^n; (h_i)_{i=1}^r)$ be the output of Definition 1 for Q. We define the k-vector space A_1, A_2, and A_3 by

$$A_1 = \bigoplus_{i=1}^l \tilde{D}_{f_i}, \qquad A_2 = \bigoplus_{i=1}^n \tilde{D}_{g_i}, \text{ and} \qquad A_3 = \bigoplus_{i=1}^r \tilde{D}_{h_i}.$$

Let (V, W) be the output of Definition 2 for T. Let v_j and w_j be the j-th column vector of V and W, respectively. The vectors v_j and w_j are elements of $\bigoplus_{i=1}^{l+n+r} k\mathcal{C}$. We define the k-vector spaces $\bar{V}$ and $\bar{W}$ by

$$\bar{V} = \langle v_j n_{a_j} \mid n_{a_j} \in \mathrm{id}_{a_j} \cdot N, 1 \leq j \leq l \rangle,$$
$$\bar{W} = \langle w_j n_{b_j} \mid n_{b_j} \in \mathrm{id}_{b_j} \cdot N, 1 \leq j \leq m \rangle.$$

Theorem 1. *The first Baues–Wirsching cohomology* $\mathrm{H}_{BW}^1(\mathcal{C}, \tilde{D})$ *is isomorphic to*

$$(A_1 \oplus A_2 \oplus A_3)/(\bar{V} + \bar{W})$$

as k-vector spaces.

Proof. According to Baues and Wirsching [1], if $\mathcal{C}$ is freely generated by $S \subset \mathrm{mor}(\mathcal{C})$, then we can identify $\mathrm{Der}(\mathcal{C}, D)$ with $\prod_{\alpha \in S} D_\alpha$. Via the identification, $\mathrm{Ider}(\mathcal{C}, D)$ is the k-vector space

$$\left\{ (\alpha_*(n_{s(\alpha)}) - \alpha^*(n_{t(\alpha)}))_\alpha \in \prod_{\alpha \in S} D_\alpha \,\middle|\, (n_x)_x \in \prod_{x \in \mathrm{ob}(\mathcal{C})} D_{\mathrm{id}_x} \right\}.$$

Let

$$Q = \{ f_i \mid 1 \leq i \leq l \} \cup \{ g_i \mid 1 \leq i \leq n \} \cup \{ h_i \mid 1 \leq i \leq r \}, \text{ and}$$
$$P = \{ a_i \mid 1 \leq i \leq l \} \cup \{ b_i \mid 1 \leq i \leq m \}.$$

It follows that $\mathrm{Der}(\mathcal{C}, \tilde{D}) \cong A_1 \oplus A_2 \oplus A_3$. Hence $\mathrm{Ider}(\mathcal{C}, \tilde{D})$ is isomorphic to the k-vector space

$$B = \left\{ (\alpha n_{s(\alpha)} - n_{t(\alpha)})_{\alpha \in Q} \in A \,\middle|\, (n_x)_x \in \bigoplus_{x \in P} \mathrm{id}_x \cdot N \right\}.$$

For $x \in P$ and $m \in \mathrm{id}_x \cdot N$, we define $r_x m = (r_{x,\alpha} m)_{\alpha \in Q} \in A$ by

$$r_{x,\alpha} m = \begin{cases} -\alpha m & (s(\alpha) = x) \\ m & (t(\alpha) = x) \\ 0 & (\text{otherwise}). \end{cases}$$

It is clear that the k-vector space B is equal to

$$\langle r_x m \mid x \in P, m \in \mathrm{id}_x \cdot N \rangle.$$

For $j = 1, \ldots, l$ and $n_{a_j} \in \mathrm{id}_{a_j} \cdot N$, we define $\overline{r_{a_j}} n_{a_j}$ to be $r_{a_j} n_{a_j} + \sum_{k=1}^{i-1} \overline{r_{a_k}} f_k n_{a_j}$. For $j = 1, \ldots, m$ and $n_{b_j} \in \mathrm{id}_{b_j} \cdot N$, we define $\overline{r_{b_j}} n_{b_j}$ to be $r_{b_j} n_{b_j} + \sum_{k=1}^{l} \overline{r_{a_k}} f_k n_{b_j}$. It follows from the direct calculation that $\overline{r_{a_j}} n_{a_j}$ and $\overline{r_{b_j}} n_{b_j}$ are equal to $v_j n_{a_j}$ and $w_j n_{b_j}$, respectively. Hence we have Theorem 1.

References

1. Baues, H.J., Wirsching, G.: Cohomology of small categories. J. Pure Appl. Algebra 38(2-3), 187–211 (1985)
2. Buchweitz, R.-O., Green, E.L., Snashall, N., Solberg, Ø.: Multiplicative structures for Koszul algebras. Q. J. Math. 59(4), 441–454 (2008)
3. Green, E.L., Solberg, Ø.: An algorithmic approach to resolutions. J. Symbolic Comput. 42(11-12), 1012–1033 (2007)
4. Han, Y.: Hochschild (co)homology dimension. J. London Math. Soc. (2) 73(3), 657–668 (2006)
5. Mitchell, B.: Rings with several objects. Advances in Math. 8, 1–161 (1972)
6. Momose, Y., Numata, Y.: On the Baues–Wirsching cohomology of B_2-free posets (in preparation)

Codes over a Non Chain Ring
with Some Applications

Aysegul Bayram,[1] Elif Segah Oztas[2], and Irfan Siap[3]

Department of Mathematics, Yildiz Technical University, Istanbul, Turkey
{aaysegulbayram,elifsegahoztas}@gmail.com, isiap@yildiz.edu.tr

Abstract. In this work, we study the structure of skew constacyclic codes over the ring $R = F_4[v]/\langle v^2 - v \rangle$ which is a non chain ring with 16 elements where F_4 denotes the field with 4 elements and v an indeterminate. We relate linear codes over R to codes over F_4 by defining a Gray map between R and F_4^2. Next, the structure of all skew constacyclic codes is completely determined. Furthermore, we construct DNA codes over R.

Keywords: Non chain rings, Linear codes, Skew codes, DNA codes.

1   Introduction

Many papers have been written on linear codes over fields especially over binary fields due to their direct application nature. The paper written by Hammons at al. [1] is recognized to be a significant milestone study in coding theory. This paper gave an important connection between non binary (quaternary) linear codes and some well known binary non linear codes. Most of the studies are concentrated on the case with codes over finite chain rings. However, optimal codes over non-chain rings exist (e.g see [2]). But the case over non-chain structure is more complicated. In [10], the algebraic structure of cyclic codes over $F_2 + vF_2$, where $v^2 = v$ are studied. Zhu and Wang studied a class of constacyclic codes in [9]. Recently, D. Boucher et al. considered codes over non commutative rings. Leonard Adleman [7] pioneered the studies on DNA computing by solving NP-complete problem by DNA molecules. DNA codes are algebraically studied over $GF(4)$ in [11]. Also DNA codes are considered over $F_2[u]/\langle u^2 - 1 \rangle$ which is a ring with four elements [5]. Later in [3], DNA double pairs are matched with the elements of F_{16} by a family of special polynomials to solve the reversibility problem.

2   Preliminaries

Throughout this paper, R denotes the commutative ring $F_4 + vF_4 = \{a + vb | a, b \in F_4\}$ with $v^2 = v$. We take $F_4 = \{0, 1, w, w + 1\}$ where $w^2 = w + 1$. The ring R is a finite non-chain ring with 16 elements. Any element of R can be uniquely expressed as $c = a + vb$, where $a, b \in F_4$. The Gray map from R to $F_4 \times F_4$ is

H. Hong and C. Yap (Eds.): ICMS 2014, LNCS 8592, pp. 106–110, 2014.

given by $\phi(c) = (a + b, a)$. It is routine to check that ϕ is a ring isomorphism, so R is a finite semi-local, Frobenius ring with maximal ideals $\langle v \rangle$ and $\langle 1 + v \rangle$.

A linear code C over R of length n is an R-submodule of R^n. For any linear code C of length n over R, the dual code is defined by $C^{\perp} = \{u \in R^n | \langle c, u \rangle = 0$ for all $c \in C\}$ where $\langle c, u \rangle$ denotes the standard Euclidean inner product of c and u in R^n. Let R be a ring and $a \in R$. The Hamming weight of a nonzero element of R equals to 1 or 0 otherwise. If $a = (a_1, a_2, \ldots, a_n) \in R^n$, then the Hamming weight of a is $w(a) = \sum_{i=1}^{n} w(a_i)$. The Hamming distance between $a, b \in R^n$ is $d(a, b) = w(a - b)$. Hamming distance is a metric on R^n. The Lee weight of any element of R is $w_L(c) = w_H(\phi(c))$. Let A and B be two linear codes. The operations $\otimes$ and $\oplus$ are defined as; $A \otimes B = \{(a, b) \mid a \in A, \ b \in B\}$ and $A \oplus B = \{a + b \mid a \in A, \ b \in B\}$. A generator matrix for a nonzero linear code C over R can be put into the following form:

$$G = \begin{pmatrix} I_{k_1} & (1+v)B_1 & vA_1 & (1+v)A_2 + vB_2 & (1+v)A_3 + vB_3 \\ 0 & vI_{k_2} & 0 & vA_4 & 0 \\ 0 & 0 & (1+v)I_{k_3} & 0 & (1+v)B_4 \end{pmatrix}$$

where A_i and B_j are F_4 matrix for all $1 \leq i, j \leq 4$. Let C be a linear code over R. The following result is presented in [4]. Let;

$$C_1 = \{x + y \in F_4^n \mid (x + y)v + x(v + 1) \in C, \ for \ some \ y \in F_4^n\} \quad (1)$$

$$C_2 = \{x \in F_4^n \mid (x + y)v + x(v + 1) \in C\} \quad (2)$$

Then, C_1 and C_2 are linear codes over F_4. Consequently, $C = vC_1 \oplus (1 + v)C_2$ and $|C| = 16^{k_1} 4^{k_2} 4^{k_3}$.

3 Skew Constacyclic Codes over R

In this section we define skew constacyclic codes for R, which is a finite non chain ring and we mention the algebraic structure skew constacyclic codes over $F_4 + vF_4$ and classify them with respect to their Gray image. Two types of nontrivial automorphisms can be defined over $F_4 + vF_4$:

$$\begin{array}{ll} \psi : F_4 + vF_4 \rightarrow F_4 + vF_4 & \theta : F_4 + vF_4 \rightarrow F_4 + vF_4 \\ \quad a + bv \rightarrow a + (1 + v)b & \quad a + bv \rightarrow a^2 + vb^2 \end{array} \quad (3)$$

Definition 1. *[8] Let θ be an automorphism over R and λ be a unit element of R. Then, a linear code C of length n is called skew constacyclic or $\theta_\lambda-$ constacyclic over R if C satisfies the property that $\sigma_{\theta_\lambda}(c) = (\theta(\lambda c_{n-1}), \theta(c_0), \ldots, \theta(c_{n-2}))$ $\in C$ whenever $c = (c_0, c_1, \ldots, c_{n-1}) \in C$ for all $c \in C$ where $\sigma_{\theta_\lambda}(c)$ is called the $\theta_\lambda-$constacyclic shift of c over R.*

In polynomial representation, a linear code of length n over F_4 is a left $F_4[x; \theta]$-submodule of $F_4[x; \theta]/(x^n - \lambda)$ where $F_4[x; \theta]$ is skew polynomial ring with the non-commutative multiplication defined as $x^i a = \theta^i(a)x^i$, $a \in F_4$ if and only if it

Table 1. The Relation of θ_λ−constacyclic C over R and the codes: C_1 and C_2 over F_4

C	C_1	C_2
θ_w	θ_w	θ_w
$\theta_{(1+w)}$	$\theta_{(1+w)}$	$\theta_{(1+w)}$
$\theta_{(v+w)}$	$\theta_{(1+w)}$	θ_w
$\theta_{(1+v+w)}$	θ_w	$\theta_{(1+w)}$
$\theta_{(1+vw)}$	$\theta_{(1+w)}$	skew cyclic
$\theta_{(1+w+vw)}$	skew cyclic	$\theta_{(1+w)}$
$\theta_{(1+v+vw)}$	θ_w	skew cyclic
$\theta_{(v+w+vw)}$	skew cyclic	θ_w

is a θ_λ−constacyclic code. Furthermore if C is a left submodule of $F_4[x;\theta]/(x^n − \lambda)$ then C is generated by a monic polynomial $g(x)$ which is a right divisor of $x^n − \lambda$ in $F_4[x;\theta]$ [6].

In particular, if λ is equal to 1 and $−1$ then C is called skew cyclic and skew negacyclic code, respectively.

Theorem 1. Let C be a linear code of length n over R and $C = (v)C_1 \oplus (1+v)C_2$. Then, C is a θ_λ-constacyclic code if and only if C_1 and C_2 are $\theta_{\pi_1(\phi(\lambda))}$−constacyclic and $\theta_{\pi_2(\phi(\lambda))}$−constacyclic codes of length n over F_4, respectively.

Corollary 1. The Table 1 summarizes all skew constacyclic codes over R and explores their relation to the Gray images.

4   Codes over R and Their DNA Code Applications

In this section, we obtain the DNA codes over R that correspond to DNA double pairs (bases) and we create a new approach to solve the reversibility problem for DNA double pairs (bases) apart from [3]. Let $S_{D_4} = \{A, T, G, C\}$ represent the set of DNA alphabet. We define the set

$$S_{D_{16}} = \{AA, AT, AG, AC, TT, TA, TG, TC, GG, GA, GC, GT, CC, CA, CG, CT\} \tag{4}$$

and define a correspondence between the elements of R and DNA double pairs in Table 2. We take the matching between elements of F_4 and S_{D_4} originally given in [11] such as $A \to 0$, $T \to 1$, $C \to w$, $G \to w + 1 = w^2$. The Watson-Crick complement is given by $A^c = T, T^c = A, C^c = G, G^c = C$. Hence in $S_{D_{16}}$ we have $(AA)^c = TT, ..., (TC)^c = AG$. If c is a codeword such as $c = (c_0, ..., c_{n-1})$, then the complement of c is $c^c = (c_0^c, ..., c_{n-1}^c)$ and the reverse-complement of c is $c^{rc} = (c_{n-1}^c, ..., c_0^c)$.

Let C be a code over R of length n and $c \in C$ be a codeword where $c = (c_0, c_1, ..., c_{n-1})$, $c_i \in R$ and $\Theta(c) : C \to S_{D_4}^{2n}$ where $(c_0, c_1, ..., c_{n-1}) \to (b_0, b_1, ..., b_{2n-1})$ given in Table 2. Each c_i is mapped to coordinate pairs b_{2i}, b_{2i+1} where $i = \{0, 1, ..., n-1\}$. $\Theta(c) = (b_0, b_1, ..., b_{2n-1})$ is a DNA codeword of $\Theta(C)$, $b_j \in$

Table 2. ψ-table for DNA correspondence

Element	Gray image	DNA double pairs
0	(0,0)	AA
1	(1,1)	TT
w	(w,w)	CC
$1+w$	$(1+w,1+w)$	GG
v	$(1,0)$	TA
$1+v$	$(0,1)$	AT
$v+w$	$(1+w,w)$	GC
$1+v+w$	$(w,1+w)$	CG
vw	$(w,0)$	CA
$1+vw$	$(1+w,1)$	GT
$w+vw$	$(0,w)$	AC
$1+w+vw$	$(1,1+w)$	TG
$v+vw$	$(1+w,0)$	GA
$1+v+vw$	$(w,1)$	CT
$w+v+vw$	$(1,w)$	TC
$1+w+v+vw$	$(0,1+w)$	AG

S_{D_4}, $j \in \{0,1,...,2n-1\}$. For instance, $(c_0,c_1,c_2,c_3) = (1,v,w,w+v) \to$ is mapped to $(TTTACCGC) = (b_0,b_1,b_2,b_3,b_4,b_5,b_6,b_7)$.

Definition 2. Let $g_1(x)$ and $g_2(x)$ be polynomials with $\deg g_1(x) = t_1$, $\deg g_2(x) = t_2$ and both dividing $x^n - 1$ over F_4. Let $\ell = \min\{n-t_1, n-t_2\}$, and $g(x) = vg_1(x) + (v+1)g_2(x)$ over R. C is a linear code over R, generated by the set E_g. The set E_g is called a ψ set.

$$E_g = \{E_0, E_1, ..., E_{\ell-1}\}$$

where

$$E_i = \begin{cases} x^i g(x) & \text{if } i \text{ is even} \\ x^i \psi(g(x)) & \text{if } i \text{ is odd.} \end{cases}$$

Theorem 2. Let $g_1(x)$ and $g_2(x)$ be self reciprocal polynomials dividing $x^n - 1$ over F_4 and $2 \le \ell$ be an even integer.

1. Let $\deg g_1(x) = \deg g_2(x)$ or one of $g_1(x)$ or $g_2(x)$ be equal zero. If we take $g(x) = vg_1(x) + (v+1)g_2(x)$, then $|\langle E_g \rangle| = 16^\ell$ and $|\langle E_g \rangle| = 4^\ell$, respectively.
2. Let $\deg g_1(x) \ne \deg g_2(x)$ and $2 \le s = |t_1 - t_2|$ be even. If we take $g(x) = vg_1(x) + (v+1)x^{s/2}g_2(x)$ for $\deg g_1(x) > \deg g_2(x)$, or $g(x) = vx^{s/2}g_1(x) + (v+1)g_2(x)$ for $\deg g_1(x) < \deg g_2(x)$ then $|\langle E_g \rangle| = 16^\ell$.

In both cases, $C = \langle E_g \rangle$ is a linear code over R and $\Theta(C)$ is a reversible DNA code.

Theorem 3. Let $g(x)$ be a self reciprocal polynomial dividing $x^n - 1$ but $x - 1 \nmid g(x)$ over F_4. If $C = \langle E_g \rangle$, then $\Theta(C)$ is a reversible complement DNA code.

Acknowledgements. We thank to the referee(s) for their valuable remarks.

References

1. Hammons Jr., A.R., Kumar, P.V., Calderbank, J.A., Sloane, N.J.A., Sole, P.: The Z-linearity of Kerdock, Preparata, Goethals, and related codes. IEEE Trans. Inf. Theory. 40(2), 301–319 (1994)
2. Yildiz, B., Karadeniz, S.: Linear codes over $F_2 + uF_2 + vF_2 + uvF_2$. Des. Codes Crypt. 54, 61–81 (2010)
3. Oztas, E.S., Siap, I.: Lifted Polynomials Over F_{16} and Their Applications to DNA Codes. Filomat 27(3), 459–466 (2013)
4. Gursoy, F., Siap, I., Yildiz, B.: Construction of Skew Cylic Codes Over $F_q + vF_q$. Advances in Mathematics of Communications (accepted, 2014)
5. Siap, I., Abualrub, T., Ghrayeb, A.: Cyclic DNA codes over the ring $F_2[u]/(u^2 - 1)$ based on the deletion distance. Journal of the Franklin Ins. 346, 731–740 (2006)
6. Siap, I., Abualrub, T., Aydin, N., Seneviratne, P.: Skew cyclic codes of arbitrary length. Int. J. Inf. Coding Theory 2, 10–20 (2011)
7. Adleman, L.: Molecular computation of solutions to combinatorial problems. Science 266, 1021–1024 (1994)
8. Jitman, S., Ling, S., Udomkavanich, P.: Skew constacyclic codes over finite chain rings. Advances in Mathematics of Communications 6, 39–63 (2012)
9. Zhu, S., Wang, Y.: A class of constacyclic codes over $F_p + vFp$ and their Gray image Discrete. Math. Theory 311, 2677–2682 (2011)
10. Zhu, S., Wang, Y., Shi, M.: Some result on cyclic codes over $F_2 + vF_2$. IEEE Trans. Inform. Theory 56(4), 1680–1684 (2010)
11. Abualrub, T., Ghrayeb, A., Zeng, X.N.: Consruction of cyclic codes over F_4 for DNA computing. Journal of the Franklin Ins. 343, 448–457 (2006)

On the Weight Enumerators of the Projections of the 2-adic Golay Code of Length 24 to $\mathbb{Z}_{2^e}$

Sunghyu Han

School of Liberal Arts, KoreaTech, Republic of Korea
sunghyu@koreatech.ac.kr

Abstract. Dougherty et al. calculated some of the weight enumerators of the projections of the 2-adic Golay code of length 24 to finite rings [Lifted codes and their weight enumerators, Discrete Mathematics, 305 (2005) 123–135]. In this paper, we calculate the missing values so that we complete the calculation of the weight enumerators of the projections of the 2-adic Golay code of length 24 to all the finite rings.

Keywords: Golay codes, Lifted codes, p-Adic codes, Weight enumerators.

1 Introduction

Let p be a prime. $\mathbb{Z}_{p^e}$ is the ring of integers modulo p^e, $(1 \leq e < \infty)$. $\mathbb{Z}_{p^\infty}$ is the ring of p-adic integers. For undefined notations, refer to [2,3].

The factorization of $x^{23} - 1$ over $\mathbb{Z}_{2^\infty}$ is

$$x^{23} - 1 = (x - 1)\pi_1(x)\pi_2(x),$$

where, $\pi_1(x) = x^{11} + ax^{10} + (a-3)x^9 - 4x^8 - (a+3)x^7 - (2a+1)x^6 - (2a-3)x^5 - (a-4)x^4 + 4x^3 + (a+2)x^2 + (a-1)x - 1$ and $a = 0 + 2 + 8 + 32 + 64 + 128 + \cdots$ is a 2-adic integer satisfying $a^2 - a + 6 = 0$. We make a cyclic code generated by $\pi_1(x)$, and extend this code by appending 1 to the generators. Then we have the $[24, 12, 13]$ self-dual 2-adic Golay code $\mathcal{G}$.

Let $\mathcal{G}^e = \Psi_e(\mathcal{G})$ which is the projection of the 2-adic Golay code $\mathcal{G}$ to $\mathbb{Z}_{p^e}$. We have the following series

$$\mathcal{G}^1 \prec \mathcal{G}^2 \prec \cdots \prec \mathcal{G}^e \prec \cdots \prec \mathcal{G} \tag{1}$$

of lifts of codes [3]. Since $\mathcal{G}^1$ is the well known binary Golay code,

$$W_{\mathcal{G}^1}(x, y) = x^{24} + 759x^{16}y^8 + 2576x^{12}y^{12} + 759x^8y^{16} + y^{24}.$$

In [3], they computed $W_{\mathcal{G}^2}(x, y)$. Since $\mathcal{G}^e$ is self-dual, we have

$$W_{\mathcal{G}^e}(x, y) = \sum_{j=0}^{12} c_i \left(x^2 + (2^e - 1)y^2\right)^j (xy - y^2)^{24-2j}. \tag{2}$$

H. Hong and C. Yap (Eds.): ICMS 2014, LNCS 8592, pp. 111–114, 2014.

There are 13 unknowns $c_0, c_1, c_2, \ldots, c_{12}$ in (2). Let $(A_0^e, A_1^e, \ldots, A_{24}^e)$ be the weight distribution of $\mathcal{G}^e$. From [2], we know the minimum distance of $\mathcal{G}^e$ is eight for all $e \geq 1$, and from [3] we know $A_8^e = 759$ and $A_9^e = 0$ for all $e \geq 1$. So that if we know A_{10}^e, A_{11}^e, and A_{12}^e, then we can determine $W_{\mathcal{G}^e}(x, y)$. From [3], we also know that $A_{10}^e, A_{11}^e, A_{12}^e$ remain constant for $e \geq 7$. In summary, we only have to calculate A_{10}^e, A_{11}^e, and A_{12}^e for $e = 3, 4, 5, 6, 7$ to complete the weight distribution of $\mathcal{G}^e$ for all $e \geq 1$. The object of this paper is to compute these values. All the computation of this paper was made using Magma [1] with a 2.3GHz and 3.0GB RAM PC.

2   Main Results

The calculation times of $W_{\mathcal{G}^1}(x, y)$, $W_{\mathcal{G}^2}(x, y)$, and $W_{\mathcal{G}^3}(x, y)$ using the Magma function, "WeightDistribution", are $0.000(\text{sec}), 0.983(\text{sec}), 14653.704(\text{sec})(\approx \text{four}$ hours), respectively. We expect that the running time of $W_{\mathcal{G}^4}$ is more than two years. Therefore this naive method can not be applied to our problem.

2.1   Calculation of A_{10}^e, A_{11}^e, and A_{12}^e

We only have to calculate A_{10}^e, A_{11}^e, and $A_{12}^e, (e = 3, 4, 5, 6, 7)$. There are Magma functions, "NumberOfWords", "PartialWeightDistribution". But these functions can be applied only for codes over finite fields. So, we have to make a program.

In the following, we state the algorithm in [4, p. 404]. Let $\mathbb{F}_q$ be the finite field of order q. Let $\mathcal{C}$ be an $[n, k]$ linear code over $\mathbb{F}_q$ with $n = 2k$ and two generator matrices $G' = (I, A')$ and $G'' = (A'', I)$ for $\mathcal{C}$, where I is the $k \times k$ identity matrix. Let G_i' and G_i'' be the i-th row of G' and G'', respectively. In order to determine the number of codewords of weight w in $\mathcal{C}$ we count

(i) for all values of t with $1 \leq t \leq w/2$ and all possible choices of $a_i \in (\mathbb{F}_q - \{0\})$ the number of codewords c of weight w which are the linear combination of t rows of G'

$$c = a_1 G_{i_1}' + a_2 G_{i_2}' + \cdots + a_t G_{i_t}'; \tag{3}$$

(ii) for all values of t with $1 \leq t < w/2$ and all possible choices of $a_i \in (\mathbb{F}_q - \{0\})$ the number of codewords c of weight w which are the linear combination of t rows of G''

$$c = a_1 G_{i_1}'' + a_2 G_{i_2}'' + \cdots + a_t G_{i_t}''. \tag{4}$$

We can easily apply this algorithm to a linear code over $\mathbb{Z}_{p^e}$ with some minor improvement by the following. Let $\mathcal{C}$ be an $[n, k]$ linear code over $\mathbb{Z}_{p^e}$ with $n = 2k$ and two generator matrices $G' = (I, A')$ and $G'' = (A'', I)$ for $\mathcal{C}$, where I is the $k \times k$ identity matrix. Let A_i' and A_i'' be the i-th row of A' and A'', respectively. In order to determine the number of codewords of weight w in $\mathcal{C}$ we count

(i) for all values of t with $1 \leq t \leq w/2$ and all possible choices of $a_i \in (\mathbb{Z}_{p^e} - \{0\})$ the number of codewords c of weight $w - t$ which are the linear combination of t rows of A'

$$c = a_1 A_{i_1}' + a_2 A_{i_2}' + \cdots + a_t A_{i_t}'; \tag{5}$$

(ii) for all values of t with $1 \leq t < w/2$ and all possible choices of $a_i \in (\mathbb{Z}_{p^e} - \{0\})$ the number of codewords c of weight $w - t$ which are the linear combination of t rows of A''

$$c = a_1 A''_{i_1} + a_2 A''_{i_2} + \cdots + a_t A''_{i_t}. \tag{6}$$

Let $G = (I, A)$ be a generator matrix of $\mathcal{G}^e, (1 \leq e < \infty)$. Then its parity check matrix $H = (A^T, I)$ is also a generator matrix of $\mathcal{G}^e$. So, we can apply the above algorithm to $\mathcal{G}^e$. The number of the calculation of Eqn. (5) and Eqn. (6) is

$$\sum_{1 \leq t \leq w/2} \binom{12}{t} (2^e - 1)^t + \sum_{1 \leq t < w/2} \binom{12}{t} (2^e - 1)^t. \tag{7}$$

Expression (7) has a factor 2^{et} and we know that the complexity grows exponentially as e and w grow. If $e = 7$, $w = 12$, and $t = 6$, then 2^{et} is 2^{42}. Here is our calculation results. For $w = 8$, the running times of $e = 1, 2, 3, 4$ are $0.016, 0.609, 17.109, 340.344$ seconds, respectively. So, we expect that it is computationally impossible to calculate the number of codewords of weight 12 in $W_{\mathcal{G}^7}$.

2.2 Our Method

We state our algorithm. Let $\mathcal{C}$ be an $[n, k]$ linear code over $\mathbb{Z}_{p^e}$ with $n = 2k$ and two generator matrices $G' = (I, A')$ and $G'' = (A'', I)$ of $\mathcal{C}$, where I is the $k \times k$ identity matrix. For a given $w \geq 1$, our algorithm computes the number of codewords of $\mathcal{C}$ with weight w, i.e., A^e_w.

We still consider all values of t with $1 \leq t \leq w/2$ (or $1 \leq t < w/2$) as in the algorithm of the previous subsection. But we avoid "all possible choices of $a_i \in (\mathbb{Z}_{p^e} - \{0\})$". Let $LC(A', t)$ be the set of all choices of t rows of A'. Therefore $|LC(A', t)| = \binom{k}{t}$. Let $i_1, i_2, \ldots, i_t$ be an element of $LC(A', t)$ which represent some t rows. Let $M' = (m'_{ij})$ be the submatrix of A' which consists of i_1-th row, i_2-th row, ..., and i_t-th row of A'. We define

$$Z'_j = \{(x_1, x_2, \ldots, x_t) \in (\mathbb{Z}_{p^e} - \{0\})^t | m'_{1j} x_1 + m'_{2j} x_2 + \cdots + m'_{tj} x_t = 0\} \tag{8}$$

for $j = 1, 2, \ldots, k$, and

$$f(A', \{i_1, i_2, \ldots, i_t\}) = \sum_{I \subseteq \{1,2,\ldots,k\}, |I|=k-(w-t)} \left| \bigcap_{j \in I} Z'_j - \bigcup_{j \notin I} Z'_j \right|.$$

Then the number of codewords of weight w in $\mathcal{C}$ is

$$A^e_w = \sum_{1 \leq t \leq w/2} \sum_{\{i_1, i_2, \ldots, i_t\} \in LC(A', t)} f(A', \{i_1, i_2, \ldots, i_t\})$$

$$+ \sum_{1 \leq t < w/2} \sum_{\{i_1, i_2, \ldots, i_t\} \in LC(A'', t)} f(A'', \{i_1, i_2, \ldots, i_t\}). \tag{9}$$

We explain Eqn. (9). Let $\mathbf{c} = (c_1, \ldots, c_k, c_{k+1}, \ldots, c_{2k})$ be a codeword of $\mathcal{C}$ and $wt(\mathbf{c}) = w$. Since we consider t non-zero coefficient linear combination of

$G' = (I, A')$, $wt(c_1, \ldots, c_k) = t$ and $wt(c_{k+1}, \ldots, c_{2k}) = w - t$. Therefore the number of zero components of $(c_{k+1}, \ldots, c_{2k})$ is $k - (w - t)$ which corresponds to $\bigcap_{j \in I} Z'_j$, and the number of nonzero components $(c_{k+1}, \ldots, c_{2k})$ is $w - t$ which corresponds to $\bigcup_{j \notin I} Z'_j$.

We can quickly calculate $\bigcap_{j \in I} Z'_j$ since we can view $\bigcap_{j \in I} Z'_j$ as a homogeneous system of linear equations with $|I|$ equations and t unknowns. So that we avoid the calculation in Eqn. (5). Our algorithm is very fast for $\mathcal{G}^e$. For example, the running time for $e = 7$ with $w = 8, 9, 10, 11, 12$ are $15, 27, 43, 60, 135$ seconds, respectively. Our result is in Table 1.

Table 1. A^e_w : the number of codewords of weight w in $\mathcal{G}^e$

	$e = 3$	$e = 4$	$e = 5$	$e = 6$	$e = 7$
$w = 8$	759	759	759	759	759
$w = 9$	0	0	0	0	0
$w = 10$	12144	12144	12144	12144	12144
$w = 11$	48576	48576	48576	48576	48576
$w = 12$	658352	1629872	2504240	3281456	3281456

Remark 1. We state our method for a linear code over $\mathbb{Z}_{p^e}$. But our method can be described in the same way for a linear code over an arbitrary commutative ring with unity.

Acknowledgment. The author would like to thank Young Ho Park (Kangwon National University) for his helpful discussions on this research.

References

1. Bosma, W., Cannon, J., Playoust, C.: The Magma Algebra System I: The User Language. J. Symbolic Comput. 24, 235–265 (1997)
2. Calderbank, A.R., Sloane, N.J.A.: Modular and p-adic Cyclic Codes. Designs Codes Cryptogr. 6, 21–35 (1995)
3. Dougherty, S.T., Kim, S.Y., Park, Y.H.: Lifted codes and their weight enumerators. Discr. Math. 305, 123–135 (2005)
4. Gaborit, P., Nedeloaia, C.-S., Wassermann, A.: On the Weight Enumerators of Duadic and Quadratic Residue Codes. IEEE Trans. Inf. Theory 51, 402–407 (2005)

Computer Based Reconstruction of Binary Extremal Self-dual Codes of Length 32

Jon-Lark Kim

Department of Mathematics, Sogang University,
Republic of Korea
jlkim@sogang.ac.kr

Abstract. It is well known that there are exactly five inequivalent doubly-even binary self-dual codes of length 32 and minimum distance 8. The first proof was done by Conway and Pless in 1980. The second proof was given by H. Koch in 1989 by using the balance principle. Both proofs require nontrivial mathematical arguments. In this talk, we give a computer-aided proof of this fact.

Keywords: balance principle, self-dual codes, classification of linear codes.

1 Introduction

Self-dual codes have been of great interest since the beginning of coding theory around 1948. For example, the extended binary Hamming $[8, 4, 4]$ code is self-dual and the extended binary Golay $[24, 12, 8]$ code is also self-dual. Furthermore, each code is unique up to isomorphism. Hence it is a natural question to classify all binary self-dual codes for a given length n.

A binary linear code C is called *self-dual* if C is equal to its dual $C^\perp$. A self-dual code is called *Type I* (or *singly-even*) if there is a codeword with weight $\equiv 2 \pmod 4$, and called *Type II* (or *doubly-even*) if all codewords have weight $\equiv 0 \pmod 4$. Two binary codes are said to be *equivalent* if they differ only by a permutation of the coordinates. Let C be a binary self-dual code of length n and minimum distance $d(C)$. Rains gave the following upper bound on the minimum distance $d(C)$.

$$d(C) \leq \begin{cases} 4\left\lfloor \frac{n}{24} \right\rfloor + 4, & \text{if } n \neq 22 \pmod{24}, \\ 4\left\lfloor \frac{n}{24} \right\rfloor + 6, & \text{if } n = 22 \pmod{24}. \end{cases}$$

A self-dual code meeting one of the above bounds is called *extremal*. A code is called *optimal* if it has the highest possible minimum distance for its length and dimension.

It is well known from the Gleason theorem that if C is a Type II code of length n then $n \equiv 0 \pmod 8$. V. Pless has asked me if there is a purely combinatorial proof of this fact and I have not found one yet. I hope that someone can find such a proof.

H. Hong and C. Yap (Eds.): ICMS 2014, LNCS 8592, pp. 115–118, 2014.

In the following table, we give a current status of the classification of binary self-dual codes of lengths up to 40. Here d_I (d_{II} respectively) denotes the optimal (O) or extremal (E) minimum distance of Type I (II respectively) codes. The symbol num$_I$ (num$_{II}$ respectively) denotes the number of Type I (II respectively) codes with minimum distance d_I (d_{II} respectively).

Table 1. Type I and II codes of length $2 \le n \le 40$

n	d_I	num$_I$	d_{II}	num$_{II}$	n	d_I	num$_I$	d_{II}	num$_{II}$
2	2^O	1			22	6^E	1		
4	2^O	1			24	6^E	1	8^E	1
6	2^O	1			26	6°	1		
8	2^O	1	4^E	1	28	6°	3		
10	2^O	2			30	6°	13		
12	4^E	1			32	8^E	3	8^E	5
14	4^E	1			34	6°	938		
16	4^E	1	4^E	2	36	8^E	41		
18	4^E	2			38	8^E	2744		
20	4^E	7			40	8^E	10200655	8^E	16470

In particular, it is well known that there are exactly five inequivalent doubly-even binary self-dual codes of length 32 and minimum distance 8. The first proof was done by Conway and Pless [3] in 1980. The second proof was given by H. Koch [4] in 1989 by using the balance principle. Both proofs require nontrivial mathematical arguments.

In this talk, we give a computer-aided proof of this fact. We also use the balance principle by classifying all [24,9,8] doubly-even codes. We show that there are exactly 22 such codes by the program called Q-extension made by Iliya Bouyukliev [1]. Using this result, we reprove that there are exactly five doubly-even self-dual [32,16,8] codes by writing Magma [2] programs implementing a restricted equivalence test. This experiment shows that the combination of Q-extension and Magma can be used for classification problems. We will discuss the difficulty of this approach and some possible reduction of the running complexity.

2   Balance Principle

We recall the balance principle below.

Theorem 1 ([4,5]). *Let C be a self-dual $[n, \frac{n}{2}]$ code. Choose a set of coordinate positions P_{n_1} of size n_1 and let P_{n_2} be the complementary set of coordinate positions of size $n_2 = n - n_1$. Let C_i be the subcode of C all of whose codewords have support in P_{n_i}. The following hold.*

(i) *(Balance Principle)*

$$\dim C_1 - \frac{n_1}{2} = \dim C_2 - \frac{n_2}{2}.$$

(ii) *If we reorder coordinates so that P_{n_1} is the left-most n_1 coordinates and P_{n_2} is the right-most n_2 coordinates, then C has a generator matrix of the form*

$$G = \begin{bmatrix} A & O \\ O & B \\ D & E \end{bmatrix} \tag{1}$$

where $[A \ \ O]$ is a generator matrix of C_1 and $[O \ \ B]$ is a generator matrix of C_2, O being the appropriate size zero matrix. We also have the following.
(a) *If $k_i = \dim C_i$, then D and E each have rank $\frac{n}{2} - k_1 - k_2$.*
(b) *Let $\mathcal{A}$ be the code of length n_1 generated by A, $\mathcal{A}_D$ the code of length n_1 generated by the rows of A and D, $\mathcal{B}$ the code of length n_2 generated by B, and $\mathcal{B}_E$ the code of length n_2 generated by the rows of B and E. Then $\mathcal{A}^\perp = \mathcal{A}_D$ and $\mathcal{B}^\perp = \mathcal{B}_E$.*

3 Some Computational Results

In this section, we describe how to apply the balance principle to Type II $[32, 16, 8]$ codes. Let C be a Type II $[32, 16, 8]$ code. We let $n = 32$. We need to choose n_1 in Theorem 1. The most common way to choose n_1 is to take the weight of a codeword with minimum weight in a Type II $[32, 16, 8]$ code, that is, $n_1 = 8$. Hence we may take $A = [1, \cdots, 1, 0, \cdots, 0]$. Now we have to compute the rank of the matrix B or the dimension of the code C_2 in the notation of Theorem 1. Note that

$$\dim C_1 - \frac{n_1}{2} = \dim C_2 - \frac{n_2}{2}$$
$$1 - \frac{8}{2} = \dim C_2 - \frac{24}{2}$$
$$\dim C_2 = 9$$

Therefore, the code C_2 generated by B has parameters $[24, 9, 8]$ and should be doubly-even since C is doubly-even. It may be possible to generate all doubly-even self-orthogonal $[24, 9, 8]$ codes in Magma. However, we have found that there is a very efficient program called Q-extension made by Iliya Bouyukliev [1]. A good thing about this program is that without much programming it can construct all linear $[n + n', k, d + d']$ codes for $n_1 = 1, 2, \cdots, d_1 = 0, 1, 2, \cdots$, from given parameters $[n, k]$. Below is the command of Q-extension to construct all $[24, 9, 8]$ doubly-even codes with all one-vector from the $[9, 9, 1]$ full code.

```
Extension:
(1) [9,9,1] to [24,9,8]
    1. Start
*  2. Restrictions on weights
    3. Column multiplicity restrictions
    4. Change inp --> outp
2  5. Dual distance
o  6. Form of the output matrices (c-convenient for extension, o-ordinary)
o  7. Self-orthogonal
```

```
 8. Number of ones adding in first row (Now :1)
 9. Show input
 10. Show output
 11. Start with all even weights: 1 to 5
 12. Restrictions on even weights
y 13. Self-complementary
 14. Help
 15. About Q-Extension
 16. Exit
Choose:
```

As a result, we have obtained exactly 22 inequivalent $[24, 9, 8]$ doubly-even codes. For each such code, we must fill in the two matrices B and E in Theorem 1. Since A is the all one-vector, we can choose D to be the subcode of the even code in $\mathbb{F}_2^8$ which does not contain the all one-vector. Therefore all we have to do is to find all possible matrices E such that

1. The vectors of E should be orthogonal to those of B.
2. The code generated by $[D \ E]$ must have minimum weight 8.

This task cannot be done by Q-extension and so we have used Magma. Since E is a 6×24 matrix and each row of E should have vectors of weight $6, 10, 14, 18, 22$, there are about $(\sum_{i=1}^{5} \binom{24}{4i+2})^6 = 4191980^6 \approx 5.426 * 10^{39}$ possible matrices for E. Hence it is infeasible to find E this way. Instead, in each step when we add a row to E, we apply an equivalence test to the subcode coming from A, B, D, and that part of E such that the equivalence fixes the first 6 coordinates. This method reduces the time to construct the matrices C. Finally, we have reproved the following fact using computer algebra systems.

Theorem 2 ([3],[4]). *There are exactly five Type II $[32, 16, 8]$ codes.*

References

1. Bouyukliev, I.: Q-extension, http://www.moi.math.bas.bg/~iliya/Q_ext.htm
2. Bosma, W., Cannon, J., Playoust, C.: The Magma Algebra System I: The User Language. J. Symbolic Comput. 24, 235–265 (1997)
3. Conway, J.H., Pless, V.: On the enumeration of self-dual codes. J. Combin. Theory Ser. A 28, 26–53 (1980)
4. Koch, H.: On self-dual, doubly even codes of length 32. J. Combin. Theory Ser. A 51, 63–76 (1989)
5. Pless, V.: Introduction to the Theory of Error-Correcting Codes, 3rd edn. J. Wiley and Sons, New York (1998)

Magma Implementation of Decoding Algorithms for General Algebraic Geometry Codes

Kwankyu Lee

Department of Mathematics and Education, Chosun University, Korea
kwankyu@chosun.ac.kr

Abstract. Goppa's codes on algebraic curves defined over finite fields, called AG codes, are usually regarded as the most successful class of error correcting codes in theory as well as in practice. Despite the splendid history of theoretic achievements though, an efficient algorithm decoding general AG codes appeared only recently. The decoding algorithm requires some precomputed data about the Riemann-Roch spaces of functions or differentials of the given curve of positive genus. As Magma is particularly good at computing with these spaces, the algorithm was implemented on Magma. We present its Magma implementation and describe certain details of the implementation.

Keywords: Decoding algorithm, Algebraic Geometry coodes, Algebraic curves.

1 Introduction

Let X be a smooth geometrically irreducible projective curve defined over a finite field $\mathbb{F}$ of genus g. Let $\mathbb{F}(X)$ and Ω_X denote the function field and the module of differentials of X respectively. Let $P_1, P_2, \ldots, P_n$ be distinct rational points on X, and $D = P_1 + P_2 + \cdots + P_n$. Let G be an arbitrary divisor on X, whose support does not contain the rational points. Recall that $\mathcal{L}(G) = \{ f \in \mathbb{F}(X) \mid (f) + G \geq 0 \}$ and $\Omega(G) = \{ \omega \in \Omega_X \mid (\omega) \geq G \}$. Goppa [4] defined two kinds of error correcting codes

$$C_{\mathcal{L}}(D, G) = \{ (f(P_1), f(P_2), \ldots, f(P_n)) \mid f \in \mathcal{L}(G) \} \tag{1}$$

and

$$C_{\Omega}(D, G) = \{ (\operatorname{res}_{P_1}(\omega), \operatorname{res}_{P_2}(\omega), \ldots, \operatorname{res}_{P_n}(\omega)) \mid \omega \in \Omega(-D + G) \}, \tag{2}$$

which are respectively called an evaluation AG code and a differential AG code. As well known, they are dual to each other.

The class of AG codes provides series of codes with large minimum distances surpassing the Gilbert-Varshamov bound [9]. Research on the dimensions and the minimum distances of AG codes has deep connections with important problems of discrete mathematics, number theory and algebraic geometry [8]. Moreover,

H. Hong and C. Yap (Eds.): ICMS 2014, LNCS 8592, pp. 119–123, 2014.

with the projective line taken for X, the class of AG codes includes Reed-Solomon codes and BCH codes that have been used in various communication and storage devices. Generalizing decoding algorithms for Reed-Solomon and BCH codes, efficient decoding algorithms for a subclass of AG codes have been devised. Some AG codes are strong candidates to replace Reed-Solomon codes. Furthermore AG codes and their subfield subcodes are also considered for the McEliece cryptosystem and for secret sharing schemes. For research on and practice of these applications, availability of efficient decoding algorithms for AG codes is essential.

Then it is unfortunate to find that decoding algorithms for AG codes are currently very poorly available in computer algebra systems. In Singular and Magma, Skorobogatov and Vladut's basic algorithm for general differential AG codes [9] is implemented, but this is almost all one can find in public. The basic algorithm requires as input another divisor G_1 such that the supports of G_1 and D are disjoint and

$$\deg(G_1) < \deg(G) - 2g + 2 - t \quad \text{and} \quad \dim_{\mathbb{F}}(\mathcal{L}(G_1)) > t \tag{3}$$

to correct all errors of weight $\leq t$, and it is known that G_1 exists for all $t \leq (d^* - 1 - g)/2$, where $d^* = \deg G - 2g + 2$ is the Goppa bound on the minimum distance or the designed distance of $C_{\Omega}(D, G)$.

However, in the research literature, actually there have been much more advances on the problem of decoding AG codes. By extensive works of Feng, Rao, Sakata and many others, a fast decoding algorithm that corrects errors of weight less than half of d^* for one-point differential AG codes, in which $G = mQ$, was already established in 1990's [3,6]. Moreover, decoding algorithms for multipoint differential and evaluation AG codes have been proposed by Duursma [2], Beelen and Høholdt [1], and Sakata and Fujisawa [7]. Unfortunately these algorithms were never available in public on computer algebra systems. It seems that implementations of these algorithms are only kept in the researchers' private computers. A more important reason of this situation may be that none of these algorithms are fast, simple to implement, and easy to apply for general AG codes.

To improve on the current situation, recently, there appeared a fast decoding algorithm for general evaluation and differential AG codes [5] that can correct up to half of their designed distances.[1] This algorithm is very simple and easy to implement. Indeed all heavy computations are just to provide required initial data to the algorithm, and for each received vector, the algorithm performs very simple iterative procedure to recover message symbols. Computations of the initial data are all ultimately based on the computation of a basis of the Riemann-Roch space $\mathcal{L}(G)$ or $\Omega(G)$ for any divisor G. As Magma has very nice facilities for computing with these spaces, the decoding algorithm can be most easily implemented in Magma. In the next section, we show an example session on Magma with a decoding example from the Magma documentation.

---

[1] Precisely speaking, general AG codes mean multi-point AG codes, which allows arbitrary G but requires a rational point Q not in the support of D.

2   Example Session

The following Magma scripts construct $[23, 14, 7]$ differential AG code C on the Klein quartic, of genus $g = 3$, defined over $\mathbb{F}_8$ on the projective plane.

```
> F<a> := GF(8);
> PS<x,y,z> := ProjectiveSpace(F, 2);
> Cv := Curve(PS, x^3*y + y^3*z + x*z^3);
> FF<X,Y> := FunctionField(Cv);
> Pl := Places(Cv, 1); // rational points
> Q := Place(Cv![0,1,0]);
> P := [Pl[i]: i in [1..#Pl] | Pl[i] ne Q];
> G := 11*Q;
> C := AGDualCode(P, G);
```

With the rational point $Q = [0 : 1 : 0]$ on the curve, the divisor $G_1 = 4Q$ satisfies the condition (3) for $t = 1$. Thus the Magma intrinsic AGDecode, which implements Skorobogatov and Vladut's basic algorithm, can correct arbitrary errors of weight 1.

```
> v := Random(C);
> rec_vec := v;
> rec_vec[Random(1,Length(C))] +:= Random(F);
> res := AGDecode(C, rec_vec, 4*Q);
> res eq v;
true
```

However, observe that as the minimum distance of C is 7, the code has capability of correcting unambiguously errors of weight at most 3. So the decoding algorithm in Magma fails to exert the full potential of the code.

Now we turn to the new Magma intrinsic DifferentialAGCode, which implements the interpolation-based decoding algorithm from [5] for differential AG codes (2).

```
> D:=&+P;
> code:=DifferentialAGCode(D,G,Q);
> code eq C;
true
> code'DecodingRadius;
3
```

The last output indicates that the code is capable of correcting errors of weight up to half of the actual minimum distance of C.

```
> res := DecodeAGCode(code,rec_vec);
> res eq v;
true
> rec_vec[Random(1,Length(C))] +:= Random(F);
```

```
> rec_vec[Random(1,Length(C))] +:= Random(F);
> Distance(rec_vec, v);
3
> res := DecodeAGCode(code,rec_vec);
> res eq v;
true
```

There is also new Magama intrinsic `EvaluationAGCode`, which implements the new decoding algorithm for evaluation AG code (1).

```
> ecode := EvaluationAGCode(D,G,Q);
> Dual(ecode) eq code;
true
> ecode'DecodingRadius;
5
> MinimumDistance(ecode);
12
```

Note that `ecode` is a $[23, 9, 12]$ evaluation AG code, dual of `code`, and can correct errors of weight up to 5.

3   Implementation Details

The details of the decoding algorithm implemented in the Magma intrinsic `EvaluationAGCode` is fully described in [5]. The decoding algorithm for differential AG codes is not much different from the companion algorithm. The efficiency and simplicity of the decoding algorithm are partially due to the initial data precomputed before actual decoding procedure. Moreover, the data are just polynomials over the finite field $\mathbb{F}$ representing elements of the Riemann-Roch spaces. This polynomial description is made possible by so-called Apéry systems on the function field. In this section, we explain how to compute the Apéry systems and the initial data for the decoding algorithm.

First we recall some definitions from [5]. Let

$$R = \bigcup_{s=0}^{\infty} \mathcal{L}(sQ) \subset \mathbb{F}(X).$$

For $f \in R$, let $\rho(f) = -v_Q(f)$. The Weierstrass semigroup at Q is then

$$\Lambda = \{\rho(f) \mid f \in R\} = \{\lambda_0, \lambda_1, \lambda_2, \ldots\} \subset \mathbb{Z}_{\geq 0}.$$

It is well known that Λ is a numerical semigroup whose number of gaps is the genus g of X. The nonnegative integers in Λ are called nongaps. Let γ be the smallest positive nongap, and let $\rho(x) = \gamma$ for some $x \in R$. For each $0 \leq i < \gamma$, let a_i be the smallest nongap such that $a_i \equiv i \pmod{\gamma}$ and $\rho(y_i) = a_i$ for some $y_i \in R$. Then every element of R can be written as a unique $\mathbb{F}$-linear combination of the monomials $\{x^k y_i \mid k \geq 0, 0 \leq i < \gamma\}$. The set $\{y_i \mid 0 \leq i < \gamma\}$ is called the Apéry system of R.

Recall that for divisor G, the Magma command `Basis(G)` computes a basis of $\mathcal{L}(G)$ and `Dimension(G)` computes the dimension of $\mathcal{L}(G)$. These commands are used to compute γ and x with $\rho(x) = \gamma$ and the Apéry system of R. Further details of the computations will be dealt in the full paper.

Acknowledgments. This work was supported by Basic Science Research Program through the National Research Foundation of Korea (NRF) funded by the Ministry of Education, Science and Technology (2013R1A1A2009714).

References

1. Beelen, P., Høholdt, T.: The decoding of algebraic geometry codes. In: Advances in Algebraic Geometry Codes. Ser. Coding Theory Cryptol., vol. 5, pp. 49–98. World Sci. Publ., Hackensack (2008)
2. Duursma, I.M.: Majority coset decoding. IEEE Trans. Inf. Theory 39(3), 1067–1070 (1993)
3. Feng, G.L., Rao, T.T.N.: Decoding algebraic-geometric codes up to the designed minimum distance. IEEE Trans. Inf. Theory 39(1), 37–45 (1993)
4. Goppa, V.D.: Codes on algebraic curves. Sov. Math. Dokl. 24(1), 170–172 (1981)
5. Lee, K., Bras-Amorós, M., O'Sullivan, M.E.: Unique decoding of general AG codes. IEEE Trans. Inf. Theory 60(4), 2038–2053 (2014)
6. Sakata, S., Jensen, H.E., Høholdt, T.: Generalized Berlekamp-Massey decoding of algebraic-geometric codes up to half the Feng-Rao bound. IEEE Trans. Inf. Theory 41(6), 1762–1768 (1995)
7. Sakata, S., Fujisawa, M.: Fast decoding of multipoint codes from algebraic curves. IEEE Trans. Inf. Theory 60(4), 2054–2063 (2014)
8. Stepanov, S.A.: Codes on Algebraic Curves. Springer (1999)
9. Stichtenoth, H.: Algebraic Function Fields and Codes, 2nd edn. Springer (2009)

Reversible Codes and Applications to DNA*

Elif Segah Oztas[1], Irfan Siap[1], and Bahattin Yildiz[2]

[1] Yildiz Technical University, Turkey
elifsegahoztas@gmail.com, isiap@yildiz.edu.tr
[2] Fatih University, Turkey
byildiz@fatih.edu.tr

Abstract. In this study we focus on codes over a special family of commutative rings where we are able to construct a map that gives a correspondence between k-bases (k-letter words) of DNA with elements of the ring. By making use of so called coterm polynomials, we are able to solve the reversibility and complement problems in DNA codes and construct DNA codes over this ring.

Keywords: Coterm polynomials, Reversible-complement codes, DNA codes.

1 Introduction

The first and remarkable application of DNA was demonstrated by Adleman [2] where DNA molecules are used in an experiment to solve the famous directed salesman problem. The solution of this problem relies on the Watson Crick Complement (WCC) property of DNA. The salesman problem is well known to be an NP problem. Hence, the interest of DNA in computing science attracted many researchers. Later, in [4], D. Boneh, et al., and independently Adleman, et al. in [3] developed a molecular program based on DNA where they showed that Data Encryption *(DES)* can be broken. Due to the structure of DNA, DNA molecules are proposed to be used for data storage systems [5]. DNA sequences can be viewed as strings of a four alphabet set which consists of adenine (A), guanine (G), thymine (T) and cytosine (C). The DNA replicates by taking advantage of its Watson-Crick property which helps to detect errors and hence avoid any misconstructions which cause diseases or deformations in cells. The WCC property makes sure that A and T bound to each other and G and C bound to each other in the process of replications. A and G are called the complements of T and C respectively or vice versa. In this work, the complement of a base say X will be denoted by X^c, for instance, the complement of A is $A^c = T$. Since DNA enjoys error detection and correction while the cells reproduce and this occurs in a huge numbers in the living bodies, the structure of DNA can be used as a model for error correcting codes. The first attempts naturally are done for codes

---

* This study is supported by The Scientific and Technological Research Council of Turkey (TÜBİTAK) (No: 113F071).

H. Hong and C. Yap (Eds.): ICMS 2014, LNCS 8592, pp. 124–128, 2014.

over four element alphabets. One such study is presented in [1] by Abualrub et al. where DNA codes are studied over four element finite fields. Also, Siap et al. studied DNA codes over the finite ring $\mathbb{F}_2[u]/(u^2 - 1)$ in [8]. Later, for the first time pairs of nucleotides (two letter strings) were identified with elements of the ring $\mathbb{F}_2[u]/(u^4 - 1)$. Also in [6], DNA-double pairs are used with the field F_{16} and optimal codes were obtained. These studies encouraged further considerations for k-letter words and algebraic structures that support such considerations.

Here, being inspired by the work [9] we study DNA codes over an extension ring of $\mathbb{F}_2 + u\mathbb{F}_2$, by means of a notion so called "coterm polynomials". In doing so, we identify sequences of DNA bases with elements from the extension ring $\mathcal{R}_{2k} = \mathbb{F}_2[u]/(u^{2k} - 1)$. With the introduction of coterm polynomials, we construct linear codes that enjoy the WCC property of DNA i.e linear codes that have the reverse-complement property. While the matching between the rings elements and DNA strings are to be found, so called reversibility problem arises. Basically each codeword (DNA string) under the ring operations and the linearity of the code has to fall inside the code itself. This has to be true especially for the reversibility property. To clarify the reversibility problem, we let (u_1, u_2, u_3) be a codeword that corresponds to the DNA string AGTTCCGTC where $u_1 = 1 + u + u^2$, $u_2 = u + u^3 + u^4 + u^5$, $u_3 = u + u^2 + u^3 + u^4 \in \mathcal{R}_6$. The reverse of (u_1, u_2, u_3) is (u_3, u_2, u_1) and (u_3, u_2, u_1) corresponds to GTCTCCAGT. But GTCTCCAGT is not the reverse of AGTTCCGTC. As seen from this example one needs to make sure that the DNA code is closed under such operations in order to enjoy the DNA property.

2   Reversible Codes over R with Coterm Polynomials

In this section, we suppose that the ring R is a commutative ring with identity.

Definition 1. *Let* $f(x) = a_0 + a_1 x + \cdots + a_{n-1}x^{n-1} \in R[x]/(x^n - 1)$ *be a polynomial, with* $a_i \in R$. *If for all* $1 \leq i \leq \lfloor \frac{n}{2} \rfloor$, *we have* $a_i = a_{n-i}$, *then* $f(x)$ *is said to be a coterm polynomial over* R.

Identifying the elements $\Phi(c(x)) = \bar{c} = (c_0, c_1, \ldots c_{n-1})$ in R^n with polynomials $c(x) = c_0 + c_1 x + \cdots + c_{n-1}x^{n-1} \in R[x]$.

$\langle S_g \rangle$ is a set in R^n generated by the right and the left shifts of coterm $g(x)$, respectively. For any integer i, by τ^i we mean shifts on n coordinates given by the composition of τ i times where we let i to be negative as well. In all these shifts, the indices are shifted modulo n to the right if i positive or to the left if i is negative.

Example 1. Let $g(x) = 1 + w^2 x^2 + w x^3 + x^4 + x^5 + w x^6 + w^2 x^7$ be a coterm polynomial over $\mathbb{F}_4$ and $C = \langle S_g \rangle$ be a linear code of length 9. For $t = 2$ (t is chosen to be 2 here for this example, i.e the right and the left shifts up to 2 are taken here) ($k = 2t + 2$) the $[9, 6, 3]$ optimal reversible code is obtained and the details of the construction are presented in the sequel. The corresponding generator matrix is as follows, with the row

$$\bar{c} = (1, 0, w^2, w, 1, 1, w, w^2, 0)$$

corresponding to the polynomial $g(x)$ we have:

$$\langle S_g^2 \rangle = \{\tau^{-3}(\overline{c}), \tau^{-2}(\overline{c}), \tau^{-1}(\overline{c}), \tau^0(\overline{c}), \tau^1(\overline{c}), \tau^2(\overline{c})\} =$$

$$\begin{pmatrix} w & 1 & 1 & w & w^2 & 0 & 1 & 0 & w^2 \\ w^2 & w & 1 & 1 & w & w^2 & 0 & 1 & 0 \\ 0 & w^2 & w & 1 & 1 & w & w^2 & 0 & 1 \\ 1 & 0 & w^2 & w & 1 & 1 & w & w^2 & 0 \\ 0 & 1 & 0 & w^2 & w & 1 & 1 & w & w^2 \\ w^2 & 0 & 1 & 0 & w^2 & w & 1 & 1 & w \end{pmatrix}.$$

3   Reversible DNA Codes over $\mathbb{F}_2[u]/(u^{2k} - 1)$

In order to match the DNA strings with the ring elements, first we borrow the function η given in [8] which gives a correspondence between DNA single bases and the ring $\mathbb{F}_2[u]/(u^2 - 1)$.

$$\eta(A) = 0, \eta(T) = 1 + u, \eta(G) = 1, \eta(C) = u.$$

Definition 2. *Elements of the ring $\mathcal{R}_{2k}$ can be expressed uniquely by the digits $1s$ (units), u^2s, u^4s, u^6s, u^8s, ... , $u^{2(k-1)}s$ over $\mathcal{R}_{2k}$. We call this system the u^2-adic system. In other words any element in $\mathcal{R}_{2k}$ can be expressed as a linear combination of $1, u^2, \ldots, u^{2k-2}$, where the coefficients are from $\mathbb{F}_2[u]/(u^2 - 1)$.*

Definition 3. *Let $b_1 b_2 ... b_k$ be a DNA k-base (k-leter word) where $b_i \in \{A, T, G, C\}$. The corresponding elements of DNA k-bases in R_{2k} based on the u^2-adic systems are obtained as follows:*

$$\zeta(b_1 b_2 ... b_k) = \alpha \in \mathcal{R}_{2k}, \tag{1}$$

where

$$\alpha = \eta(b_k)1 + \eta(b_{k-1})u^2 + \eta(b_{k-2})u^4 + ... + \eta(b_1)u^{2(k-1)} = \sum_{t=1}^{k} \eta(b_t)u^{2(k-t)}.$$

Example 2. Consider $TCGCAT$ as a DNA 6-bases. The corresponding element of $\mathcal{R}_{12} = \mathbb{Z}_2[u]/(u^{12} - 1)$ that represents this base-sequence is given by

$$TCGCAT \to \zeta(TCGCAT) = \eta(T)1 + \eta(A)u^2 + \eta(C)u^4 + \eta(G)u^6 + \eta(C)u^8 + \eta(T)u^{10}.$$

So we have

$$TCGCAT \to \zeta(TCGCAT) = (1+u)1 + (0)u^2 + (u)u^4 + (1)u^6 + (u)u^8 + (1+u)u^{10}$$

hence

$$TCGCAT \to \zeta(TCGCAT) = 1 + u + u^5 + u^6 + u^9 + u^{10} + u^{11}.$$

Definition 4. *Let $\langle r \rangle$ be a subring of R and generated by an element r. Let C be a code generated by $g(x)$ in $R[x]/(x^n - 1)$ as follows. If $S = \{c_1, c_2, ..., c_z\}$ is a subset of R^n, then C is a code generated by S with $\langle r \rangle$ over R^n, i.e.,*

$$C = \{r_0 c_1 + r_1 c_2 + ... + r_z c_z \mid r_i \in \langle r \rangle\}.$$

In both cases C is called an r-module code.

Definition 5. *Let $\mathfrak{B}^k = b_1 b_2 ... b_k$ be a DNA k-base. Then $(\mathfrak{B}_1^k, \mathfrak{B}_2^k, ..., \mathfrak{B}_n^k)$ is called an n-tuple of DNA k-bases. Thus, we have the following notations:*

1. $(\mathfrak{B}^k)^r = b_k b_{k-1} ... b_1$ *is the reverse of $\mathfrak{B}^k$.*
2. $(\mathfrak{B}^k)^c = b_1^c b_2^c ... b_k^c$ *is the complement of $\mathfrak{B}^k$.*
3. $(\mathfrak{B}^k)^{rc} = b_k^c b_{k-1}^c ... b_1^c$ *is the reverse complement of $\mathfrak{B}^k$.*
4. $\zeta(\mathfrak{B}_1^k, \mathfrak{B}_2^k, ..., \mathfrak{B}_n^k) = (\zeta(\mathfrak{B}_1^k), \zeta(\mathfrak{B}_2^k), ..., \zeta(\mathfrak{B}_n^k))$.

Definition 6. *We define rC_{DNA_k} to be the following n-tuple of DNA k-bases: If n is odd,*

$$rC_{DNA_k} = (\mathfrak{B}_{sr}^k, \mathfrak{B}_1^k, \mathfrak{B}_2^k, ..., \mathfrak{B}_{\frac{n-1}{2}}^k, (\mathfrak{B}_{\frac{n-1}{2}}^k)^r, ...(\mathfrak{B}_2^k)^r, (\mathfrak{B}_1^k)^r) \qquad (2)$$

if n is even,

$$rC_{DNA_k} = (\mathfrak{B}_{sr}^k, \mathfrak{B}_1^k, \mathfrak{B}_2^k, ..., \mathfrak{B}_{\frac{n}{2}-1}^k, \mathfrak{B}_{sr\frac{n}{2}}^k, (\mathfrak{B}_{\frac{n}{2}-1}^k)^r, ...(\mathfrak{B}_2^k)^r, (\mathfrak{B}_1^k)^r), \qquad (3)$$

where $\mathfrak{B}_{sr}^k = (\mathfrak{B}_{sr}^k)^r$ and $\mathfrak{B}_{sr\frac{n}{2}}^k = (\mathfrak{B}_{sr\frac{n}{2}}^k)^r$, i.e. they are self-reversible. These n-tuples are called r-coterm n-tuples of DNA k-bases. $\Phi^{-1}\zeta(rC_{DNA_k})$ is called an r-coterm polynomial.

Example 3. Suppose we have a 7-tuple of DNA 3-bases given by $rC_{DNA_k} = (TTT, GAC, TCA, CGT, TGC, ACT, CAG)$. Then $\zeta(rC_{DNA_k}) = (1 + u + u^2 + u^3 + u^4 + u^5, u + u^4, u^3 + u^4 + u^5, 1 + u + u^2 + u^5, u + u^2 + u^4 + u^5, 1 + u + u^3, 1 + u^5)$. Let $t = 1$, hence $k = 4$. $C = \langle S_g \rangle$ is a u^2-module code of length 7 and $\zeta^{-1}(C)$ is a reversible DNA code of length 21.

Example 4. Let us consider

$$rC_{DNA_3} = (ATA, AGG, TGC, CGT, GGA)$$

or

$$\zeta(rC_{DNA_3}) = (u^2 + u^3, 1 + u^2, u + u^2 + u^4 + u^5, 1 + u + u^2 + u^5, u^2 + u^4).$$

The r-coterm poynomial is $g(x) = \Phi^{-1}(\zeta(rC_{DNA_3})) = u^2 + u^3 + (1 + u^2)x + (u + u^2 + u^4 + u^5)x^2 + (1 + u + u^2 + u^5)x^3 + (u^2 + u^4)x^4$.

The generator set for $t = 0$ is given by $S_g = \{\tau^{-1}(\bar{c}), \tau(\bar{c})^0\} = \{(u^2 + u^3, 1 + u^2, u + u^2 + u^4 + u^5, 1 + u + u^2 + u^5, u^2 + u^4), (1 + u^2, u + u^2 + u^4 + u^5, 1 + u + u^2 + u^5, u^2 + u^4, u^2 + u^3)\}$

This can be considered as a matrix as follows:

$$\begin{pmatrix} \tau^{-1}(\overline{c}) \\ \tau(\overline{c})^0 \end{pmatrix} =$$

$$\begin{pmatrix} 1+u^2 & u+u^2+u^4+u^5 & 1+u+u^2+u^5 & u^2+u^4 & u^2+u^3 \\ u^2+u^3 & 1+u^2 & u+u^2+u^4+u^5 & 1+u+u^2+u^5 & u^2+u^4 \end{pmatrix} .$$

4 Conclusion

In this work, we introduce coterm and r-coterm polynomials to construct reversible codes and hence DNA codes. We solve reversibility problem for DNA k bases by means of coterm polynomials. The advantage of using these polynomials is evident from providing simple constructions and providing a rich source of DNA codes. Further, some optimal reversible codes are obtained by coterm polynomials.

Acknowledgment. We thank to the referee(s) for his/her valuable remarks.

References

1. Abulraub, T., Ghrayeb, A., Zeng, X.N.: Construction of cyclic codes over $GF(4)$ for DNA computing. Journal of the Franklin Institute 343(4-5), 448–457 (2006)
2. Adleman, L.: Molecular computation of solutions to combinatorial problems. Science 266, 1021–1024 (1994)
3. Adleman, L., Rothemund, P.W.K., Roweis, S., Winfree, E.: On applying molecular computation to the Data Encryption Standard. Journal of Computational Biology 6, 53–63 (1999)
4. Boneh, D., Dunworth, C., Lipton, R.: Breaking DES using molecular computer, Princeton CS Tech-Report, Number CS-TR-489-95 (1995)
5. Mansuripur, M., Khulbe, P.K., Kuebler, S.M., Perry, J.W., Giridhar, M.S., Peyghambarian, N.: Information Storage and retrieval using macromolecules as storage media, University of Arizona Technical Report (2003)
6. Oztas, E.S., Siap, I.: Lifted Polynomials Over F_{16} and Their Applications to DNA Codes. Filomat 27(3), 459–466 (2013)
7. Siap, I., Abulraub, T., Ghayreb, A.: Similarity Cyclic DNA Codes over Rings. International Conference on Bioinformatics and Biomedical Engineering, Shanghai, PRC, May 16–18 (2008)
8. Siap, I., Abulraub, T., Ghrayeb, A.: Cyclic DNA codes over the ring $\mathbb{F}_2[u]/(u^2-1)$ based on the deletion distance. Journal of the Franklin Institute 346(8), 731–740 (2009)
9. Yildiz, B., Siap, I.: Cyclic Codes over $\mathbb{F}_2[u]/(u^4-1)$ and applications to DNA codes. Computers and Mathematics with Applications 63(7), 1169–1176 (2012)

javaPlex: A Research Software Package for Persistent (Co)Homology

Henry Adams[1], Andrew Tausz[2], and Mikael Vejdemo-Johansson[3,4]

[1] Institute of Mathematics and its Applications, USA
[2] Stanford University, USA
[3] KTH Royal Institute of Technology, Sweden
`mvj@kth.se`
[4] Institut Jozef Stefan, Slovenia

Abstract. The computation of persistent homology has proven a fundamental component of the nascent field of topological data analysis and computational topology. We describe a new software package for topological computation, with design focus on needs of the research community. This tool, replacing previous jPlex and Plex, enables researchers to access state of the art algorithms for persistent homology, cohomology, hom complexes, filtered simplicial complexes, filtered cell complexes, witness complex constructions, and many more essential components of computational topology.

We describe, herewithin, the design goals we have chosen, as well as the resulting software package, and some of its more novel capabilities.

Keywords: persistent homology, topological data analysis, computational topology.

1 Motivation and Design Goals

The main reason for the existence of javaPlex is to provide researchers in the area of topological data analysis a unified software library to support their investigations. With this in mind, the design goals for it are as follows:

- **Support for new directions for research:** The main goal of the javaPlex package is to provide an extensible base to support new avenues for research in computational homology and data analysis. While its predecesor jPlex was very well suited towards computing simplicial homology, its design made extension difficult.
- **Interoperability:** Since javaPlex is a java package, it is accessible from anything that runs in the Java runtime environment: in a Java or Scala application, or called from Matlab or Mathematica, or as a library loaded into beanshell, jython for a scripting interface.
- **Adherence to generally accepted software engineering practices:** As a means to realizing the first goal, the javaPlex software package was designed and implemented with software engineering best-practices. Emphasis was placed on maintainability, modularity, and reusability of the different parts of the code.

H. Hong and C. Yap (Eds.): ICMS 2014, LNCS 8592, pp. 129–136, 2014.

We refer the reader to [Car09] for a very readable introduction to the field of topological data analysis as well as the computational tasks involved.

2 Previous Work

The javaPlex package is the fourth version in the Plex family. These programs have been developed over the past decade by members of the computational topology research group at Stanford University. Each successive version incorporated the results of new advances in the relatively quickly developing fields of computational topology and topological data analysis.

Like javaPlex, its predecessor jPlex was also written in the Java language. However, it differed in that the main goal of jPlex was the computation of *simplicial* homology. Recent research topics in topological data analysis have required practitioners to move beyond conventional simplicial homology to more general scenarios.

3 Persistent Homology and Topological Data Analysis

The javaPlex library is focused on computing persistent homology and enabling research and use of topological data analysis methods. At the core of these tasks is the ability to compute the homology indicated by a point cloud. To this end, [ELZ02] introduced persistent homology, refined by [ZC05]. Using one of a whole family of methods, the point cloud induces a filtered simplicial complex, where the filtration encodes distance data as increasing "closeness" data for the data points in the point cloud. From a simplicial complex, we can generate a *chain complex*: a vector space with vectors representing the geometry and with an operator ∂ capturing what it means to be on the boundary of a cell. The homology is defined as $\ker \partial / \operatorname{im} \partial$, and can be considered to represent *essential cycles* or *bubbles* in arbitrary high dimensions as witnessed in the data itself.

Persistence captures the notion of computing this homology for a range of parameter values, connecting the local results to extract a *barcode* that represents each such essential feature as a discrete component coupled with a spread of parameter values at which the feature exists in the dataset. For more details, we refer an interested reader to [Car09].

4 Filtered Complex Generation

As mentioned in the abstract, the primary function of javaPlex is the construction of filtered chain complexes of vector spaces associated to actual point cloud datasets. The motivation for such constructions is that they provide a persistent model of the dataset in question across all scales. javaPlex currently supports the construction of two main types of filtered simplicial complexes: the Vietoris-Rips and lazy-witness constructions. To begin, suppose that we have a finite metric space $(\mathcal{X}, d)$. In practice, it is possible that $\mathcal{X}$ is a set of points in Euclidean space, although this is not necessary.

4.1   The Vietoris-Rips Construction

We define the filtered complex $VR(\mathcal{X}, r)$ as follows. Suppose that the points of $\mathcal{X}$ are $\{x_1, ... x_N\}$, where $N = |\mathcal{X}|$. The Vietoris-Rips complex is constructed as follows:

- **Add points:** For all points $x \in \mathcal{X}$, $x \in VR_0(\mathcal{X}, 0)$.
- **Add 1-skeleton:** The 1-simplex $[x_i, x_j]$ is in $VR_1(\mathcal{X}, r)$ iff $d(x_i, x_j) \leq r$.
- **Expansion:** We define $VR(\mathcal{X}, r)$ to be the maximal simplicial complex with 1-skeleton $VR_1(\mathcal{X}, r)$. That is, a simplex $[x_0, .. x_k]$ is in $VR(\mathcal{X}, r)$ if and only if all of its edges are in $VR_1(\mathcal{X}, r)$.

An extensive discussion on algorithms for computing the Vietoris-Rips complex can be found in [Zom10]. The javaPlex implementation is based on the results of this paper.

4.2   The Lazy-Witness Construction

The fundamental idea behind the lazy-witness construction is that a relatively small subset of a point cloud can accurately describe the shape of the dataset. This construction has the advantage of being more resistant to noise than the Vietoris-Rips construction. An extensive discussion about it can be found in [dSC04].

The lazy-witness construction starts with a selection of landmark points, $\mathcal{L} \subset \mathcal{X}$ with $|\mathcal{L}| = L$. One possibility is to simply choose a random subset of $\mathcal{X}$. Another possibility is to perform a sequential max-min selection: An initial point l_0 is selected, and then we inductively select the point l_k which maximizes the minimum distance to all previously generated points. This max-min construction tends to produce more evenly spaced points than the random selection. Again we refer the reader to [dSC04] for a more detailed discussion, as well as empirical results supporting these claims.

This construction is parameterized by a value ν, which most commonly takes the values 0, 1, or 2. We also define the distance matrix D to contain the pairwise distances between the points in $\mathcal{X}$.

- **Define m_i:** If $\nu = 0$, let $m_i = 0$, otherwise, define m_i to be the ν-th smallest entry in the i-th column of D.
- **Add points:** For all points $l \in \mathcal{L}$, $l \in LW_0(\mathcal{X}, 0, \nu)$.
- **Add 1-skeleton:** The 1-simplex $[l_i, l_j]$ is in $LW_1(\mathcal{X}, r, \nu)$ iff there exists an $x \in \mathcal{X}$ such that $\max(d(l_i, x), d(l_j, x)) \leq r + m_i$.
- **Expansion:** We define $LW(\mathcal{X}, r, \nu)$ to be the maximal simplicial complex with 1-skeleton $LW_1(\mathcal{X}, r, \nu)$.

5   Homology Computation

At the core of the javaPlex library is the set of algorithms that actually compute the homology of a filtered chain complex. Key references to background material regarding these algorithms can be found in [ZC05, dSMVJ10]. Although we do not describe them in detail here, we note that the algorithms for computing persistent absolute/relative (co)homology can be formulated as matrix decomposition problems. The fundamental reason for this is the equivalence of the category of persistent vector spaces of finite type, and the category of finitely generated graded modules over $\mathbb{F}[t]$. This correspondence is described in [ZC05].

The homology algorithms are built in a way that is optimized for chain complexes implemented as *streams*. By this we mean that a filtered chain complex is represented by a sequence of basis elements that are produced in increasing order of their filtration indices. Enforcing the constraint that all complexes must be implemented this way allows javaPlex to perform the matrix decomposition operations in an efficient online fashion.

6   Applications

Although in principle javaPlex can compute the persistent homology of arbitrary chain complexes of vector spaces, almost always these complexes arise from some sort of topological construction. Below we outline these different situations.

6.1   Simplicial Homology

Computing simplicial homology of a filtered sequence of complexes is performed by generating a `SimplexStream` and running it through the persistent homology algorithm. A sample invocation would look like follows:

```
ExplicitSimplexStream stream = new ExplicitSimplexStream();
// add vertices with stream.addVertex; simplices with
// stream.addElement
stream.finalizeStream();
AbstractPersistenceAlgorithm<Simplex> pA =
  Plex4.getDefaultSimplicialAlgorithm(d + 1);
BarcodeCollection<Double> intervals = pA.computeIntervals(stream);
```

6.2   Simplicial Cohomology

As described in [dSMVJ10], persistent cohomology can be computed by consuming simplices in the opposite order, computing coboundaries instead of boundaries, and reversing the order of simplices when picking out leading terms. This is supported in javaPlex through the `DualStream` class that transforms an existing simplex stream to a reversed version, together with the Java utility method `java.util.Collections.reverseOrder` which can reverse the simplex order declaration instantiating the homology algorithm:

```
AbstractPersistenceAlgorithm<Simplex> pA =
  new IntAbsoluteHomology(ModularIntField.getInstance(prime),
        Collections.reverseOrder(SimplexComparator.getInstance()),
        0, d+1);
BarcodeCollection<Double> intervals = pA.computeIntervals(stream);
```

6.3   Novel Operations and Types

javaPlex also supports arbitrary cell complexes, where the gluing maps are given explicitly rather than generating them implicitly from the simplices. In addition to this, there is also support for computing with **tensor products** and **Hom complexes** of chain complexes of any type in the system. The particular case of dualizing a chain complex to get cochains is handled by `DualStream`, while the general homomorphism complexes are handled by `HomStream` instead.

In a recent preprint [TC11], the hom-complex was used to compute a parameterization for the space of homotopy classes of chain maps between simplicial complexes.

7   Examples

7.1   Simplicial Homology

In Figure 1, one can see an example of a filtered simplicial complex generated from points on a torus. As one moves from left to right, the filtration parameter r is increased, yielding a more connected complex. In Figure 2 we show the persistence barcodes for the same shape. Note that the significant intervals correspond to homological features that last for a long time in the filtration.

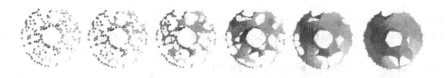

Fig. 1. Example of a lazy-witness complex generated from randomly sampled points on a torus

7.2   Matlab Scripting Example - Cellular Homology

In this section we show a brief Matlab session in which the cellular homology is computed for a Klein bottle over different coefficient fields. Essentially, this code example constructs a cellular Klein bottle, initializes persistence algorithm objects over the fields $\mathbb{Z}/2\mathbb{Z}$, $\mathbb{Z}/3\mathbb{Z}$, and $\mathbb{Q}$, and computes the persistence intervals.

```
% get the cellular Klein bottle
stream = examples.CellStreamExamples.getCellularKleinBottle();

% get cellular homology algorithm over Z/2Z
Z2_persistence = api.Plex4.getModularCellularAlgorithm(3, 2);
% get cellular homology algorithm over Z/3Z
Z3_persistence = api.Plex4.getModularCellularAlgorithm(3, 3);
```

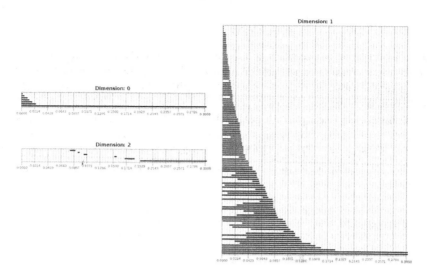

Fig. 2. Persistence barcodes for a lazy-witness filtration of random points on a torus. The parameters used were: $N = 1000$, $L = 300$, $r_{max} = 0.3$. The inner and outer radii of the torus were 0.5 and 1, respectively. The max-min selection procedure was used to create the landmark set. Note that long intervals correspond to significant homological features, and short ones are most likely the result of noise. We can see that the number of significant intervals in each dimension equals the expected Betti number.

```
% get cellular homology algorithm over Q
Q_persistence = api.Plex4.getRationalCellularAlgorithm(3);

% compute over Z/2Z - should give (1, 2, 1)
Z2_intervals = Z2_persistence.computeIntervals(stream)

% compute over Z/3Z - should give (1, 1, 0)
Z3_intervals = Z3_persistence.computeIntervals(stream)

% compute over Q - should give (1, 1, 0)
Q_intervals = Q_persistence.computeIntervals(stream)
```

The output of this example is:

```
Z2_intervals =
Dimension: 2 [0, infinity)
Dimension: 1 [0, infinity), [0, infinity)
Dimension: 0 [0, infinity)

Z3_intervals =
Dimension: 1 [0, infinity)
Dimension: 0 [0, infinity)
```

```
Q_intervals =
Dimension: 1 [0, infinity)
Dimension: 0 [0, infinity)
```

This is exactly what we expect, due to the presence of 2-torsion in the Klein bottle.

7.3   Hom Complex Examples

As mentioned earlier, the hom complex is another homological construction that is useful in algebraic topology. A key result is that the 0-dimensional homology classes of $\mathrm{Hom}(A, B)_*$ correspond exactly with homotopy classes of chain maps between A and B. Thus, by computing homology (with field coefficients in our case), we can obtain an explicit parameterization of the affine space of homotopy classes of chain maps for simplicial complexes. Additionally, a practitioner can also optimize over this space to select a particular map that optimizes some sort of geometric objective.

In Figures 3 and 4 we show examples of the computation of homotopy representatives of chain maps between two simplicial complexes. The specific maps were computed by minimizing the maximum ℓ_1 norm of the images and adjoint images. The maps are represented by composing the color of the domain with the computed map. In the first example we can see that the larger circle is essentially partitioned into different segments of constant color. The second example shows the mapping of a trefoil knot to a square.

Fig. 3. Example of a homotopy representative from the class of chain maps that induce isomorphisms on both 0 and 1 dimensional homology

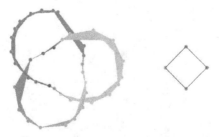

Fig. 4. Example of a chain map on a lazy-witness filtration of a trefoil knot

References

[Car09]   Carlsson, G.: Topology and data. Bulletin of the American Mathematical
             Society 46(2), 255–308 (2009)

[dSC04]   de Silva, V., Carlsson, G.: Topological estimation using witness com-
             plexes. In: Alexa, M., Rusinkiewicz, S. (eds.) Eurographics Symposium
             on Point-Based Graphics. The Eurographics Association (2004)

[dSMVJ10]   De Silva, V., Morozov, D., Vejdemo-Johansson, M.: Dualities in persistent
             (co) homology. Inverse Problems 27(12), 124003 (2011)

[ELZ02]   Edelsbrunner, H., Letscher, D., Zomorodian, A.: Topological persistence
             and simplification. Discrete Comput. Geom. 28(4), 511–533 (2002), Dis-
             crete and computational geometry and graph drawing (Columbia, SC,
             2001). MR 1949898 (2003m:52019)

[TC11]   Tausz, A., Carlsson, G.: Homological Coordinatization (2011) (preprint)

[ZC05]   Zomorodian, A., Carlsson, G.: Computing persistent homology. Discrete
             Comput. Geom. 33, 249–274 (2005)

[Zom10]   Zomorodian, A.: Fast construction of the Vietoris-Rips complex. Com-
             puters & Graphics 34(3), 263–271 (2010)

PHAT – Persistent Homology Algorithms Toolbox

Ulrich Bauer[1], Michael Kerber[2], Jan Reininghaus[1], and Hubert Wagner[3]

[1] IST Austria, Klosterneuburg, Austria
[2] Max-Planck-Institut für Informatik, Saarbrücken, Germany
[3] Jagiellonian University, Poland

Abstract. PHAT is a C++ library for the computation of persistent homology by matrix reduction. We aim for a simple generic design that decouples algorithms from data structures without sacrificing efficiency or user-friendliness. This makes PHAT a versatile platform for experimenting with algorithmic ideas and comparing them to state of the art implementations.

1 Introduction

Persistent homology is one of the most widely applicable tools in the emerging field of computational topology. Intuitively, persistent homology tracks the topological features in a growing sequence of shapes. For a comprehensive introduction to the theory and some applications, see [4,5].

Computing persistent homology for a given data set requires the construction of a *filtered cell complex*, i.e., an ordered list of cells such that every prefix forms a subcomplex. We represent a filtered cell complex by its *boundary matrix*, a square matrix whose indices correspond to the ordering of the cells, and whose entries encode the boundary relation of the complex. Since we only consider homology with coefficients in $\mathbb{Z}_2$, the entries of the boundary matrix are in $\mathbb{Z}_2$. Given a boundary matrix, computing persistent homology amounts to transforming it into a *reduced form* using certain elementary matrix operations, similar to Gaussian elimination. A boundary matrix is called *reduced* iff the column pivots, i.e., the maximal indices of the nonzero column entries, are disjoint.

The purpose of PHAT[1] is to provide a platform for comparative evaluation of new and existing algorithms and data structures for matrix reduction. PHAT is among the fastest codes for computing persistent homology currently available [1] and can be obtained under the GNU Lesser General Public License.

While the worst case computational complexity is cubic for all combinations of algorithms and data structures, the actual running time for real world data sets can differ drastically. We demonstrate that there are three key ideas that can have a tremendous impact on the running time: the *clearing* optimization suggested in [2], the computation of persistent cohomology as proposed in [3], and the use of an efficient data structure for column additions.

---

[1] http://phat.googlecode.com

H. Hong and C. Yap (Eds.): ICMS 2014, LNCS 8592, pp. 137–143, 2014.
© Springer-Verlag Berlin Heidelberg 2014

2   Design

PHAT is a C++ library consisting of about 3200 lines of code. Its main purpose
is the computation of the persistent homology of a boundary matrix in an extensible, simple, and efficient way. A boundary matrix can be passed to PHAT
through an input file that encodes the boundary in a compact form; we refer
to the README file of the library for details on the file format. Internally, the
matrix is accessed through a template class called boundary_matrix with the
following interface:

```
template < typename Representation >
class boundary_matrix   {
    Representation rep;
    int get_max_index ( int idx );
    void add_to ( int source , int target );
    . . .
};
```

A boundary_matrix stores an internal object of the supplied Representation
type and forwards all matrix access and manipulation requests to this object.
Furthermore, boundary_matrix implements several functions independent of the
representation, such as reading from and writing to a file. This way, the required
interface of the representation class is kept as small as possible. We remark that
dynamic polymorphism would give similar advantages (collecting the interface
of boundary_matrix in an abstract base class and implementing it in several
subclasses). We decided for the templated version for efficiency reasons: in a
polymorphic implementation, every call to a matrix operation requires a lookup
in the virtual function table, and most of the execution time is in fact spent for
such low-level matrix operations.

The main function of the library has the following signature:

```
template < typename ReductionAlgorithm , typename Representation >
void compute_persistence_pairs ( persistence_pairs& pairs ,
    boundary_matrix < Representation >& boundary_matrix );
```

It takes a boundary_matrix as input and transforms it into reduced form. Moreover, it computes the persistence pairs (as pairs of indices) from the reduced
matrix and stores them in the container pairs. The function has two template
parameters: Representation defines the data structure for the boundary matrix (see above), and ReductionAlgorithm specifies which method is used for
matrix reduction. The template class ReductionAlgorithm must yield *function
objects*, i.e., it must implement the () operator. This operator is assumed to take
a boundary matrix as an argument and transform it into reduced form. The template parameter ReductionAlgorithm thereby specifies *in which order* columns
of the matrix are accessed, while the template parameter Representation specifies *how* columns of the matrix are accessed. As shown in Section 5, both aspects
are equally important for an efficient implementation, and we describe the options provided by the PHAT library below.

3   Algorithms

We use the following notation throughout this paper. The *pivot index* of a column in the matrix is the largest index of any nonzero entry. All our reduction algorithms perform left-to-right column additions until no two columns have the same pivot index. A matrix with this property is called *reduced*.

PHAT 1.4 provides five different choices of reduction strategies, two of which have a parallel implementation using OpenMP.

Sequential algorithms. The `standard` algorithm for reducing boundary matrices [6] traverses the columns from left to right and reduces a column completely (through left-to-right column additions) before proceeding with the next one. The reduction of a column is complete once either its pivot index does not appear as a pivot index in a previous column or the column becomes zero.

The algorithm `twist` [2] is based on the standard algorithm and exploits the observation that a column will eventually be reduced to an empty column if its index appears as the pivot of another column. By reducing columns in decreasing order of the dimensions of the corresponding cells, we can explicitly *clear* the columns corresponding to pivot indices. Since the omitted column operations often constitute the bulk of the column operations in the standard algorithm, the clearing optimization can have a tremendous impact on practical performance. It is therefore also used in all other algorithms described below. Due to its simplicity and efficiency, the `twist` algorithm is the default in PHAT.

The idea behind the `row` algorithm [3] is to traverse the columns from right to left. Whenever the pivot of a newly inspected column A equals the pivot of a column B to its right, we add A to B.

Parallel algorithms. The algorithm `spectral_sequence` decomposes the boundary matrix into blocks and processes them in diagonals from the main diagonal outward. The reduction of the individual blocks in a single diagonal is then independent and can be performed in parallel.

The `chunk` algorithm [1] begins with the reduction of two diagonals of blocks, as in the spectral sequence algorithm. In a second step, it simplifies the partially reduced boundary matrix by eliminating the indices of the already found pairs from the matrix. Finally, the simplified matrix is reduced using the twist algorithm. The chunk algorithm is a generalized version of the approach in [8] to compute persistence using discrete Morse theory [7]. The first two steps of the algorithm be run in parallel.

Dualization. Every algorithm for persistent homology also yields an algorithm for persistent cohomology by applying it to the corresponding coboundary matrix. This matrix is given by the *anti-transpose* of the boundary matrix D, obtained by swapping $D_{i,j}$ with $D_{n+1-j, n+1-i}$. Reducing the coboundary matrix yields the same persistence pairs, up to reindexing. In some cases, this *dualization* yields significant speed-ups [3], in particular for Vietoris–Rips complexes. PHAT contains a method to anti-transpose a boundary matrix, thus providing the option to dualize every algorithm proposed in this section.

4   Data Structures

All boundary matrix data structures ·currently implemented in PHAT use a
vector containing the individual columns. The column type is defined by the
representation. There are two kinds of representations in PHAT, direct and accelerated. A direct representation makes use of a single column type for storage
and computation, whereas an accelerated representation provides an additional
type that is optimized for fast column additions. To simplify notation, we refer
to the number of nonzero entries of a column as its *length*.

Direct representations. The class vector_list represents a column as a doubly-
linked list (list<int>) storing the indices of nonzero entries in increasing order,
as suggested in [5]. Adding two columns of lengths k and m can therefore be
performed in time $O(k + m)$ by computing the symmetric difference of the lists.
The pivot of a column can be found in $O(1)$ by querying the last element of
the list. The representation vector_vector is analogous, using a dynamically
growing array (vector<int>) instead of a linked list. This representation is more
machine friendly, since it makes use of a contiguous memory region. However,
both representations have the disadvantage that column additions are expensive
when a small column is added to a large column.

An alternative representation is vector_set, where columns are stored as
balanced binary search trees (set<int>). Adding a column A of length k to
a column B of length m can be performed as follows. We iterate through the
entries of A, removing the entry from B if it is already present, and inserting it
otherwise. The complexity of such an addition is $O(k \log(k + m))$, which can be
much better than $O(k + m)$ when $k \ll m$. The pivot of a column can be found
in $O(1)$.

The representation vector_heap combines the advantages of contiguous storage and efficient column addition. Columns are again stored as vector<int>,
but the indices are now arranged in heap order. Adding a column A of length k
to a column B of length m can be lazily performed by inserting the indices of A
into B in amortized time $O(k \log(k + m))$. This implies that an index may temporarily appear multiple times in the heap. The symmetric difference operation
is delayed until a certain number of insertions is exceeded or the content of the
column is queried. This allows for the pivot of a column of length k to be found
in amortized time $O(1)$.

Accelerated representations. Similar to vector_heap, accelerated representations
also aim at combining contiguous storage with efficient column additions. An
accelerated representation is a subclass derived from vector_vector that contains an instance of a specialized column type, supporting fast column additions
and pivot queries. This column is used as a cache for the *active column*, i.e.,
the last column modified by the algorithm. For a net gain in performance, efficient conversion between vector<int> and cache type, column additions, and
pivot queries are required. Moreover, the employed algorithm needs to exhibit a

cache-friendly access pattern. This is the case for all algorithms of Section 3 except from the `row` algorithm.

A simple yet efficient choice for the column cache is `set<int>`. The resulting representation is called `sparse_pivot_column`. Another option is the use of a heap as explained in the description of `vector_heap`. The corresponding representation is called `heap_pivot_column`.

Alternatively, `full_pivot_column` uses a bit array of size n to store the entries of the column explicitly. Adding a column of length k to this representation still takes time $O(k \log(k + m))$ due to some additional structure required to quickly access the pivot of a column. We store all modified indices in a heap called *history*, and extract elements from the history until we find an index with nonzero column entry in the bit array. To ensure that no element is inserted into the history multiple times, we also maintain a bit array representing its content. When converting the bit array back into a `vector<int>`, we repeatedly extract and remove the pivot in order to re-initialize it to zero for further use.

The `bit_tree_pivot_column` representation stores the column explicitly in a tree structure. Internally, an 64-ary tree is used, which is encoded implicitly in a bit array. It not only supports fast insertions, deletions and lookup of entries, but also maximum, minimum, successor, and predecessor queries, all in time $O(\log n)$. Furthermore, the structure can be traversed and cleared in time proportional to the number of nonzero entries.

5   Experiments

To evaluate the performance of the different algorithms and data structures, we perform computational experiments using two data sets. The running times are measured on a workstation with two Intel Xeon E5645 CPUs using the integrated benchmark utility of PHAT v1.4. All data sets are available on the project homepage.

The first data set is the 3-skeleton of the Vietoris-Rips filtration of a point cloud generated by a uniform random sample of the 2-sphere. Using 64 points, the resulting boundary matrix has 679,120 columns and 2,670,528 nonzero entries. The running times in seconds for the matrix reduction of this data set are shown in Table 1, where we denote the dualized algorithms by $(\cdot)^*$. It can be observed that the combination of dualization with algorithms employing the clearing optimization leads to drastically shorter running times compared to other choices. To admit a meaningful comparison for these fast algorithms, we repeat the experiment using 192 points, resulting in a boundary matrix with 56,050,288 columns and 223,002,432 nonzero entries. The running times in Table 2 show that the more sophisticated data structures significantly improve the running time. Moreover, the simple sequential `twist` algorithm is about as fast as the parallel algorithms on this example.

The second data set is a lower-star Morse filtration of a cubical complex generated from a 3D image that indicates separation behavior in a vector field [9]. Using a 64^3 sub-region of the image, we get a boundary matrix with 2,048,383

Table 1. Running times (in seconds) for the 3-skeleton of the Vietoris-Rips filtration of 64 points on a 2-sphere, using different combinations of algorithms and data structures. The prefix "A-" refers to accelerated representations, while $(\cdot)^*$ denotes dualization.

	List	Vector	Set	Heap	A-Heap	A-Set	A-Full	A-Bit-Tree
standard	15.5	2.7	6.5	5.6	5.2	7.7	2.3	1.6
standard*	2353.4	160.3	15.9	13.4	13.5	15.1	4.1	0.6
twist	15.1	2.4	6.4	5.7	5.3	7.1	2.0	1.4
twist*	0.2	0.0	0.0	0.0	0.0	0.0	0.0	0.0
row	44.1	4.5	19.0	7.4	20.7	34.3	15.4	13.4
row*	0.2	0.1	0.1	0.1	0.1	0.1	0.1	0.0
chunk	0.6	0.2	0.5	0.5	0.3	0.5	0.2	0.2
chunk*	2.8	0.3	0.1	0.1	0.1	0.1	0.1	0.0
spectral sequence	9.7	1.7	4.1	3.3	3.4	4.2	1.5	1.1
spectral sequence*	0.3	0.0	0.0	0.0	0.0	0.0	0.0	0.0

Table 2. Running times for the 3-skeleton of the Vietoris-Rips filtration of 192 points on a 2-sphere. See Table 1 for details.

	List	Vector	Set	Heap	A-Heap	A-Set	A-Full	A-Bit-Tree
twist*	2635.4	339.9	4.9	2.0	2.5	6.1	2.1	1.0
row*	2842.3	434.6	5.6	4.0	59.4	107.9	45.9	17.3
chunk*	24391.6	3276.2	25.2	14.2	14.0	20.7	8.7	4.0
spectral sequence*	2644.8	349.2	5.2	1.9	3.3	6.6	3.1	1.0

Table 3. Running times for a 64^3 image data set. See Table 1 for details.

	List	Vector	Set	Heap	A-Heap	A-Set	A-Full	A-Bit-Tree
standard	144.8	15.2	27.9	19.5	16.6	18.6	13.6	9.5
standard*	461.0	39.1	33.1	22.3	22.1	21.4	12.4	14.9
twist	9.6	0.6	0.4	0.1	0.1	0.1	0.1	0.1
twist*	343.0	19.1	1.0	0.5	0.6	0.7	0.2	0.2
row	10.3	1.1	0.4	0.4	1.5	2.1	1.2	0.7
row*	339.9	32.9	1.1	1.0	27.0	44.6	17.8	7.5
chunk	1.8	0.2	0.2	0.1	0.1	0.1	0.1	0.1
chunk*	5.7	0.5	0.3	0.2	0.2	0.2	0.2	0.2
spectral sequence	9.6	0.8	0.2	0.1	0.1	0.1	0.1	0.1
spectral sequence*	338.1	21.4	0.9	0.7	0.8	0.8	0.3	0.1

Table 4. Running times for a 256^3 image data set. See Table 1 for details.

	List	Vector	Set	Heap	A-Heap	A-Set	A-Full	A-Bit-Tree
twist	2080.2	101.7	26.4	11.3	11.1	12.3	10.4	8.8
row	3201.8	210.4	52.1	32.5	1778.7	3437.1	1244.1	734.3
chunk	894.5	156.1	9.8	6.5	6.3	6.2	5.6	4.7
spectral sequence	1197.7	261.7	11.9	8.5	8.5	8.4	7.1	6.1

columns and 6,096,762 nonzero entries. The running times for this data set are shown in Table 3. We observe that homology computation is generally faster than cohomology computation for this data set, and the clearing optimization is again crucial for a fast algorithm. To investigate the performance behavior further, we also apply a subset of the algorithms to the full data set consisting of 256^3 voxels – the corresponding boundary matrix has 133,432,831 columns and 399,515,130 nonzero entries. The results in Table 4 again demonstrate the usefulness of the complex data structures introduced in Section 4. In contrast to the first data set, the parallel algorithms can outperform the sequential methods in this case.

Acknowledgements. This research is partially supported by the TOPOSYS project FP7-ICT-318493-STREP, the Max Planck Center for Visual Computing and Communication, and the Foundation for Polish Science IPP Programme *Geometry and Topology in Physical Models*.

References

1. Bauer, U., Kerber, M., Reininghaus, J.: Clear and compress: Computing persistent homology in chunks. In: Topological Methods in Data Analysis and Visualization III. Mathematics and Visualization, pp. 103–117. Springer (2014)
2. Chen, C., Kerber, M.: Persistent homology computation with a twist. In: 27th European Workshop on Computational Geometry (EuroCG), pp. 197–200 (2011)
3. de Silva, V., Morozov, D., Vejdemo-Johansson, M.: Dualities in persistent (co)homology. Inverse Problems 27(12), 124003+ (2011)
4. Edelsbrunner, H., Harer, J.: Persistent homology — a survey. In: Surveys on Discrete and Computational Geometry: Twenty Years Later, Contemporary Mathematics, pp. 257–282 (2008)
5. Edelsbrunner, H., Harer, J.: Computational Topology. An Introduction. American Mathematical Society (2010)
6. Edelsbrunner, H., Letscher, D., Zomorodian, A.: Topological persistence and simplification. Discrete & Computational Geometry 28(4), 511–533 (2002)
7. Forman, R.: Morse theory for cell complexes. Advances in Mathematics 134(1), 90–145 (1998)
8. Günther, D., Reininghaus, J., Wagner, H., Hotz, I.: Efficient computation of 3D Morse –Smale complexes and persistent homology using discrete Morse theory. The Visual Computer 28(10), 959–969 (2012)
9. Kasten, J., Reininghaus, J., Reich, W., Scheuermann, G.: Toward the extraction of saddle periodic orbits. In: Topological Methods in Data Analysis and Visualization III. Mathematics and Visualization, pp. 55–69. Springer (2014)

Computing Persistence Modules
on Commutative Ladders of Finite Type

Emerson G. Escolar[1] and Yasuaki Hiraoka[2]

[1] Graduate School of Mathematics, Kyushu University, Japan
`eescolar@math.kyushu-u.ac.jp`
[2] Institute of Mathematics for Industry, Kyushu University, Japan
`hiraoka@imi.kyushu-u.ac.jp`

Abstract. Persistence modules on commutative ladders naturally arise in topological data analysis. It is known that all isomorphism classes of indecomposable modules, which are the counterparts to persistence intervals in the standard setting of persistent homology, can be derived for persistence modules on commutative ladders of finite type. Furthermore, the concept of persistence diagrams can be naturally generalized as functions defined on the Auslander-Reiten quivers of commutative ladders. A previous paper [4] presents an algorithm to compute persistence diagrams by inductively applying echelon form reductions to a given persistence module. In this work, we show that discrete Morse reduction can be generalized to this setting. Given a quiver complex $\mathbb{X}$, we show that its persistence module $H_q(\mathbb{X})$ is isomorphic to the persistence module $H_q(\mathbb{A})$ of its Morse quiver complex $\mathbb{A}$. With this preprocessing step, we reduce the computation time by computing $H_q(\mathbb{A})$ instead, since $\mathbb{A}$ is generally smaller in size. We also provide an algorithm to obtain such Morse quiver complexes.

Keywords: Homology groups, Representation theory of quivers, Discrete Morse theory.

1 Introduction

1.1 Motivations

Suppose that we have two two-step filtrations of complexes $X^1 \subset X^2$ and $Y^1 \subset Y^2$ and that we would like to study the robust and common topological features between them. Standard persistence [3] allows us to study the persistent features in X and Y independently of each other, while zigzag persistence [2] allows us to extract common features through the diagrams $X^i \hookrightarrow X^i \cup Y^i \hookleftarrow Y^i$ for $i = 1, 2$. To extract robust *and* common features however, we use the diagram

$$\mathbb{X} = \begin{array}{ccccc} X^2 & \longhookrightarrow & X^2 \cup Y^2 & \longhookleftarrow & Y^2 \\ \uparrow & & \uparrow & & \uparrow \\ X^1 & \longhookrightarrow & X^1 \cup Y^1 & \longhookleftarrow & Y^1 \end{array} \qquad (1)$$

H. Hong and C. Yap (Eds.): ICMS 2014, LNCS 8592, pp. 144–151, 2014.

of complexes and study the induced diagram

$$H_q(\mathbb{X}) = \begin{array}{ccccc} H_q(X^2) & \longrightarrow & H_q(X^2 \cup Y^2) & \longleftarrow & H_q(Y^2) \\ \uparrow & & \uparrow & & \uparrow \\ H_q(X^1) & \longrightarrow & H_q(X^1 \cup Y^1) & \longleftarrow & H_q(Y^1) \end{array} \qquad (2)$$

for some dimension q. This diagram can be thought of as a representation on the quiver

$$\begin{array}{ccccc} \overset{4}{\circ} & \longrightarrow & \overset{5}{\circ} & \longleftarrow & \overset{6}{\circ} \\ \uparrow & & \uparrow & & \uparrow \\ \underset{1}{\circ} & \longrightarrow & \underset{2}{\circ} & \longleftarrow & \underset{3}{\circ} \end{array} \qquad (3)$$

with commutative relations. In general, define an orientation τ_n to be an $n-1$ sequence of symbols f or b meaning "forwards" or "backwards," respectively. The commutative ladder $CL(\tau_n)$ is defined as the path algebra of the quiver

$$\begin{array}{ccccccccc} \overset{1}{\circ} & \longleftrightarrow & \overset{2}{\circ} & \longleftrightarrow & \overset{3}{\circ} & \longleftrightarrow & \cdots & \longleftrightarrow & \overset{n}{\circ} \\ \uparrow & & \uparrow & & \uparrow & & & & \uparrow \\ \underset{1'}{\circ} & \longleftrightarrow & \underset{2'}{\circ} & \longleftrightarrow & \underset{3'}{\circ} & \longleftrightarrow & \cdots & \longleftrightarrow & \underset{n'}{\circ} \end{array}, $$

where the orientation of each pair of ith horizontal arrows corresponds to the ith term of τ_n, taken together with commutative relations. We give precise definitions for these concepts in the next section. For a more detailed treatment, we refer the reader to [1], for example.

For now, we consider the diagram (2). The Krull-Remak-Schmidt theorem guarantees that it can be written uniquely as an indecomposable decomposition

$$H_q(\mathbb{X}) \cong \bigoplus_{[I] \in \Gamma_0} I^{k_I}.$$

We recall the following theorem from the paper [4].

Theorem 1. *The commutative ladders $CL(\tau_n)$ of length $n \leq 4$ are representation-finite for arbitrary orientations τ_n.*

In particular, Theorem 1 tells us that Γ_0 can be taken as a fixed finite list of isomorphism classes of indecomposable representations. The paper [4] provides Γ_0 using the so-called Auslander-Reiten quiver. With this decomposition, an extended definition of persistence diagrams has been given for this setting, from which we can study the robust common topological features.

Moreover, the paper [4] provides an algorithm to compute the multiplicities k_I for $H_q(\mathbb{X})$. In this work, we apply the technique of Morse reductions to our settings in order to reduce the time taken for computing $H_q(\mathbb{X})$. We generalize the results of [7] concerning the use of discrete Morse theory for filtrations and those of [5] for zigzag complexes to what we call quiver complexes. A particular example of a quiver complex is the diagram (1).

1.2 Background

Let K be a field. A complex [9] is a pair (X, κ) of a graded set $X = \sqcup_{q \geq 0} X_q$ (a disjoint union) together with an incidence map $\kappa : X \times X \to K$. The elements of X are called cells, and each cell $\sigma \in X_q$ is assigned a dimension $\dim \sigma = q$. An incidence map satisfies the properties that $\kappa(\sigma, \tau) \neq 0$ implies $\dim \sigma = \dim \tau + 1$ and that $\sum_{\tau \in X} \kappa(\sigma, \tau) \kappa(\tau, \rho) = 0$ for any $\sigma, \rho \in X$.

We use complexes to model our geometric objects. In this work, we consider only complexes with X finite. In particular, a simplicial complex is an example of a complex by setting $\kappa(\sigma, \tau) = \pm 1$ for all faces τ of σ with $\dim \tau = \dim \sigma - 1$, with sign determined by the orientations of σ and τ.

A subcomplex of (X, κ) is a complex (X', κ') such that $X' \subset X$, $\kappa|_{X' \times X'} = \kappa'$, and for any $\sigma \in X'$, $\tau \in X \setminus X'$ implies $\kappa(\sigma, \tau) = 0$. We define a relation $<$ by setting $\tau < \sigma$ if and only if $\kappa(\sigma, \tau) \neq 0$. The face order $<'$ of X is defined by transitive extension of $<$. When $\tau <' \sigma$, τ is said to be a face of σ. Then, the third condition for subcomplexes can be rephrased as that all faces of cells in X' are also in X'.

A quiver $G = (G_0, G_1)$ is a directed graph with vertices G_0 and arrows G_1. Given an arrow $\alpha : a \to b$, we write $s(\alpha) = a$ and $t(\alpha) = b$, called its source and target, respectively. A directed path $p = (a|\alpha_1, \ldots, \alpha_l|b)$ from a to b is a sequence of arrows α_i such that $s(\alpha_{i+1}) = t(\alpha_i)$ for $i = 1, \ldots, l - 1$, $s(\alpha_1) = a$, and $t(\alpha_l) = b$. The path algebra of G, denoted KG, is the K-algebra generated by the paths in G with multiplication defined by composing paths. A representation (V^a, f^α) of KG is a set of K-vector spaces $\{V^a\}_{a \in G_0}$, together with linear maps $f^\alpha : V^a \to V^b$ for every arrow $\alpha : a \to b$. This forms an object of the category $\mathrm{rep}(KG)$ with each morphism $\psi : (V^a, f^\alpha) \to (W^a, g^\alpha)$ defined as a set of linear maps $\psi^a : V^a \to W^a$ for $a \in G_0$ satisfying the commutativity $g^\alpha \psi^a = \psi^b f^\alpha$ for each arrow $\alpha : a \to b$.

A relation $w = \sum_p c_p p \in KG$ is a K-linear combination of paths with the same source and target and with length at least 2. Given a representation (V^a, f^α), we define ϕ_w for relations w by first setting $\phi_p = f^{\alpha_l} \circ \cdots \circ f^{\alpha_1}$ for a path $p = (a|\alpha_1, \ldots, \alpha_l|b)$ and extending linearly to $\phi_w = \sum_p c_p \phi_p$. Let $I = \langle w_1, \ldots, w_n \rangle$ be a two-sided ideal of KG generated by relations. We define $\mathrm{rep}(KG/I)$ as the full subcategory of $\mathrm{rep}(KG)$ of objects such that $\phi_w = 0$ for all relations $w \in I$.

Let $G = (G_0, G_1)$ be a finite acyclic connected quiver. We restrict our attention to what we call *quiver complexes*. A quiver complex $\mathbb{X}$ is a complex (X^a, κ^a) for each $a \in G_0$ satisfying the condition that if there is an arrow $\alpha : a \to b$ in G_1, then we have a subcomplex inclusion $\iota_\alpha : (X^a, \kappa^a) \hookrightarrow (X^b, \kappa^b)$. We adopt the term *slice* to refer to the individual complexes (X^a, κ^a) in contrast to the whole quiver complex $\mathbb{X}$. For example, the filtrations of complexes studied in [7] are quiver complexes over $G = \circ \to \circ \to \circ \cdots \to \circ$.

Given a quiver complex $\mathbb{X}$, we define its qth chain complex $C_q(\mathbb{X})$ with coefficients in K by taking the K-vector space $C_q(X^a)$ of q-chains of X^a with coefficients in K for every $a \in G_0$, and for every arrow $\alpha : a \to b$, the induced inclusion $\iota_\alpha : C_q(X^a) \to C_q(X^b)$. It is easy to see that each $C_q(\mathbb{X})$, for $q \geq 0$, is an object of $\mathrm{rep}(KG)$. Now, for every $a \in G_0$, we define the boundary maps

$\partial_q^a : C_q(X^a) \to C_{q-1}(X^a)$ by $\partial_q^a(\sigma) = \sum_{\tau \in X^a} \kappa^a(\sigma, \tau)\tau$. These boundary maps for each $C_q(X^a)$ define a morphism $\partial_q : C_q(\mathbb{X}) \to C_{q-1}(\mathbb{X})$ in rep(KG).

This allows us to construct the homology module of $\mathbb{X}$ by taking $H_q(\mathbb{X}) = \frac{\ker \partial_q}{\operatorname{im} \partial_{q+1}}$ in rep(KG). Equivalently, one can construct $H_q(\mathbb{X})$ by taking $H_q(X^a)$ of each slice (X^a, κ^a) and using the maps $\iota_\alpha^* : H_q(X^a) \to H_q(X^b)$ induced from inclusions for every arrow $\alpha : a \to b$.

In fact, this construction gives us representations of KG with added commutative relations in the following sense. Let I be the two-sided ideal of KG generated by relations of the form $p - p'$, where $p = (a|\alpha_1, \ldots, \alpha_l|b)$ and $p' = (a|\alpha_1', \ldots, \alpha_k'|b)$ are any two nontrivial and unequal paths from a to b for all pairs $a, b \in G_0$. Then, it can be shown that $C_q(\mathbb{X})$ and $H_q(\mathbb{X})$ are objects of rep(KG/I). In particular, the diagram (2) is an object of rep($CL(fb)$).

Once we obtain $H_q(\mathbb{X})$, we can use the algorithm in [4] to compute its indecomposable decomposition and thus its persistence diagram. To go from a quiver complex $\mathbb{X}$ to $H_q(\mathbb{X})$, one may apply classical computations for homology groups and induced maps.

In this work, we generalize the technique of [7] to quiver complexes. That is, we replace $\mathbb{X}$ by a related and smaller quiver complex $\mathbb{A}$, called its Morse quiver complex. We show that $H_q(\mathbb{X}) \cong H_q(\mathbb{A})$, which guarantees that this replacement preserves homological information. Even though we are particularly interested in the quiver (3), our discussions will apply for any quiver complex over any finite, acylic, connected quiver.

2   Underlying Theory

We recall some ideas found in [7] for Morse reductions, but we extend the definitions for use in our more general setting. Moreover, these ideas are very much related to discrete Morse theory [6]. The results here are further generalizations of those in [5].

Let (X, κ) be a complex. A matching of (X, κ) is a partition of X into the sets $\mathcal{A}, \mathcal{Q}, \mathcal{K}$, together with a bijection $w : \mathcal{Q} \to \mathcal{K}$ such that $\kappa(w(Q), Q)$ is invertible (nonzero, since we are working over a field) for all $Q \in \mathcal{Q}$. Define the relation $\lhd$ on $\mathcal{Q}$ by $Q' \lhd Q$ if and only if $Q' < w(Q)$. Then, the matching above is said to be an acyclic matching if and only if the transitive closure of $\lhd$ is a partial order. For $\sigma, \tau \in \mathcal{A}$, a gradient path from σ to τ is a sequence $p = (Q_1, \ldots, Q_l)$ of elements of $\mathcal{Q}$ such that $Q_i \neq Q_{i+1} \lhd Q_i$ for $i = 1, \ldots, l-1$, and $Q_1 < \sigma$, $\tau < w(Q_l)$. The multiplicity of p is defined as $m(p) = \kappa(\sigma, Q_1) \frac{\prod_{i=1}^{l-1} \kappa(w(Q_i), Q_{i+1})}{\prod_{i=1}^{l} -\kappa(w(Q_i), Q_i)} \kappa(w(Q_l), \tau)$ We define the incidence map $\tilde{\kappa} : \mathcal{A} \times \mathcal{A} \to K$ by $\tilde{\kappa}(\sigma, \tau) = \kappa(\sigma, \tau) + \sum_p m(p)$, where the sum is taken over all gradient paths p from σ to τ, for $\sigma, \tau \in \mathcal{A}$.

It is known that $(\mathcal{A}, \tilde{\kappa})$, called the Morse complex of (X, κ) induced by the above acyclic matching, is also a complex and that for $q \geq 0$, $H_q(X) \cong H_q(\mathcal{A})$ [6,7]. The paper [7] extends this fact to the case where we have a filtration of a complex. Here, we extend it to our more general setting.

Definition 1. An acylic matching of a quiver complex $\mathbb{X}$ is a set of acylic matchings $\{(\mathcal{A}^a, w^a : \mathcal{Q}^a \to \mathcal{K}^a)\}_{a \in G_0}$ such that for every arrow $\alpha : a \to b$, we have $\mathcal{A}^a \subset \mathcal{A}^b$, $\mathcal{Q}^a \subset \mathcal{Q}^b$, $\mathcal{K}^a \subset \mathcal{K}^b$, and $w^b(Q) = w^a(Q)$ for all $Q \in \mathcal{Q}^a$.

Given an acyclic matching of $\mathbb{X}$, we have a Morse complex $(\mathcal{A}^a, \tilde{\kappa}^a)$ in each slice. It can be shown that there is a subcomplex inclusion $(\mathcal{A}^a, \tilde{\kappa}^a) \hookrightarrow (\mathcal{A}^b, \tilde{\kappa}^b)$ for every arrow $\alpha : a \to b$. These data define what we call the *Morse quiver complex* $\mathbb{A}$ of $\mathbb{X}$ induced by the above acyclic matching.

We take $C(\mathbb{X})$ to be the sequence $\{C_q(\mathbb{X})\}_{q \geq 0}$ in $\mathrm{rep}(KG/I)$ together with the boundary maps $\partial_q : C_q(\mathbb{X}) \to C_{q-1}(\mathbb{X})$. A chain map $\phi : C(\mathbb{X}) \to C(\mathbb{Y})$ is a sequence of morphisms $\{\phi_q : C_q(\mathbb{X}) \to C_q(\mathbb{Y})\}$ that commute with the boundary maps. Two chain maps $\phi, \psi : C(\mathbb{X}) \to C(\mathbb{Y})$ are said to be chain homotopic if for every $q \geq 0$, there is a map $\theta_q : C_q(\mathbb{X}) \to C_{q+1}(\mathbb{Y})$ such that $\partial^Y_{q+1} \theta_q + \theta_{q-1} \partial^X_q = \phi_q - \psi_q$. Two chain complexes $C(\mathbb{X})$ and $C(\mathbb{Y})$ are said to be chain equivalent if there exists chain maps $\phi : C(\mathbb{X}) \to C(\mathbb{Y})$ and $\psi : C(\mathbb{Y}) \to C(\mathbb{X})$ such that $\psi\phi$ and $\phi\psi$ are chain homotopic to the appropriate identity chain maps.

It can be shown, through a simple generalization of the standard proof, that if $C(\mathbb{X})$ and $C(\mathbb{Y})$ are chain equivalent, then $H_q(\mathbb{X}) \cong H_q(\mathbb{Y})$ for all $q \geq 0$. With this background, we can state our main theorem.

Theorem 2. *Let $\mathbb{X}$ be a quiver complex and $\{(\mathcal{A}^a, w^a : \mathcal{Q}^a \to \mathcal{K}^a)\}_{a \in G_0}$ an acyclic matching of $\mathbb{X}$ that induces the Morse quiver complex $\mathbb{A}$. Then, $C(\mathbb{X})$ and $C(\mathbb{A})$ are chain equivalent and thus $H_q(\mathbb{X}) \cong H_q(\mathbb{A})$ for all $q \geq 0$.*

The proof, which we briefly summarize below, is to construct the required chain equivalences between $C(\mathbb{X})$ and $C(\mathbb{A})$. The chain equivalences are defined by an inductive process. Let (X^a_Q, κ^a_Q) be the complex obtained from (X^a, κ^a) by removing the pair $Q, w(Q)$, for some $Q \in \mathcal{Q}^a$, and appropriately modifying the incidence map. We have chain equivalences $\psi^a_Q : C_q(X^a) \to C_q(X^a_Q)$ and $\phi^a_Q : C_q(X^a_Q) \to C_q(X^a)$ of the same form as in [7]. By taking the compositions of ψ^a_Q over all $Q \in \mathcal{Q}^a$, we obtain $\psi^a : C_q(X^a) \to C_q(\mathcal{A}^a)$, and similarly ϕ^a in the opposite direction. One condition that needs to be satisfied is the commutativity of for every arrow $\alpha : a \to b$, with a similar statement for ϕ^a.

$$
\begin{array}{ccc}
C_q(X^a) & \xrightarrow{\ \psi^a\ } & C_q(\mathcal{A}^a) \\
\downarrow & & \downarrow \\
C_q(X^b) & \xrightarrow{\ \psi^b\ } & C_q(\mathcal{A}^b)
\end{array}
\tag{4}
$$

To that end, we set an order on the elements of $\mathcal{Q}^a$ for all a such that if Q appears before Q' in $\mathcal{Q}^a$ and there is an arrow $\alpha : a \to b$, then Q appears before Q' in $\mathcal{Q}^b$. In the inductive process above, we remove the pairs $Q, w(Q)$ in the given order. This consistent ordering of the elements of $\mathcal{Q}^a$ for $a \in G_0$ is important for satisfying the commutativity of (4). In contrast, the paper [7] deals with the quiver $G = \circ \to \circ \to \cdots \to \circ$, and the consistent ordering follows from the nesting of $\mathcal{Q}^a$ for $a \in G_0$.

For every arrow $\alpha : a \to b$, we have the cases where Q is an element of both $\mathcal{Q}^a$ and $\mathcal{Q}^b$, and where Q is an element of $\mathcal{Q}^b$ but not $\mathcal{Q}^a$. It can be shown that we have the commutative diagrams

$$
\begin{array}{ccc}
C_q(X^a) & \xrightarrow{\psi_Q^a} & C_q(X_Q^a) \\
\downarrow & & \downarrow \\
C_q(X^b) & \xrightarrow{\psi_Q^b} & C_q(X_Q^b)
\end{array}
\quad , \text{ and } \quad
\begin{array}{ccc}
C_q(X^a) & \xrightarrow{1} & C_q(X^a) \\
\downarrow & & \downarrow \\
C_q(X^b) & \xrightarrow{\psi_Q^b} & C_q(X_Q^b)
\end{array}
\tag{5}
$$

respectively. Recall that ψ^a and ψ^b are defined by composing ψ_Q^a for $a \in \mathcal{Q}^a$ and ψ_Q^b for $b \in \mathcal{Q}^b$, respectively. Every time we encounter a Q in $\mathcal{Q}^a$ and $\mathcal{Q}^b$, we use the diagram on the left, while if Q is in $\mathcal{Q}^b$ but not $\mathcal{Q}^a$, we use the diagram on the right. By consistent ordering, composing diagrams of the forms given in (5) indeed gives us ψ^a and ψ^b on the top and bottom rows, and leads to the commutative diagram (4). The arguments for ϕ^a proceed similarly.

3   Algorithm

We say that σ is an element of a quiver complex $\mathbb{X}$, denoted $\sigma \in \mathbb{X}$, if $\sigma \in X^a$ for some slice X^a of $\mathbb{X}$. For every $\sigma \in \mathbb{X}$, we define the support of σ, denoted $b(\sigma)$, as a function $b(\sigma) : G_0 \to \{0,1\}$ with $b(\sigma)(a) = 1$ if $\sigma \in X^a$ and 0 otherwise.

We assume that $\kappa^a(\sigma, \tau) = \kappa^b(\sigma, \tau)$ whenever σ, τ are elements of X^a and X^b. With this condition, we can define $\kappa(\sigma, \tau)$ by

$$
\kappa(\sigma, \tau) = \begin{cases} \kappa^a(\sigma, \tau) & \text{if } \sigma, \tau \in X^a \text{ for some } a, \\ 0 & \text{otherwise.} \end{cases}
$$

Let us show that the assumption above holds without loss of generality, by a suitable renaming of elements of $\mathbb{X}$. Let $\sqcup_{a \in G_0} X^a \doteq \{(\sigma, a) | \sigma \in X^a, a \in G_0\}$ be the disjoint union of the slices. We define an equivalence relation $\sim$ on $\sqcup_{a \in G_0} X^a$ as the transitive closure of $\approx$, where $(\sigma, a) \approx (\sigma', b)$ if and only if $\sigma = \sigma'$ and there is a path from a to b or b to a (or trivially, $a = b$). Under this equivalence relation, we denote the class of (σ, a) by $[\sigma, a]$. Furthermore, it is easy to see that if $(\sigma, a) \sim (\sigma, b)$ and $(\tau, a) \sim (\tau, b)$, then $\kappa^a(\sigma, \tau) = \kappa^b(\sigma, \tau)$.

We rename σ as $[\sigma, a]$ for every $\sigma \in X^a$ and $a \in G_0$ to obtain a quiver complex $\hat{\mathbb{X}}$ defined by $\hat{X}^a = \{[\sigma, a] | \sigma \in X^a\}$ and $\hat{\kappa}^a([\sigma, a], [\tau, a]) = \kappa^a(\sigma, \tau)$. Now, suppose that $\pi, \rho \in \hat{X}^a, \hat{X}^b$. Then, $\pi = [\sigma, a] = [\sigma, b]$ and $\rho = [\tau, a] = [\tau, b]$ for some σ, τ in the original complex. Thus, $\kappa^a(\sigma, \tau) = \kappa^b(\sigma, \tau)$ and we conclude $\hat{\kappa}^a(\pi, \rho) = \hat{\kappa}^b(\pi, \rho)$. This shows that $\hat{\mathbb{X}}$ satisfies the assumption above.

We adapt the algorithm of [7], making changes as required. The input to MORSEREDUCE is the cells of $\mathbb{X}$, together with incidence map κ and the support functions b. Let D be the maximum of the dimensions of the cells. We place all cells in a container U of unprocessed cells. As we go through the algorithm, we remove cells from U either by MAKECRITICAL, which moves cells to $\mathcal{A}$, or by REMOVEPAIR, which moves pairs of cells to $\mathcal{Q}$ and $\mathcal{K}$.

The boundary of σ relative to the remaining cells is $\partial^U \sigma = \sum_{\tau \in U} \kappa(\sigma, \tau)$, while the coboundary of σ is $\mathrm{cb}_U(\sigma) = \{\rho \in U | \kappa(\rho, \sigma) \neq 0\}$. An elementary coreduction pair [8] is a pair of cells ξ, η such that $\partial^U \xi = u\eta$, where $0 \neq u \in K$. The algorithm identifies such elementary coreduction pairs and removes them

---

procedure MORSEREDUCE(U, κ, b)
    for $d \in \{0, 1, \ldots, D\}$ **do**
        while $\{\sigma \in U | \dim \sigma = d\} \neq \emptyset$ **do**
            $A \leftarrow$ MAKECRITICAL(d)
            Que $\leftarrow$ **new** Queue
            enqueue: $\mathrm{cb}_U(A)$ in Que
            while Que $\neq \emptyset$ **do**
                dequeue: ξ from Que
                if $\partial^U \xi = 0$ **then**
                    enqueue: $\mathrm{cb}_U(\xi)$ in Que
                else if $\partial^U(\xi) = u \cdot \eta$ with $b(\eta) = b(\xi)$, $0 \neq u \in K$ **then**
                    REMOVEPAIR(ξ, η, d)
    return $\mathbb{A}$

---

via REMOVEPAIR. As we remove cells by MAKECRITICAL or REMOVEPAIR, new elementary coreduction pairs are possibly created. We queue candidates and check for the existence of any new elementary coreduction pairs.

Require: $\partial^U K = uQ$, $b(K) = b(Q)$
    procedure REMOVEPAIR(K, Q, d)
        remove: K from U
        enqueue: $\mathrm{cb}_U(Q)$ in Que
        if $\dim Q = d$ **then**
            $g(Q) \leftarrow -\frac{g(K)}{u}$
            UPDATEGRADIENTCHAIN(Q)
        remove: Q from U

Require: d: $\{\sigma \in \mathbb{X} | \dim \sigma = d\} \neq \emptyset$
    procedure MAKECRITICAL(d)
        choose: $A \in U$ of dimension d
        add: A to $\mathbb{A}$
        UPDATEGRADIENTCHAIN(A)
        remove: A from U
        $\partial^{\mathbb{A}} A \leftarrow g(A)$
        return A

For every cell σ, we store a variable $g(\sigma)$ that keeps track of how the removed cells affect the boundary of σ. At the time the algorithm terminates, we have $\tilde{\kappa}(\sigma, \tau) = \langle g(\sigma), \tau \rangle$ for $\sigma, \tau \in \mathbb{A}$. Here, we take the inner product $\langle \cdot, \cdot \rangle$ relative to the standard basis consisting of all cells of $\mathbb{A}$. We set $\mathcal{A}^a = \{\sigma \in \mathbb{A} | b(\sigma)(a) = 1\}$

---

procedure UPDATEGRADIENTCHAIN(ξ)
    for $\zeta \in \mathrm{cb}_U(\xi)$ **do**
        if $\xi \in \mathbb{A}$ **then**
            $g(\zeta) \leftarrow g(\zeta) + \kappa(\zeta, \xi)\xi$
        else
            $g(\zeta) \leftarrow g(\zeta) + \kappa(\zeta, \xi)g(\xi)$

---

and define $\tilde{\kappa}^a(\sigma, \tau) = \langle \partial^{\mathbb{A}} \sigma, \tau \rangle$, for every $\sigma, \tau \in \mathcal{A}^a$ and $a \in G_0$. This gives a Morse quiver complex of the input $\mathbb{X}$ induced from the acyclic matching $\{(\mathcal{A}^a, w^a : \mathcal{Q}^a \to \mathcal{K}^a)\}_{a \in G_0}$, where $w^a(\eta) = \xi$ for every ξ, η sent to REMOVEPAIR.

Given an input quiver complex $\mathbb{X}$, an existing method to obtain its persistence diagram is to compute $H_q(\mathbb{X})$ and then use the algorithm in [4]. Here, we introduce MORSEREDUCE as a preprocessing step and use the following flowchart.

1. Compute a Morse quiver complex $\mathbb{A}$ of $\mathbb{X}$ by MORSEREDUCE.
2. Compute $H_q(\mathbb{A})$ by the process sketched in the introduction.
3. Apply the algorithm in [4] on $V = H_q(\mathbb{A})$ to obtain its persistence diagram.

By Theorem 2, $H_q(\mathbb{X}) \cong H_q(\mathbb{A})$ and this flowchart gives the same output as working with $H_q(\mathbb{X})$ instead. Since the number of cells of $\mathbb{A}$ is at most that of $\mathbb{X}$, subsequent computations tend to complete faster with the preprocessing step.

In the following table, we summarize the time taken in seconds for computations on three sample quiver complexes $\mathbb{X}$ over the quiver (3) with $K = \mathbb{Z}_2$ and $q = 1$, together with sizes $|\mathbb{X}|$ and $|\mathbb{A}|$. The column under t_{without} contains the total times taken for computing without using Morse reductions and working on $H_q(\mathbb{X})$ directly, while t_{with} lists the total times taken for following the flowchart above.

#	$|\mathbb{X}|$	$|\mathbb{A}|$	t_{without}	t_{with}
1	15,341	2,777	903.31	39.53
2	17,626	7,164	3497.55	143.41
3	32,540	7,834	5162.12	42.34

For every input, the only difference between the computations performed is whether or not we perform preprocessing by MORSEREDUCE, with other implementation details kept the same. Since the preprocessing technique is essentially independent of the subsequent steps, if one were to use a more efficient algorithm for steps 2 and 3, we expect to observe similar improvements in performance.

References

1. Assem, I., Simson, D., Skowroński, A.: Elements of the Representation Theory of Associative Algebras 1: Techniques of Representation Theory. Cambridge University Press, Cambridge (2006)
2. Carlsson, G., de Silva, V.: Zigzag Persistence. Found. Comput. Math. 10(4), 367–405 (2010)
3. Edelsbrunner, H., Letscher, D., Zomorodian, A.: Topological Persistence and Simplification. Discrete Comput. Geom. 28(4), 511–533 (2002)
4. Escolar, E., Hiraoka, Y.: Persistence Modules on Commutative Ladders of Finite Type, http://arxiv.org/abs/1404.7588v1
5. Escolar, E., Hiraoka, Y.: Morse Reduction for Zigzag Complexes. J. Indones. Math. Soc. 20(1), 47–75 (2014)
6. Forman, R.: Morse Theory for Cell Complexes. Adv. Math. 134, 90–145 (1998)
7. Mischaikow, K., Nanda, V.: Morse Theory for Filtrations and Efficient Computation of Persistent Homology. Discrete Comput. Geom. 50(2), 330–353 (2013)
8. Mrozek, M., Batko, B.: Coreduction Homology Algorithm. Discrete Comput. Geom. 41(1), 96–118 (2009)
9. Tucker, A.W.: Cell Spaces. Ann. of Math. (2) 37(1), 92–100 (1936)

Heuristics for Sphere Recognition

Michael Joswig*, Frank H. Lutz**, and Mimi Tsuruga***

Institut für Mathematik, TU Berlin, Straße des 17, Juni 136, 10623 Berlin, Germany
{joswig,lutz,tsuruga}@math.tu-berlin.de

Abstract. The problem of determining whether a given (finite abstract) simplicial complex is homeomorphic to a sphere is undecidable. Still, the task naturally appears in a number of practical applications and can often be solved, even for huge instances, with the use of appropriate heuristics. We report on the current status of suitable techniques and their limitations. We also present implementations in **polymake** and relevant test examples.

Keywords: sphere recognition, combinatorial manifolds, discrete Morse theory, presentations of fundamental groups, bistellar flips.

1 Introduction

The sphere recognition problem often arises in the guise of *manifold recognition*, that is, deciding whether a given finite abstract simplicial complex triangulates some manifold then determining the type of the manifold it triangulates. In the piecewise linear (PL) category, recognizing whether a given complex triangulates a PL manifold can be reduced to *PL sphere recognition* since the links of all vertices of the given complex need to be PL spheres. The following is a (very incomplete) list of scenarios where manifold recognition can be used:

1. Enumeration. When enumerating triangulations of manifolds of a given dimension with a fixed number of vertices or facets, we want to ensure that the objects produced are indeed manifolds and discard all others [6,7,31].
2. Topological Constructions. Various topological manifold constructions can be discretized so that the objects of interest can be studied with the help of a computer. To ensure the discretization has been carried out correctly, we want to confirm the manifold property. In practice, this test effectively detects the majority of construction errors [1,33,30].
3. Meshing. The goal here is to obtain a triangulation of a hypersurface in some (higher-dimensional) Euclidean space by sampling; see, e.g., [27]. As in the case of the topological constructions, we want to verify that the triangulation is non-degenerate.

---

* Partially supported by the Einstein Foundation Berlin and DFG within the Priority Program 1489.
** Partially supported by VILLUM FONDEN through the Experimental Mathematics Network and by the Danish National Research Foundation (DNRF) through the Centre for Symmetry and Deformation.
*** Supported by the Berlin Mathematical School (BMS).

H. Hong and C. Yap (Eds.): ICMS 2014, LNCS 8592, pp. 152–159, 2014.

Throughout this extended abstract we will only consider closed manifolds encoded as finite abstract simplicial complexes. However, our methods can easily be modified to deal with manifolds with boundary or more general cell complexes.

2   An Integrated Recognition Procedure

Let K be a d-dimensional (finite abstract simplicial) complex with n vertices and m facets. A *facet* is a face that is maximal with respect to inclusion. A d-dimensional complex is *pure* if each facet has exactly $d + 1$ vertices. A codimension-1-face in a pure complex is called a *ridge*.

To verify whether K is a PL d-sphere, there are three elementary combinatorial checks that are useful to perform first. These checks are fast; their running time is bounded by a low-degree polynomial in the parameters d, m and n. If one of the checks fails, this will serve as the certificate that K is not a sphere.

(1) Check if K is pure.

(2) Check if each ridge is contained in exactly two facets.

Success in these two tests will ascertain that K is a *pseudo-manifold* (without boundary). A pseudo-manifold K of dimension $d = 0$ is the 0-dimensional sphere $\mathbb{S}^0$; it consists of two isolated vertices.

(3) If $d \geq 1$, check if the 1-skeleton of K is a connected graph.

A connected pseudo-manifold K of dimension $d = 1$ is a polygon, and thus triangulates the 1-dimensional sphere $\mathbb{S}^1$.

The pseudo-manifold property of a simplicial complex is inherited by all face links. In particular, a connected pseudo-manifold of dimension 2 is a triangulation of a closed surface or of a closed surface with pinch points. A *pinch point* has multiple disjoint cycles as its vertex link.

A d-dimensional pseudo-manifold is a *combinatorial d-manifold* if all vertex links are PL homeomorphic to the boundary of the d-simplex. In particular, a combinatorial d-manifold is a triangulation of a PL d-manifold.

A (connected) 2-dimensional pseudo-manifold K with the additional property that all vertex links are single cycles is a combinatorial 2-manifold and triangulates a closed surface. If the Euler characteristic of K is 2, then K is $\mathbb{S}^2$.

The sphere recognition problem becomes more interesting from $d \geq 3$ and requires additional steps; see also the discussion in Section 3 below.

We begin by computing the Hasse diagram of the complex K; this is a directed graph with one node per face and a directed edge for each pair of incident faces whose dimensions differ by one. How to orient the edges is merely a matter of convention; here we assume they point towards the higher-dimensional faces. We use a method introduced by Kaibel and Pfetsch [18] which is output sensitive in the sense that it is linear in ϕ, the total number of faces of K. More precisely, this algorithm's worst-case complexity is in $O(d \cdot m^2 \cdot \phi)$.

After the initial Tests (1), (2), and (3), our next test for verifying whether K is a PL d-sphere is to first verify that K is a combinatorial manifold. A pure d-dimensional simplicial complex K is a combinatorial d-manifold iff for any proper i-face F of K, $0 \leq i \leq d-1$, the link of F in K is a PL $(d-i-1)$-sphere. Notice that if all links of i-faces are PL spheres, then all links of $(i-1)$-faces are combinatorial $(d-i)$-manifolds, which is a necessary condition to verify that the links of $(i-1)$-faces are PL spheres. In this way, this property is recursive. In practice, however, a recursive method is likely to encounter repetitions so a level-wise approach is preferred.

According to Whitehead [35], any combinatorial d-manifold that becomes collapsible after the removal of one facet is a PL d-sphere. This statement is equivalent to the existence of an *acyclic matching* in the Hasse diagram with exactly two *critical cells*. That is, a matching given by the pairings induced by the performed elementary collapses such that if the edges of the matching are reversed, the resulting directed graph is acyclic and has precisely two unmatched nodes: one representing a facet and one representing a vertex. In the language of discrete Morse theory, as developed by Forman [12,13], to every acyclic matching in the Hasse diagram there is a corresponding discrete Morse function with the same number of critical cells. For a general discrete Morse function on a d-dimensional simplicial complex, the *discrete Morse vector* $(c_0, c_1, \ldots, c_d)$ counts the critical cells per dimension.

A randomized search for small discrete Morse vectors was introduced in [3]. This approach proceeds level-wise from top to bottom. The free faces for elementary collapses are chosen uniformly at random; if there are no free faces, a face of the current maximal dimension is chosen uniformly at random, marked critical, and removed.

A combinatorial manifold K is a PL sphere iff some subdivision of K admits $(1, 0, \ldots, 0, 1)$ as its discrete Morse vector. We say that $(1, 0, \ldots, 0, 1)$ is a *spherical discrete Morse vector* and recursively (or level-wise) check the links of all faces to see whether they admit such a vector. However, we need to overcome four major difficulties:

▷ Computing an optimal discrete Morse vector for a simplicial complex is NP-hard [17,20].

▷ There are combinatorial d-spheres that do not admit a spherical discrete Morse vector [2,4].

▷ For $d \geq 5$, the question of whether a given simplicial d-complex is a (combinatorial) d-sphere is undecidable [34]; in particular, there is no bound on the number of, say, barycentric subdivisions needed to permit the discrete Morse vector $(1, 0, \ldots, 0, 1)$.

▷ In iterated barycentric subdivisions, finding the vector $(1, 0, \ldots, 0, 1)$ quickly becomes unlikely [1]; see Section 3 for experimental results.

In Section 3, we will demonstrate that despite these drawbacks, finding optimal discrete Morse functions (within some 'horizon') is often surprisingly easy, even for huge complexes; see also [1,3].

(4) Heuristic: Search for a spherical discrete Morse vector using the random discrete Morse algorithm [3].

Brehm and Kühnel [6] introduced a basic version of (4) to show that some 8-dimensional simplicial complex with 15 vertices is a combinatorial 8-manifold.

Algorithm 1. Sphere recognition heuristics

Input: K: triangulation of connected closed PL d-manifold, where $d \geq 3$
Output: Decision: Is K PL homeomorphic to $\mathbb{S}^d$?

1 compute Hasse diagram
2 **for** N *rounds* **do**
3   compute random discrete Morse vector
4   **if** *discrete Morse vector spherical* **then** **return** YES

5 compute homology
6 **if** *homology not spherical* **then** **return** NO
7
8 compute and simplify presentation of fundamental group π_1
9 **if** *presentation of π_1 trivial* **then**
10   **if** $d \neq 4$ **then**
11   **return** YES
12   **else**
13   **for** N' *rounds* **do**
14   perform random bistellar flip
15   **if** *boundary of simplex reached* **then** **return** YES

16 **else**
17   **if** $\pi_1 \cong \mathbb{Z}_2, \mathbb{Z}, \ldots$ **then**
18   **return** NO
19   **else**
20   **for** N' *rounds* **do**
21   perform random bistellar flip
22   **if** *boundary of simplex reached* **then** **return** YES

23 **return** UNDECIDED

A necessary condition for K to be homeomorphic to the d-sphere is that the homology, say with integer coefficients, is trivial. If it is not, we know that K is *not* homeomorphic to the d-sphere. Writing down the (simplicial) boundary matrices from the Hasse diagram is a straightforward procedure. Computing the homology then amounts to determining the Smith Normal Forms of these matrices; see e.g. Munkres [23, §11]. To improve the running time, reduction of the initial complex is essential; see CHomP [11], RedHom [9], or polymake's application topaz [15] for implementations.

(5) Check the homology.

Let K be a simply connected combinatorial d-manifold, $d \neq 4$, with the homology of the d-dimensional sphere. Then K is PL homeomorphic to the boundary of the $(d + 1)$-simplex (PL Poincaré conjecture [25,29]).

To determine whether K is simply connected, we compute a finite presentation of the fundamental group $\pi_1(K)$ [28] and use heuristics to simplify the presentation, e.g., as implemented in the software package GAP. Deciding whether a finitely presented group is trivial is again an undecidable problem [24], but GAP often finds a simplification of the presentation quickly.

(6) Heuristic: Compute and simplify a presentation of the fundamental group.

Since the 4-dimensional PL Poincaré conjecture is open, *exotic 4-spheres* may exist. Exotic 4-spheres are simply-connected combinatorial 4-manifolds with the homology of $\mathbb{S}^4$ (and are thus homeomorphic to $\mathbb{S}^4$ by the 4-dimensional topological Poincaré conjecture [14]), but not PL homeomorphic to the boundary of the 5-simplex. This means that there are two situations in which our algorithm may return UNDECIDED. When a combinatorial manifold has the homology of a sphere, no spherical discrete Morse vector is found and

▷ the combinatorial manifold is 4-dimensional and is found to be simply connected; or

▷ no decision was possible on the fundamental group.

As a last and final resort for these two cases, attempt to reach the boundary of the $(d+1)$-simplex by applying random bistellar flips as suggested by Björner and Lutz [5].

(7) Heuristic: Simplify using random bistellar flips.

If Test (7) is inconclusive, the decision problem remains unsolved.

We indicate how our method can be further refined.

Remark 1. Depending on how likely K is homeomorphic to $\mathbb{S}^d$ we can first check its homology before we try the random discrete Morse algorithm.

Remark 2. In the case of non-PL triangulations of $\mathbb{S}^d$, our approach may be used to show that a given triangulation is isomorphic to the double-suspension of some homology sphere or has some other not too complicated PL singular set.

Remark 3. In special cases of non-spherical manifolds computing other invariants of K may be worthwhile; for a survey, e.g., see [16]. In particular, if $d = 4$ and $\pi_1(K) = 1$, computing the intersection form of K decides the homeomorphism type of K, however, it does not settle the PL type; see Freedman [14].

3   Computational Results and Horizon for Computations

S.P. Novikov (cf. [10,34]) proved that recognizing the d-sphere is undecidable for $d \geq 5$. For the 4-sphere it is unknown whether it can be recognized algorithmically. Rubinstein [26] used normal surface theory to provide an algorithm to recognize the 3-dimensional sphere; see also Thompson [32] and Matveev [21]. Unfortunately, the known deterministic 3-sphere recognition algorithms have exponential running time or worse. King [19] showed that two triangulations (as pseudo-simplicial complexes) of S^3 with t tetrahedra are related by a sequence of less than 2^{201t^2} edge contractions and expansions. A similar result was obtained by Mijatovic [22], who proved that two pseudo-simplicial triangulations of S^3 with t tetrahedra are related by at most $6 \cdot 10^6 t^2 2^{2 \cdot 10^4 t^2}$ bistellar flips. Burton implemented the Rubinstein–Thompson–Matveev–Casson algorithm in his 3-manifold software package Regina [8].

While our Algorithm 1 cannot guarantee success, we are not aware of a single triangulation of the 3-sphere for which the recognition of π_1 via Test (6) fails, and substantial work was required to provide such examples in dimension 4 [33].

The smallest known triangulated 3-sphere for which the discrete Morse Test (4) fails is the example triple_trefoil (constructed via a complicated knot in the 1-skeleton with 18 vertices) from [2,3]; it does not admit a spherical discrete Morse vector. For 'more standard' triangulations of S^3, even with up to $5 \cdot 10^5$ vertices and 10^7 tetrahedra, it is very likely that Test (4) is positive; see [2] and the examples below. For triangulations of an even larger size, Test (4) will eventually fail [1]. While for smaller examples computations will be successful most of the time, there is a *horizon* beyond which computing good discrete Morse vectors becomes hopeless, and — for the first time — we actually see this horizon in the experiments on iterated barycentric subdivisions reported below. Test (6) is successful on the 3-sphere triple_trefoil.

For the 4-sphere, there is a triangulation with 30 vertices [33] on which the GAP implementation of Test (6) fails as well as our current implementation of Test (7). This example is constructed via a non-trivial presentation of the trivial group, and precisely this presentation with two generators and two relators is obtained for the simplified fundamental group. The example is obtained after applying bistellar flips to the $r = 4$ case of a series of triangulations of the Akbulut–Kirby spheres with face vectors $f = (176 + 64r, 2390 + 1120r, 7820 + 3840r, 9340 + 4640r, 3736 + 1856r)$ for $r \geq 3$. Test (4) and (6) fail on all small examples of this series. With Test (7), we were successful only for $r = 3$. The flipping algorithm for Test (7) retains the complex having the (lexicographically) smallest f-vector found during the search; Test (4) is positive for a few of those simplified complexes.

In fact, finding interesting and challenging test examples for our recognition procedure is non-trivial. Most examples from the literature are tiny, easily fit into memory, and can be recognized instantaneously. A recent example of larger size is contractible_non_5_ball [1], a non-PL triangulation of a contractible and collapsible 5-manifold, different from the 5-ball, with f-vector $f = (5013, 72300, 290944, 495912, 383136, 110880)$. The example was shown to be collapsible [1]; the vector $(1, 0, 0, 0, 0, 0)$ was obtained after only a single random discrete Morse vector search in a running time of 82 hours. Our new implementation in the polymake system [15] of the search for random discrete Morse vectors produced the same result in about 9 seconds; the computations ran on a standard desktop computer with AMD Phenom II X6 1090T CPU (3.2 GHz, 6422 bogomips) and 8 GB RAM.

The boundary of contractible_non_5_ball is a combinatorial 4-manifold with $f = (5010, 65520, 212000, 252480, 100992)$. We used Algorithm 1 to confirm that this example is indeed a combinatorial manifold. For all face links a spherical discrete Morse vector was found immediately. In total, recognition of all face links took about 7.5 hours. The example itself is a homology 4-sphere that has the binary icosahedral group as its fundamental group, as was confirmed computationally in [1].

We ran our implementation on higher barycentric subdivisions of boundaries of simplices. For the 3rd barycentric subdivision sd_3_bd_delta_4 of the boundary of the 4-simplex with $f = (12600, 81720, 138240, 69120)$ the optimal discrete Morse vector $(1, 0, 0, 1)$ was found in 994 out of 1000 runs of the random-revlex version [1] of the random discrete Morse search. For the 4th barycentric subdivision sd_4_bd_delta_4 of the boundary of the 4-simplex with face vector $f = (301680, 1960560, 3317760, 1658880)$ the optimal discrete Morse vector $(1, 0, 0, 1)$ was found in only 844 out of 1000 runs, which may indicate that the horizon for computations lies near the 5th barycentric subdivision.

4 Conclusion

For 'standard' triangulations of the d-sphere, Test (4) provides a reliable and extremely fast tool for recognition. The test is essentially linear in the number of faces, and thus can be repeated several times if the first try does not produce a spherical discrete Morse vector. Test (6) depends only on the 2-skeleton of a complex, so higher-dimensional complexes of considerable size can be processed. Some of our experiments show that Test (4) can be run for complexes with 10^7 or more faces, Test (6) for complexes with $5 \cdot 10^5$ or more triangles, while Test (7) is efficient in dimension 3 for examples with up to 10^4 vertices, but slow in higher dimensions.

References

1. Adiprasito, K., Benedetti, B., Lutz, F.H.: Extremal examples of collapsible complexes and random discrete morse theory. arXiv:1404.4239, 20 pages (2014)
2. Benedetti, B., Lutz, F.H.: Knots in collapsible and non-collapsible balls. Electron. J. Comb. 20(3), Research Paper P31, 29 p. (2013)
3. Benedetti, B., Lutz, F.H.: Random discrete Morse theory and a new library of triangulations. Exp. Math. 23, 66–94 (2014)
4. Benedetti, B., Ziegler, G.M.: On locally constructible spheres and balls. Acta Math. 206, 205–243 (2011)
5. Björner, A., Lutz, F.H.: Simplicial manifolds, bistellar flips and a 16-vertex triangulation of the Poincaré homology 3-sphere. Exp. Math. 9, 275–289 (2000)
6. Brehm, U., Kühnel, W.: 15-vertex triangulations of an 8-manifold. Math. Ann. 294, 167–193 (1992)
7. Burton, B.A.: Simplification paths in the Pachner graphs of closed orientable 3-manifold triangulations. arXiv:1110.6080, 39 pages (2011)
8. Burton, B.A.: Regina: A normal surface theory calculator, version 4.95 (1999-2013), http://regina.sourceforge.net
9. CAPD::RedHom. Redhom software library, http://redhom.ii.uj.edu.pl
10. Chernavsky, A.V., Leksine, V.P.: Unrecognizability of manifolds. Ann. Pure Appl. Logic 141, 325–335 (2006)
11. CHomP. Computational Homology Project, http://chomp.rutgers.edu
12. Forman, R.: Morse theory for cell complexes. Adv. Math. 134, 90–145 (1998)
13. Forman, R.: A user's guide to discrete Morse theory. Sémin. Lothar. Comb 48(B48c), 35 p., electronic only. (2002)

14. Freedman, M.H.: The topology of four-dimensional manifolds. J. Differ. Geom. 17, 357–453 (1982)
15. Gawrilow, E., Joswig, M.: Polymake: a framework for analyzing convex polytopes. In: Polytopes—Combinatorics and Computation (Oberwolfach, 1997). DMV Sem., vol. 29, pp. 43–73. Birkhäuser, Basel (2000)
16. Joswig, M.: Computing invariants of simplicial manifolds, preprint arXiv:math/0401176 (2004)
17. Joswig, M., Pfetsch, M.E.: Computing optimal Morse matchings. SIAM J. Discrete Math. 20, 11–25 (2006)
18. Kaibel, V., Pfetsch, M.E.: Computing the face lattice of a polytope from its vertex-facet incidences. Comput. Geom. 23, 281–290 (2002)
19. King, S.A.: How to make a triangulation of S^3 polytopal. Trans. Am. Math. Soc. 356, 4519–4542 (2004)
20. Lewiner, T., Lopes, H., Tavares, G.: Optimal discrete Morse functions for 2-manifolds. Comput. Geom. 26, 221–233 (2003)
21. Matveev, S.V.: An algorithm for the recognition of 3-spheres (according to Thompson). Sb. Math. 186, 695–710 (1995), Translation from Mat. Sb. 186, 69–84 (1995)
22. Mijatović, A.: Simplifying triangulations of S^3. Pac. J. Math. 208, 291–324 (2003)
23. Munkres, J.R.: Elements of Algebraic Topology. Addison-Wesley Publishing Company, Menlo Park (1984)
24. Novikov, P.S.: On the algorithmic unsolvability of the word problem in group theory. Trudy Matematicheskogo Instituta imeni V. A. Steklova, vol. 44. Izdatel'stvo Akademii Nauk SSSR, Moscow (1955)
25. Perelman, G.: The entropy formula for the Ricci flow and its geometric applications. arXiv:math.DG/0211159, 39 pages (2002)
26. Rubinstein, J.H.: An algorithm to recognize the 3-sphere. In: Chatterji, S.D. (ed.) Proc. Internat. Congr. of Mathematicians, ICM 1994, Zürich, vol. 1, pp. 601–611. Birkhäuser Verlag, Basel (1995)
27. Saucan, E., Appleboim, E., Zeevi, Y.Y.: Sampling and reconstruction of surfaces and higher dimensional manifolds. J. Math. Imaging Vision 30(1), 105–123 (2008)
28. Seifert, H., Threlfall, W.: Lehrbuch der Topologie. B. G. Teubner, Leipzig (1934)
29. Smale, S.: On the structure of manifolds. Am. J. Math. 84, 387–399 (1962)
30. Spreer, J., Kühnel, W.: Combinatorial properties of the $K3$ surface: Simplicial blowups and slicings. Exp. Math. 20, 201–216 (2011)
31. Sulanke, T., Lutz, F.H.: Isomorphism free lexicographic enumeration of triangulated surfaces and 3-manifolds. Eur. J. Comb. 30, 1965–1979 (2009)
32. Thompson, A.: Thin position and the recognition problem for S^3. Math. Res. Lett. 1, 613–630 (1994)
33. Tsuruga, M., Lutz, F.H.: Constructing complicated spheres. arXiv:1302.6856, EuroCG 2013, 4 pages (2013)
34. Volodin, I.A., Kuznetsov, V.E., Fomenko, A.T.: The problem of discriminating algorithmically the standard three-dimensional sphere. Russ. Math. Surv. 29(5), 71–172 (1974)
35. Whitehead, J.H.C.: Simplicial spaces, nuclei and m-groups. Proc. Lond. Math. Soc., II. Ser. 45, 243–327 (1939)

CAPD::RedHom v2 - Homology Software Based on Reduction Algorithms

Mateusz Juda and Marian Mrozek*

Institute of Computer Science and Computational Mathematics,
Jagiellonian University, Poland
mateusz.juda@uj.edu.pl
http://redhom.ii.uj.edu.pl/

Abstract. We present an efficient software package for computing homology of sets, maps and filtrations represented as cubical, simplicial and regular CW complexes. The core homology computation is based on classical Smith diagonalization, but the efficiency of our approach comes from applying several geometric and algebraic reduction techniques combined with smart implementation.

Keywords: Homology software, homology algorithms, Betti numbers, homology groups, homology generators, homology maps, persistent homology, cubical sets, simplicial complexes, CW complexes.

1 Introduction

In 1995 M. Mrozek and K. Mischaikow presented a computer assisted proof of the existence of chaotic dynamics in Lorenz equations [10,11]. The computer programs needed for the proof became the seed of the software package developed by members of the CAPD (Computer Assisted Proofs in Dynamics) group [21]. Throughout the years the package became a reach collection of software libraries and tools for rigorous numerics of dynamical systems (see [9] for the description of the mainstream CAPD package).

An important ingredient of the mentioned proof is Conley index, a homological invariant of dynamical systems. The computer assisted proofs based on Conley index brought interest in cubical homology theory [7] and stimulated the development of the homology package for the needs of computer assisted proofs. Since 2005 the homology software for CAPD has been developed jointly with the Computational Homology Project (CHomP) [23].

After having implemented the classical algorithm based on Smith diagonalization it became clear that it is much too slow for the needs of computer assisted proofs. This originated the search for faster homology algorithms.

CAPD::RedHom is a software package for efficient homology computations of cubical and simplicial complexes as well as some special cases of regular CW

* This research is partially supported by the Polish National Science Center under grant 2012/05/N/ST6/03621 and by the TOPOSYS project FP7-ICT-318493-STREP.

H. Hong and C. Yap (Eds.): ICMS 2014, LNCS 8592, pp. 160–166, 2014.

complexes. Originally, the software was designed for applications in rigorous numerics of Topological Dynamics. Such applications, based on interval arithmetic, lead in a natural way to cubical sets. They may be represented very efficiently as bitmaps. The cubical sets arising from the algorithms in dynamics usually are strongly inflated in the sense the sets which much smaller representation have the same topology or homotopy type. Such small representations may be found in linear time by various geometric reduction techniques. The algebraic invariants of topology, in particular homology, are then computed for the small representation. This leads to a very significant speed up. In particular, the expensive, linear algebra computations, such as Smith diagonalization, are performed on small data.

The package was developed by: P. Brendel, P. Dlotko, G. Jablonski, M. Juda, A. Krajniak, M. Mrozek, P. Pilarczyk, H. Wagner, N. Zelazna.

2 Functionality

The CAPD::RedHom software package, which is currently under intensive development, constitutes a redesign of the CAPD homology software. It is based on the already mentioned as well as the very recent reduction ideas proposed in [1,4,5,6,14,17]. It is designed to meet the needs of various areas of applications, to apply to cubical and simplicial sets as well as CW complexes and at the same time to maintain the efficiency of the original CAPD software for cubical sets. This is achieved by applying the techniques of static polymorphism based on C++ templates so that the reduction algorithms may be applied to various representations of sets without any overhead run-time costs. An unwanted side effect is that this makes the code very hard to use as a library or a plug-in. For this reason recently we put a lot of effort to make the efficient C++ code accessible in external, commonly used libraries. Presently, the code is available as a plug-in for *GAP* [24], *Python*, and *Sage* [25].

The package is intended both for users who are interested in stand-alone programs as well as programmers who want to use the library in their programs. The ultimate goal is that the package will provide Betti numbers, torsion coefficients, homology generators and matrices of maps induced in homology. Moreover, for filtered sets the package will provide persistence intervals [3].

3 Applications

The original CAPD homology software was written for applications in rigorous numerics of dynamical systems. However, the range of applicability of homology software encompasses several other areas: electromagnetism, image analysis, visualization, data mining, sensor networks, robotics and many others. Although the general goal is the same, these areas differ in details of input and output. The cubical representation of sets is convenient in dynamics, because interval arithmetic used in computer assisted proofs in the theory of dynamical systems leads in a natural way to such sets. It is also convenient in the analysis of raster images. However, in many situations the simplicial representation is more natural. In electromagnetics and in all cases when sets exhibit fractal structure a general CW complex representation is most convenient.

Apart from the original applications in rigorous numerics of dynamical systems [13], so far the package has found applications in image analysis [18], material science [17,1], electromagnetism [2], and group representation theory [20].

4   Underlying Theory

One way of avoiding the supercubical cost of the classical homology algorithm is decreasing the size of the input to Smith algorithm without changing the homology. Such an approach was first proposed in [8] by means of a linear time reduction of chain complexes. The reduction process considered in that paper is purely algebraic and may be viewed as a method of limiting the fill-in process in the Smith diagonalization.

However, reductions may be performed directly on the level of the topological space. At first, this may look like acting against the fundamentals of algebraic topology. Algebraic topology solves problems in topology by translating them to the ground of much simpler algebra. But, experiments indicate that in many applications doing geometric reductions directly at the topological level instead of algebraic reductions after translating the problem from topology to algebra may significantly speed up the computations. Also, such an approach often uses significantly less memory.

The first implemented algorithm of this type is based on the observation that for a cube $Q \subset X$, if $Q \cap \mathrm{cl}(X \setminus Q)$ is acyclic then X can be replaced by $\mathrm{cl}(X \setminus Q)$ without affecting the homology (see [19]). This fact was used in the reduction techniques proposed in [12] and motivated the Acyclic Subspace Homology Algorithm (see [15]), based on the construction of a possibly large acyclic subspace A of the topological space X. The computation of the homology groups $H(X)$ reduces then to the computation of $H(X \setminus A)$ in the sense of one space homology theory (see [16]). The method is particularly useful for cubical subsets of $\mathbb{R}^n$ with $n \in \{2, 3\}$, because in these dimensions the acyclic subspace may be constructed extremely fast due to the possibility of storing all possible neighborhood configurations and using them as look-up tables for testing the acyclicity.

The simplest example of reductions on the topological level are free face collapses proposed in [7]. Unfortunately, in many situations free faces are quickly exhausted and the remaining set is still large. Significantly deeper reductions in low dimensions may be achieved by means of the dual concept of free cofaces. This idea leads to the Coreduction Homology Algorithm [16].

The Acyclic Subspace Homology Algorithm and the Coreduction Homology Algorithm together with Discrete Morse Theory [4] seem to be the fastest homology algorithms for inflated cubical and simplicial sets available so far. In particular, they outperform algebraic homology algorithms just because they run in a fraction of time needed to translate the problem to algebra.

5   Technical Contribution

Algorithms implemented in the CAPD::RedHom package behave incredibly well for inflated data sets. We see such sets especially in applications, where a continuous problem is translated into a combinatorial problem. Among many examples, there is a common pattern: to achieve sufficient theoretical conditions for the discretization, we need to subdivide our space. That operation do not change homology, but increase data size.

We compared CAPD::RedHom with latest CHomP [23] (programs *homsimpl* and *chomp-simplicial*) and Linbox [26] (program *homology_gap* 1.4.3 used in GAP [24] - we cannot use latest version, GPC compiler removed from Ubuntu/Debian in 2011). For the comparison we generated simplicial complexes using Sage [25]. For classical examples available in module `sage.simplicial_complexes` we generate their subdivisions with `subdivide()` routine. Using various parameters we generated 380 input files. We will present detailed list of examples in the full paper. For the purpose of this article, on the Figure 1 we show CPU usage for following complexes:

- `Torus()` with 4 subdivisions, 18144 2-dimensional simplices on input;
- `KleinBottle()` with 4 subdivisions, 20736 2-dimensional simplices on input;
- `MooreSpace(9)` with 3 subdivisions, 13176 2-dimensional simplices on input;
- `ProjectivePlane()` with 4 subdivisions, 12960 2-dimensional simplices on input;
- `MatchingComplex(7)` with 3 subdivisions, 22680 2-dimensional simplices on input;
- `ChessboardComplex(5,5)` with 1 subdivision, 14400 4-dimensional simplices on input;
- `RandomComplex(11,5)` with 1 subdivision, 172680 5-dimensional simplices on input;

The number of subdivisions in each case is big enough to force non-instant computations. The case `RandomComplex(11,5)` emphasize benefits from our approach: CAPD::RedHom is almost three times faster than CHomP and Linbox cannot finish computations in one hour. On the chart *CumulativeCPU* we presents total CPU usage by each program in the experiment.

During development of the CAPD::RedHom package we faced a lot of interesting technical and theoretical problems. The most important challenge in our applications is in data set size. The biggest set computed so far contains 10^9 simplices in dimension $0 - 3$ ($600 \cdot 10^6$ facets) [20]. The set required 3 days of computations on a machine with 512 GB of RAM. Big data sets in applications convinced us to start implementations of our algorithms for parallel and distributed computations. This is a big challenge in the area of computational homology. In the full paper we will show our progress in this subject.

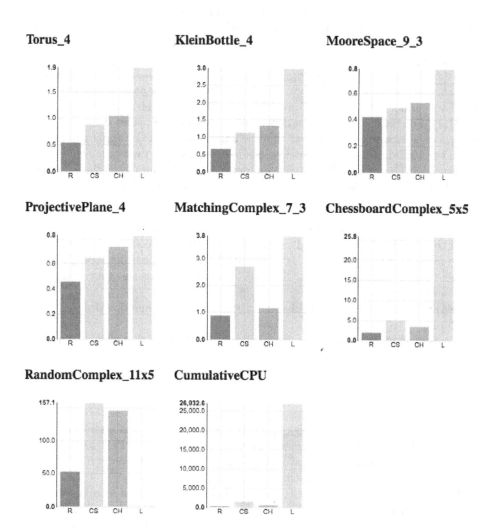

Fig. 1. CPU usage charts. On each picture from left: R (CAPD::RedHom), CS (CHomP - chomp-simplicial), CH (CHomP - homsimpl), L (Linbox - homology_gap).

References

1. Dłotko, P., Kaczynski, T., Mrozek, M., Wanner, T.: Coreduction Homology Algorithm for Regular CW-Complexes. Discrete and Computational Geometry 46, 361–388 (2011), doi:10.1007/s00454-010-9303-y
2. Dłotko, P., Specogna, R., Trevisan, F.: Automatic generation of cuts on large-sized meshes for the T-Omega geometric eddy-current formulation. Computer Methods in Applied Mechanics and Engineering 198, 3765–3781 (2009)
3. Edelsbrunner, H., Letscher, D., Zomorodian, A.: Topological Persistence and Simplification. Discrete and Computational Geometry 28, 511–533 (2002)
4. Harker, S., Mischaikow, K., Mrozek, M., Nanda, V., Wagner, H., Juda, M., Dłotko, P.: The Efficiency of a Homology Algorithm based on Discrete Morse Theory and Coreductions. In: Diaz, R.G., Jurado, P.R. (eds.) Proceedings of the 3rd International Workshop on Computational Topology in Image Context, Chipiona, Spain, Image A, vol. 1, pp. 41–47 (November 2010) ISSN: 1885-4508
5. Juda, M., Mrozek, M.: Z_2-Homology of weak $(p-2)$-faceless p-manifolds may be computed in $O(n)$ time. Topological Methods in Nonlinear Analysis 40, 137–159 (2012)
6. Kaczynski, T., Dłotko, P., Mrozek, M.: Computing the Cubical Cohomology Ring. In: Diaz, R.G., Jurado, P.R. (eds.) Proceedings of the 3rd International Workshop on Computational Topology in Image Context, Chipiona, Spain, Image A, vol. 3, pp. 137–142 (2010) ISSN: 1885-4508
7. Kaczynski, T., Mischaikow, M., Mrozek, M.: Computational Homology. Applied Mathematical Sciences, vol. 157. Springer-Verlag (2004)
8. Kaczynski, T., Mrozek, M., Ślusarek, M.: Homology computation by reduction of chain complexes. Computers and Math. Appl. 35, 59–70 (1998)
9. Kapela, T., Mrozek, M., Pilarczyk, P., Wilczak, D., Zgliczyński, P.: CAPD - a Rigorous Toolbox for Computer Assisted Proofs in Dynamics, technical report, Jagiellonian University (2010)
10. Mischaikow, K., Mrozek, M.: Chaos in Lorenz equations: a computer assisted proof. Bull. AMS (N.S.) 33, 66–72 (1995)
11. Mischaikow, K., Mrozek, M.: Chaos in the Lorenz equations: a computer assisted proof. Part II: details. Mathematics of Computation 67, 1023–1046 (1998)
12. Mischaikow, K., Mrozek, M., Pilarczyk, P.: Graph approach to the computation of the homology of continuous maps. Found. Comp. Mathematics 5, 199–229 (2005)
13. Mrozek, M.: Index Pairs Algorithms. Found. Comp. Mathematics 6, 457–493 (2006)
14. Mrozek, M.: Čech Type Approach to Computing Homology of Maps. Discrete and Computational Geometry 44(3), 546–576 (2010), doi:10.1007/s00454-010-9255-2.
15. Mrozek, M., Pilarczyk, P., Żelazna, N.: Homology algorithm based on acyclic subspace. Computers and Mathematics with Applications 55, 2395–2412 (2008)
16. Mrozek, M., Batko, B.: Coreduction homology algorithm. Discrete and Computational Geometry 41, 96–118 (2009)
17. Mrozek, M., Wanner, T.: Coreduction homology algorithm for inclusions and persistent homology. Computers and Mathematics with Applications 60(10), 2812–2833 (2010), doi:10.1016/j.camwa.2010.09.036
18. Mrozek, M., Żelawski, M., Gryglewski, A., Han, S., Krajniak, A.: Homological methods for extraction and analysis of linear features in multidimensional images. Pattern Recognition 45, 285–298 (2012)
19. Pilarczyk, P.: Computer assisted method for proving existence of periodic orbits. Topological Methods in Nonlinear Analysis 13, 365–377 (1999)

20. Koonin, J.: Topology of eigenspace posets for imprimitive reflection groups, http://arxiv.org/abs/1208.4435
21. Computer Assisted Proofs in Dynamics, http://capd.ii.uj.edu.pl/
22. Reduction Homology Algorithms, http://redhom.ii.uj.edu.pl/
23. Computational Homology Project, http://chomp.rutgers.edu/
24. GAP System, http://www.gap-system.org/
25. Sage, http://www.sagemath.org/
26. Linbox, http://www.eecis.udel.edu/~dumas/Homology/

The Gudhi Library: Simplicial Complexes and Persistent Homology

Clément Maria, Jean-Daniel Boissonnat, Marc Glisse, and Mariette Yvinec

INRIA, France
{clement.maria,jean-daniel.boissonnat,marc.glisse,
mariette.yvinec}@inria.fr

Abstract. We present the main algorithmic and design choices that have been made to represent complexes and compute persistent homology in the Gudhi library. The Gudhi library (Geometric Understanding in Higher Dimensions) is a generic C++ library for computational topology. Its goal is to provide robust, efficient, flexible and easy to use implementations of state-of-the-art algorithms and data structures for computational topology. We present the different components of the software, their interaction and the user interface. We justify the algorithmic and design decisions made in Gudhi and provide benchmarks for the code. The software, which has been developed by the first author, will be available soon at `project.inria.fr/gudhi/software/`.

Keywords: persistent homology, simplicial complex, software library, computational topology, generic programming.

1 Introduction

The principle of algebraic topology is to attach algebraic invariants to topological spaces in order to classify them up to homeomorphism. One can consequently study the property of a discrete algebraic structure (a sequence of homology groups in our case) instead of studying a continuous domain directly, which would be hard to handle algorithmically. Persistent homology [14, 16] may be considered as a "dynamic version" of this principle: given a sequence of topological spaces connected by continuous maps, we study the corresponding sequence of homology groups connected by group homomorphisms, induced by the topological space maps. The whole sequence of groups together with their homomorphisms form an algebraic structure (specifically a module) that we study. Very efficient methods have been developed for computing persistent homology [1, 14] and its dual, persistent cohomology [2, 11, 13]. The generality and stability with regard to noise [9] of persistence have made it a widely used tool in practice.

An application of interest for computational topology is topological data analysis, where one is interested in learning topological invariants of a shape, sampled by a point cloud. A popular approach is to construct, at different scales, an approximation of the shape using complexes built on top of the points, and then

H. Hong and C. Yap (Eds.): ICMS 2014, LNCS 8592, pp. 167–174, 2014.
© Springer-Verlag Berlin Heidelberg 2014

compute the persistent homology of these complexes. This approach has been successfully used in various areas of science and engineering, as for example in sensor networks [10], image analysis [6], and data analysis [8], where one typically needs to deal with big data sets in high dimensions and with general metrics. The simplicial complex and persistent homology packages in Gudhi provide all software components for this approach.

The challenge is twofold. On the one hand we need to design a generic library in computational topology, in order to adapt to the various configurations of the problem: nature of the complexes (simplicial, cubical, etc) and their representation, nature of the maps between them (inclusions, edge contractions, etc), ordering of the maps (linear, zigzag, etc) and types of algorithm for persistence (homology, cohomology). On the other hand, we need to implement a high-performance library to handle complex practical examples.

We recall in Section 2 the definition of homology and persistent homology constructed from simplicial complexes. In Section 3, we describe the design of the Gudhi library. In Section 4, we discuss the implementation choices and the user interface. Specifically, simplicial complexes are implemented with a simplex tree data structure [4]. The simplex tree is an efficient and flexible data structure for representing general (filtered) simplicial complexes. The persistent homology of a filtered simplicial complex is computed by means of the persistent cohomology algorithm [11, 13], implemented with a compressed annotation matrix [2]. The persistent homology package provides the computation of persistence with different coefficient fields, including the implementation of the multi-field persistence algorithm of [3], i.e. the simultaneous computation of persistence with various coefficient fields. Finally, in Section 5 we discuss the future components of the library and their integration in the design.

2   Theoretical Foundation of Persistent Homology

The theory of homology consists in attaching to a topological space a sequence of (homology) groups, capturing global topological features like connected components, holes, cavities, etc. Persistent homology studies the evolution – birth, life and death – of these features when the topological space is changing. Consequently, the theory is essentially composed of three elements: topological spaces, their homology groups and an evolution scheme.

Simplicial Complexes: In computer science, topological spaces are represented by their discrete counterpart: (cell) complexes. On the following, we focus on simplicial complexes, but our approach applies to all kinds of cell complexes. Let $V = \{1, \cdots, |V|\}$ be a set of *vertices*. A *simplex* σ is a subset of vertices $\sigma \subseteq V$. A *simplicial complex* $\mathbf{K}$ on V is a collection of *simplices* $\{\sigma\}$, $\sigma \subseteq V$, such that $\tau \subseteq \sigma \in \mathbf{K} \Rightarrow \tau \in \mathbf{K}$. The dimension $n = |\sigma| - 1$ of σ is its number of elements minus 1.

A *simplicial map* $f : \mathbf{K} \to \mathbf{K}'$ between simplicial complexes $\mathbf{K}$ and $\mathbf{K}'$, with respective vertex sets V and V', is a map that sends every vertex $v \in V$ to a vertex

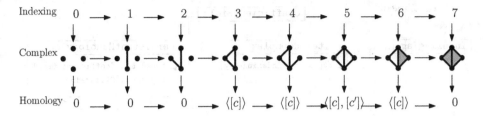

Fig. 1. Indexing of eight simplicial complexes and corresponding sequence of homology groups in dimension 1

$f(v) \in V'$, and every simplex $[v_0, \cdots, v_n] \in \mathbf{K}$ to a simplex $[f(v_0), \cdots, f(v_n)] \in \mathbf{K}'$. Note that they may be redundancy in the set $\{f(v_0), \cdots, f(v_n)\}$, in which case the simplex image has lower dimension that its pre-image. In the following, we focus on *inclusions*, which are a particular case of simplicial maps, and discuss the case of general simplicial maps in Section 5.

Homology: For a ring $\mathcal{R}$, the group of n-chains, denoted $\mathbf{C}_n(\mathbf{K}, \mathcal{R})$, of $\mathbf{K}$ is the group of formal sums of n-simplices with $\mathcal{R}$ coefficients. The *boundary operator* is a linear operator $\partial_n : \mathbf{C}_n(\mathbf{K}, \mathcal{R}) \rightarrow \mathbf{C}_{n-1}(\mathbf{K}, \mathcal{R})$ such that $\partial_n \sigma = \partial_n[v_0, \cdots, v_n] = \sum_{i=0}^{n}(-1)^i[v_0, \cdots, \widehat{v_i}, \cdots, v_n]$, where $\widehat{v_i}$ means v_i is omitted from the list. The chain groups form a sequence:

$$\cdots \; \mathbf{C}_n(\mathbf{K}, \mathcal{R}) \xrightarrow{\partial_n} \mathbf{C}_{n-1}(\mathbf{K}, \mathcal{R}) \xrightarrow{\partial_{n-1}} \cdots \xrightarrow{\partial_2} \mathbf{C}_1(\mathbf{K}, \mathcal{R}) \xrightarrow{\partial_1} \mathbf{C}_0(\mathbf{K}, \mathcal{R})$$

of finitely many groups $\mathbf{C}_n(\mathbf{K}, \mathcal{R})$ and homomorphisms ∂_n, indexed by the dimension $n \geq 0$. The boundary operators satisfy the property $\partial_n \circ \partial_{n+1} = 0$ for every $n > 0$ and we define the homology groups:

$$\mathbf{H}_n(\mathbf{K}, \mathcal{R}) = \ker \partial_n / \mathrm{im} \, \partial_{n+1}$$

We refer to [15] for an introduction to homology theory and to [14] for an introduction to persistent homology.

Indexing Scheme: "Changing" a simplicial complex consists in applying a simplicial map. An *indexing scheme* is a directed graph together with a traversal order, such that two consecutive nodes in the graph are connected by an arrow (either forward or backward). The nodes represent simplicial complexes and the directed edges simplicial maps.

From the computational point of view, there are two types of indexing schemes of interest in persistent homology [1]: *linear* ones $\bullet \longrightarrow \bullet \longrightarrow \cdots \longrightarrow \bullet \longrightarrow \bullet$ in persistent homology [16], and *zigzag* ones $\bullet \longrightarrow \bullet \longleftarrow \cdots \longrightarrow \bullet \longleftarrow \bullet$ in

---

[1] i.e. from which an interval decomposition of the persistence module exists: Gabriel's theorem [12] in quiver theory classifies these graphs.

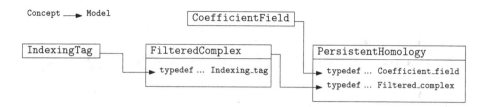

Fig. 2. Overview of the design of the library

zigzag persistent homology [7]. These indexing schemes have a natural left-to-right traversal order, and we describe them with ranges and iterators. We focus in the following on the linear case, and discuss the zigzag case in Section 5.

In the following, we consider the case where the indexing scheme is induced by a filtration. A *filtration* of a simplicial complex is a function $f : \mathbf{K} \to \mathbb{R}$ satisfying $f(\tau) \leq f(\sigma)$ whenever $\tau \subseteq \sigma$. Ordering the simplices by increasing filtration values (breaking ties so as a simplex appears after its subsimplices of same filtration value) provides an indexing scheme.

We refer to Figure 1 for an illustration of the three components of the theory and their connections. The figure pictures the linear indexing of eight simplicial complexes connected by inclusions, and the corresponding sequence of homology groups in dimension 1. Every inclusion induces a group homomorphism at the homology level. Persistent homology studies this sequence of homology groups connected by homomorphisms. Specifically, computing persistent homology consists in computing a primary decomposition of this sequence of homology groups (forming a module); the decomposition is usually represented by means of a *persistence diagram* [14].

Remark: The reader may have found a category theory taste to this presentation of persistent homology. In particular, the vertical arrows in Figure 1 represent functors of categories. We refer to [5] for more details on the categorification of persistent homology.

3   Design of the Library

A *concept* is a set of requirements (valid expression, associated types, etc) for a type. If a type satisfies these requirements, it is a *model* of the concept. The general idea under our design is to associate a concept per component presented in Section 2: the three components of the theory (indexing, complex and homology) are illustrated in Figure 1. Given two components related by a vertical arrow in Figure 1, and two models A and B of their respective associated concepts, we connect B with A through a template argument B<A>.

IndexingTag Concept: In order to describe the indexing scheme, we use a tag IndexingTag that is either linear_indexing_tag or zigzag_indexing_tag,

```
void compute_persistent_homology( FilteredComplex cpx ) {
    for( Simplex_handle sh : cpx.filtration_simplex_range() ) {
        int dim = cpx.dimension(sh);
        update_cohomology_groups( dim, sh, cpx );
            //inside update_cohomology_groups
            for( Simplex_handle b_sh : cpx.boundary_simplex_range(sh) )
                {...}
            //out
} } }
```

Fig. 3. Sample code for the computation of persistence, illustrating the use of a model of concept `FilteredComplex`

corresponding to the two indexing schemes of interest mentioned above. The tag is passed as template argument to a model of the concept `FilteredComplex` (described below and representing filtered cell complexes).

FilteredComplex Concept: We define the concept `FilteredComplex` that describes the requirement for a type to implement a filtered cell complex. We use the vocabulary of simplicial complexes, but the concept is valid for any type of cell complex. The main requirements are the definition of:

1. type `Indexing_tag`, which is a model of the concept `IndexingTag`, describing the nature of the indexing scheme,
2. type `Simplex_handle` to manipulate simplices,
3. method `int dimension(Simplex_handle)` returning the dimension of a simplex,
4. type and method `Boundary_simplex_range boundary_simplex_range( Simplex_handle)` that returns a range giving access to the codimension 1 subsimplices of the input simplex, as-well-as the coefficients $(-1)^i$ in the definition of the operator ∂. The iterators have value type `Simplex_handle`,
5. type and method `Filtration_simplex_range filtration_simplex_range()` that returns a range giving access to all the simplices of the complex read in the order assigned by the indexing scheme,
6. type and method `Filtration_value filtration(Simplex_handle)` that returns the value of the filtration on the simplex represented by the handle.

Figure 3 illustrates the use of a model of the concept `FilteredComplex`. It sketches the algorithm used for computing persistent homology via the approach of [11, 13].

PersistentHomology Concept: The concept `PersistentHomology` describes the requirement for a type to compute the persistent homology of a filtered complex. The requirement are the definition of:

1. a type `Filtered_complex`, which is a model of `FilteredComplex` and provides the type of complex on which persistence is computed,

2. a type `Coefficient_field`, which is a model of `CoefficientField` and provides the coefficient field on which homology is computed.

The requirements of the concept `CoefficientField` are essentially the definition of field operations (addition, multiplication, inversion, etc). We refer to Figure 2 for a presentation of the concepts and their connections.

4   Implementation

In this section we describe how these concepts are implemented. The code will be available soon at `project.inria.fr/gudhi/software/`.

Simplicial Complex: We use a *Simplex Tree* [4] to represent simplicial complexes. The class `Simplex_tree` is a model of `FilteredComplex` and hence furnishes all requirements of the concept. Moreover, it furnishes algorithms to construct efficiently simplicial complexes, and in particular flag complexes [14]. Details on the implementation of the algorithms may be found in [4].

Persistent Homology: We use the *Compressed Annotation Matrix* [2] to implement the persistent cohomology algorithm [11, 13] for persistence. This leads to the class `Persistent_cohomology`, which is a model of `PersistentHomology`. The class `Persistent_cohomology` allows the computation of the persistence diagram of a filtered complex, using the method `compute_persistent_homology` (see Figure 3).

The coefficient fields available as models of `CoefficientField` are `Field_Zp` for $\mathbb{Z}_p$ (for any prime p) and `Multi_field` for the multi-field persistence algorithm – computing persistence simultaneously in various coefficient fields – described in [3].

Example of Use of the Library: Figure 4 illustrates the user interface for constructing a flag complex [14] from a graph and computing its persistent homology with various coefficient fields.

```
Graph g; ...  //compute the graph
Simplex_tree< linear_indexing_tag > st; //linear ordering
st.insert(g); //insert the graph as 1-skeleton of the complex
st.expand(5); //construct the 5-skeleton of the associated flag complex
Persistent_cohomology< Simplex_tree<linear_indexing_tag>, Multi_field >
    pcoh; //persistence with "multi field coefficients" defined on a
    simplex tree
pcoh.compute_persistent_homology(st,2,1223); //compute persistent
    homology of st in all fields Zp for p prime between 2 and 1223
```

Fig. 4. User interface for the construction of a filtered flag complex with a simplex tree and the computation of its persistent homology

Data	$\|\mathcal{P}\|$	D	d	r	$\|\mathcal{K}\|$	T_{st}	$T_{\mathbb{Z}_2}^{\text{ph}}$	$T_{\mathbb{Z}_{1223}}^{\text{ph}}$	$T_{\mathbb{Z}_{1223}^2}^{\text{ph}}$
Bud	49,990	3	2	0.09	$127 \cdot 10^6$	5.7	161	161	252
Bro	15,000	25	?	0.04	$142 \cdot 10^6$	5.8	252	252	380
Cy8	6,040	24	2	0.8	$193 \cdot 10^6$	8.4	249	249	325
Kl	90,000	5	2	0.25	$114 \cdot 10^6$	8.3	228	227	401
S3	50,000	4	3	0.65	$134 \cdot 10^6$	7.2	176	176	310

Fig. 5. Timings in seconds for the various algorithms

Experiments: Figure 5 presents timings T_{st} for the construction of flag complexes with a simplex tree using the algorithm of [4], $T_{\mathbb{Z}_2}^{\text{ph}}$ and $T_{\mathbb{Z}_{1223}}^{\text{ph}}$ for the computation of persistent homology with coefficient is $\mathbb{Z}_2$ and $\mathbb{Z}_{1223}$ respectively, using the implementation of [2], and $T_{\mathbb{Z}_{1223}^2}^{\text{ph}}$ for the simultaneous computation of persistent homology in the 200 coefficient fields $\mathbb{Z}_p$ with p prime, for $2 \leq p \leq 1223$, using the multi-field persistent homology algorithm described in [3]. Experiments have been realized on a Linux machine with 3.00 GHz processor and 32 GB RAM, for Rips complexes [14] built on a variety of data points. Datasets are listed in Figure 5 with the size of points sets $|\mathcal{P}|$, the ambient dimension D and intrinsic dimension d of the sample points ("?" if unknown), the parameter r for the Rips complex and the size of the complex $|\mathcal{K}|$. More details about the implementation, the experimental protocol, the data sets as-well-as additional experiments can be found in [2–4].

The average timings per simplex of the various algorithms are ranging between $4.08 \cdot 10^{-8}$ and $7.28 \cdot 10^{-8}$ seconds per simplex for the construction of the simplex tree, between $1.27 \cdot 10^{-6}$ and $2.00 \cdot 10^{-6}$ seconds per simplex for the computation of persistent homology with coefficient field $\mathbb{Z}_2$ or $\mathbb{Z}_{1223}$, and between $1.68 \cdot 10^{-6}$ and $3.52 \cdot 10^{-6}$ seconds per simplex for the computation of multi-field persistent homology in all fields $\mathbb{Z}_p$ for p prime, $2 \leq p \leq 1223$. Note that most of the time for the computation of persistent homology is spent computing boundaries in the simplex tree.

5   Future Components

The library may be extended in various directions that fit naturally in the design. The first direction is to allow zigzag indexing schemes, by the creation of a tag `zigzag_indexing_tag`. In this case, the method `filtration_simplex_range` must indicate the direction of the arrows.

New implementations and models for `FilteredComplex` may be added. For example, the construction of witness complexes [4] will be added to the class `Simplex_tree`. Additionnaly, new types of complexes (like cubical complexes) and new data structures to represent them may be added to the library: in order to compute their persistent homology, they only need to satisfy the requirements of the concept `FilteredComplex`.

So far, only inclusions have been considered for simplicial maps between simplicial complexes. As explained in [13], any simplicial map may be implemented with a sequence of inclusions and edge contractions. We will consequently add edge contractions as updates in the class `Simplex_tree` and implement the induced updates in the class `Persistent_cohomology` (algorithms exist for edge contractions in a simplex tree [4] and for the corresponding updates at the cohomology level [13]). This way, we will be able to compute persistent homology of simplicial maps. In this case, the range provided by `filtration_simplex_range` must indicate the nature of the map between complexes.

Future works include also the implementation of a class `Field_Q`, model of concept `CoefficientField`, for homology with $\mathbb{Q}$ coefficients. Finally the interface between complexes and persistent homology allows us to implement more persistent homology algorithms.

References

1. Bauer, U., Kerber, M., Reininghaus, J.: Clear and compress: Computing persistent homology in chunks. In: Topological Methods in Data Analysis and Visualization III, pp. 103–117 (2014)
2. Boissonnat, J.-D., Dey, T.K., Maria, C.: The compressed annotation matrix: An efficient data structure for computing persistent cohomology. In: Bodlaender, H.L., Italiano, G.F. (eds.) ESA 2013. LNCS, vol. 8125, pp. 695–706. Springer, Heidelberg (2013)
3. Boissonnat, J.-D., Maria, C.: Computing Persistent Homology with Various Coefficient Fields in a Single Pass. Rapport de recherche RR-8436, INRIA (December 2013)
4. Boissonnat, J.-D., Maria, C.: The simplex tree: An efficient data structure for general simplicial complexes. Algorithmica, 1–22 (2014)
5. Bubenik, P., Scott, J.A.: Categorification of persistent homology. CoRR, abs/1205.3669 (2012)
6. Carlsson, G.E., de Ishkhanov, T., Silva, V., Zomorodian, A.: On the local behavior of spaces of natural images. Int. J. Comput. Vision 76, 1–12 (2008)
7. Carlsson, G.E., de Silva, V.: Zigzag persistence. Foundations of Computational Mathematics 10(4), 367–405 (2010)
8. Chazal, F., Oudot, S.: Towards persistence-based reconstruction in euclidean spaces. In: Proc. 24th. Annu. Sympos. Comput. Geom., pp. 231–241 (2008)
9. Cohen-Steiner, D., Edelsbrunner, H., Harer, J.: Stability of persistence diagrams. Discrete & Computational Geometry 37(1), 103–120 (2007)
10. de Silva, V., Ghrist, R.: Coverage in sensor network via persistent homology. Algebraic & Geometric Topology 7, 339–358 (2007)
11. de Silva, V., Morozov, D., Vejdemo-Johansson, M.: Persistent cohomology and circular coordinates. Discrete & Computational Geometry 45(4), 737–759 (2011)
12. Derksen, H., Weyman, J.: Quiver representations. Notices of the AMS 52(2), 200–206 (2005)
13. Dey, T.K., Fan, F., Wang, Y.: Computing topological persistence for simplicial maps. In: Symposium on Computational Geometry, p. 345 (2014)
14. Edelsbrunner, H., Harer, J.: Computational Topology - an Introduction. American Mathematical Society (2010)
15. Munkres, J.R.: Elements of algebraic topology. Addison-Wesley (1984)
16. Zomorodian, A., Carlsson, G.E.: Computing persistent homology. Discrete & Computational Geometry 33(2), 249–274 (2005)

Bertini_real: Software for One- and Two-Dimensional Real Algebraic Sets

Daniel A. Brake[1], Daniel J. Bates[2], Wenrui Hao[3], Jonathan D. Hauenstein[1], Andrew J. Sommese[4], and Charles W. Wampler[5]

[1] North Carolina State University, USA
danielthebrake@gmail.com, hauenstein@ncsu.edu
danielthebrake.org, www.math.ncsu.edu/~jdhauens
[2] Colorado State University, USA
bates@math.colostate.edu
www.math.colostate.edu/~bates
[3] Mathematical Biosciences Institute, USA
hao.50@osu.edu
people.mbi.ohio-state.edu/hao.50
[4] University of Notre Dame, USA
sommese@nd.edu
www.nd.edu/~sommese
[5] General Motors Research and Development, USA
charles.w.wampler@gm.com
www.nd.edu/~cwample1

Abstract. Bertini_real is a command line program for numerically decomposing the real portion of a one- or two-dimensional complex irreducible algebraic set in any reasonable number of variables. Using numerical homotopy continuation to solve a series of polynomial systems via regeneration from a witness set, a set of real vertices is computed, along with connection information and associated homotopy functions. The challenge of embedded singular curves is overcome using isosingular deflation. This decomposition captures the topological information and can be used for further computation and refinement.

Keywords: Numerical algebraic geometry, cell decomposition, algebraic surface, algebraic curve, homotopy continuation, deflation.

1 Introduction

Bertini_real seeks to automate the task of visualizing and computing on real algebraic curves and surfaces. From only a defining polynomial system, the program computes a cellular decomposition of the real portion of a one- or two-dimensional complex algebraic set. The output of Bertini_real is a set of text files, containing the set of computed vertices, the connections between them, and any associated homotopies. Using the homotopies, the decomposition can be refined to the user's desire with a supplemental program simply titled sampler. An interactive visualization suite is provided in MATLAB.

H. Hong and C. Yap (Eds.): ICMS 2014, LNCS 8592, pp. 175–182, 2014.

Bertini_real works by leveraging the power of homotopy continuation [2,3], numerical irreducible decomposition [8], regeneration [5], randomization [3], and isosingular deflation [6] to decompose the real parts of complex one- and two-dimensional components of algebraic varieties. It produces a cell decomposition, similar to the output of other decomposition methods, most notably the Cylindrical Algebraic Decomposition [1].

2   Functionality

Bertini_real is an MPI parallel-enabled compiled program called from the command line. The two necessary ingredients to run the software are: 1) a Bertini input file, and 2) a Numerical Irreducible Decomposition (NID) produced by Bertini. It further depends on MATLAB for symbolic calculations (*e.g.* deflation, symbolic derivatives and determinants), the Boost C++ support library, as well as GMP and MPFR (for multiple-precision numerics). Compilation requires a library-compiled version of Bertini [2].

The basic pattern for usage of Bertini_real is summarized below.

1. Create a NID, via Bertini. This gives a witness set for each irreducible component, as well as information on each component's degree, multiplicity, and deflation requirements.
2. Run Bertini_real on a single irreducible component. Bertini_real checks if the component is self-conjugate. If it is not, Bertini_real finds the intersection of the component with its conjugate and proceeds. The program further deflates the system [6] so that the component is reduced and properly deflated, so that we may track on it. It then finds a cell decomposition of the real points in the complex set.
3. Refine the decomposition. Bertini_real produces raw decompositions that are bare skeletons of the objects they describe. If the user wants to view a smoothed version, or use the decomposition for further calculations, they might want to refine using the program **sampler**, provided as part of the Bertini_real package.
4. Visualize. Visual interpretation of the data typically quickly reveals any problems which might have been encountered during computations. The suite of graphical software is provided through MATLAB.

3   Application

3.1   Curve

Consider a three-jointed revolute planar robot, with equal link lengths – and let the length be unity. If we fix a point in the workspace of the robot, we get a curve of solutions in terms of the joint angles such that the end effector is placed at the point. Equations are given below in (1), with $s_i = si = \sin \theta_i$, and $c_i = ci = \cos \theta_i$.

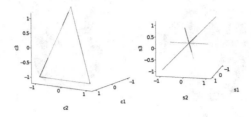

Fig. 1. Example of curve decomposition. A 3R planar robot of unit link length places its end effector at the point $(x,y) = (1,0)$. Left: projection onto the cosines of the angles. Right: projection onto sines. These two plots are simpler than viewing the joint angles directly, due to the periodic nature of trigonometric functions.

$$\begin{bmatrix} c_1 - s_3(c_1 s_2 + c_2 s_1) + c_1 c_2 - s_1 s_2 + c_3(c_1 c_2 - s_1 s_2) - 1 \\ s_1 + c_1 s_2 + c_2 s_1 + s_3(c_1 c_2 - s_1 s_2) + c_3(c_1 s_2 + c_2 s_1) \\ c_2^2 + s_1^2 - 1 \\ c_2^2 + s_2^2 - 1 \\ c_3^2 + s_3^2 - 1 \end{bmatrix} = 0 \qquad (1)$$

In Fig. 1, we present the three components of the solution curve when we grasp the point $(x,y) = (1,0)$. On the left is a projection of the set onto the cosines, and the figure on the right are the sines.

3.2 Surface

Now consider a two-joint revolute planar robot with link lengths $\ell_1 = 1, \ell_2 = 0.5$, and let the target position for the end effector be variable and denoted (x,y), as in (2). The set of points in the plane the robot can reach is realizable using a surface decomposition. The workspace ought to be an annulus, and this is indeed the result of the decomposition when projected onto (x,y) as in Fig. 2.

$$\begin{bmatrix} c_1 - x + (c_1 c_2)/2 - (s_1 s_2)/2 \\ s_1 - y + (c_1 s_2)/2 + (c_2 s_1)/2 \\ c_1^2 + s_1^2 - 1 \\ c_2^2 + s_2^2 - 1 \end{bmatrix} = 0 \qquad (2)$$

4   Underlying Theory

4.1   Curve

The implementation of curve decomposition in Bertini_real follows the algorithm laid out in [7], depicted in Fig. 3, and summarized informally below.

To begin, there is a little user set up, the foremost of which is to run Bertini with configuration setting TrackType:1 to obtain a NID. Optionally, the user

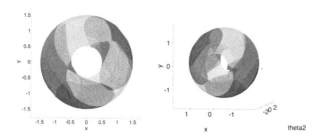

Fig. 2. Example of surface decomposition. A 2R planar robot with differing link lengths is allowed to move freely, and we decompose its workspace as a surface in terms of (x, y) and the sines and cosines of the joint variables. On the left, projection of the surface onto (x, y) gives an annulus as expected. At the right, the surface is tilted, revealing the two solutions, in terms of the arctangent of (s_2, c_2).

may write a file containing a (random) real projection and a sphere of interest. Bertini_real automatically tests for self-conjugacy. A non-self-conjugate component is intersected with its own conjugate to produce a finite set of isolated real points, which terminates the computation. Otherwise, Bertini_real carries out the following six steps.

1. Find critical points. These points will include singular points, and points such that the curve is tangent to the direction of projection, and they will satisfy the system:

$$f_{\text{crit}} = \begin{bmatrix} f(x) \\ \det \begin{pmatrix} Jf(x) \\ J\pi_1(x) \end{pmatrix} \end{bmatrix} = 0, \tag{3}$$

where J indicates the Jacobian matrix of partial derivatives and $\pi_1 : \mathbb{C}^N \to \mathbb{C}$ is the random real projection being used for the decomposition. Let $c_1, \ldots, c_n$ be the real critical points, ordered so that $\pi_1(c_1) < \pi_1(c_2) < \cdots < \pi_1(c_n)$.

2. Intersect with sphere. To cut off unbounded arcs of the curve, or to focus the view to the user's region, we intersect with a sphere of center x_0 and radius r, and solve the system (4), inserting the real intersection points into the list of ordered critical points.

$$f_{\text{sphere}} = \begin{bmatrix} f(x) \\ ||x - x_0||_2^2 - r^2 \end{bmatrix}. \tag{4}$$

3. Slice. To find what will become the midpoints of the edges of the decomposition, slice the curve between its critical points, by tracking from the single witness linear $\mathcal{L}$ to each midpoint projection value, $p_{m_i} = (\pi_1(c_i) + \pi_1(c_{i+1}))/2$, as:

$$H_{\text{midslice}} = \begin{bmatrix} f(x) \\ t\mathcal{L}(x) + (1 - t)(\pi_1(x) - p_{m_i}) \end{bmatrix}.$$

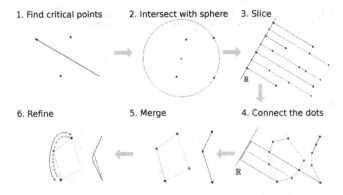

Fig. 3. The six major steps for a curve decomposition as implemented in Bertini_real. This illustration uses an elliptic curve, $x^3 - 2x + 1 - y^2 = 0$.

4. Connect the dots. Use the following homotopy to track midpoints first left and then right to the points on the curve above each critical point:

$$H_{\text{track}} = \begin{bmatrix} f(x) \\ \pi(x) - (tp_{m_i} + (1-t)p) \end{bmatrix},$$

where p is taken first as $p = \pi_1(c_i)$ and then as $p = \pi_1(c_{i+1})$.

5. Merge. Optionally, we can remove superfluous intersections which lie in the same projection fiber as critical points. These points arise when the curve has non-critical branches above a critical point, and they can be removed to produce a simpler decomposition.

6. Refine. Optionally, the user can refine the decomposition to their specification. By using the same homotopy as in Step 4, we can move the generic point in the center of each edge to any projection value p, $\pi_1(c_i) < p < \pi_1(c_{i+1})$. Two methods are available in Bertini_real: 1) a fixed-number method, where the user specifies how many points they want per edge; and 2) an adaptive method, where the user specifies a distance tolerance and a limit on the of number of refinement iterations.

4.2   Surface

The implementation of surface decomposition in Bertini_real follows the algorithm laid out in [4], depicted in Fig. 4, and summarized informally below.

Similarly to a curve decomposition, there is a small amount of user set up. Of course, one must obtain a NID, and the user may choose a projection and sphere. Self-conjugacy testing, and deflation are performed automatically. The six steps below are for self-conjugate components only. Any non-self-conjugate component is intersected with its conjugate component, producing at most a curve, which is then treated as in the curve case above. The decomposition of a

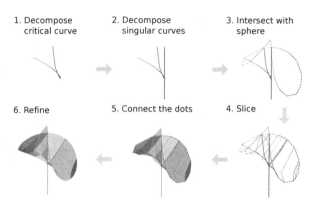

Fig. 4. The six major steps for a surface decomposition as implemented in `Bertini_real`. This example uses the Whitney Umbrella, $x^2 - y^2 z = 0$, a degree 3 surface in three variables, which is unbounded and contains a curve of singularity (around part of which, the surface is one-real dimensional.

surface is found with respect to two random real projections, $\pi_1, \pi_2 : \mathbb{C}^N \to \mathbb{C}$, as follows.

1. Decompose the critical curve. The critical curve is analogous to the outline of an object when viewed in an image plane and is also the set where the tangent is parallel to the *two* directions of projection, π_1, π_2. The curve is defined by the system:

$$f_{\text{critcurve}} = \left[\det \begin{pmatrix} f \\ Jf(x) \\ J\pi_1(x) \\ J\pi_2(x) \end{pmatrix} \right] = 0. \tag{5}$$

Witness sets for the components of the critical curve are obtained via regeneration from the witness set, using a left-nullspace approach, and these are passed the curve method for decomposition with respect to π_1.

2. Decompose singular curves. As a matter of course from computing the witness set for the critical curve, we also obtain witness points for singular curves, since every singular curve will also satisfy (5). We use isosingular deflation [6] to deflate the input system these witness points, thereby producing full witness sets. These are then decomposed with respect to π_1 exactly as for any other component of the critical curve.

3. Intersect with sphere. The intersection of the surface with a sphere of radius r and center x_0 will result in a curve, defined by (4). The intersection curve is treated as part of the critical curve, so it too is decomposed with respect to π_1.

4. Slice. We perform a curve decomposition at each of two sets of π_1 projection values – *at* each critical π_1-value, and *halfway between* each pair, coming from

the critical points of the critical curve, singular curves, and the sphere curve. Call these *critical slices* and *mid-slices*, respectively. Each of these slices has a constant π_1 value and is decomposed with respect to π_2.

5. Connect the dots. The midpoints of each edge of each mid-slice become the center point for a face of the decomposition. The decompositions of the mid-slices reveal how the midpoint is connected to the top and bottom edges of its face, each coming from the critical curve, the sphere curve, or a singular curve. The description of the face is completed by finding which edges in the adjacent left and right critical slices connect to the midpoint. This is determined using a homotopy that keeps the midpoint from crossing its top and bottom edges as it is moved to the left and right critical projection values: see [4].

6. Refine. The decomposition to this point is coarse, in that it provides a coarse triangulation of the surface. A refinement method is provided in the separate executable `sampler`, which refines each edge and face in the decomposition, to contain a number of points of the user's choice. Adaptive and eventually optimal sampling for surfaces is a matter of ongoing development.

5   Technical Contribution

5.1   Advances

Bertini_real allows a non-expert access to the algorithms of [7,4] for decomposing the real points of complex algebraic curves and surfaces, whereas the previous prototype codes required expertise and worked only on sets of low degree. Importantly, Bertini_real is the first implementation that removes the restriction to almost-smooth surfaces that was needed in [4] — non-smooth surfaces can now be treated in any number of variables. The largest curve we have decomposed so far is a 3-3 Burmester curve [9] in 14 variables of degree 630.

5.2   Challenges

The main algorithms as implemented in Bertini_real are all for affine varieties. One can decompose any projective variety one wants, by considering patch equations and the transformation into an affine space. However, the Bertini tracker loops used by Bertini_real expect there to be a single homogenizing variable for a single non-homogeneous variable group. Furthermore, Bertini as written was not intended to be called as a library as we do with Bertini_real, so linking into the loops required a great deal of finesse. This experience is helping the setting up of specifications for the next version of Bertini.

While curves are comparatively easy to decompose, Bertini_real's surface decomposer is currently capable of dealing with only moderately sized systems — surfaces involving no randomization and six variables are generally currently tractable. However, we have encountered difficulty decomposing a particular Burmester surface, involving eight polynomials in ten variables. While we can readily obtain the witness points for the critical curve, computing the critical points of the critical curve remains a barrier for this problem. The code uses a determinantal formulation of the criticality condition, wherein we compute a symbolic determinant involving a Jacobian matrix. `MATLAB` struggles with this, eventually spitting out a system over 25 MB in size. Worse, `Bertini` must then parse this input file to create procedures for evaluating the function and its Jacobian, which overwhelms the available computing resource.

The major obstacle to running large problems through the surface decomposer is therefore the elimination of the determinant. Alternate methods that avoid the determinant are the subject of further research.

Acknowledgments. All authors were partially supported by AFOSR grant FA8650-13-1-7317. DJB was partially supported by NSF grant DMS-1025564. DAB and JDH were additionally supported by DARPA YFA.

References

1. Arnon, D.S., Collins, G.E., McCallum, S.: Cylindrical algebraic decomposition I: The basic algorithm. SIAM Journal on Computing 13(4), 865–877 (1984)
2. Bates, D.J., Hauenstein, J.D., Sommese, A.J., Wampler, C.W.: Bertini: Software for Numerical Algebraic Geometry, http://bertini.nd.edu
3. Bates, D.J., Hauenstein, J.D., Sommese, A.J., Wampler, C.W.: Numerically Solving Polynomial Systems with Bertini, vol. 25. SIAM (2013)
4. Besana, G.M., Di Rocco, S., Hauenstein, J.D., Sommese, A.J., Wampler, C.W.: Cell decomposition of almost smooth real algebraic surfaces. Numerical Algorithms 63(4), 645–678 (2013)
5. Hauenstein, J.D., Sommese, A.J., Wampler, C.W.: Regeneration homotopies for solving systems of polynomials. Mathematics of Computation 80(273), 345–377 (2011)
6. Hauenstein, J.D., Wampler, C.W.: Isosingular sets and deflation. Foundations of Computational Mathematics 13(3), 371–403 (2013)
7. Lu, Y., Bates, D.J., Sommese, A.J., Wampler, C.W.: Finding all real points of a complex curve. Contemporary Mathematics 448, 183–205 (2007)
8. Sommese, A.J., Verschelde, J., Wampler, C.W.: Numerical decomposition of the solution sets of polynomial systems into irreducible components. SIAM Journal on Numerical Analysis 38(6), 2022–2046 (2001)
9. Tong, Y., Myszka, D.H., Murray, A.P.: Four-bar linkage synthesis for a combination of motion and path-point generation. In: Proc. ASME IDETC/CIE 2013, Portland, OR, August 4-7 (2013)

Hom4PS-3: A Parallel Numerical Solver for Systems of Polynomial Equations Based on Polyhedral Homotopy Continuation Methods

Tianran Chen[1], Tsung-Lin Lee[2], and Tien-Yien Li[1]

[1] Michigan State University, USA
chentia1@msu.edu, li@math.msu.edu
http://www.math.msu.edu/~chentia1/
http://www.math.msu.edu/~li/
[2] National Sun Yat-sen University, Taiwan ROC
leetsung@math.nsysu.edu.tw
http://www.math.nsysu.edu.tw/~leetsung/

Abstract. Homotopy continuation methods have been proved to be an efficient and reliable class of numerical methods for solving systems of polynomial equations which occur frequently in various fields of mathematics, science, and engineering. Based on the successful software package Hom4PS-2.0 for solving such polynomial systems, Hom4PS-3 has a new fully modular design which allows it to be easily extended. It implements many different numerical homotopy methods including the Polyhedral Homotopy continuation method. Furthermore, it is capable of carrying out computation in parallel on a wide range of hardware architectures including multi-core systems, computer clusters, distributed environments, and GPUs with great efficiency and scalability. Designed to be user-friendly, it includes interfaces to a variety of existing mathematical software and programming languages such as Python, Ruby, Octave, Sage and Matlab.

Keywords: polynomial systems, homotopy continuation, polyhedral homotopy, binomial system.

1 Introduction

The problem of solving systems of polynomial equations, or polynomial systems, has been, and will continue to be, one of the most important subjects in both pure and applied mathematics. The need to solve polynomial systems arises naturally and frequently in various fields of science and engineering as documented in [1,13,18]. The *homotopy continuation method* has been established, in recent years, as one of the most reliable and efficient class of numerical methods for finding the full set of isolated solutions to a general polynomial system. There are many mature software implementing this method, including Bertini[3], HOMPACK[20], NAG4M2[12], and etc. See [2,13,16,18] for a survey. One important branch among them is the *polyhedral homotopy* method initiated in [9]. The

H. Hong and C. Yap (Eds.): ICMS 2014, LNCS 8592, pp. 183–190, 2014.

method has been successfully implemented in software packages PHCpack [19] developed by J. VERSCHELDE at University of Illinois at Chicago Circle, PHoM [8] developed by a group led by M. KOJIMA at Tokyo Institute of Technology in Japan, and Hom4PS-2.0 [10] developed by a group led by the authors. The efficiency and reliability in real world applications of Hom4PS-2.0 is documented in [10,11,13]. Based Hom4PS-2.0, a new numerical solver for polynomial systems Hom4PS-3[5] is created around the same core mathematical algorithms. Written in the C++ programming language and taking advantage of the object-oriented programming paradigm, Hom4PS-3 has a fully modular structure following modern design principles that allows it to be easily extended by individual "modules". Designed to be user-friendly from the ground up, it includes interfaces to a variety of existing mathematical software and programming languages such as Sage, Python, Ruby, Octave, and Matlab.

In real world applications from science and engineering, there is no shortage in the demand of solving larger and larger polynomial systems. Homotopy continuation methods are particularly suited to handle these large polynomial systems due to its *pleasantly parallel* nature: each isolated solution is computed independently of the others. Hom4PS-3 is designed to take advantage of a variety of parallel hardware architectures including *multi-core* systems, *NUMA* systems, *computer clusters*, *distributed environments*, and *GPUs*. Using parallel computation techniques tailored for each architecture (a symmetric model using Intel TBB and OpenMP on multi-core architectures, a hierarchical model for NUMA architectures, a master-worker model using MPI on clusters, an asynchronous message passing model for distributed environments, and a hybrid *"single-thread-multiple-data"* model for GPUs), excellent efficiency and scalability have been achieved on these systems.

2   Functionality

Given an input polynomial system, which can be represented in a number of different formats, Hom4PS-3 solves the polynomial system and outputs a list of complex solutions. This list includes *all* isolated nonsingular solutions of the given system in $\mathbb{C}^n$ as well as isolated singular solutions together with their multiplicity information. Optionally, Hom4PS-3 can also produce "solutions at infinity" by carrying out computation in the complex projective space or weighted projective spaces. For polynomial systems having solution components of positive dimensions, an included module posdim can be used to compute sample or *"witness"* points on solution components of any given dimension. Furthermore, the number of components and their degrees can be computed via the "witness" points.

On a Unix/Linux or similar operating system, one can solve a polynomial system with Hom4PS-3 simply by invoking the command

```
hom4ps-easy FILE
```

on a terminal, where FILE is the path of the file that contains the representation of the input polynomial system. This command runs Hom4PS-3 in its "easy mode" in which a predetermined set of parameters for controlling the behavior of the program that is likely appropriate for most situations is used.

The behavior of Hom4PS-3 can be controlled via a long list of switches and parameters given either on the command line or as a configuration file. They control the usage of certain modules, the precision to be used for floating point arithmetic, the strategy for adjusting "step sizes" in the procedure of tracking homotopy paths (Section 3), and many other aspects of the program. A complete list can be found in its reference manual. Both the downloadable packages and complete documentation can be found on the website http://www.hom4ps3.org.

2.1   Parallel Computation Capabilities

Multi-core Systems. A multi-core processor contains multiple processing units, called "cores", each capable of executing program instructions and carrying out computation independently. On a multi-core system, Hom4PS-3 automatically spreads work load across all available processor cores on the system via a multi-thread model. The implementation supports both Intel TBB and OpenMP, two of the most popular programming frameworks for multi-core systems.

Computer Cluster. A computer cluster is a group of computers, connected via high speed network, that work together on a single task and can be viewed as a single computer system. Using MPI, the *de facto* standard for communication on clusters, Hom4PS-3 can distribute work load among nodes in the cluster.

GPU Computing. GPU computing is the use of graphics processing units, or GPUs, which are originally designed for rendering graphics, to perform general purpose computation in a highly parallel fashion. On platforms where one or more GPU devices are available, Hom4PS-3 can take advantage of these highly parallel hardware on specific tasks involving intensive floating point matrix and vector manipulations such as polynomial and derivative evaluation and "mixed volume" computation (Section 3). The current implementation is built on top CUDA, a popular proprietary GPU programming framework developed by NVidia. Experimental supports for OpenCL, the dominant open standard developed by multiple vendors, are under active development.

2.2   Interfaces with Existing Mathematical Software and Programming Languages

Sage Interface. Sage is a free open-source mathematical software with features covering many aspects of mathematics, including algebra, combinatorics, numerical mathematics, number theory, and calculus. The Sage interface is one of the easiest way to use Hom4PS-3. For example, one can use Hom4PS-3 to solve a extremely simple polynomial system in Sage by using the following commands:

```
import hom4pspy
R.<x,y> = CC['x,y']
f = x^2 - 3*x + 2
g = y^2 - 4*y + 3
hom4pspy.solve_real ( [f,g] )
```

In this example, the first line imports the Hom4PS-3 interface. The next three lines creates a polynomial ring in two variables over the complex (floating point) field and two polynomials using the Sage syntax. The last line solves the polynomial system for the real solutions via Hom4PS-3 and returns a list of dict each describing a solution. Of course it is typically used to handle much more complicated and larger systems than this simple example, and with the power of Sage one can perform complicated algebraic construction to build the input system for Hom4PS-3, bridging the world of symbolic and numerical computation.

Python Interface. Python is a popular programming language (and the solid foundation on top of which Sage was built). The hom4pspy module used in the Sage can also be used separately as a Python interface. The commands

```
import hom4pspy
hom4pspy.solve_real ( ["x^2 - 3*x + 2", "y^2 - 4*y + 3"] )
```

solves the same simple polynomial system, now represented as strings.

Octave Interface. (GNU) Octave is a software and programming language designed for numerical computations that is mostly compatible with Matlab. In Octave, with the Hom4PS-3 interface, the commands

```
hom4psoct
hom4psoct.solve_real ( "x^2 - 3*x + 2, y^2 - 4*y + 3" )
```

solves the same simple polynomial system, represented as a single string (due to the lack of sophisticated symbolic manipulation capabilities in Octave).

3   Underlying Theory

In the 90's, a considerable research effort in Europe had been directed to the problem of solving polynomial systems in two consecutive major projects, PoSSo (**Po**lynomial **S**ystem **So**lving) and FRISCO (**FR**amework for **I**ntegrated **S**ymbolic and numerical **CO**mputation), supported by the European Commission. Those research projects focused on the development of the well-established Gröbner basis methods within the framework of computer algebra. Their reliance on symbolic manipulation makes those methods seem somewhat limited to relatively small problems. In 1977, GARCIA and ZANGWILL [7] and DREXLER [6] independently discovered that the *homotopy continuation method* could be used to find the full set of isolated solutions to a polynomial system numerically. In the last several decades, the method has been quite well developed and proved to be reliable and efficient. Note that *continuation methods* are the method of choice to deal with general nonlinear systems of equations numerically and globally as illustrated by the extensive bibliography listed in [1] where general ideas of the method were discussed.

One of the most important branches of the homotopy continuation method for solving general polynomial systems is the *polyhedral* homotopy method initiated by B. HUBER and B. STURMFELS [9]. For an $n \times n$ square polynomial systems

$$
P(x_1, \ldots, x_n) = P(\mathbf{x}) = \begin{cases} p_1(\mathbf{x}) = \sum_{\mathbf{a} \in S_1} c_{1,\mathbf{a}} \mathbf{x^a} \\ \quad \vdots \\ p_n(\mathbf{x}) = \sum_{\mathbf{a} \in S_n} c_{n,\mathbf{a}} \mathbf{x^a} \end{cases} \tag{1}
$$

where $\mathbf{x} = (x_1, \ldots, x_n)$, $\mathbf{a} = (a_1, \ldots, a_n)^\top \in \mathbb{N}_0^n$, and $\mathbf{x^a} = x_1^{a_1} \cdots x_n^{a_n}$. Here S_j, a finite subset of $\mathbb{N}_0^n$, is called the *support* of $p_j(\mathbf{x})$. For fixed supports $S_1, \ldots, S_n$, it is a basic fact in algebraic geometry that for generic choices of the complex coefficients $c_{j,\mathbf{a}} \in \mathbb{C}^*$ the number of isolated solutions of the system $P(\mathbf{x}) = 0$ in $(\mathbb{C}^*)^n$ is a fixed number. The word "generic" here can be understood as "randomly chosen". Its precise meaning can be found in [4], [9] and [13]. This fixed number also serves as an upper bound on the number of isolated solutions $P(\mathbf{x}) = 0$ can have in $(\mathbb{C}^*)^n$ among all choices of coefficients. In [4], this upper bound, now commonly known as the *BKK bound*, is formulated in terms of *mixed volume*: For convex polytopes $\mathcal{Q}_1, \ldots, \mathcal{Q}_k \subset \mathbb{R}^k$, let $\lambda_1 \mathcal{Q}_1, \ldots, \lambda_k \mathcal{Q}_k$ represent their scaled version, by factors of positive $\lambda_1, \ldots, \lambda_k$ respectively. Then the *Minkowski sum* $\lambda_1 \mathcal{Q}_1 + \cdots + \lambda_k \mathcal{Q}_k$ is also a convex polytope. It can be shown that the volume $\mathrm{Vol}_k(\lambda_1 \mathcal{Q}_1 + \cdots + \lambda_k \mathcal{Q}_k)$ in $\mathbb{R}^k$ is a homogeneous polynomial in $\lambda_1, \ldots, \lambda_k$. The *mixed volume*, denoted by $\mathrm{MVol}(\mathcal{Q}_1, \ldots, \mathcal{Q}_k)$, is defined to be the coefficient of $\lambda_1 \times \lambda_2 \times \cdots \times \lambda_k$ in this polynomial. The theory of BKK bound [4] states that the number of isolated solutions of the system $P(\mathbf{x}) = 0$ in $(\mathbb{C}^*)^n$ for generic choices of the coefficients is the mixed volume of the convex hull of the supports of $p_1, \ldots, p_n$, i.e.

$$
\mathrm{MVol}(\mathrm{conv} S_1, \ldots, \mathrm{conv} S_n).
$$

We shall restrict our focus on solving a polynomial system $P(\mathbf{x}) = 0$ in (1) with "generic" (nonzero) complex coefficients $c_{j,\mathbf{a}} \in \mathbb{C}^*$. When the system with generic coefficients is solved, one can always use it to solve the system with specifically given coefficients with the same supports by the *Cheater's* homotopy [14] (or [17]).

To solve $P(\mathbf{x}) = 0$ in (1), consider, with a new variable t, the homotopy

$$
H(x_1, \ldots, x_n, t) = H(\mathbf{x}, t) = \begin{cases} h_1(\mathbf{x}, t) = \sum_{\mathbf{a} \in S_1} c_{1,\mathbf{a}} \mathbf{x^a} t^{\omega_1(\mathbf{a})} \\ \quad \vdots \\ h_n(\mathbf{x}, t) = \sum_{\mathbf{a} \in S_n} c_{n,\mathbf{a}} \mathbf{x^a} t^{\omega_n(\mathbf{a})} \end{cases} \tag{2}
$$

with "lifting" functions $\omega_1, \ldots, \omega_n$, where each $\omega_k : S_k \to \mathbb{Q}$ has randomly chosen images. Note that when $t = 1$, $H(\mathbf{x}, 1) = P(\mathbf{x})$. For $\mathbf{a} \in S_k$, write

$\hat{\mathbf{a}} = (\mathbf{a}, \omega_k(\mathbf{a}))$. In [9], it was shown that if the system $P(\mathbf{x}) = 0$ has isolated solutions in $(\mathbb{C}^*)^n$, then there exists $\hat{\alpha} = (\alpha, 1) \in \mathbb{R}^{n+1}$ with $\alpha = (\alpha_1, \ldots, \alpha_n)$ and a corresponding collection of pairs $\{\mathbf{a}_1, \mathbf{a}_1'\} \subset S_1, \ldots, \{\mathbf{a}_n, \mathbf{a}_n'\} \subset S_n$ such that for each $k = 1, \ldots, n$

$$\langle \hat{\mathbf{a}}_k, \hat{\alpha} \rangle = \langle \hat{\mathbf{a}}_k', \hat{\alpha} \rangle < \langle \hat{\mathbf{a}}, \hat{\alpha} \rangle \quad \text{for all } \mathbf{a} \in S_k \setminus \{\mathbf{a}_k, \mathbf{a}_k'\} \tag{3}$$

and

$$\kappa_\alpha := \left| \det \left[\mathbf{a}_1 - \mathbf{a}_1' \ \ldots \ \mathbf{a}_n - \mathbf{a}_n' \right] \right| > 0.$$

Here $\langle \, , \, \rangle$ stands for the standard inner product in Euclidean space. Let $\mathcal{T}$ be the collection of all such α's, then

$$\sum_{\alpha \in \mathcal{T}} \kappa_\alpha$$

is independent of the choice of the lifting functions $\omega_1, \ldots, \omega_k$. In fact, this number agrees with the number of isolated solutions of the system $P(\mathbf{x}) = 0$ in $(\mathbb{C}^*)^n$, counting multiplicities, known as the BKK bound mentioned before.

Now, for a fixed α in $\mathcal{T}$ along with its corresponding set of pairs $\{\mathbf{a}_1, \mathbf{a}_1'\} \subset S_1, \ldots, \{\mathbf{a}_n, \mathbf{a}_n'\} \subset S_n$, let $\beta_k = \langle \hat{\mathbf{a}}_k, \hat{\alpha} \rangle = \langle \hat{\mathbf{a}}_k', \hat{\alpha} \rangle = \langle \mathbf{a}_k', \alpha \rangle + \omega_k(\mathbf{a}_k')$ for $k = 1, \ldots, n$. Then by (3), for each $k = 1, \ldots, n$,

$$\beta_k < \langle \hat{\mathbf{a}}, \alpha \rangle \quad \text{for all } \mathbf{a} \in S_k \setminus \{\mathbf{a}_k, \mathbf{a}_k'\}. \tag{4}$$

By the change of variables $\mathbf{x} = t^\alpha \bullet \mathbf{y}$, i.e., for $\mathbf{y} = (y_1, \ldots, y_n)$

$$\begin{cases} x_1 = t^{\alpha_1} y_1 \\ \quad \vdots \\ x_n = t^{\alpha_n} y_n, \end{cases} \tag{5}$$

we have, for $\mathbf{a} = (a_1, \ldots, a_n) \in S_k$ and $\hat{\mathbf{a}} = (\mathbf{a}, \omega_k(\mathbf{a}))$

$$\begin{aligned} \mathbf{x}^{\mathbf{a}} t^{\omega_k(\mathbf{a})} &= x_1^{a_1} \ldots x_n^{a_n} t^{\omega_k(\mathbf{a})} \\ &= (t^{\alpha_1} y_1)^{a_1} \ldots (t^{\alpha_n} y_n)^{a_n} t^{\omega_k(\mathbf{a})} \\ &= y_1^{a_1} \ldots y_n^{a_n} t^{a_1 \alpha_1 + \cdots + a_n \alpha_n + \omega_k(\mathbf{a})} \\ &= \mathbf{y}^{\mathbf{a}} t^{\langle (\mathbf{a}, \omega_k(\mathbf{a})), (\alpha, 1) \rangle} \\ &= \mathbf{y}^{\mathbf{a}} t^{\langle \hat{\mathbf{a}}, \hat{\alpha} \rangle} \end{aligned}$$

with $\hat{\alpha} = (\alpha, 1)$. Substituting the result into $H(\mathbf{x}, t)$ in (2), it follows that

$$\bar{H}^\alpha(\mathbf{y}, t) := H(t^\alpha \bullet \mathbf{y}, t) = \begin{cases} \bar{h}_1^\alpha(\mathbf{y}, t) := h_1(t^\alpha \bullet \mathbf{y}, t) = \sum_{\mathbf{a} \in S_1} c_{1,\mathbf{a}} \, \mathbf{y}^{\mathbf{a}} \, t^{\langle \hat{\mathbf{a}}, \hat{\alpha} \rangle} \\ \quad \vdots \\ \bar{h}_n^\alpha(\mathbf{y}, t) := h_n(t^\alpha \bullet \mathbf{y}, t) = \sum_{\mathbf{a} \in S_n} c_{n,\mathbf{a}} \, \mathbf{y}^{\mathbf{a}} \, t^{\langle \hat{\mathbf{a}}, \hat{\alpha} \rangle} \end{cases}.$$

Though the above expression may contain positive or negative powers of t, the minimum exponents of t in each $\bar{h}_k^\alpha$ is actually given by β_k. Therefore, if

$$H^\alpha(\mathbf{y}, t) := \begin{cases} t^{-\beta_1}\bar{h}_1(\mathbf{y}, t) = \sum_{\mathbf{a} \in S_1} c_{1,\mathbf{a}} \, \mathbf{y}^\mathbf{a} \, t^{\langle \hat{\mathbf{a}}, \hat{\alpha} \rangle - \beta_1} \\ \qquad \vdots \\ t^{-\beta_n}\bar{h}_n(\mathbf{y}, t) = \sum_{\mathbf{a} \in S_n} c_{n,\mathbf{a}} \, \mathbf{y}^\mathbf{a} \, t^{\langle \hat{\mathbf{a}}, \hat{\alpha} \rangle - \beta_n} \end{cases}, \qquad (6)$$

then, by (4), each component of H^α has exactly two terms having no powers of t while all other terms have positive powers of t. Hence, when $t = 0$, $H^\alpha(\mathbf{y}, 0) = 0$ is the "binomial system of equations"

$$\begin{cases} c_{1,\mathbf{a}_1} \mathbf{y}^{\mathbf{a}_1} + c_{1,\mathbf{a}_1'} \mathbf{y}^{\mathbf{a}_1'} = 0 \\ \qquad \vdots \\ c_{n,\mathbf{a}_n} \mathbf{y}^{\mathbf{a}_n} + c_{n,\mathbf{a}_n'} \mathbf{y}^{\mathbf{a}_n'} = 0 \end{cases} \qquad (7)$$

with κ_α nonsingular isolated solutions in $(\mathbb{C}^*)^n$. It is known that such binomial systems can be solved accurately and efficiently via numerical methods [13]. After (7) is solved, these nonsingular solutions obtained can be used as the starting points for following the homotopy paths $\mathbf{y}(t)$ of $H^\alpha(\mathbf{y}, t) = 0$, for which $H^\alpha(\mathbf{y}(t), t) = 0$ from $t = 0$ to $t = 1$. Note that the change of variables $\mathbf{x} = t^\alpha \bullet \mathbf{y}$ in (5) yields $\mathbf{x} \equiv \mathbf{y}$ at $t = 1$. Therefore, each end point $\mathbf{y}(1)$ at $t = 1$ of the homotopy path $\mathbf{y}(t)$ of $H^\alpha(\mathbf{y}, t) = 0$ is also an end point $\mathbf{x}(1)$ of the homotopy path $\mathbf{x}(t)$ of the homotopy $H(\mathbf{x}, t) = 0$ given in (2) which, in turn, provides a solution of the target system $P(\mathbf{x}) = 0$ in (1). Altogether it yields κ_α of the isolated solutions of $P(\mathbf{x}) = 0$ in $(\mathbb{C}^*)^n$ along this route. In [9], it was shown that as one follows the homotopy paths defined by $H^\alpha(\mathbf{y}, t) = 0$ for all individual $\alpha \in \mathcal{T}$, one obtains all (isolated) solutions of $P(\mathbf{x}) = 0$ in $(\mathbb{C}^*)^n$, justifying, in fact, the BKK bound agrees with $\sum_{\alpha \in \mathcal{T}} \kappa_\alpha$.

Even though the above procedure only addresses the solution set in $(\mathbb{C}^*)^n$ of the target system $P(\mathbf{x}) = 0$ in (1), this method has been extended in [15] so that all isolated zeros of the target system $P(\mathbf{x})$ in $\mathbb{C}^n$ can be obtained. Since its inception, this general method has achieved a great success. It is widely considered to be one of the most efficient, robust and reliable numerical methods for solving systems of polynomial equations. Hom4PS-3 implements this method as its primary tool.

References

1. Allgower, E., Georg, K.: Introduction to numerical continuation methods, vol. 45. Society for Industrial and Applied Mathematics (2003)
2. Attardi, G., Traverso, C.: The PoSSo library for polynomial system solving. In: Proc. of AIHENP 1995 (1995)
3. Bates, D.J., Hauenstein, J.D., Sommese, A.J., Wampler, C.W.: Numerically Solving Polynomial Systems with Bertini. Society for Industrial and Applied Mathematics (2013)

4. Bernshtein, D.N.: The number of roots of a system of equations. Functional Analysis and its Applications 9(3), 183–185 (1975)
5. Chen, T.R., Lee, T.L., Li, T.Y.: Hom4PS-3: an numerical solver for polynomial systems using homotopy continuation methods, http://www.hom4ps3.org
6. Drexler, F.-J.: Eine methode zur berechnung sämtlicher lösungen von polynomgleichungssystemen. Numerische Mathematik 29(1), 45–58 (1977)
7. Garcia, C.B., Zangwill, W.I.: Finding all solutions to polynomial systems and other systems of equations. Mathematical Programming 16(1), 159–176 (1979)
8. Gunji, T., Kim, S., Kojima, M., Takeda, A., Fujisawa, K., Mizutani, T.: PHoM– a polyhedral homotopy continuation method for polynomial systems. Computing 73(1), 57–77 (2004)
9. Huber, B., Sturmfels, B.: A polyhedral method for solving sparse polynomial systems. Mathematics of Computation 64(212), 1541–1555 (1995)
10. Lee, T.L., Li, T.Y., Tsai, C.H.: HOM4PS-2.0: a software package for solving polynomial systems by the polyhedral homotopy continuation method. Computing 83(2), 109–133 (2008)
11. Lee, T.L., Li, T.Y., Tsai, C.H.: HOM4PS-2.0 para: Parallelization of HOM4PS-2.0 for solving polynomial systems. Parallel Computing 35(4), 226–238 (2009)
12. Leykin, A.: NAG4M2: Numerical algebraic geometry for Macaulay2, http://people.math.gatech.edu/~aleykin3/NAG4M2/
13. Li, T.Y.: Numerical solution of polynomial systems by homotopy continuation methods. In: Ciarlet, P.G. (ed.) Handbook of Numerical Analysis, vol. 11, pp. 209–304. North-Holland (2003)
14. Li, T.Y., Sauer, T., Yorke, J.: The cheater's homotopy: an efficient procedure for solving systems of polynomial equations. SIAM Journal on Numerical Analysis, 1241–1251 (1989)
15. Li, T.Y., Wang, X.: The BKK root count in $\mathbb{C}^n$. Mathematics of Computation of the American Mathematical Society 65(216), 1477–1484 (1996)
16. Morgan, A.P.: Solving polynomial systems using continuation for engineering and scientific problems. Classics in Applied Mathematics, vol. 57. Society for Industrial and Applied Mathematics (2009)
17. Morgan, A.P., Sommese, A.J.: Coefficient-parameter polynomial continuation. Applied Mathematics and Computation 29(2), 123–160 (1989)
18. Sommese, A.J., Wampler, C.W.: The Numerical solution of systems of polynomials arising in engineering and science. World Scientific Pub. Co. Inc. (2005)
19. Verschelde, J.: Algorithm 795: PHCpack: A general-purpose solver for polynomial systems by homotopy continuation. ACM Transactions on Mathematical Software (TOMS) 25(2), 251–276 (1999)
20. Watson, L.T., Billups, S.C., Morgan, A.P.: Algorithm 652: Hompack: A suite of codes for globally convergent homotopy algorithms. ACM Transactions on Mathematical Software (TOMS) 13(3), 281–310 (1987)

CGAL – Reliable Geometric Computing for Academia and Industry

Eric Berberich

Max-Planck-Institut für Informatik, Germany
eric@mpi-inf.mpg.de
http://people.mpi-inf.mpg.de/~eric

Abstract. CGAL, the Computational Geometry Algorithms Library, provides easy access to efficient and reliable geometric algorithms. Since its first release in 1997, CGAL is reducing the the gap between theoretical algorithms and data structure and implementations that can be used in practical scenarios. CGAL's philosophy dictates correct results for any given input, even if intermediate round-off errors occur. This is achieved by its design, which separates numerical constructions and predicates from combinatorial algorithms and data structures. A naive implementation of the predicates and constructions still leads to wrong results, but reliable versions are shipped with the library.

CGAL is successful in academic prototypical development and widely spread among industrial users. It follows the design principles of C++'s Standard Template Library. CGAL, now available in version 4.4, is already quite comprehensive. Nevertheless, it is still growing and improving.

We first introduce the library, its design and basics and then present major packages, such as triangulations and arrangements. We also illustrate showcases of how CGAL is used for real world applications asking for reliable geometric computing. The second part covers recent additions and contributions to the project: We discuss periodic and hyperbolic triangulations. The arrangement package has seen improvements for point location, rational functions and multi-part curves. It has also been extended with support for algebraic curves. This relies on several new packages that enable operations on (multivariate) polynomials and topology computation of such curves. Geometric objects in higher dimensions can now be represented using combinatorial maps; the instance for linear objects, namely the linear cell complex, is of particular interest. CGAL also provides data structures and algorithms for geometric sorting in arbitrary dimensions. Finally we present CGAL's achievements to replace serial implementations with versions that support up-to-date multi-core architectures.

Keywords: CGAL, library, computational geometry, arrangements, curves, triangulations, high-dimensional geometry.

H. Hong and C. Yap (Eds.): ICMS 2014, LNCS 8592, pp. 191–197, 2014.
© Springer-Verlag Berlin Heidelberg 2014

1 Introduction

 is an open source software library for reliable and efficient geometric computing. It serves both academic and industrial users that demand such implementations from various domains such as computer graphics, scientific visualization, motion planning, geographic information systems, modeling, computer aided design, bioinformatics and many more.

The origin of CGAL dates back into the 90s, when several research groups joint efforts to combine several geometry libraries into one. Eventually, transforming the advanced theoretical results into robust software was set as goal and recommended by the Computational Geometry Impact Task Force Report [6].

CGAL is distributed under the GPL license (with few basic parts under the LGPL license). As such, it is immediately and freely available to academic users and other open source projects. Industrial users that are in favor of closed source development of their products can purchase commercial licenses from GEOMETRYFACTORY.[1]

CGAL is a C++ library whose design follows the generic programming paradigm in the style of the well-known *Standard Template Library* (STL) or BOOST. This way, it abstracts combinatorial layers from numerical and geometrical operations, and allows users to instantiate CGAL's algorithms and data structures with their own geometric objects – as long as they model the well-documented concepts. Of course, CGAL also ships default kernels that offer various geometric objects and hundreds of reliable predicates and constructions on them – if instantiated with robust software number types. Machine types will work, but are doomed – by design – to result in crashes, infinite loops or wrong results for many ill-conditioned inputs. CGAL, however, follows the *Exact Geometric Computing Paradigm*: As long as all basic operations are computed correctly, the overall result will be correct. The library comes with an extensive sets of examples and demos.

CGAL uses state-of-the-art development tools: A GIT repository for version control, CMAKE as build system, and DOXYGEN for the huge user and reference manual. Each package undergoes a daily function and regression testing to ensure the quality of the library and the support for various compilers on the three major platforms: Windows, Linux, Mac OS X. The overall quality assurance is steered by CGAL's *editorial board* via reviews of new additions to the library. The board is also responsible for technical decisions and the promotion of CGAL.

2 Packages

Besides the kernels that are central to the library, CGAL consists of many *packages* serving various goals and application areas. We are highlighting next the library's main packages and some that have recently been added or significantly extended:

---

[1] http://www.geometryfactory.com

2.1   Triangulations

CGAL's triangulations in 2D and 3D were among the first packages added to the library [17,14]. The packages cover methods to build and handle various triangulations for point sets in two and three dimensions. Each triangulation covers the convex hull of the input points. Triangulations are build incrementally and can be modified by insertion and deletion of (ranges of) vertices. Besides modifications, point location queries can be performed. Beyond *plain* triangulations, whose result depends on the insertion order, the packages provide *Delaunay* and *regular* triangulations, the latter also for weighted points. Delaunay and regular triangulations allow nearest neighbor queries and cover methods to build the dual Voronoi and power diagrams. In 2D, certain edges can be forced to appear in a triangulation, resulting in a *constrained* (Delaunay) triangulation.

The packages have constantly seen improvements, in particular the API has been adapted and performance boosts have been implemented. Beyond, the successful mesh generation packages of CGAL depend on its triangulations.

A major recent addition to CGAL have been *periodic* triangulations in 2D and 3D [12,5], which allow to build and handle triangulations of the points contained in the flat torus, that is, within a periodic domain. Other than that, these triangulations offer the same capabilities as their non-periodic counterparts: point location, nearest neighbor queries, dual Voronoi construction. Periodic triangulations have applications in astronomy, material engineering, crystallography, modeling of foams, structural biology and many more. Work in progress extends periodic triangulation to *periodic meshes.*

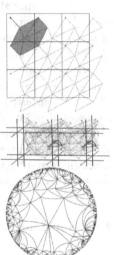

CGAL's triangulations also get enhanced with respect to the embedding space. Recent work [4] implements hyperbolic Delaunay complexes and Voronoi diagrams. This software for 2D is expected to be part of one of CGAL's next releases.

Similarly, a near release will contain triangulations in arbitrary dimensions. The new package will deal with pure manifold simplicial complexes which are connected and have no boundaries. Again, the convex hull of the points is part of the triangulation. As for the low-dimensional triangulations, insertions and point location queries will be supported – and in the case of d-dimensional Delaunay triangulations also deletions of vertices.

2.2   Arrangements

The arrangement package supports subdivisions of a two-dimensional space into maximal cells induced by a set of curves. Cells can be of dimension 0 (vertices), 1 (edges) and 2 (faces). The arrangement can be constructed with the well-known plane-sweep algorithm or in an incremental way using a zone-computation. The implementation uses generic programming techniques to separate combinatorial algorithms and the representation of the arrangement as DCEL (doubly-connected-edge-list) on one side from geometric predicates and constructions on the other side. For each task a certain set of such operations is required. Examples are decomposing a curve into x-monotone subpieces, lexicographic comparison of points, vertical alignment of x-monotone curves and the construction of the intersection of x-monotone curves. The package also supports to overlay two arrangements (while having the chance to combine data that is attached to cells), to perform (batched) point location queries and to compute Boolean set operations.

Many families of curves are shipped with CGAL; we list the non-linear ones in Section 2.3. In order to support a new family, the user only needs to provide a *traits class* that models the required concept. All details about CGAL's arrangements can be found in [9].

Recently the arrangement packages has seen several feature additions: Originally, only bounded arrangements were allowed, that is, there always must be a (large enough) rectangle that contains all input curves. This restriction has been removed. Arrangements of unbounded curves are supported, and all traits classes of unbounded curves have been adapted. Support for spherical arrangements is on the way, and we like to mention geodesic arcs on a sphere. This class of arrangement allows to compute power diagrams (using the `Envelope_3` package) of point sites on a sphere.

The random incremental point location using a trapezoidal decomposition has also been redesigned to support unbounded arrangements, and now features a heuristic to reduce the number of rebuilds [11]. Finally, the poly-line meta-traits has been redesigned to eventually support poly-curves.

2.3 Non-Linear Curves and the Algebraic Kernel

Slightly more than a decade ago it was considered an impossible challenge to reliably and efficiently compute with non-linear objects. The first break-through was to support arrangements of circular arcs [1] and conic segments [16] in CGAL. The next step comprised rational functions (that have recently seen a performance boost [15]) and Bézier curves [10].

The latest addition models the most general setting, as it subsumes all the previous cases: Arrangements of algebraic curves of arbitrary degree [8]. This implementation even outperforms the conic and Bézier implementation, as it relies less on explicit representation of low-degree algebraic numbers. Algebraic curves heavily rely on CGAL's newer packages for arithmetic and multivariate polynomials. Uni- and bivariate *algebraic kernels* form a middle layer that can be used independently to solve and work with uni- and bivariate polynomial systems [2,3].

2.4   Combinatorial Maps

CGAL now also offers data structures to represent objects in higher dimensions. *Combinatorial Maps* are an edge-centered structure to store orientable subdivided objects in *d*-dimensional space. It describes all cells of the subdivision and all the incidence and adjacency information between them. In 3D it consists of vertices, edges, faces, and volumes. To each cell, information can be added in terms of attributes. The package provides basic creation 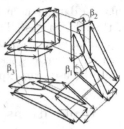 and modification operations, as well as iterators allowing to trace through some specific part of the object.

A particular instance of a combinatorial map is the *linear cell complex* that allows to represent an orientable subdivided *d*-dimensional object having linear geometry: Each vertex refers to a point, an edge stores a segment, the geometry of a 2-cell is obtained from the segments associated with the edges describing the boundary of the 2-cell and so on. A 2D combinatorial map that uses 3D points results in a linear cell complex that is equivalent to a polyhedron in 3D. That is, the dimension of the combinatorial map is not required to match the dimension of the ambient space.

2.5   Spatial Sorting

Many geometric algorithms are of incremental nature and thus, their performance can depend on the order the input is given in. The spatial sorting package [7] allows to pre-sort the input along a *space-filling curve* such that geometrically close objects are sorted close to each other. This should improve overall performance by having less cache misses and better memory locality. While the first implementation only allowed sorting in 2D and 3D, CGAL 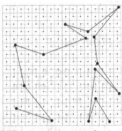 now supports spatial sorting in arbitrary dimensions and Hilbert sorting admits two policies: splitting at the median or splitting at the middle.

2.6   Supporting Multi-core Architectures

Modern hardware architectures feature multiple cores and shared RAM. Exploiting these resources is a desired goal for most software projects. Supporting multi-threaded software is a non-trivial task, in particular for geometric algorithms where often parallel access to underlying data destroys the combinatorial invariants. However, CGAL has recently evaluated guidelines to encourage concurrent implementations, and recommends to rely on Intel's *Threading Building Blocks* (TBB).[2] This library supports concurrency that follows the generic programming paradigm and thus is perfectly suited for use within CGAL.

---

[2] https://www.threadingbuildingblocks.org

The upcoming release 4.5 of CGAL will feature first concurrency-enabled packages: the ability to create and delete vertices and cells in parallel to the triangulation data structure [13] has been added. The user can control the number of threads. In case TBB is not available, CGAL automatically falls back to a sequential implementation. The experiments have shown that the insertion and deletion of vertices scales linearly with the number of cores available: For eight cores, the user can expect a factor 6 speedup.

3 Applications

CGAL is widely used in academic and industrial projects. An (incomplete) list of industrial users is available online[3] which comprises companies from various areas such as geomodeling, geographical information systems, computer aided design, image processing, telecommunication, and many more. GEOMETRYFACTORY also lists details to projects run by professional users.[4] About 100 projects form various research institutes are listed on CGAL's web-site[5] – they range from architecture, astronomy via shape matching and medical imaging to sensor networks and particle physics. CGAL's triangulation package has also been integrated in Matlab. C++-bindings into other programming language, such as Java and Python, exist and are constantly extended.

Acknowledgments. Figures taken from CGAL's manual available online: http://doc.cgal.org

References

1. de Castro, P.M.M., Pion, S., Teillaud, M.: 2D circular geometry kernel. In: CGAL User and Reference Manual, 4.4 edn., CGAL Editorial Board (2014)
2. Berberich, E., Hemmer, M., Kerber, M.: A generic algebraic kernel for non-linear geometric applications. In: Hurtado, F., van Kreveld, M.J. (eds.) Symposium on Computational Geometry, pp. 179–186. ACM (2011)
3. Berberich, E., Hemmer, M., Kerber, M., Lazard, S., Peñaranda, L., Teillaud, M.: Algebraic kernel. In: CGAL User and Reference Manual, 4.4 edn., CGAL Editorial Board (2014)
4. Bogdanov, M., Devillers, O., Teillaud, M.: Hyperbolic delaunay complexes and voronoi diagrams made practical. In: da Fonseca, G.D., Lewiner, T., Peñaranda, L.M., Chan, T., Klein, R. (eds.) Symposium on Computational Geometry, pp. 67–76. ACM (2013)
5. Caroli, M., Teillaud, M.: 3D periodic triangulations. In: CGAL User and Reference Manual, 4.4 edn., CGAL Editorial Board (2014)
6. Chazelle, B.: The computational geometry impact task force report: An executive summary. In: Lin, M.C., Manocha, D. (eds.) FCRC-WS 1996 and WACG 1996. LNCS, vol. 1148, pp. 59–65. Springer, Heidelberg (1996)

---

[3] http://www.geometryfactory.com/markets-and-customers/
[4] http://www.geometryfactory.com/portfolio/
[5] http://www.cgal.org/projects.html

7. Delage, C., Devillers, O.: Spatial sorting. In: CGAL User and Reference Manual, 4.4 edn., CGAL Editorial Board (2014)
8. Eigenwillig, A., Kerber, M.: Exact and efficient 2d-arrangements of arbitrary algebraic curves. In: Teng, S.-H. (ed.) SODA, pp. 122–131. SIAM (2008)
9. Fogel, E., Halperin, D., Wein, R.: CGAL Arrangements and Their Applications - A Step-by-Step Guide. Geometry and computing, vol. 7. Springer (2012)
10. Hanniel, I., Wein, R.: An exact, complete and efficient computation of arrangements of Bézier curves. IEEE T. Automation Science and Engineering 6(3), 399–408 (2009)
11. Hemmer, M., Kleinbort, M., Halperin, D.: Improved implementation of point location in general two-dimensional subdivisions. In: Epstein, L., Ferragina, P. (eds.) ESA 2012. LNCS, vol. 7501, pp. 611–623. Springer, Heidelberg (2012)
12. Kruithof, N.: 2D periodic triangulations. In: CGAL User and Reference Manual, 4.4 edn., CGAL Editorial Board (2014)
13. Pion, S., Teillaud, M.: 3D triangulation data structure. In: CGAL User and Reference Manual, 4.4 edn., CGAL Editorial Board (2014)
14. Pion, S., Teillaud, M.: 3D triangulations. In: CGAL User and Reference Manual, 4.4 edn., CGAL Editorial Board (2014)
15. Salzman, O., Hemmer, M., Raveh, B., Halperin, D.: Motion planning via manifold samples. Algorithmica 67(4), 547–565 (2013)
16. Wein, R.: High-level filtering for arrangements of conic arcs. In: Möhring, R.H., Raman, R. (eds.) ESA 2002. LNCS, vol. 2461, pp. 884–895. Springer, Heidelberg (2002)
17. Yvinec, M.: 2D triangulations. In: CGAL User and Reference Manual, 4.4 edn., CGAL Editorial Board (2014)

Implementing the L_∞ Segment Voronoi Diagram in CGAL and Applying in VLSI Pattern Analysis

Panagiotis Cheilaris[1], Sandeep Kumar Dey[1], Maria Gabrani[2], and Evanthia Papadopoulou[1*]

[1] Faculty of Informatics, Università della Svizzera Italiana, Switzerland
[2] IBM Zurich Research Laboratory, Switzerland

Abstract. In this work we present a CGAL (Computational Geometry Algorithm Library) implementation of the line segment Voronoi diagram under the L_∞ metric, building on top of the existing line segment Voronoi diagram under the Euclidean (L_2) metric in CGAL. CGAL is an open-source collection of geometric algorithms implemented in C++, used in both academia and industry. We also discuss an application of the L_∞ segment Voronoi diagram in the area of VLSI pattern analysis. In particular, we identify potentially critical locations in VLSI design patterns, where a pattern, when printed, may differ substantially from the original intended VLSI design, improving on existing methods.

1 Introduction

Let S be a set of n sites in the plane (simple geometric shapes, such as points, line segments, or circular arcs). The (nearest-neighbor) *Voronoi diagram* [3] of S is a subdivision of the plane into regions such that the region of a site $s \in S$ is the locus of points closer to s than to any other site in S. The *distance* of a site s from a point q in the plane is defined as $d(s, q) = \min_{p \in s} d(p, q)$, where the the interpoint distance $d(p, q)$ can be the Euclidean (L_2) distance or any other metric. In this paper, we focus on the Voronoi diagram of line segments in the plane, under the L_∞ metric (or maximum norm): $d(p, q) = d_\infty(p, q) = \max(|p_x - q_x|, |p_y - q_y|)$. Fig. 1a shows in red the Voronoi diagram of the same set of segments under the L_2 and the L_∞ metric (interiors of segments and their endpoints are considered different sites).

The Voronoi diagram of segments under the L_∞ metric has some nice properties, compared to the corresponding L_2 diagram: (1) The L_∞ diagram consists solely of *straight line segments* [9], whereas the L_2 diagram contains also *parabolic arcs*. (2) If the coordinates of the endpoints of the input segments (sites) are rational, then the L_∞ Voronoi vertices are also on rational coordinates. (3) The *degree* of an algorithm [7] is a complexity measure capturing its potential for

---

* Supported in part by the Swiss National Science Foundation grant 134355, under the auspices of the ESF EUROCORES program EuroGIGA/VORONOI.

H. Hong and C. Yap (Eds.): ICMS 2014, LNCS 8592, pp. 198–205, 2014.

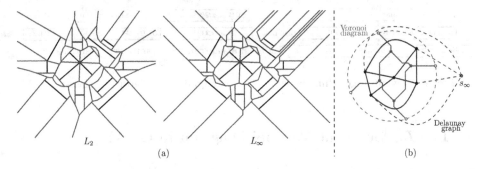

Fig. 1. (a) Segment Voronoi diagram under L_2 and L_∞ metric, with distinct sites for interiors of segments and their endpoints, (b) L_∞ Voronoi diagram and Delaunay graph of five point sites (black) with additional site s_∞ at infinity (infinite edges: dashed; Voronoi vertices: red, finite ones filled and infinite ones unfilled)

robust implementation. An algorithm has degree d if its test computations involve the evaluation of multivariate polynomials of arithmetic degree at most d. The degree captures the precision to which arithmetic calculations need to be executed in the exact computation paradigm [11], for a robust implementation of the algorithm. A crucial predicate for a Voronoi algorithm is the *in-circle* test, which checks whether a new input segment is altering or erasing an existing vertex of the diagram. The L_2 in-circle test for arbitrary segments can be implemented with degree 40 [4], whereas the corresponding L_∞ test only with degree 5 [9]. (4) The Voronoi diagram in L_∞ coincides with *straight skeleton* [2] when the input consists of axis-parallel segments, which is predominant in VLSI designs.

Segment Voronoi diagrams encode proximity information between polygonal objects. In many applications proximity is most naturally expressed with the L_2 distance, but there are applications, particularly in VLSI CAD, for which the L_∞ distance is a good and simpler alternative, see, e.g., [9,8] and references therein. Thus, a robust implementation of the L_∞ segment Voronoi diagram is desirable, but, as far as we know, there is none freely available (except the proprietary [10]). Instead of building such software from scratch, we decided to develop it in the CGAL framework, on top of the existing L_2 segment Voronoi diagram of CGAL [6]. CGAL is an open-source collection of a wide range of geometric algorithms implemented in C++. CGAL is built in a modular way and there is provision for code reuse. We exploit this provision by using a significant part of the L_2 segment Voronoi diagram incremental construction code in CGAL [6].

We also use our code for the L_∞ diagram for an application in VLSI pattern analysis. With the increase in miniaturization of current VLSI patterns, there is a significant rise in the pattern printability problems. We discuss the potential of using the L_∞ segment Voronoi diagram to identify critical locations in a VLSI pattern.

Fig. 2. Previous algorithm classes   **Fig. 3.** New algorithm classes

2   The L_∞ Segment Voronoi Diagram in CGAL

The *2D segment Delaunay graph package* of CGAL provides a randomized incremental construction of the L_2 *segment Delaunay graph* [6] and its *dual* graph, the segment Voronoi diagram. It is typical to always include an additional site s_∞ at infinity, as it simplifies the construction algorithms (see Fig. 1b). The SDG L_2 package contains two algorithm template classes (see Fig. 2) to construct the SDG: (1) The *segment Delaunay graph* class Segment_Delaunay_graph_2 (abbreviation: SDGL2), which is derived from a CGAL triangulation class. Among other things, it contains the functionality to maintain and update the arrangement of the input sites. It also contains functions to construct duals of edges of the SDG, i.e., edges of the Voronoi diagram. (2) The *segment Delaunay (graph) hierarchy* class Segment_Delaunay_graph_hierarchy_2 (abbreviation: SDH) is derived from the SDG class. It builds a hierarchy of SDGs and uses it to achieve faster insertion of a new site in the segment Delaunay graph. This is an implementation with better worst-case complexity than the SDG class (for details, see [5,6]). Both template classes have a mandatory template argument (denoted by GT in Fig. 2), that must be instantiated with one of four geometric traits classes with geometric predicates related to the L_2 diagram (like the in-circle test).

We implement the segment Delaunay graph under the L_∞ metric (SDG L_∞) in CGAL, trying to reuse as much code as possible from the SDG L_2 package of CGAL. Ideally, we would like the situation to be as follows: We only write a geometric traits class containing the L_∞-related predicates (and constructions) and supply it as the GT template argument of the SDG algorithm template classes of Fig. 2. In any case, the most significant part of the algorithm, like the maintenance of the arrangement of input sites and the high-level incremental construction of the Delaunay graph is the same under both the L_2 and the L_∞ metric. Unfortunately, since the SDG L_2 algorithm classes were not designed with provision for other metrics except L_2, there is some L_2-specific code in them, the most significant being the code for drawing dual edges for the Voronoi diagram. Fortunately, these hard-coded L_2-specific functions in the algorithms are few; most of the functionality is indeed in the L_2 geometric traits class.

We make the following design decisions related to the SDG L_2 implementation: (1) Keep the same interface for users of the SDG L_2 package. (2) Change SDG L_2 code as little as possible. (3) Preserve the efficiency of the SDG L_2 algorithms. Therefore, we implement a few local changes in the L_2 code (mostly in the SDGL2 class) that allow us to implement Segment_Delaunay_graph_Linf_2 (abbreviation: SDGLinf) as a class derived from SDGL2 (see Fig. 3).

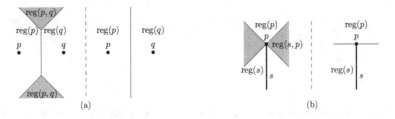

Fig. 4. L_∞ bisector and its dimensionalization: (a) between two points with the same y coordinate and (b) between a vertical segment and one of its endpoints

Since the existing hierarchy class SDH is hard-coded to use only instances of SDGL2 at its levels, we alter SDH so that it has an additional optional template parameter SDGLx (with default value SDGL2), which is the segment Delaunay graph class that is used in every level of the hierarchy (and from which SDH is derived). In Fig. 3, the altered classes SDGL2 and SDH are shown with gray, together with the new class SDGLinf. Since SDGLx is an optional parameter with default value SDGL2, there is no change for old user code of the L_2 segment Delaunay hierarchy. By setting SDGLx to SDGLinf in the SDH template, we obtain the segment Delaunay hierarchy under the L_∞ metric, for which we also create an alias template class Segment_Delaunay_graph_hierarchy_Linf_2 (abbreviation: SDHLinf) for easy access to the user (see Fig. 3).

A user of the SDG L_∞ package must instantiate the above two algorithm classes with one of the four L_∞ geometric traits classes, which are analogous to the corresponding L_2 geometric traits classes. Apart from the classes, we also provide a GUI demo, examples, and an ipelet for the L_∞ segment Voronoi diagram. Our package is currently under review for inclusion in the CGAL library.

One important difference in the L_∞ setting (in comparison to the L_2 setting) is that in some special non-general position cases the L_∞ bisector between two sites can be 2-dimensional. We resort to a 1-*dimensionalization* of these bisectors, by assigning portions of 2-dimensional regions of a bisector to the two sites of the bisector. This way it is also easier to draw the Voronoi diagram. We remark that this simplification of the diagram is acceptable in the VLSI applications, where the L_∞ diagram is employed [9]. We show examples of 2-dimensional bisectors and their 1-dimensionalization in Fig. 4.

The L_∞ parabola is the geometric locus of points equidistant under the L_∞ distance from a line ℓ (the directrix) and a (focus) point $p \notin \ell$. In contrast with the standard L_2 parabola, the L_∞ parabola consists of a constant number of linear segments and rays [9]. Only bounded parabolic arcs appear as edges of the L_2 segment Voronoi diagram and never a complete parabola. On the other hand, unbounded L_∞ parabolic arcs can survive in the corresponding L_∞ diagram (see Fig. 5). The existing SDGL2 code is not ready to support the peculiarities of the L_∞ parabolas. For example, the Voronoi region of any segment is expected to have 0, 1, or 2 infinite edges (these are edges with the infinite site s_∞). While this is true in the L_2 setting, it is not true in the L_∞ setting, where the

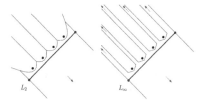

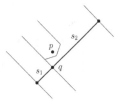

Fig. 5. Only bounded parabolic arcs survive in the L_2 diagram, whereas even complete L_∞ parabolas can survive in the L_∞ diagram. Arrows point to distinct infinite edges of the diagrams.

Fig. 6. The bisector through q touches the parabolic arc at the parabolic arc's portion which is parallel to this bisector

aforementioned number of infinite edges is unbounded in general. For example, in the L_∞ diagram of Fig. 5, there are six distinct infinite edges neighboring with the region of the open segment.

Several problems may occur when a new point site q is inserted in the interior of an existing segment s. This operation is needed when, for example, a newly inserted segment crosses an existing segment. The algorithm checks the neighbors of s in the segment Delaunay graph, splits the site of s to two sites s_1 and s_2 and adds the site q to the diagram. In the L_∞ setting this has to be done more carefully than in the L_2 setting. For example, when the site q shares a coordinate with a point p for which there is an L_∞ parabolic arc in the diagram, we have to be careful, because the bisector that passes through q might touch a portion of the L_∞ parabolic arc that is parallel to this bisector (see Fig. 6). Our solution is to derive SDGLinf from SDGL2 and override some SDGL2 member functions in SDGLinf, in particular the ones that insert a point in the interior of a segment.

In the old L_2 code, when there are two points in the diagram and a third one is inserted, the resulting Delaunay graph construction is based on the *orientation* test for three points p, q, r (i.e., whether the three points make a left, a right turn, or they are collinear), which is very specific to the L_2 case. To make the code work for both L_2 and L_∞, we substitute the orientation test with a call to an in-circle predicate from the corresponding L_2 or L_∞ geometric traits class.

Like in L_2, the L_∞ traits contain predicates resolving whether a new site conflicts with an existing Voronoi vertex (vertex conflict) or an existing edge (edge conflict). There are also special conflict-like predicates used when a new point site is inserted in the interior of a segment. The vertex conflict predicates are also known as in-circle tests. The in-circle test in L_2 is analog to an "in-square" test in L_∞. A new site is tested for containment in the minimum shape (circle or axis-parallel square) that touches the sites associated with an existing Voronoi vertex. In L_2, the circle that touches three non-collinear points is unique and its center corresponds to the Voronoi vertex. In L_∞, however, the analog axis-parallel square might not be unique (see Fig. 7). Again our 1-dimensionalization comes to the rescue, since we can define the Voronoi vertex to be the intersection of L_∞ bisectors of these three points and then the square becomes unique.

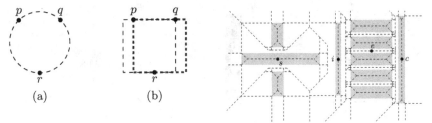

Fig. 7. p, q, and r define a unique circle (a),
but not a unique axis-parallel square (b)

Fig. 8. Different gauge types

3   Application in VLSI Pattern Analysis

As VLSI patterns keep on shrinking in size, their error-free printing is increasingly challenging the chip manufacturing industry. There are two kind of faults that occur during printing: (1) a *pinch*, which corresponds to an open fault and occurs due to incomplete printing or due to discontinuity in the printing of a shape; (2) a *bridge*, which corresponds to a short fault and occurs when two printed shapes are touching each other. The identification of fault prone patterns, also known as *patterns of interest* (POI), in a complete layout is a difficult and very time-consuming task. The actual location of faults within a POI is known as a *hotspot*. The measurement location in each pattern is called a *gauge*, which is generally represented by a line. Current gauge suggestion techniques are rule-based or they are done manually by VLSI designers. The suggested gauges very often miss the location of critical distance or hotspots on the clip.

We propose to use the L_∞ segment Voronoi diagram to suggest good gauge locations based on the proximity information of the shapes of a pattern. We suggest four types of gauges (see Fig. 8): (1) *Internal gauge* G_i (inside a shape): It lies on the center of the Voronoi edge inside the polygonal shape of minimum width in the pattern. The position of G_i is the most probable for a pinch. (2) *External gauge* G_e (between different neighboring shapes): It lies on the center of the Voronoi edge between the two shapes that are closest in the pattern. The position of G_e is the most probable for the formation of a bridge between the two corresponding shapes. (3) *Sandwich gauge,* G_s: It lies on the center of the Voronoi edge inside a polygonal shape P_1 that is "sandwiched" between two other shapes P_2 and P_3 for which the distance between P_2 and P_3 is the minimum in the pattern. There is a probability of a pinch happening at P_1 around G_s because of the influence of P_2 and P_3. (4) *Comb gauge,* G_c: It lies on the center of the Voronoi edge inside a long polygonal shape P_1 (the base of the comb) that has close to it and on one side other polygonal shapes (the teeth of the comb). We report the gauge for the configuration where the base of the comb shape is closer to the teeth in the pattern. The position of G_c is dangerous for a pinch, when printing the pattern.

We have performed experiments on patterns provided by IBM Zurich Research Laboratory, in order to assess the quality of our gauges. These patterns were hand-picked by engineers at IBM and for most of them the previous gauge

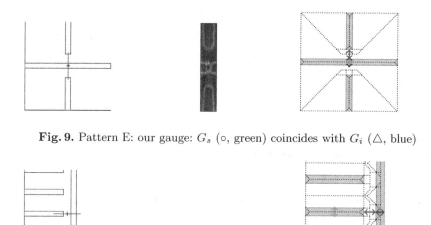

Fig. 9. Pattern E: our gauge: G_s (○, green) coincides with G_i (△, blue)

Fig. 10. Pattern H: our gauge: G_c (◇, purple) detects a pinch, improves on old gauge

suggestion missed the critical location in each pattern. Each pattern is a representative of a wide set of patterns with similar behavior. We run experiments on ten such patterns: A, B, C, D, E, F, G, H, I, J. Due to lack of space, we only show patterns E (Fig. 9) and H (Fig 10) as follows: (a) in the left, the pattern and the old gauge suggestion; (b) in the center, the SEM image around the old suggested gauge and the location of the critical distance measurement with a cyan arrow; (c) in the right, the pattern together with the gauge suggestions provided by our tool based on the L_∞ segment Voronoi diagram. Figures with explanation for all patterns and detailed analysis can be found in http://compgeom.inf.usi.ch/papers/linfcgal.pdf. For each pattern we measure the distance in pixels in the corresponding SEM image for each of our suggested gauges and we take the minimum. We show the comparison with the old measurements in Table 1. Our experiment shows that the L_∞ segment Voronoi diagram can be used effectively to identify potentially critical locations of VLSI layouts.

Future work. A *model-based OPC* (MB-OPC) simulator [1] is essential for advanced lithography processes. Optimization for MB-OPC requires selection of optimal test patterns that would cover the whole layout. We intend to use the L_∞ segment Voronoi diagram for obtaining a good set of patterns for a VLSI layout that would in turn enhance MB-OPC. We would also like to analyze bigger layout clips on the order of tens of thousands of shapes. The main advantage of our proposed methodology is that it provides gauges that are pattern dependent, include context information, and at the same time are orders of magnitude faster to compute.

Table 1. Comparison of CD measurement at different gauge locations

Pattern	old measurement	type of our gauge	our measurement	improvement
A	17	internal	17	(same) 0%
B	58	external	10	83%
C	52	internal	18	65%
D	31	external	21	32%
E	16	sandwich	16	(same) 0%
F	27	external	18	33%
G	86	external	13	85%
H	35	comb	(pinch) 0	100%
I	47	external	10	79%
J	28	external	8	71%

Acknowledgment. We thank Menelaos Karavelas for providing us insight into the L_2 segment Voronoi diagram implementation in CGAL.

References

1. Abdo, A., Viswanathan, R.: The feasibility of using image parameters for test pattern selection during OPC model calibration. In: Proc. SPIE, vol. 7640, p. 76401E (2010)
2. Aichholzer, O., Aurenhammer, F.: Straight skeletons for general polygonal figures in the plane. In: Cai, J.-Y., Wong, C.K. (eds.) COCOON 1996. LNCS, vol. 1090, pp. 117–126. Springer, Heidelberg (1996)
3. Aurenhammer, F., Klein, R., Lee, D.T.: Voronoi Diagrams and Delaunay Triangulations. World Scientific Publishing Company, Singapore (2013)
4. Burnikel, C., Mehlhorn, K., Schirra, S.: How to compute the Voronoi diagram of line segments: Theoretical and experimental results. In: van Leeuwen, J. (ed.) ESA 1994. LNCS, vol. 855, pp. 227–239. Springer, Heidelberg (1994)
5. Devillers, O.: The Delaunay hierarchy. International Journal of Foundations of Computer Science 13, 163–180 (2002)
6. Karavelas, M.: A robust and efficient implementation for the segment Voronoi diagram. In: International Symposium on Voronoi Diagrams in Science and Engineering, pp. 51–62 (2004)
7. Liotta, G., Preparata, F.P., Tamassia, R.: Robust proximity queries: An illustration of degree-driven algorithm design. SIAM Journal on Computing 28(3), 864–889 (1998)
8. Papadopoulou, E.: Net-aware critical area extraction for opens in VLSI circuits via higher-order Voronoi diagrams. IEEE Transactions on Computer-Aided Design of Integrated Circuits and Systems 30(5), 704–716 (2011)
9. Papadopoulou, E., Lee, D.T.: The L_∞ Voronoi diagram of segments and VLSI applications. International Journal of Computational Geometry and Application 11(5), 502–528 (2001)
10. Voronoi CAA: Voronoi Critical Area Analysis. IBM CAD Tool, Department of Electronic Design Automation, IBM Microelectronics Division, Burlington, VT, initial patents: US6178539, US6317859
11. Yap, C., Dubé, T.: The exact computation paradigm. In: Computing in Euclidean Geometry, pp. 452–492 (1994)

BULL! - The Molecular Geometry Engine Based on Voronoi Diagram, Quasi-Triangulation, and Beta-Complex

Deok-Soo Kim[1,2,*], Youngsong Cho[2], Jae-Kwan Kim[2], Joonghyun Ryu[2], Mokwon Lee[1], Jehyun Cha[1], and Chanyoung Song[1]

[1] Department of Mechanical Engineering, Hanyang University, Seoul, Korea
dskim@hanyang.ac.kr
http://voronoi.hanyang.ac.kr
[2] Voronoi Diagram Research Center, Hanyang University, Seoul, Korea
{ycho,jkkim,jhryu,mwlee,jhcha,cysong}@voronoi.hanyang.ac.kr

Abstract. Libraries are available for the power diagram and the ordinary Voronoi diagram of points upon which application programs can be easily built. However, its counterpart for the Voronoi diagram of spheres does not exist despite of enormous applications, particularly those in molecular worlds. In this paper, we present the BULL! library which abbreviates "Beta Universe Library Liberandam!" for computing the Voronoi diagram of spheres, transforming it to the quasi-triangulation, and extracting the beta-complex. Being an engine library implemented in the standard C++, application programmers can simply call API-functions of BULL! to build application programs correctly, efficiently, and easily. The BULL! engine is designed so that the application programs developed by embedding API-functions are completely independent of the future modifications of the engine.

Keywords: application program interface, engine, molecular structure, computational geometry, geometric modeling, C++.

1 Introduction

Molecular structure determines molecular function. While the meaning of "structure" varies in molecular worlds, "geometry" is always central to molecular structure and there exist rich prior studies on the geometry of molecules. However, the studies were mostly conducted in ad hoc manner depending on discipline or even depending on a particular aspect of a problem at hand; There have been no unified framework of theory to deal with the geometry of molecular structure.

Authors' group at the Voronoi Diagram Research Center (VDRC) [1], Hanyang University, has been developing the **Molecular Geometry (MG)** theory during the past decade based on the Voronoi diagram, in particular the Voronoi diagram of spheres, and its derivative structures [6,11,4,8,7,10]. Suppose that P

---

* Corresponding author.

H. Hong and C. Yap (Eds.): ICMS 2014, LNCS 8592, pp. 206–213, 2014.

is a molecular structure problem of interest at hand. Let S(P) be its solution to be found. In the MG paradigm, the problem P is first transformed to a corresponding geometry problem G of three-dimensional spheres whose solution S(G) can be easily found via a geometry engine. Then, S(G) is inverse-transformed to get $S^{-1}(G)$ which is expected to be close to S(P). Assuming a geometric engine for transforming G to S(G), $S^{-1}(G)$ can converge to S(P) with a sufficient number of iterations if the forward- and backward-transformations are well-defined.

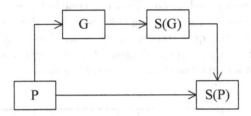

Fig. 1. The **Molecular Geometry** paradigm for solving all geometry problems in molecular worlds

BULL! is the engine for the MG paradigm based on the Voronoi diagram of spherical atoms and its two derivative structures: the quasi-triangulation and the beta-complex. This paper is the initial proposal of the BULL! library which abbreviates "Beta Universe Library Liberandam!" meaning that the library liberates researchers who are working on molecular structure from the hard and tedious job of developing accurate and efficient geometric algorithms and their implementation. The BULL! library is implemented in the standard C++ language and will be freely available from VDRC (http://voronoi.hanyang.ac.kr).

2   Three Fundamental Constructs in BULL!

There are three fundamental computational constructs in the MG theory: the primal, the dual, and the interested subset of the dual which correspond to the Voronoi diagram, the quasi-triangulation, and the beta-complex in BULL!, respectively. Any, perhaps all, geometry problems in molecular worlds can be correctly, efficiently, and easily solved using either one of these constructs or their combinations.

Among various types of Voronoi diagram, the Voronoi diagram of spheres is the key construct. The Voronoi diagram of spheres is the generalization of the power diagram [2] which is a generalization of the ordinary Voronoi diagram of points from the Euclidean distance point of view [3,13]. By the same token, in the dual space, the quasi-triangulation [9,7] is the generalization of the regular triangulation and the Delaunay triangulation and the beta-complex [8] is the generalization of the (weighted) alpha-complex [5]. Note that the Voronoi diagram of spheres is also called the additively-weighted Voronoi diagram. The quasi-, regular, and the Delaunay triangulations are the dual of the Voronoi diagram of spheres, the power diagram, and the ordinary Voronoi diagram of points, respectively.

The Voronoi diagram $\mathcal{VD}$ of three-dimensional spheres can be computed by the edge-tracing algorithm taking $O(n^3)$ time in the worst case but $O(n)$ time on average for molecules [6]. The quasi-triangulation $\mathcal{QT}$ is obtained by transforming $\mathcal{VD}$ in $O(n)$ time in the worst case. Then, the beta-complex is extracted from $\mathcal{QT}$ using a binary search in $O(n \log n + k)$ time in the worst case where k is the number of simplexes in the resulting beta-complex. More powerful approach to general queries on the quasi-triangulation is available [10]. Fig. 2 summarizes the process from the Voronoi diagram to the quasi-triangulation to the beta-complex, given an input of the arrangement of atoms.

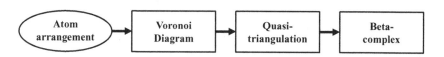

Fig. 2. An atom arrangement is given as an input. The Voronoi diagram is first computed and is transformed to the quasi-triangulation which is then used to extract the beta-complex.

Fig. 3 shows an example of this process in the plane. Fig. 3(a), (b), and (c) show the Voronoi diagram of circular disks, the quasi-triangulation, and the beta-complexes corresponding to the probe of a certain radius. The beta-complex defines the neighborhood information between atom pairs within the boundary of the disk set defined by the probe whereas the quasi-triangulation defines that for all disks. The figures are all created using the `BetaConcept` program [12] freely available at VDRC.

3   Data Structures in BULL!

Data structure is one of the key issues in BULL!. *Radial-edge data structure* (REDS) stores the topology of the Voronoi diagram because Voronoi diagrams have a cell structure. Fig. 4 shows the schematic diagram of the REDS used in BULL!. In the following, "V-" denotes "Voronoi." Each V-cell of the Voronoi diagram has a direct pointer to each of its V-faces and thus has $|f|$ V-face pointers where $|f|$ is the number of V-faces of the V-cell. Each V-face has two pointers to the incident V-cells. Each V-face has a pointer to each one of its bounding loops and each loop points to the V-face that it lies on. A V-face has one or more loops where the first one is external and the others are for interior holes. Thus a V-face has $|l|$ V-loop pointers. Each loop has a pointer to the V-face that it belongs to. Each loop points to one of the partial edges that belongs to the loop

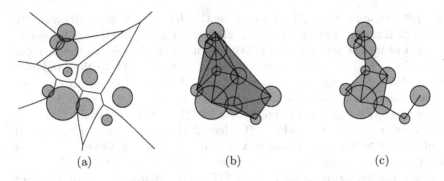

Fig. 3. (a) The Voronoi diagram of disks, (b) the quasi-triangulation, and (c) the beta-complexes corresponding to a circular probe. Figures created using the `BetaConcept` program [12].

and each partial edge points to the loop that it belongs to. Each partial edge has two more types of pointers for the two types of cycles on a V-face: one for the *radial cycle* which consists of a single pointer; the other for the *loop cycle* which consists of two pointers for its predecessor and successor. Each V-edge has two pointers to its V-vertices and each V-vertex has four pointers where each points to a V-edge incident to it.

V-cell, V-face, V-edge, and V-vertex may be associated with corresponding geometry. Each V-cell has the coordinates and radius of the spherical atom generator corresponding to itself and each V-vertex has its coordinate data. Each V-face and V-edge may or may not have its surface and curve equation, respectively, depending on application. If the geometry part of V-face and V-edge is not explicitly stored, they can be easily computed if the topology of the Voronoi diagram is available.

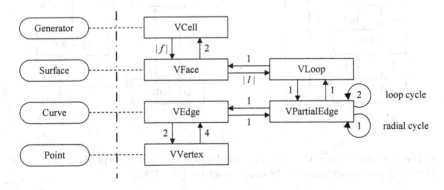

Fig. 4. Radial edge data structure (REDS) for Voronoi diagram of spherical atoms

The quasi-triangulation $\mathcal{QT}$ is stored in the *Inter-world data structure* [9] (IWDS) which is schematically shown in Fig. 5(a). Each QT-cell has four pointers to each of its QT-faces and each QT-face has two pointers to the incident QT-cells. Each QT-cell has also four pointers to each of its QT-vertices and each QT-vertex has a pointer to the incident QT-cells. Each QT-face has three pointers to each one of its QT-edges and each QT-edge has $|w|$ pointers to the incident QT-faces. Each QT-edge has two pointers to its QT-vertices and each QT-vertex has a pointer to one of the incident QT-edges. If there are a fixed number m of incident QT-simplexes to a QT-simplex σ, σ has m pointers. Otherwise (i.e., if there are an arbitrary number of QT-simplexes incident to a QT-simplex σ), σ has only one pointer to one of the incident QT-simplexes from which all the other QT-simplexes can be traversed. The number $|w|$ is to connect small-worlds. The explicit representation of QT-faces and QT-edges are necessary for the extraction of beta-complexes from the quasi-triangulation. Otherwise, if it is not necessary to store QT-faces and QT-edges explicitly, then a more compact data structure can be devised as shown in Fig. 5(b) which is now we call a compact IWDS, abbreviated cIWDS. Note that these two data structures were called differently in our earlier papers.

It is important for us to state the following: BULL! uses REDS to store all the three types of Voronoi diagrams in 3D: The Voronoi diagram of spheres, the power diagram, and the ordinary Voronoi diagram of points. This is possible because REDS can store the most general one: the Voronoi diagram of spheres. By the same token, BULL! uses IWDS to store all the three types of triangulations in a compact form: the quasi-, the regular, and Delaunay triangulations. Note that this observation is critically used in the design of the classes in BULL!.

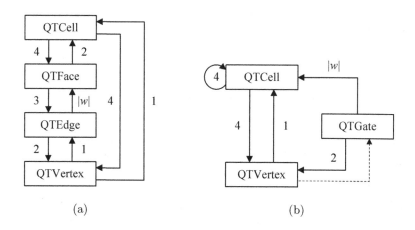

(a)   (b)

Fig. 5. Data structure for the quasi-triangulation. (a) Inter-world data structure (IWDS) and (b) compact inter-world data structure (cIWDS).

4 Architecture of BULL! and an Example Application Program

The architecture of BULL! is designed so that programmers can create application programs easily and conveniently through out entire software life cycle. To achieve this goal, we have designed BULL! so that application program is completely separated from the internal functions which may be modified as the development of BULL! goes on.

BULL! has a three-tier architecture as shown in Fig. 6: **API-tier**, **Core-tier**, and **Geometry-tier**. The API-tier is only visible to and the other two are completely hidden from application programmers. Thus, application program interacts only with the API-tier by including the related head files and embedding the API-functions. The API-tier interacts with Core-tier which implements the application neutral data structure of the primal and dual structures. The Core-tier interacts with the Geometry-tier which actually contains the Voronoi diagram construction codes, possibly implementing more than one algorithms for the Voronoi diagram construction. Currently, the Geometry-tier contains the implementations of the edge-tracing algorithm and the region-expansion algorithm for three-dimensional spheres. In this architecture, the modifications to be made in the Geometry-tier in future does not cause any change in the codes of already-existing application programs.

In principle, the API-tier currently contains three main classes: `AtomSetVoronoiDiagram`, `QuasiTriangulation`, and `BetaComplex`. Each contains API-functions that can be embedded in application programs to perform various computations. There are their respective counterparts in the Core-tier: `BallSetVoronoiDiagramCore`, `QuasiTriangulationCore`, and `BetaComplexCore`. Both the transformation between the Voronoi diagram and the quasi-triangulation and the extraction of the beta-complexes from the quasi-triangulation are in fact all performed in the Core-tier. Each of the API-tier classes communicates its counterpart in the Core-tier. In the Geometry-tier, there are currently two classes for the Voronoi diagram of spheres: `SphereSetVoronoiDiagramByEdgeTracing` and `SphereSetVoronoiDiagramByRegionExpansion` where both communicate with `BallSetVoronoiDiagramCore`. Note that we try to exclusively use the words "atom," "ball," and "sphere" for API-, Core-, and Geometry-tier, respectively.

Fig. 7 illustrates an example of simple application program which calls a few API-functions of BULL!. Assume that necessary header files are included. Given

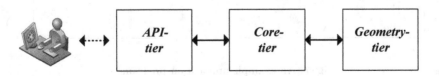

Fig. 6. 3-tier architecture of BULL!

a set of 100 input disks (ie, atoms) generated by the line 2, the simple API-functions in the lines 3 and 4 creates the Voronoi diagram `atomVD` and stores in `atomlist`. The lines 5 and 6 retrieves all the Voronoi cells in the Voronoi diagram. The lines 7 through 11 counts the number of adjacent Voronoi cells of each Voronoi cell in the Voronoi diagram and accumulates these numbers for all Voronoi cells in the Voronoi diagram. Then, the lines 12 and 13 prints out the average number of adjacent Voronoi cells in the entire Voronoi diagram. The lines 14 through 17 perform similar operation for the quasi-triangulation which is transformed from the Voronoi diagram in the line 15. The lines 18 through 22 perform similar operation for the beta-complex which is extracted from the quasi-triangulation by the line 20. As this code shows, the creation of such an application program is very straightforward once a user understands the basics of the theory and learns a few API-functions which are necessary to build an application program. Application programs performing more complicated tasks can be similarly created by using API-functions of BULL!.

```
1    int main()
     {
2        list<Sphere3d> atomlist = generateRandomAtoms(100);

3        AtomSetVoronoiDiagram atomVD;
4        atomVD.compute( atomlist );

5        list<VCellHandle> all_cells;
6        atomVD.getAllVCells( all_cells );

7        int countAdjacentCells = 0;
8        list<VCellHandle>::iterator i_cell;
9        for (i_cell=all_cells.begin(); i_cell!=all_cells.end(); i_cell++ ) {
10           countAdjacentCells += atomVD.countNumberOfAdjacentCells( *i_cell );
11       }
12       cout << "The average number of adjacent cells is ";
13       cout << (double)countAdjacentCells/atomVD.countNumberOfCells() << endl;

14       QuasiTriangulation QT;
15       QT.transform( atomVD );
16       cout << "The number of 2-adjacency anomaly is ";
17       cout << QT.countNumberOf2AdjacencyAnomaly() << endl;

18       double betaValue = 10.0;
19       BetaComplex BC;
20       BC.extract( QT, betaValue );
21       cout << "The number of atoms on the boundary of BC(beta=50) is ";
22       cout << BC.countNumberOfBCVerticesOnBoundary() << endl;

23       return 0;
     };
```

Fig. 7. An example application program

5   Conclusion

The Voronoi diagram of spherical atoms has many important applications, particularly for molecular worlds. While there are libraries for the ordinary Voronoi diagram and power diagram, its counterpart for spheres is not available. In this paper, we present the BULL! engine library for the Voronoi diagram of spheres, the quasi-triangulation, and the beta-complex with which application programmers can easily build application programs dealing with molecules. The BULL! engine is designed so that the application programs developed based on the API-tier is completely independent of future modifications. BULL! will be available from VDRC (http://voronoi.hanyang.ac.kr) at Hanyang University.

Acknowledgement. This work was supported by the National Research Foundation of Korea (NRF) grant funded by the Korea government (MSIP) (No. 2012R1A2A1A05026395).

References

1. Voronoi Diagram Research Center, http://voronoi.hanyang.ac.kr/
2. Aurenhammer, F.: Power diagrams: Properties, algorithms and applications. SIAM Journal on Computing 16, 78–96 (1987)
3. Aurenhammer, F.: Voronoi diagrams – a survey of a fundamental geometric data structure. ACM Computing Surveys 23(3), 345–405 (1991)
4. Cho, Y., Kim, D., Kim, D.S.: Topology representation for the Voronoi diagram of 3D spheres. International Journal of CAD/CAM 5(1), 59–68 (2005), http://www.ijcc.org
5. Edelsbrunner, H., Mücke, E.P.: Three-dimensional alpha shapes. ACM Transactions on Graphics 13(1), 43–72 (1994)
6. Kim, D.S., Cho, Y., Kim, D.: Euclidean Voronoi diagram of 3D balls and its computation via tracing edges. Computer-Aided Design 37(13), 1412–1424 (2005)
7. Kim, D.S., Cho, Y., Sugihara, K.: Quasi-worlds and quasi-operators on quasi-triangulations. Computer-Aided Design 42(10), 874–888 (2010)
8. Kim, D.S., Cho, Y., Sugihara, K., Ryu, J., Kim, D.: Three-dimensional beta-shapes and beta-complexes via quasi-triangulation. Computer-Aided Design 42(10), 911–929 (2010)
9. Kim, D.S., Kim, D., Cho, Y., Sugihara, K.: Quasi-triangulation and interworld data structure in three dimensions. Computer-Aided Design 38(7), 808–819 (2006)
10. Kim, D.S., Kim, J.K., Cho, Y., Kim, C.M.: Querying simplexes in quasi-triangulation. Computer-Aided Design 44(2), 85–98 (2012)
11. Kim, D., Kim, D.S.: Region-expansion for the Voronoi diagram of 3D spheres. Computer-Aided Design 38(5), 417–430 (2006)
12. Kim, J.K., Cho, Y., Kim, D., Kim, D.S.: Voronoi diagrams, quasi-triangulations, and beta-complexes for disks in $\mathbb{R}^2$: The theory and implementation in BetaConcept. Journal of Computational Design and Engineering 1(2), 78–86 (2014)
13. Okabe, A., Boots, B., Sugihara, K., Chiu, S.N.: Spatial Tessellations: Concepts and Applications of Voronoi Diagrams, 2nd edn. John Wiley & Sons, Chichester (1999)

Integrating Circumradius and Area Formulae for Cyclic Pentagons*

Shuichi Moritsugu

University of Tsukuba, Japan
moritsug@slis.tsukuba.ac.jp

Abstract. This paper describes computations of the relations between circumradius R and area S of cyclic polygons given by the lengths of the sides. Classic results by Heron and Brahmagupta clearly show the relation of circumradius and area for triangles and cyclic quadrilaterals. In contrast, formulae for the circumradius and the area of cyclic pentagons have been studied separately. D.P.Robbins obtained the area formula in 1994, which is a polynomial equation in $(4S)^2$ with degree 7. The circumradius formula was given by P.Pech in 2006, which is also a polynomial equation in R^2 with degree 7. In this study, we succeeded in computing the *integrated formula* for the circumradius and the area of cyclic pentagons. It is found to be a polynomial equation in $(4SR)$ with degree 7. This equation is easily transformed into the equation in $(4SR)^2$ with degree 7, hence both the expressions are meaningful. The existence of the latter form of formula was pointed out by D.Svrtan et al. in 2004, but somehow their result seems to contain typographical errors. Therefore, we believe that our results correspond to the correction and extension of already known formulae.

Keywords: cyclic polygons, circumradius formula, area formula.

1 Introduction

In this study, we consider a classic problem in Euclidean geometry for cyclic polygons; that is, n-gons inscribed in a circle, given by the length of sides $a_1, a_2, \ldots, a_n$. In particular, we focus on the relation between the circumradius R and the area S of cyclic pentagons.

Firstly, for a triangle with side length a_1, a_2, and a_3, the classic formula by Heron gives its circumradius and area as follows:

$$
\begin{cases}
R = \dfrac{a_1 a_2 a_3}{\sqrt{(a_1+a_2+a_3)(-a_1+a_2+a_3)(a_1-a_2+a_3)(a_1+a_2-a_3)}} \\
S = \dfrac{\sqrt{(a_1+a_2+a_3)(-a_1+a_2+a_3)(a_1-a_2+a_3)(a_1+a_2-a_3)}}{4}.
\end{cases}
\tag{1}
$$

---

* This work was supported by a Grant-in-Aid for Scientific Research (25330006) from the Japan Society for the Promotion of Science (JSPS).

H. Hong and C. Yap (Eds.): ICMS 2014, LNCS 8592, pp. 214–221, 2014.

It is straightforward to combine these equations, and we obtain the relation

$$4SR = a_1 a_2 a_3. \tag{2}$$

Secondly, Brahmagupta's formula gives the circumradius and the area of a cyclic quadrilateral as

$$
\begin{cases}
R = \sqrt{\dfrac{(a_1 a_2 + a_3 a_4)(a_1 a_3 + a_2 a_4)(a_1 a_4 + a_2 a_3)}{(-a_1 + a_2 + a_3 + a_4)(a_1 - a_2 + a_3 + a_4)(a_1 + a_2 - a_3 + a_4)(a_1 + a_2 + a_3 - a_4)}} \\
S = \dfrac{\sqrt{(-a_1 + a_2 + a_3 + a_4)(a_1 - a_2 + a_3 + a_4)(a_1 + a_2 - a_3 + a_4)(a_1 + a_2 + a_3 - a_4)}}{4}.
\end{cases}
\tag{3}
$$

It is direct again to integrate Equation (3) into

$$16S^2 R^2 = (a_1 a_2 + a_3 a_4)(a_1 a_3 + a_2 a_4)(a_1 a_4 + a_2 a_3). \tag{4}$$

We should note that Equation (3) represents the case of convex quadrilaterals, and the other case of non-convex, crossing figures is given by

$$
\begin{cases}
R = \sqrt{\dfrac{-(a_1 a_2 - a_3 a_4)(a_1 a_3 - a_2 a_4)(a_1 a_4 - a_2 a_3)}{(a_1 + a_2 + a_3 + a_4)(a_1 + a_2 - a_3 - a_4)(a_1 - a_2 - a_3 + a_4)(-a_1 + a_2 - a_3 + a_4)}} \\
S = \dfrac{\sqrt{(a_1 + a_2 + a_3 + a_4)(a_1 + a_2 - a_3 - a_4)(a_1 - a_2 - a_3 + a_4)(-a_1 + a_2 - a_3 + a_4)}}{4}.
\end{cases}
\tag{5}
$$

Hence, the latter case is expressed by the relation

$$16S^2 R^2 = -(a_1 a_2 - a_3 a_4)(a_1 a_3 - a_2 a_4)(a_1 a_4 - a_2 a_3). \tag{6}$$

If we let $Z = (4SR)^2$, then the above results for triangles and cyclic quadrilaterals are summarized as follows:

$$
\begin{cases}
Z - a_1^2 a_2^2 a_3^2 = 0, \\
(Z - (a_1 a_2 + a_3 a_4)(a_1 a_3 + a_2 a_4)(a_1 a_4 + a_2 a_3)) \\
\quad \times \ (Z + (a_1 a_2 - a_3 a_4)(a_1 a_3 - a_2 a_4)(a_1 a_4 - a_2 a_3)) = 0.
\end{cases}
\tag{7}
$$

The goal of this paper is to find such *integrated formula* for cyclic pentagons as Equation (7). Since D.P.Robbins [4] discovered the area formula for cyclic pentagons in 1994, the following facts are confirmed by several authors [1][3]:

- The area formula is a polynomial equation in $(4S)^2$ with degree 7,
- The circumradius formula is a polynomial equation in R^2 with degree 7.

Therefore, we can speculate that the relation between S and R is also expressed by a polynomial in $Z = (4SR)^2$ with degree 7, analogously from Equation (7). As a result, we succeeded in computing such formula exactly as speculated.

It might sound strange that the relation between the area and circumradius has been seldom discussed, and Pech [3] describes that "it is still missing". We have found that D.Svrtan et al. [5] shows a likely formula with degree 7, but their result seems to contain typographical errors or something, and their proof is too much abbreviated to follow.

In contrast, we show two ways of computation and confirm the correctness of both results. Hence, we believe that our result gives the correction and extension to that of D.Svrtan [5], and we have succeeded in specifying the structure of the integrated formula in detail.

2   Brute Force Algorithm

2.1   Expression by Elementary Symmetric Functions

Since the coefficients in the area and circumradius formulae are symmetric with a_i^2's, such expressions as Equation (7) can be reduced if the coefficients are expressed by elementary symmetric functions.

The conversion is processed by the following algorithm. First, we consider the polynomial ideal with elementary symmetric functions of n-th order:

$$I = \left\{ s_1 - (a_1^2 + \cdots + a_n^2), \ \ldots, \ s_n - (a_1^2 \cdots a_n^2) \right\}, \tag{8}$$

and compute its Gröbner basis using a group ordering ("lexdeg" in Maple computer algebra system) as

$$G := \text{Basis}(I, \{a_1, \ldots, a_n\} \succ \{s_1, \ldots, s_n\}). \tag{9}$$

Next, computing $p := \text{NormalForm}(f, G)$ for a symmetric function f, we get the expression p by elementary symmetric functions. For example, Equation (7), formulae for triangles and cyclic quadrilaterals are converted into

$$\begin{cases} Z - s_3 = 0, \\ Z^2 - 2s_3 Z + \left(s_3^2 - s_1^2 s_4\right) = \left(Z - s_3 + s_1\sqrt{s_4}\right)\left(Z - s_3 - s_1\sqrt{s_4}\right) = 0. \end{cases} \tag{10}$$

We should note that it gives a good insight into the structure of the formulae to introduce an auxiliary expression $\sqrt{s_n} = a_1 \cdots a_n$. Therefore, the formula for triangles given by Equation (2) is rewritten as

$$z - \sqrt{s_3} = 0 \qquad (z = 4SR = \sqrt{Z}, \quad \sqrt{s_3} = a_1 a_2 a_3), \tag{11}$$

which should be simpler than that in Equation (10).

2.2   Integrated Formula for Cyclic Pentagons

We assume that, according to [4][3][2], we have already computed the circumradius formula with 2,922 terms for cyclic pentagons:

$$\begin{aligned} \Phi_R(y) &= B_7 y^7 + B_6 y^6 + B_5 y^5 + B_4 y^4 + B_3 y^3 + B_2 y^2 + B_1 y + B_0 \\ &= 0 \qquad\qquad \left(y = R^2, \quad B_i \in \mathbf{Z}[a_1, \ldots a_5]\right), \end{aligned} \tag{12}$$

and the area formula with 6,672 terms:

$$\begin{aligned} \Phi_S(x) &= x^7 + C_6 x^6 + C_5 x^5 + C_4 x^4 + C_3 x^3 + C_2 x^2 + C_1 x + C_0 \\ &= 0 \qquad\qquad \left(x = (4S)^2, \quad C_i \in \mathbf{Z}[a_1, \ldots a_5]\right). \end{aligned} \tag{13}$$

Using elementary symmetric functions $s_1 = a_1^2 + \cdots + a_5^2, \ \ldots, \ s_5 = a_1^2 \cdots a_5^2$, we rewrite Equations (12) and (13) into simpler forms:

$$\begin{cases} \tilde{\Phi}_R(y) = \tilde{B}_7 y^7 + \tilde{B}_6 y^6 + \cdots + \tilde{B}_1 y + \tilde{B}_0 = 0 & \text{(81 terms)} \\ \tilde{\Phi}_S(x) = x^7 + \tilde{C}_6 x^6 + \cdots + \tilde{C}_1 x + \tilde{C}_0 \ \ = 0 & \text{(153 terms)}, \end{cases} \tag{14}$$

where $\tilde{B}_i, \tilde{C}_i \in \mathbf{Z}[s_1, \ldots, s_5]$. In order to combine $\tilde{\Phi}_R(y)$ and $\tilde{\Phi}_S(x)$ in Equation (14), first we substitute $y = Z/x$ into the radius formula, and we obtain

$$\tilde{\Phi}'_R(x, Z) = x^7 \tilde{\Phi}_R(Z/x) = \tilde{B}_7 Z^7 + \tilde{B}_6 Z^6 x + \cdots + \tilde{B}_1 Z x^6 + \tilde{B}_0 x^7. \quad (15)$$

Next, eliminating x by the resultant of $\tilde{\Phi}'_R(x, Z)$ and $\tilde{\Phi}_S(x)$, we obtain

$$\begin{aligned}
\tilde{\Psi}(Z) &= \mathrm{Res}_x(\tilde{\Phi}'_R(x, Z), \tilde{\Phi}_S(x)) \\
&= \tilde{A}_{49} Z^{49} + \cdots + \tilde{A}_0 \quad \left(\tilde{A}_i \in \mathbf{Z}[s_1, \ldots s_5]\right) \quad (2{,}093{,}279 \text{ terms}).
\end{aligned}$$
$$(16)$$

This computation required about 15 minutes of CPU time in the following environment: Maple14 on Win64, Xeon(2.93GHz)$\times$2, 192GB RAM.

Finally, we factorize this polynomial with degree 49. Together with the other factor of degree 42, we obtain the polynomial in Z with degree 7:

$$\begin{aligned}
\psi(Z) = {}& Z^7 - 4s_3 Z^6 + \left(-28 s_1 s_5 - 2 s_1^2 s_4 + 6 s_3^2\right) Z^5 \\
& + \left((-s_1^4 - 10 s_1^2 s_2 - 8 s_2^2 + 52 s_1 s_3 - 32 s_4) s_5 + 4 s_3 (s_1^2 s_4 - s_3^2)\right) Z^4 \\
& + \big(4(37 s_1^2 - 48 s_2) s_5^2 + (-2 s_1^4 s_3 + 12 s_1^3 s_4 + 64 s_3 s_4 + 4 s_1^2 s_2 s_3 + 16 s_2^2 s_3 \\
& \qquad\qquad - 20 s_1 s_3^2 - 64 s_1 s_2 s_4) s_5 + (s_1^2 s_4 - s_3^2)^2\big) Z^3 \\
& + \big(-576 s_5^3 + (64 s_1 s_4 - 80 s_1 s_2^2 + 28 s_1^3 s_2 - 8 s_1^2 s_3 + 128 s_2 s_3 - 2 s_1^5) s_5^2 \\
& \qquad + (2 s_1^4 s_2 s_4 - 12 s_1^3 s_3 s_4 - 8 s_1^2 s_2^2 s_4 - s_1^4 s_3^2 - 4 s_1 s_3^3 - 32 s_1^2 s_4^2 - 32 s_2^2 s_4 \\
& \qquad\qquad + 64 s_1 s_2 s_3 s_4 - 8 s_2^2 s_3^2 + 6 s_1^2 s_2 s_3^2) s_5\big) Z^2 \\
& + \big(-48(s_1^3 - 4 s_1 s_2 + 8 s_3) s_5^3 \\
& \qquad (-8 s_1^2 s_3^2 + 256 s_4^2 + s_1^4 s_2^2 - 64 s_1 s_3 s_4 - 128 s_2^2 s_4 + 64 s_2 s_3^2 + 16 s_2^4 \\
& \qquad + 96 s_1^2 s_2 s_4 - 12 s_1^4 s_3^2 - 48 s_1 s_2^2 s_3 - 2 s_1^5 s_3 - 16 s_1^4 s_4 + 20 s_1^3 s_2 s_3) s_5^2\big) Z \\
& - s_5^3 (s_1^3 - 4 s_1 s_2 + 8 s_3)^2 \quad = \quad 0 \qquad (63 \text{ terms}).
\end{aligned}$$
$$(17)$$

This factorization required nearly 80 hours of CPU time in the same environment shown above. If we put $\forall a_i := 1$, Equation (17) is reduced to

$$(Z^2 - 35Z + 25)(Z - 1)^5 = 0, \qquad (18)$$

which respectively correspond to the cases of regular pentagon/pentagram and (five degenerated) regular triangles. Therefore, we believe that Equation (17) is the *integrated formula* for cyclic pentagons which is the goal of this study.

[Remark]

The notion of the formula in $Z = (4SR)^2$ is already proposed by D.Svrtan et al. [5], and their Equation (35) is supposed to correspond to Equation (17) above. Unfortunately, their result does not coincide with ours, and theirs is not factored when we let $\forall a_i := 1$. Therefore, D.Svrtan's result seems to contain typographical errors or something, which might be caused in the process of converting into the expression by elementary symmetric functions.

3 Stepwise Algorithm

In this algorithm, first we divide a cyclic pentagon with side length $\{a_1, \ldots, a_5\}$, by a diagonal of length d, into a triangle of sides $\{a_1, a_2, d\}$ and a quadrilateral of sides $\{a_3, a_4, a_5, d\}$. Next, using the common circumradius R, we consider the sum of areas of the triangle and the quadrilateral.

Firstly, the circumradius formula of Heron gives the defining polynomial in $y = R^2$ as follows:

$$H_3(a_1, a_2, d;\ y) := (a_1 + a_2 + d)(-a_1 + a_2 + d)(a_1 - a_2 + d)(a_1 + a_2 - d)y + a_1^2 a_2^2 d^2.$$
(19)

Similarly, the formula of Brahmagupta gives the following polynomials for convex and non-convex cases respectively:

$$\begin{cases} H_4^{(+)}(a_3, a_4, a_5, d;\ y) = \\ \quad (-a_3 + a_4 + a_5 + d)(a_3 - a_4 + a_5 + d)(a_3 + a_4 - a_5 + d)(a_3 + a_4 + a_5 - d)y \\ \qquad\qquad\qquad\qquad\qquad\qquad -(a_3 a_4 + a_5 d)(a_3 a_5 + a_4 d)(a_3 d + a_4 a_5) \\ H_4^{(-)}(a_3, a_4, a_5, d;\ y) = \\ \quad (a_3 + a_4 + a_5 + d)(a_3 - a_4 - a_5 + d)(a_3 - a_4 + a_5 - d)(a_3 + a_4 - a_5 - d)y \\ \qquad\qquad\qquad\qquad\qquad\qquad -(a_3 a_4 - a_5 d)(a_3 a_5 - a_4 d)(a_3 d - a_4 a_5) \end{cases}$$
(20)

Since the circumradius R is common to these triangle and quadrilateral, we eliminate y using resultant and obtain the defining polynomials in the diagonal d with degree 7:

$$\begin{cases} F^{(+)}(d) := \operatorname{Res}_y(H_4^{(+)}(a_3, a_4, a_5, d;\ y), H_3(a_1, a_2, d;\ y)) \\ \qquad = a_3 a_4 a_5 d^7 + (a_3^2 a_4^2 + a_3^2 a_5^2 + a_4^2 a_5^2 - a_1^2 a_2^2)d^6 + \cdots \\ F^{(-)}(d) := \operatorname{Res}_y(H_4^{(-)}(a_3, a_4, a_5, d;\ y), H_3(a_1, a_2, d;\ y)) \\ \qquad = a_3 a_4 a_5 d^7 - (a_3^2 a_4^2 + a_3^2 a_5^2 + a_4^2 a_5^2 - a_1^2 a_2^2)d^6 + \cdots \end{cases}$$
(21)

We should note that we have $F^{(-)}(-d) = -F^{(+)}(d)$, which means the roots of both polynomials are equivalent up to signs.

Secondly, we let S_3, S_4, and S_5 be the area for the triangle, cyclic quadrilateral, and cyclic pentagon respectively given in the above. Since we have $S_5 = S_3 + S_4$, we get $4S_4 R = 4S_5 R - 4S_3 R$, where R is the common circumradius. For the triangle, we have $4S_3 R = a_1 a_2 d$ from Equation (2). If we let $z = 4S_5 R$ and substitute $4S_4 R = z - a_1 a_2 d$ into Equations (4) and (6), we obtain the following polynomial equations in z and d, for the cases of convex and non-convex quadrilaterals respectively:

$$\begin{cases} f^{(+)}(z, d) := (z - a_1 a_2 d)^2 - (a_3 a_4 + a_5 d)(a_3 a_5 + a_4 d)(a_3 d + a_4 a_5) = 0 \\ f^{(-)}(z, d) := (z - a_1 a_2 d)^2 + (a_3 a_4 - a_5 d)(a_3 a_5 - a_4 d)(a_3 d - a_4 a_5) = 0. \end{cases}$$
(22)

Finally, we eliminate the diagonal d from $F^{(+)}(d)$ and $f^{(+)}(z, d)$ by computing the resultant:

$$\begin{aligned} P^{(+)}(z) &:= \operatorname{Res}_d(F^{(+)}(d), f^{(+)}(z, d)) \\ &= (z^7 + \cdots)(a_3 a_4 a_5 z^7 + \cdots). \end{aligned}$$
(23)

This computation took about 2.5 minutes of CPU time in total, which is drastically reduced from the brute force algorithm shown in the previous section. Since the latter factor in Equation (23) is asymmetry with a_i's, we adopt the former one as the defining polynomial in $z = 4SR$ for cyclic pentagons. Converting it into the expression by elementary symmetric functions, with $\sqrt{s_5} = a_1 a_2 a_3 a_4 a_5$, we obtain the final result:

$$
\begin{aligned}
\varphi^{(+)}(z) = {} & z^7 - 2s_3 z^5 - (s_1^2 + 4s_2)\sqrt{s_5}z^4 + (s_3^2 - s_1^2 s_4 - 14s_1 s_5)z^3 \\
& - (s_1^2 s_3 + 8s_1 s_4 - 4s_2 s_3 + 24s_5)\sqrt{s_5}z^2 \\
& - (s_1^2 s_2 - 4s_2^2 + 2s_1 s_3 + 16s_4)s_5 z \\
& - (s_1^3 - 4s_1 s_2 + 8s_3)s_5\sqrt{s_5} = 0 \qquad \text{(18 terms)}.
\end{aligned}
\tag{24}
$$

If we put $\forall a_i := 1$, Equation (24) is reduced to

$$
(z^2 - 5z - 5)(z + 1)^5 = 0,
\tag{25}
$$

which respectively correspond to the cases of regular pentagon/pentagram and (five degenerated) regular triangles, similarly to Equation (18).

For the case of non-convex quadrilateral, we compute the resultant of another pair of $F^{(-)}(d)$ and $f^{(-)}(z, d)$, and obtain a similar result $\varphi^{(-)}(z)$, where we have $\varphi^{(-)}(-z) = -\varphi^{(+)}(z)$ and we can regard the roots of $\varphi^{(+)}(z) = 0$ and $\varphi^{(-)}(z) = 0$ as equivalent up to signs of S_5. If we put $\forall a_i := 1$, the equation $\varphi^{(-)}(z) = 0$ is reduced to

$$
(z^2 + 5z - 5)(z - 1)^5 = 0.
\tag{26}
$$

We should note that, in our formulation, the area of the triangle between $\overrightarrow{OA} = [x_1, y_1]$ and $\overrightarrow{OB} = [x_2, y_2]$ is defined as the determinant

$$
S = \frac{1}{2}\begin{vmatrix} x_1 & y_1 \\ x_2 & y_2 \end{vmatrix}.
\tag{27}
$$

Hence, the sign of area S of polygons should be essentially discarded. Therefore, we do not need to distinguish $\varphi^{(+)}(z)$ from $\varphi^{(-)}(z)$, and we conclude that Equation (24) also shows the relation between circumradius and area for cyclic pentagons.

It is straightforward to rewrite Equation (24) as a polynomial equation in $Z = z^2 = (4SR)^2$. We separate the equation $\varphi^{(+)}(z) = 0$ into the terms with even degrees and odd degrees:

$$
z\left(z^6 - 2s_3 z^4 + \cdots\right) = (s_1^2 + 4s_2)\sqrt{s_5}z^4 + \cdots + (s_1^3 - 4s_1 s_2 + 8s_3)s_5\sqrt{s_5}.
\tag{28}
$$

Squaring both sides and substituting $z^2 = Z$, we obtain the same result as Equation (17), which is a polynomial in $Z = (4SR)^2$ with degree 7. Another equation $\varphi^{(-)}(z) = 0$ gives the same result by similar computation. Hence, we believe that the correctness of two algorithms we propose in this paper is confirmed by those results.

[Remark]

The paper by D.Svrtan et al. [5] is also based on this approach, but it does not specify the final step of the computation of resultant, which should correspond to Equation (23), or does not refer the existence of the defining polynomial in $z = 4SR$. Therefore, we consider that our computation makes corrections to theirs, and it contains new results $\varphi^{(\pm)}(z)$ unknown so far.

4   Application to a Numerical Problem

We consider the problem where the length of sides are given as $a_1 = 5, a_2 = 6, a_3 = 7, a_4 = 8, a_5 = 9$. The area ($x = 16S^2$) and the circumradius ($y = R^2$) formula leads to the following equations:

$$\begin{cases} x^7 - 145377x^6 + \cdots - 1360512306447018480615234375 &= 0 \\ 5810802381759375y^7 - \cdots - 119484273420824739840000000 = 0. \end{cases} \tag{29}$$

These equations are easily solved by conventional algorithms for numerical computation, and we obtain 5 real roots of each equation. However, the correspondence between area and circumradius is unknown yet, or we cannot even point out which the case of convex pentagon is. Hence, we apply the set $\{5, 6, 7, 8, 9\}$ to the integrated formula (24):

$$z^7 - 2356490z^5 - \cdots - 1672586387136000000 = 0, \tag{30}$$

which yields also 5 real roots.

Since we have $|z| = \sqrt{xy} = 4|S|R$, we check all $125(= 5^3)$ combinations of roots $\{x_i, y_j, z_k\}$, and pick up those satisfy the condition

$$\left| |S_i| \cdot R_j - \frac{|z_k|}{4} \right| < 10^{-6}. \tag{31}$$

As a result, the following pairs are selected:

$$\begin{aligned} [|S_i|, R_j] = &[10.47633365, \ 6.035515309], \\ &[16.78535280, \ 4.505907128], \\ &[23.09053708, \ 4.602116876], \\ &[30.69973405, \ 4.802909240], \\ &[82.47639518, \ 6.019756631], \end{aligned} \tag{32}$$

which shows that the last pair implies the convex pentagon figure.

5   Concluding Remarks

We investigated closely the relations among several quantities of cyclic pentagons, and succeeded in computing the *integrated formula* of area and circumradius, that is, Equations (24) and (17). We compared two algorithms and confirmed the coincidence of both results, where the stepwise algorithm turned out to be more efficient than the brute force algorithm,

A polynomial in $(4SR)^2$ like Equation (17) was shown before by D.Svrtan et al. [5], but their result seems to contain errors somehow. To the best of our knowledge, there exist no other reports in which defining polynomial in $z = 4SR$ like Equation (24) is discussed. In addition to Equation (17), we believe that our results contain original formulae $\varphi^{(+)}(z)$ in Equation (24).

The degrees of defining polynomials of area and circumradius for cyclic n-gons are given by the theorem of Robbins [4]. It is also speculated that the integrated formula has the same degree as $(4S)^2$ and R^2. Therefore, on the analogy of formulae for quadrilaterals in Equation (10), the integrated formula in $Z = (4SR)^2$ for cyclic hexagons should have the following structure:

$$
\begin{aligned}
Z^{14} &+ \cdots && (\mathbf{Z}\,[s_1, \ldots, s_5, s_6;\ Z]) \\
&= (Z^7 + \cdots)(Z^7 + \cdots) && (\mathbf{Z}\,[s_1, \ldots, s_5, \sqrt{s_6};\ Z])
\end{aligned}
\tag{33}
$$

where $s_1 = a_1^2 + a_2^2 + a_3^2 + a_4^2 + a_5^2 + a_6^2$, ..., $\sqrt{s_6} = a_1 a_2 a_3 a_4 a_5 a_6$.

The present status of computations by two algorithms is as follows, and neither has succeeded in the computation yet, because of the exploding size of polynomials. The Area formula is straightforwardly obtained by Maley [1]:

$$
\tilde{\Phi}_S^{(+)}(s_1, \ldots, s_5, \sqrt{s_6};\ x) = x^7 + \tilde{C}_6 x^6 + \cdots + \tilde{C}_1 x + \tilde{C}_0 \qquad (282\text{ terms}). \tag{34}
$$

The radius formula was computed before by the author [2]:

$$
\begin{aligned}
\Phi_R(a_1, \ldots, a_6;\ y) \\
&:= B_{14} y^{14} + \cdots + B_1 y + B_0 \quad (497{,}417\text{ terms}) \quad (B_i \in \mathbf{Z}[a_1, \ldots a_6]) \\
&= \Phi_R^{(+)}(a_i;\ y) \cdot \Phi_R^{(-)}(a_i;\ y) \quad (\text{each has degree 7 and 19,449 terms}),
\end{aligned}
\tag{35}
$$

which is rewritten into the expressions by elementary symmetric functions:

$$
\tilde{\Phi}_R^{(+)}(s_1, \ldots, s_5, \sqrt{s_6};\ y) = \tilde{B}_7 y^7 + \tilde{B}_6 y^6 + \cdots + \tilde{B}_1 y + \tilde{B}_0 \qquad (224\text{ terms}). \tag{36}
$$

We discard the non-convex cases, because they are easily obtained by replacing $\sqrt{s_6}$ in the convex cases by $-\sqrt{s_6}$.

[Brute force algorithm]

Computing the resultant from $\tilde{\Phi}_S^{(+)}$ and $\tilde{\Phi}_R^{(+)}$, we have obtained a polynomial of degree 49 with $52{,}490{,}772$ terms (nearly 3.0GB), which corresponds to Equation (16). This polynomial should also be factored into two polynomials of degrees 42 and 7, but it is hopeless to execute it actually.

[Stepwise algorithm]

The resultant that corresponds to Equation (23) is not finished yet, because its computation proceeds in the coefficient $\mathbf{Z}\,[a_1, \ldots, a_6]$, and it cannot take the advantage of the simpler expression by elementary symmetric functions.

References

1. Maley, F.M., Robbins, D.P., Roskies, J.: On the Areas of Cyclic and Semicyclic Polygons. Advances in Applied Mathematics 34(4), 669–689 (2005)
2. Moritsugu, S.: Computing Explicit Formulae for the Radius of Cyclic Hexagons and Heptagons. Bulletin of Japan Soc. Symbolic and Algebraic Computation 18(1), 3–9 (2011)
3. Pech, P.: Computations of the Area and Radius of Cyclic Polygons Given by the Lengths of Sides. In: Hong, H., Wang, D. (eds.) ADG 2004. LNCS (LNAI), vol. 3763, pp. 44–58. Springer, Heidelberg (2006)
4. Robbins, D.P.: Areas of Polygons Inscribed in a Circle. Discrete & Computational Geometry 12(1), 223–236 (1994)
5. Svrtan, D., Veljan, D., Volenec, V.: Geometry of Pentagons: from Gauss to Robbins. arXiv:math.MG/0403503 v1 (2004)

Computer Aided Geometry

Douglas Navarro Guevara[1] and Adrian Navarro Alvarez[2]

[1] Universidad Nacional, Costa Rica
navarro.douglas@gmail.com
[2] Universidad de Costa Rica, Costa Rica
adrnavarro@gmail.com

Abstract. This paper presents a software to work with 3D dynamic geometry and multivariate calculus. It provides many resources to define and manipulate diverse 0D, 1D, 2D and 3D objects. Functions are defined explicitly or as the result of operations. Functions can (for example) be associated to 3D objects to calculate an iterated or a surface integral. The embedded CAS uses a novel and efficient scheme of representation for the common transcendental functions. Applications range from mathematical education to scientific simulation events passing by banal or utilities applications.

Keywords: dynamic geometry, computer algebra, function's representation, primitive transcendental functions, symbolic computation.

1 Introduction

The purpose of the software presented here is to facilitate several applications of mathematics related to 3D analytic geometry and symbolic/numeric calculations. It has many resources for definition and manipulation of diverse 0D, 1D, 2D, 3D objects. 3D objects are represented as solid objects and may come from primitives, user libraries, 3D scanner files, mechanical engineering software as **SolidWorks** or **Inventor**, or, be the result of operations as symmetries, extrusions, Boolean operations, involutions and so on. It also includes a Computer Algebra System kernel that uses a novel and efficient scheme of representation for the "common transcendental functions". Such representation is based on a few types of power series characterized by a periodic sequence of numbers. This representation scheme allows orders of execution similar to those of arithmetic calculations. The software will be available from December 2014.

2 Functionality

In this software, objects are organized as: 3D, 2D, 1D, 0D geometric objects and functions. These objects can come from various sources, for example, an object can be recovered from a previous work session, can be imported from an external software, it may be a primitive, it can be explicitly declared, can be retrieved from a file of points to interpolate or to interpret as a b-spline. They can also

H. Hong and C. Yap (Eds.): ICMS 2014, LNCS 8592, pp. 222–229, 2014.

come from active operations as a difference of solids, the rotation of a curve segment about an axis, they can also be the value of a definite integral or the solution of a differential equation, etc.

Operations are executed from a *context sensitive* calculator, so if the operands are two solid objects some possible operations would be: intersection or difference; but if operands are a solid object and a real function f of three real variables the calculator proposes operations like: calculation of the surface integral or calculation of the gravity center (with f as the density function).

The functionality of operators is dynamic, but the reliance can be released so the new object becomes independent of the operation and the operands. Geometric objects can be associated with rigid transformations that can operate together. They can be executed selectively, forward and backward, one step at a time or continuously. The total configuration of objects is conceptualized as a *"scenario"* and can be saved and retrieved as such. Solids can also be saved and retrieved individually in proprietary format. The currently supported input formats are PLY (Polygon File Format or the Stanford Triangle Format) and STL (Standard Tessellation Language or STereoLithography file format). Solids can be exported in STL format, allowing lift in external software and 3D printers. The 3D-view manipulation includes functionalities such as the selective projection, zoom in, zoom out, light source manipulation, rotations, etc. The real functions of real variable can be plotted in a specialized window.

Possible applications for this software include: a banal construction for a desired 3D-printing; construction corresponding to a calculus optimization problem; mathematical *practical work* support; calculation of a surface integral over a certain 3D-object, calculation of Euler Characteristic for a double torus, simulation of a laser propagation between two concave mirrors, finite element simulation, etc. Currently the software runs on Windows but soon will be available for MAC. The graphical output is based on OpenGL.

3   Application Examples

Example 1: elementary geometry
Calculation of the sphere volume [10]: A *Cavalieri Principle* illustration.

Example 2: elementary calculus
An optimization problem [11]: Given a cube, find the inscribed cylinder of maximum volume such that its axis coincides with one of the diagonals of this cube.

First, a dynamic construction and then, the corresponding trace of the cylinder volume function for manipulation (interpolation, derivation, etc.).

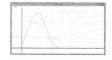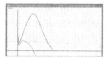

Example 3: elementary calculus, visualization
Problem 1 [12]: Find the volume of the intersection between two cylinders of radio r whose axes are perpendicular.

Example 4: elementary calculus, visualization
Problem 2 [1]: Calculate the work done by the force field $f(x, y, z) = y^2 i + z^2 j + x^2 k$ along the curve of intersection of the sphere $x^2 + y^2 + z^2 = r^2$ and the cylinder $x^2 + y^2 = r^2$ where $z > 0$, $r > 0$. The path is traversed in a direction ...

Example 5: computational geometry
Convex hull [13]: Calculation of convex hull for two convex polyhedrons based on the involution of the intersection of the involutions of the original intersecting polyhedrons (in this case of a cube and a five sides prism).

Example 6: algebraic topology
Nine torus Euler characteristic $\chi()$: Exploration of certain invariants.

Label:	Vertices-Edges+Faces (9 torus)
Value:	-16

Example 7: simulation
Laser reflection: Simulation of a perfect laser reflection between concave mirrors.

Example 8: discrete function
Discrete function: Discrete input from a file, interpolation, solid of revolution, parameters calculation, etc.

| Label: | Surface (solid of revolution) |
| Value: | 154.572677612305 |

Example 9: differential equation
Differential equation $y''-4y'=e^{2x}$ [1]: Input from a file, solution (variation of parameters method).

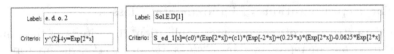

Example 10: others examples
Boolean operators applications:

Edited inputs:

Special operators [1]:
Incremental union:

Example 11: output examples
Output examples: 3D printer output.

---

[1] For the construction of more elaborate geometries.

4   Function Representation

The Computer Algebra System kernel embedded in this software uses a novel and efficient scheme of representation for the criteria of a subset of functions here named: *primitive transcendental functions*. This subset intersects the set of holonomic functions [20] and includes the "common transcendental functions" [2]. Such representation is based on a few types of power series characterized by a periodic sequence of numbers. The induced representation allows to define a natural isomorphism between some subsets of functions and the n-dimensional space R^n that allows the implementation of diverse analytical operators as well as the decoding of the results of such operators.

4.1   The Representations of Fundamental Classes

For the data structure of the computer application, the criteria of the complex variable functions are represented by a *"classic binary tree"* structure (operands in the leaves and operators in the interior nodes) [3]. The basic types of functions considered (named: *Exponential, Geometric, Arc*) consist of the functions whose power series expansion at the origin are of the form [3] :

$$\sum_{i=0}^{\infty} b_i \frac{x^i}{i!}, \quad \sum_{i=max\{0,-k\}}^{\infty} b_i(i+k)!\frac{x^i}{i!}, \quad \sum_{i=0}^{\infty} b_i(\lfloor i+k \rfloor!)^2 \frac{x^i}{i!} \tag{1}$$

where $\{b_i\}$ is a periodic sequence of period n, $k \in Z$, and $\lfloor m \rfloor! = \{1$ if $m \leqslant 0$; $1.3.5....m$ if m is odd; $1.3.5....(m-1)$ if m is even.

Coding. Coding of a function of the X *Type* is made by $F[X; k; b_0, b_1, \cdots, b_{n-1}](x)$.

Decoding. Decoding (calculation of the sum in terms of usual functions) is based upon the fact that, for a given length n, a given k and a fixed type, the defined set of functions forms an n-dimensional vector space. As an example, for the *Exponential Type*, the isomorphism is given by the function $\varphi : R^n \longrightarrow B_n$:

$$\varphi(b_0, b_1, ..., b_{n-1}) = \sum_{i=0}^{\infty} b_i \frac{x^i}{i!} \quad where \quad B_n = \left\{ \sum_{i=0}^{\infty} b_i \frac{x^i}{i!} \text{ with } b_i = b_{i+n} \right\}. \tag{2}$$

Fundamental Properties. The sum and scalar multiplication in B_n correspond to the usual vector operations in R^n. The differentiation operator (integration with constant b_n) corresponds to a circular left (right) shift of the generating sequence and the unitary increase (decrease) of k (if any). Other properties of R^n are also transferred by the isomorphism φ.

Below, there are some application examples of this approach. In certain examples they are compared with the results obtained from a leading commercial CAS (**Mathematica**).

---

[2] *e.g.:* $\exp(x)$, $\cos(x)$, $\sin(x)$, $\cosh(x)$, $\sinh(x)$, $\arctan(x)$, $\operatorname{arctanh}(x)$, $\ln(1\text{-}x)$, $\arcsin(x)$, $\operatorname{arcsinh}(x)$, etc.

[3] They are extended to include certain *Bessel* and *Bernoulli* functions and this extension process can be continued.

Example 12: Decoding of the series $F[Geo; -1; 1, 2, 3, 4]$:

$$2\frac{x}{1} + 3\frac{x^2}{2} + 4\frac{x^3}{3} + \frac{x^4}{4} + 2\frac{x^5}{5} + 3\frac{x^6}{6} + 4\frac{x^7}{7} + \frac{x^8}{8} + \dots \tag{3}$$

Solution. In the *Geometric Type* the functions corresponding to the base: $\{[-1; 1,0,1,0], [-1; 0,1,0,1], [-1; 1,0,-1,0], [-1; 0,1,0,-1]\}$ are: $\{-\ln\left(\sqrt{1-x^2}\right),$ arctanh(x), $-\ln\left(\sqrt{1+x^2}\right),$ arctan(x)$\}$. Thus, the sum in question corresponds to:

$$[-1; 1, 2, 3, 4] = 2[-1; 1, 0, 1, 0] + 3[-1; 0, 1, 0, 1] - [-1; 1, 0, -1, 0] - [-1; 0, 1, 0 - 1]$$

$$= -2\ln\left(\sqrt{1-x^2}\right) + 3arctanh(x) + \ln\left(\sqrt{1+x^2}\right) - \arctan(x) \tag{4}$$

With **Mathematica**

The corresponding command is:

```
Sum[x^(4n)/((4n))+2x^(4n+1)/((4n+1))+3x^(4n+2)/((4n+2))
    +4x^(4n+3)/((4n+3)),{n,1,Infinity}]
```

Resulting in:

```
1/12 (-24x-18x^2-16x^3+18ArcTanh[x^2]+24x
Hypergeometric2F1[1/4,1,5/4,x^4]+16x^3
Hypergeometric2F1[3/4,1,7/4,x^4]-3Log[1-x^4])
```

Example 13: Reduction of the identity [4] $\sinh^2(x) + \cosh^2(x) - \cosh(2x) = 0$:
Solution. The simplification is immediate from the mechanical treatment of products and the respective sums [5]:

$$[Exp; 0, 1]^2(x) + [Exp; 1, 0]^2(x) - [Exp; 1, 0](2x)$$

$$= -\frac{1}{2} + \frac{1}{2}[Exp; 1, 0,](2x) + \frac{1}{2} + \frac{1}{2}[Exp; 1, 0](2x) - [Exp; 1, 0](2x)$$

$$= [Exp; 1, 0](2x) - [Exp; 1, 0](2x) = 0 * [Exp; 1, 0](2x) = 0 \tag{5}$$

With **Mathematica**

Consider the **Mathematica** commands:

```
f(x_)=Sinh(x)Sinh(x)+Cosh(x)Cosh(x)-Cosh(2x)
g(x_)=f(x)^2
```

Producing respectively:

```
:Cosh[x]^2-Cosh[2x]+Sinh[x]^2
:(Cosh[x]^2 - Cosh[2x] + Sinh[x]^2)^2
```

---

[4] It is worth noting that the problem of solving equivalences is, in general, recursively undecidable [14].

[5] All products of the elements of the basis $A = \{\cosh(x), \sinh(x), \cos(x), \sin(x)\}$ of B_4 (*i.e.* $\{[1,0,1,0], [0,1,0,1], [1,0,-1,0], [0,1,0,-1]\}$) are explicitly calculable.

Example 14: Non polynomial solution of $y'' + y' = e^x + \sin(x)$:
Coding. The function $e^x + \sin(x)$ is of the Exp type, it corresponds to $(1,1,1,1) + (0,1,0,-1) = (1,2,1,0)$. If $y(x)$ corresponds to a vector of the form (a,b,c,d). Then, $y'(x)$ corresponds to (b,c,d,a) and $y''(x)$ corresponds to (c,d,a,b), so the equation in question would be written $(c,d,a,b) + (b,c,d,a) = (1,2,1,0)$. *i.e.*

$$
\begin{bmatrix} 0 & 1 & 1 & 0 \\ 0 & 0 & 1 & 1 \\ 1 & 0 & 0 & 1 \\ 1 & 1 & 0 & 0 \end{bmatrix} \begin{bmatrix} a \\ b \\ c \\ d \end{bmatrix} = \begin{bmatrix} 1 \\ 2 \\ 1 \\ 0 \end{bmatrix} \tag{6}
$$

Solution. The solution is then: $(1,-1,2,0) + d(-1,1,-1,1)$.
Decoding. The solution corresponds to (changing the base to the base A of B_4):

$$
3\cosh(x)/2 - \sinh(x)/2 - \cos(x)/2 - \sin(x)/2 + d(\sinh(x) - \cosh(x)) \tag{7}
$$

Example 15: The non polynomial solution of $y'' = \ln(1 - x)$:

$$
\int^{(2)} [Geo; -1; -1](x) = [Geo; -3; -1](x) \tag{8}
$$

The decoding is automatic: $[Geo;-3;-1](x) = (1 - x)^2 (\ln(1 - x) - 3/2)/2$
With **Mathematica**

```
DSolve[y''[x] == Ln[1 - x], y[x], x]
```

$$
y(x) \rightarrow \int_1^x \left(\int_1^{K[2]} Ln(1 - K[1]) \, dK[1] \right) dK[2] + c_2 x + c_1 \tag{9}
$$

Noteworthy that the examples show the difficulty found by this type of CAS to solve relatively simple problems.

5   General Remarks

In regards to mathematics education, the expectation is to offer a mathematical exercise that can provide a stimulating experience. One can imagine, for example, assigning a *practical work* that consists in the construction of a model that eventually will be 3D printed. An activity that can exploit diverse mathematical concepts like: vectors, planes, symmetries, solids of revolution, cylinders, cones, Boolean operations, center of mass, surface calculation, etc.

Regarding scientific and engineering simulations, currently the system has direct and iterative methods to solve systems of equations of higher orders. It operates sparse matrices with real entries and sparse matrices with functional entries. The implementation of finite element methods is in preparation.

The function representation utilized and the definition of fundamental analytical calculations allows the definition of the basic *classes* and *methods* to develop an efficient CAS under the *object-oriented programming paradigm*. It is appropriate noting that this is very important in the implementation of software projects related with subjects such as dynamic geometry or finite element.

Dual calculations (calculations made with the coordinates) pass analytical calculations to the domain of *computable functions* [5]. The approach of this representation scheme consists of three stages (encoding - computing - decoding). It allows to produce effective results where other CAS produce useless results. In fact, this is a good example of the significance of the paradigm change that occurs in the search for discrete and algorithmic solutions [9].

References

1. Apostol, T.: Calculus, vol. 2. Blaisdell Publishing, Massachusetts (1967)
2. Atkinson, K.E.: An Introduction to Numerical Analysis. John Wiley & Sons, New York (1978)
3. Cormen, T., Leiserson, C., Rivest, R.: Introduction to algorithms. The MIT Press, Cambridge (1990)
4. Davenport, J., Siret, Y., Tournier, E.: Calcul formel: systèmes et algorithmes de manipulations algébriques. Masson, Paris (1987)
5. Hennie, F.: Introduction to Computability, Massachusetts Institute of Technology, University of California, Addison-Wesley, Massachusetts (1977)
6. Hoffman, K., Kunze, R.: Linear Algebra, 2nd edn. Prentice-Hall, Inc., New Jersey (1971)
7. Lang, S.: Complex Analysis. Yale University, New Haven (1977)
8. Navarro, D.: Sur l'utilisation des outils informatiques dans l'enseignement des mathématiques; doctoral thèse. Université Paul Sabatier, Toulouse (2006)
9. Navarro, D.: Des changements de paradigme dans le developpement de logiciels. Revista Brasileira de Ensino de Ciência e Tecnologia 6(1) (2013) ISSN - 1982-873X
10. Moise, E.: Elementary Geometry from an Advanced Standpoint. Addison-Wesley, Massachusetts (1962)
11. Piskunov, N.: Calculo Diferencial e Integral, tomo I. Editorial MIR. Moscu (1969)
12. Simmons, G.: Calculus with Analytic Geometry, 2nd edn. McGraw-Hill, USA (1996)
13. Preparata, F., Shamos, M.: Computational Geometry, an introduction. Springer (1985)
14. Richardson, D.: Some undecidable Problems Involving Elementary Functions of a Real Variable. The Journal of Symbolic Logic 33, 514–520 (1968)
15. Risch, R.: The problem of integration in finite terms. Transactions of American Mathematical Society 139, 167–189 (1969)
16. Wolff, D.: OpenGL 4.0 Shading Language Cookbook. Packt Publishing Ltd., Birmingham (2011)
17. Wolfram, S.: Mathematica Book. Cambridge University Press, USA (1999)
18. Wright, R., Haemel, N., Sellers, G., Lipchak, B.: OpenGL SuperBible, 5th edn. Adisson Wesley, Person Education, Inc., Boston, MA (2011)
19. Zeid, I.: CAD/CAM Theory and Practice. McGraw-Hill (1991)
20. Zeilberger, D.: A holonomic systems approach to special functions identities. Journal of Computational and Applied Mathematics 32, 321–368 (1990)

The Sustainability
of Digital Educational Resources[*]

Yongsheng Rao[1], Ying Wang[2,**], Yu Zou[1], and Jingzhong Zhang[1]

[1] School of Computer Science and Educational Software,
Guangzhou University, Guangzhou, China
{rysheng,zouyu020,zjz2271}@163.com
[2] South China Institute of Software Engineering,
Guangzhou University, Guangzhou, China
waniny@163.com

Abstract. Among a large number of digital educational resources, high-quality resources are very scarce. For quite some time, the shortage of high-quality resources is the bottleneck problem of education informatization. In this paper, we will introduce an innovational feature of digital educational resources – sustainability, which includes interactivity, transparency, and openness. The transparency means that users can obtain the original creating process of educational resources by their contents. The openness means that users can conveniently edit the contents of educational resources. Sustainable educational resources will become better and better in quality through continuous optimization. The sustainable optimization of educational resources needs intelligent subject knowledge platform. We will demonstrate several cases of sustainable optimization based on Super Sketchpad (SSP) which is an excellent intelligent subject knowledge platform.

Keywords: High-quality educational resources, sustainability, transparency, openness.

1 Introduction

The positive role of informatization in education is indubitable; however the educational goal that people expected has not been achieved. An investigation report by the U.S. congress shows that using computer and educational software is helpless to improve student achievement[1]. In China, government has spent much efforts on education informatization, but lots of facts show that the practical effects are not satisfactory and improvement of student achievement is not significant.

In fact, most of the existing educational resources are made by common software. Users hardly know how the resources are made. Therefore, it is very hard for the users to edit or recreate the resources appropriately according to their

---

[*] This work is supported by NSFC-GD project U1201252.
[**] Corresponding author.

H. Hong and C. Yap (Eds.): ICMS 2014, LNCS 8592, pp. 230–234, 2014.

personal demands. That is to say that the making process of the resources is not transparent and the resources are not open.

For example, it is difficult for users to know the exact process of making the simple geometric diagram shown in Fig. 1. If it is a piece of courseware of format PPT, WORD or FLASH, it would be very difficult to add a foot point from point O to segment BC into the diagram.

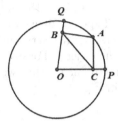

Fig. 1. A Simple Geometric Diagram

Therefore, although there have been heavy investments in educational informatization, high-quality educational resources are still scarce. The main reason is that most current educational resources are not made for sustainable development. Creation, application, and sharing of high-quality educational resources are the bottleneck problems of education informatization that have not been satisfactorily resolved, which need to be urgently addressed nowadays[2].

Intelligent subject knowledge platform can resolve the bottleneck problems of making high-quality educational resources [3]. So sustainable optimization of educational resources based on intelligent subject knowledge platform is feasible to deal with the problems of application and sharing of high-quality educational resources.

2   Sustainability

Sustainable optimization of educational resources means that users can conveniently edit or recreate the resources to satisfy their personal demands. Thus, the resources first should be editable by users; that is to say the resources are open. Second, the original process of making the resources should be easy to be understood and the way of editing resources should be convenient for users, which means that the resource is transparent. If an interactive educational resource has the features of openness and transparency, then we say it has the feature of sustainability. Sustainable educational resources will become better and better in quality through continuous optimization, their utilization rate can be improved and their useful lives can be prolonged, which will promote the applications and sharing of the resources.

2.1 Transparency

The transparency of an educational resource means the original making process of the resource is shown explicitly.

The diagram in Fig. 2 is created by Super Sketchpad (SSP) which is a mature and excellent intelligent subject knowledge platform [4]. Object information listed on the left side of the figure is generated automatically by the system according to construction steps. The list includes key information about the geometric diagram: (1) Object Index. It records the index of constructing geometric objects in diagram; (2) Object Type. It shows of which type (e.g. point, line, circle, and curve, etc.) the objects are; (3) Structural conditions. This indicates the conditions for constructing the objects. For example, according to the object information, we easily know that point C, the twelfth object, is the foot from point A on a circle to line PO in Fig. 2.

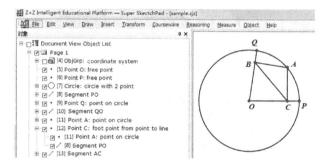

Fig. 2. Transparency

2.2 Openness

The openness of an educational resource means that the resource can be conveniently edited or recreated by users according to their personal demands.

In SSP, it is very easy to create the foot point from point O to line BC. One only needs to draw one line segment from O to BC with an intelligent pencil tool in one step, and then SSP will generate a tip automatically when the mouse approaches to the foot point. One can simply do more constructions, such as tracing line segment BC and point D and making a logical animation [5] for point A running along the circle O (Fig. 3(a)). Fig. 3(b) shows the results after the logical animation runs once.

3 Sustainable Optimization

If a resource is sustainable, it should be optimized easily. There are several ways of optimizing sustainable resources. For example, editing or changing the

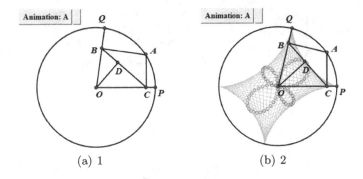

(a) 1   (b) 2

Fig. 3. Openness

properties of objects, deleting or adding some objects, decomposing one resource into different parts, combining some parts of different resources into a new one, and so on.

Here we show an optimization example by editing the properties of a resource which involves the number of intersection points of exponential and logarithmic functions.

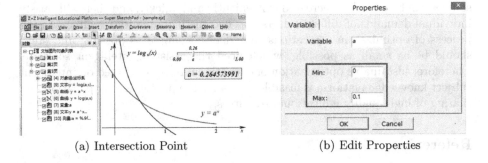

(a) Intersection Point   (b) Edit Properties

Fig. 4. Exponential and Logarithmic Functions

It seems easy to discuss the intersection of exponential function $y = a^x$ and logarithmic function $y = log(a, x)$. We can create the figure as shown in Fig. 4(a). Obviously, when $0 < a < 1$, there is an intersection point on $y = x$. Most people think that there should be only one intersection point.

However, this is not the fact. Through checking the object information listed in Fig. 4(a), we can make a simple optimization to the resource as follows: Change the range of variable a from $(0, 1)$ to $(0, 0.1)$ by opening the property dialog for variable a and inputting new data as shown in Fig. 4(b). By sliding variable scale continuously, we can find that there are three intersection points when a is less than about 0.06 (Fig. 6). In fact, it can be proved that the two curves have

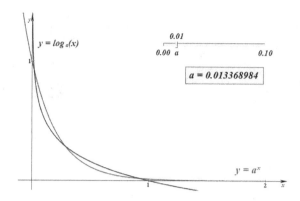

Fig. 5. Three Intersection Points

three intersection points when $0 < a < e^{-e} \approx 0.065988$. It is indeed difficult to imagine and understand the result if we don't make such an optimization.

4 Conclusion

With the features of interactivity, transparency, and openness, educational resources can be edited or recreated conveniently and quickly, and then will satisfy individual demands of different users and become high-quality resources. In the process of optimization, operations should be as simple as possible and steps should be as small as possible, which need the assist of intelligent software. Therefore, sustainable optimization of educational resources based on intelligent subject knowledge platform is feasible to resolve the problems of application and sharing of high-quality educational resources.

References

1. Wang, P.: Is educational software helpless to improve student achievement. The Chinese Journal of ICT in Education (May 2007)
2. Zhu, Z.: Ten Years of Chinese Educational Informatization. China Educational Technology (January 2011)
3. Zhang, J., Peng, X.: Design Idea of Super Sketchpad. The Monthly Journal of High School Mathematics (October 2012)
4. Zhang, J., Li, C.: Super Sketchpad (software). Beijing Normal University Press (2004-2014)
5. Zhang, J., Li, C.: An Introduction to Logical Animation. In: Li, H., J. Olver, P., Sommer, G. (eds.) IWMM-GIAE 2004. LNCS, vol. 3519, pp. 418–428. Springer, Heidelberg (2005)

A Touch-Operation-Based Dynamic Geometry System: Design and Implementation

Wei Su[1], Paul S. Wang[2], Chuan Cai[1], and Lian Li[1]

[1] School of Information Science & Engineering, Lanzhou University, China
[2] Department of Computer Science, Kent State University, USA
caichuan@lzu.edu.cn

Abstract. GeometryTouch is a dynamic geometry software system with touch operation. Developed by JavaScript and SVG, GeometryTouch is a Web-based application which can run on browsers of mobile devices. When using GeometryTouch, users can draw geometric figures and create or modify geometry-based interactive manipulatives. A virtual cursor is designed to implement precise operations in GeometryTouch. Geometric operations consist of 4 continuous actions and it is a challenge to implement the action of "the first point confirming". Three methods have been investigated to tackle the problem in the paper.

1 Introduction

A dynamic geometry software (DGS) system is a computer program for interactive creation and manipulation of geometric constructions. It enables users to construct geometric objects such as points, lines, and circles, and together with the dependencies that may relate the objects to each other. It is a dynamic construction, demonstration, and exploration tool for the study of mathematics. Teachers and students can use a DGS system to build and investigate mathematical models, objects, figures, diagrams, and graphs. In the past 20 years, many dynamic geometry software systems such as Cabri [2], Geometers SketchPad [10], Cinderella [3], Euclides [4], C.a.R. [1], GeoGebra [5], Kig [6], and KSeg [7], have been widely used in schools and colleges all over the world. Most of the DGS systems were designed as desktop-based systems without considering any usage on the Web or touch-based mobile devices. There are many difficulties or limitations to transplant them to the Web or a touch-orientated system.

Developed by the WME Group of Kent State University, GeometryEditor [11,9,8] is an attempt to build a totally Web-based DGS system. The main functionalities of GeometryEditor are as follows.

- Basic geometric object drawing. Drawing basic geometric shapes such as points, segments, rays, lines, circles, ellipses, and polygons.
- Geometric object construction. Constructing a new geometric object subject to mathematical relations with the constructed objects.
- Measurement. Measuring length, slope, radius, distance, area, circumference, perimeter, angle, coordinates, parallel, perpendicular, and tangent relations.

H. Hong and C. Yap (Eds.): ICMS 2014, LNCS 8592, pp. 235–239, 2014.

- – Loci and Envelopes. Constructing loci of moving points and envelopes of moving lines.
- – Animation. Moving and changing objects for illustration and demonstration.
- – Calculation. Creating and evaluating mathematical expressions based on the existing measurements and calculations.
- – Graphing. Plotting points and function graphs in coordinate systems.
- – Geometric transforms. Translation, reflection, dilation, and rotation of objects.
- – Defining macros. Grouping several steps of a construction into one command.

GeometryEditor defines a set of APIs to access constructions by other Web-based systems. GeoSite (http://wme.lzu.edu.cn/geosite), a Web site for hosting the GeometryEditor and the created geometry manipulatives by GeometryEditor's users, is also created by the WME Group.

2   GeometryTouch

GeometryTouch is a touch version of GemetryEditor. Developed by JavaScript and SVG, GeometryTouch is a Web-based application which can run on browsers of mobile devices such as iPad and Android Tablets. When using Geometry-Touch, users including specialists, teachers, and students may draw geometry figures and create or modify geometry-based interactive manipulatives. Figure 1 shows a screen shot of GeometryTouch.[1]

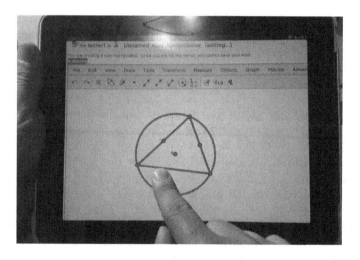

Fig. 1. The virtual cursor of GeometryTouch

In order to provide a touch-based user-friendly interface, touch events offer the ability to interpret finger activities on touch screens. Multi-touch gestures are predefined to interact with multi-touch devices. They can be recognized

---

[1] The GeometryTouch can be accessed on http://wme.lzu.edu.cn/geositeipad.

by detecting one or more touch events. In GeometryTouch, we define 9 basic gestures including 6 single-touch and 3 multi-touch operations. When using GeometryTouch, we call each operation like choosing, drawing, moving, or editing an object as a geometric operation. Different gestures can be mapped to different geometric operations. Each geometric operation also corresponds to one or more gestures. Figure 2 shows the relationship of gestures and some general geometric operations.

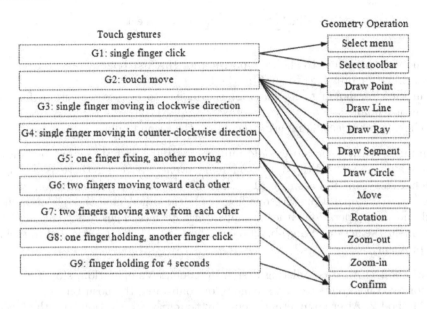

Fig. 2. The relationships between touch gestures and geometric operations

Most of geometric operations of a DGS system need accurate operations. However the size of human fingers and the limitation of sensing precision make precise interactions of touch on the screen difficult. A virtual cursor is designed in GeometryTouch to indicate the current focus. As shown in Figure 1, a small red cross will appear on the top of a touch finger and follow with the finger. It will change to a point when users begin to draw geometric objects. The virtual cursor makes users easily and conveniently implement the precise operations, such as selecting one point from several adjacent points in GeometryTouch.

In a DGS system, many geometric operations, such as drawing a circle, a ray, or a segment, consist of four continuous actions (see Figure 3):

- Step-a: locating the first point;
- Step-b: confirming the position of the first point;
- Step-c: locating the second point;
- Step-d: confirming the position of the second point and stop.

In the mouse and keyboard environment, the four actions can be performed as the following steps:

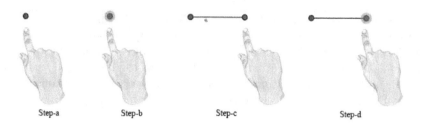

Fig. 3. Steps of drawing a segment

1. moving the mouse to choose a position for the first point;
2. clicking (or pressing) to confirm the first point;
3. moving (or holding and moving) the mouse to choose a position for the second point;
4. clicking (or releasing mouse) to confirm the second point.

In multi-touch devices, we could implement Step-a and Step-c by `touchmove` action and complete Step-d by `touchend` event. Although there is a single `touchmove` action and a single `click` action, one cannot perform a continuous move-and-click operation with one finger. On the other hand there is no mouse hold-and-move operation either in multi-touch devices. Thus it is a challenge to implement the action of Step-b. Three methods have been investigated to tackle the problem.

- Method 1: After completing Step-a by `touchmove` operation, stop and hold for several seconds (e.g., 4 seconds) for confirming the position.
- Method 2: After completing Step-a by `touchmove` operation, rise the finger for confirming the position.
- Method 3: After completing Step-a by one finger `touchmove` operation, click with another finger for confirming the position.

For Method 1, the novice users can explore and learn the operation easily even though there is no guide. However, it may cost more time to wait during the holding operation. If we set a shorter interval in the system, it may cause a wrong operation. For Method 2, the novice users may meet difficulties if there are no pre-instructions. But the experienced users can quickly draw the objects through the method. The operation of Method 3 looks like a mouse-click action. However it needs a second finger to assist. Both the novice and experienced users may feel tired if they use the system for a long time through this method.

3   Conclusion

GeometryTouch, a Web- and touch-based dynamic geometry software, is introduced. We use a virtual cursor to make touch interactions on the screen precisely and easily. For some continuous operations on touch device, it is hard to know whether users have completed the previous operation. In the paper, we provide three solutions for the problem. The three solutions may be useful instructions for designing a program on touch device. Some experiments on evaluating the three solutions are needed in the further work.

Acknowledgments. We'd like to thank Dr. Lai Xun for making available his GeometryEditor for further development.

This work is supported by Natural Science Foundation of China (61003139, 60903102), Innovation Found For Technology Based Firms(12C26216206998) of China, Fundamental Research Funds for Central Universities (lzujbky-2013-39, lzujbky-2013-188, lzujbky-2013-187)the MOE-Intel Joint Research Fund (MOE-INTEL-11-03). Any opinions, findings, and conclusions or recommendations expressed in this material are those of the authors and do not necessarily reflect the views of the funding agencies.

References

1. C.a.R., http://mathsrv.ku-eichstaett.de/MGF/homes/grothmann/java/zirkel/index.html
2. Cabri Geometry II, http://www-cabri.imag.fr/cabri2/accueil-e.php
3. Cinderella, http://www.cinderella.de/tiki-index.php
4. Euclides, http://www.euklides.hu/eng/euklides.htm
5. GeoGebra, http://www.geogebra.org
6. Kig, http://edu.kde.org/kig/
7. Kseg, http://www.mit.edu/ibaran/kseg.html
8. Xun, L., Wang, P.: An SVG Based Tool for Plane Geometry and Mathematics Education. In: IEEE Southeast Con. [s.n.], Fort Lauderdale (2005)
9. Xun, L., Wang, P.: GeoSVG:A Web-based Interactive Plane Geometry System for Mathematics Education. In: Proc.of the 2nd LASTED International Conference on Education and Technology, pp. 5–10 [s.n.], Puerto Vallarta (2006)
10. Jackiw, N.: The Geometers Sketchpad [Computer software]. Key Curriculum Press, Emeryville (1995)
11. Lai, X.: GeometryEditor: A Web-Based System for Authoring, Sharing and Support of Plane Geometry Manipulative for Mathematics Education. AmericaKent State University (2010)

OpenGeo: An Open Geometric Knowledge Base

Dongming Wang[1,2], Xiaoyu Chen[2], Wenya An[1], Lei Jiang[1], and Dan Song[1]

[1] LMIB - School of Mathematics and Systems Science, Beihang University,
Beijing 100191, China
[2] SKLSDE - School of Computer Science and Engineering, Beihang University,
Beijing 100191, China

Abstract. OpenGeo is an enhanced version of the geometric knowledge base developed by Chen, Huang, and Wang, which is equipped with web-based interfaces and new management facilities and made open and online. The kernel of the knowledge base consists of typical geometric knowledge objects such as definitions, theorems, and proofs. Several tools have been developed to support users to manage the knowledge objects contained in OpenGeo. Users can create new knowledge objects and add them to OpenGeo.

Keywords: Data management, geometry software, knowledge object, open database.

1 Introduction

There is a large amount of geometric knowledge resources created by researchers and educators and accessible in different ways (see for example [13,8,10]). The knowledge data in such resources are represented with different structures in different formats. To facilitate data exchange among those resources, one has to study how to standardize, formalize, and structure geometric knowledge data. This question has been partially answered by Chen and others in [1] along with the development of a geometric knowledge base. Following the work of [1], we present in this paper an enhanced geometric knowledge base, called OpenGeo, which is equipped with web-based interfaces and new management facilities and will be open and online. The kernel of OpenGeo consists of typical geometric knowledge objects such as definitions, theorems, and proofs. Based on the methods proposed in [3], several tools have been developed to support users to manage the knowledge objects contained in OpenGeo. For example, using the developed tools, (1) knowledge objects can be edited or deleted; (2) meta-information (e.g., language, format, and keyword) can be annotated for organizing and classifying knowledge objects; (3) revisions of knowledge objects can be recorded; (4) knowledge objects can be retrieved in meta-information-based ways; (5) knowledge objects can be rated and commented for screening high-quality versions. In addition, users can create new knowledge objects (containing texts, images, diagrams, files, videos, and audios) and add them to OpenGeo.

H. Hong and C. Yap (Eds.): ICMS 2014, LNCS 8592, pp. 240–245, 2014.

OpenGeo is created for the purpose of research and education. Creative Commons Attribution-ShareAlike license is adopted as its main content license. OpenGeo may serve as a public resource for users to test, for instance, geometric theorem provers and problem solvers and as an infrastructure for developing new educational applications (e.g., generation of textbooks and courses) in online learning environments.

2   The Knowledge Base

2.1   Ontologies for Geometric Knowledge Objects

By *knowledge object*, we mean an individual knowledge unit that can be recognized, differentiated, understood, and manipulated in the process of management. We adopt ontology [4] to formally specify geometric knowledge objects and intrinsic relations among them. Ontology is often used to represent models composed of sets of objects, attributes, and relations. In most examples, classes, individuals, attributes, and properties are used as ontology terminologies to explicitly specify domain concepts. They may be expressed by using, e.g., the Web Ontology Language (OWL [4]). OWL is recognized by the World Wide Web Consortium as a formal language for the representation of ontology statements. Protégé [5] is one of the most widely used ontology editors which allows users to export ontologies of the OWL form. The structure of the knowledge base, OpenGeo, presented in this paper is built up by using Protégé.

Geometric knowledge objects can be categorized into specific classes, such as definition, proposition, problem, proof, solution, and method. Each class may contain several data items, such as natural language expression, formal expression, algebraic expression, nondegeneracy condition, diagram, and keyword. An ontology class may be constructed for each class O of geometric knowledge objects or each data item of O. Ontology classes constructed for data items of O can be divided into resource classes and annotation classes. The resource classes specify different types of media (such as text, audio, and video) to interpret and/or illustrate the knowledge objects in O, while the annotation classes are used to annotate the knowledge objects in O as well as the media specified by the resource classes.

The constructed ontology classes may satisfy certain properties. Both logical relations among geometric knowledge objects and semantic relations between geometric knowledge objects and data items need be considered. There are three kinds of logical relations among geometric knowledge objects: inheritance relations, dependence relations, and incidence relations. These logical relations may be described by introducing ontology properties like `hasFather`, `deriveFrom`, and `hasProof`. Semantic relations may be represented by means of the relationships "geometric knowledge objects contain media which are individuals of the corresponding resource classes" and "geometric knowledge objects and media have annotations which are individuals of the corresponding annotation classes." They may be depicted by ontology properties like `hasVideo`, `hasKeyword`, and `hasFormat`.

We have been constructing ontologies for geometric knowledge objects. For each specific object, its annotation and resource attributes and relations with other objects can be easily obtained on the basis of the ontology properties. For example, from the `hasDiagram` property, one may obtain the diagram to which the object is related; the `hasProof` and `forTheorem` properties may indicate the logical relations between a theorem object and a proof object.

2.2   The Database Schema

Geometric knowledge objects are stored in the database of OpenGeo and are managed through interfaces with management facilities and external software tools. Without using entity-relation diagrams [6], relational data tables for the database are created automatically from the ontologies for geometric knowledge objects and their modifications can be automated. We use Protégé to edit ontologies for geometric knowledge objects and export them to OWL files. A parser is developed to process the contents of the OWL files and to generate SQL statements for automated creation of relational data tables. When some ontology is modified, the parser would re-generate new SQL statements automatically for updating the relational data tables. The paradigm of automated database schema generation explained above is also used for the form generation of the user interface of OpenGeo.

Geometric knowledge objects may be modified or refined gradually. One object may exist in different versions created by different users. Therefore, database schema need be designed to meet the management requirements of multi-users and multi-versions. In OpenGeo, relational data tables are created to store data about users and about which geometric knowledge objects the users have handled.

Figure 1 illustrates the structure of the relational data tables Knowledge, Process, Image, Users, and Know2res we have constructed, where an arrow $A \to B$ indicates the link from the foreign key B in one table to the primary key A in

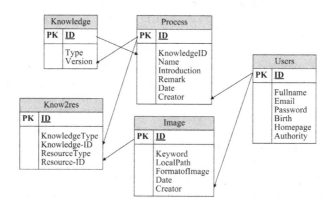

Fig. 1. Part of the database schema for OpenGeo

another table. The table Process records the results of processing for geometric knowledge objects (e.g., proofs of theorems or steps of drawing diagrams for theorems). Name, Introduction, and Remark in Process are the data items and ID, the primary key of Process, identifies the version of a knowledge object of type Process. The table Image contains data items such as Keyword, LocalPath, and ID for images uploaded by users. The table Know2res is used to store relations between knowledge objects stored in Process and images stored in Image. The table Users records some basic information about users. Its primary key ID is the identity of the creator of the knowledge objects in Process and the images in Image. Since different users may create different versions of the same knowledge object, Type and Version are introduced to indicate of which type (Definition, Theorem, or Process, ...) and of which version the knowledge objects are, respectively, in the Knowledge table. The current version corresponds to the ID in the Process table. Other types of knowledge objects and media are structured similarly.

3   The User Interface

Geometric knowledge sharing can be realized by applying network techniques. The LAMP [7] (Linux Apache MySQL PHP/Perl/Python) is a typical framework under which web servers, databases, and scripting techniques can be integrated into powerful web application platforms. OpenGeo is developed by using LAMP techniques. It also integrates tools developed by the third party to facilitate users to create data of special format. In particular, MathEdit [9] and Sketchometry [14] are used for editing formatted formulas and demo scripts. Users can edit complicated formulas in the manner of WYSIWYG in MathEdit that runs in the browser. Sketchometry is a browser-based interactive tool for drawing dynamic geometric diagrams by clicking and dragging with mouse.

We have developed a user-friendly interface for low-level management of geometric knowledge objects in OpenGeo.[1] Through this interface, the user can easily edit, modify, and delete geometric knowledge objects. To create a geometric knowledge object, such as a definition, a theorem, or a proof, one can upload media and fill in the form of the user interface generated according to the ontology of the knowledge class to which the object belongs. To modify a created object of a certain version, one can revise the data displayed in the form of the user interface; the resulting object is stored as a new version. To delete a created object, one needs to simply remove the ID of its current version; in this case, the media for the object remain stored in the knowledge base.

In OpenGeo, geometric knowledge objects are stored in multi-versions with multi-users. One can browse any version of an object and check who created and edited which version of the object. Mathematical formulas and dynamic

---

[1] For more advanced management of geometric knowledge objects, OpenGeo interfaces with other sophisticated software tools such as GeoGebra (a dynamic geometry software system [10]), GEOTHER (a geometry theorem prover [12]), and GeoText (a dynamic geometry textbook [11]).

diagrams can be displayed or animated, while media such as images, audios, and videos can be viewed or played online. Geometric knowledge objects can be retrieved by searching keywords or according to classifications based on ontology classes.

OpenGeo users are allowed to reuse and evaluate geometric knowledge objects stored in the knowledge base. Before creating a geometric knowledge object, the user is urged to search the existing versions of relevant objects in the knowledge base. The version which is first found to be appropriate by the user is regarded as an initial version. The user can keep the reusable part and modify other parts of the initial version and then save the revised version as a new version of the object. This permits each user to contribute data to the knowledge base and the contributed data are subject to evaluation and further revisions by other users. It is expected that in this way the knowledge objects in OpenGeo will be improved gradually by the users.

Media for knowledge objects can be easily reused and be shared by many versions of the objects. Users can evaluate the media and rate the rationality and quality of the interpretations for the objects by using scores. For each object, the provisionally best version of it can be determined through interactions between creators and viewers using collaborative filtering techniques.

The user interface also includes other auxiliary modules. One of them is the user module, through which users can modify their profiles and view the histories of their creations, modifications, and evaluations.

4   Ongoing and Future Work

Currently, we are formalizing geometric theorems in the OpenGeo collection and developing semantic querying tools based on images of diagrams and formalized geometric statements. We expect to complete these tasks and release a preliminary version of OpenGeo in early 2015.

Very recently, we have proposed a general approach for discovering geometric theorems automatically from images of diagrams [2]. Preliminary experiments have shown the feasibility and efficiency of our approach. We plan to implement and apply this approach to deal with different kinds of images, including photographed images of hand-drawn diagrams. We will develop software tools to help generate geometric theorems for OpenGeo from images of diagrams taken from the Internet or hand-sketched on mobile devices.

References

1. Chen, X., Huang, Y., Wang, D.: On the design and implementation of a geometric knowledge base. In: Sturm, T., Zengler, C. (eds.) ADG 2008. LNCS, vol. 6301, pp. 22–41. Springer, Heidelberg (2011)
2. Chen, X., Song, D., Wang, D.: Automated generation of geometric theorems from images of diagrams. Submitted to Geometric Reasoning — Special issue of Annals of Mathematics and Articial Intelligence (with the editor after minor revisions). Available at arXiv:1406.1638 (2014)

3. Chen, X., Wang, D.: Management of geometric knowledge in textbooks. Data & Knowledge Engineering 73, 43–57 (2012)
4. McGuinness, D.L., van Harmelen, F.: OWL web ontology language overview. W3C Recommendation, February 10 (2004)
5. Gennari, J.H., Musen, M.A., Fergerson, R.W., Grosso, W.E., Crubézy, M., Eriksson, H., Noy, N.F., Tu, S.W.: The evolution of Protégé: an environment for knowledge-based systems development. International Journal of Human-Computer Studies 58(1), 89–123 (2003)
6. Batini, C., Ceri, S., Navathe, S.B.: Conceptual Database Design: An Entity-Relationship Approach. Benjamin/Cummings (1992)
7. Gerner, J., Owens, M., Naramore, E., Warden, M.: Professional Lamp: Linux, Apache, MySQL and PHP5 Web Development. John Wiley & Sons (2006)
8. Quaresma, P.: Thousands of geometric problems for geometric theorem provers (TGTP). In: Schreck, P., Narboux, J., Richter-Gebert, J. (eds.) ADG 2010. LNCS (LNAI), vol. 6877, pp. 169–181. Springer, Heidelberg (2011)
9. Su, W., Wang, P., Li, L., Li, G., Zhao, Y.: MathEdit, A browser-based visual mathematics expression editor. In: Electronic Proceedings of the 11th Asian Technology Conference in Mathematics, Hong Kong, December 12-16, pp. 271–279. ATCM, Inc. (2006)
10. GeoGebra: http://www.geogebra.org/ (accessed May 18, 2014)
11. GeoText: http://geo.cc4cm.org/text/ (accessed May 18, 2014)
12. GEOTHER: http://www-salsa.lip6.fr/~wang/GEOTHER/ (accessed May 18, 2014)
13. Intergeo: http://i2geo.net/ (accessed May 18, 2014)
14. JSXGraph: http://jsxgraph.uni-bayreuth.de/wp/ (accessed May 18, 2014)

On Computing a Cell Decomposition of a Real Surface Containing Infinitely Many Singularities

Daniel J. Bates[1], Daniel A. Brake[2], Jonathan D. Hauenstein[2],
Andrew J. Sommese[3], and Charles W. Wampler[4]

[1] Colorado State University, USA
bates@math.colostate.edu
www.math.colostate.edu/~bates
[2] North Carolina State University, USA
danielthebrake@gmail.com, hauenstein@ncsu.edu
danielthebrake.org
www.math.ncsu.edu/~jdhauens
[3] University of Notre Dame, USA
sommese@nd.edu
www.nd.edu/~sommese
[4] General Motors Research and Development, USA
charles.w.wampler@gm.com
www.nd.edu/~cwample1

Abstract. Numerical algorithms for decomposing the real points of a complex curve or surface in any number of variables have been developed and implemented in the new software package `Bertini_real`. These algorithms use homotopy continuation to produce a cell decomposition. The previously existing algorithm for surfaces is restricted to the "almost smooth" case, i.e., the given surface must contain only finitely many singular points. We describe the use of isosingular deflation to remove this almost smooth condition and describe an implementation of deflation via `Bertini` with `MATLAB`.

Keywords: Real decomposition, real algebraic set, numerical algebraic geometry, isosingular deflation, homotopy continuation.

1   Introduction

Polynomial systems appear throughout the sciences, engineering, and mathematics. Given a polynomial system, $f(z)$, with N polynomials in n variables, a common problem is to find all solutions $\widehat{z}$ in $\mathbb{C}$ or in $\mathbb{R}$ such that $f(\widehat{z}) = 0$, i.e., the *solution set* of $f(z) = 0$, also denoted $V(f)$. Such a solution set (for either $\mathbb{C}$ or $\mathbb{R}$) may consist of points, curves, surfaces, and/or higher-dimensional components.

In *numerical algebraic geometry*, there are now several numerical methods to produce the *numerical irreducible decomposition* over $\mathbb{C}$ of $V(f)$. It is a fundamental fact from algebraic geometry that a degree d irreducible algebraic set meets a general linear space of complementary dimension in d distinct points.

H. Hong and C. Yap (Eds.): ICMS 2014, LNCS 8592, pp. 246–252, 2014.
© Springer-Verlag Berlin Heidelberg 2014

For each irreducible component $A \subset V(f)$ of dimension m, the numerical irreducible decomposition contains a *witness set*, which is the triplet of $f(z)$, $L(z)$, and W, where L is a general set of m linear equations, and W consists of numerical approximations to the set of d points $A \cap V(L)$. The books [3,10] discuss these concepts and the associated algorithms, and many of these methods are implemented in [4].

Working over $\mathbb{R}$ is significantly different than working over $\mathbb{C}$, reflecting a more complicated geometry. Real slices of real algebraic sets do not behave so uniformly as their complex counterparts, so there is no simple real analog to the witness sets that suffice when working over $\mathbb{C}$. Instead, as described below, we break the real sets into a finite number of pieces, each having a uniform behavior within. The decomposition of the real subsets within complex curves was accomplished in [9], while the decomposition of the real subsets within an adequately nice complex surface was achieved in [5]. In particular, the surface was required to be "almost smooth," meaning that it could contain at most a finite number of singularities. In this article, we remove the almost smooth condition from the surface case by incorporating *isosingular deflation* [6] into the approach of [5]. The resulting surface method and the method for curves are both implemented in the software package `Bertini_real` [2].

A fundamental problem with real solution sets of polynomial systems is the choice of a data type. We have opted for a topological description, a *cell decomposition*, of real curves and surfaces, dependent on the (typically random) choice of two linear projections. The next section provides some basic details on this data type and the previously known numerical algebraic geometry method for computing it. Section 3 illustrates the need for isosingular deflation, which is then described in §4. The inclusion of isosingular deflation is finally briefly described and illustrated in §5.

2   Cell Decomposition

The cell decomposition of an algebraic surface [5] breaks it into a finite number of regions over which the implicit function theorem holds. The construction is related to Morse theory and similar in essence to the Cylindrical Algebraic Decomposition [1], although the specifics of the data structure and the algorithms for computing it are quite different. The decomposition consists of '2-cells' or *faces*, which are bounded by '1-cells' or *edges*, which are themselves bounded by vertices. Each face and edge is equipped with a generic point in the middle and a homotopy such that the generic point can be tracked along the face.

This decomposition is computed with respect to two real linear projections, $\pi_1(x)$ and $\pi_2(x)$, typically chosen randomly, which give rise to the *implicit* parameterization of the surface. Each face describes some portion of the surface with boundary either a curve over which the generic point cannot be tracked or part of an artificially imposed edge.

The process for decomposing an irreducible algebraic surface is illustrated in Fig. 1 for a surface given by the Zitrus system [7]:

$$f(x, y, z) = x^2 + z^2 + y^3(y - 1)^3. \tag{1}$$

Letting S denote the surface to be decomposed, this process is loosely given as follows. Given a witness set for S, the techniques of numerical algebraic geometry allow us to restrict all of the following computations to S, even in the case where $V(f)$ contains other irreducible components.

1. **Compute the *critical set* C of S with respect to π_1, π_2.** C consists of points $\widehat{x}$ on S that are either singular or include the direction of the projection in the tangent space at $\widehat{x}$. The points of C are solutions of the system

$$\left[\det \begin{pmatrix} f(x) \\ \begin{pmatrix} Jf \\ J\pi_1 \\ J\pi_2 \end{pmatrix} \end{pmatrix} \right] = 0,$$

where J means the Jacobian matrix of partial derivatives. The top and bottom edges of the faces in the eventual surface decomposition will be edges coming from the curve decomposition of C. See the top left of Fig. 1, where for the particular projection we illustrate, the critical curve consists of a ring around the surface and the two singular points at its extremities. The critical curve is itself decomposed with respect to π_1.

2. **Intersect with a suitably chosen sphere.** Because the surface might be noncompact, with parts that extend to infinity, we consider only the compact part of it lying within a suitably chosen sphere. In particular, after computing the critical curve, the locations of all topologically interesting parts of the surface are known, so we may choose a sphere containing all critical points of the critical curve, and intersect it with S. In the Zitrus example, the sphere intersection curve, i.e., the intersection of S with a sphere, is empty because the surface is compact.

3. **Slice at all critical points of π_1, and halfway between.** The boundary of a face is a graph of edges of curve decompositions, the right and left of which are slices of the surface at critical points under the first projection, π_1. In contrast, the midpoint of each face is the midpoint of an edge of a *midslice*, i.e., a slice of the surface at a point halfway between two critical points under the first projection, π_1. Each slice is the intersection of the surface with a plane corresponding to fixing π_1 at a projection value, and decomposing with respect to π_2. This step is the top right in Fig. 1.

4. **Connect midpoints to build faces.** For each edge of each midslice, track its midpoint to each candidate edge of each left- and right-bounding critical slice. Using a specially crafted homotopy which couples the midpoint, top, and bottom points, as in [5], we establish the network of connections between midpoints. This step corresponds to the bottom left in Fig. 1, where each color corresponds to an individual face. After this step is complete we have a topologically correct triangulation of the surface.

Compute critical set   Slice

Connect   Refine

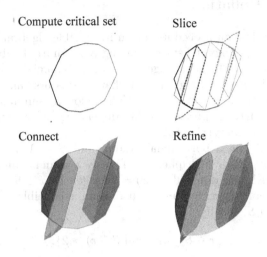

Fig. 1. Computing a cell decomposition of the Zitrus

5. **Refine and smooth.** The initially computed decomposition is rough, containing only the bare skeleton of the surface. Since each decomposition is equipped not only with a graph of connecting points, but also with a homotopy and generic point, we can refine the decomposition to obtain a more accurate geometric representation of S. The lower right figure of Fig. 1 is a moderately fine smoothing of the Zitrus.

3 Singular Curves on Surfaces

The Zitrus surface described in §2 is almost smooth since it has only two singular points. In the almost smooth case, the singular points are simply part of the critical set. In particular, numerical tracking does not need to be performed starting from such singular points.

In contrast, when the surface contains a curve of singularities, one needs the ability to numerically track along these singular curves to compute the cell decomposition. An example of such a curve is the "handle" of the Whitney umbrella [7], i.e., the z-axis, $x = y = 0$, in the surface implicity defined by $x^2 - y^2z = 0$. As another example, consider the Solitude surface [7] defined by the vanishing of

$$f(x,y,z) = x^2yz + xy^2 + y^3 + y^3z - x^2z^2. \tag{2}$$

There are two singular lines on this surface, one is defined by $x = y = 0$ while the other is defined by $y = z = 0$. In order to perform tracking on such singular curves, we use isosingular deflation [6], which is summarized in the next section.

4 Isosingular Deflation

Deflation is a regularization procedure for an irreducible algebraic set $X \subset \mathbb{C}^N$ which produces a new polynomial system having X as an irreducible component of generic multiplicity 1. The advantage of such a polynomial system is that it facilitates numerical path tracking on X. Deflation was first introduced in the specific setting of polynomial systems in [8]. The following summarizes the more recent isosingular deflation approach of [6], depending on determinants, as is currently being used in `Bertini_real`.

Let $f : \mathbb{C}^N \to \mathbb{C}^n$ be a polynomial system and $S \subset \mathcal{V}(f) \subset \mathbb{C}^N$ be an irreducible surface of generic multiplicity 1. That is, S is an irreducible algebraic set of dimension 2 such that $\dim \operatorname{null} Jf(x) = 2$ for generic $x \in S$, where $Jf(x)$ is the Jacobian matrix of f at x. Suppose that C is an irreducible curve contained in the singular set of S, that is,

$$C \subset \{x \in S \mid \dim \operatorname{null} Jf(x) > 2\}.$$

Isosingular deflation results in a polynomial system $g(z)$ such that C is an irreducible component of the solution set of $g(z) = 0$ of generic multiplicity 1. Letting c be a generic point of curve C, isosingular deflation proceeds as follows:

1. Initialize $g := f$.
2. Loop until $\dim \operatorname{null} Jg(c) = 1$:
   (a) Set $r := \operatorname{rank} Jg(c)$.
   (b) Append to g the $(r+1) \times (r+1)$ determinants of $Jg(x)$.

This loop will terminate and produce a polynomial system that can be used to perform computations on C. If the surface S was of multiplicity greater than 1, a minor modification to the stopping criterion would give a procedure for deflating S.

The following example illustrates the deflation of the singular curves on the Solitude surface.

Example 1. Let f be as in (2) and consider $C = \{(a,0,0) \mid a \in \mathbb{C}\}$. For simplicity of presentation, we take $c = (1,0,0)$. Since all first order partial derivatives of f vanish at c, $r = 0$ and we add all first partial derivatives to f, yielding

$$g(x,y,z) = \begin{bmatrix} x^2yz + xy^2 + y^3 + y^3z - x^2z^2 \\ y^2 + 2xyz - 2xz^2 \\ 2xy + x^2z + 3y^2z + 3y^2 \\ x^2y - 2x^2z + y^3 \end{bmatrix}$$

It is easy to verify that $\dim \operatorname{null} Jg(c) = 1$ so g has deflated C.

Now, we consider the other curve $C' = \{(0,0,a) \mid a \in \mathbb{C}\}$ with $c' = (0,0,1)$. The first iteration of isosingular deflation again produces g as above, but since $\dim \operatorname{null} Jg(c') = 2$, we need to perform another iteration. Adding in the 2×2 determinants of $Jg(x,y,z)$ produces a polynomial system $g' : \mathbb{C}^3 \to \mathbb{C}^{22}$ such that $\dim \operatorname{null} Jg'(c') = 1$.

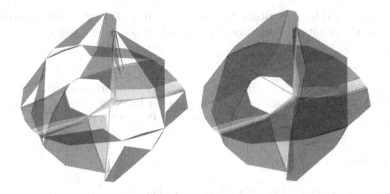

Fig. 2. Comparison of results from decomposition without deflation (left) and with deflation (right). This is the raw decomposition before refinement.

In the procedure above, the required null space dimension was known *a priori* The required determinants are computed via MATLAB with the rank r computed in multiple precision using Bertini. For deflating at points for which the corresponding dimension may not be known, we use as stopping criterion the isosingular stabilization test described in [6], as implemented in Bertini.

5   Decomposing Surfaces

With isosingular deflation [6], we have now removed the almost smooth restriction from [5] so that this new algorithm can produce a cell decomposition of the set of real points on a complex surface regardless of the presence of singular curves on the surface. To demonstrate, Fig. 2 presents the Solitude surface defined by (2). The figure on the left shows the decomposition where the presence of the singular curves is ignored, demonstrating the failure of the decomposition method without using isosingular deflation. The figure on the right uses isosingular deflation to track along the singular curves, yielding a complete decomposition. In this figure, part of the singular line corresponding to the x-axis is isolated in that it is not an edge of any face, similar to the "handle" of the Whitney umbrella [7].

6   Conclusion

The use of isosingular deflation permits numerical path tracking to be performed on singular sets. We have applied this technique to remove the almost smooth assumption for the algorithm presented in [5] to allow one to compute a cell decomposition of the real points of any complex surface. There is no theoretical limitation on the number of variables. The drawback of using the determinantal formulation of isosingular deflation is the potentially large number of additional polynomials added to the system. We are currently exploring various approaches for limiting the number of additional polynomials needed to deflate the components of interest.

Acknowledgments. All authors were partially supported by AFOSR grant FA8650-13-1-7317. DJB was partially supported by NSF grant DMS-1025564. DAB and JDH were additionally supported by DARPA YFA.

References

1. Arnon, D.S., Collins, G.E., McCallum, S.: Cylindrical algebraic decomposition I: The basic algorithm. SIAM J. Computing 13(4), 865–877 (1984)
2. Brake, D.A., Bates, D.J., Hao, W., Hauenstein, J.D., Sommese, A.J., Wampler, C.W.: Bertini_real: software for real algebraic sets, `bertinireal.com`
3. Bates, D.J., Hauenstein, J.D., Sommese, A.J., Wampler, C.W.: Numerically Solving Polynomial Systems with Bertini. SIAM, Philadelphia (2013)
4. Bates, D.J., Hauenstein, J.D., Sommese, A.J., Wampler, C.W.: Bertini: software for numerical algebraic geometry, `bertini.nd.edu`
5. Besana, G.M., Di Rocco, S., Hauenstein, J.D., Sommese, A.J., Wampler, C.W.: Cell decomposition of almost smooth real algebraic surfaces. Numer. Algorithms 63(4), 645–678 (2013)
6. Hauenstein, J.D., Wampler, C.W.: Isosingular sets and deflation. Found. Comp. Math. 13(3), 371–403 (2013)
7. Hauser, H., Schicho, J.: Algebraic Surfaces, `www1-c703.uibk.ac.at/mathematik/project/bildergalerie/gallery.html` (accessed May 19, 2014)
8. Leykin, A., Verschelde, J., Zhao, A.: Newton's method with deflation for isolated singularities of polynomial systems. Theor. Comp. Sci. 359(1-3), 111–122 (2006)
9. Lu, Y., Bates, D.J., Sommese, A.J., Wampler, C.W.: Finding all real points of a complex curve. Contemp. Math. 448, 183–205 (2007)
10. Sommese, A.J., Wampler, C.W.: The Numerical Solution to Systems of Polynomials Arising in Engineering and Science. World Scientific, Singapore (2005)

Robustly and Efficiently Computing Algebraic Curves and Surfaces

Eric Berberich

Max-Planck-Institut für Informatik, Germany
eric@mpi-inf.mpg.de
http://people.mpi-inf.mpg.de/~eric

Abstract. Computing with curved geometric objects forms the basis for many algorithms in areas such as geometric modeling, computer aided design and robot motion planning. In general, such computations cannot be carried out reliably with standard machine precision arithmetic.

Slightly more than a decade ago robustly and efficiently dealing with conics and Bézier curves in 2D and quadrics and splines in 3D was considered an enormous challenge. This picture has changed. Our first successes were achieved for conics and quadrics, mainly relying on properties of the involved low-degree polynomials. In a second step, to tackle general algebraic curves and surfaces, we exploited more involved mathematical tools such as subresultants. In addition with clever filtering techniques, these methods already beat the previous specialized solutions. The most recent *drastical* success in performance gain for algebraic curves is due to several ingredients: The central one consists of a cylindrical algebraic decomposition with a revised lifting step. Using results from algebraic geometry we avoid any change of coordinates and replace the costly symbolic operations by numerical tools. The new algorithms for curve topology computation only need to compute the resultant and the gcd of bivariate polynomials and to perform numerical root finding. For the symbolic operations, we can rely on implementations exploiting graphics hardware, which is several magnitudes faster than corresponding CPU implementations.

All algorithms have been implemented as contributions to the C++ project CGAL. Excellent practical behavior of our algorithms has been shown in exhaustive sets of experiments, where we compared them with our previous and recent competing approaches. Beyond, the algorithms are also proven to be efficient in theory. Recent work shows that our implemented and practical algorithm needs $\tilde{O}(d^6 + d^5\tau)$ bit operations (d degree, τ bitsize of coefficients) to compute the topology of an algebraic curve and for solving bivariate systems.

Joint work with Pavel Emeliyanenko, Michael Kerber, Kurt Mehlhorn, Michael Sagraloff, Alexander Kobel, and Pengming Wang.

Keywords: algebraic curves, algebraic surfaces, geometric computing, symbolic operations, arrangements.

H. Hong and C. Yap (Eds.): ICMS 2014, LNCS 8592, pp. 253–260, 2014.

1   Computing with Algebraic Curves and Surfaces

Computing with geometric objects has always been a challenge. The first implementations in the 90s, that only dealt with linear objects, basically line segments, were facing robustness issues due to rounding errors of built-in machine arithmetic. It was needed to implement reliable number types. Such types (e.g. software rational numbers) solved the robustness issue, but introduced a performance penalty. They are much slower than machine arithmetic, due to memory allocations. Software libraries, like CGAL, the Computational Geometry Algorithms Library[1] [14], overcome this disadvantage by relying on various filter techniques, such as static, dynamic and combinatorial ones. The state-of-the-art implementations of many algorithms loose only a few percent of runtime, compared to non-reliable versions based on hardware arithmetic. Besides the numerical problems geometric computing is constantly handling with degenerate situations.

Dealing with more general algebraic curves and surfaces complicates both aspects. Real algebraic curves and surfaces (of degree d) are defined as vanishing sets of bivariate and trivariate polynomials:

$$C_f := \{(x, y) \in \mathbb{R}^2 \mid 0 = f(x, y) \in \mathbb{Z}[x, y]\}$$

and

$$S_f := \{(x, y, z) \in \mathbb{R}^3 \mid 0 = f(x, y, z) \in \mathbb{Z}[x, y, z]\}.$$

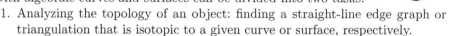

The simplest algebraic curves and surfaces are lines in 2D and planes in 3D, the ones with $d = 1$. Going beyond that limit has been considered practically impossible slightly more than a decade ago. Meanwhile, software that breaks through the linear limit has been presented. Computing with algebraic curves and surfaces can be divided into two tasks:

1. Analyzing the topology of an object: finding a straight-line edge graph or triangulation that is isotopic to a given curve or surface, respectively.
2. Computing the arrangement induced by a set of n objects: decomposing the ambient space into maximal cells such that all points of a cell are attained by the same set of objects.

The later is basically achievable for algebraic curves in terms of the former using CGAL's arrangement package [16]. The main technique to construct a planar arrangement is the plane-sweep algorithm. In order to use CGAL's implementation, it is required to provide a set of geometric predicates and constructions: for instance, decomposing a curve into x-monotone subcurves, comparing two points lexicographically, determining the relative vertical order of a point and an x-monotone subcurve, or computing the intersections of two x-monotone subcurves. With basically the same needs, CGAL provides tools to overlay two arrangements, perform point location queries, or create curved polygons. Computing arrangements in 3D while handling all degeneracies is still not fully achieved, even if only linear objects (like planes or triangles) play a role.

---

[1] http://www.cgal.org

Algebraic Curves. The general technique how we analyze a single algebraic curve follows the classical *cylindrical algebraic decomposition* scheme [6], which exploits the fact that algebraic curves are *delineable*. That is, a curve can be split, over disjoint intervals, into arcs, where each arc can be represented by an implicit function. It consists of three steps: Projection, Lifting and Connection. The *projection* step consists of first computing the resultant $\mathrm{res}(f, f_y, y)$, whose roots, as known from elimination theory, form a superset of the x-coordinates of a curve's C_f x-critical and singular points and then isolating the real roots of this univariate polynomial. It is known that for a curve of degree d, the resultant has degree $O(d^2)$. Next, these coordinates are *lifted*: For a given root x_0, this is done by isolating the real roots of $f(x_0, y)$, which is the main difficulty, in particular for high-degree algebraic curves: The univariate *fiber* polynomial $f(x_0, y)$ is usually formed by non-rational coefficients and has multiple roots. In addition, we lift rational values in all open intervals induced by the x-critical coordinates. This is done by choosing for each interval I a rational q_I and isolate the real roots of the univariate polynomial $f(q_I, y) \in Q[y]$. It has rational coefficients and only simple roots. The result of the lifting phase is the vertex set of the desired straight-line graph. The final *connection* step inserts segments in between these vertices to capture the topology of the curve.

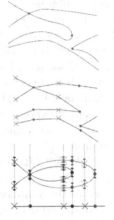

The analysis of a single curve is required to split a curve into its x-monotone subcurves, the other needed predicates can be expressed in terms of the analysis of pairs of curves (C_f, C_g). That task is similar: Project and compute x-critical coordinates by additionally isolating the real roots of another resultant, namely $\mathrm{res}(f, g, y)$. Next, these values (and rational coordinates for each induced open interval) are lifted: We obtain a vertical order of the two curves over the intervals and over each x-critical coordinate. Only for the latter, the order can contain discrete intersection information (if f and g are coprime, which can be assumed by applying a pre-processing; otherwise, two curves share a common part).

Our achievements for low-degree and arbitrary-degree curves mainly differ in the technique to lift x-critical points. For conics, details are in [1], the resultant of a single curve is a polynomial of degree 2, which has either two simple solutions or a multiple rational one. As C_{f_y} is a line, the critical points can be expressed by *one-root numbers*, that is, they are of the form $a + b\sqrt{c}$, $a, b, c \in \mathbb{Q}, c \geq 0$. Similarly, the y-coordinates of a conic can be parameterized in x with using at most one square-root. The coordinates of intersection points are roots of a polynomial of degree 4, and thus are either simple roots, or one-root numbers. There exists number types in software (`CORE::expr`, `leda::real`) that allow exact arithmetic with such numbers, which enables to implement the needed predicates.

For cubics, algebraic curves of degree 3, the analyses of curves and pair of such exploit geometric properties of the curves in order to avoid arithmetic on irrationals as far as possible [9]. This goal is basically achieved by a careful case analysis, for which the approach often relies on evaluations of f_y and f_{yy}, the detection of sign changes, and signs of discriminants. For the analysis of pairs of cubics, the approach already needs for the most general case the first subresultant: $sres_1(f, g, y)$ in order to decide an intersection over a critical x_i. The overall distinction is lengthy and also imposes various conditions on the input curves, which often requires the approach to transform into a generic coordinate system (shearing) and back (shearing-back), which comes at additional cost.

The work for cubics also builds the basis for general algebraic curves. Both follow the abstraction that the analysis of single curves and pairs of them are sufficient to compute arrangements. The work on single curves [8] introduces a uniform numerical lifting (bitstream-Descartes method), which however needs additional counts obtained from sign evaluations of the principal Sturm-Habicht coefficients (which are based on subresultants). The approach still requires shearing, but detects non-generic positions along the way, namely when the bitstream-Descartes method fails to isolate $f(x_i, y)$ in case of multiple roots. If so, the triggered shearing and shearing back requires refineable approximations to match critical events (and the connecting arcs) in the original and sheared coordinate system. These ideas have been extended to analyze pairs of curves in [7]; that is, to reach completeness, expensive subresultants and resource-consuming shearing must be used.

Both disadvantages for arbitrary algebraic curves disappeared in our newest approach [2], which only deviates in the lifting phase(s) from the previous one. The default lifting method is based on a numerical but certified (complex) root solver which, in iterations, lower-bounds the number of "points" along a fiber. As soon as this number matches an upper bound the fiber is lifted. The upper bound is computed using a previous result that relates the intersection multiplicities of the curves C_f and C_{f_x} and C_{f_y} to the multiplicities of roots of two resultants. We have shown that this match is almost always attained. In particular, it holds for curves in generic position. Only for a few special cases, this highly efficient lifting fails and triggers a more expensive, but complete lifting method, based on BISOLVE, a solver for bivariate polynomial systems. It is used to refine the roots of $f(x_i, y)$ with respect to roots of derivatives $f^{(k)}(x_i, y)$ to guide a bitstream-Descartes method in his decisions. BISOLVE, which only needs resultant and bivariate gcd computations, is also used to eventually compute the intersections of a pair of curves along a fiber, and thus replacing the costly subresultants and shearing. As a performance booster, the two remaining symbolic operations (resultants and gcd) are now computable in a highly-parallel fashion on graphics hardware, and thus do not form any longer bottlenecks.

The connection phase for arbitrary algebraic curves is performed either with a simple enumeration, in case there is only one critical vertex along a fiber, or with a continuation argument of the arcs in the neighborhood of the lifted vertices. Obtaining the connection in the later case only requires to isolate real roots of rational polynomials and to compare them.

Algebraic Surfaces. Our work on algebraic surfaces is also based on elimination theory. In [4] we consider arrangements on one quadric q_1 (an algebraic surface of degree 2) induced by its intersection with $n - 1$ other quadrics $q_2 \ldots q_n$. We project the *silhouette* of q_1 and the *intersection curves* $q_1 \cap q_i$ into the plane, which results in degree 2 and 4 curves. The analysis of the projected curves is shaped as a lengthy case distinction that heavily relies on their low degree. However, it allows the curves to be split into x-monotone pieces where each can - again exploiting low-degrees - be lifted uniquely onto the lower or upper part of q_1 using ray-shooting techniques.

As for a general curve, the work to stratify a general algebraic surface [5] also uses general techniques. The first step is to obtain a special planar arrangement $\mathcal{A}$: The points of each of its cells are invariant with respect to the local degree n_p of $f(p_x, p_y, z)$, the local gcd degree k_p and, as shown, also the local real degree m_p (i.e. the number of lifted cells over a planar cell). $\mathcal{A}$ is constructed using the presented planar approach by considering curves of degree at most $d(d - 1)$ in an iterative way: the projected silhouette of S_f, the coefficients of f and the principal Sturm-Habicht coefficients. Neighboring cells that will result in the same (n, k)-signature are merged immediately. The final $\mathcal{A}$ consists of at most $O(d^4)$ cells. Lifting these cells requires to isolate the real roots of $f(p_x, p_y, z)$, for (p_x, p_y) being a representative point of a given cell of $\mathcal{A}$. Using the bitstream-Descartes method for this task finally yields to describe the topology of S_f with $O(d^5)$ items. The approach never assumes generic position, however suffers from the use of expensive subresultants. The adjacency between lifted items is mainly computed from a bucketing technique based on continuation arguments, with a complication if a vertical line is contained in S_f. The stratification has been extended to obtain an isotopic triangulation of the surface.

2   Software

Our software for general algebraic curves and surfaces is implemented in C++ in the scope of CGAL.[2] It relies on the library's provided functionality for arithmetic and polynomials. Honestly, lots of tools in these packages have been achieved while working on the described goals.

Throughout the library, we follow the generic programming paradigm. This is used to separate combinatorial tasks from geometric and numerical operations. This allows, for instance, to parameterize our layer with different sets of number types, like the ones provided by GMP, LEDA or CORE. We can also exchange the essential method to isolate real roots of univariate polynomials. Options here are isolators from CGAL, as well as the method provided by RS.[3] We also provide a bivariate algebraic kernel [3] and a traits class to use algebraic curves within CGAL's arrangement package. Arcs of algebraic curves can also be visualized in a reliable and efficient way by combining the computed topological information with a tracking algorithm [10]. The analysis of single algebraic curves and the computation of arrangements of such is also demonstrated in a web-demo[4].

---

[2] The general implementations outperform the specialized ones, thus we skip their exposition here.

[3] http://vegas.loria.fr/rs/

[4] http://exacus.mpi-inf.mpg.de/cgi-bin/xalci.cgi

Performance. For comparison, we first want to mention older running times. When conics became available in 2002, it took (on a reasonable machine at that time) about 19s to compute the arrangement of 30 random conics and 49s for 60 random ones. It was claimed that filtering can improve this performance. The software for cubics needed for random 30 (or 60) cubics only 6.1s (25.1s); degenerated ones were slightly slower. Already the first implementation for general algebraic curves outperformed both approaches. The speedup is mainly due to relying on numerical decisions more than on explicit representation for low-degree algebraic numbers. An arrangement of 50 random degree 6 curves became computable in about 50s. Finally, the biggest performance gain was observed with the revised analysis of single algebraic curves [2]. This approach avoids subresultants, exploits graphics hardware for resultants and gcd, relies on Rs for real root solving and usually lifts with a very cheap numerical root solver. Figure 1 shows large improvement factors between the previous and the newest approach for general algebraic curves:

It is important to mention now that the current algorithms are not only faster than previous ones, but also capable of handling geometric difficult instances (singular) at least as fast as seemingly easy ones (random ones). The theoretical performance is also analyzed. [12] uses amortized analysis to obtain the best known deterministic bounds for computing the topology of algebraic curves: The new bound of $\tilde{O}(d^6 + d^5\tau)$ bit operations (d degree, τ bitsize of coefficients) improves the state of the art by four magnitudes.

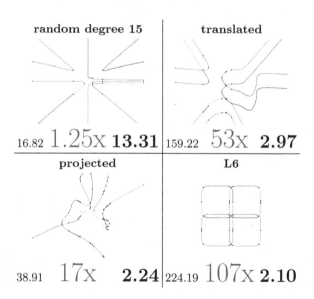

Fig. 1. Speedups for analyzing different kinds of algebraic curves. Left time: seconds for previous approach with CPU-subresultants. Right time: seconds for approach with numerical lifting filter. Both use resultants on the graphics card!

Stratifying an algebraic surface with the presented approach needs for many well-studied instance of moderate degrees often about one second. However, higher degree instances (such as random, interpolated or projected surfaces) require significantly more time [5]. We foresee that running times should drastically improve when eventually switching to the newest numerical lifting to compute $\mathcal{A}$ and for lifting $\mathcal{A}$'s a cells into 3D.

3 Applications

Computing with algebraic curves and surfaces has many applications, or is essential in many as subtasks. We list a few of them with details:

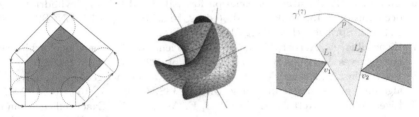

- Offsets of polygons need conics as exact representations [15]. Beyond polygonal shapes offsets are defined as parts of higher-degree algebraic curves, whereas to combinatorially pick the right parts constitutes a major difficulty. (left picture)
- Voronoi diagrams subdivide the ambient space of n objects into n cells, one for each input object that represents the influence of an object with respect to all other. Voronoi diagrams exist in all kind of dimensions, for all kind of objects and with various distance measures. The Voronoi diagram for lines in Euclidean 3D needs support by arrangement of algebraic curves. Namely, each cell is represented as the lower envelope of a minimization diagram of the trisectors in the parameter space of a cylinder; see [11]. (middle picture)
- The main task in robot motion planning is to decide whether there exists a collision free path from the current position of the robot to the desired destination. The robot's configurations are modeled in a (high-dimensional) configuration space, which is partitioned by a set of obstacles into a free and a forbidden space. It remains to compute the cell of the starting configuration and to decide whether the goal configuration belongs to the same cell. For polygonal robots in the plane this task is known as the Piano Mover's problem, where the mentioned partitioning is induced by algebraic curves [13]. (right picture)

References

1. Berberich, E., Eigenwillig, A., Hemmer, M., Hert, S., Mehlhorn, K., Schömer, E.: A computational basis for conic arcs and boolean operations on conic polygons. In: Möhring, R.H., Raman, R. (eds.) ESA 2002. LNCS, vol. 2461, pp. 174–186. Springer, Heidelberg (2002)

2. Berberich, E., Emeliyanenko, P., Kobel, A., Sagraloff, M.: Exact symbolic-numeric computation of planar algebraic curves. Theoretical Computer Science 491, 1–32 (2013)
3. Berberich, E., Hemmer, M., Kerber, M.: A generic algebraic kernel for non-linear geometric applications. In: Hurtado, F., van Kreveld, M.J. (eds.) Symposium on Computational Geometry, pp. 179–186. ACM (2011)
4. Berberich, E., Hemmer, M., Kettner, L., Schömer, E., Wolpert, N.: An exact, complete and efficient implementation for computing planar maps of quadric intersection curves. In: Mitchell, J.S.B., Rote, G. (eds.) Symposium on Computational Geometry, pp. 99–106. ACM (2005)
5. Berberich, E., Kerber, M., Sagraloff, M.: An efficient algorithm for the stratification and triangulation of an algebraic surface. Comput. Geom. 43(3), 257–278 (2010)
6. Collins, G.E.: Quantifier elimination for real closed fields by cylindrical algebraic decompostion. In: Brakhage, H. (ed.) GI-Fachtagung 1975. LNCS, vol. 33, pp. 134–183. Springer, Heidelberg (1975)
7. Eigenwillig, A., Kerber, M.: Exact and efficient 2d-arrangements of arbitrary algebraic curves. In: Teng, S.-H. (ed.) SODA, pp. 122–131. SIAM (2008)
8. Eigenwillig, A., Kerber, M., Wolpert, N.: Fast and exact geometric analysis of real algebraic plane curves. In: Wang, D. (ed.) ISSAC, pp. 151–158. ACM (2007)
9. Eigenwillig, A., Kettner, L., Schömer, E., Wolpert, N.: Exact, efficient, and complete arrangement computation for cubic curves. Comput. Geom. 35(1-2), 36–73 (2006)
10. Emeliyanenko, P., Berberich, E., Sagraloff, M.: Visualizing arcs of implicit algebraic curves, exactly and fast. In: Bebis, G., et al. (eds.) ISVC 2009, Part I. LNCS, vol. 5875, pp. 608–619. Springer, Heidelberg (2009)
11. Hemmer, M., Setter, O., Halperin, D.: Constructing the exact voronoi diagram of arbitrary lines in three-dimensional space - with fast point-location. In: de Berg, M., Meyer, U. (eds.) ESA 2010, Part I. LNCS, vol. 6346, pp. 398–409. Springer, Heidelberg (2010)
12. Kobel, A., Sagraloff, M.: Improved complexity bounds for computing with planar algebraic curves. CoRR, abs/1401.5690 (2014)
13. Schwartz, J.T., Sharir, M.: On the piano movers' problem i. The case of a two-dimensional rigid polygonal body moving amidst polygonal barriers. Communications on Pure and Applied Mathematics 36(3), 345–398 (1983)
14. The CGAL Project. CGAL User and Reference Manual, 4.4 edn. CGAL Editorial Board (2014)
15. Wein, R.: 2D Minkowski sums. CGAL User and Reference Manual, 4.4 edn., CGAL Editorial Board (2014)
16. Wein, R., Berberich, E., Fogel, E., Halperin, D., Hemmer, M., Salzman, O., Zukerman, B.: 2D arrangements. CGAL User and Reference Manual, 4.4 edn., CGAL Editorial Board (2014)

Computing the Orthogonal Projection of Rational Curves onto Rational Parameterized Surface by Symbolic Methods

Zhiwang Gan[1] and Meng Zhou[2]

[1] School of Mathematics and System Science, Beihang University, Beijing, China
ganzw@smss.buaa.edu.cn
[2] School of Mathematics and System Science, Beihang University, Beijing, China
zm1613@sina.com

Abstract. This paper presents three algorithms to compute orthogonal projection of rational curves onto rational parameterized surface. One of them, based on regular systems, is able to compute the exact parametric loci of projection. The one based on Gröbner basis can compute the minimal variety that contains the parametric loci. The rest one computes a variety that contains the parametric loci via resultant. Examples show that our algorithms are efficient and valuable.

Keywords: Orthogonal projection, Point projection,Curve projection, Rational curve, Rational surface.

1 Introduction

Computing the projection of a point onto a surface is to find a closest point on the surface, and projection of a curve onto a surface is the locus of all points on the curve project onto the surface. The orthogonal projection problem attracted great interest in minimal distance computation[7,3], calculating the intersection of curves and surfaces[11], surface curve design[16,6], curve or surface selecting[13], shape registration[18]. And many algorithms have been developed[8,4,9,12,15]. Among these methods,the commonly steps are to find the approach projective point in normed space by iteration techniques,which rely on good initial values, and then determine the approximate parameters in parametric space, which is called a *point inversion* problem.

Numerical methods above are efficient and stable in computing orthogonal projection,and are easy applied. However, there exist common drawbacks: The computation relies on samplings and the step size determines the accuracy of the result. The projective locus might be invisible while the locus is smaller than the step size. And the curve is always assumed to keep close enough to the surface so that a single solution is guaranteed. Symbolic methods would be necessary to overcome the shortcomings. Previous applications of symbolic methods in CAGD could be seen in [2,21,10]. In order to apply symbolic methods, we only concern about curves and surfaces that have rational parametric representations.

H. Hong and C. Yap (Eds.): ICMS 2014, LNCS 8592, pp. 261–268, 2014.

As known to all that a commonly used representation of surface and curves is NURBS[17], which is formed by rational patches. And since the parametric locus could uniquely determine the projection in 3D space, we focus on the parametric locus of orthogonal projection. Moreover, the range of surfaces and curves are restricted in $\mathbb{R}^3$.

Classical symbolic tools applied in this paper are regular systems[19], Gröbner basis[1] and resultant(see [14,5]). With the rational assumptions of curves and surfaces,the orthogonal condition would be transformed into a simple polynomial system. Then the orthogonal projection problem equals to determine the real solution of the polynomial system, which can be solved by symbolic or mix symbolic-numeric techniques.

In this paper,three algorithms are presented to compute the orthogonal projection of a rational parameterized curve onto a rational parameterized surface. The algorithm based on regular systems is able to compute the exact loci of orthogonal projection, and the false points will be detected. By means of Gröbner bases, we can get the minimal variety that contains the projective loci. And the resultant method efficiently computes a variety that contains the projective loci. The former two algorithms can particularly be used to compute point projections.

Compared with numerical algorithms,our algorithms have distinct advantages:

1. We generate the exact results without numerical errors.
2. Both point projection and curve projection are included.
3. There's no point inversion problem involved since we directly concern about the parametric loci.

In addition,the decomposition method in [20] would generate duplicate zeros between different regular systems and Huang Yanli[10] proposed a method to simplify the result. We improve Huang's method and directly consider the symbolic representation of zeros. Once the redundancy of zeros is judged, the corresponding regular system could be deleted without changing the zeros.

The rest of the paper is organized as follows. Section 2 presents the main theorems that calculate the orthogonal projections. And section 3 describes the algorithms based on the theorems in section 2. In section 4, we demonstrate non-trivial examples and experiment results.

2 The Main Theorems

In this section, we consider the orthogonal projection of a rational parameterized curve onto a rational parameterized surface. In the rest of the paper, let $\Phi(t) = (\frac{\Phi_1(t)}{\Phi_0(t)}, \frac{\Phi_2(t)}{\Phi_0(t)}, \frac{\Phi_3(t)}{\Phi_0(t)})$ be the parametric equation of rational curve C, and

$$\Psi(u, v) = (\frac{\Psi_1(u, v)}{\Psi_0(u, v)}, \frac{\Psi_2(u, v)}{\Psi_0(u, v)}, \frac{\Psi_3(u, v)}{\Psi_0(u, v)})$$

be the parametric mapping of surface S.

Given a rational parameterized curve C with parametric equation $\Phi(t)$ and a rational parameterized surface S with parametric equation $\Psi(u, v)$, the orthogonal projection of C onto S is defined to be the set Γ_{CS} of points (u, v, t) that satisfying the following condition:

$$(\Psi(u, v) - \Phi(t)) \times N(u, v) = 0, \tag{1}$$

where $N(u, v)$ stands for the normal vector of $\Psi(u, v)$ of S at (u, v), and "$\times$" denotes the cross product of two vectors. Since $N(u, v)$ is parallel with $\Psi_u(u, v) \times \Psi_v(u, v)$, the above condition can be written as:

$$\begin{cases} (\Psi(u, v) - \Phi(t)) \cdot \Psi_u(u, v) = 0, \\ (\Psi(u, v) - \Phi(t)) \cdot \Psi_v(u, v) = 0. \end{cases} \tag{2}$$

Here "." is the operator of the scalar product. The problem of orthogonal projection is to find the solution of system (2). And note that (2) can be treated as polynomial systems when Φ, Ψ are rational mappings.

To study the locus of orthogonal projection in three dimension space, we can equivalently discuss the parametric locus of orthogonal projection. We denote $\Gamma_{CS}(u, v) = \{(u, v) | \exists t, s.t.(u, v, t) \in \Gamma_{CS}\}$.

Denote

$$\begin{array}{l} P_{CS1} = \sum_{i=1}^{3} (\Psi_i \Phi_0 - \Phi_i \Psi_0)(\Psi_{iu} \Psi_0 - \Psi_i \Psi_{0u}), \\ P_{CS2} = \sum_{i=1}^{3} (\Psi_i \Phi_0 - \Phi_i \Psi_0)(\Psi_{iv} \Psi_0 - \Psi_i \Psi_{0v}), \\ \mathbb{P}_{CS} = \{P_{CS1}, P_{CS2}\}, Q_{CS} = \Psi_0 \Phi_0, \mathbb{Q}_{CS} = \{\Psi_0, \Phi_0\}. \end{array}$$

Theorem 1. $[\mathbb{T}_1, \mathbb{U}_1], \cdots, [\mathbb{T}_k, \mathbb{U}_k]$ *are regular systems with the variable order* $u < v < t$, *such that* $Zero([\mathbb{P}_{CS}, \mathbb{Q}_{CS}]) = \bigcup_{i=1}^{k} Zero([\mathbb{T}_i, \mathbb{U}_i])$. *Then*

$$\Gamma_{CS} = \bigcup_{i=1}^{k} Zero([\mathbb{T}_i, \mathbb{U}_i]),$$

And $\Gamma_{CS}(u, v) = \bigcup_{i=1}^{k} Zero([\mathbb{T}_i^{(2)}, \mathbb{U}_i^{(2)}])$.

Remark 2. *For the polynomial system* $[\mathbb{P}_{CS}, \mathbb{Q}_{CS}]$, *an algorithm*

$$RegSer([\mathbb{P}_{CS}, \mathbb{Q}_{CS}], [u, v, t]) = \{[\mathbb{T}_1, \mathbb{U}_1], \cdots, [\mathbb{T}_k, \mathbb{U}_k]\}$$

such that $Zero([\mathbb{P}_{CS}, \mathbb{Q}_{CS}]) = \bigcup_{i=1}^{k} Zero([\mathbb{T}_i, \mathbb{U}_i])$ *had been established[19], where* $[u, v, t]$ *means the variable order is* $u < v < t$.

Theorem 3. $\mathbb{G}$ *is a Gröbner basis of*

$$I = \langle P_{CS1}, P_{CS2}, zQ_{CS} - 1 \rangle$$

under a variable order $u < v < t < z$. *Then*

$$\overline{\Gamma_{CS}} = Zero(\mathbb{G} \cap \mathbb{R}[u, v, t]).$$

Furthermore, $\overline{\Gamma_{CS}(u, v)} = Zero(\mathbb{G} \cap \mathbb{R}[u, v])$.

Theorem 4

$$Zero(Res(P_{CS1}, P_{CS2}, t))$$
$$\supseteq Zero(Res(P_{CS1}, P_{CS2}, t)) \setminus Zero(\Psi_0)$$
$$\supseteq \Gamma_{CS}(u, v).$$

Furthermore, if

1. $Zero(Res(P_{CS1}, P_{CS2}, t)) \cap Zero(lc(P_{CS1}, t)) \cap Zero(lc(P_{CS2}, t)) = \emptyset$;
2. $Zero(Res(P_{CS1}, P_{CS2}, t)) \cap Zero(Cof(P_{CS1}, t)) = \emptyset$,
   and $Zero(Res(P_{CS1}, P_{CS2}, t)) \cap Zero(Cof(P_{CS2}, t)) = \emptyset$,
3. $Zero(\Phi_0) \cap Zero(\mathbb{P}_{CS}) = \emptyset$.

then

$$Zero(Res(P_{CS1}, P_{CS2}, t)) \setminus Zero(\Psi_0) = \Gamma_{CS}(u, v).$$

Remark 5. *In Theorem 4, $Res(P_{CS1}, P_{CS2}, t)$ is the resultant of P_{CS1} and P_{CS2} in variable t. Theorem 4 show the relationship between the loci of the projection and the resultant method.*

3   Algorithms

In this section, a series of algorithms will be concisely introduced according to the theorems discussed above.

For a polynomial set $\mathbb{P}$ and a set $M \subseteq \widehat{\mathcal{K}}^2$, we denote $Zero([\mathbb{P}, M]) = Zero(\mathbb{P}) - M$. Then for a polynomial system $[\mathbb{P}, \mathbb{Q}]$, we have

$$Zero([\mathbb{P}, \mathbb{Q}]) = Zero(\mathbb{P}) \setminus \bigcup_{Q \in \mathbb{Q}} Zero(Q)$$
$$= Zero(\mathbb{P}) \setminus \bigcup_{Q \in \mathbb{Q}} Zero(\mathbb{P} \cup \{Q\})$$
$$= Zero([\mathbb{P}, \bigcup_{Q \in \mathbb{Q}} Zero(\mathbb{P} \cup \{Q\})]).$$

Let

$$\Omega = \{[\mathbb{T}_1, M_1], \cdots, [\mathbb{T}_k, M_k]\},$$

we define $Zero(\Omega) = \bigcup_{i=1}^{k} Zero([\mathbb{T}_i, M_i])$.

Theorem 1 induces that the exact loci of projection could be decomposed into the union of zeros of regular systems, which could be in a complex form. In order to analyze the result easier, we developed an algorithm,denoted by **Simplify**(Ω),which is improved from **SIM**[10], to simplify regular systems.

Given a rational curve C and a rational surface S, **Algorithm 2** computes the exact parametric loci of the orthogonal projection of C onto S basic on theorem 1.

According to theorem 3, **Algorithm 3** returns the minimal variety that contains $\Gamma_{CS}(u, v)$. And **Algorithm 4**, deduced from theorem 4, calculates a variety that contains $\Gamma_{CS}(u, v)$.

4   Examples and Comparision

Example 1. *We consider a simple case with an algebraic surface* $S : \Psi(u, v) = (v^2 + u, 4uv + 2, u^2 + 3)$ *and an algebraic curve* $C : \Phi(t) = (t + 3, -2t, 5t + 5)$.
Firstly,

$$P_{CS1}(u, v, t) = (8v - 10u - 1)t + v^2 + u + 2(u^2 - 2)u - 3 + 4(4uv + 2)v,$$
$$P_{CS2}(u, v, t) = (8u - 2v)t + 2(v^2 + u - 3)v + 4(4uv + 2)u.$$

In step 3 of **Algorithm 2,** $\Omega = \{[\mathbb{T}_1, \mathbb{U}_1], [\mathbb{T}_2, \mathbb{U}_2], [\mathbb{T}_3, \mathbb{U}_3]\}$, *where*

$$\mathbb{T}_1 = \{-13uv + 9u^2v - 4u - 4v^4 - 2uv^2 + 8v^2 - 3uv^2 + 39u^3v + 14u^2 + 4u^4\},$$
$$\mathbb{U}_1 = \{-3278u + 10228u^2 + 38016u^4 + 7512u^3 + 127\},$$
$$\mathbb{T}_2 = \{1 - 8v + 10u, -492v^2 - 1351 + 2824v + 6912v^3\},$$
$$\mathbb{T}_3 = \{-3278u + 10228u^2 + 38016u^4 + 7512u^3 + 127, 3392u^3 + 412u^2 - 1280u^2$$
$$v - 464uv - 452u - 1024uv^2 + 1016v + 127 - 64v^2 - 512v^3\},$$
$$\mathbb{U}_2 = \emptyset. \mathbb{U}_3 = \emptyset.$$

And Huang's[10] algorithm **SIM** *outputs* $\{[\mathbb{T}_1, \mathbb{U}], [\mathbb{T}_3, \emptyset]\}$, *where*

$$\mathbb{U} = \{(\tfrac{1}{22}, \tfrac{2}{11}), (\tfrac{1}{22}RootOf(4Z^3 + 19Z^2 - 3752Z - 8953), \tfrac{1}{22}), (\tfrac{1}{2}\alpha, \beta)\},$$
$$\alpha = RootOf(216Z^3 + 105Z^2 + 242Z - 127),$$
$$\beta = RootOf(-1728Z^3 + (1728 + 216)Z^2 + (1080 - 3429)Z - 1270$$
$$+348\alpha^2 + 2366\alpha).$$

While algorithm **Simplify**(Ω) *yields* $\{[\mathbb{T}_1, \{(\tfrac{1}{22}, \tfrac{2}{11})\}]\}$. *Compared with algorithm* **SIM***, algorithm* **Simplify** *returns a more laconic result by directly computing the zero sets.*
Algorithm 3 *returns* $Zero(\mathbb{T}_1)$, *while* **Algorithm 4** *returns the same variety. And* $Genus(\mathbb{T}_1[1]) = 3$, *so the variety couldn't be rational parameterized. Furthermore,* $(\tfrac{1}{22}, \tfrac{2}{11}) \in Zero(lc(P_{CS1}, t)) \cap Zero(lc(P_{CS2}, t))$, *so* $(\tfrac{1}{22}, \tfrac{2}{11})$ *is not in the exact parametric of the loci.*

Example 2. *consider the algebraic surface* S *with:*

$$\Psi_0(u, v) = 1, \Psi_1(u, v) = -94.4 + 88.9v + 5.6v^2,$$
$$\Psi_2(u, v) = -131.3u + 28.1u^2,$$
$$\Psi_3(u, v) = 5.9(u^2v^2 + u^2v) - 3.9v^2u + 76.2u^2 + 6.7v^2 - 27.3uv - 50.8u$$
$$+25v + 12.1,$$

We randomly pick a curve C: $\Phi(t) = (\tfrac{\Phi_1(t)}{\Phi_0(t)}, \tfrac{\Phi_2(t)}{\Phi_0(t)}, \tfrac{\Phi_3(t)}{\Phi_0(t)})$ *passing over* S, *where*

$$\Phi_1(t) = (-90t - 1)(t + 5), \Phi_2(t) = -4t - 200,$$
$$\Phi_3(t) = (-5t + 30)(t + 5), \Phi_0(t) = t + 5.$$

S *is a common surface in mold industry[22], and is also a popular test surface for CNC machining methods. And note that* C *is a rational curve.*

Algorithm 2 *yields* $\{[\mathbb{T}_1, \mathbb{U}_1], [\mathbb{T}_2, \emptyset], [\mathbb{T}_3, \emptyset], [\mathbb{T}_4, \emptyset]\}$, *where*

$$\mathbb{T}_1 = \{M\}, \mathbb{U}_1 = \{(\alpha_1, \beta_1), (\alpha_2, \beta_2)\}.$$
$$M = 129735449752u^9v^4 + \cdots (82terms),$$
$$|\mathbb{T}_2| = |\mathbb{T}_3| = |\mathbb{T}_4| = 2.$$

Since $\mathbb{T}_2, \mathbb{T}_3, \mathbb{T}_4$ *are triangular systems with two elements,* $\bigcup_{i=2}^{4} Zero(\mathbb{T}_i)$ *consisted of finitely points. Furthermore, one could check that* $\bigcup_{i=2}^{4} Zero(\mathbb{T}_i) \subseteq Zero([\mathbb{T}_1, \mathbb{U}_1])$. *So the exact parametric locus is* $Zero([\mathbb{T}_1, \mathbb{U}_1])$. *And moreover,* $(\alpha_1, \beta_1, -5)$ *and* $(\alpha_2, \beta_2, -5)$ *are in* $Zero(\mathbb{P}_{CS})$, *while* $t = -5$ *is not in the domain of* $\Phi(t)$.

Algorithm 3 *returns* $Z(M)$. *And one can obtain* $Z(M) \cup Z(u - \frac{1313}{562})$ *by means of* **Algorithm 4**. *Compared with* **Alorithm 3**, $u - \frac{1313}{562}$ *is a redundant branch of the projective loci. As a matter of fact,* $(\frac{1313}{562}, v, -5)$ *is a zero of* $\mathbb{P}_{CS}$, *while* $t = -5$ *is not in the domain of* $\Phi(t)$.

More examples have been computed with a 3GHz CPU and 2GB memories. And the cost of time for each algorithm have been demonstrated in table 1. And table 2 records the number of regular systems before simplified, simplified via **SIM** and simplified using **Simplify**. "Y" in the chart induces that **Algorithm 4** have redundant branches with respect to the result of **Algorithm 3**, and "N" for no. "NA" means the result is not available in 3600s or the memory reached the hareware limit. In each example, $Degree(a, b, c, d)$ means $\max_{1 \leq i \leq 3} degree(\Psi_i) = a$, $degree(\Psi_0) = b$, $\max_{1 \leq i \leq 3} degree(\Phi_i) = c$, $degree(\Phi_0) = d$.

Table 1. Time cost of the algorithms

Degree	Algorithm 2	Algorithm 3	Algorithm 4	Redundancy
EX1 (1,0,1,0)	0.016s	0.016s	<0.001s	N
EX2 (2,0,1,0)	0.188s	0.063s	<0.001s	N
EX3 (3,0,1,0)	0.032s	0.468s	0.016s	N
EX4 (4,0,1,0)	0.734s	2.671s	<0.001s	N
EX5 (2,0,2,0)	0.203s	0.031s	<0.001s	N
EX6 (3,0,2,0)	596.797s	0.047s	<0.001s	N
EX7 (1,1,1,1)	0.078s	0.016s	<0.001s	Y
EX8 (2,1,2,1)	0.891s	0.297s	0.016s	Y
EX9 (2,2,2,1)	0.078s	0.015s	0.032s	Y
EX10 (2,2,2,2)	1.515s	0.984s	0.016s	Y
EX11 (3,1,2,1)	NA	113.938s	0.016s	Y

Table 1 illustrates that **Algorithm 2** performs well in low degree case , but the time cost increasing fast while the degree of surface and curve increasing, this is because the degree and the number of output regular systems in step 3 are getting enormous(see EX6 and EX11). Since most commonly used surfaces and curves have low degree, **Algorithm 2** is valuable for engineering practice.

Table 2. Comparison on simplification methods(number of elements)

	RegSer	SIM	Simplify
EX1	1	1	1
EX2	3	3	3
EX3	4	2	2
EX4	6	3	2
EX5	4	1	1
EX6	8	2	2
EX7	2	1	1
EX8	7	2	1
EX9	2	1	1
EX10	6	2	1
EX11	NA	NA	NA

Algorithm 3 is significantly better than **Algorithm 2** at time cost, and works fine while the input degree grows.

Algorithm 4 performs steady and excellent at the computation cost, but it always generates redundant branches when the inputs are rational.

Table 2 induces that the algorithm **Simplify** could reduce the number of regular systems for the parametric loci in most circumstances. And compared with **SIM**, our algorithm has a more concise output.

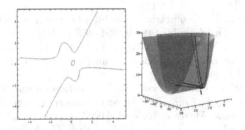

Fig. 1. Parametric loci and 3-D curve of orthogonal projection for Example 1

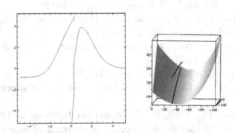

Fig. 2. Parametric loci and 3-D curve of orthogonal projection for Example 2

References

1. Buchberger, B.: Gröbner Bases: A Short Introduction for Systems Theorists. In: Moreno-Díaz Jr., R., Buchberger, B., Freire, J.-L. (eds.) EUROCAST 2001. LNCS, vol. 2178, pp. 1–19. Springer, Heidelberg (2001)
2. Busé, L., Elkadi, M., Galligo, A.: Intersection and self-intersection of surfaces by means of bezoutian matrices. Computer Aided Geometric Design 25, 53–68 (2008)
3. Chen, X.-D., Yong, J.-H., Wang, G., Paul, J.-C., Xu, G.: Computing the minimum distance between a point and a nurbs curve. Computer Aided Design 40(10-11), 1051–1054 (2008)
4. Chernov, N., Wijewickrema, S.: Algorithms for projecting points onto conics. Journal of Computational and Applied Mathematics 251, 8–21 (2013)
5. Cox, D., Little, J., O'Shea, D.: Ideals, varieties, and algorithms: An Introduction to Computational Algebraic Geometry and Commutative Algebra. Springer (2007)
6. Dietz, R., Hoschek, J., Jüttler, B.: An algebraic approach to curves and surfaces on the sphere and on other quadrics. Computer Aided Geometric Design 10(3-4), 211–229 (1993)
7. Gilbert, E., Johnson, D., Keerthi, S.: A fast procedure for computing the distance between complex objects in three-dimensional space. IEEE Transactions on Robotics and Automation 4(2), 193–203 (1988)
8. Song, H.C., Yong, J.H., Yang, Y.J., Liu, X.M.: Algorithm for orthogonal projection of parametric curves onto b-spline surfaces. Computer Aided Design 43, 381–393 (2011)
9. Hu, S.M., Wallner, J.: A second order algorithm for orthogonal projection onto curves and surfaces. Computer Aided Geometric Design 22, 251–260 (2005)
10. Huang, Y., Wang, D.: Computing intersection and self-intersection loci of parametrized surfaces using regular systems and Gröbner bases. Computer Aided Geometric Design 28, 566–581 (2011)
11. Limaiem, A., Trochu, F.: Geometric algorithms for the intersection of curves and surfaces. Computers and Graphics 19(3), 391–403 (1995)
12. Liu, X.-M., Yang, L., Yong, J.-H., Gu, H.-J., Sun, J.-G.: A torus patch approximation approach for point projection on surfaces. Computer Aided Geometric Design 26, 593–598 (2009)
13. Ma, Y.L., Hewitt, W.T.: Point inversion and projection for nurbs curve and surface: control polygon approach. Computer Aided Geometric Design 20, 79–99 (2003)
14. Mishra, B.: Algorithmic Algebra. Springer (1993)
15. Oh, Y.-T., Kim, Y.-J., Lee, J., Kim, M.-S., Elber, G.: Efficient point-projection to freeform curves and surfaces. Computer Aided Geometric Design 29, 242–254 (2012)
16. Pegna, J., Wolter, F.-E.: Surface curve design by orthogonal projection of space curves onto free-form surfaces. Journal of Mechanical Design 118, 45–52 (1996)
17. Piegl, L., Tiller, W.: The NURBS book. Springer (2012)
18. Pottmann, H., Leopoldseder, S., Hofer, M.: Registration without icp. Computer Vision and Image Understanding 95(1), 54–71 (2004)
19. Wang, D.: Computing triangular systems and regular systems. Journal of Symbolic Computation 30, 221–236 (2000)
20. Wang, D.: Elimination Methods. Springer (2001)
21. Wang, D.: Elimination Practice: Software Tools and Applications. Imperial College Press (2004)
22. Warkentin, A., Ismail, F., Bedi, S.: Comparison between multi-point and other 5-axis tool positioning strategies. International Journal of Machine Tools & Manufacture 40, 185–208 (2000)

Isotopic ϵ-Approximation of Algebraic Curves

(Extended Abstract)

Kai Jin

Key Lab of Mathematics Mechanization
Institute of Systems Science, AMSS, Chinese Academy of Sciences
jinkai@amss.ac.cn

Abstract. In this paper, we will introduce our implementation for isotopic approximation of plane and space algebraic curves. The important basic algorithm used in our implementation is real solving of zero-dimensional polynomial systems, especially for bivariate polynomial systems. For the topology computation of plane curves, compared to other symbolic methods, the novelty of our method is that we can get many simple roots on the fibers when computing the critical points of the plane curve, which greatly improves the lifting step. After the topology is computed, we also give a certified approximation for the plane curve, which is a basic operation for approximating a space curve and further for an algebraic surface. As to space curve case, the topology and approximation are recovered from that of their projection plane curves. We implemented our algorithms in Maple 15. The benchmarks show the high efficiency of the implementation.

Keywords: Bivariate polynomial system, real roots isolation, plane (space) algebraic curves, topology, isotopic approximation.

1 Introduction

To determine the topology of a given algebraic curves and to use line segments to approximately represent the curves are basic operations in computer graphics and geometric modeling. A mesh of a plane curve or an algebraic space curve could be used to display the corresponding curve correctly, and is the foundation of further displaying space surfaces. We consider an algebraic plane (space) curve defined by $f(x, y) = 0$ (correp:$(f(x, y, z) = g(x, y, z) = 0)$), denoted by $\mathcal{C}_f = \{(x, y) \in \mathbb{R}^2 \big| f(x, y) = 0\}$ ($\mathcal{C}_{f,g} = \{(x, y, z) \in \mathbb{R}^3 \big| f(x, y, z) = g(x, y, z) = 0\}$), where $f, g \in \mathbb{Z}[x, y, z]$. Typically, the topology of $\mathcal{C}_f$ is given in terms of a planar graph $\mathcal{G}$ embedded in $\mathbb{R}^2$ that is isotopic to $\mathcal{C}_f$. For a geometric-topological analysis, we further require the vertices of $\mathcal{G}$ to be located on $\mathcal{C}_f$ (ignoring the representation error). As there are many papers study the problem of computing the topology of plane curves, we omit the corresponding references in this extended abstract for the sake of space limitation, and we will list them in the full paper. For the space algebraic curve, though there are some papers which

H. Hong and C. Yap (Eds.): ICMS 2014, LNCS 8592, pp. 269–276, 2014.

studied it [1, 4–7], the problem still needs to be explored. For example, the efficient generic position checking method, efficient and complete implementation of the algorithms for the problem are welcome.

The paper is organized as follows. In the next section, we give some brief introduction on the functions of the algorithms. The underlying theory and technique contributions are presented in section 3. Applications and Experiments are shown in the fourth section.

2   Functionality

First we give some notations. Let $\mathcal{C}_h$ denote the plane algebraic curve with defining polynomial $h(x, y) \in \mathbb{Z}[x, y]$. Let $\mathbf{p} = (x_0, y_0)$ be a point on $\mathcal{C}_f$. We call $\mathbf{p}$ as an **x-critical point** (**y-critical point**) if $\mathbf{p}$ satisfies $\mathbf{p} \in \mathcal{C}_f, \frac{\partial f}{\partial y}(\mathbf{p}) = 0$ ($\mathbf{p} \in \mathcal{C}_f, \frac{\partial f}{\partial x}(\mathbf{p}) = 0$), and a **singular point** if $\mathbf{p} \in \mathcal{C}_f, \frac{\partial f}{\partial y}(\mathbf{p}) = \frac{\partial f}{\partial x}(\mathbf{p}) = 0$. We briefly introduce some main functions in our softwares.

Bivariate System Solving: This function is an efficient solver for bivariate polynomial systems. The input of the function are a zero dimensional polynomial system $\{f, g\}$ and a rational number ϵ, while the output are two sets. The first one is the isolating intervals of the real roots for the system with the length of each interval smaller than ϵ, the second one is the set of multiplicities of each real roots of the system. We can also get the linear univariate representation (LUR) from the output of the algorithm. For more details please see [3].

LUR-Top: This function computes the topology graph of a plane algebraic curve. The input of this function is a square free polynomial $f(x, y) \in \mathbb{Z}[x, y]$ while the output is the topology graph of the plane curve $\mathcal{C}_f = \{(x, y) \in \mathbb{R} | f(x, y) = 0\}$. We use a modified version of **Bivariate system solving** to compute the system $\{f, f_y\}$. The difference is we can get the isolating intervals for each simple points as soon as they are not in the isolating intervals of real roots of $\{f, f_y\}$ on each fiber. Hence, it greatly improve the efficiency of the algorithm.

Curve-app: This function is to used to approximate a regular plane curve. It mainly use a modified Newton's method to approximate a regular plane curve segment, which consist of two ingredients. The first one is Newton's method, the second one is a simple check which makes sure that the approximation is reliable.

For the detailed information about functions **LUR-top** and **Curve-app**, please refer to [9].

LUR-3top: This function computes the topology graph and approximation of an algebraic space curve. It mainly call the function **LUR-top** to compute the topology of two projection plane curves. The topology and the approximation of the space curve are recovered from the topology and approximation of the two projection plane curves.

3 Underlying Theories and Technical Contributions

In this section, we mainly introduce the underlying theories for computing the topology graph of algebraic curves, for the bivariate polynomial systems solving, please see [3]. For the sake of space, we just list the outline of algorithms for computing the isotopic meshing of algebraic curves.

Isotopic Approximation for Algebraic Plane Curves

Assume $f \in \mathbb{Z}[x, y]$ be a square free polynomial. The plane curve $\mathcal{C}_f$ has no vertical lines parallel to y axis. This conditions can be easily satisfied if we do a pretreatment to the polynomial f.

Overview of the Algorithm

1. Solving the system $\{f, f_y\}$:
   - Project the x-critical points of the plane curve onto the x-axis and isolate the real roots of $R(x)$, where $R(x) = \mathrm{sqrfree}(\mathrm{Res}_y(f, f_y))$.
   - Lifting: Lift the real zeros of $R(x)$ to obtain the candidates of the x-critical points of $\mathcal{C}_h$.
   - Certification: Certify the candidates above to make sure that each candidate contains exactly one x-critical point.
2. Computing on the fiber: Separating the simple roots apart from the multiple roots of $f(\alpha, y)$ on each fiber $x = \alpha$.
3. Connection: Compute the branch numbers for each x-critical points and connect the plane key points appropriately. Hence we get the topology of the plane curve.
4. ϵ-Meshing: Reuse line segments to approximate the plane curve such that the error is bounded by ϵ.

We should indicate that the algorithm of solving the system $\{f, f_y\}$ has a little difference with the function **Bivariate system solving**. Explicitly, we can get the left simple roots of $f(\alpha, y)$ when we compute the system $\{f(\alpha, y) = \frac{\partial f}{\partial y}(\alpha, y)\}$ on the α fiber, which highly improves the lifting step since we need not compute simple roots of $f(\alpha, y)$ any more, so it speedup the algorithm. From the outline of the algorithm, we know the algorithm mainly involves resultant computation and real roots isolation for univariate polynomials, this is the another reason for the high efficiency of the alsorithm. For more details, please refer to [9].

Isotopic Approximation for Algebraic Space Curves

We consider the topology computation and approximation of an algebraic space curve. Assume $\mathcal{C} = \{(x, y, z) \in \mathbb{R}^3 \mid f(x, y, z) = g(x, y, z) = 0, f, g \in \mathbb{Z}[x, y, z]\}$.

Since we also consider the ϵ-meshing of the projection plane curve, we have the following assumptions for the input algebraic space curve.

- For any $x_0 \in \mathbb{R}$, $f(x_0, y, z) = g(x_0, y, z) = 0$ has a finite number of solutions; and
- the leading coefficients of f, g w.r.t. z have no common factors only in x.

The main steps to obtain the topology of $\mathcal{C}$ are similar as that presented in [4]. But we revise some steps of the original method to compute the points on the space curve. Now we give an outline of our method below:

Overview of the Algorithm

1. Projection: Project the space onto the XY plane and compute the topology of the projection plane curve. This step we can use the function **LUR-top**.
2. Lifting: Lift the points on the $\mathcal{C}_h$ to obtain the space point candidates. This is a crucial step for computing the topology of algebraic space curves. Many algorithms use subresultant sequence to do the lifting which has been turned out to be very time-consuming, since the process needs a large amount of symbolic computation. We use interval polynomials to get the candidates of the space key points which involves only real roots isolation for univariate polynomials. Hence, the lifting process is efficient.
3. Certification: To determine which candidates contain points of the space curve.
   - We compute a rational number s with small bitsize such that the sheared space curve $\overline{\mathcal{C}} = \{(x, y, z) \in \mathbb{R}^3 \mid F = f(x, y + sz, z) = 0, G = g(x, y + sz, z) = 0\}$ is in a weak generic position w.r.t z.
   - Compute the topology of $\mathcal{C}_{\bar{h}}$, where $\bar{h} = \mathrm{sqrfree}(\mathrm{Res}_z(F, G))$.
   - Certify the space root candidates by comparing the space root candidates of $\mathcal{C}$ and the points on the plane curve $\mathcal{C}_{\bar{h}}$.
4. Connection: Connect the space points using line segments by comparing the topology of the the plane curves of $\mathcal{C}_h$ and $\mathcal{C}_{\bar{h}}$. Hence we get the topology of the space curve.
5. ϵ-Meshing: Reuse line segments to approximate the space curve such that the error is bounded by ϵ.

Compared to some existing methods [1, 4–7] computing generic position for the algebraic space curves, we provide a criterion to check a weak generic position for an algebraic space curves, and the criterion is simple and efficient. Using This criterion, the topology computation of an algebraic space curve is transformed into the topology computation of projection plane curves. This greatly improves the efficiency of the algorithm. For more details, please refer to [8].

4   Application and Experiments

In this section, we main consider topology computation for algebraic curves, As to the performance of **Bivariate system solving**, please refer to [3]. We implemented our algorithm in Maple and test examples in Maple 15 on a PC with Inter(R)Core(TM)i3-2100 CPU @3.10GHz 3.10GHz, 2G memory and Windows 7 operating system.

In the Table 1, we compare our algorithm with the algorithms in [2] indirectly. Consider that their algorithms are implemented in C++ language and they use the GPU acceleration technique, our algorithm is competitive.

Table 1. Total running times for analysing the topology of five random curves from [2]. GEOTOP-BS and GEOTOP are two two algorithms in [2] for computing the topology of the curve, the corresponding columns are the times in their paper (their code is not available). Moreover they outsourced many symbolic computations to the graphics hardware to reduce the computing time, For the GPU-part of the algorithm, they used the GeForce GTX580 Graphics card (Fermi Core). We implemented our code in maple and illustrate the times for the same benchmarks, where T_1 denotes the total time for computing $\mathrm{Res}_y(f, f_y)$, and T_2 denotes total time for isolating the real roots of $\mathrm{Res}_y(f, f_y)$. T denotes the total time for computing the topology of the curves.

Sets of five random dense curves					
degree, bits	GEOTOP-BS	GEOTOP	LUR-top		
			T_1	T_2	T
Machine	Linux platform on a 2.8 GHz 8-Core Inter Xeon W3530 with 8MB of L2 cache		Win 7 on 3.1GHz dual Core i3-2100 CPU with 256KB of L2 cache		
Code language	C++		Maple language		
GPU speedup	YES		NO		
06, 10	0.71	0.14	0.046	0.016	0.655
06, 512	0.15	0.29	0.219	0.047	0.874
09, 10	1.50	0.23	0.124	0.016	1.155
09, 512	2.38	0.57	2.012	0.156	3.166
12, 10	4.54	0.65	0.281	0.187	2.854
12, 512	7.37	1.49	5.553	0.468	8.628
15, 10	5.81	0.92	0.686	0.312	3.931
15, 512	11.16	2.46	13.011	0.982	17.161
Sets of five random sparse curves					
06, 10	0.25	0.07	0.016	0	0.265
06, 512	0.42	0.13	0.032	0.031	0.235
09, 10	0.54	0.11	0.031	0.015	0.265
09, 512	0.78	0.20	0.499	0.157	0.899
12, 10	0.88	0.17	0.062	0.047	0.359
12, 512	1.73	0.42	1.311	0.888	2.543
15, 10	3.03	0.59	0.201	0.204	0.890
15, 512	5.88	1.22	5.054	3.401	9.126

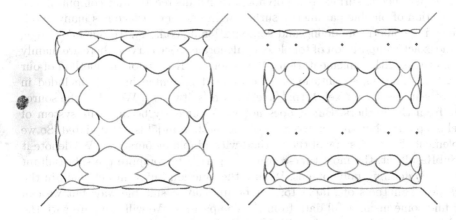

Fig. 1. The visualization of $FTT^2_{3,5}$

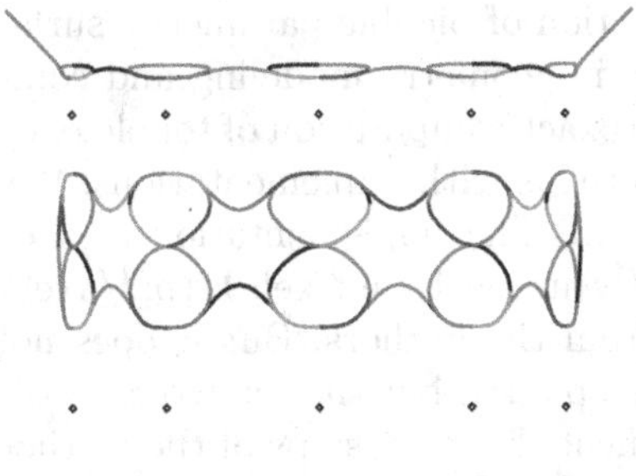

Fig. 2. The visualization of $FTT^2_{4,5}$

In Figure 1,2, we list the approximations for two complicated plane curves which are taken from [10], for detailed information of this two curves, please refer to [10]. In Figure 3 the defining polynomial is $F^* = \mathrm{Res}_z(f, g)$, and $f = -y^3 - y^2z + x^4z + x^2yz^2 - yz^5 - x^5z^2 + x^4yz^2 - x^3y^3z - x^7y + x^7z^2 + x^5z^4 + x^3y^6 + y^{10} + z^{10}$, $g = x - y - z + z^2 + x^2y - x^2z - xyz - x^4 + x^2yz - xy^3 - yz^3 + y^5 + z^5$.

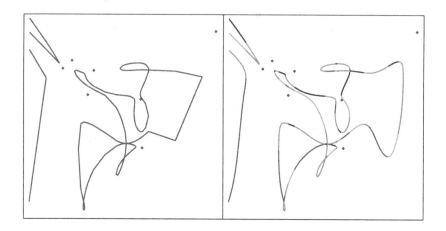

Fig. 3. The topology and visualization of curve F^*, from the above picture, we can see, it has 11 singular points, of which 7 are isolated singular points. The time is 261.83 seconds for the topology computation while the visualization time is 97.02 second.

For the performance of space curve, we can compute the intersection of two surfaces with degree 2 (3) and 18. As we know, the implicitization equation of bicubic parametric surface is a polynomial with degree 18, and computing the intersection of bicubic parametric surface with another surface has many applications in geometry modeling and computational geometry. To our knowledge, for the exact computation of topology of algebraic space curves, there are mainly two methods with implementations. We denote LUR-3top for the method of our algorithm. The implementation of the method presented in [5] is included in the software package Axel: http://axel.inria.fr/softwares/. We asked the source code from the authors. But it does not work currently because the system of Axel is updated but the related code of the method in [5] is not updated. So we implement the main steps of the method with Maple by ourselves. We denote it as SubResultant, the most time-consuming part is to compute the subresultant of the two defining polynomials. We use the function SubresultantChain in the package ChainTools of Maple 15. Maybe it is not a suitable way, but we can still find some meaningful data from the comparison. We will compare with the implementation of the method in [5] after they revise their version in our full version paper.

In Table 2, we compare two methods for some random and dense benchmarks. We can find that SubResultant runs faster than LUR-3top when the degrees of

Table 2. The time (in sec) for the topology computation of space curve of random generated dense polynomials with coefficients between −5 to 5. The time is an averaged one for five random space curves.

degree	LUR-3top	SubResultant
3-3	0.918	0.125
4-4	5.9	2.303
5-5	27.172	92.421
6-6	114.648	>3000
7-7	581.849	>3000

Table 3. The timing for the topology computation of four given examples

Ex & degree	LUR-3top	SubResultant
Example 1: (4, 4)	0.827	0.281
Example 2: (6, 7)	324.466	>3000
Example 3: (2, 18)	250.195	1226.043
Example 4: (3, 18)	1724.560	>3000

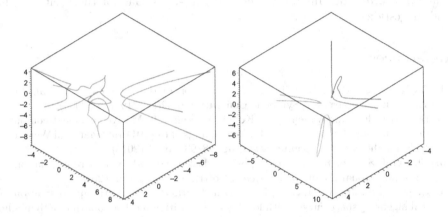

Fig. 4. Example for degree 6 and 7   **Fig. 5.** Example for degree 2 and 18

defining polynomials of the space curves are low. But when the degree become higher, it runs much slower than LUR-3top.

In Table 3, we test four explicit examples. Example 1 is taken from [5]. Example 2 is shown in Figure 4, where f is a polynomial in three variables with degree 6 and 77 terms, g is with degree 7 and 102 terms. Their expressions are listed below.

$f = 2 - 11\,y^2 + 2\,y^5 + 2\,z^5 - 18\,y^3 - 2\,z^3 + 2\,z^2 + 2\,z^6 + y^6 + z^4 + 2\,x + z + 23\,y + x^4yz + x^3yz - 2\,x^4 - 20\,yz + yz^2 - y^2z^2 - yz^3 - 2\,y^3z^2 - y^2z^3 + yz^4 + y^5z + y^2z^4 + 2\,xy^3z - 2\,xyz^4 - xy^2z^3 + 2\,xy^2z + x^2yz^3 - x^2yz^2 - x^2yz - 2\,x^2y^2z + 2\,x^3 - 2\,x^4y^2 + 2\,x^4z^2 - 2\,x^4z + 2\,x^3y^3 + x^3y^2 + 2\,x^3z^2 - x^2y^3 + x^2y^2 - 2\,x^2z^3 + 2\,xy^4 - xy^3 + 2\,xy + 56\,xz^2 - yz^5 + 2\,x^5y + 2\,x^3y - 2\,x^3z^3 - x^3z + 2\,x^2y^4 - x^2y + x^2z^4 - x^2z - xy^5 - 72\,xy^2 - xz^4 + xz^3 - x^5 + x^6 + 2\,x^2 + 2\,x^3yz^2 - 2\,x^3y^2z + 2\,xy^4z - 2\,xy^3z^2 + 2\,xyz^3 + 2\,x^2z^2 - 2\,x^4y + y^3z^3 + y^4z - xz^5 + x^5z + 2\,x^2y^3z - xy^2z^2 - 2\,xyz,$

$g = -2 - 2\,y^2 - 2\,y^5 - 2\,z^5 - 2\,z^3 + 2\,z^2 + z^7 - y^6 + z^4 - x - 84\,z + y - 2\,x^2y^3z^2 + x^5yz - 2\,x^4yz - 2\,x^3y^3z - x^3y^2z^2 + 2\,x^3yz - x^2y^2z^2 + 2\,x^2yz^4 + x^4 - xyz^2 - 95\,yz - yz^2 - 2\,y^3z - y^2z^2 + yz^3 + y^3z^2 - 2\,yz^4 + 2\,y^5z + y^2z^4 - y^6z - xy^4z^2 - 2\,xy^3z^3 -$

$2\,xy^3z - 2\,xy^5z - 2\,xyz^4 - 2\,xy^2z^4 + xy^2z^3 - 2\,xy^2z - 2\,x^2yz^3 + x^2yz^2 + 2\,x^2yz +$
$2\,x^2y^2z^3 + x^2y^2z + x^3 + 2\,x^5y^2 + x^5z^2 + x^4y^2 + 2\,x^4z^3 - x^4z^2 - 2\,x^4z - 2\,x^6z +$
$x^3y^3 - x^3y^2 + 2\,x^3z^2 + 2\,x^2y^5 + 2\,x^2z^3 - 2\,xy^6 - xy^4 + xy^3 - 17\,xz^2 - 81\,xz -$
$12\,y^2z + y^4z^2 - yz^5 - 2\,y^4z^3 - 2\,y^3z^4 - 2\,x^5y - x^3y^4 + x^3z^4 - 2\,x^3z^3 + 2\,x^3z -$
$x^2y^4 - 74\,x^2y - x^2z^4 + 2\,x^2z + xy^5 + xy^2 + xz^6 + xz^4 - x^5 - x^4y^2z - 2\,y^7 - 2\,x^6 +$
$2\,x^3yz^2 - 2\,x^2y^4z - x^4yz^2 + 2\,x^3y^2z - xy^4z - xy^3z^2 + 2\,xyz^5 - x^2z^2 - 2\,x^4y -$
$2\,y^3z^3 - y^4z + xz^5 - x^5z - 2\,x^2y^3z - 2\,xy^2z^2 - 2\,xyz + x^6y.$

From the performance of the above experiments, we can find that our algorithm is efficient and stable.

Acknowledgement. The work is partially supported by NKBRPC (2011CB302400), NSFC Grants (11001258, 60821002, 11371356). I am deeply grateful to professors Jin-San Cheng and Xiao-Shan Gao for their selfless help and suggestions.

References

1. Alcázar, J.G., Rafael~Sendra, J.: Computation of the topology of real algebraic space curves. Journal of Symbolic Computation 39(6), 719–744 (2005)
2. Berberich, E., Emeliyanenko, P., Kobel, A., Sagraloff, M.: Arrangement computation for planar algebraic curves. In: Proceedings of the 2011 International Workshop on Symbolic-Numeric Computation, pp. 88–98. ACM (2012)
3. Cheng, J.-S., Jin, K.: A generic position based method for real root isolation of zero-dimensional polynomial systems. CoRR, abs/1312.0462 (2013)
4. Cheng, J.-S., Jin, K., Lazard, D.: Certified rational parametric approximation of real algebraic space curves with local generic position method. Journal of Symbolic Computation 58, 18–40 (2013)
5. Daouda, D.N., Mourrain, B., Ruatta, O.: On the computation of the topology of a non-reduced implicit space curve. In: Proceedings of the Twenty-first International Symposium on Symbolic and Algebraic Computation, pp. 47–54. ACM (2008)
6. El Kahoui, M.: Topology of real algebraic space curves. Journal of Symbolic Computation 43(4), 235–258 (2008)
7. Gatellier, G., Labrouzy, A., Mourrain, B., Técourt, J.-P.: Computing the topology of three-dimensional algebraic curves. In: Computational Methods for Algebraic Spline Surfaces, pp. 27–43. Springer (2005)
8. Jin, K., Cheng, J.-S.: Isotopic ε-meshing of real algebraic space curves. Submitted to SNC 2014 (2014)
9. Jin, K., Cheng, J.-S., Gao, X.-S.: On the topology and certified approximation of plane algebraic curves (2013) (preprint)
10. Labs, O.: A list of challenges for real algebraic plane curve visualization software. In: Nonlinear Computational Geometry, pp. 137–164 (2010)

Isotopic Arrangement of Simple Curves:
An Exact Numerical Approach
Based on Subdivision

Jyh-Ming Lien[1], Vikram Sharma[2], Gert Vegter[3], and Chee Yap[4]

[1] Department of Computer Science, George Mason University, USA
`jmlien@cs.gmu.edu`
`www.cs.gmu.edu/~jmlien/`
[2] Institute of Mathematical Sciences, India
`vikram@imsc.res.in`
`www.imsc.res.in/~vikram/`
[3] Johann Bernoulli Institute for Mathematics and Computer Science, Netherlands
`gert@rug.nl`
`www.cs.rug.nl/gert/`
[4] Department of Computer Science, New York University, USA
`chee@cs.nyu.edu`
`www.cs.nyu.edu/yap/`

Abstract. We present a purely numerical (i.e., non-algebraic) subdivision algorithm for computing an isotopic approximation of a simple arrangement of curves. The arrangement is "simple" in the sense that any three curves have no common intersection, any two curves intersect transversally, and each curve is non-singular. A curve is given as the zero set of an analytic function on the plane, along with effective interval forms of the function and its partial derivatives. Our solution generalizes the isotopic curve approximation algorithms of Plantinga-Vegter (2004) and Lin-Yap (2009). We use certified numerical primitives based on interval methods. Such algorithms have many favorable properties: they are practical, easy to implement, suffer no implementation gaps, integrate topological with geometric computation, and have adaptive as well as local complexity. A preliminary implementation is available in Core Library.

Keywords: Isotopy, arrangement of curves, interval arithmetic, subdivision algorithms, marching-cube.

1 Introduction

Computing arrangements of curves and surfaces is a fundamental problem in computational geometry. Current algorithms that guarantee the topology of the arrangement rely on algebraic tools, such as resultants. This limits their applicability to algebraic curves and surfaces. However, in many situations we need to work with curves and surfaces defined as the zero set of an analytic function. In such setting, we are necessarily restricted to numerical primitives, such as evaluating the function at some set of points. In this weaker setting, therefore, it

H. Hong and C. Yap (Eds.): ICMS 2014, LNCS 8592, pp. 277–282, 2014.

is very hard to provide topological guarantees. We describe a new approach for computing curve arrangements based on purely numerical (i.e., non-algebraic) primitives.

By a **simple curve arrangement** we mean a collection of non-singular curves such that no three of them intersect, and any two of them intersect transversally. The most fundamental case in handling the simple arrangement of three or more curves can is the case of two curves. Let $F : \mathbb{R}^2 \to \mathbb{R}^2$, where $F(x, y) = (f(x, y), g(x, y))$ be a pair of analytic functions. It generically defines two planar curves $S = f^{-1}(0) \subseteq \mathbb{R}^2$ and $T = g^{-1}(0)$. The concept of hyperplane arrangement is classical in computational geometry [4]; recent interest focuses on nonlinear arrangements [2].

Our basic problem is the following: suppose we are given an $\epsilon > 0$ and a region $B_0 \subseteq \mathbb{R}^2$, called the **region-of-interest** or ROI, which is usually in the shape of an axes-aligned box. We want to compute *an ϵ-approximation to the arrangement of the pair (S, T) of curves restricted to B_0*. This will be a planar straight-line graph $G = (V, E)$ where V is a finite set of points in B_0 and E is a set of polygonal paths in B_0. Each path $e \in E$ connects a pair of points in V, and no path intersects another path or any point in V (except at endpoints). Moreover, E is partitioned into two sets $E = E_S \cup E_T$ such that $\cup E_S$ (resp., $\cup E_T$) is an approximation of S (resp., T). The correctness of this graph G has two aspects: topological and geometric. The latter is easy to formulate: we require that the set $\cup E_S \subseteq B_0$ is ϵ-close to S in the sense of Hausdorff distance; similarly, the set $\cup E_T$ is ϵ-close to T. If we specify $\epsilon = \infty$, then we are basically unconcerned about geometric closeness. Topological correctness is harder to formulate; we use the notion of isotopy of arrangements, which is stated in the next section.

Compared to algebraic approaches (e.g., those based on resultant computation), the algorithm presented here has following advantages:

1. In the algebraic approach, one computes the overall topology of the arrangement first, followed by an ϵ-approximation to the curves; the second step is usually not fully addressed in these approaches. In contrast to such a "decoupled" approach, our algorithm provides an integrated approach, whereby we can commence to compute the geometric approximation, even before we determine the global topology. Ultimately, we would be able to determine the topology exactly using zero bounds as in [13,3]. Moreover, in this integrated approach we can cut off the computation at any desired resolution, without fully resolving all aspects of the topology, which is useful in applications such as visualization.

2. Our use of analytic (numerical) primitives means that our approach is applicable to the much larger class of analytic curves.

3. Numerical algorithms have adaptive as well as "local" complexity. Adaptive means that the worst case complexity does not characterize the complexity for most inputs, and local means the computational effort is restricted to ROI.

One disadvantage of our current method is that it places some strong restrictions on the class of curve arrangements: the curves must be non-singular with pairwise transversal intersections in the ROI. In practice, these restrictions can be ameliorated in different ways.

Our algorithm falls under the popular literature on Marching-cube type algorithms [9]. The results we present are a generalization of the algorithms of Plantinga-Vegter [11,10] and Lin and Yap [6] for computing isotopic approximation of a single non-singular curve or surface. We give an extension of their results to simple curve arrangements.

2   Underlying Theory and Algorithm

Two closed sets $S, T \subseteq \mathbb{R}^2$ are (ambient) **isotopic**, denoted by $S \simeq T$ if there exists a continuous mapping $\gamma : [0, 1] \times \mathbb{R}^2 \to \mathbb{R}^2$ such that for each $t \in [0, 1]$, the function $\gamma_t : \mathbb{R}^2 \to \mathbb{R}^2$ (with $\gamma_t(x, y) = \gamma(t, x, y)$) is a homeomorphism, γ_0 is the identity map, and $\gamma_1(S) = T$. This can be generalized to arrangement of sets. Let $\overline{S} = (S_1, \ldots, S_m)$ and $\overline{T} = (T_1, \ldots, T_m)$ be two sequences of m closed sets. For each non-empty subset $J \subseteq [m]$, let $\overline{S}_J$ denote the intersection $\cap_{i \in J} S_i$; similarly for $\overline{T}_J$. We say that $\overline{S}$ and $\overline{T}$ are **isotopic** if there exists a continuous mapping of the form γ such that for each non-empty subset $J \subseteq [m]$, we have $\overline{S}_J \simeq \overline{T}_J$ under γ. For simple curve arrangements, the critical problem to solve is when $m = 2$. We assume that the two curves S_1, S_2 are restricted to a region or box B. Our basic problem is to compute a pair of piecewise linear curves (T_1, T_2) such that $(T_1, T_2) \simeq (S_1 \cap B, S_2 \cap B)$. See [1] for a general discussion of isotopy of the case $m = 1$.

Interval arithmetic [7,12] is central to our computational toolkit. Let $\square\mathbb{R}$ be the set of closed and bounded intervals in $\mathbb{R}$, and $\square\mathbb{R}^2$ be the set of axis aligned boxes in the Euclidean plane. For a function $f : \mathbb{R}^2 \to \mathbb{R}$, a **box function** is of the form $\square f : \square\mathbb{R}^2 \to \square\mathbb{R}$ such that

1. for all $B \in \square\mathbb{R}^2$, $\square f(B)$ contains the range of f on B, and
2. for all $\{B_i \in \square\mathbb{R}^2 : i \in \mathbb{N}\}$, if B_i converges monotonically (i.e., $B_1 \supset B_2 \supset B_3 \supset \cdots$) to a point $p \in \mathbb{R}^2$ then $\square f(B_i)$ converges monotonically to $f(p)$.

Note that box functions are easy to construct for polynomials, and most of the real functions commonly used, such as $\sin, \cos, \tan$, etc.

We will use a variety of box predicates. These predicates will determine the subdivision process. Typically, we will keep subdividing boxes until some Boolean combination of these box predicates hold. The following pair of box predicates is crucially used in our algorithm:

$$C_0^f(B) \equiv 0 \notin \square f(B),$$
$$C_1^f(B) \equiv 0 \notin (\square f_x(B))^2 + (\square f_y(B))^2.$$

Note that C_1^f is taken from [11], where the interval operation I^2 is defined as $\{xy : x, y \in I\}$; if the predicate holds then the gradient of f in B does not change by more than $\pi/2$. These predicates will help us approximate the isotopy of individual curves correctly. We next consider two predicates that will allow us to detect boxes containing a common root of two functions f, g. First is the **Jacobian condition**,

$$0 \notin \det(\Box J_{f,g}(B)),$$

where $\Box J_F(B)$ is the interval evaluation on B of the entries in the Jacobian matrix of (f, g). It can be shown that if $JC(B)$ holds, then B has at most one root of $f = g = 0$. The second predicate is the **Moore-Kioustelidis condition** (MK-condition, for short) [8], which can be viewed as a preconditioned form of the famous Miranda Test [5]: If f is sign determinate on the left and right segments of B and the two signs are different, whereas g is sign determinate on the top and bottom segments and the two signs are different, then f and g have a common root inside B (in fact, an odd number of roots). This is a generalization to higher dimensions of the following property in one dimension: if a continuous function changes sign on an interval then it has an odd number of roots in the interval. Thus, if the MK-condition holds, then B has at least one root of $f = g = 0$. Therefore, if both the Jacobian-condition and MK-condition hold for a box B, then we know that B contains a unique common root of f, g.

Our algorithm uses a quadtree based subdivision of the input box B_0. Let S and T be the planar curves defined by f and g, respectively. We want to compute a pair of piecewise linear curves that are isotopic to $(S_1 \cap B_0, S_2 \cap B_0)$. We use the predicates given above to decide when to subdivide a box. Initially, we subdivide a box until one of the following conditions hold:

1. either C_0^f or C_0^g holds (i.e., we can exclude one of S or T from the box), or
2. at least one of C_1^f or C_1^g holds (i.e., change in the gradient for at least one of them is $< \pi/2$).

So far our algorithm is similar to the algorithm in [11]. Boxes for which we cannot exclude S and T, but which satisfy both the predicates in the second condition are the interesting boxes as they can contain common roots of f and g. To detect the presence of a root, we first subdivide such a box until the Jacobian-condition holds (or we have excluded one of S or T from the box), and then until the MK-condition holds. If both the conditions hold for a box, we have isolated a common root of f and g. Typically, at this point we can deduce the arrangement of the curves inside the box from the sign of the curves on the boundary of this box. However, if the common intersection points are located on the boundary of a box generated in the subdivision process then the MK-condition will not terminate. To circumvent this problem, we introduce the notion of non-aligned boxes; roughly speaking, these are boxes obtained by scaling a box in the subdivision tree (which is an aligned box) by some constant; see Figure 1, where a non-aligned box isolating a common root is highlighted. The roots are then isolated inside these non-aligned boxes using essentially the same predicates as above (but on larger boxes).

We now refine the subdivision tree further to aid in the construction of an isotopic approximation to S and T inside B_0. A crucial requirement of the subdivision tree for constructing an isotopic arrangement is that the subdivision should be balanced, i.e., the levels of two neighbors in the tree differ by at most one. This is easily attained for aligned boxes. However, for non-aligned boxes we need an additional step to balance their interior and also to balance them with respect to the neighboring aligned boxes. Once we have a balanced subdivision, we can construct an isotopic arrangement by evaluating the signs of f and g at the endpoints of the boxes; this is where the similarity with marching-cube like methods appears. This suffices for most boxes, but there might be some boxes for which this is not possible locally (e.g., those having a branch of S and T and having no roots in the interior but nearby). In these cases, we need to propagate the arrangement from neighboring boxes where the arrangement has been resolved.

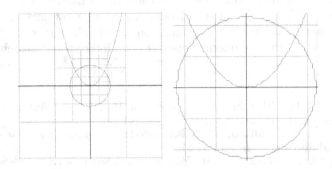

Fig. 1. An isotopic approximation to the arrangement of the circle $x^2 + y^2 = 2$ and the parabola $y = x^2$. The right hand side figure is a zoom-in of the figure on the left, and clearly shows the planar straight-line nature of the approximation. The highlighted square on the left is a non-aligned box containing a common root. Also, note the balanced nature of the subdivision.

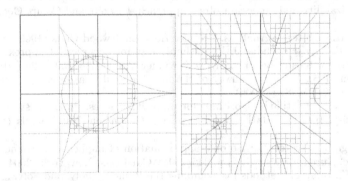

Fig. 2. The figure on the left is the arrangement of the circle $x^2 + y^2 = 2$ and the curve $(x^2 + y^2 + 12x + 9)^2 - 4(2x+3)^3$. The RHS figure corresponds to the arrangement of the curves F and G corresponding to the real and imaginary parts of a degree 5 univariate polynomial $f(z)$, i.e., $f(x+iy) = F(x,y) + iG(x,y)$. Note that the algorithm computes the correct arrangement in a region not containing singularities of the curves.

3 Experimental Results

In this section we show some results of running our algorithm to compute the arrangement of two curves. A preliminary version of the algorithm is implemented in **Core Library** [14]. We are unaware of any other purely numerical approach that computes an isotopic approximation to arrangement of curves, and hence could not compare our results with other software.

4 Conclusion

The algorithm presented here is work in progress. Our initial implementation has been promising. To truly exploit the advantages of the algorithm, we have to test it on non-algebraic curves. The current algorithm handles a pair of curves. The approach has to be extended to handle arrangement of curves, in general. Moreover, removing the strong constraints on the simple nature of curves that we can handle in a purely numerical manner is an interesting research problem.

References

1. Boissonnat, J.-D., Cohen-Steiner, D., Mourrain, B., Rote, G., Vegter, G.: Meshing of surfaces. In: Boissonnat, Teillaud (eds.) [2], ch. 5
2. Boissonnat, J.-D., Teillaud, M. (eds.): Effective Computational Geometry for Curves and Surfaces. Springer (2006)
3. Burr, M., Choi, S., Galehouse, B., Yap, C.: Complete subdivision algorithms, II: Isotopic meshing of singular algebraic curves. In: 33rd Int'l Symp. Symb. and Alge. Comp., pp. 87–94 (2008); Special Issue of JSC 47(2), 131–152 (2012)
4. de Berg, M., van Kreveld, M., Overmars, M., Schwarzkopf, O.: Computational Geometry: Algorithms and Applications. Springer, Berlin (1997)
5. Kulpa, W.: The Poincaré-Miranda theorem. The American Mathematical Monthly 104(6), 545–550 (1997)
6. Lin, L., Yap, C.: Adaptive isotopic approximation of nonsingular curves: the parameterizability and nonlocal isotopy approach. Discrete and Comp. Geom. 45(4), 760–795 (2011)
7. Moore, R.E.: Interval Analysis. Prentice Hall, Englewood Cliffs (1966)
8. Moore, R.E., Kioustelidis, J.B.: A simple test for accuracy of approx. solns. to nonlinear (or linear) systems. SIAM J. Numer. Anal. 17(4), 521–529 (1980)
9. Newman, T.S., Yi, H.: A survey of the marching cubes algorithm. Computers & Graphics 30, 854–879 (2006)
10. Plantinga, S.: Certified Algorithms for Implicit Surfaces. Ph.D. thesis, Groningen University, Inst. for Math. and Comp. Sc., Groningen, Netherlands (December 2006)
11. Plantinga, S., Vegter, G.: Isotopic approximation of implicit curves and surfaces. In: Proc. Eurographics SGP, pp. 245–254. ACM Press, New York (2004)
12. Stahl, V.: Interval Methods for Bounding the Range of Poly. and Solving Systems of Nonlin. Eqns. Ph.D. thesis, Johannes Kepler University, Linz (1995)
13. Yap, C.K.: Complete subdivision algorithms, I: Intersection of Bezier curves. In: 22nd ACM Symp. on Comp. Geom., pp. 217–226 (July 2006)
14. Yu, J., Yap, C., Du, Z., Pion, S., Brönnimann, H.: The Design of Core 2: A Library for Exact Numeric Computation in Geometry and Algebra. In: Fukuda, K., van der Hoeven, J., Joswig, M., Takayama, N. (eds.) ICMS 2010. LNCS, vol. 6327, pp. 121–141. Springer, Heidelberg (2010)

Real Quantifier Elimination
in the RegularChains Library

Changbo Chen[1] and Marc Moreno Maza[2]

[1] Chongqing Key Laboratory of Automated Reasoning and Cognition,
Chongqing Institute of Green and Intelligent Technology,
Chinese Academy of Sciences, China
changbo.chen@hotmail.com
http://www.orcca.on.ca/~cchen
[2] ORCCA, University of Western Ontario, Canada
moreno@csd.uwo.ca
www.csd.uwo.ca/~moreno

Abstract. Quantifier elimination (QE) over real closed fields has found numerous applications. Cylindrical algebraic decomposition (CAD) is one of the main tools for handling quantifier elimination of nonlinear input formulas. Despite of its worst case doubly exponential complexity, CAD-based quantifier elimination remains interesting for handling general quantified formulas and producing simple quantifier-free formulas.

In this paper, we report on the implementation of a QE procedure, called QuantifierElimination, based on the CAD implementations in the RegularChains library. This command supports both standard quantifier-free formula and extended Tarski formula in the output. The use of the QE procedure is illustrated by solving examples from different applications.

Keywords: Quantifier elimination, cylindrical algebraic decomposition, triangular decomposition, RegularChains.

1 Introduction

In the 1930's, A. Tarski [11] proved that quantifier elimination over the reals is possible and provided the first algorithm for real quantifier elimination, although the complexity of his algorithm is not even elementary recursive. In 1975, G. E. Collins [7] invented cylindrical algebraic decomposition, which opens the door for solving quantifier elimination practically. The worst case complexity for solving QE by means of CAD is doubly exponential in the number of variables. In the 1990's, QE algorithms, whose worst complexity are doubly exponential in the number of alternative quantifier blocks instead of variables, emerged [1]. Although QE based on CAD is not favorable in terms of complexity, it remains a practical tool for solving general QE problems and obtaining simple quantifier free formula.

Many authors have improved the practical efficiency of CAD based on the original projection-lifting scheme proposed by Collins. In [6], with B. Xia and L.

H. Hong and C. Yap (Eds.): ICMS 2014, LNCS 8592, pp. 283–290, 2014.

Yang, we introduced an alternative way of computing CADs based on triangular decompositions. In this new method, one first computes a complex cylindrical decomposition (CCD), which partitions the complex space into cylindrically arranged cells, each of which is the complex zero set of a regular system. In a second stage, the real connected components of each regular system are computed, which all together form a CAD of the real space. A CAD computed in this way is called an RC-CAD. The efficiency of RC-CAD was substantially improved in [5], where an incremental algorithm was proposed to compute CCDs. Moreover, in the same paper, a systematic way for making use of equational constraints is presented.

In [4], an RC-CAD based quantifier elimination algorithm was proposed. A preliminary implementation of it in the **RegularChains** library is available through the function QuantifierElimination. The goal of this paper is to present the implementation details of such an algorithm. Several important optimizations are also discussed. The paper is organized as follows. In Section 2, we illustrate the user interface of QuantifierElimination by some simple examples. In Section 3, we present some non-trivial applications of it. In Section 4, we explain the underlying theory and algorithm as well as some optimizations realized in the implementation.

2   Functionality

In this section, we explain the usage of QuantifierElimination by some simple examples.

In Figure 1, the user interface of QuantifierElimination is illustrated by the famous Davenport-Heintz example.

Solve the Davenport–Heintz problem by QuantifierElimination.

$$(\exists\ c, \forall\ b, a)\ ((a = d \wedge b = c) \vee (a{=}c \wedge b = 1)) \Rightarrow a^2 = b.$$

> f := &E([c]), &A([b, a]), ((a=d) &and (b=c)) &or
        ((a=c) &and (b=1)) &implies (a^2=b);

$f := \&E([c]), \&A([b, a]), a = d$ &and $b = c$ &or $a = c$ &and $b = 1$ &implies $a^2 = b$

> out := QuantifierElimination(f);

$$out := d - 1 = 0 \text{ &or } d + 1 = 0$$

Fig. 1. The user interface of QuantifierElimination

The user interface of QuantifierElimination is implemented on top of the Logic package of MAPLE. This package supports usual logical operators, such as $\wedge$, $\vee$, $\neg$, $\Longrightarrow$, $\Longleftrightarrow$, and represent them respectively by &and, &or, &not, &implies, &iff. There is also a function called Normalize, which can convert a given logical formula into its disjunctive normal form or conjunctive normal form. However, the quantifiers are missing in the Logic package. We create the symbol $\&E$ and $\&A$ to represent respectively the existential quantifier $\exists$ and the universal quantifier $\forall$. To use them, the quantified variables have to been put in a list, as shown in Figure 1. Note that all operators in the Logic package have the same precedence. Parentheses should be used to correctly specify the precedence.

In Figure 1, the order of variables is not specified. In such case, QuantifierElimination calls the function SuggestVariableOrder of RegularChains library to pick a "good" order by some heuristic strategy. It is also possible for the user to choose her favorable order, as shown in Figure 2, where the variables supplied to the function PolynomialRing are in descending order.

```
> R := PolynomialRing([x, a, b, c]);
  f := &E([x]), a*x^2+b*x+c=0;
  out := QuantifierElimination(f, R);
```

$$R := polynomial\_ring$$

$$f := \&E([x]), x^2\,a + x\,b + c = 0$$

$$out := \left(\left(4\,a\,c - b^2 < 0 \text{ \&or } 4\,a\,c - b^2 = 0 \text{ \&and } a < 0\right) \text{ \&or } 4\,a\,c - b^2\right.$$
$$= 0 \text{ \&and } 0 < a\left.\right) \text{ \&or } \left(4\,a\,c - b^2 = 0 \text{ \&and } a = 0\right) \text{ \&and } c = 0$$

Fig. 2. The default output of QuantifierElimination is quantifier free formula

The default output of QuantifierElimination is a quantifier free formula formed by polynomial constraints and logical connectives, which is the same as the default output of QEPCAD. Such formulas are called Tarski formulas. An alternative output format, called extended Tarski formula, is also available, when the option 'output'='rootof' is specified. An extended Tarski formula extends Tarski formula by allowing indexed roots of polynomials to appear. This is illustrated by Figure 3 and Figure 4. Such an output format is the same as the default output of Mathematica.

```
> f := &E([x]), a*x^2+b*x+c=0;
  out := QuantifierElimination(f,'output'='rootof');
```

$$f := \&E([x]), a\,x^2 + b\,x + c = 0$$

$$out := \left(\left(\left(a < 0 \text{ \&and } \frac{1}{4}\frac{b^2}{a} \le c \text{ \&or } a = 0 \text{ \&and } b < 0\right) \text{ \&or } (a = 0 \text{ \&and } b\right.\right.$$
$$= 0) \text{ \&and } c = 0\left.\right) \text{ \&or } a = 0 \text{ \&and } 0 < b\left.\right) \text{ \&or } 0 < a \text{ \&and } c \le \frac{1}{4}\frac{b^2}{a}$$

Fig. 3. The output of QuantifierElimination in extended Tarski formula

The users who are familiar with MAPLE's RootOf may be surprised to see the real index there. Indeed, it is a new feature we added to Rootof, which is currently supported by the evalf function of MAPLE as shown in Figure 4.

```
> f := &E([y]), y^2+x^2=2;
  out := QuantifierElimination(f, output=rootof);
```
$$f := \&E([y]), x^2 + y^2 = 2$$
$$out := -\sqrt{2} \le x \text{ \&and } x \le \sqrt{2}$$

```
> f := &E([y]), y^4+x^4=2;
  out := QuantifierElimination(f, output=rootof);
```
$$f := \&E([y]), x^4 + y^4 = 2$$
$$out := RootOf\left(\_Z^4 - 2, index = real_1\right) \le x \text{ \&and } x \le RootOf\left(\_Z^4 - 2, index\right.$$
$$\left. = real_2\right)$$

```
> evalf(op(1, out)); evalf(op(2, out));
```
$$-1.189207115 \le x$$
$$x \le 1.189207115$$

Fig. 4. Solve QuantifierElimination in extended Tarski formula

3   Application

In this section, we present how QuantifierElimination is applied to solve several applications.

The first application is on the verification and synthesis of switched and hybrid dynamical systems [10]. A common problem studied in this field is to determine if a system remains in the safe state if it starts in an initial safe state. A typical approach to solve this problem is to find a certificate, or an invariant set, such that the following are satisfied simultaneously:

- the initial states satisfy the invariant set
- any states that satisfy the invariant set are safe
- the system dynamics cannot force the system to leave the invariant set

Finding such a certificate can be casted into a real quantifier elimination problem.

In Figure 5, we show how to use QuantifierElimination to solve the quantifier elimination problem casted from an 1-D robot model [10], where the details of the casting are explained. This problem was originally solved in [10] by a combination of Reduce and QEPCAD.

The second application is on computing control Lyapunov function. Suppose we are given a dynamical system $\dot{x} = f(x, u)$, where $x \in \mathbb{R}^n$ and $u \in \mathbb{R}$ are respectively the state variables and the control input implicitly depending on t.

```
> phi1 := ( ( 74 <= x ) &and ( x <= 76 ) &and ( v = 0 )
  &implies ( -v^2 - a * (x-75)^2 + b >= 0 ) ):

> phi2 := ( ( -v^2 - a * (x-75)^2 + b >= 0 )
  &implies (( 80 >= x ) &and ( x >= 70 )) ):

> phi3 := ( ( -v^2 - a * (x-75)^2 + b = 0 )
  &implies (( -2*v - a * 2 * (x-75)* v >= 0 ) &or ( 2*v - a
  * 2 * (x-75)* v >= 0 )) ):

> phi := phi1 &and phi2 &and phi3:
> t0 := time():
  psi := QuantifierElimination(&A([x,v]),phi,output=rootof);
  t1 := time() - t0;
```

$$\psi := ((0 < a \,\&and\, a \le 1) \,\&and\, a \le b) \,\&and\, b \le \min\left(\frac{1}{a}, 25\,a\right)$$

$$t1 := 15.094$$

Fig. 5. Solve a QE problem related to 1-D robot model

Definition 1. *A function* $V(x) : \mathbb{R}^n \to \mathbb{R}$ *is called a* control Lyapunov function *of the dynamical system* $\dot{x} = f(x, u)$ *if the following are satisfied:*

- $V(x)$ *is positive definite, that is* $V(0) = 0$ *and* $\forall x \neq 0$, $V(x) > 0$.
- $\dot{V}(0) = 0$ *and* $\forall x \neq 0$, $\exists u$, *such that* $\dot{V} < 0$, *where* $\dot{V} = \nabla V(x) \cdot f(x, u)$.
- V *is radically unbounded, that is* $\|x\| \to \infty$ *implies that* $V \to \infty$.

Suppose one wants to know if there exists a control Lyapunov function of a given template $V(a, x)$, where a are parameters. The equivalent QE problem is:

$$(\forall x, \exists u)(x \neq 0) \implies (V > 0 \land \nabla V(x) \cdot f(x, u) < 0).$$

If one also wants to find out if there exists u of a given template $g(b, x)$, then the equivalent QE problem is:

$$(\forall x, \exists u)\left((u = g(b, x)) \land (x \neq 0 \implies (V > 0 \land \nabla V(x) \cdot f(x, u) < 0))\right).$$

Let's illustrate this application by a bivariate dynamical system.

$$\begin{cases} \frac{dx_1}{dt} = -x_1 + u \\ \frac{dx_2}{dt} = -x_1 - x_2^3 \end{cases}$$

We aim to find control Lyapunov function of the form $V := a_1 x_1^2 + a_2 x_2^2$ and control input of the form $u := b_1 x_1 + b_2 x_2$. Figure 6 shows how to call QuantifierElimination to find parameters a_1, a_2, b_1, b_2 such that control Lyapunov function exists. The computation takes several seconds. To verify the result, let $a_1 = a_2 = b_2 = 1$ and $b_1 = 0$, we obtain $u = x_2$, $V = x_1^2 + x_2^2$ and $\dot{V} = -2x_1^2 - 4x_2^2$. Clearly V is a control Lyapunov function.

```
> f1 := -x_1+u: f2 := -x_1-x_2^3:
  V := a_1*x_1^2+a_2*x_2^2;
  Vt := diff(V, x_1)*f1 + diff(V, x_2)*f2;
```
$$V := x\_2^2\, a\_2 + x\_1^2\, a\_1$$
$$Vt := 2\, a\_1\, x\_1\, (u - x\_1) + 2\, a\_2\, x\_2 \left(-x\_2^3 - x\_1\right)$$

```
> QuantifierElimination( &A([x_1,x_2]), &E([u]), (x_1<>0)
  &or (x_2<>0) &implies ((V>0) &and (Vt<0)) );
```
$$0 < a\_1 \,\&\text{and}\, 0 < a\_2$$

```
> QuantifierElimination( &A([x_1, x_2]), &E([u]), (u=b_1*
  x_1+b_2*x_2) &and (a_1>0) &and (a_2>0) &and ((x_1<>0) &or
  (x_2<>0) &implies ((Vt<0))) );
```
$$((b\_2\, a\_1 - a\_2 = 0 \,\&\text{and}\, 0 < a\_1) \,\&\text{and}\, 0 < a\_2) \,\&\text{and}\, b\_1 < 1$$

Fig. 6. Compute control Layapunov function

4   Underlying Theory

Let $PF := (Q_{k+1}x_{k+1}, \ldots, Q_n x_n) FF(x_1, \ldots, x_n)$, where FF is a logical formula formed by polynomial constraints with real number coefficients and logical connectives and each Q_i, $k+1 \leq i \leq n$, is an existential or universal quantifier. The problem of quantifier elimination looks for an equivalent quantifier free formula SF involving only the free variables $x_1, \cdots, x_k$. Let F be the set of polynomials appearing in FF.

The QE algorithm based on RC-CAD consists of the following steps:

1. Compute an F-sign invariant CCD of $\mathbb{C}^n$, that is a CCD such that above any given cell of it, each polynomial in F either vanishes at all points of the cell or no points of the cell.
2. Produce an F-invariant CAD of $\mathbb{R}^n$ from the CCD by real root isolation.
3. For each cell c of the CAD, evaluate FF at a sample point of c and attach the resulting truth value to c.
4. Propagate the truth value according to the quantifiers until each cell in the the induced CAD of $\mathbb{R}^k$ is attached with a truth value, see Figure 7.
5. Each true cell has a defining extended Tarski formula representation. If only extended Tarski formula output is required, then the disjunction of the representation of all true cells, with possible simplification, gives the solution formula SF. If Tarski formula is required, one tests if the signs of polynomials in the CCD are enough to distinguish true and false cells of the CAD. If yes, a representation of the true cells by the signs of these polynomials gives SF. If no, the CCD is refined and the algorithm resumes from Step 2.

It was proved in [4] that the above process terminates in finitely many steps.

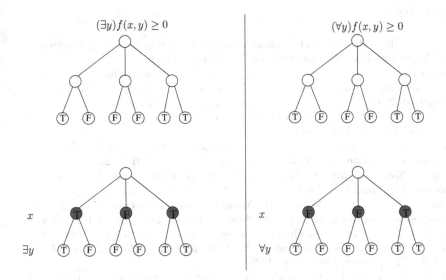

Fig. 7. Propagate truth values

We explain now briefly a few optimizations that have been implemented in QuantifierElimination. Let $PF := (Q_{k+1}x_{k+1}, \ldots, Q_n x_n) FF(x_1, \ldots, x_n)$. If FF is a conjunctive formula having equational constraints, then truth-invariant CCDs and CADs are computed instead of sign-invariant ones using techniques proposed in [5]. If FF is in disjunctive normal form and has equational constraints, then truth table invariant CCDs and CADs are computed using algorithm presented in [2]. If there exists m, $k+1 \leq m \leq n$, such that $Q_m = \cdots = Q_n = \forall$, then PF is converted to its equivalent form

$$(Q_{k+1}x_{k+1}, \ldots, Q_{m-1}x_{m-1}, \neg \exists x_m, \exists x_{m+1}, \ldots, \exists x_n) \neg FF(x_1, \ldots, x_n).$$

This trick is particular useful if FF is of the form $A \implies B$, where A has equational constraints, as $\neg FF$ is equivalent to $A \wedge \neg B$, which can benefit from the techniques for making use of equational constraints in [5,2].

We have also implemented some simple partial lifting techniques when FF is a conjunctive formula. Exploiting systematically the partial lifting techniques as in [8] is working in progress. In [4], some simplification strategies for the Tarski formula output of QuantifierElimination was proposed. The simplification remains to be enhanced by integrating techniques as in [9,3]. For the extended Tarski formula, a better technique for merging true cells is working in progress.

Acknowledgments. This work was supported by the NSFC (11301524) and the CSTC (cstc2013jjys0002).

References

1. Basu, S., Pollack, R., Roy, M.-F.: Algorithms in real algebraic geometry. Algorithms and Computations in Mathematics, vol. 10. Springer (2006)
2. Bradford, R., Chen, C., Davenport, J.H., England, M., Moreno Maza, M., Wilson, D.: Truth table invariant cylindrical algebraic decomposition by regular chains (submitted, 2014), Preprint: http://opus.bath.ac.uk/38344/
3. Brown, C.W.: Solution Formula Construction for Truth Invariant CAD's. PhD thesis, University of Delaware (1999)
4. Chen, C., Moreno Maza, M.: Quantifier elimination by cylindrical algebraic decomposition based on regular chains (Accepted for ISSAC 2014) (2014)
5. Chen, C., Moreno Maza, M.: An incremental algorithm for computing cylindrical algebraic decompositions. In: Proc. ASCM 2012 (October 2012)
6. Chen, C., Moreno Maza, M., Xia, B., Yang, L.: Computing cylindrical algebraic decomposition via triangular decomposition. In: ISSAC 2009, pp. 95–102 (2009)
7. Collins, G.E.: Quantifier elimination for real closed fields by cylindrical algebraic decomposition. In: Brakhage, H. (ed.) GI-Fachtagung 1975. LNCS, vol. 33, pp. 515–532. Springer, Heidelberg (1975)
8. Collins, G.E., Hong, H.: Partial cylindrical algebraic decomposition. Journal of Symbolic Computation 12(3), 299–328 (1991)
9. Hong, H.: Simple solution formula construction in cylindrical algebraic decomposition based quantifier elimination. In: ISSAC, pp. 177–188 (1992)
10. Sturm, T., Tiwari, A.: Verification and synthesis using real quantifier elimination. In: ISSAC, pp. 329–336 (2011)
11. Tarski, A.: A decision method for elementary algebra and geometry. University of California Press (1951)

Software for Quantifier Elimination in Propositional Logic

Eugene Goldberg[1] and Panagiotis Manolios[2]

[1] Northeastern University, USA
eigold@ccs.neu.edu
http://www.ccs.neu.edu/home/eigold
[2] Northeastern University, USA
pete@ccs.neu.edu
http://www.ccs.neu.edu/home/pete

Abstract. We consider the following problem of Quantifier Elimination (QE). Given a Boolean CNF formula F where some variables are existentially quantified, find a logically equivalent CNF formula that is free of quantifiers. Solving this problem comes down to finding a set of clauses depending only on free variables that has the following property: adding the clauses of this set to F makes all the clauses of F with quantified variables redundant. To solve the QE problem we developed a tool meant for handling a more general problem called partial QE. This tool builds a set of clauses adding which to F renders a specified subset of clauses with quantified variables redundant. In particular, if the specified subset contains all the clauses with quantified variables, our tool performs QE.

Keywords: Propositional logic, quantifier elimination, dependency sequents.

1 Introduction

In this extended abstract, we describe software for solving the problem of Quantifier Elimination (QE) and that of Partial QE (PQE). Let $H(X, Y)$ be a Boolean formula in Conjunctive Normal Form (CNF). Given a formula $\exists X[H]$, the **QE problem** is to find a CNF formula $H^*(Y)$ such that $H^* \equiv \exists X[H]$.

Let $F(X, Y)$ and $G(X, Y)$ be CNF formulas. Given a formula $\exists X[F \wedge G]$, the **PQE problem** is to find a CNF formula $F^*(Y)$ such that $F^* \wedge \exists X[G] \equiv \exists X[F \wedge G]$. We will say that formula F^* is obtained by taking F out of the scope of quantifiers. Obviously, QE is a special case of PQE where the entire formula is taken out of the scope of quantifiers.

QE has numerous applications in verification. For instance, to find if a system specified by a transition relation $T(S, S')$ can reach a bad state, one needs to perform reachability analysis. Here S, S' specify current and next state variables. The set of states reachable in one transition from states specified by Boolean formula $G(S')$ is described by $\exists S[G \wedge T]$. To represent this set of states in a quantifier-free form one needs to find a quantifier-free formula logically equivalent to $\exists S[G \wedge T]$ i.e. to solve the QE problem.

H. Hong and C. Yap (Eds.): ICMS 2014, LNCS 8592, pp. 291–294, 2014.
© Springer-Verlag Berlin Heidelberg 2014

Unfortunately, the "straightforward" methods of QE seem to be very time-consuming even in propositional logic. This is one reason that many successful theorem proving methods such as interpolation [4] and IC3 [1] avoid QE and use SAT-based reasoning instead. This motivates our interest in studying variations of QE that can be solved efficiently. PQE is one of such variations. A detailed description of our algorithm for solving the PQE problem is given in [3].

2   Application of PQE: Solving SAT by PQE

In [3], we list some applications of PQE to verification problems. In this section, we give one more application not mentioned in [3]. Namely, we show how PQE can be used to solve a version of SAT called Circuit-SAT. We give two methods of reducing Circuit-SAT to PQE that are complementary to each other.

2.1   Circuit-SAT

Let $N(X, Y, z)$ be a single-output combinational circuit. Here X and Y are the sets of input and internal variables respectively and z specifies the output of N. Suppose that one needs to check the satisfiability of N i.e. whether N ever evaluates to 1. We will refer to this problem as Circuit-SAT (as opposed to SAT that is the problem of checking the satisfiability of arbitrary Boolean formulas). A common way of solving Circuit-SAT is to represent N as a CNF formula $H(X, Y, z)$ obtained by Tseitsin transformations and then check if $H \land z$ is satisfiable.

2.2   Reducing Circuit-SAT to PQE: First Method

One can reduce checking the satisfiability of formula $H \land z$ above to PQE as follows. Let F be the set of all clauses of H with literal $\overline{z}$. We will refer to such clauses as $\overline{z}$-**clauses** of H. Let $G = H \setminus F$. Checking the satisfiability of $H \land z$ is equivalent to solving the PQE problem of finding formula $F^*(z)$ such that $F^*(z) \land \exists W[G] \equiv \exists W[F \land G]$ where $W = X \cup Y$. If $F^*(z) \equiv 1$, i.e. if F^* consists of an empty set of clauses, formula $H \land z$ is satisfiable. If $F^*(z) = \overline{z}$, then $H \land z$ is unsatisfiable. In other words, if all $\overline{z}$-clauses are redundant in $\exists W[H]$, then $H \land z$ is satisfiable. However, if making the original $\overline{z}$-clauses of H redundant requires derivation and adding to H clause $\overline{z}$, then $H \land z$ is unsatisfiable.

Indeed, if clause $\overline{z}$ is derived from H it can be resolved with clause z of $H \land z$ to produce an empty clause. This proves the unsatisfiability of $H \land z$. If $\overline{z}$-clauses are redundant in $\exists W[H]$ without derivation of $\overline{z}$, the fact that H is satisfiable, implies that assignment $z = 1$ can be extended to an assignment satisfying H. This assignment obviously satisfies $H \land z$.

Note that in case $H \land z$ is unsatisfiable, the final goal of a PQE-algorithm solving the PQE problem above is the same as that of a SAT-solver. The PQE-algorithm aims at deriving clause $\overline{z}$ that is only one resolution operation away from producing an empty clause. However, in case $H \land z$ is satisfiable, there is an important difference: a PQE-algorithm can prove satisfiability just by showing redundancy of $\overline{z}$-clauses of H i.e. *without finding a satisfying assignment*.

2.3 Reducing Circuit-SAT to PQE: Second Method

Here we give a different method of reducing Circuit-SAT to PQE. We will refer to the methods of the previous and current subsections as first and second method respectively. The second method is to solve the PQE problem of finding a CNF formula $K(X)$ such that $K \wedge \exists V[H] \equiv \exists V[\overline{z} \wedge H]$ where $V = Y \cup \{z\}$. That is K is obtained by taking clause $\overline{z}$ out of the scope of quantifiers. It is not hard to show that $H \wedge z$ is satisfiable if and only if formula K contains at least one clause. Every complete assignment to X falsifying K specifies an input for which N evaluates to 1, i.e. a counterexample. So if finding one counterexample suffices, one can stop as soon as a clause is added to K.

Notice that the first and second methods are, in a sense, *complementary*. To prove unsatisfiability of $H \wedge z$ by the first method, one needs to produce an explicit derivation of an empty clause. However, proving satisfiability of $H \wedge z$ does not require finding an explicit satisfying assignment. In the second method, the situation is the opposite. Proving satisfiability requires generating at least one clause of H and hence finding at least one counterexample. On the other hand, the fact that clause $\overline{z}$ is redundant in $\exists V[\overline{z} \wedge H]$ means that $H \wedge z$ is unsatisfiable. However, the second method does not give an explicit proof of this fact (like generation of an empty clause).

An interesting feature of the second method is that it provides a *derivation* of a counterexample. Usually a counterexample is a result of guesswork even in a formal verification tool. For example, finding a satisfying assignment by a SAT-solver requires *guessing* the decision assignments. (Implied assignments are derived from learned clauses and do not need guesswork.) This makes it hard to measure the *complexity* of finding a counterexample. In the second method, a counterexample x is a complete assignment falsifying a clause C of K. This clause is *derived* from $\overline{z} \wedge H$ and the length of this derivation can be used to measure the complexity of finding counterexample x.

3 Quantifier Elimination by Dependency Sequents

In this section, we give the high-level view of our algorithms for QE and PQE.

Suppose that one needs to eliminate quantifiers from formula $\exists X[H]$. In [2], we developed a QE algorithm based on the notion of a Dependency Sequent (D-sequent). This algorithm is called $DCDS$ (Derivation of Clause D-sequents). $DCDS$ is based on the following two ideas. First, if one adds to H a "sufficient" number of resolvent clauses, all X-clauses (i.e. clauses containing variables of X) will become redundant. Second, proving clause redundancy globally is hard. So it makes sense to use branching to prove redundancy of X-clauses in subspaces first and then merge the results of different branches. Proving redundancy of X-clauses of H in subspaces, in general, requires adding resolvent clauses to H.

Let q be an assignment to variables of H. A record $(\exists X[H], q) \rightarrow R$ called **D-sequent** is used by $DCDS$ to store the fact that a set R of X-clauses is redundant in $\exists X[H]$ in subspace q. Assignment q is called the conditional part of the D-sequent. When $DCDS$ merges results of branching on a variable v, it "merges" D-sequents obtained in subspaces $v = 0$ and $v = 1$ using a resolution-like operation called *join*. This results in producing new D-sequents that do not have an assignment to variable v in their conditional parts. The objective of $DCDS$ is to derive D-sequent $(\exists X[H], \emptyset) \rightarrow H^X$ where H^X is the set of all X-clauses of H. This D-sequents states unconditional redundancy of X-clauses in $\exists X[H]$. Once this D-sequent is derived, a solution to the QE problem is obtained by removing all X-clauses from H.

We have developed a PQE algorithm based on $DCDS$. This algorithm is called **DS-PQE** (DS stands for D-Sequents). Suppose one needs to solve the PQE problem of taking F out of the scope of quantifiers in $\exists X[F \wedge G]$. $DS\text{-}PQE$ is based on the same two ideas as above. The main difference of $DS\text{-}PQE$ from $DCDS$, is that the former needs to prove only the redundancy of X-clauses of F. So the objective of $DS\text{-}PQE$ is to derive D-sequent $(\exists X[F \wedge G], \emptyset) \rightarrow F^X$ where F^X is the set of all X-clauses of F. In $DS\text{-}PQE$, new resolvent clauses are assumed to be added to F while G stays unchanged. So after the final D-sequent above is derived, a solution to the PQE problem is obtained from F by discarding the clauses of F^X.

4   Software Description

$DS\text{-}PQE$ is implemented as a stand-alone program written in C++. $DS\text{-}PQE$ accepts formula $\exists X[F \wedge G]$ and returns formula F^* such that $F^* \wedge \exists X[G] \equiv \exists X[F \wedge G]$. Formula $\exists X[F \wedge G]$ is specified by three files. The first file describes CNF formula $F \wedge G$ in the DIMACS format. The second file lists the free variables of $\exists X[F \wedge G]$. So if a variable is not mentioned in this file it is assumed to be quantified. The third file lists the clauses of F. The resulting CNF formula F^* is returned in a file in the DIMACS format.

Acknowledgments. This research was supported in part by DARPA under AFRL Cooperative Agreement No. FA8750-10-2-0233 and by NSF grants CCF-1117184 and CCF-1319580.

References

1. Bradley, A.R.: SAT-based model checking without unrolling. In: Jhala, R., Schmidt, D. (eds.) VMCAI 2011. LNCS, vol. 6538, pp. 70–87. Springer, Heidelberg (2011)
2. Goldberg, E., Manolios, P.: Quantifier elimination via clause redundancy. In: FM-CAD 2013, pp. 85–92 (2013)
3. Goldberg, E., Manolios, P.: Partial quantifier elimination, submitted to a conference (2014)
4. McMillan, K.L.: Interpolation and SAT-based model checking. In: Hunt Jr., W.A., Somenzi, F. (eds.) CAV 2003. LNCS, vol. 2725, pp. 1–13. Springer, Heidelberg (2003)

Quantifier Elimination
for Linear Modular Constraints

Ajith K. John[1] and Supratik Chakraborty[2]

[1] Homi Bhabha National Institute, BARC, Mumbai, India
[2] Dept. of Computer Sc. & Engg., IIT Bombay, India

Abstract. Linear equalities, disequalities and inequalities on fixed-width bit-vectors, collectively called linear modular constraints, form an important fragment of the theory of fixed-width bit-vectors. We present an efficient and bit-precise algorithm for quantifier elimination from conjunctions of linear modular constraints. Our algorithm uses a layered approach, whereby sound but incomplete and cheaper layers are invoked first, and expensive but complete layers are called only when required. We have extended the above algorithm to work with boolean combinations of linear modular constraints as well. Experiments on an extensive set of benchmarks demonstrate that our techniques significantly outperform alternative quantifier elimination techniques based on bit-blasting and Presburger Arithmetic.

Keywords: Quantifier Elimination, Linear Modular Arithmetic.

1 Introduction

A first-order theory T is said to admit quantifier elimination (henceforth called QE) if every quantified formula φ in the theory is T-equivalent to a quantifier-free formula ψ. The theory admits *effective* QE if there exists an algorithm that computes ψ on input φ. An example of a theory admitting effective QE is the theory of fixed-width bit-vectors. This theory is extremely important in the context of word-level verification and analysis of hardware and software systems. QE is a key operation in such verification and analysis tasks.

For ease of analysis, words in hardware and software systems are often abstracted as unbounded integers, and QE techniques for integers [5,6] are used by verification and analysis tools. However the results of verification and analysis using QE for unbounded integers may not be sound [3,9] if the underlying implementation uses fixed-width bit-vectors. Therefore, bit-precise QE techniques from fixed-width bit-vector constraints is an important problem.

Boolean combinations of linear equalities, disequalities and inequalities on fixed-width bit-vectors, collectively called linear modular constraints, form an important fragment of the theory of fixed-width bit-vectors. Let p be a positive integer constant, $x_1, \ldots, x_n$ be p-bit non-negative integer variables, and $a_0, \ldots, a_n$ be integer constants in $\{0, \ldots, 2^p - 1\}$. A linear term over $x_1, \ldots, x_n$ is a term of the form $a_1 \cdot x_1 + \cdots a_n \cdot x_n + a_0$. A linear modular equality (LME)

H. Hong and C. Yap (Eds.): ICMS 2014, LNCS 8592, pp. 295–302, 2014.
© Springer-Verlag Berlin Heidelberg 2014

is a formula of the form $t_1 = t_2 \pmod{2^p}$, where t_1 and t_2 are linear terms over $x_1, \ldots, x_n$. Similarly, a linear modular disequality (LMD) is a formula of the form $t_1 \neq t_2 \pmod{2^p}$, and a linear modular inequality (LMI) is a formula of the form $t_1 \bowtie t_2 \pmod{2^p}$, where $\bowtie \in \{<, \leq\}$. We will use linear modular constraint (LMC) when the distinction between LME, LMD and LMI is not important. Conventionally 2^p is called the modulus of the LMC. Since every variable in an LMC with modulus 2^p represents a p-bit integer, we will assume without loss of generality that whenever we consider a conjunction of LMCs sharing a variable, all the LMCs have the same modulus.

The most dominant technique used in practice for eliminating quantifiers from LMCs is conversion to bit-level constraints (also called bit-blasting [4]), followed by bit-level QE. However this technique scales poorly as the width of bit-vectors increases. In addition, the quantifier-eliminated formula appears more like a propositional logic formula on blasted bits instead of being a bit-vector formula. This reduces the scope for word-level reasoning on the quantifier-eliminated formula if it is used in further reasoning. Since LMCs can be expressed as formulae in Presburger Arithmetic (PA), QE techniques for PA such as Omega Test [6] can also be used to eliminate quantifiers from LMCs. However using PA-based techniques for QE from LMCs scales poorly in practice [4]. Moreover, these techniques destroy the word-level structure of the problem.

We present efficient and bit-precise techniques for QE from LMCs that overcome the above drawbacks in practice. In contrast to bit-blasting and PA-based techniques, our techniques keep the quantifier-eliminated formula in linear modular arithmetic, so that it is amenable to further bit-vector level reasoning.

Our techniques have applications in model checking, program analysis and counterexample guided abstraction refinement (CEGAR) of word-level RTL designs and embedded programs. Symbolic transition relations of word-level RTL designs and embedded programs involve boolean combinations of LMCs. LMEs arise from the assignment statements, whereas LMDs and LMIs arise primarily from branch and loop conditions that compare words/registers. QE from formulae involving symbolic transition relation is the key operation during image computation, computation of strongest post-conditions and computation of predicate abstractions in the verification of such word-level RTL designs and embedded programs. In a CEGAR framework, our techniques can be used to compute abstraction of symbolic transition relation by existentially quantifying out a selected set of variables from the transition relation, and to compute Craig interpolants from spurious counterexamples.

There are two fundamental technical contributions of this work. First, we present a practically efficient and bit-precise algorithm for QE from conjunctions of LMCs called *Project*. Secondly, we extend *Project* for eliminating quantifiers from boolean combinations of LMCs. The work presented here is a collation of our earlier works in [1] and [2]. We have skipped the details of the algorithms and proofs due to lack of space. For interested reader, these details can be found in [1,2].

2 *Project*: Algorithm for QE from Conjunctions of LMCs

Project uses a layered approach to eliminate quantifiers from a conjunction of LMCs. Sound but incomplete, cheaper layers are invoked first, and expensive but complete layers are called only when required.

2.1 Layer1: Simplifications Using LMEs

Layer1 is an extension of the work by Ganesh and Dill [8]. It involves simplification of the given conjunction of LMCs using LMEs present in the conjunction.

For example, consider the problem of computing $\exists x.\,((6x+y=4)\wedge(2x+z\neq 0)\wedge(4x+y\leq 3))$, where all LMCs have modulus 8. Note that $(6x+y=4)$ can be equivalently expressed as $(2\cdot 3x=7y+4)$. Multiplying both the sides of $(2\cdot 3x=7y+4)$ by the multiplicative inverse of 3 modulo 8, i.e. 3, we get $(2x=5y+4)$. Replacing the occurrences of $2x$ by $5y+4$, the original problem can be equivalently expressed as $\exists x.\,((2x=5y+4)\wedge(5y+4+z\neq 0)\wedge(2\cdot(5y+4)+y\leq 3))$. Simplifying modulo 8, we get $(5y+z+4\neq 0)\wedge(3y\leq 3)\wedge\exists x.\,(2x=5y+4)$. Note that $\exists x.\,(2x=5y+4)$ is equivalent to $(4y=0)$. Hence the result of QE is $(5y+z+4\neq 0)\wedge(3y\leq 3)\wedge(4y=0)$.

Simplifications as above using LMEs present in the conjunction forms the crux of Layer1. It can be observed that Layer1 may not always eliminate the quantifier. For example, consider the problem of computing $\exists x.\,((2x+3y=4)\wedge(x+y\leq 3))$ with modulus 8. Note that simplifications in Layer1 cannot eliminate the quantifier in this case. Such cases are handled by the following layers which are more expensive.

2.2 Layer2: Dropping Unconstraining LMIs and LMDs

Consider the problem of computing $\exists x.\,A$ obtained after Layer1, where A is a conjunction of LMCs. Let $A\equiv C\wedge D\wedge I$, where (i) D is a conjunction of (zero or more) LMDs in A, (ii) I is a conjunction of (zero or more) LMIs in A, (iii) C is the conjunction of the remaining LMCs in A, and (iv) $\exists x.\,(C)\Rightarrow\exists x.\,(C\wedge D\wedge I)$. Since $\exists x.\,(C\wedge D\wedge I)\Rightarrow\exists x.\,(C)$ always holds, this would mean that $\exists x.\,(C\wedge D\wedge I)$ is equivalent to $\exists x.\,C$. We say that D and I are "unconstraining" in such cases.

Given $\exists x.\,(C\wedge D\wedge I)$ satisfying conditions (i), (ii) and (iii) above, Layer2 uses efficiently computable conditions sufficient for condition (iv) to hold. Let $x[i]$ denote the ith bit of the bit-vector x, where $x[0]$ denotes the least significant bit of x. For $i\leq j$, let $x[i:j]$ denote the slice of bit-vector x consisting of bits $x[i]$ through $x[j]$. Let each LMI in I be of the form $s_i\bowtie t_i$, where $\bowtie\in\{\leq,\geq\}$, s_i is a linear term with x in its support, and t_i is a linear term free of x. Let $s_1,\ldots,s_r$ be the distinct linear terms in I with x in their support. We assume without loss of generality that I contains the trivial LMIs $s_i\geq 0$ and $s_i\leq 2^p-1$ for each linear term s_i. Suppose the LMIs in I are grouped into inequalities of the form $Z_i:u_i\leq s_i\leq v_i$, where u_i denotes the maximum among the lower bounds of s_i in I and v_i denotes the minimum among the upper bounds of s_i in I. Let $k_1,\ldots,k_r$ be the highest powers of 2 in the coefficients of x in $s_1,\ldots,s_r$.

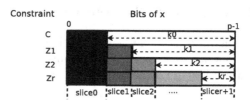

Fig. 1. Slicing of bits of x by $k_0, \ldots, k_r$

Similarly, let k_0 and k_D be the highest powers of 2 in the coefficients of x in C and D respectively. Suppose further that $k_1 > \ldots > k_r$ and $k_0 > \max(k_D, k_1)$.

We can partition the bits of x into $r + 2$ slices as shown in Fig. 1, where $slice_0$ represents $x[0 : p - k_0 - 1]$, $slice_i$ represents $x[p - k_{i-1} : p - k_i - 1]$ for $1 \le i \le r$, and $slice_{r+1}$ represents $x[p - k_r : p - 1]$. Note that the value of $slice_0$ potentially affects the satisfaction of C as well as that of Z_1 through Z_r, the value of $slice_i$ potentially affects the satisfaction of Z_i through Z_r for $1 \le i \le r$, and the value of $slice_{r+1}$ does not affect the satisfaction of any Z_i or C.

Suppose, given a solution θ_1 of C, there exists a solution θ_2 of $C \wedge Z_1$ that matches θ_1 except possibly in the bits of $slice_1$. In such cases, we say that the solution θ_1 of C can be "engineered" w.r.t. $slice_1$ to satisfy $C \wedge Z_1$. Suppose an arbitrary solution of C can be engineered w.r.t. $slice_1$ to satisfy $C \wedge Z_1$. This would mean that $\exists x. (C \wedge Z_1)$ is equivalent to $\exists x. C$. Following this argument, if an arbitrary solution of C can be engineered w.r.t. $slice_1$ through $slice_r$ to satisfy $C \wedge Z_1 \wedge \ldots \wedge Z_r$, then $\exists x. (C \wedge I)$ is equivalent to $\exists x. C$, and I is unconstraining. A similar argument as above can be used to identify unconstraining LMDs. Layer2 computes an efficiently computable *under-approximation* η of the number of ways in which an arbitrary solution of C can be engineered w.r.t. $slice_1$ through $slice_{r+1}$ to satisfy $C \wedge D \wedge I$. If $\eta \ge 1$, then D and I are unconstraining.

For example, consider the problem of computing $\exists x. ((z = 4x + y) \wedge (6x + y \le 4) \wedge (x \ne z))$ with modulus 8. Suppose $C \equiv (z = 4x + y)$, $D \equiv (x \ne z)$, and $I \equiv (6x + y \le 4)$. Note that the bits of x can be partitioned into $slice_0$, $slice_1$ and $slice_2$, where $slice_0$ represents $x[0 : 0]$, $slice_1$ represents $x[1 : 1]$ and $slice_2$ represents $x[2 : 2]$. $Slice_1$ and $slice_2$ do not affect the satisfaction of C. Moreover, it can be observed that an arbitrary solution of C can be engineered w.r.t. $slice_1$ through $slice_2$ to satisfy $C \wedge D \wedge I$. Layer2 computes η as 1 in this case, and thus identifies that $\exists x. (C \wedge D \wedge I)$ is equivalent to $\exists x. (z = 4x + y)$. Note that $\exists x. (z = 4x + y)$ is equivalent to $(4y + 4z = 0)$. Hence the result of QE is $(4y + 4z = 0)$.

2.3 Layer3: Fourier-Motzkin Elimination for LMIs

There are two fundamental problems when trying to apply FM elimination for reals [4] to a conjunction of LMIs. The first step in FM elimination is "normalization" of each inequality l w.r.t. the variable x being quantified. This involves expressing l in an equivalent form $x \bowtie t$, where $\bowtie \in \{\leq, \geq\}$ and t is a term free of x. However, normalizing an LMI w.r.t. a variable is much more difficult than normalizing in the case for reals, since standard equivalences used for normalizing inequalities over reals do not hold in modular arithmetic [3]. Moreover, even if we could normalize LMIs w.r.t. the variable being quantified, due to the lack of density of integers, FM elimination cannot be directly lifted to integers.

Layer3 makes use of a weak normal form for LMIs. We say that an LMI l with x in its support is *normalized w.r.t.* x if it is of the form $a \cdot x \bowtie t$ (*first normal form*), or of the form $a \cdot x \bowtie b \cdot x$ (*second normal form*), where $\bowtie \in \{\leq, \geq\}$, and t is a linear term free of x. A boolean combination of LMCs φ is said to be normalized w.r.t. x if every LMI in φ with x in its support is normalized w.r.t. x.

Given $\exists x. I$, where I is a conjunction of LMIs, Layer3 converts I to an equivalent boolean combination of LMCs normalized w.r.t. x. For example, suppose we wish to normalize $x + 2 \leq y$ modulo 8 w.r.t. x. Consider adding the additive inverse of 2 modulo 8, i.e. 6 to both sides of $x + 2 \leq y$. Let ω_1 be the condition under which the addition of $x + 2$ with 6 overflows the 3-bit representation. Similarly, let ω_2 be the condition under which the addition of y with 6 overflows 3-bit representation. Note that if $\omega_1 \equiv \omega_2$, then $(x + 2 \leq y) \equiv (x \leq y + 6)$ holds; otherwise $(x + 2 \leq y) \equiv (x > y + 6)$ holds. $\omega_1 \equiv \omega_2$ can be equivalently expressed as $(x \leq 5) \equiv (y \geq 2)$. Hence, $(x + 2 \leq y)$ can be equivalently expressed in the normalized form $\text{ite}(\varphi, (x \leq y + 6), (x > y + 6))$, where φ denotes $(x \leq 5) \equiv (y \geq 2)$, and $\text{ite}(\alpha, \beta, \gamma)$ denotes $(\alpha \wedge \beta) \vee (\neg\alpha \wedge \gamma)$.

Layer3 applies a variant of FM elimination to achieve QE from the normalized LMIs. We illustrate the idea with help of an example. Consider the problem of computing $\exists x. C$, where $C \equiv (y \leq 4x) \wedge (4x \leq z)$ with modulus 16. Observe that $\exists x. C$ is "the condition under which there exists a multiple of 4 between y and z, where $y \leq z$". It can be shown that $\exists x. C$ is equivalent to the disjunction of the following three conditions: (i) $(y \leq z)$, and y is a multiple of 4, i.e., $(y \leq z) \wedge (4y = 0)$, (ii) $(y \leq z) \wedge (y \leq 12) \wedge (z \geq y + 3)$, (iii) $(y \leq z)$, $(z < y + 3)$, and $(y > z \pmod 4)$, i.e., $(y \leq z) \wedge (z < y + 3) \wedge (4y > 4z)$. In general, suppose we wish to compute $\exists x. (l_1 \wedge l_2)$, where $l_1 : (t_1 \leq a \cdot x)$ and $l_2 : (a \cdot x \leq t_2)$ are LMIs in the *first normal form* w.r.t. x. Let k be the highest power of 2 in the coefficient a of x. Then, $\exists x. (l_1 \wedge l_2)$ is equivalent to $(t_1 \leq t_2) \wedge \varphi$, where φ is the disjunction of the formulas: (i) $(2^{p-k} \cdot t_1 = 0)$, (ii) $(t_2 \geq t_1 + 2^k - 1) \wedge (t_1 \leq 2^p - 2^k)$, and (iii) $(t_2 < t_1 + 2^k - 1) \wedge (2^{p-k} \cdot t_1 > 2^{p-k} \cdot t_2)$.

The conjunction of LMIs such as $(l_1 \wedge l_2)$ above, where all LMIs are in the *first normal form* w.r.t. x, and have the same coefficient of x are said to be "unified" w.r.t. x. Unfortunately, unifying a conjunction of LMIs I w.r.t. x is inefficient in general. Hence we unify I w.r.t. x and apply FM elimination only in the cases where the unification can be done efficiently (the details of unification can be found in [2]). In the other cases, we compute $\exists x. I$ using *model enumeration*, i.e., by expressing $\exists x. I$ in the equivalent form $I|_{x \leftarrow 0} \vee \ldots \vee I|_{x \leftarrow 2^p - 1}$ where $I|_{x \leftarrow i}$ denotes I with x replaced by the constant i.

3   QE from Boolean Combinations of LMCs

We extend *Project* to work with boolean combinations of LMCs using three approaches - a decision diagram (DD) based approach, an SMT-solving based approach and a hybrid approach that combines the strengths of the DD based and the SMT-solving based approaches.

The DD based approach makes use of a data structure called Linear Modular Decision Diagram (LMDD). LMDDs are BDDs [10] with nodes labeled with LMEs or LMIs. They represent boolean combinations of LMCs. Suppose we wish to compute $\exists X.f$, where f is an LMDD over a set of variables V and $X \subseteq V$. A naive algorithm to compute $\exists X.f$ is to apply *Project* to each path in f. However, this algorithm, similar to the Black-box QE algorithm [5] for Linear Decision Diagrams, has running time linear in the number of paths in f. We use an alternate algorithm *QE_LMDD* to compute $\exists X.f$, which is motivated by the White-box QE approach suggested in [5]. *QE_LMDD* makes use of a procedure *QE1_LMDD* that eliminates a single variable x from f. *QE1_LMDD* performs a recursive traversal of the LMDD f. In each recursive call, *QE1_LMDD* computes the LMDD for $\exists x.(g \wedge C_x)$, where g is the LMDD encountered during the traversal and C_x is the conjunction of LMCs containing x encountered in the path from the root node of f to the root node of g. If g is a 1-terminal, then *QE1_LMDD* computes $\exists x.(g \wedge C_x)$ by calling *Project* on $\exists x.C_x$. If the root node of g is a non-terminal, then *QE1_LMDD* first simplifies g using the LMEs in C_x and then traverses g recursively. The single variable elimination strategy gives opportunities for reuse of results through dynamic programming, and in practice significantly outperforms the Black-box QE algorithm.

The SMT-solving based approach is a straightforward extension of the work in [7] for QE from boolean combinations of linear inequalities on reals. Suppose we wish to compute $\exists X.f$, where f is a boolean combination of LMCs over a set of variables V and $X \subseteq V$. A naive way of computing this is by converting f to DNF by enumerating all satisfying assignments, and by invoking *Project* on each conjunction of LMCs in the DNF. We use an algorithm *QE_SMT* which generalizes a satisfying assignment to obtain a conjunction of LMCs, and projects the conjunction of LMCs on the variables in $V \setminus X$. The complement of the projected conjunction of LMCs is conjoined with f before further satisfying assignments are obtained. The interleaving of projection and model enumeration in *QE_SMT* helps in significant pruning of the solution space.

The hybrid approach tries to combine the strengths of the DD based and the SMT-solving based approaches. Suppose we wish to compute $\exists X.f$, where f is an LMDD over a set of variables V and $X \subseteq V$. The hybrid algorithm *QE_Combined* splits $\exists X.f$ into an equivalent disjunction of sub-problems $\bigvee_{i=1}^{n}(\exists X.(f_i \wedge C_i))$, where f_i denotes an internal LMDD node in f and C_i denotes the conjunction of LMCs in the path from the root node of f to f_i. *QE_Combined* now computes $g \equiv \bigvee_{i=1}^{n}(\exists X.(f_i \wedge C_i))$ in the following manner: if $f_i \wedge C_i \wedge \neg g$ is satisfiable, then $h \equiv \exists X.(f_i \wedge C_i)$ is computed using the DD-based approach, and then h is disjoined with g. Computing the sub-problems using the DD-based approach helps in achieving reuse of results through dynamic programming. Unlike

QE_SMT, *QE_Combined* does not explicitly interleave projections inside model enumeration. However disjoining the result of $\exists X. (f_i \wedge C_i)$ with g, and computing $\exists X. (f_i \wedge C_i)$ only if $f_i \wedge C_i \wedge \neg g$ is satisfiable helps in pruning the solution space of the problem, as achieved in *QE_SMT*.

4   Experiments and Comparison with Existing Software

In order to evaluate the performance of our algorithms and compare them with alternate QE techniques, we used a benchmark suite consisting of a set of *lindd* benchmarks from [5] and a set of *vhdl* benchmarks. Each benchmark is a boolean combination of LMCs with a subset of the variables in their support existentially quantified. The *lindd* benchmarks are boolean combinations of octagonal constraints over integers. These benchmarks are converted to boolean combinations of LMCs by assuming the size of integer as 16 bits. The *vhdl* benchmarks are obtained from transition relation abstraction. We derived the symbolic transition relations of a set of VHDL designs. All the internal variables in these symbolic transition relations are quantified out, which gives abstract transition relations of the vhdl designs.

We measured the time taken by *QE_LMDD*, *QE_SMT*, and *QE_Combined* for QE from each benchmark. We observed that (i) each approach performs better than the others for some benchmarks, (ii) *DD* and *SMT* based approaches are incomparable, and (iii) hybrid approach inherits the strengths of both *DD* and *SMT* based approaches. We also measured the contributions and costs of different layers of *Project* in performing QE from the benchmarks. Layer1 and Layer2 together eliminated 95% of the quantifiers in *lindd* benchmarks and 99.5% of the quantifiers in *vhdl* benchmarks. The remaining quantifiers were eliminated by Layer3. However, none of the benchmarks required model enumeration. Layer1 and Layer2 were cheap (on average, took 1-6 milliseconds per quantifier eliminated). Layer3 was comparatively expensive. On average, Layer3 took 13 seconds per quantifier eliminated for *lindd* benchmarks and 161 milliseconds per quantifier eliminated for *vhdl* benchmarks.

We compared the performance of *Project* with alternate QE techniques. This included comparison of *Project* with PA based QE using Omega Test [6] and with bit-level QE using BDDs [11]. Since Layer1 is a simple extension of the work in [8], we applied Layer1 as a pre-processing step before applying the PA based/ bit-level QE. The procedure that first quantifies out the variables using *Layer1*, and then uses conversion to PA and Omega Test for the remaining variables is called *Layer1_OT*. Similarly, the procedure that first quantifies out the variables using *Layer1*, and then uses bit-blasting and bit-level BDD based QE for the remaining variables is called *Layer1_Blast*. The instances of QE problem for conjunctions of LMCs arising from *QE_SMT* when QE is performed on each benchmark were collected. The procedures *Project*, *Layer1_Blast* and *Layer1_OT* were applied on these instances of the QE problem for conjunctions of LMCs. The results demonstrated that (see Fig.2) *Project* outperforms the alternative QE techniques.

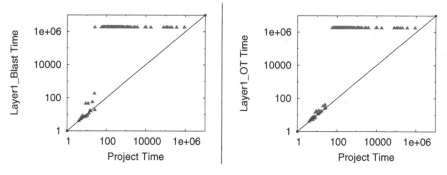

Fig. 2. Plots comparing (a) *Project* and *Layer1_Blast* and (b) *Project* and *Layer1_OT* (All times are in milliseconds)

5   Conclusion

We presented practically efficient and bit-precise techniques for QE from LMCs. Our experiments demonstrate that modular arithmetic based techniques for QE outperform PA and bit-blasting based QE techniques and keep the final result in modular arithmetic.

References

1. John, A.K., Chakraborty, S.: A quantifier elimination algorithm for linear modular equations and disequations. In: Gopalakrishnan, G., Qadeer, S. (eds.) CAV 2011. LNCS, vol. 6806, pp. 486–503. Springer, Heidelberg (2011)
2. John, A.K., Chakraborty, S.: Extending quantifier elimination to linear inequalities on bit-vectors. In: Piterman, N., Smolka, S.A. (eds.) TACAS 2013. LNCS, vol. 7795, pp. 78–92. Springer, Heidelberg (2013)
3. Bjørner, N., Blass, A., Gurevich, Y., Musuvathi, M.: Modular difference logic is hard. CoRR abs/0811.0987 (2008)
4. Kroening, D., Strichman, O.: Decision procedures: an algorithmic point of view. Texts in Theoretical Computer Science. Springer (2008)
5. Chaki, S., Gurfinkel, A., Strichman, O.: Decision diagrams for linear arithmetic. In: FMCAD 2009 (2009)
6. Pugh, W.: The Omega Test: A fast and practical integer programming algorithm for dependence analysis. Communications of the ACM, 102–114 (1992)
7. Monniaux, D.: A quantifier elimination algorithm for linear real arithmetic. In: Cervesato, I., Veith, H., Voronkov, A. (eds.) LPAR 2008. LNCS (LNAI), vol. 5330, pp. 243–257. Springer, Heidelberg (2008)
8. Ganesh, V., Dill, D.L.: A decision procedure for bit-vectors and arrays. In: Damm, W., Hermanns, H. (eds.) CAV 2007. LNCS, vol. 4590, pp. 519–531. Springer, Heidelberg (2007)
9. Muller-Olm, M., Seidl, H.: Analysis of modular arithmetic. ACM Transactions on Programming Languages and Systems 29(5), 29 (2007)
10. Bryant, R.E.: Graph-based algorithms for boolean function manipulation. IEEE Transactions on Computers C-35(8), 677–691 (1986)
11. CUDD release 2.4.2 website, vlsi.colorado.edu/fabio/CUDD

Skolemization Modulo Theories

Konstantin Korovin[1] and Margus Veanes[2]

[1] The University of Manchester, UK
korovin@cs.man.ac.uk
[2] Microsoft Research, USA
margus@microsoft.com

Abstract. Combining classical automated theorem proving techniques with theory based reasoning, such as satisfiability modulo theories, is a new approach to first-order reasoning modulo theories. Skolemization is a classical technique used to transform first-order formulas into *equisatisfiable* form. We show how Skolemization can benefit from a new satisfiability modulo theories based simplification technique of formulas called *monadic decomposition*. The technique can be used to transform a theory dependent formula over multiple variables into an equivalent form as a Boolean combination of unary formulas, where a unary formula depends on a single variable. In this way, theory specific variable dependencies can be eliminated and consequently, Skolemization can be refined by minimizing variable scopes in the decomposed formula in order to yield simpler Skolem terms.

1 The Role of Skolemization

In classical automated theorem proving, Skolemization [9,2,4,8] is a technique used to transform formulas into *equisatisfiable* form by replacing existentially quantified variables by Skolem terms. In resolution based methods using clausal normal form (CNF) this is a necessary preprocessing step of the input formula. A CNF represents a universally quantified conjunction of clauses, where each clause is a disjunction of literals, a literal being an atom or a negated atom. The arguments of the atoms are terms, some of which may contain Skolem terms as subterms where a Skolem term has the form $f(\bar{x})$ for some Skolem function symbol f and a sequence $\bar{x}$ of variables; f may also be a constant ($\bar{x}$ is empty). The input to Skolemization is a formula ψ in *prenex normal form*: $Q_1 x_1 Q_2 x_2 \ldots Q_n x_n \varphi$ where $Q_i \in \{\exists, \forall\}$ and φ is quantifier free and the free variables of φ, $FV(\varphi)$, form a subset of $\{x_i\}_{i=1}^{n}$; φ is called the *matrix* of ψ. In its most basic form, one Skolemization step *Skolemize1* is a transformation that is applied to the outermost prefix of the given prenex formula, $Skolemize1(\forall \bar{x} \exists y \chi(\bar{x}, y)) \overset{\text{def}}{=} \forall \bar{x} \chi(\bar{x}, f_{\chi, \bar{x}}(\bar{x}))$, whose output is another prenex formula with one less existential quantifier and where $f_{\chi, \bar{x}}$ is a new function symbol called a *Skolem function* (or a *Skolem constant* when $n = 0$). *Skolemize1* is applied repeatedly, denoted here by *Skolemize*, until no more existential quantifiers remain. *Skolem Normal Form Theorem* [4, Corollary 3.1.3] implies that *Skolemize*(ψ) is satisfiable if and only if ψ is satisfiable.

H. Hong and C. Yap (Eds.): ICMS 2014, LNCS 8592, pp. 303–306, 2014.

In the context of theorem proving it is assumed that the Skolem functions are uninterpreted.

There are several important techniques related to Skolemization. The main objective is to *minimize the arity* of the Skolem functions. *Mini-scoping* [1], also called *antiprenexing* or creating an *antiprenex normal form* [2,8], is the main Skolemization technique that is used in theorem proving as a method to minimize quantifier scopes by shifting quantifiers from the prenex back into the formula. Mini-scoping can be seen as a separate preprocessing step prior to Skolemization, consisting of the following rewrite steps that correspond to standard equivalence preserving laws of logic. The formula is first transformed into an equivalent negation normal form (NNF), so that all quantifiers occur in a positive context.

$$Qx(\varphi \diamond \psi) \overset{x \notin FV(\psi), x \in FV(\varphi)}{\Longrightarrow}_{mini\text{-}scope} Qx\varphi \diamond \psi, \quad \diamond \in \{\vee, \wedge\}$$
$$\forall x(\varphi \wedge \psi) \Longrightarrow_{mini\text{-}scope} \forall x\varphi \wedge \forall x\psi$$
$$\exists x(\varphi \vee \psi) \Longrightarrow_{mini\text{-}scope} \exists x\varphi \vee \exists x\psi$$

After mini-scoping, all quantified variables are renamed apart. Finally, *standard Skolemization* [2] is applied to the resulting formula by replacing a subformula $\exists y \chi$ that occurs in the context of universal variables $\bar{x}$, by the formula $\chi\{y \mapsto f(\bar{x})\}$. We refer to the full procedure as mini-scoping. Without theory based reasoning, mini-scoping results in the lowest possible arities of the Skolem functions, and is thus optimal in that sense. In theory based reasoning this is not always true, one can do better, we discuss this below.

2 Working Modulo a Theory

Theory based automated reasoning is a new area of automated reasoning that combines techniques from propositional satisfiability (SAT) and satisfiability modulo theories (SMT) area into the expressive power of first-order reasoning with quantifiers [5,6]. Skolemization is one piece of the big picture, it has been considered as a solved problem, much due to the *Skolemization Theorem* [4, Theorem 3.1.2]. Skolemization Theorem implies a much stronger property than equisatisfiability, that allows the use of Skolemization modulo *arbitrary* theories.

Let L be the language of the theory and let Σ be the *Skolem Theory* consisting of axioms $\forall \bar{x}(\exists y \, \chi(\bar{x}, y) \rightarrow \chi(\bar{x}, f_{\chi, \bar{x}}(\bar{x})))$ for all L-formulas $\chi(\bar{x}, y)$ and new function symbols $f_{\chi, \bar{x}}$. In other words, the Skolem theory axiomatizes the intended interpretations of the Skolem functions. Skolemization Theorem says that any L-structure A can be expanded to be a model A^{Σ} of Σ. Therefore, if we work with uninterpreted function symbols, i.e., without assuming Σ, and A is a model of the original formula, then some expansion of A models the Skolemized one: just pick the intended interpretations from A^{Σ} for the uninterpreted Skolem functions. In the other direction, the Skolemized formula always entails the original formula. In particular if the Skolemized formula is satisfiable then so is the original one.

Often the starting point in theory based reasoning is a formula which presumes Skolemization. For example, assume the theory of integer linear arithmetic and consider the following (true) sentence:

$$\forall x \exists y (0 \leq x \leq 1 \to (0 \leq y \land x + y \leq 1)) \tag{1}$$

It is already in prenex form and mini-scoping produces the equisatisfiable formula where f is a Skolem function:

$$\forall x (0 \leq x \leq 1 \to (0 \leq f(x) \land x + f(x) \leq 1)) \tag{2}$$

We will see below how introduction of f can be avoided completely in this case.

3   Using Monadic Decomposition

We consider theories that satisfy the following conditions. More general theories fall outside the scope of this paper. Let A be a recursively enumerable (re) L-structure with an re universe so that all elements can be named by L-terms. As the theory we take the *theory of A*.

Moreover, let Ψ be an re set of L-formulas that is closed under Boolean operations, and if a is an element, x a variable, and $\psi \in \Psi$ then $\psi\{x \mapsto a\} \in \Psi$. Furthermore, satisfiability of formulas in Ψ is assumed *decidable*: it is decidable, for $\psi(\bar{x}) \in \Psi$, if $A \models \exists \bar{x} \psi(\bar{x})$. It follows from A being re that concrete witnesses can also be generated for satisfiable formulas. Examples of A are: standard integers or standard rational numbers (or A may be multi-sorted), and an example of Ψ is quantifier free L-formulas where all variables have a fixed sort. These conditions are very natural from the standpoint of modern SMT solvers, because Ψ embodies the basic properties supported by any state-of-the-art SMT solver [3].

We need some additional notions before defining monadic decomposition formally. A *unary* formula is a formula with at most one free variable. An *explicitly monadic* formula is a Boolean combination of unary formulas. A *monadic* formula is a formula for which there exists an *equivalent* explicitly monadic formula. Now, *monadic decomposition* (for Ψ) is the following problem: given a monadic formula $\psi \in \Psi$, construct and explicitly monadic formula that is equivalent to ψ. It is shown in [10] that this problem is solvable, the given algorithm **mondec** relies solely on the assumptions of Ψ as stated above. *Deciding* if a formula is monadic is shown decidable in two cases but is an open problem in general.

Now, monadic decomposition can be applied as a preprocessing step to miniscoping. This can happen in several different ways. First, several variables can be grouped together and viewed as a single variable by using tuples; the structure A as well as Ψ can, without loss of generality, be extended with tuples. Second, the decomposition can be applied selectively to some subformulas only. Finally, if the formula is not known to be monadic then **mondec** might not terminate and thus heuristics need to be developed to decide when to abandon the decomposition attempt.

As an example assume that the theory is linear arithmetic and pick the matrix of the prenex formula (1). This formula is monadic (which can also be decided [10] with a Presburger formula but there is currently no particular implementation for this decision procedure). If we apply **mondec** to the matrix of (1) we get the following concrete output as the result of running the python script from [10]:

```
And(Or(Not(And(x >= 0, x <= 1)), x <= 1),
    Or(Not(And(x >= 0, x <= 1)), x <= -1,
        And(y >= 0, y <= 1, Or(Not(And(x >= 0, x <= 1)), x <= 0)),
        And(y >= 0, y <= 0, Or(Not(And(x >= 0, x <= 1)), x <= 1))))
```

This formula is explicitly monadic and equivalent to the matrix of (1). Thus, we can replace the matrix of (1) by this formula. Now mini-scoping will produce a formula where y is not in the scope of x any more. So the final Skolemized formula will use *Skolem constants* for y.

In general, if all maximal quantifier free subformulas are monadic (as is the case with formula (1)) and mini-scoping is slightly modified and applied so that quantifiers are pushed all the way to the unary sub-formulas, then the quantifiers are effectively eliminated and the final formula will be a Boolean combination of $\forall x \varphi(x)$ or $\exists x \varphi(x)$ where $\varphi(x)$ is a formula in Ψ and thus decidable. So the final formula is essentially propositional (modulo the theory of A). Overall, this implies that the full first-order fragment over monadic formulas is decidable, as an extension of of the Löwenheim class [7].

References

1. Andrews, P.B.: An Introduction to Mathematical Logic and Type Theory: To Truth through Proof. Academic Press (1986)
2. Baaz, M., Egly, U., Leitsch, A.: Normal form transformations. In: Robinson, A., Voronkov, A. (eds.) Handbook of Automated Reasoning, vol. I, ch. 5, pp. 273–333. North Holland (2001)
3. De Moura, L., Bjørner, N.: Satisfiability modulo theories: introduction and applications. Commun. ACM 54(9), 69–77 (2011)
4. Hodges, W.: Model theory. Cambridge Univ. Press (1995)
5. Korovin, K.: Instantiation-based automated reasoning: From theory to practice. In: Schmidt, R.A. (ed.) CADE 2009. LNCS (LNAI), vol. 5663, pp. 163–166. Springer, Heidelberg (2009)
6. Korovin, K.: Inst-gen – a modular approach to instantiation-based automated reasoning. In: Voronkov, A., Weidenbach, C. (eds.) Programming Logics. LNCS, vol. 7797, pp. 239–270. Springer, Heidelberg (2013)
7. Löwenheim, L.: Über Möglichkeiten im Relativkalkül. Math. Annalen 76, 447–470 (1915)
8. Nonnengart, A., Weidenbach, C.: Computing small clause normal forms. In: Robinson, A., Voronkov, A. (eds.) Handbook of Automated Reasoning, vol. I, ch. 6, pp. 335–367. North Holland (2001)
9. Skolem, T.: Logisch-kombinatorische Untersuchungen über die Erfüllbarkeit oder Beweisbarkeit mathematischer Sätze nebst einem Theoreme über dichte Mengen. In: Skrifter utgitt av Videnskapsselskapet i Kristiania (1920)
10. Veanes, M., Bjørner, N., Nachmanson, L., Bereg, S.: Monadic decomposition. In: Biere, A., Bloem, R. (eds.) CAV 2014. LNCS, vol. 8559, pp. 628–645. Springer, Heidelberg (2014)

Incremental QBF Solving by DepQBF[*]

Florian Lonsing and Uwe Egly

Vienna University of Technology
Institute of Information Systems
Knowledge-Based Systems Group
http://www.kr.tuwien.ac.at/

Abstract. The logic of quantified Boolean formulae (QBF) extends propositional logic by explicit existential and universal quantification of the variables. We present the search-based QBF solver DepQBF which allows to solve a sequence of QBFs incrementally. The goal is to exploit information which was learned when solving previous formulae in the process of solving the next formula in a sequence. We illustrate incremental QBF solving and potential usage scenarios by examples. Incremental QBF solving has the potential to considerably improve QBF-based workflows in many application domains.

Keywords: quantified Boolean formulae, QBF, search-based solving, Q-resolution, clause learning, cube learning, incremental solving.

1 Introduction

Propositional logic (SAT) has been widely applied to encode problems from model checking, formal verification, and synthesis. In these practical applications, an instance of a given problem is encoded as a formula. The satisfiability of this formula is checked using a SAT solver. The result of the satisfiability check is then mapped back and interpreted on the level of the problem instance.

Encodings of problems often give rise to sequences of closely related formulae to be solved, in contrast to one single formula. A prominent example is SAT-based bounded model checking (BMC) [1]. Rather than solving each formula in the sequence individually, *incremental solving* [6] aims at employing information that was learned when solving one formula for solving the next formulae. The overall goal is to speed up the solving process of the entire sequence of formulae.

We consider the problem of incrementally solving a sequence of quantified Boolean formulae (QBF). The decision problem of QBF is PSPACE-complete. Existential and universal quantification together with possible quantifier alternations in QBF potentially allow for exponentially more succinct encodings of problems than propositional logic [2]. This property makes QBF an interesting modelling language for practical applications.

Incremental QBF solving was first applied in the context of QBF-based bounded model checking of partial designs [14]. We extended our QBF solver DepQBF [11,12]

---

[*] Supported by the Austrian Science Fund (FWF) under grant S11409-N23.

H. Hong and C. Yap (Eds.): ICMS 2014, LNCS 8592, pp. 307–314, 2014.

by *general-purpose* incremental solving capabilities. Our approach adopts ideas from incremental SAT solving, it is *application-independent* and hence applicable to QBF encodings of *arbitrary* problems. Furthermore, our implementation is publicly available, it features APIs in the C and Java languages and thus facilitates the use of incremental QBF solving in practice.[1]

We present incremental QBF solving from a general perspective. During the solving process, QBF solvers learn information about a QBF in terms of restricted inferences in the Q-resolution calculus. Information learned from previous QBFs must be maintained to prevent unsound inferences. Regarding practical applications, we illustrate the API of our incremental QBF solver DepQBF by means of examples to make its use more accessible. Incremental QBF solving has the potential to improve QBF-based workflows in many applications.

2   Quantified Boolean Formulae

A QBF $\psi := \hat{Q}.\phi$ in prenex conjunctive normal form (PCNF) consists of a quantifier-free propositional formula ϕ in CNF containing the variables V and a quantifier prefix $\hat{Q}$. The prefix $\hat{Q} := Q_1 B_1 \ldots Q_n B_n$ contains sets B_i of propositional variables and quantifiers $Q_i \in \{\forall, \exists\}$. We assume that $B_i \neq \emptyset$, $\bigcup B_i = V$ and $B_i \cap B_j = \emptyset$ for $i \neq j$. The sequence of sets B_i introduces a linear ordering of the variables: given two variables x, y, we define $x < y$ if and only if $x \in B_i$, $y \in B_j$ and $i < j$. In the following, we consider QBFs in PCNF.

An *assignment* $A : V \to \{\mathbf{t}, \mathbf{f}\}$ is a (partial) mapping from the set of all propositional variables V to truth values *true* ($\mathbf{t}$) and *false* ($\mathbf{f}$). To allow for simple notation, we represent an assignment A as a set $\{l_1, \ldots, l_k\}$ of literals where, for a variable x assigned by A, we have $l_i = x$ ($l_i = \neg x$) if x is mapped to $\mathbf{t}$ ($\mathbf{f}$). Given a QBF ψ, a variable $x \in B_i$ and the assignment $A = \{l\}$ with $l = x$ ($l = \neg x$), the QBF $\psi[A]$ *under the assignment* A is obtained from ψ by replacing every occurrence of x in ψ with the syntactic truth constant $\top$ ($\bot$) denoting *true* (*false*), deleting x from the prefix (along with $Q_i B_i$ if $B_i = \emptyset$) and applying simplifications using the annihilator and identity properties of $\wedge, \vee, \top$ and $\bot$ of Boolean algebra.

The semantics of QBF is defined recursively based on the syntactic structure. The QBF $\psi = \top$ ($\psi = \bot$), which consists of the syntactic truth constant *true* (*false*), is satisfiable (unsatisfiable). The QBF $\psi = \exists(\{x\} \cup B_1) \ldots Q_n B_n. \phi$ is satisfiable if and only if $\psi[x]$ or $\psi[\neg x]$ is satisfiable. The QBF $\psi = \forall(\{x\} \cup B_1) \ldots Q_n B_n. \phi$ is satisfiable if and only if $\psi[x]$ and $\psi[\neg x]$ is satisfiable.

Example 1. The QBF $\forall x \exists y. (x \vee \neg y) \wedge (\neg x \vee y)$ is satisfiable. We assign the variables in the ordering of the prefix. Both $\psi[x] = \exists y. (y)$ and $\psi[\neg x] = \exists y. (\neg y)$ are satisfiable, since $\psi[x, y] = \top$ and $\psi[\neg x, \neg y] = \top$, respectively.

---

[1] DepQBF is free software: `http://lonsing.github.io/depqbf/`

3   Search-Based QBF Solving with Learning

Modern search-based QBF solvers are based on an extension of the conflict-driven clause learning approach (CDCL), which is applied in SAT solving [15]. In this QBF-specific approach called QCDCL [8,10,13,16], a backtracking search procedure related to the DPLL algorithm [4,5] is used to generate assignments to control the application of the inference rules in the *Q-resolution calculus* [3,8]. New learned clauses and cubes are inferred and added to the formula.

In the following, we present the rules of the Q-resolution calculus to derive learned clauses and cubes, called *constraints*, in QCDCL-based QBF solvers. Given a QBF $Q_1 B_1 \ldots Q_n B_n . \phi$ and a literal l of a variable $x \in B_i$, the quantifier type of the variable x of l is denoted by $q(l)$ where $q(l) = \forall$ ($q(l) = \exists$) if $Q_i = \forall$ ($Q_i = \exists$). To allow for a uniform presentation of the rules to derive clauses and cubes in the calculus, we represent clauses and cubes as sets of literals.

$$\frac{C_1 \cup \{p\} \qquad C_2 \cup \{\neg p\}}{C_1 \cup C_2} \quad \begin{array}{l} \text{if } \{x, \neg x\} \nsubseteq (C_1 \cup C_2), \neg p \notin C_1, p \notin C_2 \text{ and} \\ \text{either (1) } C_1, C_2 \text{ are clauses and } q(p) = \exists \\ \text{or (2) } C_1, C_2 \text{ are cubes and } q(p) = \forall \end{array} \quad (res)$$

$$\frac{C \cup \{l\}}{C} \quad \begin{array}{l} \text{if } \{x, \neg x\} \nsubseteq (C \cup \{l\}) \text{ and either} \\ (1) \ C \text{ is a clause, } q(l) = \forall \text{ and } \forall l' \in C : q(l') = \exists \to l' < l \text{ or} \\ (2) \ C \text{ is a cube, } q(l) = \exists \text{ and } \forall l' \in C : q(l') = \forall \to l' < l \end{array} \quad (red)$$

$$\frac{}{C} \quad \begin{array}{l} \text{if } \{x, \neg x\} \nsubseteq C \text{ and either (1) } C \in \phi \text{ with } \psi = \hat{Q}. \phi \text{ or} \\ (2) \ \psi[A] = \top \text{ for an assignment } A \text{ and } C = (\bigwedge_{l \in A} l) \end{array} \quad (init)$$

The rule *res* defines *Q-resolution* with a *pivot variable* p. The constraints $C_1 \cup \{p\}$ and $C_2 \cup \{\neg p\}$ and the resolvent $C_1 \cup C_2$ must not contain complementary literals and the quantifier type $q(p)$ of the pivot variable is restricted to $\exists$ [3].

The rule *red* defines *constraint reduction* [3,8], which deletes universal (existential) literals from a clause (cube) C which are maximal among the literals in C with respect to the ordering of the quantifier prefix.

The rule *init* defines the axioms. Any clause $C \in \phi$ of a QBF $\psi = \hat{Q}. \phi$ can be used as a start point of a resolution derivation. Given an assignment A such that $\psi[A] = \top$, that is $C'[A] = \top$ for all $C' \in \phi$, the cube $C = (\bigwedge_{l \in A} l)$ can be inferred as a start point of a cube resolution derivation. The application of the rule *init* to infer cubes is also called *model generation* [8,10,16].

Due to the soundness of the calculus, a derived learned clause C' is added conjunctively to ψ and has the property that $\hat{Q}. \phi \equiv \hat{Q}. (\phi \wedge C')$. A derived learned cube C' is added disjunctively to ψ and has the property that $\hat{Q}. \phi \equiv \hat{Q}. (\phi \vee C')$.

A QBF is unsatisfiable (satisfiable) if and only if the empty clause (cube) can be derived using the rules *res*, *red* and *init*. In this case, the steps in the derivations of the learned clauses (cubes) up to $\emptyset$ correspond to a Q-resolution proof of the unsatisfiability (satisfiability) of ψ. We write $\psi \vdash C$ if the clause (cube) C can be derived from C using the rules of the Q-resolution calculus.

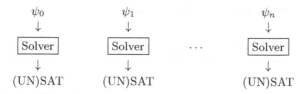

Fig. 1. Solving a sequence $S := \langle \psi_1, \ldots, \psi_n \rangle$ of PCNFs non-incrementally

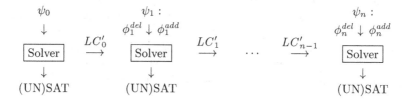

Fig. 2. Solving a sequence $S := \langle \psi_1, \ldots, \psi_n \rangle$ of PCNFs incrementally

Example 2. Given the satisfiable QBF $\psi = \forall x \exists y. (x \vee \neg y) \wedge (\neg x \vee y)$ from Example 1. By the rule *init*, we generate the cubes $C_1 := (x \wedge y)$ and $C_2 := (\neg x \wedge \neg y)$ using the assignments $A_1 = \{x, y\}$ and $A_2 = \{\neg x, \neg y\}$. Constraint reduction of C_1 and C_2 by rule *red* produces the cubes $C_3 = (x)$ and $C_4 = (\neg x)$, respectively. Finally, resolution by rule *res* of C_3 and C_4 produces the empty cube.

Example 3. Given the unsatisfiable QBF $\psi = \forall x \exists y. (x \vee \neg y) \wedge (\neg x \vee y) \wedge (x \vee y)$. Resolution of the clauses $(x \vee \neg y)$ and $(x \vee y)$ by the rule *res* produces the clause $C_1 := (x)$. Finally, constraint reduction by rule *red* results in the empty clause.

4   Incremental QBF Solving

Let $S := \langle \psi_1, \ldots, \psi_n \rangle$ be a sequence of QBFs to be solved where $\psi_i = \hat{Q}_i . \phi_i$. The QBF $\psi_{i+1} = \hat{Q}_{i+1} . \phi_{i+1}$ is obtained from the previous QBF ψ_i by adding and deleting the sets ϕ_{i+1}^{add} and ϕ_{i+1}^{del} of clauses, respectively: $\phi_{i+1} = (\phi_i \setminus \phi_{i+1}^{del}) \cup \phi_{i+1}^{add}$. Similarly, the quantifier prefix $\hat{Q}_{i+1}$ of ψ_{i+1} is obtained from $\hat{Q}_i$ by adding and deleting quantifiers, provided that in ψ_{i+1} still all the variables are quantified.

In *non-incremental* solving (Fig. 1), the solver tackles each QBF ψ_i in S from scratch. The entire formula is parsed and solving starts without using any learned constraint that was inferred when solving previous QBFs ψ_j with $j < i$.

In *incremental* solving (Fig. 2), the solver retains in a correctness preserving way a subset LC'_{i-1} of the constraints that were learned from previously solved QBFs in S in order to solve the current QBF ψ_i. The constraints in LC'_{i-1} can be used for inferences by the Q-resolution calculus. The choice of the set LC'_{i-1} depends on the sets ϕ_i^{add} and ϕ_i^{del} of clauses that were added to and deleted from the previous QBF ψ_{i-1}, respectively. For all constraints $C \in LC'_{i-1}$, it must hold that C can be derived from ψ_i and hence $\psi_i \vdash C$. Due to the soundness of Q-resolution, in this case we have that (1) $\hat{Q}_i.\phi_i \equiv \hat{Q}_i.(\phi_i \wedge C)$ if $C \in LC'_{i-1}$ is a clause and (2) $\hat{Q}_i.\phi_i \equiv \hat{Q}_i.(\phi_i \vee C)$ if $C \in LC'_{i-1}$ is a cube.

Compared to non-incremental solving, incremental solving has several advantages. First, the solver has to parse only the clauses ϕ_i^{add} which are added to ψ_{i-1} to obtain the new QBF ψ_i rather than the entire QBF ψ_i. In practice, an incremental solver typically is called as a library from another application program which generates the sequence $S := \langle \psi_1, \ldots, \psi_n \rangle$ of QBFs to be solved and retrieves the result returned by the solver. The solver is configured to solve the next QBF in S through its API. In contrast to that, a non-incremental solver is called as a standalone program to solve each QBF in S, where the QBFs are first written to hard disk, accessed by the solver and then parsed. This may result in I/O overhead, which is avoided in incremental solving.

The addition of ϕ_i^{add} to the previous QBF ψ_{i-1} can make the derivations of cubes learned from ψ_{i-1} invalid with respect to the current QBF ψ_i. Similarly, the deletion of ϕ_i^{del} from the previous QBF ψ_{i-1} can make the derivations of learned clauses invalid. The reason is that the side conditions of the rule *init*, which held with respect to ψ_{i-1}, might no longer hold with respect to ψ_i.

Example 4. Let $\psi' = \forall x \exists y.\, (x \vee \neg y) \wedge (\neg x \vee y) \wedge (x \vee y)$ be obtained from the satisfiable QBF ψ in Example 2 by adding the clause $(x \vee y)$. The QBF ψ' is unsatisfiable. Consider the cubes C_1 to C_4 inferred from ψ as shown in Example 2. We have $\psi' \nvdash C_2$ and $\psi' \nvdash C_4$ but $\psi' \vdash C_1$ and $\psi' \vdash C_3$ because the assignment $A_1 = \{x, y\}$ is still a model of ψ' whereas $A_2 = \{\neg x, \neg y\}$ is not.

Example 5. Let $\psi' = \forall x \exists y.\, (x \vee \neg y) \wedge (\neg x \vee y)$ be obtained from the unsatisfiable ψ in Example 3 by deleting the clause $(x \vee y)$. The QBF ψ' is satisfiable. We have $\psi' \nvdash C_1$ and no clauses can be derived from ψ' by the rules *red* and *res*.

If the set LC'_{i-1} of constraints which is retained contains constraints C for which $\psi_i \nvdash C$ then the solver might perform unsound inferences on ψ_i by the Q-resolution calculus, using the constraints in LC'_{i-1}.

Keeping learned constraints in incremental solving might give speedups in the solving time, as illustrated in the following experiment.

Proposition 1 ([9,10]). $\psi_n^C := \forall x_1 \exists y_2 \ldots \forall x_{2n-1} \exists y_{2n}.\, \bigwedge_{i=0}^{n-1} [(x_{2i+1} \vee \neg y_{2i+2}) \wedge (\neg x_{2i+1} \vee y_{2i+2})]$ *is a class of satisfiable QBFs. For each QBF ψ_n^C, the length of the shortest cube resolution proof of satisfiability of ψ_n^C is exponential in n.*

Let $S := \langle \psi_{10}^C, \ldots, \psi_{20}^C \rangle$ be the sequence of QBFs by Proposition 1 for $n = 10, 11, \ldots, 20$. The left part of the plot on the right shows that the number of learned cubes (y-axis, in millions) carried out by DepQBF when solving S incrementally ("inc") and non-incrementally ("noninc") scales exponentially with the size parameter n ($10 \ldots 20$ on the x-axis). Consider the reversed sequence $S' = \langle \psi_{20}^C, \ldots, \psi_{10}^C \rangle$ and the right part of the plot ($20 \ldots 10$ on the x-axis). When solving S' incrementally, then *all* the cubes learned when solving ψ_i^C in S' can be fully retained and used to solve the next QBF ψ_j^C with $j < i$. No new cubes are inferred. This is possible because clauses are only deleted from ψ_i^C to obtain ψ_j^C for $j <$ i in S' but not added. Therefore, for all cubes C' derived from ψ_i^C, it holds that $\psi_j^C \vdash C''$ for a subcube $C'' \subseteq C'$. The subcube C'' is obtained from C' by removing any literals which no longer occur in ψ_j^C. For further experiments, we refer to the technical reports related to DepQBF [12] and QBF-based conformant planning by incremental QBF solving [7].

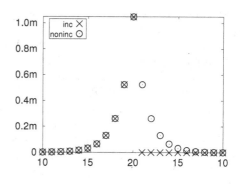

5   Implementation of DepQBF

In incremental solving, the set of learned constraints must be maintained across different calls of the solver. Regarding the learned clauses, the implementation of DepQBF [12] is based on the idea of *selector variables* from incremental SAT solving [6]. Thereby, a *fresh* variable v is added to *each* clause in the QBF $\psi = Q_1 B_1 \ldots Q_n B_n . \phi$ so that the clause $C \cup \{v\}$ is added to ψ instead of C. The selector variables are existentially quantified in a separate block V' at the left end of the quantifier prefix, which has the form $\exists V' Q_1 B_1 \ldots Q_n B_n . \phi$. If new clauses are derived using the rule *res* of the Q-resolution calculus, then the selector variables are always transferred to the derived clauses. In order to remove a clause C from the CNF of ψ including all learned clauses derived from C, the solver assigns the selector variable $v \in C$ to *true*. This causes v to be replaced by $\top$ in every clause, which effectively removes the clauses. They can no longer be used to make inferences by the Q-resolution calculus.

Regarding the learned cubes, we keep only cubes derived by applications of the rule *init*. For every cube C which is kept, the side condition of this rule with respect to the *current* QBF ψ' must hold: $\psi'[A] = \top$ for the assignment A which was used to derive $C = (\bigwedge_{l \in A} l)$.

The API of DepQBF provides the user with functions to manipulate the input formula by incrementally adding and deleting clauses and variables. As an additional API feature, the user can add and delete sets of clauses by means of push and pop operations. This way, the set of clauses of the input formula is organized as a sequence of frames on a stack. The same selector variable is added to all clauses of a particular stack frame. As a unique feature, DepQBF maintains the selector variables internally, which are invisible to the user. This design increases the usability of the solver from a user's perspective.

We illustrate the API of DepQBF by the code example in Fig. 3. The source release of DepQBF comes with further examples.[2] A solver object is created using the function `qdpll_create`. We create the quantifier prefix $\forall x \exists y$ by calling `qdpll_new_scope_at_nesting` followed by `qdpll_add` to add the variables x

---

[2] DepQBF tutorial: `http://lonsing.github.io/depqbf/depqbf-in-practice.pdf`

```
int main (int argc, char ** argv) {      ...//continued from left column.
  QDPLL *s = qdpll_create();              qdpll_add(s,2); qdpll_add(s,0);
  ...
  qdpll_new_scope_at_nesting             QDPLLResult res = qdpll_sat(s);
    (s,QDPLL_QTYPE_FORALL,1);            assert(res == QDPLL_RESULT_UNSAT);
  qdpll_add(s,1); qdpll_add(s,0);         assert(qdpll_get_value (s,1) ==
  qdpll_new_scope_at_nesting                      QDPLL_ASSIGNMENT_FALSE);
    (s,QDPLL_QTYPE_EXISTS,2);
  qdpll_add(s,2); qdpll_add(s,0);         qdpll_reset(s);
                                          qdpll_pop(s);
  qdpll_add(s,1); qdpll_add(s,-2);
    qdpll_add(s,0);                       res = qdpll_sat(s);
                                          assert(res == QDPLL_RESULT_SAT);
  qdpll_push(s);
  ...//continues on right column.         qdpll_delete (s); }
```

Fig. 3. DepQBF API usage example. Some configuration code was omitted for brevity

and y which we encode by the unsigned integers 1 and 2, respectively.[3] Then we add the clauses $(x \lor \neg y)$ by `qdpll_add` where the negative integer -2 encodes the negative literal $\neg y$ and `qdpll_add(s,0)` closes the clause. A new frame of clauses is allocated by `qdpll_push`. We add the clause (y) to the new frame (right column in Fig. 3). The call of `qdpll_sat` starts the solver given the current QBF $\psi = \forall x \exists y.\,(x \lor \neg y) \land (y)$. The function `qdpll_get_value` returns a partial countermodel of the QBF: x was assigned to *false* which explains the unsatisfiability of ψ since $\psi[\neg x] = (\neg y) \land (y)$. By calling `qdpll_pop` we remove the clauses of the most recently added frame, which contains only (y). Thus the new QBF is $\psi = \forall x \exists y.\,(x \lor \neg y)$, which is satisfiable as found out by `qdpll_sat`.

The API of DepQBF allows to add new variables at any position in the prefix and provides functions to inspect the prefix. Variables and clauses can not explicitly deleted. Instead, a garbage collection phase can be triggered through the API which deletes all the clauses, variables and quantifiers which have been effectively removed by previous calls of `qdpll_pop`. The push/pop functionality of DepQBF is particularly useful for sequences S of QBFs where a large part of the CNFs is shared between the individual QBFs in S.

Originally, DepQBF is written in C. The release of version 3.03 (or later) comes with the Java API *DepQBF4J*, which allows to call DepQBF as a library from Java programs and thus makes incremental QBF solving more accessible.

6   Conclusion

We presented an overview of incremental QBF solving and our incremental QBF solver DepQBF. Incremental solving is useful to solve sequences of related formulae. Information learned from previously solved formulae in terms of derived clauses and cubes can be employed to solve the next formulae in the sequence.

---

[3] DepQBF takes input in QDIMACS format: http://www.qbflib.org/qdimacs.html

We implemented a simple approach to keep only particular cubes derived by model generation across incremental solver calls. As future work, we consider more sophisticated approaches to also keep cubes derived by Q-resolution.

Another important direction is the combination of incremental QBF solving with advanced techniques such as preprocessing and the generation of proofs and certificates. Currently, these techniques are implemented in separate tools. It is necessary to efficiently integrate them into a uniform framework to leverage the full power of the state of the art of QBF reasoning in practical applications.

References

1. Biere, A., Cimatti, A., Clarke, E., Zhu, Y.: Symbolic Model Checking without BDDs. In: Cleaveland, W.R. (ed.) TACAS 1999. LNCS, vol. 1579, pp. 193–207. Springer, Heidelberg (1999)
2. Bubeck, U., Kleine Büning, H.: Encoding Nested Boolean Functions as Quantified Boolean Formulas. JSAT 8(1/2), 101–116 (2012)
3. Kleine Büning, H., Karpinski, M., Flögel, A.: Resolution for Quantified Boolean Formulas. Inf. Comput. 117(1), 12–18 (1995)
4. Cadoli, M., Schaerf, M., Giovanardi, A., Giovanardi, M.: An Algorithm to Evaluate Quantified Boolean Formulae and Its Experimental Evaluation. J. Autom. Reasoning 28(2), 101–142 (2002)
5. Davis, M., Logemann, G., Loveland, D.: A Machine Program for Theorem-proving. Commun. ACM 5(7), 394–397 (1962)
6. Eén, N., Sörensson, N.: Temporal Induction by Incremental SAT Solving. Electr. Notes Theor. Comput. Sci. 89(4), 543–560 (2003)
7. Egly, U., Kronegger, M., Lonsing, F., Pfandler, A.: Conformant Planning as a Case Study of Incremental QBF Solving. CoRR (submitted, 2014)
8. Giunchiglia, E., Narizzano, M., Tacchella, A.: Clause/Term Resolution and Learning in the Evaluation of Quantified Boolean Formulas. J. Artif. Intell. Res (JAIR) 26, 371–416 (2006)
9. Janota, M., Grigore, R., Marques-Silva, J.: On QBF Proofs and Preprocessing. In: McMillan, K., Middeldorp, A., Voronkov, A. (eds.) LPAR-19. LNCS, vol. 8312, pp. 473–489. Springer, Heidelberg (2013)
10. Letz, R.: Lemma and Model Caching in Decision Procedures for Quantified Boolean Formulas. In: Egly, U., Fermüller, C. (eds.) TABLEAUX 2002. LNCS (LNAI), vol. 2381, pp. 160–175. Springer, Heidelberg (2002)
11. Lonsing, F., Biere, A.: DepQBF: A Dependency-Aware QBF Solver. JSAT 7(2-3), 71–76 (2010)
12. Lonsing, F., Egly, U.: Incremental QBF Solving. CoRR, abs/1402.2410 (2014)
13. Lonsing, F., Egly, U., Van Gelder, A.: Efficient Clause Learning for Quantified Boolean Formulas via QBF Pseudo Unit Propagation. In: Järvisalo, M., Van Gelder, A. (eds.) SAT 2013. LNCS, vol. 7962, pp. 100–115. Springer, Heidelberg (2013)
14. Marin, P., Miller, C., Lewis, M.D.T., Becker, B.: Verification of Partial Designs using Incremental QBF Solving. In: Proc. DATE. IEEE (2012)
15. Marques Silva, J.P., Lynce, I., Malik, S.: Conflict-Driven Clause Learning SAT Solvers. In: Handbook of Satisfiability. FAIA, vol. 185. IOS Press (2009)
16. Zhang, L., Malik, S.: Towards a Symmetric Treatment of Satisfaction and Conflicts in Quantified Boolean Formula Evaluation. In: Van Hentenryck, P. (ed.) CP 2002. LNCS, vol. 2470, pp. 200–215. Springer, Heidelberg (2002)

NLCertify: A Tool for Formal Nonlinear Optimization

Victor Magron

LAAS-CNRS, 7 avenue du colonel Roche, F-31400 Toulouse, France
magron@laas.fr
http://homepages.laas.fr/vmagron

Abstract. NLCertify is a software package for handling formal certification of nonlinear inequalities involving transcendental multivariate functions. The tool exploits sparse semialgebraic optimization techniques with approximation methods for transcendental functions, as well as formal features. Given a box and a transcendental multivariate function as input, NLCertify provides OCAMLlibraries that produce nonnegativity certificates for the function over the box, which can be ultimately proved correct inside the CoQ proof assistant.

Keywords: Formal Nonlinear Optimization, Hybrid Symbolic-Numeric Certification, Proof Assistant, Sparse SOS, Maxplus Approximation.

1 Introduction

A variety of tools for solving nonlinear systems are being adapted for the field of formal reasoning. One way to import the technology available inside an *informal* tool is the *skeptical* approach: the tool yields a form of certificate which can be verified on the *formal* side, i.e. inside a theoretical prover such as CoQ [7]. A recent illustration [4] is the integration of the computational features of SAT/SMT solvers in CoQ. The NLCertify[1] tool has informal nonlinear optimization features and enables formal verification in a skeptical way. An ambitious application is to automatically verify real numbers inequalities occurring by thousands in Thomas Hales' proof of Kepler's conjecture [8]. In the present article, nonlinear functions include polynomials, semialgebraic functions obtained by composition of polynomials with some basic operations (including the square root, sup, inf, $+, \times, -, /$, etc.) and composition of semialgebraic functions with transcendental functions (arctan, cos, exp, etc.) or basic operations.

Polynomial inequalities over a finite set of polynomial constraints can be certified using a hierarchy of sums of squares (SOS) relaxations [10]. Several variants of these relaxations are implemented in some MATLAB toolboxes: GLOPTIPOLY 3 [9] solves the Generalized Problem of Moments while SPARSE-POP takes sparsity into account [16], YALMIP [12] is a high-level parser for

---

[1] The source code is available at https://forge.ocamlcore.org/frs/?group_id=351.
See also the documentation at http://nl-certify.forge.ocamlcore.org

H. Hong and C. Yap (Eds.): ICMS 2014, LNCS 8592, pp. 315–320, 2014.

nonlinear optimization problems and has a built-in module for SOS calculations. These toolboxes rely on external SOS solvers for solving the relaxations (e.g. SDPA [17]) . However, the validity of the bounds obtained with these numerical tools can be compromise, due to the rounding error of the SOS solver. The tool[2] mentioned in [13] allows to handle some degenerate situations. For a more general class of problems (when the functions are not restricted to polynomials), one can combine SOS software with frameworks that approximate transcendental functions. Sollya [6] returns safe tight bounds for the approximation error obtained when computing minimax estimators of nonlinear univariate functions.

On the formal side, recent efforts have been done to verify nonlinear inequalities with theorem provers. A tool[3] in HOL-LIGHT combines formal interval arithmetic computation and quadratic Taylor approximation [15]. The features of the METITARSKI [1] theorem prover include continued fractions expansions of univariate transcendental functions such as log, arctan, etc. PVS incorporates nonlinear optimization libraries relying on Bernstein polynomial approximation [14]. The interval[4] tactic can assert the validity of interval enclosures of nonlinear functions over a finite set of box constraints inside COQ. The micromega tactic returns emptiness certificates for basic semialgebraic sets [5].

One specific challenge of the field of formal nonlinear optimization is to develop adaptive techniques to produce certificates with a reduced complexity. NLCertify provides efficient informal libraries by implementing the nonlinear maxplus method [2], which combines sparse SOS relaxations with maxplus quadratic approximation. In addition, the tool offers a secure certification framework for the bounds obtained with these semialgebraic relaxations [3]. These various features are placed in a unified framework extending to about 15000 lines of OCAML code and 3600 lines of COQ code. The NLCertify package can solve successfully non-trivial inequalities from the Flyspeck project (essentially tight inequalities, involving both semialgebraic and transcendental expressions of 6 to 12 variables) as well as significant global optimization benchmarks. The running tests for the verification of polynomial inequalities (Section 2) and transcendental inequalities (Section 3) are performed on Intel Core i5 CPU (2.40 GHz)[5].

2   Certified Polynomial Optimization

One particular problem among certification of nonlinear problems is to verify the inequality $\forall \mathbf{x} \in \mathbf{K}, f_{\mathrm{pop}}(\mathbf{x}) \geq 0$, where f_{pop} is an n-variate positive polynomial, $\mathbf{K} := \{\mathbf{x} \in \mathbb{R}^n : g_1(\mathbf{x}) \geq 0, \dots, g_m(\mathbf{x}) \geq 0\}$ is a semialgebraic set obtained with polynomials $g_1, \dots, g_m$. One way to convexify this polynomial optimization problem is to find sums of squares of polynomials $\sigma_0, \sigma_1, \dots, \sigma_m$ satisfying $f_{\mathrm{pop}}(\mathbf{x}) = \sigma_0 + \sum_{j=1}^{m} \sigma_j(\mathbf{x})g_j(\mathbf{x})$ and $\deg \sigma_0 \leq 2k, \deg(\sigma_1 g_1) \dots, \deg(\sigma_m g_m) \leq 2k$, for a fixed positive integer k (called the *relaxation* order). When k increases,

---

[2] Available from the pages http://bit.ly/fBNLhR and bit.ly/gPXNF8
[3] http://flyspeck.googlecode.com/files/FormalVerifier.zip
[4] https://www.lri.fr/~melquion/soft/coq-interval/
[5] With OCAML 4.01.0, COQ 8.4pl2, SSREFLECT 1.4, SDPA 7.3.6 and Sollya 3.0.

one obtains progressively stronger relaxations. In this way, it is always possible to certify the inequality $\forall \mathbf{x} \in \mathbf{K}, f_{\mathrm{pop}}(\mathbf{x}) \geq 0$ for a sufficiently large order (under certain assumptions [10]). These relaxations are implemented in NLCertify and numerically solved with the SOS solver SDPA. The tool performs a rational extraction from the SOS solver output with the LACAML[6] library. Then the corresponding remainder ϵ_{pop} (the difference between the objective polynomial f_{pop} and the SOS representation) can be bounded on a box which contains $\mathbf{K}$. More details can be found in [3].

Example 1. (caprasse) Here, we consider the degree 4 polynomial inequality $\forall \mathbf{x} \in [-0.5, 0.5]^4, -x_1 x_3^3 + 4x_2 x_3^2 x_4 + 4x_1 x_3 x_4^2 + 2x_2 x_4^3 + 4x_1 x_3 + 4x_3^2 - 10x_2 x_4 - 10x_4^2 + 5.1801 \geq 0$. The inequality is scaled on $[0, 1]^4$ with the solver option scale_pol = true and one adds the redundant constraints $x_1^2 \leq 1, \ldots, x_4^2 \leq 1$ by setting bound_squares_variables = true. The inequality can be solved numerically at the second SOS relaxation order (relax_order = 2). The correctness of the SOS representation is verified inside COQ (via the reflexive tactic ring) by setting check_certif_coq = true. Then the execution of NLCertify returns the following output:

```
% ./nlcertify caprasse
Proving that - x1 * x3 * x3 * x3 + 4 * x1 * x3 * x4 * x4 + 4 * x1 * x3 + 4 *
x2 *x3 * x3 * x4 + 2 * x2 * x4 * x4 * x4 - 10 * x2 * x4 + 4 * x3 * x3 - 10 *
x4 * x4 + 5.1801 >= 0 over the box [(-0.5, 0.5); (-0.5, 0.5); (-0.5, 0.5); (-0.5, 0.5)]
...
Computing lower bound ...
 SOS numerical computation in 0.045087 secs
Proving non-negativity inside Coq
    = true
    : bool
Finished transaction in  1. secs (0.813333u,0.s)
Lower Bound with SOS of degree at most 4 = 0.0000021671
...
0.0000021642 >= 0.0000000000
Inequality caprasse verified
```

Here, the caprasse inequality is formally proved in less than 1 second which is 8 times faster than the verification procedure in HOL-LIGHT with the framework described in [15] and 10 times faster than the tool based on Bernstein approximation in PVS [14].

3   Certificates for Nonlinear Transcendental Inequalities

Now, we consider a more general goal $\forall \mathbf{x} \in \mathbf{K}, f(\mathbf{x}) \geq 0$, where f is an n-variate transcendental function and $\mathbf{K} \subset \mathbb{R}^n$ is a box. NLCertify implements the nonlinear maxplus method [2], which can be summarized as follows. The tool builds first the abstract syntax tree t of f (see Figure 1 for an illustration). The leaves of t are semialgebraic functions. The other nodes can be either univariate transcendental functions or basic operations. NLCertify approximates t with means of semialgebraic estimators and provides lower and upper bounds of t over $\mathbf{K}$. When t represents a polynomial, the tool computes lower and upper bounds of t using a

---

[6] Linear Algebra with OCAML, this library implements BLAS/LAPACK routines.

hierarchy of sparse SOS relaxations, as outlined in Section 2. The extension to the semialgebraic case is straightforward through the implementation of the Lasserre-Putinar lifting-strategy [11]. The user can choose to approximate transcendental functions with maxplus estimators as well as best uniform (or minimax) polynomials. The maxplus method derives lower (resp. upper) estimators using concave maxima (resp. convex infima) of quadratic forms (see Figure 2 for an example). Alternatively, univariate minimax polynomials are provided with an interface to the Sollya environment, in which the Remez iterative algorithm is implemented. In this way, NLCertify computes certified global estimators from approximations of primitive functions by induction over the syntax tree t.

Example 2 (from LEMMA 9922699028 Flyspeck [7]). Let define the polynomial $\Delta \mathbf{x} := x_1 x_4 (-x_1 + x_2 + x_3 - x_4 + x_5 + x_6) + x_2 x_5 (x_1 - x_2 + x_3 + x_4 - x_5 + x_6) + x_3 x_6 (x_1 + x_2 - x_3 + x_4 + x_5 - x_6) - x_2 x_3 x_4 - x_1 x_3 x_5 - x_1 x_2 x_6 - x_4 x_5 x_6$, the semialgebraic functions $r(\mathbf{x}) := \partial_4 \Delta \mathbf{x} / \sqrt{4 x_1 \Delta \mathbf{x}}$ and $l(\mathbf{x}) := 1.6294 - \pi/2 - 0.2213(\sqrt{x_2} + \sqrt{x_3} + \sqrt{x_5} + \sqrt{x_6} - 8.0) + 0.913(\sqrt{x_4} - 2.52) + 0.728(\sqrt{x_1} - 2.0)$, as well as the box $\mathbf{K} := [4, 2.1^2]^3 \times [2.65^2, 8] \times [4, 2.1^2]^2$. Note that for illustration purpose, the inequality has been modified by taking a sub-box of the original Flyspeck inequality box $[4, 2.52^2]^3 \times [2.52^2, 8] \times [4, 2.52^2]^2$.

Here we display and comment the output of NLCertify for the inequality $\forall \mathbf{x} \in \mathbf{K}, l(\mathbf{x}) + \arctan(r(\mathbf{x})) \geq 0$. The total computation time is about 20 seconds when running the informal algorithm (check_certif_coq = false) with 3 iterations (samp_iters = 3), no box subdivisions (bb = false) and the additional setting xconvert_variables = true.

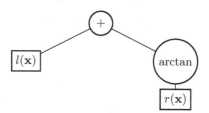

Fig. 1. The abstract syntax tree of the function f from LEMMA 9922699028 Flyspeck

```
% ./nlcertify 9922699028_modified
Proving that - 1.5708 + atan ... >= 0 over the box  [(4, 4.41);
(4, 4.41); (4, 4.41); (7.0225, 8); (4, 4.41); (4, 4.41)] ...
Bounding semialgebraic components
 Computing approximation of atan on [0.0297, 0.4165]
 Minimizer candidate x = [4; 4; 4; 8; 4; 4]
  Control points set: [0.3535] ...
Semialgebraic components bounded

Iteration 1
 Lower bound = -0.00463
 Minimizer candidate x = [4; 4; 4; 7.0225; 4; 4]
Iteration 2
```

---

[7] See the file available at
http://code.google.com/p/flyspeck/source/browse/
trunk/text_formalization/nonlinear/ineq.hl

```
Control points set: [0.1729; 0.3535] ...
Lower bound = -0.00006025
Minimizer candidate x = [4; 4; 4; 7.6622; 4; 4]
Iteration 3
  Control points set: [0.1729; 0.2884; 0.3535] ...
  Lower bound = 0.000004662
Minimizer candidate x = [4; 4; 4; 7.8083; 4; 4]
```

Lower and upper bounds for the semialgebraic components (i.e. r and l) are computed using SOS relaxations. An interval enclosure for r is $[m, M]$, with $m := 0.0297$ and $M := 0.4165$. Multiple evaluations of f return a set of values and we obtain a first minimizer guess $\mathbf{x}_1 := (4, 4, 4, 8, 4, 4)$ of f over $\mathbf{K}$, which corresponds to the minimal value of the set. Then, the solver performs three iterations of the nonlinear maxplus algorithm.

1. The tool returns an underestimator par_{a_1} of arctan over $[m, M]$, with $a_1 := r(\mathbf{x}_1) = 0.3535$. Then, it computes $m_1 \leq \min_{\mathbf{x} \in \mathbf{K}} \{l(\mathbf{x}) + \mathrm{par}_{a_1}(r(\mathbf{x}))\}$. It yields $m_1 = -4.63 \times 10^{-3} < 0$ and $\mathbf{x}_2 := (4, 4, 4, 7.0225, 4, 4)$.
2. From the second control point, we get $a_2 := r(\mathbf{x}_2) = 0.1729$ and a tighter bound $m_2 \leq \min_{\mathbf{x} \in \mathbf{K}} \{l(\mathbf{x}) + \max_{1 \leq i \leq 2} \{\mathrm{par}_{a_i}(r(\mathbf{x}))\}\}$. We get $m_2 = -6.025 \times 10^{-5} < 0$ and $\mathbf{x}_3 := (4, 4, 4, 7.6622, 4, 4)$.
3. From the third control point, we get $a_3 := r(\mathbf{x}_3) = 0.2884$ and $m_3 \leq \min_{\mathbf{x} \in \mathbf{K}} \{l(\mathbf{x}) + \max_{1 \leq i \leq 3} \{\mathrm{par}_{a_i}(r(\mathbf{x}))\}\}$. We obtain $m_3 = 4.662 \times 10^{-6} > 0$. Thus, the inequality is solved.

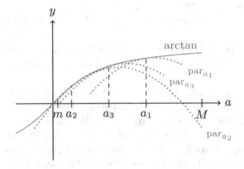

Fig. 2. A hierarchy of maxplus quadratic underestimators for arctan

4   Conclusion

NLCertify aims at combining the safety of the COQ proof assistant with the efficiency of informal optimization algorithms, based on low degree maxplus estimators and sparse semialgebraic relaxations. This could allow to derive safe solutions for challenging problems that require to certify both approximation of transcendental functions and bounds for polynomial programs such as impulsive Rendezvous problems. Further developments on the formal side include the integration of a new reflexive tactic inside the COQ standard library. Adding faster arithmetic for the polynomial coefficients ring would speedup the computation of the SOS checker. The current features could also be extended to handle non-commutative SOS certificates as well as discrete combinatorial optimization.

References

1. Akbarpour, B., Paulson, L.C.: Metitarski: An automatic theorem prover for real-valued special functions. J. Autom. Reason. 44(3), 175–205 (2010)
2. Allamigeon, X., Gaubert, S., Magron, V., Werner, B.: Certification of real inequalities – templates and sums of squares (submitted for publication, March 2014), http://arxiv.org/abs/1403.5899
3. Allamigeon, X., Gaubert, S., Magron, V., Werner, B.: Formal proofs for nonlinear optimization (submitted for publication, April 2014), http://arxiv.org/abs/1404.7282
4. Armand, M., Faure, G., Grégoire, B., Keller, C., Théry, L., Werner, B.: A Modular Integration of SAT/SMT Solvers to Coq through Proof Witnesses. In: Jouannaud, J.-P., Shao, Z. (eds.) CPP 2011. LNCS, vol. 7086, pp. 135–150. Springer, Heidelberg (2011)
5. Besson, F.: Fast reflexive arithmetic tactics the linear case and beyond. In: Altenkirch, T., McBride, C. (eds.) TYPES 2006. LNCS, vol. 4502, pp. 48–62. Springer, Heidelberg (2007)
6. Chevillard, S., Joldeş, M., Lauter, C.: Sollya: An environment for the development of numerical codes. In: Fukuda, K., van der Hoeven, J., Joswig, M., Takayama, N. (eds.) ICMS 2010. LNCS, vol. 6327, pp. 28–31. Springer, Heidelberg (2010)
7. The Coq Proof Assistant, http://coq.inria.fr/
8. Hales, T.C.: A proof of the Kepler conjecture. Ann. of Math (2) 162(3), 1065–1185 (2005)
9. Henrion, D., Lasserre, J.-B., Löfberg, J.: GloptiPoly 3: moments, optimization and semidefinite programming. Optimization Methods and Software 24(4-5), 761–779 (2009)
10. Lasserre, J.B.: Global optimization with polynomials and the problem of moments. SIAM Journal on Optimization 11(3), 796–817 (2001)
11. Lasserre, J.B., Putinar, M.: Positivity and optimization for semi-algebraic functions. SIAM Journal on Optimization 20(6), 3364–3383 (2010)
12. Löfberg, J.: Yalmip: A toolbox for modeling and optimization in MATLAB. In: Proceedings of the CACSD Conference, Taipei, Taiwan (2004)
13. Monniaux, D., Corbineau, P.: On the Generation of Positivstellensatz Witnesses in Degenerate Cases. In: van Eekelen, M., Geuvers, H., Schmaltz, J., Wiedijk, F. (eds.) ITP 2011. LNCS, vol. 6898, pp. 249–264. Springer, Heidelberg (2011)
14. Muoz, C., Narkawicz, A.: Formalization of bernstein polynomials and applications to global optimization. Journal of Automated Reasoning 51(2), 151–196 (2013)
15. Solovyev, A., Hales, T.C.: Formal verification of nonlinear inequalities with taylor interval approximations. CoRR, abs/1301.1702 (2013)
16. Waki, H., Kim, S., Kojima, M., Muramatsu, M., Sugimoto, H.: Sparsepop—a sparse semidefinite programming relaxation of polynomial optimization problems. ACM Trans. Math. Softw. 35(2) (2008)
17. Yamashita, M., Fujisawa, K., Nakata, K., Nakata, M., Fukuda, M., Kobayashi, K., Goto, K.: A high-performance software package for semidefinite programs: Sdpa7. Technical report, Dept. of Information Sciences, Tokyo Institute of Technology, Tokyo, Japan (2010)

Developing Linear Algebra Packages on Risa/Asir for Eigenproblems

Katsuyoshi Ohara[1], Shinichi Tajima[2], and Akira Terui[3]

[1] Faculty of Mathematics and Physics, Kanazawa University, Japan
ohara@air.s.kanazawa-u.ac.jp
http://air.s.kanazawa-u.ac.jp/~ohara/
[2] Faculty of Pure and Applied Sciences, University of Tsukuba, Japan
tajima@math.tsukuba.ac.jp
[3] Faculty of Pure and Applied Sciences, University of Tsukuba, Japan
terui@math.tsukuba.ac.jp
http://researchmap.jp/aterui

Abstract. We are developing linear algebra packages on Risa/Asir, a computer algebra system. The aim is to provide programs for efficiently and exactly solving eigenproblems on the computer algebra system for large scale square matrices over integers or rational numbers. The software package consists of some programs. The followings are currently prepared for solving eigenproblems: computing eigenspaces, the spectral decomposition, Jordan chains and minimal annihilating polynomials.

Keywords: Linear Algebra, Eigenproblems, Residue Calculus.

1 Introduction

We are developing linear algebra packages on Risa/Asir, a computer algebra system. The aim is to provide programs for efficiently and exactly solving eigenproblems on the computer algebra system for large scale square matrices over integers or rational numbers.

For a square matrix with rational number entries, in general, eigenvalues are algebraic numbers. A square matrix can be regarded as a linear transformation on a finite-dimensional complex vector space. Any invariant subspace has a basis which can be expressed by polynomials of the corresponding eigenvalue with rational number coefficients. Here each polynomial has smaller degree than the defining polynomial of the eigenvalue. Moreover the projection matrix to the invariant subspace, which appears in the spectral decomposition, can be also represented by polynomials of the eigenvalue. Since invariant subspaces have common expression for conjugate eigenvalues, it depends on only the defining polynomial of the eigenvalue.

Some mathematical software give verbose expressions as solutions of eigenproblems. For example, "Eigenvectors" of Maple 17 returns vectors whose entries are represented by rationals of eigenvalues. In this case it is hard to analyze eigenproblems for large scale matrices because of high costs of normalization for

H. Hong and C. Yap (Eds.): ICMS 2014, LNCS 8592, pp. 321–324, 2014.

algebraic numbers. It is our motivation to develop and implement efficient algorithms for large scale problems. We also pay attention to parallel computation from the beginning of the research.

Our approach is based on minimal (or pseudo-)annihilating polynomials and residue calculus of resolvents. The resolvent of square matrix is a matrix-valued rational function of a complex variable and its denominator agrees with the minimal polynomial of the matrix. Theoretically we have the spectral decomposition by using Laurent expansion of the resolvent. In order to execute residue calculus of the resolvent we use algebraic local cohomology groups. On the other hand a monic polynomial is called an annihilating polynomial for a vector if the polynomial of the matrix annihilates the vector. The minimal polynomial is given by the LCM of minimal annihilating polynomials for the standard basis of euclidean vector space. Minimal annihilating polynomials are also applied to parallel computation.

2   Residue Calculus for Resolvents

In this section, we explain one of our algorithms developed for eigenproblems. Let A be a square matrix with rational number entries. A matrix-valued rational function $R(z) = (zE - A)^{-1}$ with complex argument is called the *resolvent* and has poles at eigenvalues of A. The following theorem holds:

Theorem 1 ([2], pp.40–43). *Let $\lambda_1, \ldots, \lambda_m$ be eigenvalues of A. If $R(z)$ has the Laurent expansion $\cdots + \frac{D_i}{(z-\lambda_i)^2} + \frac{P_i}{z-\lambda_i} + \cdots$ at $z = \lambda_i$, then the spectral decomposition of A can be expressed by*

$$A = \lambda_1 P_1 + \cdots + \lambda_m P_m + \sum_{i=1}^{m} D_i, \qquad E = P_1 + \cdots + P_m,$$

where P_i is the projection matrix to the invariant subspace corresponding to λ_i and all D_i are nilpotent.

Example 1. Let $A = \begin{pmatrix} 0 & 4 & 0 \\ -1 & 4 & 0 \\ 0 & 0 & 3 \end{pmatrix}$ with the minimal polynomial $(x-2)^2(x-3)$.

The resolvent has an expression $R(z) = \dfrac{1}{(z-2)^2}D_1 + \dfrac{1}{z-2}P_1 + \dfrac{1}{z-3}P_2$. Here $P_1 = -3E + 4A - A^2$, $P_2 = 4E - 4A + A^2$ and $D_1 = -6E + 5A - A^2$. Then the spectral decomposition is given by $A = 2P_1 + 3P_2 + D_1$.

Example 2. Let $A = \begin{pmatrix} 0 & 2 & 0 & 1 \\ 1 & 0 & 0 & 0 \\ 0 & 0 & 0 & 2 \\ 0 & 0 & 1 & 0 \end{pmatrix}$ with the minimal polynomial $\pi_A(x) = (x^2 - 2)^2$.

For conjugate eigenvalues $\lambda_1, \lambda_2 = \pm\sqrt{2}$, we put $P_i = \frac{1}{16}(8E + 6\lambda_i A - \lambda_i A^3)$ and $D_i = \frac{1}{8}(-2\lambda_i E - 2A + \lambda_i A^2 + A^3)$. Then the spectral decomposition is expressed by $A = \lambda_1 P_1 + \lambda_2 P_2 + (D_1 + D_2)$.

From these examples, we can observe some properties for spectral decomposition: (1) P_i and D_i can be written as polynomials of λ_i and A, (2) these have common expression for conjugate eigenvalues.

We can give explicit algebraic representations for P_i and D_i by using residue calculus for resolvents. We consider a polynomial ideal $\mathrm{Ann}_{\mathbf{Q}[x]}(A) = \{f(x) \in \mathbf{Q}[x] \mid f(A) = O\}$ generated by the minimal polynomial of A. There exists a symmetric polynomial $\Psi_f \in \mathbf{Q}[x, y]$ which satisfies the relation $f(x) - f(y) = (x - y)\Psi_f(x, y)$ for given $f \in \mathrm{Ann}_{\mathbf{Q}[x]}(A)$. By assigning zE and A to $\Psi_f(x, y)$, we have an expression

$$R(z) = \frac{1}{f(z)}\Psi_f(zE, A), \qquad f \in \mathrm{Ann}_{\mathbf{Q}[x]}(A).$$

Let f_i be the defining polynomial of an eigenvalue λ_i. Any $f \in \mathrm{Ann}_{\mathbf{Q}[x]}(A)$ has a decomposition $f(z) = f_i(z)^{\ell_i} g_i(z)$ with $\gcd(f_i, g_i) = 1$. Let us introduce an algebraic local cohomology group $H_{[Z_i]}(\mathbf{Q}[x])$ with support $Z_i = \{x \mid f_i(x) = 0\}$. We can compute an explicit Noetherian differential operator

$$T_f = (-\partial)^{\ell_i - 1}t_0(x) + (-\partial)^{\ell_i - 1}t_1(x) + \cdots + t_{\ell_i - 1}(x) \in \mathbf{Q}[x]\langle\partial\rangle, \quad \partial = \frac{d}{dx},$$

which satisfies the relation $[1/f(x)] \equiv T_f[f_i'(x)/f_i(x)]$ in $H_{[Z_i]}(\mathbf{Q}[x])$. (see [3,5])

From the Laurent expansion of $R(z)$, projection matrices can be expressed by contour integrals around corresponding eigenvalues. By applying T_f, we have an algebraic expression of P_i as follows:

$$P_i = \frac{1}{2\pi\sqrt{-1}}\int_C R(z)dz = \frac{1}{2\pi\sqrt{-1}}\int_C \Psi_f(zE, A) T_f\left[\frac{f_i'(z)}{f_i(z)}\right] dz$$

$$= \frac{1}{2\pi\sqrt{-1}}\int_C (T_f^* \bullet \Psi_f(zE, A)) \frac{f_i'(z)}{f_i(z)} dz = F(\lambda_i E, A),$$

where T_f^* is the adjoint operator of T_f and $F(x, y) \equiv T_f^* \bullet \Psi_f(x, y)$ (mod $\mathbf{Q}[x, y]f_i(x)$). The polynomial $F(x, y)$ also satisfies $\deg_x F(x, y) < \deg f_i(x)$ and $\deg_y F(x, y) < \deg f(x)$. Nilpotent matrices are also calculated by similar formulas. So these expressions are optimized for algebraic numbers and the complexity of the method depends on the degree of $f(x)$. Thus we should choose the minimal polynomial as f, that is the generator of $\mathrm{Ann}_{\mathbf{Q}[x]}(A)$.

Next we consider a column vector of P_i. For a vector $\mathbf{p}$, let $\mathrm{Ann}_{\mathbf{Q}[x]}(A, \mathbf{p}) = \{f \in \mathbf{Q}[x] \mid f(A)\mathbf{p} = \mathbf{0}\}$. The monic generator of the ideal is called the *minimal annihilating polynomial* for $\mathbf{p}$. By an argument similar to discussion above, we have the following expressions:

$$P_i\mathbf{p} = F_\mathbf{p}(\lambda_i E, A), \quad F_\mathbf{p}(x, y) \equiv T_g^* \bullet \Psi_g(x, y), \quad g \in \mathrm{Ann}_{\mathbf{Q}[x]}(A, \mathbf{p}).$$

The projection matrix is written as $P_i = (P_i\mathbf{e}_1, P_i\mathbf{e}_2, \ldots)$ by the standard basis $\{\mathbf{e}_1, \mathbf{e}_2, \ldots\}$ of the euclidean space. Since each minimal annihilating polynomial has smaller degree than the minimal polynomial in general, the column vectors of the spectral decomposition should be calculated separately by using minimal annihilating polynomials for efficient computation. Moreover, the separated method can be executed on parallel computers.

3   Implementations

Our software package taji_mat.rr is implemented on Risa/Asir, a computer algebra system and consists of some programs. Users input a square matrix and its characteristic polynomial for invoking each function. The followings are currently prepared for solving eigenproblems:

- computing bases of eigenspaces for corresponding eigenvalues, (taji_mat.eigenspace)
- the spectral decomposition, (taji_mat.spec)
- Jordan chains for corresponding eigenvalues, (taji_mat.jordan_chain)
- minimal or pseudo-annihilating polynomials, (taji_mat.annih)
- algebraic local cohomologies and residue calculation. (taji_alc.snoether)

In several functions, minimal annihilating polynomials are automatically computed and used.

Example 3. For executing Example 2, we may call taji_mat.spec as follows:

```
[100] load("taji_mat.rr");
[101] A=newmat(4,4,[[0,2,0,1],[1,0,0,0],[0,0,0,2],[0,0,1,0]]);
[102] F=(x^2-2)^2;
[103] taji_mat.spec(A,F);
```

Finally, we show elapsed time for invoking taji_mat.spec on parallel computer. We used an integer matrix of order 48 with the minimal polynomial $\prod_{i=1}^{4}(\text{cubic})^4$. The experiment was executed on Linux/amd64 2.6.39 with dual Xeon E5645 (6-core, 2.4GHz) and 96GB memory.

Nodes	4	8	12	16
Time (sec.)	76.56	44.38	35.46	30.78

References

1. Gantmacher, F.R.: The Theory of Matrices, vol. 1. AMS Chelsea Publishing (1977)
2. Kato, T.: A Short Introduction to Perturbation Theory for Linear Operators. Springer (1982)
3. Nakamura, Y., Tajima, S.: Residue calculus with differential operators. Kyushu J. of Math. 54, 127–138 (2000)
4. Ohara, K., Tajima, S.: Spectral Decomposition and Eigenvectors of Matrices by Residue Calculus. In: Proceedings of the Joint Conference of ASCM 2009 and MACIS 2009, COE Lecture Note 22, Kyushu University, pp. 137–140 (2009)
5. Tajima, S.: An algorithm for computing the Noetherian operator representations and its applications to constant coefficients holonomic PDE's. In: Tools for Mathematical Modellings, St. Petersbourg, pp. 154–160 (2001)

Mathematical Software for Modified Bessel Functions*

Juri Rappoport

Institute for Computer Aided Design,
Russian Academy of Sciences, Russia
jmrap@landau.ac.ru
http://www.icad.org.ru

Abstract. The high-quality mathematical software for the computation of modified BESSEL functions of the second kind with integer, imaginary and complex order and real argument is elaborated. The value of function may be evaluated with high precision for given value of the independent argument x and order r. These codes are addressed to the wide audience of scientists, engineers and technical specialists. The tables of these functions are published. This software improves significantly the capability of computer libraries. These functions arise naturally in boundary value problems involving wedge-shaped or cone-shaped geometries. They are fundamental to mathematical modeling activities in applied science and engineering. Methods of mathematical and numerical analysis are adapted for the creation of appropriate algorithms for these functions, computer codes are written and tested. Power series, Tau method and numerical quadratures of trapezoidal kind are used for the construction of subroutines. New realization of the Lanczos Tau method with minimal residue is proposed for the constructive approximation of the second order differential equations solutions with polynomial coefficients. A Tau method computational scheme is applied for the constructive approximation of a system of differential equations solutions related to the differential equation of hypergeometric type. Various vector perturbations are discussed. Our choice of the perturbation term is a shifted CHEBYSHEV polynomial with a special form of selected transition and normalization. The minimality conditions for the perturbation term are found for one equation. They are sufficiently simple for verification in a number of important cases. Tau method's approach gives a big advantage in the economy of computer's time. The mathematical software for kernels of LEBEDEV type index transforms – modified BESSEL functions of the second kind with complex order is elaborated in detail. The software for new applications of LEBEDEV type integral transforms and related dual integral equations for the numerical solution of problems of mathematical physics is constructed. The algorithm of numerical solution of some mixed boundary value problems for the Helmholtz equation in wedge domains is developed. Observed examples admitting complete analytical solution demonstrate the efficiency of this approach for applied problems.

* This work was partially supported by Thematic Programme on Inverse Problems in Imaging of the FIELDS Institute.

H. Hong and C. Yap (Eds.): ICMS 2014, LNCS 8592, pp. 325–332, 2014.

Keywords: mathematical software, modified BESSEL functions, LANC-ZOS tau method, CHEBYSHEV polynomials, KONTOROVICH-LEBEDEV integral transforms.

1 Introduction

The possibility to find the expression of the solution from known functions allows one to make qualitative conclusions, estimate relations of unknown quantities from prescribed parameters and obtain required numerical values without big expenditure of time and means. It's important very much for the investigations in the field of theoretical and mathematical physics, mechanics, and for the technical and engineering applications. The mathematical tables were the principal source for numerical values of mathematical special functions a time ago. The universal character of use, high standard quality and longevity were typical for them. The purpose of the tables was to perform laborious calculations once and forever. We'd like to mention the big series of mathematical tables from the Computing Centre of the USSR Academy of Sciences. We'd like to mark out two tables from this series devoted to the modified BESSEL functions of pure imaginary and complex order [23,9].

The tables lose their value with the development of personal computers, note-books, netbooks and informatics without paper. It's inconvenient to input the tables into the computer's memory and then use and interpolate them. The effective algorithms and codes became useful for the computation of function values on widespread computers. It's important to find the numerical values of functions and their zeros, simplify such mathematical expressions as transforms and integrals, construct methods for numerical solution of differential and integral equations with these functions. So the analytical expressions on orthogonal polynomials, rational approximations and recurrent relations may be helpful very much for the creation of the mathematical software. It's possible to mark the reliable and robust properties of good mathematical software. It's possible to mention a number of packages and libraries of mathematical software for special functions. We can mark the FUNPACK package from NATS project and NAG project for example. The Internet gave the new possibilities for the mathematical software. The NIST Digital Library of Mathematical functions http://www.dlmf.nist.gov was created by great international collective of scientists. It was accompanied by NIST Handbook of Mathematical Functions [8] published in 2010 in Cambridge University Press. It may be efficient to use the cloud computing and mobile devices for the mathematical software of special functions.

2 Functionality

The program complex for evaluation of MACDONALD'S function [10] of arbitrary real argument and real and some complex values of order is described. Approximation method, quadrature formulas, decomposition on CHEBYSHEV polynomials, recurrent relations and others are used for computations. The programs of complex are used in the numerical solution of mixed boundary value problems by means of the method of integral transforms.

The complex of programs is designed for the computation of values of the MACDONALD function (modified BESSEL function of the second kind)$K_\nu(x)$ of real argument $x > 0$ for real and some complex values of order ν.

Programs $K0, K1$ compute the values of the modified BESSEL functions $K_0(x), K_1(x)$ for real argument $x > 0$. The program KN computes the values of the modified BESSEL function $K_p(x)$ for real argument $x > 0$ and for the sequence of integer values of order p from 0 to n. For the computation of the modified BESSEL function $K_\nu(x)$ for real argument $x > 0$ and arbitrary real order ν we recommend the program of N.M.TEMME [20].

The programs KIR and KIK generate values of the modified BESSEL functions $K_{i\beta}(x)$ and $K_{1/2+i\beta}(x)$, correspondingly, for real argument x and real parameter β $(0.1 \leq x \leq 10, 0 \leq \beta \leq 10)$. The results of the computations by means of the program KIR coincide with EHRENMARKS'S results [1]. It's shown [2] that programs KIR and KIK give a good accuracy in greater domain of x and β.

The methods used in the computations [2,5,7,9,10,12,13,15,16,19] were following: decompositions in CHEBYSHEV polynomials in programs $K0$ and $K1$, recurrent relations in the program KN, power series, Miller's method and other refined techniques in the program [4,20], a special approximating method (an integral form of the LANCZOS tau-method), quadrature formulas of trapezoidal kind with optimal choice of the step and power series in the programs KIR and KIK.

The Unified Library of Numerical Analysis of Moscow State University contains about 30 codes for computation of basic special functions of mathematical physics joined as a general library. All codes are carefully designed documented elements of mathematical software. The majority of codes are constructed with single and double precision. The different methods are used for the computation of different functions but one general approach is used. This is the truncated CHEBYSHEV expansions and summation of FOURIER series on CHEBYSHEV polynomials. The following codes were elaborated:

SF21R.R Computation of gamma-function $\gamma(x)$ of real argument.

SF65R.RC Computation of logarithmic derivative of gamma-function.

SF63R.R Computation of integral from first kind BESSEL function of order zero.

SF64R.R Computation of integral from NEIMAN (second kind BESSEL) function of order zero.

SF67R.R Computation of second kind modified BESSEL function $K_{i\beta}(x)$ of imaginary order for real argument x and real values β.

SF70R.R Computation of second kind modified BESSEL function $K_{\frac{1}{2}+i\beta}(x)$ of complex order for real argument x and real values β.

SF74R.R Computation of first kind modified BESSEL function for real argument x and sequence of integer orders p from n to 0.

SF75R.R Computation of second kind modified BESSEL function for real argument x and sequence of integer orders from 0 to n.

The codes ZP45R.R-ZP52R.R of the computation of partial sums of FOURIER series on CHEBYSHEV polynomials were constructed also.

3   Application

The complex of programs makes possible the evaluation of the kernels of BESSEL integral transforms (K-transforms) and KONTOROVITCH-LEBEDEV integral transforms. They have value in the solution of some problems of mathematical physics, including, for example, DIRICHLET problem and other harmonic problems in wedge-shaped domains.

The modified KONTOROVITCH-LEBEDEV integral transforms are also important in the solution of some problems of mathematical physics, including, for example, mixed boundary value problems for the HELMHOLTZ equation in wedge and conical domains. The complex of programs enables the evaluation of the kernels of the "unmodified" and modified Kontorovitch-Lebedev integral transforms with the precision 7-8 significant figures sufficient and necessary for applications. These transforms were used for the numerical solution of a number of mixed boundary value problems and dual integral equations. For example, diffusion, heat structure transfer and elasticity problems were reduced to mixed boundary value problems for the plane Helmholtz equation $\Delta u - k^2 u = 0$ and these were solved in arbitrary sectorial domains. The complex of programs makes possible the evaluation of the kernels of BESSEL integral transforms (K-transforms) [21]

$$g(y,\nu) = \int_0^\infty f(x)(xy)^{1/2} K_\nu(xy)dx, y > 0, \nu = 1, \nu \in N, \nu \in R,$$

and KONTOROVITCH-LEBEDEV integral transforms [6]

$$F(\beta) = \int_0^\infty f(x)K_{i\beta}(x)dx, 0 \le \beta < \infty.$$

These have value in the solution of some problems of mathematical physics, including, for example, DIRICHLET problem and other harmonic problems in wedge-shaped domains.

The modified KONTOROVITCH-LEBEDEV integral transforms [6,11,17,18]

$$F_+(\beta) = \int_0^\infty f(x)\text{Re}K_{1/2+i\beta}(x)dx, 0 \le \beta < \infty,$$

$$F_-(\beta) = \int_0^\infty f(x)\text{Im}K_{1/2+i\beta}(x)dx, 0 \le \beta < \infty,$$

are also important in the solution of some problems of mathematical physics, including, for example, mixed boundary value problems for the Helmholtz equation in wedge and conical domains.

As another example, a model gas mixture combustion problem in a plane wedge section was solved. The boundary conditions were that the gas mixture is given as a known law on part of the boundary, and it is known how combustion is held on the other part. The numerical solution is described in detail [14,18].

4 Underlying Theory

The programs of complex are used in the numerical solution of mixed boundary value problems by means of the method of integral transforms. The using of different quadrature formulas for evaluation $K_{1/2+i\beta}(x)$ are considering in details (FILON'S quadrature formulas, GAUSS-LEGENDRE quadrature formulas, quadrature formulas of trapezoidal kind). It is shown that the best accuracy and speed are achieved on using of quadrature formulas of trapezoidal kind in computation of modified BESSEL function $K_{1/2+i\beta}(x)$. The special procedure for "optimal" step's choice is used based on asymptotic decompositions of these functions. The suggested method allows to decrease substantially the machine time which is important in computation of special functions. The values of steps, knots and errors of quadrature formulas for separate values of index and argument are analyzed.

The other methods used in the computations were the following: decompositions in CHEBYSHEV polynomials, recurrent relations, power series, MILLER'S method and other refined techniques, approximating method and LANCZOS tau-method.

The results of computations show that the quadrature formulas of trapezoidal kind with "optimal" choice of the step to shorten considerably the number of knots required for the integration. The necessary number of knots depends significantly on x and β. It decreases with increasing x and increases with increasing β (because these functions are rapidly damped as x increases and strongly oscillating for increasing β). So quadrature formulas of trapezoidal kind with "optimal" choice of the step yield 7-8 significant digits for the computation of $K_{1/2+i\beta}(x)$ except when the ratio x/β is small.

Properties of the function $K_{1/2+i\beta}(x)$ are given in the Table's book [9]. The following methods of computation were considered: power series (1), tau-method (2), approximating method (integral form of the tau-method) (3), FILON'S quadrature formulas (4), Gauss-Legendre quadrature formulas (5), quadrature formulas of trapezoidal kind with "optima" choice of the step (6).

The results of calculations show that the methods 1, 3 and 6 are more economical. They were chosen for the preparation of the program KIK. But none of the methods gives necessary precision in 7-8 significant digits in the targeted domain of x and β. Depending on x and β, the method which yields necessary precision in 7-8 significant digits for the smallest expenditure of machine time is chosen.

The use of power series, Method 1, for the computation $K_{1/2+i\beta}(x)$ is described in Tables [9].

The use of the approximating method, Method 3, for the computation $K_{1/2+i\beta}(x)$ is described in [12,15,16,19] where also its advantages are shown in the comparison to the usual form of the tau-method with regard to precision and speed. This method gives necessary precision on 7-8 significant digits and is suitable for small values β ($0 \le \beta \le 4$) for almost all values of x. Let's note that the tau-method, Methods 2 and 3, certainly are more effective then other methods based on recurrent relations because of the necessity for multiple evaluation of the function's values for fixed parameter β as under the transition to new value x it's not necessary to repeat all scheme of calculations and it's sufficient to conduct n additions and n multiplications.

The questions of the optimal choice of the residue in the approximating method and its realization on a computer are discussed in [12,15,16,19].

Yakubovich [22] introduced the integral transforms with modified BESSEL functions of arbitrary complex order. So let's consider the question of the use of different quadrature formulas for the computation $K_{\alpha+i\beta}(x)$.

The book of LUKE [7] and paper of GAUTSCHI [3] also consider the use of quadrature formulas for the computation of BESSEL functions. The integral representations of the modified Bessel function $K_{\alpha+i\beta}(x)$ take the following form [8]:

$$\mathrm{Re}K_{\alpha+i\beta}(x) = \int_0^\infty e^{-x\mathrm{ch}t}\mathrm{ch}(\alpha t)\cos(\beta t)dt, \tag{1}$$

$$\mathrm{Im}K_{\alpha+i\beta}(x) = \int_0^\infty e^{-x\mathrm{ch}t}\mathrm{sh}(\alpha t)\sin(\beta t)dt. \tag{2}$$

They are most simple and convenient for the application of different quadrature formulas [9,10,13]. Because of the rapidly decreasing integrand for increasing t, it's possible to truncate these integrals while maintaining the necessary precision. Thus we consider the approximate integrals

$$I_1 = \int_0^b e^{-x\mathrm{ch}t}\mathrm{ch}(\alpha t)\cos(\beta t)dt, \tag{3}$$

$$I_2 = \int_0^b e^{-x\mathrm{ch}t}\mathrm{sh}(\alpha t)\sin(\beta t)dt. \tag{4}$$

The upper limit of integration, b, is determined from the condition

$$e^{x(1-\mathrm{ch}b)}\mathrm{ch}(\alpha b) = 10^{-N},$$

N is taken to be 10.

The use of quadrature formulas of the trapezoidal kind, Method 6, for the computation of the integrals (3)-(4) gives

$$I_1(h) = h(0.5e^{-x} + \sum_{j=1}^k e^{-x\mathrm{ch}(jh)}\mathrm{ch}(\alpha jh)\cos(\beta jh)), \tag{5}$$

$$I_2(h) = h\sum_{j=1}^k e^{-x\mathrm{ch}(jh)}\mathrm{sh}(\alpha jh)\sin(\beta jh), \tag{6}$$

where $k = [b/h]$.

Some calculations were conducted for the comparison of the Gauss-Legendre quadrature formulas and quadrature formulas of the trapezoidal kind with respect to speed and accuracy. It was seen that, under the choice of the same number of knots, the GAUSS-LEGENDRE formulas give a few more precise values of functions $\mathrm{Re}K_{1/2+i\beta}(x)$ and $\mathrm{Im}K_{1/2+i\beta}(x)$ for some values x and β. But the GAUSS-LEGENDRE formulas have the following negative properties: it's necessary to input a big number of knots and weights, it's difficult to perform the error estimation, it's difficult to estimate the number of knots of needed as a function of x and β. These difficulties are avoided by using a procedure based on the use of quadrature formulas of trapezoidal kind with "optimal" choice of the step.

Let's use the procedure [5,10] for the "optimal" choice of the step h of the trapezoidal quadrature formulas.

The results of computations show that the quadrature formulas of trapezoidal kind with "optimal" choice of the step to shorten considerably the number of knots required for the integration. The necessary number of knots depends significantly on x and β. It decreases with increasing x and increases with increasing β because these functions are rapidly damped as x increases and strongly oscillating for increasing β. So quadrature formulas of trapezoidal kind with "optimal" choice of the step yield 7-8 significant digits for the computation of $K_{1/2+i\beta}(x)$ except when the ratio x/β is small [10,13]. It's necessary to mention that the combination with ROMBERG integration can accelerate the convergence.

5   Technical Contribution

The values of coefficients of CHEBYSHEV polynomials expansions were taken with the precision on 15 - 20 digits. The check values were prepared and used for testing. The CHEBYSHEV polynomials expansions give 20 correct digits. But it's possible to use them for smaller precision also. It's necessary to take the coefficients with $n + 1$ decimal point and to reject expansion terms which coefficients smaller then $10^{-(n+1)}$ for the obtaining of the result with n correct digits.

The different codes were tested by the comparison of the numerical results on the boundaries of the domains of their validity. The finite differences were used for the verification of the numerical results also.

References

1. Ehrenmark, U.T.: The numerical inversion of two classes of KONTOROVICH-LEBEDEV transform by direct quadrature. J. of Comp. and Appl. Math. 61, 43–72 (1995)
2. Fabijonas, B.R., Lozier, D.W., Rappoport, J.M.: Algorithms and codes for the MACDONALD function: Recent progress and comparisons. Journ. Comput. Appl. Math. 161(1), 179–192 (2003)
3. Gautschi, W.: Computing the KONTOROVICH-LEBEDEV integral transforms and their inverses. BIT Numer. Math. 46, 21–40 (2006)

4. Gil, A., Segura, J., Temme, N.M.: Numerical methods for special functions. SIAM (2007)
5. Kiyono, T., Murashima, S.: A method of evaluation of the function $K_{is}(x)$. Mem. Fac. Engr. Kyoto Univ. 35(2), 102–127 (1973)
6. Lebedev, N.N., Skalskaya, I.P.: Some integral transforms related to KONTOROVITCH– LEBEDEV transforms, The questions of the mathematical physics, Leningrad, Nauka, 68–79 (1976) (in Russian)
7. Luke, Y.L.: Mathematical functions and their approximations. Academic Press, New York (1975)
8. Olver, F., Lozier, D., Boisvert, R., Clark, C. (eds.): NIST Handbook of mathematical functions. Cambridge University Press (2010)
9. Rappoport, J.M.: Tables of modified Bessel functions $K_{1/2+i\tau}(x)$, Nauka, Moscow (1979) (in Russian)
10. Rappoport, J.M.: The programs and some methods of the Macdonald's function computation. OVM AN U.S.S.R., Moscow (1991) (in Russian)
11. Rappoport, J.M.: Some results for modified KONTOROVITCH–LEBEDEV integral transforms. In: Proceedings of the 7th International Colloquium on Finite or Infinite Dimensional Complex Analysis, pp. 473–477. Marcel Dekker Inc (2000)
12. Rappoport, J.M.: The canonical vector-polynomials at computation of the BESSEL functions of the complex order. Comput. Math. Appl. 41(3/4), 399–406 (2001)
13. Rappoport, J.M.: Some numerical quadrature algorithms for the computation of the MACDONALD function. In: Proceedings of the Third ISAAC Congress. Progress in Analysis, vol. 2, pp. 1223–1230. World Scientific Publishing (2003)
14. Rappoport, J.M.: Dual integral equations for some mixed boundary value problems. In: Proceedings of the 4th ISAAC Congress. Advances in Analysis, pp. 167–176. World Scientific Publishing (2005)
15. Rappoport, J.M.: The properties, inequalities and numerical approximation of modified BESSEL function. Electronic Transactions on Numerical Analysis 25, 454–466 (2006)
16. Rappoport, J.M.: Some numerical methods for approximation of modified BESSEL functions. Proc. Appl. Math. Mech. 7(1), 2020017–2020018 (2007)
17. Rappoport, J.M.: About modified KONTOROVITCH–LEBEDEV integral transforms and their kernels, Imperial College of Science, Technology and Medicine, Department of Mathematics, London, preprint 09P/001, 31p. (2009)
18. Rappoport, J.M.: Some integral equations with modified BESSEL function. In: Proceedings of the 5th ISAAC Congress. More Progresses in Analysis, pp. 269–278. World Scientific Publishing (2009)
19. Rappoport, J.M.: CHEBYSHEV polynomial approximations for some hypergeometric systems, Isaac Newton Institute of Mathematical Sciences, Cambridge, preprint NI09059-DIS, 8p. (2009)
20. Temme, N.M.: On the numerical evaluation of the modified bessel function of the third kind. J. of Comp. Phys. 17, 324–337 (1975)
21. Yakubovich, S.B.: Index Transforms. World Scientific Publishing, Singapore (1996)
22. Yakubovich, S.B.: BEURLING'S theorems and inversion formulas for certain index transforms. Opuscula Mathematica 29(1), 93–110 (2009)
23. Zurina, M.I., Karmazina, L.N.: Tables of modified Bessel functions with imaginary index. VC AN U.S.S.R., Moscow (1967) (in Russian)

BetaSCP2: A Program for the Optimal Prediction of Side-Chains in Proteins

Joonghyun Ryu[1], Mokwon Lee[1], Jehyun Cha[1], Chanyoung Song[1], and Deok-Soo Kim[2],*

[1] Voronoi Diagram Research Center, Hanyang Univeristy, South Korea
[2] Hanyang Univeristy, South Korea
dskim@hanyang.ac.kr
http://voronoi.hanyang.ac.kr

Abstract. The *side-chain prediction problem* (SCP-problem), is a computational problem to predict the optimal structure of proteins by finding the optimal dihedral angles. The SCP-problem is one of key computational cornerstones for many important problems such as protein design, flexible docking of proteins, homology modeling, etc. The SCP-problem can be formulated as a minimization problem of an integer linear program which is NP-hard thus inevitably invites heuristic approach to find the solution. In this paper, we report a heuristic algorithm, called BetaSCP2, which quickly finds an excellent solution of the SCP-problem. The solution process of the BetaSCP2 is facilitated by the Voronoi diagram and its dual structure called the quasi-triangulation. The BetaSCP2 is entirely implemented using the Molecular Geometry engine called BULL! which has been developed by Voronoi Diagram Research Center (VDRC) in C++ programming language. The benchmark test of the BetaSCP2 with other programs is also provided. The BetaSCP2 program is available as both a stand-alone and a web server program from VDRC.

Keywords: protein structure/function, side-chain prediction, BetaSCP2, Voronoi diagram, quasi-triangulation, beta-complex.

1 Introduction

Bio-molecules such as protein, DNA, and RNA play important biological functions in the living bodies. It is a general consensus that the functions of molecules come from their geometric structures. Hence, there have been tremendous studies which tried to figure out the relationship between the structure and functions either computationally or experimentally.

Protein consists of linearly connected amino acids by a peptide bond where a water molecule leaves during each connection. Residue, the remaining part of each amino acid in a bonded sequence, consists of two parts: backbone and side-chain. While the backbone part is common to each residue, the side-chain

---

* Corresponding author.

H. Hong and C. Yap (Eds.): ICMS 2014, LNCS 8592, pp. 333–340, 2014.

is different depending on each type of residue. Protein structure is generally determined by the dihedral angles of some rotatable bonds in each residue because other variations except the angles are relatively negligible [11].

The side-chain prediction problem, abbreviated as SCP-problem, is a computational problem to predict the optimal structure of proteins by finding the optimal dihedral angles. Assuming that a backbone structure is fixed (i.e., the coordinates of the atoms in a backbone are given), the SCP-problem finds the optimal side-chain structure by predicting the dihedral angles in side-chains so that the total energy of the structure is minimized.

While each dihedral angle, in theory, can take any value between 0 and 360 degrees, it is well-known that there exists a preferred range of dihedral angle which maps to a representative angle through statistical analysis. A combination of such representatives for each residue is called a *rotamer* which is short for rotational isomer. Example rotamer is shown in Figure 1 . Figure 1(a) shows the chemical formula of aspartic acid whose side-chain has two dihedral angles (χ_1: between CH and CH_2; χ_2: between CH_2 and C) as shown in Fig. 1(b). The backbone atoms are also shown (H_2N, CH, C, OH) at the top of Fig. 1(b). Fig. 1(c) shows the union of nine rotamers for aspartic acid with different values of χ_1 and χ_2 (Hydrogens are usually ignored in the graphical visualization). Different residues have different number of dihedral angles. Hence, there could exist different rotamer set for each type of residue. The collection of rotamer sets for all residue types is called a *rotamer library* [12,23].

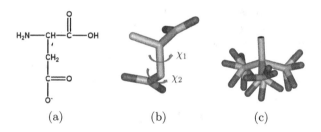

(a)   (b)   (c)

Fig. 1. Example rotamer: (a) the chemical formula of aspartic acid, (b) the two dihedral angles in the side-chain of aspartic acid, (c) the union of nine rotamers for aspartic acid

Consider a protein $\Pi = \{\rho_1, \rho_2, \ldots, \rho_n\}$ consisting of n residues. Each residue ρ has the backbone part β and the side-chain σ. Let $\mathcal{B} = \{\beta_1, \beta_2, \ldots, \beta_n\}$ and $\Sigma = \{\sigma_1, \sigma_2, \ldots, \sigma_n\}$ be the backbone parts and side-chains for the residues, respectively. Thus, β_i and σ_i constitute ρ_i and a protein can be represented as $\Pi = \mathcal{B} \cup \Sigma$. $\mathcal{B}$ is called the backbone of the protein Π. Suppose that the structure of Π (i.e. the atom coordinates of $\mathcal{B}$ and Σ) is completely defined. Then, the potential energy E_Π of Π for the SCP-problem is usually defined as follows [10]:

$$E_\Pi = E_{\mathcal{B}\Sigma} + E_{\Sigma\Sigma} \tag{1}$$

where $E_{B\Sigma}$ is the potential energy between protein backbone and the side-chain of each residue and $E_{\Sigma\Sigma}$ is the potential energy between a side-chain σ_i and another side-chain σ_j, $i \neq j$. The energy function used in this study is the van der Waals interaction between non-bonded atoms modeled by the following Lennard-Jones potential energy function

$$E = E_{B\Sigma} + E_{\Sigma\Sigma} \tag{2}$$
$$= \sum_{i\in\beta}\sum_{j\in\sigma}\{\frac{A_{ij}}{d_{ij}^{12}} - \frac{B_{ij}}{d_{ij}^{6}}\} + \sum_{i\in\sigma_i}\sum_{j\in\sigma_j}\{\frac{A_{ij}}{d_{ij}^{12}} - \frac{B_{ij}}{d_{ij}^{6}}\}$$

where d_{ij} is the Euclidean distance between the centers of a pair of atoms a_i and a_j. A_{ij} and B_{ij} are constants depending on atom types and the parameters in either AMBER [8] or CHARMM [4] could be used.

Given a rotamer library, the energy function E, and the structure of $\mathcal{B}$, the SCP-problem is to assign the optimal rotamer r^* to each residue ρ for σ so that E_Π of Eq. (1) is minimized [10]. Thus, the SCP-problem can be formulated as a minimization problem of an integer linear program[13,21,28] which is NP-hard thus necessarily invites a heuristic approach to find the solution. The NP-hardness of the SCP-problem is proved either by reducing satisfiability (SAT) problem to the decision problem of the SCP-problem [24,7] or by reducing the unconstrained quadratic 0-1 programming problem to the formulation of the SCP-problem [14]. The SCP-problem is one of key computational cornerstones for many important problems such as protein design [9,3], flexible docking of proteins [2,26], homology modeling [27], etc.

2   BetaSCP2 Algorithm

The SCP-problem can be formulated in an integer linear program (ILP) of Formulation 1 [13,6,14,21,28] where the constraints are not shown here due to space constraint. Two types of decision variables corresponding to two types of energies in Eq. (1) are defined as follows: x_{ij} decides whether rotamer j is accepted for residue i or not; x_{ijkl} decides whether the interaction between rotamer k of residue i and rotamer l of residue j is accepted or not. m_i represents the number of rotamers for a residue i in rotamer library.

Formulation 1. (ILP for SCP-problem)

$$Min. \sum_{i=1}^{n}\sum_{j=1}^{m_i} E_{B\Sigma}(i,j)x_{ij} + \sum_{i=1}^{n-1}\sum_{j=1}^{m_i}\sum_{k=i+1}^{n}\sum_{l=1}^{m_k} E_{\Sigma\Sigma}(i,j,k,l)x_{ijkl} \tag{3}$$

$$x_{ij} \in \{0,1\}, x_{ijkl} \in \{0,1\}. \tag{4}$$

While the size $\|x_{ik}\|$ of x_{ik} linearly increases with respect to $M_1 = \sum_{i=1}^{n} m_i$, the size $\|x_{ijkl}\|$ of x_{ijkl} dramatically grows according to $M_2 = \sum_{i=1}^{n-1}\sum_{k=i+1}^{n}(m_i \times m_k)$ where $M_2 \gg M_1$. Therefore, we prefer to cut down $\|x_{ijkl}\|$. BetaSCP1 algorithm, previously reported [25], decomposes the SCP-problem into subproblems and solves the ILP corresponding to each subproblem by using CPLEX solver [1]. While BetaSCP1 produces a solution very close to the global optimum, it is computationally very inefficient because it should invoke the CPLEX solver repeatedly.

BetaSCP2 algorithm decomposes the SCP-problem into small-sized subproblems and transforms each subproblem into simple geometric problem which is very efficiently solved via the theory of beta-complex, which is a derived from the Voronoi diagram of spheres (Refer to the Appendix for the brief discussion). The input of the procedure BetaSCP2 below is a protein backbone $\mathcal{B}$ from a PDB file and a rotamer library $\mathcal{R}$. In STEP 1, BetaSCP2 initially assigns a rotamer r^0 to each residue by considering the probability of the rotamer instances in $\mathcal{R}$. In STEP 2 and 3, the minimum enclosing sphere (MES) for each residue and the quasi-triangulation of MES set are computed by using BULL! engine [16]. In STEP 4, BetaSCP2 improves the rotamer r^0 initially assigned to each residue ρ by looking into the only nearby other rotamers which is defined by first-order Voronoi neighbors of Definition 1. Through STEP 4.1 and 4.2, BetaSCP2 computes the intersection volume of each candidate for ρ and choose the best one with minimum intersection volume. Note that BetaSCP2 exploits the intersection volume instead of potential energy. It can be easily proved that the rotamer with less intersection volume has the lower potential energy if there exists such an intersection. This important observation will be reported elsewhere in future.

BetaSCP2($\mathcal{B}$, $\mathcal{R}$)

1   1. assign an initial rotamer r^0 to each residue.
2   2. compute the MES of each residue.
3   3. compute the quasi-triangulation for the MES set.
4   4. **for** the first-order Voronoi neighbors $\mathcal{FN}$ of the MES for each residue ρ
5   4.1. **for** each rotamer r of ρ in $\mathcal{R}$
6   4.1.1. compute intersection volume $\mathcal{XV}(r)$ of r with
7   other rotamers in $\mathcal{FN}$.
8   4.2. find out the best rotamer r^* with minimum $\mathcal{XV}(r^*)$.

Figure 2 illustrates the BetaSCP2 algorithm: Figure 2(a) is an input protein structure; Figure 2(b) shows the backbone and the rotamer set corresponding to each residue by stick model; Figure 2(c) shows the rotamers to be initially assigned to each residue and their minimum enclosing spheres (MES) in yellow by space-filling model.

3   Benchmark Test

We have compared the BetaSCP2 algorithm with SCWRL4 [22] and CISRR [5] against 248 data from Protein Data Bank (PDB). The computational environment is as follows: Intel Core 2 Duo CPU E6850 (3.0GHz with 4GB RAM) and Windows 7. For comparing the solution quality, we evaluated Lennard-Jones potential (LJ) energy functions of the structures computed by each program. Figure 3(a) and (b) show LJ energies for three data sets from three programs.

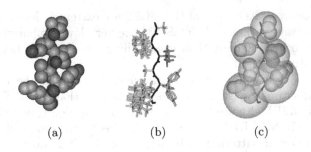

Fig. 2. Illustration of BetaSCP2 algorithm: (a) input protein structure (PDB code: 3FQP), (b) the backbone and rotamer sets of residues of (a), (c) the initially assigned rotamers to each residue and their minimum enclosing spheres

The X-axis represents the computed structures with respect to their residue sizes. The Y-axis represents LJ energies of the structures. Due to too high energies of SCWRL4, the difference between BetaSCP2 and CISRR are not clearly recognized in Figure 3(a); its zoom-up in Figure 3(b) clearly shows the powerful result of BetaSCP2.

BetaSCP2 produces energetically very stable structures, compared to both SCWRL4 and CISRR as shown in Figure 3(b). It turns out that BetaSCP2 outperforms SCWRL4 for 214 among 248 data. BetaSCP2 outperforms CISRR

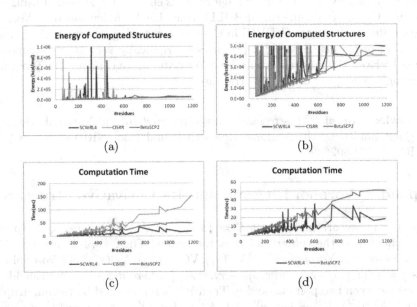

Fig. 3. Benchmark test for BetaSCP2, SCWRL4 and CISRR: (a) energies of the structures computed by BetaSCP2, SCWRL4, and CISRR, (b) the zoom-up of (a), (c) computation times for BetaSCP2, SCWRL4, and CISRR, and (d) computation times for BetaSCP2 and CISRR

for 52 data. However, the energy of BetaSCP2 are extremely lower for those 52 data. The mean and variance of BetaSCP2 are very low compared to those of CISRR. For the other 196 data, their energy differences between BetaSCP2 and CISRR are relatively small.

For comparing the computational efficiency, we counted the computation time of each program. Figure 3(c) shows the computation times of BetaSCP2, SCWRL4, and CISRR. Figure 3(d) shows the computation times of BetaSCP2 and SCWRL4 only. SCWRL4 is fastest among three programs. While BetaSCP2 shows a strongly linear pattern from both graphs, CISRR seems to have a super-linear pattern. BetaSCP2 is approximately three times faster than CISRR. The computation time taken by SCWRL4 relatively fluctuates wildly with respect to protein size.

4   Conclusion

In this paper, we reported the BetaSCP2 algorithm, and its implementation which quickly finds an excellent solution of the SCP-problem. The core idea of the BetaSCP2 algorithm is to transform the SCP-problem into a simple geometric problem whose solution process can be facilitated by the Voronoi diagram and its dual structure called the quasi-triangulation. Due to this idea, BetaSCP2 could improve the computational efficiency compared to BetaSCP1 without degrading the solution quality. The BetaSCP2 algorithm is entirely implemented using the Molecular Geometry engine called BULL! which has been developed by Voronoi Diagram Research Center (VDRC) in C++ programming language. Comparing with other programs, BetaSCP2 produces energetically very stable structures efficiently. Even though there could be other criteria, we compare the programs according to energy of computed structure.

Acknowledgement. This research was supported by the National Research Lab grant funded by the National Research Foundation (NRF) of Korea (No. 2012R1A2A1A05026395).

Appendix: Voronoi Diagram and Its Derivative Structures

Suppose that we are given a set $S = \{s_1, s_2, \ldots, s_n\}$ of spheres $s_i = (p_i, r_i)$ in $\mathbb{R}^3$ where p_i is the center and r_i is the radius. Then *Voronoi diagram* $\mathcal{VD}$ of S consists of n Voronoi cells: $\{\text{VC}(s_1), \text{VC}(s_2), \cdots, \text{VC}(s_n)\}$. A Voronoi cell $\text{VC}(s_i) = \{d(x, p_i) - r_i < d(x, p_j) - r_j, i \neq j\}$ where $d(x, y)$ is the Euclidean distance between two points x and y. Then $\mathcal{VD}$ is represented by the quadruplet $(V^\mathcal{V}, E^\mathcal{V}, F^\mathcal{V}, C^\mathcal{V})$: $V^\mathcal{V} = \{v_1^\mathcal{V}, v_2^\mathcal{V}, \ldots\}$, $E^\mathcal{V} = \{e_1^\mathcal{V}, e_2^\mathcal{V}, \ldots\}$, $F^\mathcal{V} = \{f_1^\mathcal{V}, f_2^\mathcal{V}, \ldots\}$, and $C^\mathcal{V} = \{c_1^\mathcal{V}, c_2^\mathcal{V}, \ldots c_n^\mathcal{V}\}$ are the sets of the Voronoi vertices (V-vertices), Voronoi edges (V-edges), Voronoi faces (V-faces), and Voronoi cells (V-cells) in $\mathcal{VD}$, respectively. Note that $\mathcal{VD}$ is different from the ordinary Voronoi diagram of sphere centers in many respects. One of the important properties for

$\mathcal{VD}$ is that $\mathcal{VD}$ reflects the size differences among spheres in Euclidean distance metric. For the details of $\mathcal{VD}$, refer to [15,20].

Given $\mathcal{VD}$, its dual structure *quasi-triangulation* $\mathcal{QT}$ is defined as follows: Each V-vertex maps to a tetrahedral cell simplex (q-cell); Each V-edge maps to a triangular face simplex (q-face); Each V-face maps to an edge simplex (q-edge); And each V-cell maps to a vertex simplex (q-vertex). Then $\mathcal{QT}$ is represented by the quadruplet $(V^{\mathcal{Q}}, E^{\mathcal{Q}}, F^{\mathcal{Q}}, C^{\mathcal{Q}})$: $V^{\mathcal{Q}} = \{v_1^{\mathcal{Q}}, v_2^{\mathcal{Q}}, \ldots v_n^{\mathcal{Q}}\}$, $E^{\mathcal{Q}} = \{e_1^{\mathcal{Q}}, e_2^{\mathcal{Q}}, \ldots\}$, $F^{\mathcal{Q}} = \{f_1^{\mathcal{Q}}, f_2^{\mathcal{Q}}, \ldots\}$, and $C^{\mathcal{Q}} = \{c_1^{\mathcal{Q}}, c_2^{\mathcal{Q}}, \ldots\}$ are the sets of the q-vertices, q-edges, q-faces, and q-cells in $\mathcal{QT}$, respectively.

$\mathcal{VD}$ and $\mathcal{QT}$ are equivalent to each other in mathematical and computational point of view. Given $\mathcal{VD}$, $\mathcal{QT}$ of S is computed in $O(m)$ time in the worst case where m is the number of the q-simplexes in $\mathcal{QT}$. The reverse conversion from $\mathcal{QT}$ to $\mathcal{VD}$ takes linear time in the worst case with respect to the number of the topological entities in $\mathcal{VD}$. For the details of $\mathcal{QT}$, see [19,17,18].

Definition 1 (First-order Voronoi Neighbors)
Suppose that we have $\mathcal{VD}$ of a set $S = \{s_1, s_2, \ldots, s_n\}$ of spherical balls. Let $F_i^{\mathcal{V}}$ be the V-faces bounding $VC(s_i)$ of a ball s_i. $F_j^{\mathcal{V}}$ is similarly defined. Given a spherical ball $s_i \in S$, a set $\mathcal{FN}_i = \{s_j \in S \mid F_i^{\mathcal{V}} \cap F_j^{\mathcal{V}} \neq \emptyset, i \neq j\}$ is called the first-order neighbors of s_i.

References

1. IBM ILOG CPLEX Optimizer (2013),
   http://www-01.ibm.com/software/commerce/optimization/cplex-optimizer/
2. Althaus, E., Kohlbacher, O., Lenhof, H.P., Müller, P.: A combinatorial approach to protein docking with flexible side chains. Journal of Computational Biology 9(4), 597–612 (2002)
3. Ashworth, J., Havranek, J.J., Duarte, C.M., Sussman, D., Monnat Jr., R.J., Stoddard, B.L., Baker, D.: Computational redesign of endonuclease DNA binding and cleavage specificity. Nature 441(7093), 656–659 (2006)
4. Brooks, B.R., Bruccoleri, R.E., Olafson, B.D., States, D.J., Swaminathan, S., Karplus, M.: CHARMM: A program for macromolecular energy, minimization, and dynamics calculations. Journal of Computational Chemistry 4(2), 187–217 (1983)
5. Cao, Y., Song, L., Miao, Z., Hu, Y., Tian, L., Jiang, T.: Improved side-chain modeling by coupling clash-detection guided iterative search with rotamer relaxation. Bioinformatics 27(6), 785–790 (2011)
6. Chazelle, B., Kingsford, C., Singh, M.: The side-chain positioning problem: A semidefinite programming formulation with new rounding schemes. In: Goldin, D.Q., Shvartsman, A.A., Smolka, S.A., Vitter, J.S., Zdonik, S.B. (eds.) Proceedings of the ACM International Conference Proceeding Series;Proceedings of the Paris C. Kanellakis Memorial Workshop on Pr., vol. 41, pp. 86–94 (2003)
7. Chazelle, B., Kingsford, C., Singh, M.: The inapproximability of side-chain positioning. Tech. rep., Princeton University (2004)
8. Cornell, W.D., Cieplak, P., Bayly, C.I., Gould, I.R., Merz Jr., K.M., Ferguson, D.M., Spellmeyer, D.C., Fox, T., Caldwell, J.W., Kollman, P.A.: A second generation force field for the simulation of proteins, nucleic acids, and organic molecules. Journal of the American Chemical Society 117, 5179–5197 (1995)

9. Dahiyat, B.I., Mayo, S.L.: De novo protein design: Fully automated sequence selection. Science 278(3), 82–87 (1997)
10. Desmet, J., Maeyer, M.D., Hazes, B., Lasters, I.: The dead-end elimination theorem and its use in protein side-chain positioning. Nature 356, 539–542 (1992)
11. Dunbrack Jr., R.L.: Rotamer libraries in the 21st century. Current Opinion in Structural Biology 12(4), 431–440 (2002)
12. Dunbrack Jr., R.L., Cohen, F.E.: Bayesian statistical analysis of protein side-chain rotamer preferences. Protein Science 6(8), 1661–1681 (1997)
13. Eriksson, O., Zhou, Y., Elofsson, A.: Side chain-positioning as an integer programming problem. In: Gascuel, O., Moret, B.M.E. (eds.) WABI 2001. LNCS, vol. 2149, pp. 128–141. Springer, Heidelberg (2001)
14. Fung, H., Rao, S., Floudas, C., Prokopyev, O., Pardalos, P., Rendl, F.: Computational comparison studies of quadratic assignment like formulations for the In silico sequence selection problem in De Novo protein design. Journal of Combinatorial Optimization 10(1), 41–60 (2005)
15. Kim, D.S., Cho, Y., Kim, D.: Euclidean Voronoi diagram of 3D balls and its computation via tracing edges. Computer-Aided Design 37(13), 1412–1424 (2005)
16. Kim, D.S., Cho, Y., Kim, J.K., Ryu, J., Lee, M., Cha, J., Song, C.: Bull! - the molecular geometry engine based on voronoi diagram, quasi-triangulation, and beta-complex. In: The 4th International Congress on Mathematical Software (2014)
17. Kim, D.S., Cho, Y., Sugihara, K.: Quasi-worlds and quasi-operators on quasi-triangulations. Computer-Aided Design 42(10), 874–888 (2010)
18. Kim, D.S., Cho, Y., Sugihara, K., Ryu, J., Kim, D.: Three-dimensional beta-shapes and beta-complexes via quasi-triangulation. Computer-Aided Design 42(10), 911–929 (2010)
19. Kim, D.S., Kim, D., Cho, Y., Sugihara, K.: Quasi-triangulation and interworld data structure in three dimensions. Computer-Aided Design 38(7), 808–819 (2006)
20. Kim, D., Kim, D.S.: Region-expansion for the Voronoi diagram of 3D spheres. Computer-Aided Design 38(5), 417–430 (2006)
21. Kingsford, C.L., Chazelle, B., Singh, M.: Solving and analyzing side-chain positioning problems using linear and integer programming. Bioinformatics 21(7), 1028–1036 (2005)
22. Krivov, G.G., Shapovalov, M.V., Dunbrack Jr., R.L.: Improved prediction of protein side-chain conformations with SCWRL4. PROTEINS: Structure, Function, and Bioinformatics 77(4), 778–795 (2009)
23. Lovell, S.C., Word, J.M., Richardson, J.S., Richardson, D.C.: The penultimate rotamer library. Proteins: Structure, Function, and Genetics 40(3), 389–408 (2000)
24. Pierce, N.A., Winfree, E.: Protein design is NP-hard. Protein Engineering 15(10), 779–782 (2002)
25. Ryu, J., Kim, D.S.: Protein structure optimization by side-chain positioning via beta-complex. Journal of Global Optimization 57(2), 217–250 (2013)
26. Schumann, M., Armen, R.S.: Systematic and efficient side chain optimization for molecular docking using a cheapest-path procedure. Journal of Computational Chemistry 34, 1258–1269 (2013)
27. Xiang, Z., Honig, B.: Extending the accuracy limits of prediction for side-chain conformations. Journal of Molecular Biology 311(2), 421–430 (2001)
28. Zhu, Y.: Mixed-integer linear programming algorithm for a computational protein design problem. Industrial and Engineering Chemistry Research 46, 839–845 (2007)

Computation of an Improved Lower Bound to Giuga's Primality Conjecture

Matthew Skerritt

CARMA, University of Newcastle, NSW, Australia
matthew.skerritt@newcastle.edu.au

Abstract. Our most recent computations tell us that any counterexample to Giuga's 1950 primality conjecture must have at least 19,908 decimal digits. Equivalently, any number which is both a Giuga and a Carmichael number must have at least 19,908 decimal digits. This bound has not been achieved through exhaustive testing of all numbers with up to 19,908 decimal digits, but rather through exploitation of the properties of Giuga and Carmichael numbers. This bound improves upon the 1996 bound of Borwein, Borwein, Borwein, and Girgensohn. We present the algorithm used, and discuss technical challenges and challenges to further computation.

Keywords: Giuga's conjecture, normality of primes, branch and bound.

1 Introduction

Giuseppe Giuga formulated his now well-known prime number conjecture (see Conjecture 1) in his 1950 paper [5]. The conjecture hypothesises a primality test, although we note that the test is inefficient when compared to commonly used modern primality tests. However, the conjecture is interesting for other mathematical reasons as it relates to Bernoulli numbers as shown in Takahashi Agoh's 1995 paper [1].

Conjecture 1 (Giuga's Conjecture). An integer, $n > 1$, is prime if and only if

$$s_n := \sum_{k=1}^{n-1} k^{n-1} \equiv -1 \,(\mathrm{mod}\, n). \tag{1}$$

Note that if p is prime, then $s_p \equiv p-1 \,(\mathrm{mod}\, p)$ is an immediate consequence of Fermat's little theorem. The interesting question then is whether there exists any composite n with $s_n \equiv n-1 \,(\mathrm{mod}\, n)$. Any such number would be a counterexample to the conjecture.

Herein we present an algorithm to compute lower bounds for a counterexample, used by Borwein et all in 1996 [2] to compute a lower bound of approximately $10^{13{,}886}$. We have extended this work by re-implementing the algorithm using modern techniques and have produced a considerably improved bound of approximately $10^{19{,}907}$ [4].

H. Hong and C. Yap (Eds.): ICMS 2014, LNCS 8592, pp. 341–345, 2014.

2   Underlying Theory

2.1   Giuga and Carmichael Numbers

It can be shown [2] that any number n which satisfies $s_n \equiv n - 1 \pmod{n}$ must satisfy two conditions.

$$p \mid \left(\frac{n}{p} - 1\right) \quad \text{for all prime divisors } p \text{ of } n \tag{2}$$

$$(p - 1) \mid \left(\frac{n}{p} - 1\right) \quad \text{for all prime divisors } p \text{ of } n \tag{3}$$

Composite numbers satisfying (2) are called *Giuga numbers*. They are square-free and satisfy the property that

$$\sum_{p \mid n} \frac{1}{p} - \prod_{p \mid n} \frac{1}{p} \in \mathbb{N} \tag{4}$$

Note that it follows, then, that $\sum_{p \mid n} \frac{1}{p} > 1$

Composite, square-free numbers satisfying (3) are called *Carmichael numbers*. The divisors of any Carmichael number are *normal* in the sense of the following definition.

Definition 1. *A set of odd primes, P, is* normal *if no $p, q \in P$ satisfies $p \mid (q - 1)$.*

Example 1. The set $\{3,5,17\}$ is normal because $4 = 5 - 1$ is not divisible 3, and $16 = 17 - 1$ is divisible by neither 3 nor 5.

Example 2. The set $\{3,5,7\}$ is *not* normal, because $6 = 7 - 1$ is divisible by 3.

It is unknown whether there is any Giuga number that is also a Carmichael number, however the existence of one would constitute a counterexample to Giuga's conjecture. Giuga's conjecture can, therefore, be re-written to say that there is no number that is both a Giuga number and a Carmichael number.

2.2   Exclusion Bounds

We compute a lower bound for a counterexample to Giuga's conjecture by exploiting the property (4) of Giuga numbers and the normal property of prime divisors of the Carmichael numbers. Considering the two properties in concert greatly reduces the amount of computation which must be performed in order to compute a lower bound.

Notation 1

- Let q_k denote the kth odd prime.
- Let $\mathcal{N}$ denote a normal set of odd primes.

A binary tree is constructed whose vertices are labelled by normal sets of primes. The root vertex, considered to be at depth 1, is labelled by the empty set. A vertex at depth d of the tree labelled by $\mathcal{N}$ will always have a child vertex labelled by $\mathcal{N}$, and will have a child vertex labelled by $\mathcal{N} \cup q_d$ if and only if $\mathcal{N} \cup q_d$ is normal.

Each vertex is assigned a score, called the j-value of that vertex. The j-value is calculated by first constructing a new set as follows. If $\mathcal{N}$ is a normal set of primes labelling a vertex at depth d of the tree, then we construct the set $T_d(\mathcal{N})$ by starting with $T_d(\mathcal{N}) = N$ and, for each $k \geq d$ we add q_k to $T_d(\mathcal{N})$ so long as $\mathcal{N} \cup \{q_k\}$ is normal. The primes are added one at a time, from smallest to largest, until

$$\sum_{q \in T_d(\mathcal{N})} \frac{1}{q} > 1$$

The reason for this stopping criteria is related to (4), above, and is explained in [2,4]. The j-value of the vertex is simply the cardinality of the set $T_d(\mathcal{N})$.

For a fixed depth d of the tree, the minimum j-value of all vertices at that depth, denoted j_d, is a lower bound for the number of prime numbers needed for a counterexample to Giuga's conjecture. The lower bound for the counterexample itself is computed as

$$\prod_{k=1}^{j_d} q_k$$

The branch and bound nature of the algorithm comes from the fact that the sequence of j-values is non-decreasing. That is, the j-value of a child vertex is always greater than or equal to the j-value of its parent. This allows us to forego computation of many sub-trees whose parent vertex already has a larger j-value than some known lower j-value. Doing so increases the speed of computation enormously for a fixed tree-depth.

Example 3. The tree, with j-values for each vertex, expanded and bounded using the branch and bound algorithm can be seen in Figure 1, from which we can see that $j_5 = 127$. Consequently, any counterexample to Giuga's conjecture must have at least 127 prime factors, and must therefore be greater than

$$\prod_{k=1}^{127} q_k \approx 4.962053073 \cdot 10^{297}$$

See [2,4] for more information regarding this algorithm.

3   Implementation Details and Technical Challenges

In 1996, Borwein et al [2] used the above algorithm to compute j_{100} using the Computer Algebra System *Maple*. Borwein reports in [3] that the computations

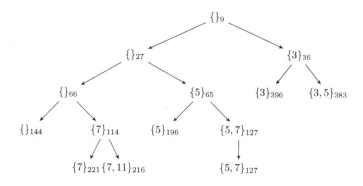

Fig. 1. Trimmed tree with j-values computed to depth 5

for j_k required "a few CPU hours for each k" when k was near 100. Using a C++ implementation (which took two months to produce) they were able to compute j_{135}, taking "303 CPU hours" [2]. The C++ implementation then crashed irrevocably before doing any extra work [3].

In 2012 we created a multi-threaded C++ implementation which was used to compute j_{311} yielding the improved bound described in Section 1. The computation ultimately took more than ten days to complete [4]. Contrary to Borwein's stated experience in 2006, the initial 2012 C++ implementation was quick and painless to write, and performed reliably. Threading was achieved through Apple's "Grand Central Dispatch" library, and was similarly quick and painless to learn and to implement.

The primary computation challenge came in the form of memory usage. Although the branch and bound algorithm significantly reduces the number of vertices of the tree to be computed, the tree still appears to grow exponentially as depth increases (see below). As available memory ran out the computation began paging, which significantly slowed the computation down. To minimise paging, vertex data was first written to disk when processing was complete, and later kept entirely on disk except when explicitly being processed. Vertex data was stored using the Berkeley database library because it already incorporated caching as well as sorting of data.

Ultimately, the paging problem was never solved and only postponed. During the final computations of j_{311} the paging problem returned in spite of the fact that only currently processing data should have been in memory. It is uncertain whether this is a limitation of the Grand Central Dispatch paradigm for such large scale computations, or whether there is a bottleneck where newly created vertex data is written to disk. Efforts to diagnose and fix the problem are ongoing, and no further computations have been performed.

The serious challenge to further computation is the fact that, although the use of the j-values allows us to avoid computing the entire binary tree, the vertices of the trimmed tree still appear to grow exponentially as the depth increases. When computing j_{311} measurements were taken for the amount of time taken to compute j_k for each $1 \leq k \leq 311$, and of the file size of the output file produced by the computation. Both measurements, when plotted on a log-linear scale (see [4]) appear to be linear, which suggests an exponential relationship.

4   Further Work

Work is ongoing to find the source of the paging problems, and to fix them so that computation can resume. The exponential growth in output file size is hoped to be able to be temporarily fixed by compressing the output. It is hoped that doing so will delay hard drive space from becoming the main barrier to more computation long enough that the apparently exponential computation times will be the primary barrier. It is hoped that a parallel implementation, either via GPU acceleration, or via grid computation might delay the inevitable computation time barrier.

References

1. Agoh, T.: On Giuga's conjecture. Manuscripta Math. 87(4), 501–510 (1995)
2. Borwein, D., Borwein, J.M., Borwein, P.B., Girgensohn, R.: Giuga's conjecture on primality. Amer. Math. Monthly 103(1), 40–50 (1996)
3. Borwein, J.M., Bailey, D.H., Girgensohn, R.: Experimentation in Mathematics. Computational Paths to Discovery. A.K. Peters Ltd. (2004)
4. Borwein, J., Maitland, C., Skerritt, M.: Computation of an improved lower bound to Giuga's primality conjecture. Integers 13, #A67 (2013)
5. Giuga, G.: Su una presumibile proprietà caratteristica dei numeri primi. Ist. Lombardo Sci. Lett. Rend. Cl. Sci. Mat. Nat (3) 14(83), 511–528 (1950)

An Extension and Efficient Calculation of the Horner's Rule for Matrices

Shinichi Tajima[1], Katsuyoshi Ohara[2], and Akira Terui[1]

[1] Faculty of Pure and Applied Sciences, University of Tsukuba, Japan
{tajima,terui}@math.tsukuba.ac.jp
http://researchmap.jp/aterui (Terui)
[2] Faculty of Mathematics and Physics, Kanazawa University, Japan
ohara@air.s.kanazawa-u.ac.jp
http://air.s.kanazawa-u.ac.jp/~ohara/

Abstract. We propose an efficient method for calculating "matrix polynomials" by extending the Horner's rule for univariate polynomials. We extend the Horner's rule by partitioning it by a given degree to reduce the number of matrix-matrix multiplications. By this extension, we show that we can calculate matrix polynomials more efficiently than by using naive Horner's rule. An implementation of our algorithm is available on the computer algebra system Risa/Asir, and our experiments have demonstrated that, at suitable degree of partitioning, our new algorithm needs significantly shorter computing time as well as much smaller amount of memory, compared to naive Horner's rule. Furthermore, we show that our new algorithm is effective for matrix polynomials not only over multiple-precision integers, but also over fixed-precision (IEEE standard) floating-point numbers by experiments.

Keywords: The Horner's rule, matrix polynomials.

1 Introduction

Give a field $\mathbb{K}$, a square matrix A of dimension n, a square matrix M or a column vector v of dimension n and a univariate polynomial $g(\lambda)$ of degree m (with all the mathematical objects defined over $\mathbb{K}$), we discuss efficient calculation of a "matrix polynomial", or a matrix $g(A)M$ or a column vector $g(A)v$. By calculating "matrix polynomials", we mean to evaluate a univariate polynomial over a field at a given value that is a square matrix over the same field by the celebrated Horner's rule.

For a given matrix, we have developed algorithms with exact arithmetic, based on residue analysis of the resolvent of the matrix, for various computations so far, including spectral decomposition, calculation of (pseudo) annihilating polynomials and eigenvectors. Calculation of matrix polynomials is at the core of these computations, thus it is important to establish an efficient algorithm for calculating matrix polynomials to increase efficiency of the overall algorithms.

While the Horner's rule has been known as an algorithm for sequential computation, several extensions have been proposed for use of the Horner's rule

H. Hong and C. Yap (Eds.): ICMS 2014, LNCS 8592, pp. 346–351, 2014.

on parallel computers ([1], [2], [4], [6]). However, our idea and/or intention for making an efficient algorithm for the Horner's rule is much different from theirs because we need to establish efficient algorithm for matrix polynomials that was not at least a main objective in previous research to the best of the authors' knowledge.

Computing time of the Horner's rule for matrices is dominated by multiplication of a matrix by another matrix, thus we aimed to reduce the number of matrix-matrix multiplications. With this strategy, we extend the Horner's rule by partitioning it by a given degree to reduce the number of matrix-matrix multiplications. By this extension, we show that we can calculate matrix polynomials more efficiently than by using naive Horner's rule.

We estimate the arithmetic time complexity of our new algorithm and derive a degree of partition of the Horner's rule which makes our extension the most efficient. This estimate has been verified with experiments with our implementation on the computer algebra system Risa/Asir, and our experiments have demonstrated that, at suitable degree of partition, our new algorithm needs significantly shorter computing time as well as much smaller amount of memory, compared to naive Horner's rule. Furthermore, we show that our new algorithm is effective for matrix polynomials not only over multiple-precision integers, but also over fixed-precision (IEEE standard) floating-point numbers by experiments.

2   The Horner's Rule for Matrices

The Horner's rule is known as an efficient algorithm for evaluating a univariate polynomial at a number [3]. Let $\mathbb{K}$ be a field, $a \in \mathbb{K}$ and $f(x) \in \mathbb{K}[x]$ be a univariate polynomial of degree m, then we see that arithmetic time complexity over $\mathbb{K}$ for calculating $f(a)$ by the Horner's rule is given as $O(m)$ (note that the number of multiplications is given as $O(m)$).

Let A and M be square matrices of dimension n over $\mathbb{K}$ and $g(\lambda) \in \mathbb{K}[\lambda]$ be a univariate polynomial of degree m, then we discuss calculating $g(A)M$ by the Horner's rule (we call it "the Horner's rule for matrices"). In this paper, we assume naive method for multiplication of two matrices over $\mathbb{K}$, in which arithmetic time complexity is given as $O(n^3)$ for multiplying square matrices of dimension n [3]. Then, in this case, we estimate arithmetic time complexity over $\mathbb{K}$ for calculating $g(\lambda)$ as $O(mn^3)$ since the number of matrix-matrix multiplications that dominates the entire calculation is given as $O(m)$. Furthermore, the arithmetic complexity becomes $O(n^4)$ if we have $m \simeq n$.

3   Efficient Calculation of Matrix Polynomials with Extension of the Horner's Rule

Our key idea for making efficient calculation of the Horner's rule is to reduce the number of matrix-matrix multiplications by partitioning the Horner's rule by a certain degree. The following example illustrates our idea.

Example 1. Let A and M be square matrices over $\mathbb{K}$ of dimension n and $g(\lambda) \in \mathbb{K}[\lambda]$ be a univariate polynomial, given as

$$g(\lambda) = a_{18}\lambda^{18} + a_{17}\lambda^{17} + \cdots + a_1\lambda + a_0, \quad a_i \in \mathbb{K}, \quad a_{18} \neq 0.$$

With the naive Horner's rule, we calculate $g(A)M$ as

$$\begin{aligned}
g(A)M &= a_{18}A^{18}M + a_{17}A^{17}M + \cdots + a_1AM + a_0M \\
&= A(A(\cdots(A(a_{18}AM + a_{17}M) + \cdots) + a_1M) + a_0M,
\end{aligned}$$

in which we employ 18 times of matrix-matrix multiplications.

Now, we carry out "partitioned" Horner's rule as follows, for example, partitioned by degree 4.

[**Step 1**]. Calculate $A^4 = (A^2)^2$ and store it in advance.
[**Step 2**]. Calculate A^3M, A^2M, AM and store them along with M in advance.
[**Step 3**]. Execute the Horner's rule with adding polynomials partitioned by degree 4 as follows.

$$\begin{aligned}
g(A)M = A^4\{A^4\{A^4\{A^4(\underline{a_{18}A^2M + a_{17}AM + a_{16}M})\} \\
+ (\underline{a_{15}A^3M + a_{14}A^2M + a_{13}AM + a_{12}M})\} \\
+ (\underline{a_{11}A^3M + a_{10}A^2M + a_9AM + a_8M})\} \\
+ (\underline{a_7A^3M + a_6A^2M + a_5AM + a_4M})\} \\
+ (\underline{a_3A^3M + a_2A^2M + a_1AM + a_0M}) \quad (1)
\end{aligned}$$

Note that we need only additions and scalar multiplications on matrices for calculating the underlined polynomials in Eq. (1) because we have already prepared A^3M, A^2M, AM and M in advance. As a consequence, we have the number of matrix-matrix multiplications in each step as follows: 2 times in Step 1, 3 times in Step 2 and 4 times in Step 3, thus 9 times in total. Note that the total number of matrix-matrix multiplications in the partitioned Horner's rule is reduced to half of that in the naive one.   □

In general, for square matrices A and M of degree n and a polynomial $g(\lambda)$ of degree m defined as $g(\lambda) = a_m\lambda^m + a_{m-1}\lambda^{m-1} + \cdots + a_1\lambda + a_0$, we present calculation of $g(A)M$ with the partitioned Horner's rule of degree $d = 2^b$ (with $d \leq m$) as Algorithm 1 below.

Algorithm 1 (A partitioned Horner's rule)

Inputs

    — A: square matrix of dimension n,
    — M: square matrix or column vector of dimension n,
    — $g(\lambda)$: univariate polynomial of degree m, given as

$$g(\lambda) = a_m\lambda^m + a_{m-1}\lambda^{m-1} + \cdots + a_1\lambda + a_0;$$

    – $d = 2^b$: degree of partition (with $d \leq m$);

Output $g(A)M$: square matrix of dimension n (if M is a matrix) or column vector of dimension n (if M is a vector);

[**Step 1**] Calculate $A^d = A^{2^b} = (\cdots(A^2)^2\cdots)^2$ and store it in advance;

[**Step 2**] Calculate $A^{d-1}M, A^{d-2}M, \ldots, AM$ and store them along with M in advance;

[**Step 3**] Execute the Horner's rule with adding polynomials partitioned by degree d as follows:

$$g(A)M = A^d\{\cdots\{A^d(a_m A^r M + \cdots + a_{qd+1}AM + a_{qd}M)\}$$
$$+ (a_{qd-1}A^{d-1}M + \cdots + a_{(q-1)d+1}AM + a_{(q-1)d}M)\}$$
$$+ \cdots$$
$$+ (a_{d-1}A^{d-1}M + \cdots + a_1 AM + a_0 M),$$

where q and r represent the quotient and the remainder of m divided by d, respectively. □

We calculate the number of matrix-matrix multiplications in Alg. 1 using b, d, m, as follows: b times in Step 1, $d - 1$ times in Step 2, $\lfloor m/d \rfloor$ times in Step 3, thus

$$b + d + \lfloor m/d \rfloor - 1 \tag{2}$$

in total, which can be represented as $T(b, m)$ with putting $d = 2^b$ as

$$T(b, m) = b + 2^b + \lfloor m/2^b \rfloor - 1. \tag{3}$$

3.1 Estimating of the Optimal Value of the Degree of Partition d

We can estimate the optimal value of the degree of partition in Alg. 1, as follows. Note that we only consider the number of matrix-matrix multiplications for the Horner's rule itself, which excludes those in the Step 1[1]. Then, from Eq. (2), we can estimate the number of matrix-matrix multiplications as

$$\frac{m}{d} + d - 1. \tag{4}$$

If we fix the degree of the polynomial m, then, by the inequality of arithmetic and geometric means, Eq. (4) attains the minimum when $m/d = d$ or $d = \sqrt{m}$. Thus, with $d = 2^b$, we estimate the optimal value for d as the maximum value of 2^b that does not exceed $\sqrt{m}$ or 2^b that is the nearest to $\sqrt{m}$.

---

[1] In calculation of minimal polynomials and/or pseudo annihilating polynomials that are among our primary intention of application of our extension of the Horner's rule, we execute the Horner's for matrices for as many times as the dimension or the degree of irreducible factors in the characteristic polynomial of the given matrix, while we calculate A^d just for once. Thus, we regard that the computing time for A^d is negligible compared to those for the Horner's rule.

4 Experiments

We have implemented our algorithm in the above on the computer algebra system Risa/Asir [5]. In this paper, we only explain the setting of our experiments because of limited space, and we show brief results of the experiments in the next section. Let A and M be square matrices of dimension n, v be a column vector of dimension n and $g(\lambda)$ be a univariate polynomial of degree m. In each experiment, we have measured computing time and the amount of memory used for computation by changing the degree of partition in the Horner's rule.

We have carried out the following experiments with computing environment as follows: Quad-core AMD Opteron 2350 at 2.0 GHz with RAM 4GB running Linux 2.6.26-amd64.

1. The Horner's rule with matrix-matrix multiplications for matrices A and M of dimension 64 and a polynomial $g(\lambda)$ of degree 64 with randomly-generated 64-bit integers to calculate $g(A)M$. We have changed the degree of partition for the Horner's rule d as 1 (without partition), 2, 4, 8 and 16. We have also tested our algorithms for matrices A and M of dimension 128 and a polynomial $g(\lambda)$ of degree 128.

2. The Horner's rule with matrix-vector multiplications for a matrix A and a column vector v of dimension 64 and a polynomial $g(\lambda)$ of degree 64 with randomly-generated 64-bit integers to calculate $g(A)v$. We have changed the degree of partition for the Horner's rule d as 1 (without partition), 2, 4, 8, 16 and 32.

3. The Horner's rule with matrix-matrix multiplications for matrices A and M of dimension 64 and a polynomial $g(\lambda)$ of degree 128 with randomly-generated IEEE double-precision standard floating-point numbers to calculate $g(A)M$. We have changed the degree of partition for the Horner's rule d as 1 (without partition), 2, 4, 8, 16, 32 and 64.

5 Conclusions

In this paper, we have proposed an extension of the Horner's rule for efficient calculation by partitioning it by a given degree. Our experiments in the previous section have shown the following results.

- Our new algorithm is especially effective for the Horner's rule with matrix-matrix multiplications. Furthermore, with appropriate degrees of partition, we can reduce not only computing time but also the amount of memory used.
- We have the above effect also in the Horner's rule with matrix-vector multiplications, although computing time for calculating matrix powers increases as the degree of partition increases.
- Our algorithm is also effective for the Horner's rule for the matrices over fixed precision (IEEE double-precision standard) floating-point numbers.

It seems that the cost for calculating matrix powers increases for larger degree of partitions, thus the overall algorithm becomes less effective in such degrees. However, we emphasis that our algorithm is still effective for our desired calculations by the following reason. In calculation of the minimal polynomials and/or the pseudo annihilating polynomials of a given matrix A, we calculate $g(A)v$ many times by changing a vector v and a polynomial $g(\lambda)$, while we calculate powers of A just once and reuse them each time we calculate $g(A)v$ for many different vs and $g(\lambda)$s. Thus, as the number of different vs increases, the relative cost for calculating powers of A becomes smaller and our algorithm will contribute to make the overall calculation more efficient.

We expect that our algorithm will be effective for matrices and/or polynomials of large dimension and/or degree, respectively, such as from hundreds to thousands, although we have actually tested our algorithm for matrices and polynomials of "moderate" degrees less than a hundred. Thus, it is among our future work to show the effectiveness of our algorithm for matrices and polynomials of large dimensions and degrees, respectively, by more experiments. Furthermore, while we give the degree d of partition in our algorithm as a power of 2 at present, it is another open problem to find the optimal value for d for matrices of larger dimensions (such as whether the power of 2 is sufficient or there exists more optimal value between 2^b and 2^{b+1}).

In the viewpoint of implementation, we have been working on incorporating practical methods for optimizing our algorithm including parallel computation, and it is important to evaluate time complexity of our algorithm properly with those methods. While we have established complexity analysis based on arithmetic complexity over the field in this paper, we will need to analyze bit complexity (with evaluating the magnitude of elements in matrices during calculation) of the algorithm, since we intend to establish efficient algorithms for matrices and polynomials over multiple-precision integers. Furthermore, with our Horner's rule, we intend to develop efficient algorithms for calculating the minimal polynomials and/or the pseudo annihilating polynomials of matrices.

References

1. Dorn, W.S.: Generalizations of Horner's Rule for Polynomial Evaluation. IBM Journal of Research and Development 6(2), 239–245 (1962)
2. Dowling, M.L.: A Fast Parallel Horner Algorithm. SIAM Journal on Computing 19(1), 133–142 (1990)
3. Knuth, D.: The Art of Computer Programming, 3rd edn. Seminumerical Algorithms, vol. 2. Addison-Wesley (1998)
4. Maruyama, K.: On the Parallel Evaluation of Polynomials. IEEE Transactions on Computers C-22(1), 2–5 (1973)
5. Noro, M.: A Computer Algebra System: Risa/Asir. In: Joswig, M., Takayama, N. (eds.) Algebra, Geometry and Software Systems, pp. 147–162. Springer (2003)
6. Munro, I., Paterson, M.: Optimal algorithms for parallel polynomial evaluation. Journal of Computer and System Sciences 7(2), 189–198 (1973)

What Is New in CoCoA?

John Abbott[1] and Anna Maria Bigatti[2]

[1] Univ. degli Studi di Genova, Italy
abbott@dima.unige.it
http://www.dima.unige.it/~abbott
[2] Univ. degli Studi di Genova, Italy
bigatti@dima.unige.it
http://www.dima.unige.it/~bigatti

Abstract. CoCoA is a well-established Computer Algebra System for Computations in Commutative Algebra, and specifically for Gröbner bases.

In the last few years CoCoA has undergone a profound change: the code has been totally re-written in C++, and includes an integral open source C++ library, called CoCoALib.

The new CoCoA-5 language still resembles the CoCoA-4 language, and maintains or improves the naturalness and ease of use for which CoCoA-4 was noted, but the clearly defined semantics of the new language make it both more robust and more flexible than CoCoA-4.

Also its C++ mathematical core, CoCoALib, focusses on ease of use and robustness, so that other software can use it as a library for multivariate polynomial computations and other Commutative Algebra operations.

Moreover the internal design makes it easy to render new extensions to the library accessible also via the interactive CoCoA-5 system.

1 Introduction

CoCoA is a well-established Computer Algebra System dating back to 1989. It was originally created as a laboratory for studying Computational Commutative Algebra, especially Gröbner bases and Buchberger's Algorithm, and still today maintains this tradition.

In the last few years CoCoA has undergone a profound change in its internal design: its "mathematical expertise" resides in a completely new, open source software library [2], a brand new interpreter grants easy, interactive access to CoCoA's capabilities [1], and an OpenMath-based server offers "remote procedure call" capabilities. All the code is in C++ and distributed under the GPL3 licence (*i.e.* free and open source). The aim of the design was to offer simple and flexible access to the mathematical operations embodied in the library.

We presented this design outline at ICMS 2010, where we showed a first prototype of CoCoA-5, which then implemented just a subset of the new CoCoA language, and was limited to some fairly basic operations on polynomials.

Four years later, 2014: what is new in CoCoA?

H. Hong and C. Yap (Eds.): ICMS 2014, LNCS 8592, pp. 352–358, 2014.

We have finished the design of the completely new CoCoALanguage, and its implementation in the new interpreter — it is simple and robust enough that it is used regularly by students.

In terms of mathematical ability, CoCoA-5 can now do (almost) everything that CoCoA-4 can, and faster. More importantly, it can also do many things that CoCoA-4 cannot.

The openness and clean design of CoCoALib and CoCoA-5 are intended to encourage contributions and extensions by users outside the main development team. This has been vindicated several times.

The internal software design makes it easy to render new extensions to Co-CoALib (whether by the authors, or by contributors) accessible via the interactive CoCoA-5 system, so there's no need to wrestle with C++ to use them.

The openness and clean design of CoCoALib and CoCoA-5 is intended to encourage extensions by users outside the main development team. For instance, (1) new contributed functions integrated in CoCoALib operations on Mayer-Vietoris trees (Saenz de Cabezon), Janet Bases (Albert, Seiler)

(2) integration of external libraries Normaliz (Bruns, Ichim, Söger) – almost "symbiotic", Frobby (Roune)

2 The New CoCoA-5 Language and Interpreter

CoCoA-4 was noted for its ease of use, and the naturalness of its language. However, it was too limited, and did have some "grey areas". As a consequence we designed a new language for CoCoA-5.

We designed the new CoCoA-5 language "from scratch" striking a balance between backward-compatibility (for existing CoCoA-4 users) and greater expressibility with a richer and more solid mathematical basis (eliminating those "grey areas").

Superficially the new CoCoA-5 language resembles that of CoCoA-4; it is largely backward compatible, and maintains or improves the naturalness and ease of use for which CoCoA-4 was noted — we are very aware that a number of CoCoA users are mathematicians with only limited programming experience. The clearly defined semantics of the new language make it both more robust and more flexible than CoCoA-4.

2.1 The Languages in CoCoA-4 and in CoCoA-5

The design of the CoCoA-5 language began by listing what we liked in CoCoA-4, what we did not like, and the new features we wished to have.

After one year of studying our three lists with F. Figari we established a good outline design of the new language which also achieved a high degree of backward compatibility, so as not to alienate existing CoCoA-4 devotees. For example, CoCoA-5 variables are dynamically typed.

Among the new features we find rings and functions are "normal objects", more properly called "first class values", and so can be assigned and passed as arguments — this avoids some of the awkward convolutions needed in CoCoA-4. Another new type of value is *ring homomorphism* which replaces the cumbersome mechanism needed previously.

The main feature we removed was "invisible multiplication" (*i.e* by juxtaposition), namely xy means x*y. While some users fiercely opposed its removal, we finally decided it was too limiting and problematic to support it in CoCoA-5 — in fact, it often caused confusion among new users too, *e.g* F (x+1) is a product but F(x+1) is a function call. Moreover this caused the constraint on the name of the indeterminates which had to be just single lower-case letters. Nevertheless, in recognition of the fact that "invisible multiplication" is sometimes handy, CoCoA-5 does accept the old CoCoA-4 syntax in expressions delimited by triple asterisks: *e.g.* you may write

```
I := *** Ideal(2x^2y-z, 3xz-5yz^3) ***;
```

Once the design of the new language was almost complete, we started a collaboration with G. Lagorio who implemented the parser and interpreter for the new language; naturally, this led to some refinements in the design! An important design goal for the new parser/interpreter was to produce genuinely helpful error messages — this was definitely a weak point of CoCoA-4. To achieve this, Lagorio's careful implementation was written entirely by hand, and is undoubtedly a vast improvement (indicating clearly what the problem was, and where it was encountered). CoCoA-5 comprises effectively Lagorio's interpreter together with CoCoALib.

Here is a typical example of an error message from CoCoA-5; note that the error was actually signalled by CoCoALib, and the interpreter caught the exception and "translated" it to human-readable form:

```
# X := isqrt(-99);
ERROR: Value must be non-negative
X := isqrt(-99);
     ^^^^^^^^^^
```

We regard good error messages (and warnings) as important assistance to both to a new user learning the language, and to a CoCoA-4 user wanting to update and clean his old code written for CoCoA-4.

2.2   The Language in CoCoALib and in CoCoA-5

We envisage that researchers and advanced users of CoCoA wishing to tackle hard computations will develop a prototype implementation in the convenient interpreted environment of CoCoA-5, and when the code is working properly, they will translate it into C++ (using CoCoALib) for better performance.

Bearing this in mind, a secondary design goal of the new CoCoA-5 language was to make it relatively easy to convert CoCoA-5 code into C++ code built upon CoCoALib (without requiring deep knowledge of advanced features of C++). That said, CoCoALib offers a richer programming environment, but also demands greater discipline from the programmer.

To facilitate the conversion into C++ we have, whenever possible, used the same function names in both CoCoA-5 and CoCoALib. We have also preferred traditional "functional" syntax in CoCoALib over object oriented "method dispatch" syntax, *e.g.* in CoCoALib we define `deg(f)` rather than `f.deg()`.

We have striven to keep the CoCoA-5 language as simple as possible (but no simpler!): for instance, there is only one type `RING` for rings, whereas CoCoALib has an inheritance hierarchy with different classes for polynomial rings, quotient rings, fraction fields, and so on.

Again for simplicity, CoCoA-5 does not regard a "power-product" as a separate type (it is just a polynomial with one term whose coefficient is 1); in contrast, CoCoALib has a special class `PPMonoidElem` which represents power-products. Thus translating into C++ a CoCoA-5 program manipulating power-products will require more effort, but the reward should be a decisive gain in speed.

3 Extending CoCoA-5

The capabilities of CoCoA-5 and CoCoALib are continually expanding as the software evolves. So we have made it easy to add new functions to CoCoA-5 — both for ourselves and for normal users.

There are several ways of extending CoCoA-5.

The easiest way to add a new function is to write it in CoCoA-5 Language. Anyone can create new CoCoA-5 functions this way, and for instance give them to students or colleagues.

Often there are several functions to be added together; in this case it is best to place them in a **CoCoA-5 package** — compared to CoCoA-4 the creation of a package has been greatly simplified. A collection of functions becomes a package by saving them in a file, then inserting as the first line "**Package** *PackageName*", and appending as the last line "**EndPackage;**". Once this has been done, any function `foo(a,b)` defined in the package can be called using the syntax *PackageName*.`foo(a,b)`. A neater solution is to *export* the function from the package: this is done by inserting the declaration `export foo;` at the start of the file, before any of the function definitions. An exported function becomes directly callable as `foo(a,b)` and is automatically protected from being accidentally overwritten.

The last way is to write the new functions in C++, and then make them "visible" to CoCoA-5. This latter stage is normally quite straightforward thanks to an ingenious combination of C++ inheritance and C macros (see [3] in this volume) — we use exactly the same mechanism for making standard CoCoALib functions accessible from an interactive CoCoA-5 session. Since it is so quick to make a C++ function (provisionally) accessible from CoCoA-5, we have also used it as a convenient way of supplying test inputs to new CoCoALib functions during development.

CoCoA and Normaliz

In [3] we explain in detail the "integration" into CoCoA-5 of numerous capabilities of the *Normaliz* library. In this instance, for technical reasons, it was most appropriate to interface CoCoALib and the *Normaliz* library directly, and then make the "new CoCoALib" functions visible to CoCoA-5 the usual way.

4   CoCoALib: The Mathematical Brain of CoCoA

From its very outset the design of CoCoA-5 was based on a software library encapsulating the mathematical knowhow of the system. It comes as no surprise then that CoCoALib is the oldest part of CoCoA. We recall briefly its salient features:

- it is well-documented, free and open source C++ code (under the GPL3 licence)
- the design is inspired by and respects the underlying mathematical structures
- the source code is clean and portable
- the function interface is natural for mathematicians
- execution speed is good with robust error detection

While most CoCoA-5 functions have now been implemented in CoCoALib, some are still awaiting migration and currently reside in CoCoA-5 packages.

Our design of CoCoALib aims to make it easy to write correct programs, and difficult to write incorrect ones or ones which produce "nasty surprises". While trying to follow this guideline we met some surprisingly tricky aspects of the design:

- function definitions in limit cases (*e.g.* determinant of 0×0 matrix)
- practical definition of a function's domain (*e.g.* what result should `IsPrime(0)` give? And `IsPrime(-2)`?)
- a choice between absolute mathematical correctness or decent computational speed (and a remote chance of a wrong answer)

4.1   An Example of Design

Finding a library interface which is easy to learn and use, mathematically correct, but also efficient at run-time often requires a delicate balance of compromise. We cite here one example from CoCoALib where the solution is untraditional but successful.

CoCoALib uses continued fractions internally in various algorithms. A *continued fraction* is an expression of the form:

$$a_0 + \cfrac{1}{a_1 + \cfrac{1}{a_2 + \cfrac{1}{a_3 + \cdots}}}$$

where a_0 is an integer, and $a_1, a_2, \ldots$ are positive integers. Every rational number has a finite continued fraction which, for compactness, is often represented as a list of integers $[a_0, a_1, a_2, \ldots, a_s]$.

The most natural implementation in CoCoALib would simply compute this list. But in many applications only the first few a_k are needed, and computing the entire list is needlessly costly. So the CoCoALib implementation produces an *iterator* (a basic concept well-known in object-oriented programming) which produces the values of the a_k one at a time.

5 Extending CoCoALib

Naturally we wrote most of the source code in CoCoALib, but the design of the library (and its openness) was chosen to facilitate and encourage "outsiders" to contribute. We distinguish two categories of contribution: code written specifically to become part of CoCoALib, and stand-alone code written without considering its integration into CoCoALib.

5.1 Specific Contributions to CoCoALib

The first outside contribution came from M. Caboara, who wrote the code for computing Gröbner bases and related operations while CoCoALib was still quite young. At that stage the detailed implementation of CoCoALib was still quite fluid, and a number of pretty radical changes in the underlying data-structures were still to occur; yet despite these upheavals Caboara's implementation of Buchberger's algorithm required virtually no changes, thus confirming the solidity and stability of the CoCoALib interface design.

Another significant outside contribution came from E. Saenz-de-Cabezon, who wrote the code for computing Mayer-Vietoris trees associated to monomial ideals. A significant aspect of this contribution is that the author worked independently (in another country) and relied entirely on the documentation of CoCoALib — thus confirming the quality of the documentation. His work has encouraged us to develop specialized, efficient handling for monomial ideals (see [7]).

A more recent contribution comes from M. Albert, who implemented an algorithm for computing Janet Bases of ideals in polynomial rings. Once a Janet Basis has been obtained, many ideal invariants can readily be determined (see [4], [5]). His code has already been incorporated into CoCoALib; we anticipate employing an automated, smart caching scheme to improve computational efficiency by avoiding recalculation of Janet Bases when computing several different invariants.

5.2 Combining with External Libraries

We have combined some of the features of various external libraries into CoCoALib. An important step in each case is the "translation" of a mathematical value from its CoCoALib representation to that of the foreign library, and *vice versa*. To make it easier to do this CoCoALib offers operations for *destructuring* the various data-structures it operates upon.

The first library we combined with CoCoALib is *Frobby* (see [8]) which is specialized for operations on monomial ideals. The experience also helped us improve the interfacibility of CoCoALib.

There is also an experimental connection to some of the functions of GSL (GNU Scientific Library [9]). This is an interesting challenge because the interface has to handle two contrasting viewpoints: the exact world of CoCoALib, and the approximate (floating-point) world of GSL.

Finally, the most advanced integration we have achieved so far is with the *Normaliz* library for computing with affine monoids or rational cones. This is part of a closer collaboration which is described in more detail in [3]. In this particular case a new data-structure was added to CoCoALib to contain the type of value (`cone`) which Normaliz computes with.

6   CoCoAServer

The CoCoA software suite includes the *CoCoAServer*, a prototype which provides a client/server mechanism for accessing CoCoALib, and accepts computation requests in an OpenMath-like language. It was developed to grant access to CoCoALib features from CoCoA-4.7 while the new CoCoA-5 system was under development.

The advantages of making a "server" are that it can be called by any other "client" software (which has an OpenMath interface), and it avoids the close integration of monolithic compilation.

Currently the server remains in prototype form as resources are directed, for the time being, at CoCoALib and CoCoA-5.

References

1. Abbott, J., Bigatti, A.M., Lagorio, G.: CoCoA-5: a system for doing Computations in Commutative Algebra, http://cocoa.dima.unige.it/cocoalib
2. Abbott, J., Bigatti, A.M.: CoCoALib: a C++ library for doing Computations in Commutative Algebra, http://cocoa.dima.unige.it/cocoalib
3. Abbott, J., Bigatti, A.M., Söger, C.: Integration of libnormaliz in CoCoALib and CoCoA 5. In: Hong, H., Yap, C. (eds.) ICMS 2014. LNCS, vol. 8592, Springer, Heidelberg (2014)
4. Albert, M.: Janet Bases in CoCoA; bachelor thesis, Institut für Mathematik, Universität Kassel (2011)
5. Albert, M.: Computing Minimal Free Resolutions of Polynomial Ideals with Pommaret Bases master thesis, Institut für Mathematik, Universität Kassel (2013)
6. Bruns, W., Ichim, B., Römer, T., Söger, C.: Normaliz, http://www.mathematik.uni-osnabrueck.de/normaliz
7. Fernàndez-Ramos, O., García-Llorente, E., Sáenz-de-Cabezón, E.: A monomial week (Spanish) Gac. R. Soc. Mat. Esp. 13(3), 515–524 (2010)
8. Roune, B.H.: Frobby, http://www.broune.com/frobby
9. Galassi, M., et al.: GNU Scientific Library Reference Manual, 3rd edn. GSL, http://www.gnu.org/software/gsl ISBN 0954612078

Maximizing Likelihood Function for Parameter Estimation in Point Clouds via Groebner Basis

Joseph Awange[1], Béla Paláncz[2], and Robert Lewis[3]

[1] Curtin University, Perth, Australia
J.awange@curtin.edu.au
https://spatial.curtin.edu.au/people/index.cfm/J.Awange
[2] Budapest Technical University of Technology and Economics, Hungary
palancz@epito.bme.hu
http://www.fmt.bme.hu/fmt/htdocs/dolgozok/
dolgozo_reszlet.php?felhasznalonev=palancz
[3] Fordham University, New York, USA
rlewis@fordham.edu
https://fordham.academia.edu/RobertLewis

Abstract. Nowadays, surface reconstruction from point clouds generated by laser scanning technology has become a fundamental task in many fields, such as robotics, computer vision, digital photogrammetry, computational geometry, digital building modeling, forest planning and operational activities. The point clouds produced by laser scanning, however, are limited due to the occurrence of occlusions, multiple reflectance and noise, and off-surface points (outliers), thus necessitating the need for robust fitting techniques. These techniques require repeated parameter estimation while eliminating outliers. Employing maximum likelihood estimation, the parameters of the model are estimated by maximizing the likelihood function, which maps the parameters to the likelihood of observing the given data. The transformation of this optimization problem into the solution of a multivariate polynomial system via computer algebra can provide two advantages. On the one hand, since all of the solutions can be computed, a single solution that provides global maximum can be selected. On the other hand, once the symbolic result has been computed, it can be used in numerical evaluations in a split second, which reduces the computation time. In our presentation, we applied Groebner basis to solve the maximization of the likelihood function in various robust techniques. A numerical example with data from a real laser scanner experiment illustrates the method. Computations have been carried out in the *Mathematica* environment.

Keywords: Groebner basis, Maximum Likelihood, Point cloud, Laser scanning, Robust estimation.

1 Introduction

Laser scanning is a cutting edge remote sensing technology that has rapidly broadened its set of applications, especially in engineering areas, e.g.,[1,5].

H. Hong and C. Yap (Eds.): ICMS 2014, LNCS 8592, pp. 359–366, 2014.

In many cases, it not only can complement the mature geodetic surveying techniques, but is able to replace them. Compared to e.g., total stations, laser scanning provides information on the entire object surface instead of discrete points. Such surface-like acquisition enables detailed surface modeling, obtaining accurate 3D products.

Laser scanning produces a point cloud, i.e., a vast amount of points reflected from the surface of objects that are to be surveyed. In order to obtain information useful for engineering purposes the data have to be processed (see, e.g., [2]); engineers require numerical data, standardized measures, cross and longitudinal sections of objects etc. Artificial, man-made objects usually have regular shapes; they often can be modeled by regular features, e.g., planes, edges, geometric primitives. Breakline detection is one major drawback of laser scanning compared to conventional geodetic survey or with photogrammetry, i.e., how to find edges and lines in a point cloud. One way (that point cloud processing software usually apply) is to fit planes on the point cloud and find the edges as intersections of planes (e.g., [4]). Obviously the reliability of edge detection highly depends on the accuracy of plane fitting in this case. Reconstructing planes can also be a primary task of the processing, many applications require 3D model of the objects (e.g., buildings) on a certain level of detail, such as 3D navigation maps, urban modeling applications, etc. In airborne laser scanning, plane fitting is used to detect roads and roofs that are usually composed of planar surfaces. Road and roof detection are key issues in airborne LiDAR (Light Detection and Ranging) segmentation and classification.

The widely used point cloud processing software uses functions that are operating as black boxes. Only a few parameters can be set but no detailed information is provided on the algorithms working in the background. Future research aims to compare the developed methods with such plane fitting functions integrated in point cloud processing software; therefore validation of existing software functions would become available.

Time consumption and reliability are of great importance to plane fitting in laser scanned point clouds. The proposed algebraic method proved to be as fast as the most popular method, Principal Component Analysis (PCA) (e.g., [3]), and provides more reliable implementation in the frequently used robust technique RANdom Sample Consensus (RANSAC), that is doubtless beneficial in case of mass data processing. Compared to integrated functions of point cloud processing software, the proposed method is open, validated by widely used techniques, and therefore its users have full control on the entire plane fitting operations.

2   Algebraic Parameter Estimation

Generally to carry out a regression procedure one needs to have a model $\mathcal{M}$ $(x, y, z : \theta = 0)$, an error definition $e_{\mathcal{M}_i}(x_i, y_i, z_i : \theta)$ as well as the probability density function of the error PDF $(e_{\mathcal{M}}(x_i, y_i, z_i : \theta))$. Now our model is linear in a plane,

$$\mathcal{M}(x, y, z : \theta) = \alpha x + \beta y + \gamma - z, \tag{1}$$

with parameters $\theta = (\alpha, \beta, \gamma)$. The error model - corresponding to the TLS - is the shortest distance of a point P_i from its perpendicular projection to the plane,

$$e_{\mathcal{M}_i(x_i, y_i, z_i : \theta)} = \frac{z_i - x_i\alpha - y_i\beta - \gamma}{\sqrt{1 + \alpha^2 + \beta^2}}. \tag{2}$$

The probability density function of the error model is considered a Gaussian - type error distribution as $\mathcal{N}(0, \sigma)$,

$$PDF(e_{\mathcal{M}}(x_i, y_i, z_i : \theta)) = \frac{e^{-\frac{(e_{\mathcal{M}})^2}{2\sigma^2}}}{\sqrt{2\pi}\sigma}. \tag{3}$$

Considering a set of $\{(x_1, y_1), (x_2, y_2) \ldots (x_N, y_N)\}$ as measurement points, the maximum likelihood approach aims at finding the parameter vector θ that maximizes the likelihood of the joint error distribution. Assuming that the measurement errors are independent, we should maximize,

$$\mathcal{L} = \prod_{i=1}^{N} \frac{e^{-\frac{(e_{\mathcal{M}_i})^2}{2\sigma^2}}}{\sqrt{2\pi}\sigma}. \tag{4}$$

In order to use sum instead of product, we can consider the logarithm of Eq. (4),

$$Log\mathcal{L} = Log\left(\prod_{i=1}^{N} PDF(e_{\mathcal{M}})\right) = -\sum_{i=1}^{N} LogPDF(e_{\mathcal{M}}). \tag{5}$$

Let us consider the Gaussian - type error distribution. Then, the function to be maximized is,

$$Log\mathcal{L}(\alpha, \beta, \gamma) = -\sum_{i=1}^{N} Log\left(\frac{e^{-\frac{(e_{\mathcal{M}})^2}{2\sigma^2}}}{\sqrt{2\pi}\sigma}\right) = N Log\left(\sqrt{2\pi}\sigma\right) + \frac{1}{2\sigma^2} \sum_{i=1}^{N} \frac{(z_i - x_i\alpha - y_i\beta - \gamma)^2}{1 + \alpha^2 + \beta^2}. \tag{6}$$

In order to avoid direct maximization and to get explicit formula for the estimated parameters, symbolic computation can be employed. We apply a SuperLog function in *Mathematica* developed by [8] that utilizes pattern-matching code that enhances *Mathematica's* ability to simplify expressions involving the natural logarithm of a product of algebraic terms, see [7]. Our log-likelihood estimator function can then be written as

$$Log\mathcal{L}(\alpha, \beta, \gamma) = -\frac{N\gamma^2}{2(1+\alpha^2+\beta^2)\sigma^2} - \frac{1}{2}NLog[2] - \frac{1}{2}NLog[\pi] - NLog[\sigma] + \sum_{i=1}^{N} -\frac{\alpha\gamma x_i}{(1+\alpha^2+\beta^2)\sigma^2} +$$

$$\sum_{i=1}^{N} -\frac{\alpha^2 x_i^2}{2(1+\alpha^2+\beta^2)\sigma^2} + \sum_{i=1}^{N} -\frac{\beta\gamma y_i}{(1+\alpha^2+\beta^2)\sigma^2} + \sum_{i=1}^{N} -\frac{\alpha\beta x_i y_i}{(1+\alpha^2+\beta^2)\sigma^2} + \sum_{i=1}^{N} -\frac{\beta^2 y_i^2}{2(1+\alpha^2+\beta^2)\sigma^2} +$$

$$\sum_{i=1}^{N} \frac{\gamma z_i}{(1+\alpha^2+\beta^2)\sigma^2} + \sum_{i=1}^{N} \frac{\alpha x_i z_i}{(1+\alpha^2+\beta^2)\sigma^2} + \sum_{i=1}^{N} \frac{\beta y_i z_i}{(1+\alpha^2+\beta^2)\sigma^2} + \sum_{i=1}^{N} -\frac{z_i^2}{2(1+\alpha^2+\beta^2)\sigma^2}.$$

From the necessary conditions of the optimum, namely

$$eq_1 = \frac{\partial Log\mathcal{L}}{\partial \alpha} = 0, \ eq_2 = \frac{\partial Log\mathcal{L}}{\partial \beta} = 0, \ eq_3 = \frac{\partial Log\mathcal{L}}{\partial \gamma} = 0, \tag{7}$$

one can obtain the following polynomial system,

$$
\begin{aligned}
eq_1 &= i - b\alpha + h\alpha - i\alpha^2 - e\beta - 2g\alpha\beta + e\alpha^2\beta + i\beta^2 - b\alpha\beta^2 + d\alpha\beta^2 - \\
&\quad e\beta^3 - a\gamma - 2f\alpha\gamma + a\alpha^2\gamma + 2c\alpha\beta\gamma - a\beta^2\gamma + \mathbb{N}\alpha\gamma^2, \\
eq_2 &= g - e\alpha + g\alpha^2 - e\alpha^3 - d\beta + h\beta - 2i\alpha\beta + b\alpha^2\beta - d\alpha^2\beta - g\beta^2 + \\
&\quad e\alpha\beta^2 - c\gamma - c\alpha^2\gamma - 2f\beta\gamma + 2a\alpha\beta\gamma + c\beta^2\gamma + \mathbb{N}\beta\gamma^2, \\
eq_3 &= f - a\alpha - c\beta - \mathbb{N}\gamma,
\end{aligned}
\tag{8}
$$

where the constants depending on the measured values, are:
$a = \sum_{i=1}^{N} x_i, \ b = \sum_{i=1}^{N} x_i^2, \ c = \sum_{i=1}^{N} y_i, \ d = \sum_{i=1}^{N} y_i^2, \ e = \sum_{i=1}^{N} x_i y_i, \ f = \sum_{i=1}^{N} z_i, \ g = \sum_{i=1}^{N} y_i z_i, \ h = \sum_{i=1}^{N} z_i^2, \ i = \sum_{i=1}^{N} x_i z_i.$

To get a symbolic solution, we reduce the multivariate polynomial system to univariate of higher order. Since the last equation in Eq. (8) is linear, γ can be solved and substituted into the other two equations (i.e., eq1 and eq2). The system is reduced to a system of two equations with two unknowns (α, β) that can be solved using reduced Groebner basis to yield

$$p_\alpha = \sum_{i=0}^{7} c_i \alpha^i = 0, \ p_\beta = \sum_{i=0}^{7} \tilde{c}_i \beta^i = 0, \tag{9}$$

where c_i and $\tilde{c}_i$ are quite complicated expressions of the constants introduced above, see [6]. If α and β are known, then γ can be computed from $eq3$. We consider the triplet $\{\alpha, \beta, \gamma\}$ as the solution of the parameter estimation problem, if it is real and provides the maximum of Eq.(6) compared to the other triplet variations of real solutions. Now, to illustrate the method, let us consider a toy example in [10]. Let us estimate the parameters of the plane on the basis of four non co-planar points. The polynomials in normalized form are
$p_\alpha = -72260849539547136 + 134691541606563840\alpha + 1622138196787200\alpha^2 - 110673082039468032\alpha^3 + 106433894218530816\alpha^4 - 146349308114141184\alpha^5 + 42586534792986624\alpha^6 + 75483497423831040\alpha^7$
and
$p_\beta = 4844786081071104 - 58829545270149120\beta + 240076453124505600\beta^2 - 336928917727346688\beta^3 + 37871519767658496\beta^4 + 4250002075582464\beta^5 - 133685815924555776\beta^6 + 75483497423831040\beta^7$,
with the real solutions as

$$\alpha = -2, \ -0.857775, \ 0.757775$$

and

$$\beta = 2, \ -1.35777, \ 0.257775.$$

Now, let us consider the log likelihood estimator in compact form using the constants. Substituting γ from Eq.(8), yields

$$
Log\mathcal{L}(\alpha, \beta) = -\frac{h}{2(1+\alpha^2+\beta^2)\sigma^2} + \frac{i\alpha}{(1+\alpha^2+\beta^2)\sigma^2} + \frac{af\alpha}{N(1+\alpha^2+\beta^2)\sigma^2} -
$$
$$
\frac{b\alpha^2}{2(1+\alpha^2+\beta^2)\sigma^2} - \frac{3a^2\alpha^2}{2N(1+\alpha^2+\beta^2)\sigma^2} + \frac{g\beta}{(1+\alpha^2+\beta^2)\sigma^2} - \frac{cf\beta}{N(1+\alpha^2+\beta^2)\sigma^2} - \frac{ea\beta}{(1+\alpha^2+\beta^2)\sigma^2} +
$$
$$
\frac{ac\alpha\beta}{N(1+\alpha^2+\beta^2)\sigma^2} - \frac{d\beta^2}{2(1+\alpha^2+\beta^2)\sigma^2} + \frac{c^2\beta^2}{2N(1+\alpha^2+\beta^2)\sigma^2} - \frac{1}{2}NLog[2] - \frac{1}{2}NLog[\pi] - NLog[\sigma].
$$

We select one combination of the real α, β pair that gives the highest value for $Log\mathcal{L}(\alpha, \beta)$ above. In our case $\alpha, \beta = (-2, 2)$. Then from Eq.(8) we get $\gamma = 15$. From a practical point of view, the best way is to carry out the computation using numerical Groebner basis solver of *Mathematica* (NSolve) since in that case the selection of the proper triplet α, β, γ is automatic. Employing *Mathematica*, the real solutions are

$$
sol\alpha\beta\gamma = \{\alpha = -0.857775, \beta = -1.35777, \gamma = 105.109\}, \{\alpha = -2, \beta = 2, \gamma = 15,
$$

$$
\{\alpha = 0.757775, \beta = 0.257775, \gamma = 9.79129\}.
$$

The values of the log-likelihood function at these solutions are,

$$
Log\mathcal{L}(\alpha, \beta, \gamma) = \{-829.275, -165.676, -393.077\}.
$$

Since the second triplet gives the highest values, its solution gives the location of the global maximum. This result represents the well known fact that log-likelihood functions may have many local maximums, therefore direct maximization is difficult and it can be successful only via global optimization techniques, which are quite time consuming. The implementation of our algebraic method is carried out in *Mathematica*.

3   Robust Parameter Estimation

Modern range sensing technologies, like laser scanners, enable detailed scans of complex objects to be made, thus generating point cloud data. The majority of point cloud data are acquired by various measurement processes using a number of sensors. The physical limitations of the sensors, boundaries between 3D features, occlusions, multiple reflectance and noise can produce off-surface points that appear to be outliers. In this study, a widespread robust technique is used for embedding the algebraic solution, namely the RANdom Sample Consensus (RANSAC) algorithm. Let us apply the RANSAC method, given in [9], which has proved to be successful for detecting outliers. The basic RANSAC algorithm is as follows:

1) Pick up a model type ($\mathcal{M}$)
2) Input data as
   - data   - data corrupted with outliers (cardinality (data) = n)
   - s - number of data elements required per subset
   - $\mathcal{N}$ - number of subsets to draw the data from

- τ - threshold, which defines if data element, $d_i \in data$ agrees with the model $\mathcal{M}$.

Remarks

In general s can be the minimal number of the data which results in a closed form system for the unknown parameters of the model. The number of subsets to draw the data from $\mathcal{N}$ is chosen high enough to ensure that at least one of the subsets of the random examples does not include an outlier (with the probability p, which is usually set to 0.99). Let u represent the probability that any selected data point is an inlier, and $v = 1 - u$ the probability of observing an outlier. Then, the iterations $\mathcal{N}$ can be computed as

$$\mathcal{N} = \frac{\log(1-p)}{\log\left(1 - (1-v)^s\right)}. \tag{10}$$

3) maximalConsensusSet $\leftarrow \emptyset$
4) Iterate $\mathcal{N}$ times:
    a) ConsensusSet $\leftarrow \emptyset$
    b) Randomly draw a subset containing s elements and estimate the parameters of the model $\mathcal{M}$ via algebraic method
    c) For each data element, $d_i \in data$:
       if $(d_i, \mathcal{M}, \tau)$ agree, ConsensusSet $\leftarrow d_i$
    d) if cardinality (maximalConsensusSet) < cardinality(ConsensusSet),
       maximalConsensusSet $\leftarrow$ ConsensusSet
5) Estimate model parameters using maximalConsensusSet.

One of the important advantage of this algorithm is that the task of step 4 can be carried out in parallel. In step 4 (b) we can employ our algebraic solution. The RANSAC implementation with the integrated algebraic solution can be undertaken in *Mathematica*.

4   Application to Real Laser Scanner Measurements

Outdoor laser scanning measurements have been carried out in a hilly park in Budapest, see Fig. (1, right). The test area is on a steep slope, covered with dense but low vegetation. The experiment was carried out with a Faro Focus 3D terrestrial laser scanner, see Fig. (1, left). The test also aimed at investigating the tie point detection capabilities of the scanner processing software; different types of spheres were deployed all over the test area. In case of multiple scanning positions, these spheres can be used for registering the point clouds, see Fig. 1. The scanning parameters were set to 1/2 resolution, which equals 3 mm/10 m point spacing. This measurement resulted in 178.8 million points that were acquired in 5 and half minutes, see Fig. (2, left). The test data set was cropped from the point cloud; moreover, further resampling was applied in order to reduce the data size. The final data set composed of 33292 points in ASCII format, and only the x, y, z coordinates were kept (no intensity values).

Employing the final data set, our algebraic method implemented in RANSAC provided the result in Fig. (2, right). In order to assess the efficiency of our

Fig. 1. The test area covered by dense, but low vegetation

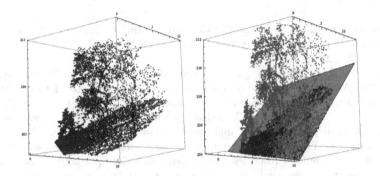

Fig. 2. Left: The test data set extracted from the laser scanner point cloud. Right: fitted plane by RANSAC algebraic technique: blue points are inliers, red points are outliers, and green is the fitted plane.

Table 1. Results of the computations for real data

Method	Computation time (sec)	Size of Inlier Set	α	β	γ	Min of error (cm)	Max of error (cm)	Mean of error (cm)	Standard deviation (cm)
RANSAC Algebraic	11.64	24 382	0.106	0.503	202.66	−22.4	28.31	0.00	6.4
Danish Algebraic	29.39	24 576	0.106	0.505	202.66	−22.0	37.0	0.00	7.0
Danish PCA	70.57	26 089	0.103	0.567	202.54	−46.0	94.6	0	18.6

algebraic method, we carried out further computations with two other robust estimation techniques, see Table 1. The values of the run time represent parallel evaluation. Undoubtedly, the RANSAC method with algebraic maximization of the likelihood equation provided the best performance, see Table 1.

5   Conclusions

This study has presented an algebraic technique that can be embedded into estimation methods such as RANSAC and Danish to offer robust solutions that adequately manage outliers. The results of the numerical tests show remarkable improvement in computational time when the algebraic method is incorporated compared to that of TLS and PCA. In addition, the algebraic method proved to have practically zero complexity considering the number of measured data points. However, if the application of the TLS error model is inevitable, and the statistical approach is not advisable, then it can be a good choice for robust estimation, since it avoids direct global maximization of the likelihood function. This is TIGeR publ. No. 571.

References

1. Armenakis, C., Gao, Y., Sohn, G.: Co-Registration of Lidar and Photogrammetric Dta for Updating Building Databases. In: ISPRS Arch., Haifa, Israel, vol. 38, pp. 96–100 (2010)
2. Fernandez, J.C., Singhania, A., Caceres, J., Slatton, K.C., Starek, M., Kumar, R.: An overview of Lidar cloud processing softwares, GEM Center Report, No. Rep-2007-12-01, Civil and Coastal Engng. Dept. Uni. of Florida (2007)
3. Hubert, M., Rousseeuw, P.J., Van den Branden, K.: ROBPCA: A new approach to robust principal component analysis. Technometrics 47(1), 64–79 (2005)
4. Lakaemper, R., Latecki, L.J.: Extended EM for Planar Approximation of 3D Data. In: IEEE Int. Conf. on Robotics and Automation (ICRa), Orlando, Florida (2006)
5. Norris-Roger, M., Behrendt, R.: From points to products - business benefits from Lidar using ArcGIS. SA Forestry Magazine (June 2013)
6. Paláncz, B., Somogyi, Lovas, T., Molnár, B.: Plane Fitting to Point Cloud via Gröbner Basis, e-publication. Wolf. Res. Inf. Center, MathSource/8491 (2013)
7. Paláncz, B.: Fitting Data with Different Error Models. Mathematica Journal 16, 1–22 (2014)
8. Rose, C., Smith, D.: Symbolic Maximum Likelihood Estimation with Mathematica. The Statistician 49, 229–240 (2000)
9. Yaniv, Z.: Random Sample Consensus (RANSAC) Algorithm, A Generic Implementation, Georgetown University Medical Center, Washington, DC, USA (2010), zivy@isis.georgetown.edu
10. Nievergelt, Y.: A tutorial history of least square with applications to astronomy and geodesy. J.Comp. and Appl.Mathematics 121, 37–72 (2000)

Groebner Basis in Geodesy and Geoinformatics

Joseph Awange[1], Béla Paláncz[2], and Robert Lewis[3]

[1] Curtin University, Perth, Australia
J.awange@curtin.edu.au
https://spatial.curtin.edu.au/people/index.cfm/J.Awange
[2] Budapest Technical University of Technology and Economics, Hungary
palancz@epito.bme.hu
http://www.fmt.bme.hu/fmt/htdocs/dolgozok/
dolgozo_reszlet.php?felhasznalonev=palancz
[3] Fordham University, New York, USA
rlewis@fordham.edu
https://fordham.academia.edu/RobertLewis

Abstract. In geodesy and geoinformatics, most problems are nonlinear in nature and often require the solution of systems of polynomial equations. Before 2002, solutions of such systems of polynomial equations, especially of higher degree remained a bottleneck, with iterative solutions being the preferred approach. With the entry of *Groebner basis* as algebraic solution to nonlinear systems of equations in geodesy and geoinformatics in the pioneering work *"Gröbner bases, multipolynomial resultants and the Gauss Jacobi combinatorial algorithms : adjustment of nonlinear GPS/LPS observations"*, the playing field changed. Most of the hitherto unsolved nonlinear problems, e.g., coordinate transformation, global navigation satellite systems (GNSS)'s pseudoranges, resection-intersection in photogrammetry, and most recently, plane fitting in point clouds in laser scanning have been solved. A comprehensive overview of such applications are captured in the first and second editions of our book *Algebraic Geodesy and Geoinformatics* published by Springer. In the coming *third edition*, an updated summary of the newest techniques and methods of combination of Groenbner basis with symbolic as well as numeric methods will be treated. To quench the appetite of the reader, this presentation considers an illustrative example of a two-dimension coordinate transformation problem solved through the combination of symbolic regression and Groebner basis.

Keywords: Groebner basis, nonlinear polynomial systems, transformation problems, GNSS, Geodesy, Geoinformatics.

1 Introduction

The solution of nonlinear systems of equations is an indispensable task in almost all geosciences, such as geodesy and geoinformatics. The rapid development of computer algebra in the last half century, with one of the most important milestones in 1965, i.e., the publication of the *Groebner Basis theory*, as well as the

H. Hong and C. Yap (Eds.): ICMS 2014, LNCS 8592, pp. 367–373, 2014.
© Springer-Verlag Berlin Heidelberg 2014

enormous improvement of computer algebra systems (CAS) such as *Mathematica* and *Maple* (just to mention two) starting 30 years ago, have provided the ability to solve many of these systems of polynomial equations in the field of geodesy and geoinformatics in analytical (exact) form.

In 2002, in a pioneering dissertation *"Gröbner bases, multipolynomial resultants and the Gauss Jacobi combinatorial algorithms : adjustment of nonlinear GPS/LPS observations"* [1], Groebner basis made its maiden entry into the fields of geodesy and geoinformatics as a powerful technique for solving systems of polynomial equations (comparable perhaps to resultant methods). Since then, problems which hitherto were unsolved, e.g., coordinate transformation, pseudoranges in global navigation satellite systems (GNSS), resection-intersection in photogrammetry, and most recently plane fitting in point clouds in laser scanning have been solved. Increased use of *Groebner basis* and other developed methods thereafter motivated the publication of our first book with Springer Verlag, e.g., *"Solving algebraic computational problems in geodesy and geoinformatics"* [2]. Wide usage of the book together with more development in computer algebraic systems (CAS) led to the writing of the second edition of the book in 2010, where both algebraic ("exact") and numerical ("approximate") methods (such as linear homotopy) were treated, see [3]. This edition was also accompanied by an electronic supplement providing the algorithms, functions, and solutions of the problems discussed in the book, in *Mathematica* code. This book, whose foreword was written by *Prof. Bruno Buchberger* (the father of Groebner basis) was a great success in the geodetic and geoinformatic society to an extend that the publisher have requested for its *third edition* currently under preparation. The *Mathematica* codes were also inspiring for many scientist and students and are stored permanently in the Wolfram Library Archive.

In the meantime, new technologies have arisen in geosciences, such as laser scanning, and new techniques have been developed in CAS software, like symbolic regression. This has led to the introduction or revision of many methods, such as robust parameter estimation, error models other than the well known least squares, Pareto optimization, parallel symbolic computation, and satellite control algorithms. Computer algebra methods can be effectively utilized in most of these techniques. Here, we present a simple illustrative example based on inverse coordinate transformation problem and refer the reader to [5] and the books mentioned above for numerous examples of application of Groebner basis to geodesy and geoinformatics.

2   The Inverse Coordinate Transformation Problem

In photogrammetry and computer vision, one is often required to perform transformation between coordinates of fiducial marks on a comparator plate and those of the corresponding points on the reseau plate during the interior orientation process, where film and lens distortions and other errors are corrected using camera calibration parameters. This process is often undertaken using two-dimension (2D) similarity, affine, or projective transformations, see e.g., [4]. Besides the solution of the 2D transformation $(x, y) \Longrightarrow (X, Y)$, where $x, y, X, Y \in \Re^2$ using the methods above, symbolic regression can also be used to find a better model where the structure as well as the coefficients of the model are computed simultaneously. To find the inverse transformation, symbolic regression can be applied in the reverse way $(X, Y) \Longrightarrow (x, y)$. The problems with this procedure are two-fold:

1. The two transformation are not really inverse of each other.
2. Symbolic regression requires a considerable computational power.

To overcome both problems, Groebner basis solution is applied. If the result of the direct symbolic regression is a multivariate polynomial, the correct inverse transformation can be computed via Greobner basis instead of using one more symbolic regression in a reversed direction. A symbolic regression combined with Groebner basis therefore may provide a mixed technique for computing a good transformation with its true inverse.

3   Two-Dimensional Transformation Models

3.1   Similarity Transformation

The advantage of this model is that it is linear in the coordinates as well as in the parameters. Consequently an iterative solution is not required and the inverse transformation is easy. This transformation can be parametrized in the following form,

$$\begin{pmatrix} x \\ y \end{pmatrix} = \begin{pmatrix} a & b \\ -b & a \end{pmatrix} \begin{pmatrix} X \\ Y \end{pmatrix} + \begin{pmatrix} c \\ d \end{pmatrix}.$$

For each observed ith point, the following pair of observation equations for the residuals (r_{x_i}) and (r_{y_i}) can be written as,

$$\begin{pmatrix} X_i & Y_i & 1 & 0 \\ Y_i & -X_i & 0 & 1 \end{pmatrix} \begin{pmatrix} a \\ b \\ c \\ d \end{pmatrix} + \begin{pmatrix} -x_i \\ -y_i \end{pmatrix} = \begin{pmatrix} r_{x_i} \\ r_{y_i} \end{pmatrix}.$$

A minimum of two fiducial marks or reseau crosses are required for a unique solution. If more observation points are available, linear least squares can be directly applied.

3.2   Affine Transformation

This is also a linear model although it needs six parameters. Therefore, a minimum of 3 fiducial marks or reseau crosses are required for a unique solution. Iterative solution is not required and the inverse transformation is easy. This model can be parametrized in the following form

$$\begin{pmatrix} x \\ y \end{pmatrix} = \begin{pmatrix} a & b \\ d & e \end{pmatrix} \begin{pmatrix} X \\ Y \end{pmatrix} + \begin{pmatrix} c \\ f \end{pmatrix}. \tag{1}$$

For each observed ith point the following pair of observation equations for the residuals (r_{x_i}) and (r_{y_i}) can be written as,

$$\begin{pmatrix} X_i & Y_i & 1 & 0 & 0 & 0 \\ 0 & 0 & 0 & X_i & Y_i & 1 \end{pmatrix} \begin{pmatrix} a \\ b \\ c \\ d \\ e \\ f \end{pmatrix} + \begin{pmatrix} -x_i \\ -y_i \end{pmatrix} = \begin{pmatrix} r_{x_i} \\ r_{y_i} \end{pmatrix}. \tag{2}$$

3.3 Projective Transformation

The projective transformation can be expressed by the following two equations,

$$x = \frac{a_1 X + a_2 Y + a_3}{c_1 X + c_2 Y + 1}, \tag{3}$$

and

$$y = \frac{b_1 X + b_2 Y + a_3}{c_1 X + c_2 Y + 1}. \tag{4}$$

These equations are a special case of the collinearity condition for mapping of 2D points from one plane onto another. There are 8 unknown parameters, so if only four fiducial marks are available, the solution is not unique. This is perhaps the main reason why it is not frequently used. However, it is relevant for systems that have been retro-fitted with a reseau plate, such as the Hasselblad camera. The least square solution is nonlinear due to the rational nature of the functions. However, an approximate linear solution can be implemented if both sides of the equations are multiplied by the denominator and partial derivatives with respect to the observable (as in the combined adjustment model) ignored. The inverse of the transformation can also be computed by solving the system of equation in symbolic form for X and Y, i.e.,

$$X = \frac{a_2\,(-y + a_3) + (x - a_3)\,b_2 + (-x + y)a_3 c_2}{-xb_2 c_1 + a_2\,(-b_1 + y c_1) + x b_1 c_2 + a_1\,(b_2 - y c_2)}, \tag{5}$$

and

$$Y = \frac{a_1\,(y - a_3) + (-x + a_3)\,b_1 + (x - y)a_3 c_1}{-xb_2 c_1 + a_2\,(-b_1 + y c_1) + x b_1 c_2 + a_1\,(b_2 - y c_2)}. \tag{6}$$

If more than 4 observation points are available, nonlinear least squares can be applied.

3.4   Application of Symbolic Regression

Per definition, neither the form nor the parameters of the model are known. However, in order to have a chance of computing the inverse of the transformation, trigonometrical functions are excluded from the functional set. Applying symbolic regression using $Eureqa$[1], the Pareto sets for $x = x(X, Y)$ and $y = y(X, Y)$ are computed and the models closest to the ideal point selected resulting into;

$$x = a_0 + Xa_1 + Ya_2 + X^2a_3 + Y^2a_4 + XYa_5 + X^3a_6 + Y^3a_7 + X^2Ya_8 + X^4a_9, \quad (7)$$

and

$$y = b_0 + Xb_1 + Yb_2 + X^2b_3 + Y^2b_4 + XYb_5 + X^3b_6 + Y^3b_7 + X^2Yb_8 + X^4b_9. \quad (8)$$

3.5   Comparison of the Different Transformation Models

In this example, a 2D transformations is performed using similarity, affine, projective, and symbolic regression transformations. 16 observed (x, y) and calibrated reseau (X, Y) coordinates are used. Table 1 summarizes the results. Although the nonlinear transformation obtained using symbolic regression gives the best fit, to compute the inverse of the transformation is not an easy task.

Table 1. Some statistical values of the different transformation models

Transformation model	Max of	Standard Deviation	RMSE
	Absolute Errors	of the Absolute Errors	of Residual Errors
	mm	mm	mm
Similarity	0.0268663	0.0088264	0.0106
Affine	0.0180314	0.00518175	0.0089
Projective	0.0128281	0.00342605	0.0066
Symbolic Regression	0.007906	0.00258456	0.0037

3.6   Computing Inverse Transformation via Groebner Basis

To get the inverse transformation, we should solve the model equations of the nonlinear transformation for values X and Y with input x and y as parameters. Namely, the following algebraic equations should be solved

$$a_0 + Xa_1 + Ya_2 + X^2a_3 + Y^2a_4 + XYa_5 + X^3a_6 + Y^3a_7 + X^2Ya_8 + X^4a_9 - x = 0, \quad (9)$$

---

[1] http://ccsl.mae.cornell.edu/eureqa

Table 2. Exponents of the variables in the polynomials of the Groebner basis

Polynomials	Exponent of X	Exponent of Y
1	0	12
2	1	11
3	1	11
4	1	11
5	1	11
6	2	11
7	2	11
8	2	11
9	3	3

and

$$b_0 + Xb_1 + Yb_2 + X^2b_3 + Y^2b_4 + XYb_5 + X^3b_6 + Y^3b_7 + X^2Yb_8 + X^4b_9 - y = 0. \quad (10)$$

Employing numerical Groebner basis solves the system with the actual numerical values of the model parameters (a_i, b_i). In order to avoid round-off errors, the actual values of the parameters should be rationalized. A Groebner basis of this system consists of 9 polynomials having the exponents of X and Y in Table 2.

In Table 2, the first polynomial of the basis is a univariate polynomial of the variable Y of order 12. The roots of this univariate polynomial provide the values of Y, which upon the selection of the proper value can be substitute into the second polynomial of the basis and solved for X. To illustrate the inverse computation, let $x = -113.767$ and $y = -107.400$. The actual univariate becomes,
$p(Y) = -1.83807 \times 10^{44} - 1.67014 \times 10^{42}Y + 7.4902 \times 10^{36}Y^2 + 7.87106 \times 10^{31}Y^3 - 5.30945 \times 10^{28}Y^4 - 7.1523 \times 10^{24}Y^5 + 1.95557 \times 10^{21}Y^6 + 1.8312 \times 10^{18}Y^7 + 4.95224 \times 10^{13}Y^8 - 8.27685 \times 10^{10}Y^9 - 8.67742 \times 10^6 Y^{10} - 360.976Y^{11} + 1.Y^{12}$.

The real roots of this polynomial are $Y = -110.001$ and $Y = 7137.73$, with the first solution being the proper one. Now employing $Y = -110.001$, we get for the second polynomial of the Groebner basis, $p(X) = 2.022687353909516 \times 10^{728} + 1.83885524266834369805 \times 10^{726}X$, leading to the corresponding X coordinate as $X = -109.997$.

4   Conclusions

Groebner basis certainly is a powerful tool in solving systems of polynomial equations in geodesy and geoinformatics besides resultants and other methods being developed. This is demonstrated in the example above. This is a TIGeR publ. No. 570.

References

1. Awange, J.L.: Gröbner bases, multipolynomial resultants and the Gauss Jacobi combinatorial algorithms: adjustment of nonlinear GPS/LPS observations (2002)
2. Awange, J.L., Grafarend, E.W.: Solving Algebraic Computational Problems in Geodesy and Geoiormatics. Springer, Berlin (2005)
3. Awange, J.L., Grafarend, E.W., Paláncz, B., Zaletnyik, P.: Algebraic Geodesy and Geoinformatics. Springer, Berlin (2010)
4. Awange, J.L., Kiema, J.B.K.: Environmental Geodesy, monitoring and management. Springer, Berlin (2013)
5. Grafarend, E.W., Awange, J.L.: Applications of Linear and Nonlinear Models. Springer, Berlin (2012)

Groebner Bases in *Theorema*[*]

Bruno Buchberger[1] and Alexander Maletzky[2]

[1] RISC, Johannes Kepler University, Linz, Austria
bruno.buchberger@risc.jku.at
http://www.risc.jku.at/home/buchberg
[2] Doctoral College "Computational Mathematics" and RISC,
Johannes Kepler University, Linz, Austria
alexander.maletzky@dk-compmath.jku.at
https://www.dk-compmath.jku.at/people/alexander-maletzky

Abstract. In this talk we show how the theory of Groebner bases can be represented in the computer system *Theorema*, a system initiated by Bruno Buchberger in the mid-nineties. The main purpose of *Theorema* is to serve mathematical theory exploration and, in particular, automated reasoning. However, it is also an essential aspect of the *Theorema* philosophy that the system also provides good facilities for carrying out computations. The main difference between *Theorema* and ordinary computer algebra systems is that in *Theorema* one can both program (and, hence, compute) and prove (generate and verify proofs of theorems and algorithms). In fact, algorithms / programs in *Theorema* are just equational (recursive) statements in predicate logic and their application to data is just a special case of simplification w. r. t. equational logic as part of predicate logic.

We present one representation of Groebner bases theory among many possible "views" on the theory. In this representation, we use functors to construct hierarchies of domains (e. g. for power products, monomials, polynomials, etc.) in a nicely structured way, which is meant to be a model for gradually more efficient implementations based on more refined and powerful theorems or at least programming tricks, data structures, etc.

Keywords: Groebner basis, Buchberger algorithm, mathematical theory exploration, Theorema.

1 Introduction

After Bruno Buchberger introduced the concept of Groebner bases and an algorithm for computing them in his 1965 PhD thesis [1,4], there has ever since been a lot of effort to implement his algorithm in various programming languages (Buchberger's thesis already contains an implementation in a version of FORTRAN and in machine language). Nowadays, there are many different computer

---

[*] This research was funded by the Austrian Science Fund (FWF): grant no. W1214-N15, project DK1.

H. Hong and C. Yap (Eds.): ICMS 2014, LNCS 8592, pp. 374–381, 2014.

systems that have either been partially inspired by or are devoted especially to the computation of Groebner bases, among them particularly successful systems such as CoCoA [8], Magma [9], Maple [10], *Mathematica* [11], Sage [12], SINGULAR [13] and many others.

In our talk we want to focus on our *Theorema* system: *Theorema* [16,7] is a system which was initiated by Bruno Buchberger and developed in his *Theorema* group at RISC since the the mid-nineties. It uses the computer algebra system *Mathematica* [11] as software frame. Its user-interface is currently re-designed and -implemented (*Theorema* Version 2.0).

The main difference between *Theorema* and other computer algebra systems (like the ones mentioned above) is that in *Theorema* one can both compute and prove within one single system, at exactly the same level: There is no need to first implement programs and then lift them to some level of abstraction for reasoning about them, or vice versa, but programs in *Theorema* are themselves just formulas. This works because the language and internal logic of *Theorema* is an elegant version of (higher-order) predicate logic, where computation is realized by repeated simplification w. r. t. equational theories.

We will present our view on how (an algorithmic treatment of) Groebner bases theory, but also mathematics in general, can be developed in a structured, generic, machine-checked, but nonetheless natural and intuitive way following the philosophy of domains, functors and categories in *Theorema*. Also, we will of course dedicate a big part to explaining how computations can effectively be carried out in *Theorema*.

2   Domains, Functors and Categories in *Theorema*

Before explaining the presentation of Groebner bases theory in *Theorema* we will briefly explain the *Theorema* view of domains, functors and categories for a hierarchical build-up of mathematics (introduced in [3], supplementary information can also be found in [17,5]).

One of many possible ways to represent domains in *Theorema* is by considering them as *interpretations of (operator) symbols*, mimicking the concept of "interpretation" in model theory. This means that, unlike in most algebra books, a domain is not characterized by a carrier set and a set of operations. Rather, a domain is simply a mapping that maps symbols to operators (i. e. functions and predicates). In *Theorema*, interpretations of operators in domains are indicated by underscripts, e. g. consider a domain D that maps the symbol f to a concrete function that is applied to arguments:

$$\underset{D}{f}[x, y]$$

After having introduced the concept of domains in *Theorema*, the role of functors can be explained in a few words: Functors, in our view, map domains to domains by defining the meaning of symbols in the new domain by formulas involving the meaning of symbols in the given domain(s). A simple example of a functor is the functor that maps a domain D to its two-fold Cartesian product, denoted by N: N consists of all pairs of elements of D, and the symbol "+" might be interpreted in N as component-wise addition in D of such pairs:

$$\langle x1, x2 \rangle \underset{N}{+} \langle y1, y2 \rangle := \langle x1 \underset{D}{+} y1, x2 \underset{D}{+} y2 \rangle$$

The significance of functors in mathematical theory exploration is:

- Building up algorithmic mathematics in a *generic* way: The formulation (programs for the operations) of a new domain has to be given only once independent of the domains from which the new domain is built.
- Constructing towers of domains in a *structured* way.
- Transforming (non-) algorithmic properties of the input domains to (non-) algorithmic properties of the output domain.

Please also note that arguments of functors are *not* restricted to domains. For instance, one can define a functor that maps a domain D and a natural number n to the polynomial ring in n indeterminates over D.

Finally, categories (in *Theorema*) describe properties of domains. Since domains are completely characterized by the interpretations they give to symbols, properties of domains are in fact properties of those interpretations: For instance, the category of Abelian groups would possibly require domains to have an interpretation of symbol "+" with all the well-known properties (associativity, commutativity, etc.).

There are many interesting interrelations between functors and categories; Some of the most important ones are so-called *conservation theorems*: If domains $D_1, \ldots, D_k$ are in categories $\mathcal{C}_1, \ldots, \mathcal{C}_k$, and F is a functor, then one may try to prove that the domain $F[D_1, \ldots, D_k]$ is in category $\mathcal{C}$. In all areas of mathematics, many of the theorems are exactly of that kind, as they describe precisely the essence of a functor.

3   Reduction- and Groebner Rings

In our approach for representing and formalizing Groebner bases in *Theorema* we strove to be as generic as possible: Neither do we want to only treat polynomial rings over fields, nor do we even want to restrict ourselves to one single *representation* of the individual components of Groebner bases theory (such as power products, monomials, polynomials, ...). Although this might sound very ambitious, thanks to the powerful concept of functors in *Theorema*, it turns out to be quite natural and intuitive.

The theoretical foundations for a generic development of Groebner bases theory were laid in [2,14,15]. Also in the present elaboration, we follow this approach. This means that the elementary domains under consideration are so-called *reduction rings*: Reduction rings are unitary commutative rings with some additional properties; In particular, they have to be equipped with

- a Noetherian partial order relation <,
- a binary function rdm (read: "reduction multiplier"), and
- a binary function lcrd (read: "least common non-trivial reducible")

that have to be related in some non-trivial way to each other.

If a domain provides all these operations (together with the usual ring operations), then all the remaining operations needed for an algorithmic treatment of Groebner bases (S-polynomial, total reduction, etc.) can be defined in terms of them[1].

3.1   The Functor-Approach to a Generic Treatment of Groebner Bases

Presenting Groebner bases theory in a generic way as described above fits nicely into the functor paradigm of *Theorema*: One basically only needs one single functor, GroebnerExtension, which maps a domain D to the *Groebner ring* of D. A Groebner ring is a reduction ring providing in addition also a function Gb for computing Groebner bases (of ideals in) D by means of the rdm- and lcrd functions. The *Theorema*-definition of function Gb according to our present work can be found in figure 1 in section 5.

Here it is important to note that in a structured development of mathematics in *Theorema*, in our view, it is *not* the task of functors to check whether their input domains satisfy all required properties; In particular, in our case of the GroebnerExtension functor, the input domain D might not even be a ring, leaving functions like $+\atop D$ undefined. If some operators are undefined, other operations defined in terms of them (like Gb) will simply not behave as expected.

The interesting cases, however, are those where the input domains do satisfy the required properties, i.e. in our case the input domains are reduction rings. Since correctness of an algorithm is always relative to the validity of the input anyway, statements of correctness of algorithms of a functor are typically conservation theorems (see section 2). In our case of the GroebnerExtension functor:

If domain D is in category ReductionRing, then domain GroebnerExtension[D] is in category GroebnerRing.

ReductionRing and GroebnerRing are the categories of reduction rings and Groebner rings, respectively. Proving statements like the one above can then be done using the automated proving facilities of *Theorema*, and after having established the validity of *one* such statement, correctness of a whole *class* of algorithms follows (one algorithm for each instantiation of the input domain).

Another integral part of a generic presentation of Groebner bases theory are conservation theorems of the form

If D is a reduction ring, then so is F[D].

---

[1] "S-polynomial" is meant to refer to the respective object in reduction rings (where the S-polynomial is not necessarily a polynomial.

where F is yet another functor, for instance a functor that maps a domain R to the univariate polynomial ring over R. Examples of functors that satisfy the above statement can be found in [15].

4   Structure of the Formalized Theory

After having described the main functor GroebnerExtension which maps reduction rings to Groebner rings, the next question is how reduction rings can be created in *Theorema*, and the answer is the same as before: Using functors. Whenever one is given some domain D which possesses all the necessary properties in order to be turned into a reduction ring, one can do so using a functor that maps the domain to a new domain where <, rdm and lcrd are interpreted properly. This does not only work for single domains, but also for whole categories: Every field K, for instance, can always be turned into a reduction ring [2].

Following our paradigm of a systematic development of Groebner bases theory, fields provide a good starting point for moving to the "next level" by considering *polynomial rings*. Constructing the polynomial ring over some domain R is again achieved by a functor, called reductionPolynomials. This functor does not only construct (one particular representation of) *univariate* polynomial rings (viewed as reduction rings), but is much more sophisticated: In addition to the coefficient domain it takes a second input domain which is meant to be the domain of power products, in arbitrarily many indeterminates. This means that for the functor it is completely irrelevant *how* power products are represented, as long as they provide operations like divisibility, multiplication, and an order relation. The advantage of such an approach is obvious: One single functor (and one single conservation theorem) is sufficient for dealing with all the infinitely many different representations of power products, and this is exactly the purpose of working with functors!

In our elaboration, we decided to represent polynomials as tuples of pairs, where each pair constitutes a monomial: The first component of each pair is a non-zero coefficient taken from the coefficient domain, and the second component is a power product taken from the domain of power products. It is clear that this representation is only one among infinitely many isomorphic ones, which are all indistinguishable from the algebraic point of view, but it proved to be quite convenient from the algorithmic point of view. Also do we provide one particular representation of power products as tuples of exponents, where $x_1^{e_1} \cdots x_n^{e_n}$ is represented as $\langle e_1, \ldots, e_n \rangle$. Still, we allow an arbitrary number of indeterminates and also provide several built-in order relations: Lexicographic, degree-lexicographic and degree-reverse-lexicographic. Other order relations and other representations of polynomials and power products, e. g. where the order relation is given by weight matrices, can easily be added as well.

It has to be mentioned that apart from fields and polynomial rings over fields there are several other reduction rings, too. Most notably, $\mathbb{Z}$ and $\mathbb{Z}_m$ (quotient ring of integers modulo m) can be made reduction rings [14], even if m is non-prime. $\mathbb{Z}$ is already included in the present state of the formalization, whereas adding $\mathbb{Z}_m$ is work in progress. Due to the properties of reduction rings and our implementation of functor `reductionPolynomials`, $\mathbb{Z}[X]$ (and, in the future, $\mathbb{Z}_m[X]$) can be dealt with as well without any further effort.

5   Computations

The computing-facility of *Theorema* builds upon the fact that (higher-order) equational predicate logic can be regarded a *rewrite mechanism*: In order to perform a computation, successively replace equals by equals (in a directed way) until no more such replacements are possible. Computations, hence, are simply transformations of syntactic expressions. The equations (and equivalences) that give rise to such rewrite rules are once again just formulas that can be entered by the user, and programs are eventually given by collections of formulas. An example can be found in figure 1, where an implementation of function Gb in Groebner rings is shown. This implementation follows Buchberger's original critical-pair/completion algorithm.

$$\text{Gb}[X] := \text{Gb}[X, \text{pairs}[X]] \qquad (145) \times$$

$$\text{Gb}[X, \langle\rangle] := X \qquad (146) \times$$

$$\text{Gb}[X, \langle\langle x, y\rangle, p \ldots\rangle] :=$$

$$\begin{array}{l} \quad \text{let} \\ h{=}\text{trd}\left[\text{cpd}[x,y],X\right] \\ \\ \left[\begin{array}{l} \text{Gb}[X, \langle p \ldots\rangle] \qquad\qquad\qquad \Longleftarrow \quad h == 0 \\ \\ \text{Gb}\left[X{\frown}h, \langle p \ldots\rangle \times \left(\langle x_k, h\rangle \quad \Big|_{k=1,-,|X|}\right)\right] \Longleftarrow \text{otherwise} \end{array} \right. \end{array} \qquad (154) \times$$

Fig. 1. Implementation of function Gb by means of predicate logic formulas

Please note the following regarding notions and notation in figure 1:

- Since the whole definition is inside a functor (`GroebnerExtension`), most of the operations that appear need to refer to the output domain; This is accomplished by adding the domain underscript N[2].
- Tuples (denoted by angle brackets) are used rather than sets for representing the input basis, the collection of critical pairs that still have to be considered, as well as the output basis. This allows us to have control over the order of elements.
- `pairs` is a function that computes all pairs of elements of a tuple.

---

[2] Further details are omitted here for the sake of simplicity.

- p... is a so-called *sequence variable*, i.e. a variable that can be instantiated by any sequence of expressions.
- trd$_N$ is an auxiliary function defined by functor GroebnerExtension, which totally reduces its first argument modulo its second argument (making use of function rdm of the underlying reduction ring).
- cpd$_N$ is an auxiliary function defined by functor GroebnerExtension, which computes the *critical pair - difference* of its arguments (making use of functions lcrd and rdm of the underlying reduction ring).
- X$_k$ refers to the k-th element of tuple X, |X| denotes the length of tuple X.
- ⌢ and ⋈ denote appending an element to a tuple and concatenating two tuples, respectively.

If the functor is applied to some concrete domain which provides (algorithmic) interpretations for the three symbols <, rdm and lcrd, then function Gb is also algorithmic in the sense that it computes *some* tuple of elements for each input tuple. If the underlying domain, in addition to giving interpretation to the aforementioned symbols, really is a reduction ring (i.e. has all the necessary properties), then the tuples computed by function Gb are indeed Groebner bases of the ideals generated by the tuples given as input to the function.

Apparently, the implementation of function Gb is certainly not the most efficient one, but it is not the purpose of our talk to present highly sophisticated, fine-tuned methods for computing Groebner bases anyway, but just to illustrate how all this can be done *in principle* in *Theorema*. Since, in *Theorema*, algorithms and theorems can be formulated within the same language and, also, proving and computing is basically the same (computing is a special case of proving), one now can proceed to prove theorems about Groebner bases automatically or semi-automatically, for example the correctness of the algorithm for computing Groebner bases under certain assumptions on the domain in which Groebner bases are considered or, for example, theorems on the complexity of Groebner bases computation or theorems on the functors that construct new domains from domains in which Groebner bases exist. Some progress on this has been made, see the companion paper "Complexity Analysis of the Bivariate Buchberger Algorithm in *Theorema*" in the session on Mathematical Theory Exploration, in which we give a completely formal and semi-automated proof of a complexity result on Groebner bases.

Properties of the polynomial functor (in particular the existence of Groebner bases in the domain generated by the polynomial functor under the existence of Groebner bases in the coefficient domain) have been proved completely formal in [2,14,15] as a preparation to what should be possible in *Theorema* in a semi-automated way. We also had a completely formal proof for the correctness of the Groebner bases algorithm quite early (see [2]), and we are now working on building up appropriate provers for this in *Theorema*.

The most significant progress along the intention of the *Theorema* project so far was the automated synthesis of the Groebner bases algorithm, see [6]. More about this will be presented in the invited talk "Soft Math / Math Soft" by Buchberger at this conference.

References

1. Buchberger, B.: Ein Algorithmus zum Auffinden der Basiselemente des Restklassenringes nach einem nulldimensionalen Polynomideal (An Algorithm for Finding the Basis Elements in the Residue Class Ring Modulo a Zero Dimensional Polynomial Ideal). PhD thesis, Mathematical Institute, University of Innsbruck, Austria (1965), English translation in J. of Symbolic Computation, Special Issue on Logic, Mathematics, and Computer Science: Interactions 41(3-4), 475–511 (2006)
2. Buchberger, B.: A Critical-Pair/Completion Algorithm in Reduction Rings. RISC Report Series 83-21, Research Institute for Symbolic Computation (RISC), University of Linz, Schloss Hagenberg, 4232 Hagenberg, Austria (1983)
3. Buchberger, B.: Mathematica as a Rewrite Language. In: Ida, T., Ohori, A., Takeichi, M. (eds.) Functional and Logic Programming (Proceedings of the 2nd Fuji International Workshop on Functional and Logic Programming, Shonan Village Center), November 1-4, pp. 1–13. World Scientific, Singapore (1996)
4. Buchberger, B.: Introduction to Groebner Bases. London Mathematical Society Lectures Notes Series, vol. 251. Cambridge University Press (April 1998)
5. Buchberger, B.: Groebner Rings in Theorema: A Case Study in Functors and Categories. Technical Report 2003-49, Johannes Kepler University Linz, Spezialforschungsbereich F013 (November 2003)
6. Buchberger, B.: Towards the Automated Synthesis of a Groebner Bases Algorithm. RACSAM - Revista de la Real Academia de Ciencias (Review of the Spanish Royal Academy of Science), Serie A: Mathematicas 98(1), 65–75 (2004)
7. Buchberger, B., Crăciun, A., Jebelean, T., Kovcs, L., Kutsia, T., Nakagawa, K., Piroi, F., Popov, N., Robu, J., Rosenkranz, M., Windsteiger, W.: Theorema: Towards Computer-Aided Mathematical Theory Exploration. Journal of Applied Logic 4(4), 470–504 (2006)
8. CoCoA system, cocoa.dima.unige.it
9. Magma Computational Algebra System, magma.maths.usyd.edu.au/magma/
10. Maple system, www.maplesoft.com/products/Maple/
11. Wolfram Mathematica, http://www.wolfram.com/mathematica/
12. Sage system, http://www.sagemath.org
13. Decker, W., Greuel, G.-M., Pfister, G., Schönemann, H.: SINGULAR 3-1-6 — A computer algebra system for polynomial computations (2012), www.singular.uni-kl.de
14. Stifter, S.: A Generalization of Reduction Rings. Journal of Symbolic Computation 4(3), 351–364 (1988)
15. Stifter, S.: The Reduction Ring Property is Hereditary. Journal of Algebra 140(89-18), 399–414 (1991)
16. Theorema system, http://www.risc.jku.at/research/theorema/description/
17. Windsteiger, W.: Building Up Hierarchical Mathematical Domains Using Functors in THEOREMA. In: Armando, A., Jebelean, T. (eds.) Electronic Notes in Theoretical Computer Science. ENTCS, vol. 23, pp. 401–419. Elsevier (1999)

Effective Computation of Radical of Ideals and Its Application to Invariant Theory[*]

Amir Hashemi[1,2]

[1] Department of Mathematical Sciences,
Isfahan University of Technology Isfahan, 84156-83111, Iran
[2] School of Mathematics,
Institute for Research in Fundamental Sciences (IPM), Tehran,
P.O. Box: 19395-5746, Iran
Amir.Hashemi@cc.iut.ac.ir
http://amirhashemi.iut.ac.ir/

Abstract. The most expensive part of the known algorithms in the calculation of primary fundamental invariants (of rings of polynomial invariants of finite linear groups over an arbitrary field) is the computation of the radicals of complete intersection ideals. Thus, in this paper, we develop effective methods for such calculation. For this purpose, we introduce first a new notion of genericity (called *D-quasi stable position*) and exhibit a novel *deterministic* algorithm to put an ideal in Nœther position (we show that this new notion of genericity is equivalent to Nœther position). Then, we use this algorithm and also the algorithm due to Krick and Logar (to compute radicals of ideals) to present an efficient algorithm to calculate the radical of a complete intersection ideal. Furthermore, we apply this algorithm, to improve the classical methods of computing primary invariants which are based on radical computation. Finally, we have implemented in MAPLE the mentioned algorithms (to put an ideal in Nœther position, to compute the radical of ideals and also primary invariants) and compare the proposed algorithms, via a set of benchmarks, with the corresponding functions in MAPLE and MAGMA. The experiments we made seem to show that these first implementations are already more efficient than the corresponding functions of MAPLE and MAGMA.

Keywords: Polynomial rings, Regular sequences, Radical of ideals, Nœther position, Deterministic algorithms.

1   Introduction

Invariant theory is a classical subject in mathematics with a long tradition which many applications to problems in diverse areas of mathematics such as algebraic geometry, combinatorics, statistics and so on. On the other hand, *algebraic* invariant theory is the study of constructive algebraic methods for finding all (polynomial) invariants under the action of a group. More precisely, let

[*] The research of the author was in part supported by a grant from IPM (No. 92550420).

H. Hong and C. Yap (Eds.): ICMS 2014, LNCS 8592, pp. 382–389, 2014.
© Springer-Verlag Berlin Heidelberg 2014

$R = K[x_1, \ldots, x_n]$ denote the polynomial ring in n variables over a field K. We consider a group G acting on R as degree preserving automorphisms. The ring of invariants, R^G, is the subalgebra in R of polynomials that is fixed by this action. Algebraic invariant theory is interested in the algebraic structure of R^G and in finding connections between properties of G and R^G. Hilbert's 14th problem (proposed at the beginning of the 20th century by Hilbert) is central in this direction: "Is R^G finitely generated as a K-algebra?". Due to some results of Hilbert, Nagata, Haboush and Popov, it is known that this question has an affirmative answer iff G is reductive. The algorithmic side of the problem is to find the generators of R^G when it is finitely generated as a K-algebra. The computational methods based on Gröbner bases (we refer to the original references [5,7,6] as well as the textbook [3] for details on Gröbner bases) give rise to more efficient algorithms in invariant theory which make many calculations now feasible. Several approaches followed for example in [27,10]. Sturmfels in his book [27, page 53] described an algorithm to compute a set of *primary invariants*, i.e., algebraically independent homogeneous invariant polynomials $p_1, \ldots, p_n$ so that R^G is a finitely generated module over $K[p_1, \ldots, p_n]$. A full generating set of R^G (as a K-algebra) is obtained by augmenting the set of primary invariants with *secondary invariants* (a generating set of R^G as a module over $K[p_1, \ldots, p_n]$). This algorithm relies on the fact that $p_1, \ldots, p_n$ is a regular sequence of homogeneous polynomials and therefore to construct p_i we shall compute the radical of the ideal $\langle p_1, \ldots, p_{i-1} \rangle$. The most expensive part of this algorithm is the computation of the radicals of polynomial ideals generated by regular sequences (see [17, page 355]). Motivated by this problem, we develop effective methods to compute the radical of an ideal generated by a regular sequence.

On the other hand, the computation of the radical of a polynomial ideal is an important problem in constructive polynomial ideal theory. The first constructive solution to the problem of computing the radical of an ideal was given by Hermann [16]. Mostly the approaches to compute radicals of (positive dimensional) ideals reduce the problem to the zero-dimensional case, see for example [1,11,21]. For a nice presentation of this method, we refer also the reader to the book [3]. Kemper in [18] presented an algorithm for the zero-dimensional case and when the base field is finitely generated over a perfect field. Further, if the characteristic of K is positive, an algorithm was proposed by Matsumoto [22] to compute the radicals of ideals. Finally, a direct method was given by Eisenbud et al. [9] by using Jacobian techniques to the problem in the case that characteristic of the ground field K is zero.

In this paper, we will adhere to the technique of reducing to the zero-dimensional case, and apply the approach of Krick and Logar in [21]. In their method the authors first proposed to put the ideal in *Nœther position* and then to reduce the computation to the zero-dimensional case. This notion of genericity has been studied by many authors in different contexts: for example by Giusti et al. to compute the dimension of a variety [12] and by Lecerf to solve a system of polynomial equations and inequations [19]. A general algorithm for the computation of a Nœther normalization was given by Vasconcelos [28]. Over the recent years, several algorithms have been proposed to put an ideal in Nœther position (see [20,23,4,14,13,24]), however, all these algorithms are probabilistic and use random changes of coordinates. Further, some of them use Gröbner basis calculations in the lexicographical monomial ordering which may lead to large coefficient growth and heavy calculations. For more details on Nœther normalization, we refer to [23,8].

In this paper, we introduce a new notion of genericity, so-called *D-quasi stable position* and using it, we present a new *deterministic* algorithm to put an ideal in Nœther position. The advantage of our algorithm with respect to the other algorithms is that, it uses sparser coordinate changes and seems more efficient. Furthermore, it computes the Gröbner bases only with respect to the degree reverse lexicographical ordering. Applying this algorithm and the one of Krick and Logar, we present an efficient algorithm to calculate the radical of a *complete intersection* ideal, i.e. an ideal generated by a regular sequence. We show that this algorithm may improve the classical methods of computing primary invariants which use radical computation. Finally, we have implemented in MAPLE the mentioned algorithms (to put an ideal in Nœther position, to compute the radical of ideals and also primary invariants) and compare the proposed algorithms, via a set of benchmarks, with the corresponding functions in MAPLE and MAGMA. The experiments we made seem to show that these first implementations are more efficient than the corresponding functions of MAPLE and MAGMA.

The paper is organized as follows. In Section 2, we introduce the notion of D-quasi stable position, and compare it with different notions of genericity. Further, we describe a deterministic algorithm to put a given ideal in Nœther position, and compare its results with those obtained by the package Involutive due to Robertz [24] and also by MAGMA function NotherNormalization. Due to space restriction, we only present the main results in this section, and the proofs are omitted. Finally, in Section 3, we discuss the functionality of the implemented package. It should be noted that we leave the description of our main results to compute the radical of complete intersection ideals and its applications in invariant theory for the full version of paper.

2   D-Quasi Stable Position

In this section, we introduce the notion of D-quasi stable position for polynomial ideals, and then compare it with some other notions of genericity like quasi-stable and D-stable position.

Let us fix the notations used throughout this paper. Let $R = K[x_1, \ldots, x_n]$ be a polynomial ring over a field K with $char(K) = 0$ and $0 \neq I \subset R$ a homogeneous ideal. We denote by $A = R/I$ the corresponding factor ring and by $D = \dim(A)$ the dimension of I. Further, we consider the monomial ordering $\prec$ on R given by the reverse degree lexicographic ordering with $x_n \prec \cdots \prec x_1$. Finally, We denote by $lt(I)$ the initial ideal (leading term ideal) of I.

Definition 1. *A monomial ideal J of dimension D is called* D-quasi stable *if for each monomial $m \in J$, for each $j = n - D + 1, \ldots, n$ and for each integer s with $x_j^s \mid m$, an integer t exists so that we have $x_i^t(m/x_j^s) \in J$ for all $i = 1, \ldots, n - D$.*

Equivalently, a monomial ideal J is in D-quasi stable position if for each $j = n - D + 1, \ldots, n$ we have $J : x_j^\infty \subset J : \langle x_1, \ldots, x_{n-D} \rangle^\infty$.

Lemma 1. *D-quasi stability for a monomial ideal J can be checked by a given generating set of J.*

Example 1. Let $J = \langle x_4^4, x_4^3 x_3, x_4^3 x_2, x_4^2 x_1, x_3^2, x_3 x_2, x_3 x_1, x_2^2, x_2 x_1, x_1^2 \rangle$ in the polynomial ring $K[x_1, \ldots, x_4]$. Then, we can see easily that J is zero-dimensional and D-quasi stable.

Based on Definition 1, we give here an algorithm to decide whether or not a monomial ideal is weakly D-stable.

Algorithm 1. DQS-test

> **Input:** A monomial ideal $J \subset R$
> **Output:** The answer to: Is J D-quasi stable?
> $G := \{m_1, \ldots, m_k\}$ a minimal system of generators for J
> $Deg := \max\{\deg(m_1), \ldots, \deg(m_k)\}$
> $D :=$ the highest integer ℓ so that $x_i^{Deg} \in J$ for $i = 1, \ldots, \ell$
> **for** each $x_1^{e_1} \cdots x_h^{e_h} \in G$ with $h > n - D$ and $e_h > 0$ **do**
>   **for** $j = 1, \ldots, n - D$ **do**
>   **if** $x_1^{e_1} \cdots x_{h-1}^{e_{h-1}} x_j^{Deg} \notin J$ **then**
>   **return**(false,x_h, x_j)
>   **end if**
>   **end for**
> **end for**
> **return**(true)

Theorem 1. *The* DQS-TEST *algorithm terminates in finitely many steps and decides whether or not the input monomial ideal is D-quasi stable.*

We recall that an ideal I is in *Nœther position* if $K[x_{n-D+1}, \ldots, x_n] \hookrightarrow R/I$ is an integral ring extension, i.e. the image in R/I of x_i for any $i = 1, \ldots, n - D$ is a root of a polynomial $X^s + g_1 X^{s-1} + \cdots + g_s = 0$ where s is an integer and $g_1, \ldots, g_s \in K[x_{n-D+1}, \ldots, x_n]$, see [8] for example.

Theorem 2. *A monomial ideal is D-quasi stable iff it is in Nœther position.*

It is worth noting that this result shows that the notion of D-quasi stable ideal is an equivalent notion of genericity to Nœther position. D-quasi stable position may be considered as a proper extension of *quasi-stable position*. The notion of quasi-stable ideals has been used by Bayer and Stillman [2] and Bermejo and Gimenez [4] to calculate the satiety and Castelnuovo-Mumford regularity of ideals. Further, Seiler in [26] studied the relation between quasi-stable ideals and Pommaret bases, and outlined a deterministic algorithm to put a given ideal in quasi-stable position.

Definition 2. *A monomial ideal J is called* quasi-stable *if for any monomial $m \in J$ and all integers i, j, s with $1 \le j < i \le n$ and $s > 0$, if $x_i^s \mid m$ there exists an integer $t \ge 0$ such that $x_j^t m / x_i^s \in J$.*

We can see easily that every quasi stable monomial ideal is D-quasi stable, however, the converse does not hold in general, as demonstrated by the following example.

Example 2. Let us consider the monomial ideal $J = \langle x_1, x_2^2, x_3 x_5 x_2, x_3 x_4^2, x_3 x_4 x_2,$
$x_3^3, x_3^2 x_2, x_3 x_5^2 x_6, x_3 x_4 x_5 x_6, x_5 x_6 x_3^2, x_4 x_6 x_3^2, x_5^3 x_3, x_4 x_5^2 x_3, x_5^2 x_3^2, x_4 x_5 x_3^2, x_3 x_6^3 x_7,$
$x_3 x_5 x_6^2 x_7, x_3 x_4 x_6^2 x_7, x_3^2 x_6^2 x_7, x_2 x_3 x_6^2 x_7, x_3 x_6^4, x_3 x_5 x_6^3, x_3 x_4 x_6^3, x_6^4 x_3^2, x_2 x_3 x_6^3,$
$x_6^2 x_3^2 h^2, x_2 x_3 x_6^2 h^2, x_3^2 x_6 x_7^2 h, x_2 x_3 x_6 x_7^2 h, x_3^2 x_5 x_7^2 h, x_3 x_6^2 x_7^3, x_3 x_5 x_6 x_7^3, x_3 x_4 x_6 x_7^3,$
$x_3^2 x_6 x_7^3, x_3^2 x_4 h^4, x_3 x_6^3 h^3, x_3 x_5 x_6^2 h^3, x_3 x_4 x_6^2 h^3, x_3 x_4 x_6 x_7^2 h^2, x_3 x_5^2 x_7^2 h^2, x_3 x_4 x_5 x_7^2 h^2,$
$x_3^2 x_4 x_7^2 h^2, x_3^2 h^5 x_5, x_3 x_6 x_7^3 h^3, x_3 x_6^2 x_7^2 h^3, x_3 x_5 x_6 x_7^2 h^3, x_3 x_6 x_7^4 h^2, x_3 x_5 x_7^3 h^4,$
$x_3 x_4 x_7^3 h^4, x_3^2 x_7^3 h^4, x_2 x_3 x_7^3 h^4, x_3 x_4 x_5 x_7 h^6, x_3 x_7^4 h^6 \rangle$ in $K[x_1, \ldots, x_7, h]$ which is the leading term ideal of the homogenization w.r.t. h of Eco7 ideal[1] after performing the linear changes $h = h - x_2, x_7 = x_7 + x_3$. We can observe that $\dim(J) = 5$ and it is D-quasi stable. However, it is not quasi stable because $x_2 x_3 x_6 x_7^2 \in J : h^\infty$ and $x_2 x_3 x_6 x_7^2 \notin J : x_7^\infty$.

It should be noted that Seiler in [25, Theorem 2.6] showed that a monomial ideal is quasi stable iff the ideal and all its primary components are in Nœther position. A weaker form of this result holds for D-quasi stable ideals.

Theorem 3. *A monomial ideal is D-quasi stable iff all its highest dimensional primary components are in Nœther position.*

Example 3. There exists a monomial ideal in D-quasi stable position so that at least one of its primary components is not in Nœther position. Let us consider the monomial ideal J given in Example 2. The associated prime ideals of J are $\langle x1, x2, x3 \rangle, \langle h, x1, x2, x3, x4, x5, x6 \rangle, \langle x1, x2, x3, x4, x5, x6, x7 \rangle, \langle h, x1, x2, x3, x4, x5, x6, x7 \rangle$. We can see that the second component has dimension one, and it is not in Nœther position.

Definition 3. *An ideal I is called* D-quasi stable *if $\mathrm{lt}(I)$ is D-quasi stable.*

Since the generic initial ideal of an ideal is (strongly stable and therefore) D-quasi stable, from Galligo's theorem we conclude that:

Proposition 1. *Let $I \subset R$ be a homogeneous ideal. Then, exists a nonempty Zariski open subset $\mathcal{U} \subset \mathrm{GL}(n, K)$ such that $A.I$ is D-quasi stable with $A \in \mathcal{U}$.*

---

[1] See http://homepages.math.uic.edu/~jan/

In the following, we describe a deterministic algorithm to transform a given ideal into a D-quasi stable ideal.

Algorithm 2. DQS-TRANSFORMATION

Input: $I \subset R$ a homogeneous ideal
Output: A linear transformation Ψ so that $\Psi(I)$ is D-quasi stable
$J := \mathrm{lt}(I)$
$D := \dim(I)$
while DQS-test(J) $\neq$ true **do**
  DQS-test(J)=(false,x_h, x_j)
  $\psi :=$ The map $x_h \longmapsto x_h + a_j x_j$ with $a_j \in K$ a random element
  $J := \mathrm{lt}(\psi(I))$
  if DQS-test(J)=true **then**
    $\Psi := \Psi \cup \{\psi\}$
    $I := \psi(I)$
  end if
end while
return(Ψ)

Theorem 4. *The* DQS-TRANSFORMATION *algorithm terminates and outputs deterministically a linear transformation Ψ so that $\Psi(I)$ is D-quasi stable (and therefore in Nœther position).*

3 Implementation

We have implemented the MAPLE package `Noether.mpl` containing a prototype implementation of the DQS-TRANSFORMATION algorithm which is available at the address `http://amirhashemi.iut.ac.ir/software.html`. In what follows we describe the functionality of the package as well as some achieved tests. After loading the package, we enter the generating set of the ideal and call the main function of the package, as follows, to compute a linear transformation to put the given ideal in D-quasi stable position.

```
>F:=[x_2x_1, x_3x_1, x_3x_2, x_3x_4, x_4x_1, x_4x_2];
>LinearChange(F, [x_1, x_2, x_3, x_4]);
Some information about the computation:
The cpu time is: 0.0468003000000863 sec
The used memory: 277613 bytes
List of variables: [x_1, x_2, x_3, x_4]
Dimension : 1
Noether position: true
Delta regularity: true
WeakDstablity : true
Dstablity : true
Borel fixed: true
```

Used change of variables: $[x_4 = x_4 + 2x_1, x_4 = x_4 + 2x_2, x4 = x_4 + 2x_3]$
These results show that after the linear change $x_4 \mapsto x_4 + 2x_1 + 2x_2 + 2x_3$, the leading term ideal of the new ideal is D-quasi stable (Nœther position), quasi stable (δ-regular), weakly D-stable, D-stable and Borel fixed. For more details on weakly D-stable and D-stable ideals we refer to [15]. It should be emphasized that our algorithm is deterministic and it finds an sparse set of linear changes to put the given ideal in Nœther position. We end this section by presenting an example illustrating the efficiency of our algorithm.

Example 4. Let us consider the Butcher example, i.e. the ideal I generated by the polynomials $a+b+c+d, u+v+w+x, 3ab+3ac+3bc+3ad+3bd+3cd, bu + cu + du + av + cv + dv + aw + bw + dw + ax + bx + cx, bcu + bdu + cdu + acv + adv + cdv + abw + adw + bdw + abx + acx + bcx, abc + abd + acd + bcd, bcdu + acdv + abdw + abcx$ in the polynomial ring $K[a, b, c, d, x, w, u, v]$. Our implementation of DQS-TRANSFORMATION algorithm (after less than one second) suggests the linear change $v \longmapsto v - d$. On the other hand, the linear transformation $w \longmapsto w - d$ is defined by the command NoetherNormalization from the package INVOLUTIVE [2] and this computation takes 24 seconds. Finally, in less than one second and using the MAGMA's command NoetherNormalisation, we get the linear changes $w \mapsto w - 2a - b - c, u \mapsto 3b + c + x, v \mapsto -3a + 4b - 2d + 2x + w + u$.

The experiments we performed seem to show that this first implementation of our new algorithm to put a given ideal in Nœther position is already very efficient. According to our experiments for about 20 examples, we observe that the new algorithm is faster than the function NoetherNormalization from the package INVOLUTIVE and the linear changes proposed by our algorithm is sparser than those proposed by the MAGMA's command NoetherNormalisation.

References

1. Alonso, M.E., Mora, T., Raimondo, M.: Local Decomposition Algorithms. In: Sakata, S. (ed.) AAECC 1990. LNCS, vol. 508, pp. 208–221. Springer, Heidelberg (1991)
2. Bayer, D., Stillman, M.: A Criterion for Detecting m-Regularity. Invent. Math. 87(1), 1–11 (1987)
3. Becker, T., Weispfenning, V.: Gröbner Bases: a Computational Approach to Commutative Algebra. Springer (1993)
4. Bermejo, I., Gimenez, P.: Saturation and Castelnuovo Mumford Regularity. J. Algebra 303, 592–617 (2006)
5. Buchberger, B.: Ein Algorithms zum Auffinden der Basiselemente des Restklassenrings nach einem nuildimensionalen Polynomideal. PhD thesis, Universität Innsbruck (1965)
6. Buchberger, B.: An Algorithm for Finding the Basis Elements in the Residue Class Ring Modulo a Zero Dimensional Polynomial Ideal. German, English translation: J. Symbolic Comput., Special Issue on Logic, Mathematics, and Computer Science: Interactions 41(3-4), 475–511 (2006)

---

[2] See http://wwwb.math.rwth-aachen.de/Janet/

7. Buchberger, B.: A Criterion for Detecting Unnecessary Reductions in the Construction of Gröbner Bases. In: Ng, K.W. (ed.) EUROSAM 1979 and ISSAC 1979. LNCS, vol. 72, pp. 3–21. Springer, Heidelberg (1979)
8. Eisenbud, D.: Commutative Algebra with a View toward Algebraic Geometry. Springer, New York (1995)
9. Eisenbud, D., Huneke, C., Vasconcelos, W.V.: Direct Methods for Primary Decomposition. Invent. Math. 110, 207–235 (1992)
10. Derksen, H., Kemper, G.: Computational Invariant Theory. Springer, Berlin (2002)
11. Gianni, P., Trager, B., Zacharias, G.: Gröbner Bases and Primary Decomposition of Polynomial Ideals. J. Symb. Comput. 6, 149–167 (1988)
12. Giusti, M., Hägele, K., Lecerf, G., Marchand, J., Salvy, B.: The Projective Nœther Maple Package: Computing the Dimension of a Projective Variety. J. Symbolic Comput. 30(3), 291–307 (2000)
13. Hashemi, A.: noether.lib. A Singular 3.0.3 distributed library for computing the Nœther normalization (2007)
14. Hashemi, A.: Efficient Algorithms for Computing Nœther Normalization. In: Kapur, D. (ed.) ASCM 2007. LNCS (LNAI), vol. 5081, pp. 97–107. Springer, Heidelberg (2008)
15. Hashemi, A., Schweinfurter, M., Seiler, W.M.: Deterministically Computing Reduction Numbers of Polynomial Ideals. arXiv:1404.1721, 16 pages (2014)
16. Hermann, D.: Die Frage der endlich vielen Schritte in der Theorie der Polynomideale. Math. Ann. 95, 736–788 (1926)
17. Kemper, G.: Calculating Invariant Rings of Finite Groups over Arbitrary Fields. J. Symb. Comput. 21, 351–366 (1996)
18. Kemper, G.: Calculating Invariant Rings of Finite Groups over Arbitrary Fields. J. Symb. Comput. 21, 351–366 (1996)
19. Lecerf, G.: Computing the Equidimensional Decomposition of an Algebraic Closed Set by Means of Lifting Fibers. J. of Complexity 19(4), 564–596 (2003)
20. Logar, A.: A Computational Proof of the Nœther Normalization Lemma. In: Mora, T. (ed.) AAECC 1988. LNCS, vol. 357, pp. 259–273. Springer, Heidelberg (1989)
21. Krick, T., Logar, A.: An Algorithm for the Computation of the Radical of an Ideal in the Ring of Polynomials. In: Mattson, H.F., Rao, T.R.N., Mora, T. (eds.) AAECC 1991. LNCS, vol. 539, pp. 195–205. Springer, Heidelberg (1991)
22. Matsumoto, R.: Computing the Radical of an Ideal in Positive Characteristic. J. Symb. Comput. 32, 263–271 (2001)
23. Greuel, G.M., Pfister, G.: A Singular Introduction to Commutative Algebra. Springer, Berlin (2002)
24. Robertz, D.: Noether Normalization Guided by Monomial Cone Decompositions. J. Symb. Comput. 44, 1359–1373 (2009)
25. Seiler, W.M.: Effective Genericity, Delta-Regularity and Strong Nœther Position. Communications in Algebra 40, 3933–3949 (2012)
26. Seiler, W.M.: A Combinatorial Approach to Involution and δ-Regularity II: Structure Analysis of Polynomial Modules with Pommaret Bases. Appl. Alg. Eng. Comm. Comp. 20, 261–338 (2009)
27. Sturmfels, B.: Algorithm in Invariant Theory. Springer (2008)
28. Vasconcelos, W.V.: Computational Methods in Commutative Algebra and Algebraic Geometry. Springer (1998)

Generic and Parallel Groebner Bases in JAS
(Extended Abstract)

Heinz Kredel

University of Mannheim, Germany
kredel@rz.uni-mannheim.de
www.uni-mannheim.de

Abstract. We present generic, type safe Groebner bases software. The implemented algorithms distinguish Groebner base computation in polynomials rings over fields, rings with pseudo division, parameter rings, regular rings, Euclidean rings, non-commutative fields in commuting, solvable and free-non-commuting main variables. The interface, class organization is described in the object-oriented programming environment of the Java Algebra System (JAS). Different critical pair selection strategies and reduction algorithms can be provided by dependency injection. Different implementations can be selected for the mentioned coefficient rings through factory classes and methods. Groebner bases algorithms can be composed according to application needs and/or hardware availability. For example, versions for shared memory sequential or parallel computation, term order optimization or fraction free coefficient ring computation can be composed. For distributed memory compute clusters there are OpenMPI and MPJ implementations of Buchberger's algorithm with optimized distributed storage of reduction polynomials.

Keywords: generic multivariate polynomials, generic Groebner bases, algorithm composition, parallel computation.

1 Introduction

As introductory example we consider the polynomial ring

$$R = E[y, z] = \mathbb{Q}(\sqrt{2})(x)(\sqrt{x})[y, z],$$

in y and z over the field E, where E is an extension of $\mathbb{Q}$ by a square-root of 2, a transcendent x and the square-root of x. In full generality the Java type of the JAS coefficients of $E = \mathbb{Q}(\sqrt{2})(x)(\sqrt{x})$ would be

```
AlgebraicNumber<Quotient<AlgebraicNumber<BigRational>>>
```

It consists of two nested algebraic number types `AlgebraicNumber` and a polynomial quotient `Quotient` over the rational numbers `BigRational`. The corresponding 'factory', a means to provide methods to create new elements of this type, has to be constructed as a Java object, referenced by `cfac`, with the type `AlgebraicNumberRing<.>`.

H. Hong and C. Yap (Eds.): ICMS 2014, LNCS 8592, pp. 390–397, 2014.

```
AlgebraicNumberRing<Quotient<AlgebraicNumber<BigRational>>> cfac = ...
```

A Gröbner base could then be constructed by first obtaining a suitable implementation via method `getImplementation(cfac)` of class `GBFactory`. `cfac` references a constructed object, representing the field extension E. The constructed algorithm is referenced by variable `bb`, which is declared as interface `Groebner-Base` with the given polynomial coefficient type. In the next step, method `GB()` of the constructed algorithm will actually do the computation. Method `isGB()` is used to test if the result polynomial list in `G` is a Gröbner base.

```
GroebnerBase<AlgebraicNumber<Quotient<AlgebraicNumber<BigRational>>>> bb;
bb = GBFactory.getImplementation(cfac);
List<
  GenPolynomial<AlgebraicNumber<Quotient<AlgebraicNumber<BigRational>>>>
> G, F = ...;
G = bb.GB(F);
System.out.println("isGB(G) = " + bb.isGB(G));
```

This example seems to be too complicated. A simpler polynomial ring could suffice, for example the polynomial ring $R' = \mathbb{Q}[w_2, x_i, x, w_x, y, z]$, over the field $\mathbb{Q}$ would do, together with adding the polynomials $w_2^2 - 2$, $w_x^2 - x$ and $x_i x - 1$ to the list of polynomials `F`. However, with ring R' it is not possible to simplify the inverse x_i of x in the result polynomials in all cases. So a better polynomial ring is $R'' = \mathbb{Q}(x)[w_2, w_x, y, z]$, over the field $\mathbb{Q}$ with transcendent extension by x. The simplification of $\sqrt{2}$ and $\sqrt{x}$ is achieved by adding the polynomials $w_2^2 - 2$ and $w_x^2 - x$ to the the list `F`. The type of the coefficients then simplifies to

```
Quotient<BigRational>
```

The corresponding factory is referenced by variable `qfac`.

```
QuotientRing<BigRational> qfac = ...;
```

The Gröbner base construction with the simpler types is then as follows.

```
GroebnerBaseAbstract<Quotient<BigRational>> bb;
bb = GBFactory.getImplementation(qfac);
List<GenPolynomial<Quotient<BigRational>>> G, F = ...;
// add w2^2 - 2 and wx^2 - x to F
G = bb.GB(F);
```

The examples show the JAS generic polynomials with different coefficient types. The generic Gröbner base implementation will, in this cases, work for different coefficient rings. As various coefficient rings are available and several Gröbner base implementations suitable for specific kinds of coefficients are available, a means to select an appropriate algorithm is mandatory.

Generic multivariate polynomials are provided by the object oriented Java computer algebra system (called JAS) as a type safe and thread safe approach to computer algebra, see [8,5,6]. JAS provides a well designed software library using generic types for algebraic computations implemented in the Java programming language thus leveraging software and hardware improvements over time. For an introduction to JAS see the cited articles.

In section 2 we will sketch some generic Gröbner base implementations and in section 3 we discuss the problem of implementation selection and composition.

2 Generic Gröbner Bases

In this section we explain the implementations behind the Gröbner base selection facilities of class GBFactory. Part of the algorithm relations and the interface and class layout is depicted in figure 1. It shows some of the Gröbner base implementations for polynomial rings over *fields*. Not shown are further implementations and also implementations for coefficients from unique factorization domains or Euclidean domains. Gröbner bases implementations for non-commutative polynomial rings or regular rings are also not discussed.

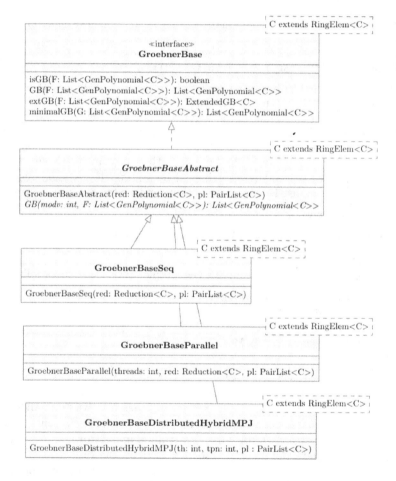

Fig. 1. UML diagram of Gröbner base classes

Figure 1 starts with the interface GroebnerBase on the top. It defines the most important methods for the usage of Gröbner base algorithms: method isGB() tests if a list of generic polynomials is in fact a Gröbner base, method GB()

computes a Gröbner base for a list of generic polynomials, method `extGB()` computes an extended Gröbner base (returning the back and forth transformations between the given polynomial list and the Gröbner base in container `ExtendedGB`) and method `minimalGB()` computes a minimal reduced Gröbner base from an arbitrary Gröbner base. The interface is parametrized by a type `C`, which is restricted to implement the `RingElem` interface. The `RingElem` interface is itself parametrized by the type `C`. This allows for recursive coefficient types. The `RingElem` interface defines all methods needed for ring arithmetic. It includes also a method `inverse()` to compute inverses of ring elements. In case of fields, all non-zero elements will have inverses, for arbitrary rings only some elements will have inverses.

The class `GroebnerBaseAbstract` implements all methods of the interface `GroebnerBase` and defines the abstract method `GB(modv: int,.)`. This method has a parameter `modv` which is used for module Gröbner bases via an embedding of a module over a polynomial ring into an polynomial ring. `modv` accounts for the number of polynomial variables to be treated as module variables. This class also defines a constructor `GroebnerBaseAbstract(red, pl)` which accepts a reduction parameter `red` of type `Reduction` and a pair-list parameter `pl` of type `PairList`. By this parameters dependencies on different implementations are injected into the algorithm. `Reduction` provides a polynomial reduction implementation consisting of methods like `normalform()` and `SPolynomial()` to compute reductions of a polynomial with respect to a list of polynomials and a S-polynomial, respectively. The `PairList` parameter provides an implementation of the basic book keeping of critical pairs during a Gröbner base construction. For example the application of the reduction avoiding strategies like the Buchberger criteria or the Gebauer-Möller criteria [2,4]. `PairList` provides methods like `put()` to add all critical pairs for a new polynomial, `removeNext()` to obtain the next critical pair according to the implemented strategy and `hasNext()` to test if there are remaining critical pairs.

The class `GroebnerBaseSeq` implements method `GB(modv,.)` as a sequential Buchberger algorithm. It inherits all other methods from `GroebnerBaseAbstract`.

A parallel implementation of the Buchberger algorithm is provided by the class `GroebnerBaseParallel`, also with method `GB(modv,.)`. All other methods are again inherited from `GroebnerBaseAbstract`. The implementation of `GB(modv,.)` uses Java `Threads` and keeps the polynomial list in shared memory. The `ReductionPar` class is tailored to tolerate asynchronous updates of the polynomial list during a reduction. For the classes implementing `PairList` it is now important that the methods `put()`, `removeNext()` and `hasNext()` are thread safe. So multiple threads will not interfere with the book keeping of the critical pairs. Some care is needed to correctly check for the termination of the algorithm, since a single remaining thread may finally produce a new non-zero polynomial, thus potentially generating a new cascade of critical pairs. Note, that since polynomials produced by the multiple threads may appear in a different sequence order than in the sequential algorithm. The constructor has additionally a parameter th which specifies the number of threads to use. So the sequential algorithm can be easily replaced by the parallel algorithm, just by using the constructor of this class. This leverages the ubiquitous multi-core computers of our time.

There are several implementations for distributed memory computers, like compute clusters consisting of multi-core nodes. One is using Java TCP/IP Sockets as transport layer, an other uses the OpenMPI Java bindings, here we sketch only one pure Java implementation based on the MPJ API, FastMPJ [9,7]. FastMPJ is thread safe if used with the newio device (OpenMPI is not thread safe at the moment, Sockets are thread safe). The class GroebnerBase-HybridMPJ extends GroebnerBaseAbstract and implements the missing method GB(modv,.). The constructor has two additional parameters: th, the number of MPJ processes to use (must match MPI.Size()), and tpn, the number of threads to use per MPJ process. The number of MPJ processes should match the number of available compute nodes. The constructor obtains the MPJ run-time system, i.e. the program must be run within the MPJ environment, and eventually calls MPI.Init(). The method GB(modv,.) now uses a case distinction whether it is running on the MPJ master node (MPI.Rank()==0), or one a worker node. The master node does the initialization and book keeping of the critical pairs. The polynomial list is replicated to all MPJ processes via class DistHashTableMPJ and updated asynchronously to the normal communication between master and worker. On the worker nodes method GB(modv,.) starts the requested number of threads and connects to the replicated polynomial list. Each thread requests a critical pair from the master node, performs the normalform() reduction and sends the reduced polynomial back to the master. On the master node the critical pairs are eventually updated. The transport of polynomials between the MPJ processes uses Java's object serialization. Critical pairs are only communicated as indexes into the replicated polynomial list. Termination detection is more complicated since a last thread on a worker node could produce a non-zero polynomial and so generate new critical pairs.

3   Implementation Selection and Composition

As there are many implemented Gröbner base algorithms it is not always evident for users which one to choose. Moreover there are further optimizations which can be composed with these basic algorithms. As solution for the first problem we provide a factory class GBFactory, which selects an implementation based on the coefficient ring of the polynomials. The composition problem is resolved with an algorithm builder class GBAlgorithmBuilder.

3.1   Selection of an Implementation

The GBFactory, see figure 2, provides static methods getImplementation() and getProxy(). These method are polymorphic for various coefficient rings, for example BigInteger, BigRational, ModInteger, QuotientRing<C> or von

GBFactory
getImplementation(f: BigInteger): GroebnerBaseAbstract<BigInteger>
getImplementation(f: BigInteger, a: GBFactory.Algo):   GroebnerBaseAbstract<BigInteger>
getImplementation(f: BigRational): GroebnerBaseAbstract<BigRational>
getImplementation(f: BigRational, a: GBFactory.Algo):   GroebnerBaseAbstract<BigRational>
getImplementation(f: ModIntegerRing): GroebnerBaseAbstract<ModInteger>
getImplementation(f: ModLongRing): GroebnerBaseAbstract<ModLong>
getImplementation(f: GenPolynomialRing<C>):   GroebnerBaseAbstract<GenPolynomial<C>>
getImplementation(f: GenPolynomialRing<C>, a: GBFactory.Algo):   GroebnerBaseAbstract<GenPolynomial<C>>
getImplementation(f: QuotientRing<C>, a: GBFactory.Algo):   GroebnerBaseAbstract<Quotient<C>>
getImplementation(f: ProductRing<C>): GroebnerBaseAbstract<Product<C>>
getImplementation(f: RingFactory<C>): GroebnerBaseAbstract<C>
getProxy(f: RingFactory<C>): GroebnerBaseAbstract<C>

Most methods also allow a `PairList<C>` parameter, which is omitted in the diagram.

Fig. 2. UML diagram of Gröbner base factory

Neumann regular rings `ProductRing<C>`. Additionally there is a general one for arbitrary `Ringfactory<C>` coefficients. In each of these methods an appropriate Gröbner base implementation is selected and returned as a `GroebnerBase-Abstract` with corresponding type parameter. In cases where the coefficient type alone is not sufficient to select an implementation there is an additional parameter `GBFactory.Algo`, see figure 3. Via this parameter, one can choose between computation with fractional coefficients (`qgb`) or fraction free coefficients (`ffgb`) (by multiplying all polynomials with the least common multiple of the denominators of the coefficients and using polynomial pseudo division algorithms). Further, for the coefficients `BigInteger` and univariate `GenPolynomial<C>`, it is possible to choose between pseudo division algorithms (`igb`) or between d- and e-Gröbner bases (`egb`, `dgb`).

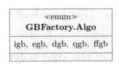

Fig. 3. Gröbner base algorithm enumeration

Method `getProxy()` in figure 2 returns a `GBProxy` object. The class `GBProxy` also extends `GroebnerBaseAbstract` and implements method `GB(modv, .)`. The constructor takes two Gröbner base implementations as parameters

```
GBProxy(GroebnerBaseAbstract<C> e1, GroebnerBaseAbstract<C> e2).
```

The GB() method then executes two Gröbner base algorithms in parallel and returns result of the first finished computation. Therefore it uses the method invokeAny() from class ExecutorService in package java.util.concurrent. In the case of getProxy(), the GBProxy is constructed with a sequential and a parallel Gröbner base implementation (as available). So, in case the sequential algorithm runs faster, its result will be returned. But if the parallel algorithm is faster, its result will be taken. The usage is as simple as with getImplementation().

```
GroebnerBaseAbstract<Quotient<BigRational>> bb;
bb = GBFactory.getProxy(qfac); // get a parallel implementation
List<GenPolynomial<Quotient<BigRational>>> G, F = ...;
G = bb.GB(F);
```

3.2 Composition of Implementations

There exist further variants of Gröbner base algorithms. One example is the FGLM algorithm [3] to compute a Gröbner base with respect to a graded term order and to construct one with respect to a lexicographic term order from it. A second example is an optimization of the variable order [1]. The selection of such variants is implemented in the class GBAlgorithmBuilder, see figure 4.

GBAlgorithmBuilder
GBAlgorithmBuilder(r: GenPolynomialRing<C>) polynomialRing(r: GenPolynomialRing<C>): GBAlgorithmBuilder<C> euclideanDomain(): GBAlgorithmBuilder<C> domainAlgorithm(a: GBFactory.Algo): GBAlgorithmBuilder<C> normalPairlist(): GBAlgorithmBuilder<C> syzygyPairlist(): GBAlgorithmBuilder<C> fractionFree(): GBAlgorithmBuilder<C> graded(): GBAlgorithmBuilder<C>// FGLM algorithm optimize(): GBAlgorithmBuilder<C>// variable ordering parallel(): GBAlgorithmBuilder<C>// using GBProxy parallel(threads: int): GBAlgorithmBuilder<C> build(): GroebnerBaseAbstract<C>// final construction

Fig. 4. UML diagram of Gröbner base algorithm builder

One starts with the definition of the used polynomial ring by method polynomialRing() (not the coefficient ring) and ends with the build() method, which returns the desired Gröbner base implementation. Method optimize() will add a variable order optimization layer, graded() will add an FGLM algorithm, the *Pairlist() methods select between critical pair selection strategies and euclideanDomain() will select a e-Gröbner base computation. For example a fraction-free and parallel algorithm using 5 threads can be selected as follows.

```
GenPolynomialRing<Quotient<BigRational>> pfac = ...;
bb = GBAlgorithmBuilder.polynomialRing(pfac)
```

```
                      .fractionFree()
                      .parallel(5)
                      .build();
List<GenPolynomial> G, F = ...;
G = bb.GB(F);
```

The composition uses the same concept as shown in class **GBProxy** above: the constructor has an additional parameter for a backing implementation of a Gröbner base algorithm. In case of the FGLM algorithm it is

```
GroebnerBaseFGLM(GroebnerBaseAbstract<C> gb).
```

Here, the **gb** implementation will be used to compute the Gröbner base with respect to the graded term order to which the FGLM algorithm is then applied.

There are various implemented applications of Gröbner bases, for example ideal constructions, syzygies, real and complex roots or primary decomposition, which can not be presented here in the available space.

Acknowledgments. I thank Thomas Becker for discussions on the implementation of a polynomial template library and Raphael Jolly for the fruitful cooperation on the generic type system suitable for a computer algebra system. Thanks also to Markus Aleksy for encouraging and supporting this work. Thanks for providing the computational resources go to the bwGRiD and bwHPC projects.

References

1. Böge, W., Gebauer, R., Kredel, H.: Some examples for solving systems of algebraic equations by calculating Gröbner bases. J. Symb. Comp. 2/1(1), 83–98 (1986)
2. Buchberger, B.: Gröbner bases: An algorithmic method in polynomial ideal theory. In: Bose, N. (ed.) Recent Trends in Multidimensional Systems Theory, Reidel, pp. 184–232 (1985)
3. Faugère, J.C., Gianni, P., Lazard, D., Mora, T.: Efficient computation of zero-dimensional Gröbner bases by change of ordering. J. Symbolic Computation 16(4), 329–344 (1994)
4. Gebauer, R., Möller, H.M.: On an installation of Buchberger's algorithm. J. Symb. Comput. 6(2/3), 275–286 (1988)
5. Kredel, H.: On a Java Computer Algebra System, its performance and applications. Science of Computer Programming 70(2-3), 185–207 (2008)
6. Kredel, H.: Unique factorization domains in the java computer algebra system. In: Sturm, T., Zengler, C. (eds.) ADG 2008. LNCS, vol. 6301, pp. 86–115. Springer, Heidelberg (2011)
7. Kredel, H.: Distributed Gröbner bases computation with MPJ. In: IEEE AINA Workshops, Barcelona, Spain, pp. 1429–1435 (2013)
8. Kredel, H.: The Java algebra system (JAS). Technical report (2000), http://krum.rz.uni-mannheim.de/jas/
9. Taboada, G.L., Ramos, S., Expósito, R.R., Touriño, J., Doallo, R.: FastMPJ a high performance Java message passing library. Technical report (2011), http://fastmpj.com/

Application of Groebner Basis Methodology to Nonlinear Mechanics Problems

Y. Jane Liu[1] and John Peddieson[2]

[1] Department of Civil and Environmental Engineering
Tennessee Technological University, Cookeville, Tennessee 38505, USA
jliu@tntech.edu
http://www.tntech.edu/people/jliu/
[2] Department of Mechanical Engineering
Tennessee Technological University, Cookeville, Tennessee 38505, USA
jpeddieson@tntech.edu
http://www.tntech.edu/people/jpeddieson/

Abstract. The application of the Groebner basis methodology to four nonlinear mechanics problems is discussed. The MAPLE software package is used in all cases to implement the Groebner basis calculation which converts a set of coupled polynomial algebraic equations into an equivalent set of uncoupled polynomial algebraic equations (the reduced Groebner basis). Observations concerning implementation of Groebner basis methodology are reported.

Keywords: Groebner basis, Computational algebraic geometry, Cable statics, Plate vibrations, Steady state vibrations, Free vibrations.

1 Introduction

With the increasing capability of symbolic computation in recent years, considerable progress has been made in the area of advanced computational algebraic geometry. One such advanced computational method is the methodology of Groebner bases which was introduced in 1965 by Buchberger [1], who was the first to provide a useful algorithm for the determination of Groebner bases. It is primarily because this algorithm has been implemented in many mathematical symbolic computational software packages that the implementation of Groebner basis methodology in science and engineering is now feasible. The purpose of this paper is to demonstrate the utility of the Groebner basis methodology in the analysis of four nonlinear mechanics problems to be described in Sections 2–5 below. Because of the variety of applications involved each of these sections employs a separate notation. This should cause no difficulty because there is no interaction between these sections. The MAPLE software package has been used in all cases to implement the conversion of a set of coupled polynomial algebraic equations into an equivalent set of uncoupled polynomial algebraic equations (the reduced Groebner basis). The details of the mathematical background underlying the methodology can be found in books such as Cox et al. [2].

H. Hong and C. Yap (Eds.): ICMS 2014, LNCS 8592, pp. 398–405, 2014.

For a fixed elimination order, the application of Groebner basis methodology produces one polynomial equation containing only one unknown (primary equation) together with additional polynomial equations (secondary equations) that can be solved sequentially to obtain the other unknowns one at a time. If the primary equation is of the fourth order or lower this process produces a closed form solution containing all the system parameters in symbolic form. If the primary equation is of the fifth order or higher the uncoupled polynomial equations must be solved numerically with specified numerical values for all system parameters. Here the potential benefit of Groebner basis methodology is the possibility that the effectiveness of iterative numerical methods could be improved by uncoupling (since iteration is performed on one equation at a time rather than on a highly coupled set).

2   Static Cable Analysis

The problem

$$(T(1+u'))'r - \frac{1}{2}r'T(1+u') + Qv'r = 0 \tag{1}$$

$$(Tv')'r - \frac{1}{2}r'Tv' - Q(1+u')r = 0 \tag{2}$$

$$r = (1+u')^2 + v'^2 \tag{3}$$

$$T = T_0 + \frac{1}{2}EA\,(r-1) \tag{4}$$

$$u(0) = u(L) = v(0) = v(L) = 0 \tag{5}$$

arises in the analysis of large plane static deflections of initially straight pretensioned linearly elastic cables fixed at both ends (see Liu et al. [3] and the references contained therein). In (1)–(5), L is the initial cable length, x is coincident with the initial cable position, y is perpendicular to x, T is the cable tension, u and v are displacements in the respective x and y directions, Q is a constant force per unit of undeformed length which initially acts the negative y direction and remains perpendicular to the deformed cable, T_0 is the initial cable pre-tension, E is the cable material Young's modulus, A is the cable cross-sectional area, and a superposed prime denotes differentiation with respect to x. The use of the Galerkin method in conjunction with the trial functions

$$u = a_0 x \left(\frac{x}{L} - 1\right) \left(\frac{x}{L} - \frac{1}{2}\right) \tag{6}$$

$$v = b_0 x \left(\frac{x}{L} - 1\right) \tag{7}$$

with a_0 and b_0 being unknown constants then leads to the system of polynomial algebraic equations

$$286QLb_0a_0^2 + 159EAa_0^5 + 10296EAa_0^3 + 48048EAa_0 + 1144QLb_0a_0$$
$$+ 1560EAa_0^4 + 13728EAa_0^2 + 32032EAb_0^2 + 25168EAa_0b_0^2$$
$$+ 12012EAa_0^2b_0^2 + 6864EAb_0^4 + 8008QLb_0 + 1768EAa_0^3b_0^2$$
$$+ 6848EAa_0b_0^4 + 9152T_0a_0b_0^2 + 3432QLb_0^3 - 16016T_0b_0^2 = 0 \quad (8)$$

$$- 3080QL + 924QLa_0 - 330QLa_0^2 + 1628EAa_0^2b_0 + 28EAa_0^4b_0$$
$$+ 11QLa_0^3 - 44QLa_0b_0^2 + 2464EAa_0b_0 + 3696EAb_0^3$$
$$+ 110EAa_0^3b_0 + 264EAa_0^2b_0^3 + 88T_0a_0^2b_0 + 1320EAa_0b_0^3$$
$$- 616QLb_0^2 + 1056EAb_0^5 - 3080T_0a_0b_0 + 6160T_0b_0 = 0 \quad (9)$$

Equations (8) and (9) (and subsequent similar equations) have been written in the un-simplified forms characteristic of MAPLE output.

The MAPLE Groebner basis module was used to uncouple (8) and (9) into one twenty-fifth order primary equation containing only b_0 and a secondary equation containing both a_0 and b_0 and linear in the former. These equations (which contain all the system parameters in symbolic form) are very lengthy and are not shown for the sake of brevity. Tables 1 and 2 contain some typical numerical results obtained from the uncoupled equations. Only a small amount of computer time was needed to generate these results, despite the large number of extraneous solutions inherent in the process.

Table 1. Coefficients appearing in (6), (7) ($QL/T_0 = 0.5$)

$T_0/EA = 0.1$		$T_0/EA = 1$		$T_0/EA = 5$	
$-a_0$	b_0	$-a_0$	b_0	$-a_0$	b_0
0.0355	0.230	0.0410	0.249	0.0406	0.251

Table 2. Coefficients appearing in (6), (7) ($QL/T_0 = 5$)

$T_0/EA = 0.1$		$T_0/EA = 1$		$T_0/EA = 5$	
$-a_0$	b_0	$-a_0$	b_0	$-a_0$	b_0
0.646	0.977	1.88	1.80	3.31	2.68

3   Forced Damped Vibration

The differential equation

$$\ddot{x} + Q\dot{x}x^2 + \omega^2 x = X_0\omega^2 \cos(\Omega t) \qquad (10)$$

arises in the analysis of harmonically forced vibration of a mass attached to a linear spring and a van der Pol damper (see Liu and Peddieson [4] and the references contained therein). In (10), x is the position of the mass, a superposed dot denotes differentiation with respect to the time t, ω is the undamped natural frequency, Ω is the forcing frequency, X_0 is the static deflection, and Q is a constant indicating the strength of the damping. Substituting the two-harmonic approximate steady state solution

$$x = C_1 \cos(\Omega t) + D_1 \sin(\Omega t) + C_3 \cos(3\Omega t) + D_3 \sin(3\Omega t) \qquad (11)$$

(with the C's and D's being constants) into (10) and performing harmonic balance (a well-known procedure for obtaining approximate harmonic solutions to nonlinear vibration problems, see Nayfeh and Mook [5]) in the usual way (using the capability of MAPLE to evaluate the required orthogonalization integrals) yields the system of polynomial algebraic equations

$$4\omega^2 C_1 - 4\Omega^2 C_1 + \Omega Q C_1^2 D_1 + \Omega Q D_1^3 - 2\Omega Q C_1 C_3 D_1 + 2\Omega Q C_3^3 D_1$$
$$+ 2\Omega Q D_1 D_3^2 - \Omega Q D_1^2 D_3 + \Omega Q C_1^2 D_3 - 4X_0\omega^2 = 0 \quad (12)$$

$$4\omega^2 D_1 - 4\Omega^2 D_1 - \Omega Q C_1 D_1^2 - \Omega Q C_1^3 - 2\Omega Q C_1 C_3 D_1 + 2\Omega Q C_3^2 D_1$$
$$+ 2\Omega Q D_1 D_3^2 - \Omega Q D_1^2 D_3 + \Omega Q C_1^2 D_3 - 4X_0\omega^2 = 0 \quad (13)$$

$$4\omega^2 C_3 - 36\Omega^2 C_3 + 3\Omega Q C_1^2 D_1 + 3\Omega Q C_3^2 D_3 - \Omega Q D_1^3 + 3\Omega Q D_3^3$$
$$+ 6\Omega Q C_1^2 D_3 + 6\Omega Q D_1^2 D_3 = 0 \quad (14)$$

$$4\omega^2 D_3 - 36\Omega^2 D_3 + 3\Omega Q C_1 D_1^2 - 3\Omega Q C_3 D_3^2 - \Omega Q C_1^3 - 3\Omega Q C_3^3$$
$$- 6\Omega Q C_1^2 C_3 - 6\Omega Q C_3 D_1^2 = 0 \quad (15)$$

to solve for the C's and D's. The MAPLE Groebner basis module produces an alternate set of four equations which are omitted from this document because of their length. The primary equation contains only C_1 and is of the ninth order. Of the secondary equations, one contains only C_1 and C_3, one only C_1 and D_1, and one only C_1 and D_3. All four contain the system parameters ω, Ω, Q, and X_0. The equation for C_1 must be solved numerically for each combination of system parameters. Some typical numerical results are presented in Table 3 for the amplitudes of the respective first and second harmonics $A_1 = (C_1^2 + D_1^2)^{1/2}$ and $A_3 = (C_3^2 + D_3^2)^{1/2}$. The computing time needed to generate these results was very small despite the large number of extraneous solutions inherent in the process. It can be seen that the amplitude of the second harmonic is small compared to the amplitude of the first harmonic for all tabulated cases.

The corresponding single harmonic approximation ($C_3 = 0$, $D_3 = 0$ in (11)) to the steady state solution of (10) can be recovered from (12)–(15) by omitting (14) and (15) and equating C_3 and D_3 to zero in (12) and (13). This produces a system of two polynomial algebraic equations to determine C_1 and D_1. The MAPLE Groebner basis module transforms these equations to a cubic primary equation containing only C_1 and a secondary equation containing C_1 and D_1 and linear in the latter. Here the known closed form solution for a cubic equation can be used to produce a solution valid for any combination of system parameters.

The Groebner bases associated with one and two harmonic solutions were easily obtained. When, however, a three harmonic solution was attempted the MAPLE Groebner basis module was unable to generate the Groebner basis because all the storage of the standard PC being used was exhausted. This would appear to signify a major limitation on the practical utility of the Groebner basis approach.

Table 3. Coefficients appearing in (11) ($QX_0^2/(4\omega) = 1.5$)

				Ω/ω				
1/10	1/2	1	3/2	2	5/2	3	7/2	4
A_1 0.98846	1.00819	0.88724	0.63555	0.33102	0.19040	0.12499	0.08889	0.06667
A_3 0.11411	0.16016	0.09776	0.02887	0.00310	0.00047	0.00011	0.00003	0.00001

4   Free Undamped Vibration

The problem
$$X'' + X + \varepsilon X^3 = 0, X(0) = 1, X'(0) = 0 \tag{16}$$

arises in the analysis of free vibration of a mass attached to a cubically non-linear spring. The differential equation is an example of a free Duffing equation (see [5]). This problem is chosen because it has a closed form solution in terms of elliptic functions that can be used as a standard of comparison for the harmonic balance approximation employed herein. In (16) X is the ratio of the mass's current position to its initial position, a superposed prime denotes differentiation with respect to the product of linear natural frequency and time T, and ε is a dimensionless measure of the importance of nonlinearity. In the previous two sections the equations were stated in dimensional forms in order to illustrate the capability of the Groebner basis methodology to deal with large numbers of parameters. The number of parameters needed to characterize a physical problem can, of course, be minimized by using a dimensionless formulation. That has been done in this section. Substituting the harmonic solution
$$X = C_1 \cos(rT) + C_3 \cos(3rT) + C_5 \cos(5rT) \tag{17}$$

(r being the ratio of the natural frequency of the nonlinear system to that of the corresponding linear system) into (16) and performing harmonic balance in the usual way yields the three polynomial equations

$$6\varepsilon C_1 C_3 C_5 + 4C_1 + 6\varepsilon C_1 C_5^2 + 6\varepsilon C_1 C_3^2 + 3\varepsilon C_3^2 C_5 + 3\varepsilon C_1^3$$
$$+ 3\varepsilon C_1^2 C_3 - 4r^2 C_1 = 0 \quad (18)$$

$$6\varepsilon C_1 C_3 C_5 + 4C_3 + 6\varepsilon C_3 C_5^2 + 6\varepsilon C_3 C_1^2 + 3\varepsilon C_1^2 C_5 + 3\varepsilon C_3^3$$
$$+ \varepsilon C_1^3 - 36r^2 C_3 = 0 \quad (19)$$

$$6\varepsilon C_1 C_3 C_5 + 4C_5 + 6\varepsilon C_5 C_1^2 + 6\varepsilon C_5 C_1^2 + 3\varepsilon C_3^2 C_1 + 3\varepsilon C_1^2 C_3$$
$$+ 3\varepsilon C_5^3 - 100r^2 C_5 = 0 \quad (20)$$

together with
$$C_1 + C_3 + C_5 - 1 = 0 \tag{21}$$

If only the first term is retained in (17) ($N = 1$) the simple closed form solution

$$C_1 = 1, r = (1 + 3\varepsilon/4)^{1/2} \tag{22}$$

results. If two terms are retained in (17) the primary equation is found to be cubic and a closed form solution exists. If all three terms are retained the MAPLE Groebner basis module is able to generate the Groebner basis with an eleventh order primary equation. No closed form solution exists but numerical results are easily obtained. A representative sample of these results is presented in Table 4, together with exact frequency ratio values based on the closed form solution in terms of elliptic functions. It can be seen that two terms of (17) appear adequate to reproduce the exact solution over the entire range of ε displayed. The value $\varepsilon = -1/3$ was chosen for inclusion in Table 4 because it is the negative value of this parameter exhibiting the largest absolute value for which the absolute value of the spring force will increase with the absolute value of the displacement in the entire permissible range of $-1 \le X \le +1$. Even the single harmonic solution produces quite accurate values with the greatest error being about 2.2%. This lack of need for the higher harmonics is indeed fortunate because when a solution was attempted using five harmonics the MAPLE Groebner basis module could not generate the Groebner basis due to insufficient storage of the standard PC being used. This was true despite the fact that only the single parameter ε was present in the equations. In addition, attempts to generate the Groebner basis using specific numerical values of ε also failed. This is a second example of the limitation mentioned earlier.

Table 4. Coefficients and frequency ratio appearing in (17)

	C_1			C_3		C_5	r			
ε	$N=1$	$N=2$	$N=3$	$N=2$	$N=3$	$N=3$	$N=1$	$N=2$	$N=3$	Exact
-1/3	1.0000	1.0139	1.0137	-0.0139	-0.0139	0.0002	0.866	0.864	0.864	0.864
1/2	1.0000	0.9886	0.9884	0.0114	0.0114	0.0001	1.173	1.171	1.171	1.171
1	1.0000	0.9820	0.9817	0.0180	0.0180	0.0003	1.323	1.318	1.318	1.318
3/2	1.0000	0.9778	0.9772	0.0222	0.0223	0.0005	1.458	1.445	1.449	1.449
2	1.0000	0.9747	0.9741	0.0253	0.0253	0.0006	1.581	1.570	1.569	1.569
20	1.0000	0.9599	0.9582	0.0401	0.0402	0.0016	4.000	3.930	3.924	3.925
100	1.0000	0.9577	0.9557	0.0423	0.0424	0.0018	8.718	8.550	8.534	8.533

5   Free Orthotropic Plate Vibrations

Equations having the form of (16) arise in many nonlinear vibration problems. One class of such problems is that of determining the fundamental frequencies of moderately large amplitude free vibrations of linearly elastic thin plate and shell structures. A specific example of such a structure selected for consideration herein is a rectangular orthotropic plate having thickness h; side lengths a and b; four independent elastic constants E_L, E_T, G_{LT}, and ν_{LT}; and initial central displacement W_0. It can then be shown that the use of the Rayleigh/Ritz method in conjunction with a particular set of trial functions leads to (22) with r representing the ratio of the plate's lowest natural frequency to its small deflection counterpart and ε being a function of the dimensionless ratios W_0/h, a/b, E_L/E_T, G_{LT}/E_T, and ν_{LT}. The details of the required analysis are omitted for the sake of brevity and the interested reader is referred to Shanmugasundaram [6]. Some typical fundamental frequency ratio results are reported in Table 5 in terms of the dimensionless ratios listed above. Predictions obtained for two types of trial functions can be seen to be in good agreement. Based on the discussion contained in Sections 3 and 4 it is to be expected that attempts to calculate higher natural frequencies of the plate will quickly exceed the capabilities of a standard PC to generate the corresponding Groebner bases using the MAPLE Groebner basis module.

Table 5. Frequency ratios for free orthotropic plate vibrations ($E_L/E_T = 10$, $G_{LT}/E_T = 0.5$, $\nu_{LT} = 0.25$, P: polynomial trial functions, T: trigonometric trial functions)

	r					
	$a/b=1$		$a/b=2$		$a/b=4$	
W_0/h	P	T	P	T	P	T
0.5	1.2286	1.2193	1.3255	1.3149	1.3546	1.3432
1	1.7430	1.7168	1.9035	1.8754	1.9620	1.9326
1.5	2.3634	2.3199	2.5688	2.5228	2.6540	2.6061
2	3.0254	2.9649	3.2685	3.2049	3.3789	3.3128

6   Conclusion

The foregoing has presented four examples of the application of Groebner basis methodology to nonlinear mechanics problems. Two of these demonstrated the apparent inability of the MAPLE Groebner basis module to generate a Groebner basis using a standard PC once a relative modest number of unknowns are exceeded. Low order Rayleigh/Ritz, Galerkin, and similar approximate methods of weighted residuals tend to produce correspondingly low order systems of polynomial algebraic equations. In such cases the Groebner basis methodology is of great value in generating closed form solutions that might not be obvious to a human analyst. In that sense there is a useful connection between methods of weighted residuals and Groebner basis methodology.

References

1. Buchberger, B.: An Algorithm for Finding a Basis for the Residue Class Ring of a Zero-Dimensional Polynomial Ideal. Ph.D. Dissertation, University of Innsbruck (1965)
2. Cox, D., Little, J., O'Shea, D.: Ideals, Varieties, and Algorithms: An Introduction to Computational Algebraic Geometry and Commutative Algebra. Springer, New York (2005)
3. Liu, J., Buchanan, G., Peddieson, J.: Application of Groebner Basis Methodology to Nonlinear Static Cable Analysis. Journal of Offshore Mechanics and Arctic Engineering 135, 041601-1–41601-6 (2013)
4. Liu, J., Peddieson, J.: Evaluation of Groebner Basis Methodology as an Aid to Harmonic Balance. Journal of Vibration and Acoustics 136, 024502-1–024502-4 (2014)
5. Nayfeh, A., Mook, D.: Nonlinear Oscillations. Wiley, New York (1979)
6. Shanmugasundaram, A.: An Application of the Method of Groebner Bases to a Geometrically Non-Linear Free Vibration Analysis of Composite Plates. M.S. Thesis, Tennessee Tech University (2009)

Software for Discussing Parametric Polynomial Systems: The GRÖBNER COVER

Antonio Montes[1] and Michael Wibmer[2]

[1] Universitat Politècnica de Catalunya, Spain
antonio.montes@upc.edu
http://www-ma2.upc.edu/montes/
[2] RWTH Aachen University, Germany
michael.wibmer@matha.rwth-aachen.de

Abstract. We present the canonical GRÖBNER COVER method for discussing parametric polynomial systems of equations. Its objective is to decompose the parameter space into subsets (*segments*) for which it exists a *generalized reduced Gröbner basis* in the whole segment with fixed set of leading power products on it. Wibmer's Theorem guarantees its existence. The GRÖBNER COVER is designed in a joint paper of the authors, and the Singular grobcov.lib library [15] implementing it, is developed by Montes. The algorithm is canonic and groups the solutions having the same kind of properties into different disjoint segments. Even if the algorithms involved have high complexity, we show how in practice it is effective in many applications of medium difficulty. An interesting application to automatic deduction of geometric theorems is roughly described here, and another one to provide a taxonomy for exact geometrical loci computations, that is experimentally implemented in a web based application using the dynamic geometry software Geogebra, is explained in another session.

Keywords: Groebner cover, parametric polynomial, canonical algorithm, automatic theorem discovering.

1 The GRÖBNER COVER

The GRÖBNER COVER algorithm for discussing parametric polynomial ideals gives a canonical description, classifying the solutions by their characteristics (number of solutions, dimension, etc.).

The GRÖBNER COVER is the analog of the *reduced Gröbner basis* of an ideal for parametric ideals. Its existence was proved by Wibmer's Theorem [14], and the method and algorithms were developed in [8]. Montes implemented in Singular the grobcov.lib library [15], whose actual version incorporates Kapur-Sun-Wang algorithm [3] for computing the initial Gröbner System used in GROBCOV algorithm, as described in [6], and recently also the LOCUS algorithm used in Dynamical Geometry software as described in [1] and in another session.

Let $\mathbf{x} = x_1, \ldots, x_n$ be the set of variables and $\mathbf{a} = a_1, \ldots, a_m$ the set of parameters. Given a generating set $F = \{f_1, \cdots, f_s\} \subset \mathbb{Q}[\mathbf{a}][\mathbf{x}]$ of the parametric

H. Hong and C. Yap (Eds.): ICMS 2014, LNCS 8592, pp. 406–413, 2014.

ideal $I = \langle F \rangle$ and a monomial order $\succ_{\mathbf{x}}$ in the variables, the GROBCOV algorithm determines

- the unique *canonical partition* of the parameter space $\mathbb{C}^m$ into locally closed sets (*segments*) with associated *generalized reduced Gröbner basis*:

$$GC = \{(S_1, B_1, \mathrm{lpp}_1), \ldots, (S_r, B_r, \mathrm{lpp}_r)\}.$$

- The segments S_i are disjoint locally closed subsets of $\mathbb{C}^m$ and $\oplus_i S_i = \mathbb{C}^m$.
- The basis B_i of a segment S_i has *fixed set of leading power products* (lpp), who ensures that the type of solutions is the same over all points of the segment, and is the *generalized reduced Gröbner basis* of $\langle F \rangle$ over the segment S_i.
- The lpp's are included in the output, even if they they are given by the basis, to characterize the segments and facilitate the applications.
- Moreover, if the ideal is homogeneous, the lpp's are characteristic of the segment as no other segment has the same lpp's.

The generalized reduced Gröbner basis B_i of a segment S_i is formed by a set of monic I-regular functions over S_i. An I-regular function, representing an element of the basis, allows a full-representation in terms of a set of polynomials that specialize for every point $\mathbf{a}_0$ of the segment, either to the corresponding element of the reduced Gröbner basis of the specialized ideal $I_{\mathbf{a}_0}$ after normalization, or to zero. It also allows a generic representation given by a single polynomial that specializes well on an open subset of the segment and to zero on the remaining points of it. Usually the generic representation is sufficient, and we can, if needed, compute the full representation from it using the EXTEND algorithm.

The segments S_i are expressed in canonical P-representation, given by a set of prime ideals of the form

$$\mathrm{Prep}(S) = \{\{\mathfrak{p}_i, \{\mathfrak{p}_{ij} : 1 \le j \le r_i\}\} : 1 \le i \le s\}$$

representing the set:

$$S = \bigcup_{i=1}^{s} \left(\mathbb{V}(\mathfrak{p}_i) \setminus \bigcup_{j=1}^{r_i} \mathbb{V}(\mathfrak{p}_{ij}) \right).$$

Each $\mathbb{V}(\mathfrak{p}_i) \setminus \bigcup_{j=1}^{r_i} \mathbb{V}(\mathfrak{p}_{ij})$ is a *component* of the segment, and its representative $\{\mathfrak{p}_i, \{\mathfrak{p}_{ij} : 1 \le j \le r_i\}\}$, by abuse of language, is also denoted a component when there is no ambiguity. $\mathfrak{p}_i$ is called the *top* of the component, and $\{\mathfrak{p}_{ij} : 1 \le j \le r_i\}$ the *holes*.

1.1 Historical Development of the Theory of Gröbner Bases for Parametric Polynomial Ideals

The first steps in the algebraic study of parametric polynomial ideals where made by V. Weispfenning (1992) in [12], who proved the existence of a Comprehensive Gröbner System (CGS) and a Comprehensive Gröbner Basis (CGB). Progress were made in two directions:

1. Improving the output: Montes (2002) [5], Weispfenning (2003) [13], Manubens & Montes (2009) [4], Montes & Wibmer (2010) [8], Montes (2012) [6].
2. Speed up the algorithms: Kapur (1995), Kalkbrenner (1997), Sato (2005), Suzuki & Sato (2006) [11], Nabeshima (2007) [9], Kapur & Sun & Wang (2010) [3].

The Gröbner Cover [8] is the final state of the research of point 1., and the actual implementation of the GC algorithm incorporates the best speed up algorithm [3] of point 2. as described in [6].

1.2 The Gröbner Cover Algorithm

The algorithm for computing the Gröbner Cover has the following steps:

1. Homogenize the input ideal wrt the variables.
2. Compute a disjoint reduced Comprehensive Gröbner System (DRCGS).[1]
3. Compute the P-representation of the segments.
4. Add together the segments with common lpp using LCUNION algorithm, knowing that the union is locally closed by Wibmer's Theorem.
5. Dehomogenize the bases.
6. For every GC-segment, compute the generic representation of the generalized reduced Gröbner basis using COMBINE algorithm.
7. Optionally, one can also compute the full representation of the bases using EXTEND algorithm after computing the generic GC

When the GC algorithm [8] was introduced in 2010, the DRCGS used for step 2. in the implementation was our own algorithm BUILDTREE [8]. But its use is not strictly necessary. We only need to compute a DRCGS. In the new 2012 implementation of the GC the DRCGS used in step 2. was KAPUR-SUN-WANG algorithm [3] because it is simpler and generally faster. This is described in [6].

1.3 Example

To fix ideas on the use of the GROBCOV algorithm of the Singular "grobcov.lib" library [15], let us consider a very simple example: the inverse kinematic problem of the robot arm of Figure 1. The problem consist of determining the angles θ_1 and θ_2 and the length ℓ to reach the point of coordinates (r, z). Setting $c_i = \cos(\theta_i)$ and $s_i = \sin(\theta_i)$ the equations are obviously:

$$F = s_1 s_2 \ell - c_1 c_2 \ell - c_1 + r, s_1 c_2 \ell - s_1 - c_1 s_2 \ell + z, s_1^2 + c_1^2 - 1, s_2^2 + c_2^2 - 1.$$

The call for solving the problem using Singular GROBVCOV is:

---

[1] A DRCGS is a CGS whose segments are dijoint and the bases specialize to the reduced Gröbner basis and have fixed lpp over the whole segment.

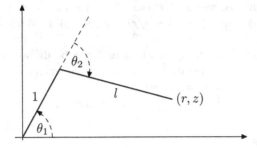

Fig. 1. Simple robot arm

Input:

```
LIB "grobcov.lib";
ring R=(0,r,z),(s1,c1,s2,c2,l),lp;
ideal F= s1*s2*l-c1*c2*l-c1+(r), s1*c2*l-s1-c1*s2*l+(z),
        s1^2+c1^2-1, s2^2+c2^2-1;
def G=grobcov(F);
"grobcov(F)=" G;
```

Output: We summarize the output in the following table

S.	lpp	Basis	Segment
1	$c_2\ell, s_2^2,$ $c_1,$ $s_1.$	$2c_2\ell + \ell^2 + (-r^2 - z^2 + 1),\ \ s_2^2 + c_2^2 - 1,$ $(2r^2 + 2z^2)c_1 + (-2z)s_2\ell + r\ell^2 + (-r^3 - rz^2 - r),$ $(2r^2 + 2z^2)s_1 + (2r)s_2\ell + (z)\ell^2 + (-r^2z - z^3 - z).$	$\mathbb{C}^2 \setminus \mathbb{V}(r^2 + z^2)$
2	$c_2\ell s_2,$ $c_1\ell^2,$ $c_1c_2,$ $s_1.$	$2c_2\ell + \ell^2 + 1, (z)s_2 + (-r)c_2 + (-r)\ell,$ $(4z^2)c_1\ell^2 + (-4z^2)c_1 + (-r)\ell^4 + (2r)\ell^2 + (4rz^2 - r),$ $(8z^2)c_1c_2 + (8z^2)c_1\ell + (-8rz^2 + 2r)c_2 + (-r)\ell^3 +$ $+(-4rz^2 + 3r)\ell,\ \ (2z)s_1 + (2r)c_1 + \ell^2 - 1.$	$\mathbb{V}(r^2 + z^2) \setminus \mathbb{V}(z, r)$
3	ℓ^2, c_2, s_2, s_1^2	$\ell^2 - 1, c_2 + 1, s_2, s_1^2 + c_1^2 - 1.$	$\mathbb{V}(z, r)$

There are 3 segments, and for each segment there are 4 arguments: 1) the lpp, 2) the basis, 3) the P-representation of the segment, 4) the lpp of the homogenized ideal. The fourth argument is purely informative to verify that each segment has a characteristic lpp of the homogenized ideal. It can be discarded, and we deleted it form the output. The output is to be read as follows:

1) The first segment represents the generic case: the solution is valid for every values of the parameters r, z, except when $r^2 + z^2 = 0$. We have one-degree of freedom in the variables. One can choose ℓ free. For each value of $\ell \neq 0$ there are two angle solutions with opposite value of θ_2. For fixed ℓ we have

$$c_2 = \frac{r^2 + z^2 - \ell^2 - 1}{2\ell}, \qquad\qquad s_2 = \pm\sqrt{1 - c_2^2},$$

$$c_1 = \frac{2zs_2\ell + r(r^2 + z^2 + 1 - \ell^2)}{2(r^2 + z^2)}, s_1 = \frac{-2rs_2\ell + z(r^2 + z^2 + 1 - \ell^2)}{2(r^2 + z^2)}.$$

As we want real solutions, we must choose ℓ such that $|c_2| \leq 1$. We set $\ell > 0$. The limits for $\cos(\theta_2)$ imply $|\ell - 1| \leq \sqrt{r^2 + z^2} \leq \ell + 1$. With this choice the angles are real.

2) The second segment is purely complex and can be discarded in practice.

3) There is only one special position $(r, z) = (0, 0)$ for which necessarily $\ell = \pm 1$ and in practice $\ell = 1$. Then $\theta_2 = \pi$, and θ_1 free.

These results correspond accurately to the geometry.

2 Applications

The GRÖBNER COVER has many applications. Let us highlight one interesting problem that can be solved using it: automatic discovering of geometrical theorems. In the "Parametric Polynomial Systems" session we show its use for determining and classifying geometrical loci that can be used by Dynamical Geometry software [1].

2.1 Automatic Deduction of Geometrical Theorems

Consider a generally false geometrical statement depending on some variable points for which we want to find the conditions in order to make the statement to hold true. Consider the coordinates of the free points of a construction as parameters and the remaining coordinates or values as variables. Then apply GROBCOV to the system defining the statement and the construction, and find the conditions over the parameters that makes the statement hold true. We show an interesting example: the generalization of the classical XIX-century known Steiner-Lehmus Theorem [10] that is described in [7]. Let us summarize here the results.

Classical Theorem states that the length of the inner bisectors of a triangle are equal if and only if the triangle is isosceles. Consider the triangle ABC of

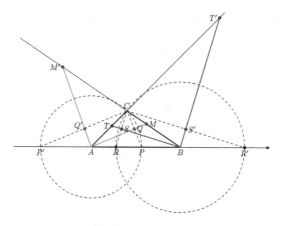

Fig. 2. Bisectors of the triangle ABC

Figure 2, and take coordinates $A(-1,0)$, $B(1,0)$ and $C(a,b)$. Trace the circles with center A and radius AC that intersects line AB (i.e the x-axis) at $P(p,0)$ and P', and the circle with center B and radius BC intersecting line AB at $R(r,0)$ and R'. The equation $(a+1)^2 + b^2 - (p+1)^2$ determining p in terms of a, b does not distinguish between P and P'. The same happens for the points R and R' and for the equations determining M and T (or T') and M (or M'), so that the statement $\overline{AM}^2 = \overline{BT}^2$, i.e. $(x_1 + 1)^2 + y_1^2 = (x_2 - 1)^2 + y_2^2$, does not distinguish between inner and outer bisectors. System F only implies that one bisector (inner or outer) of A is equal to one bisector of B. The system is

$$F = (a+1)^2 + b^2 - (p+1)^2, \ (a-1)^2 + b^2 - (r-1)^2,$$
$$ay_1 - bx_1 - y_1 + b, \ ay_2 - bx_2 + y_2 - b,$$
$$-2y_1 + bx_1 - (a+p)y_1 + b,, \ 2y_2 + bx_2 - (a+r)y_2 - b,$$
$$(x_1 + 1)^2 + y_1^2 - (x_2 - 1)^2 - y_2^2.$$

Applying GROBCOV in the ring $R=(0,a,b),(x1,y1,x2,y2,p,r),dp$ to the ideal generated by F it outputs 9 segments. Table 1 gives the 3 curves and 9 point varieties representing real and complex points in the parameter space appearing in the description of the GROBCOV(F). We do not detail the complex points as we are not interested in. Table 2 summarizes the relevant characteristics of the output of GROBCOV(F) for our purposes.

Table 1. Curves and point varieties appearing in GROBCOV(F)

Curves
$\mathcal{C}_1 = \mathbb{V}((8a^2 + 9b^2)(a^2 + b^2)^4 - 4(14a^4 + 13a^2b^2 - 3b^4)(a^2 + b^2)^2$
$\quad +2(72a^6 + 43a^4b^2 - 74a^2b^4 - 37b^6) - 4(44a^4 - 39a^2b^2 + 43b^4) + 104a^2 + 137b^2 - 24),$
$\mathcal{C}_2 = \mathbb{V}(a),$
$\mathcal{C}_3 = \mathbb{V}(b).$

Point varieties	real points	numerical values
$V_1 = \mathbb{V}(b.a + 1)$	P_1	$= (-1, 0)$
$V_2 = \mathbb{V}(b, a - 1)$	P_2	$= (1, 0)$
$V_3 = \mathbb{V}(a, b)$	P_4	$= (0, 0)$
$V_4 = \mathbb{V}(b, a^2 - 3)$	P_{42}, P_{41}	$= (\pm\sqrt{3}, 0)$
$V_5 = \mathbb{V}(3b^2 - 1, a)$	P_{52}, P_{51}	$= (0, \pm\sqrt{3}/3)$
$V_6 = \mathbb{V}(b^2 - 3, a)$	P_{62}, P_{61}	$= (0, \pm\sqrt{3})$
$V_7 = \mathbb{V}(b^4 + 5b^2 + 8, a)$	no real roots	
$V_8 = \mathbb{V}(b^4 + 44b^2 - 16, 5a + b^2 + 7)$	P_{82}, P_{81}	$= (3 - 2\sqrt{5}, \pm\sqrt{-22 + 10\sqrt{5}})$
$V_9 = \mathbb{V}(b^4 + 44b^2 - 16, 5a - b^2 - 7)$	P_{92}, P_{91}	$= (-3 + 2\sqrt{5}, \pm\sqrt{-22 + 10\sqrt{5}})$

Curves and points can be visualized on Figure 3. The fourth column in Table 2 is direct consequence of the lpp in column 3. We need the basis to determine the fifth column, who indicates which bisectors (internal i or external e) are

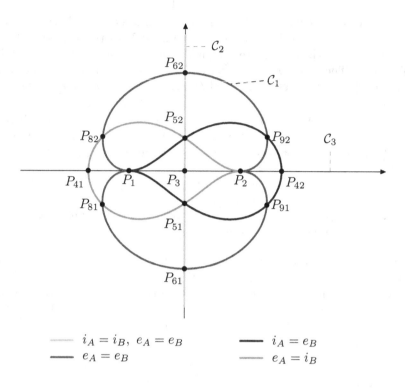

Fig. 3. Generalized Steiner-Lehmus Theorem

equal for the different solutions. From the basis we can determine the signs of $p + 1$ and of $r - 1$ for each point of the solution.

$p > -1$ corresponds to the inner bisector i_A and $p < -1$ to the external e_A,
$r < 1$ corresponds to the inner bisector i_B and $r > 1$ to the external e_B.

The line $\mathcal{C}_3$, i.e. the x-axis, corresponds to degenerate triangles, and so the segments 4,7,8 can be discarded. The remaining segments give the whole information on the generalized Theorem. The curve $\mathcal{C}_1$ has different colors, that can change only at the special self intersecting points. To determine its color it suffices to evaluate p and r on an intermediate point of the interval. The curve $\mathcal{C}_2$ corresponds to the classical Theorem and with the GRÖBNER COVER we can appreciate also more details on it. On the whole line (isosceles triangles) we have $i_A = i_B$ and also $e_A = e_B$ except for special points P_{51} and P_{52} where all bisectors are equal and special points P_{61} and P_{62} where the external bisectors become infinity. The GRÖBNER COVER reveals the generalized Theorem over the curve $\mathcal{C}_1$ with all the details.

Table 2. Segments of GROBCOV(F). (Bases are not explicitly given)

Nr.	Segment	lpp	Num. S.	Bisectors
1	$\mathbb{C}^2 \setminus (\mathcal{C}_1 \cup \mathcal{C}_2 \cup \mathcal{C}_3)$	$\{1\}$	0	-
2	$\mathcal{C}_1 \setminus ((\bigcup_{i=1}^2 V_i) \cup (\bigcup_{i=4}^9 V_i))$	$\{r,p,y_2,x_2,y_1,x_1\}$	1	depends on sector
3	$(\mathcal{C}_2 \setminus (V_3 \cup V_5 \cup V_6)) \cup V_8$	$\{p,y_2,x_2,y_1,x_1,r^2\}$	2	$i_A = i_B,\ e_A = e_B$ $e_A = e_B = i_B$
4	$\mathcal{C}_3 \setminus (V_1 \cup V_2)$	$\{y_2,y_1,r^2,p^2,x_1^2\}$	∞	
5	V_5	$\{y_2,x_2,y_1,x_1,r^2,p^2\}$	4	$i_A = i_B = e_A = e_B$
6	V_6	$\{r,p,y_2,x_2,y_1,x_1\}$	1	$i_A = i_B$
7	V_1	$\{y_1,r^2,y_2r,p^2x_1^2\}$	∞	
8	V_2	$\{y_2,r^2,p^2,y_1p,x1^2\}$	∞	
9	V_9	$\{r,y_2,x_2,y_1,x_1,p^2\}$	2	$e_A = e_B = i_A$

References

1. Abanades, M., Botana, F., Montes, A., Recio, T.: An Algebraic Taxonomy for Locus Computation in Dynamic Geometry. Computer Aided Design (submitted, 2014)
2. Gianni, P., Trager, B., Zacharia, G.: Gröbner Bases and Primary Decomposition of Polynomial Ideals. Jour. Symb. Comp. 6, 149–167 (1988)
3. Kapur, D., Sun, Y., Wang, D.: A new algorithm for computing comprehensive Gröbner systems. In: Proceedings of ISSAC 2010, pp. 29–36. ACM Press, New York (2010)
4. Manubens, M., Montes, A.: Minimal canonical comprehensive Gröbner systems. Jour. Symb. Comp. 44(5), 463–478 (2009)
5. Montes, A.: A new algorithm for discussing Gröbner bases with parameters. Jour. Symb. Comp. 22, 183–208 (2002)
6. Montes, A.: Using Kapur-Sun-Wang algorithm for the Gröbner Cover. In: Sendra, R., Villarino, C. (eds.) Proceedings of EACA 2012, pp. 135–138. Universidad de Alcalá de Henares (2012)
7. Montes, A., Recio, T.: Generalizing the Steiner-Lehmus theorem using the Gröbner Cover. Mathematics and Computers in Simulation 104, 67–81 (2014), http://dx.doi.org/10.1016/j.matcom.2013.06.006
8. Montes, A., Wibmer, M.: Gröbner bases for polynomial systems with parameters. Jour. Symb. Comp. 45, 1391–1425 (2010)
9. Nabeshima, K.: A speed-up of the algorithm for computing comprehensive Gröbner systems. In: Proceedings of ISSAC 2007, pp. 299–306. ACM Press (2007)
10. Classical Steiner-Lehmus Theorem (2010), http://www.mathematik.uni-bielefeld.de/silke/PUZZLES/steinerlehmus
11. Suzuki, A., Sato, Y.: A simple algorithm to compute comprehensive Gröbner bases. In: Proceedings of ISSAC 2006, pp. 326–331. ACM, New York (2006)
12. Weispfenning, V.: Comprehensive Gröbner bases. Jour. of Symb. Comp. 14, 1–29 (1992)
13. Weispfenning, V.: "Canonical Comprehensive Gröbner Basis". Jour. Symb. Comp. 36, 669–683 (2003)
14. Wibmer, M.: Gröbner bases for families of affine or projective schemes. Jour. Symb. Comp. 42, 803–834 (2007)
15. Singular, http://www.singular.uni-kl.de/ (last accessed February 2014)

An Algorithm for Computing Standard Bases by Change of Ordering via Algebraic Local Cohomology

Katsusuke Nabeshima[1] and Shinichi Tajima[2]

[1] Institute of Socio-Arts and Sciences, The University of Tokushima,
1-1 Minamijosanjima, Tokushima, Japan
nabeshima@tokushima-u.ac.jp

[2] Graduate School of Pure and Applied Sciences, University of Tsukuba,
1-1-1 Tennoudai, Tsukuba, Japan
tajima@math.tsukuba.ac.jp

Abstract. An algorithm is introduced for transforming a standard basis of a zero-dimensional ideal, in the formal power series ring, into another standard basis with respect to any given local ordering. The key ingredient of the proposed algorithm is an efficient method for solving membership problems for Jacobi ideals in local rings, that utilizes the Grothendieck local duality theorem. Namely, a new algorithm for computing a standard basis of a given zero-dimensional ideal with respect to any given local ordering, is derived by using algebraic local cohomology. Its implementation is introduced, too.

Keywords: standard bases, algebraic local cohomology, singularities.

1 Introduction

Let $f \in \mathbb{C}[x_1, \ldots, x_n]$ be a polynomial with an isolated singularity at the origin. In singularity theory, we sometimes need to compute standard bases of the Jacobi ideal $J = \langle \frac{\partial f}{\partial x_1}, \ldots, \frac{\partial f}{\partial x_n} \rangle$ w.r.t. a local ordering to analyze properties of singularities. In [8], we have introduced algorithms for computing a basis of algebraic local cohomology classes w.r.t. f, and a standard basis for J w.r.t. a local degree lexicographic ordering. In [2], we have considered the case where f is a semi-quasihomogeneous polynomial. In this case, we have also introduced algorithms for computing a basis of algebraic local cohomology classes w.r.t. f, and a standard basis for J w.r.t. a local weighted lexicographic ordering.

In this paper, we present an algorithm for computing the reduced standard basis of J w.r.t. "any given local ordering" by using algebraic local cohomology and its implementation. The main part of the proposed algorithm is changing of ordering from a local degree lexicographic ordering or weighted lexicographic ordering. Hence, the proposed algorithm is based on our previous algorithms [2,8]. There are two main advantages of the proposed algorithm. First, the algorithm always outputs the reduced standard bases. Other implementation does

H. Hong and C. Yap (Eds.): ICMS 2014, LNCS 8592, pp. 414–418, 2014.

not have this property. Second, the proposed algorithm is free from Mora's reduction (tangent cone algorithms). The computation consists of only linear algebra computation.

2   Functionality

Let $f \in \mathbb{C}[x_1, \ldots, x_n]$ be a polynomial with an isolated singularity at the origin. Our implementation works for computing a standard basis of the Jacobi ideal $J = \langle \frac{\partial f}{\partial x_1}, \ldots, \frac{\partial f}{\partial x_n} \rangle$ w.r.t. a local ordering, in a local ring $\mathbb{C}[[x_1, \ldots, x_n]]$. In particular, our implementation always outputs the reduced standard basis of J w.r.t. a given local ordering. The proposed algorithm is implemented in the CA system Risa/Asir[4].

Example 1. A polynomial $f = x^3 y + x y^4 + x^2 y^3 \in \mathbb{C}[x, y]$ has an isolated singularity at the origin. Set $J = \langle \frac{\partial f}{\partial x}, \frac{\partial f}{\partial y} \rangle$. Our implementation outputs a basis of H_f and the reduced standard basis of J w.r.t. the local degree lexicographic ordering $\succ$ such that $x \succ y$, where H_f will be introduced in section 3.

```
stand_coho(x^3*y+x*y^4+x^2*y^3,[x,y],1,1,1);
Basis of Algebraic cohomo.  12
[y^(-1)*x^(-1),y^(-2)*x^(-1),y^(-1)*x^(-2),y^(-3)*x^(-1),y^(-2)*x^
(-2),y^(-1)*x^(-3),y^(-4)*x^(-1),y^(-3)*x^(-2)],[y^(-5)*x^(-1)-1/3
*y^(-2)*x^(-3),y^(-4)*x^(-2)-2/3*y^(-2)*x^(-3)-4*y^(-1)*x^(-4),y^(
-6)*x^(-1)-1/3*y^(-3)*x^(-3)+y^(-1)*x^(-4),y^(-7)*x^(-1)+7/33*y^(-
5)*x^(-2)+(-14/99*y^(-3)-1/3*y^(-4))*x^(-3)+(14/33*y^(-1)+5/33*y^(
-2))*x^(-4)+4/3*y^(-1)*x^(-5)]
Standard Basis
[y*x^2+2/3*y^3*x+1/3*y^4,x^3+4*y^3*x-14/33*y^6-y^5,y^4*x-7/33*y^6,
y^7]
```

The meaning of the output is the following. A basis of the vector space H_f is
$\{[\frac{1}{yx}], [\frac{1}{y^2x}], [\frac{1}{yx^2}], [\frac{1}{y^3x}], [\frac{1}{y^2x^2}], [\frac{1}{yx^3}], [\frac{1}{y^4x}], [\frac{1}{y^3x^2}], [\frac{1}{y^5x}] - \frac{1}{3}[\frac{1}{y^2x^3}], [\frac{1}{y^4x^2}] - \frac{2}{3}[\frac{1}{y^2x^3}] - 4[\frac{1}{yx^4}], [\frac{1}{y^6x}] - \frac{1}{3}[\frac{1}{y^3x^3}] + [\frac{1}{yx^4}], [\frac{1}{y^7x}] + \frac{7}{33}[\frac{1}{y^5x^2}] - \frac{14}{99}[\frac{1}{y^3x^3}] - \frac{1}{3}[\frac{1}{y^4x^3}] + \frac{14}{33}[\frac{1}{y^1x^4}] + \frac{5}{33}[\frac{1}{y^2x^4}] + \frac{4}{3}[\frac{1}{yx^5}]\}$. The reduced standard basis of J w.r.t. $\succ$ is $\{yx^2 + \frac{2}{3}y^3x + \frac{1}{3}y^4, x^3 + 4y^3x - \frac{14}{33}y^6 - y^5, y^4x - \frac{7}{33}y^6, y^7\}$.

Example 2. Let consider $\mathbb{Z}_{12}$ singularity defined by $f = x^3 y + x y^4 + x^2 y^3$. Set $J = \langle \frac{\partial f}{\partial x}, \frac{\partial f}{\partial y} \rangle$. Our implementation also can output only the reduced standard basis w.r.t. a local lexicographic ordering such that $y \succ x$ (i.e., $1 \succ y \succ y^2 \succ \cdots \succ x \succ xy \succ \cdots$).

```
stand_coho(x^3*y+x*y^4+x^2*y^3,[x,y],2,0,0);
[y^4+3*x^2*y-3/2*x^2*y^2-1/2*x^3, x*y^3+3/4*x^*y^2+1/4*x^3, x^3*y
-5/44*x^4, x^5]
```

The CA system Singular has a command "std" which computes a standard basis of J. Let us compare the output of our implementation with Singular's one. Singular outputs the following as a standard basis of J w.r.t. $\succ$.

```
> ring A=0, (x,y),ls;
> polynomial f=x^3*y+x*y^4+x^2*y^3;
> poly f=x^3*y+x*y^4+x^2*y^3;
> ideal I=diff(f,x),diff(f,y);
> std(I);
_[1]=y4+2xy3+3x2y
_[2]=4xy3+3x2y2+x3
_[3]=44x3y-15x3y2-5x4
_[4]=x5
```

It is easy to see that the output is not the reduced standard basis w.r.t. $\succ$.

Our implementation has the command "groeb_coho" which computes the reduced Gröbner basis of a primary component at the origin of a primary ideal decomposition of J.

3   Underlying Theory

Let X be an open neighborhood of the origin O of the n-dimensional complex space $\mathbb{C}^m$ with coordinates $x = (x_1, x_2, \ldots, x_n)$ and let $\mathcal{O}_X$ be the sheaf on X of holomorphic functions. Let f be a holomorphic function defined on X with an isolated singularity at the origin O and let J denote the Jacobi ideal $J = \left\langle \frac{\partial f}{\partial x_1}, \frac{\partial f}{\partial x_2}, \ldots, \frac{\partial f}{\partial x_n} \right\rangle$ in $\mathcal{O}_{X,O}$ generated by the partial derivatives $\frac{\partial f}{\partial x_1}, \frac{\partial f}{\partial x_2}, \ldots, \frac{\partial f}{\partial x_n}$, where $\mathcal{O}_{X,O}$ is the stalk at O of the sheaf $\mathcal{O}_X$. Let $\mathcal{H}^n_{[O]}(\mathcal{O}_X)$ denote the set of algebraic local cohomology classes, defined by $\mathcal{H}^n_{[O]}(\mathcal{O}_X) = \lim_{k \to \infty} Ext^n_{\mathcal{O}_X}(\mathcal{O}_X / \langle x_1, x_2, .., x_n \rangle^k, \mathcal{O}_X)$ where $\langle x_1, x_2, \ldots, x_n \rangle$ is the maximal ideal generated by $x_1, x_2, \ldots, x_n$. We define a vector space H_f to be the set of algebraic local cohomology classes in $\mathcal{H}^n_{[O]}(\mathcal{O}_X)$ that are annihilated by the Jacobi ideal J

$$H_f = \left\{ \psi \in \mathcal{H}^n_{[O]}(\mathcal{O}_X) \,\middle|\, \frac{\partial f}{\partial x_1}(x)\psi = \frac{\partial f}{\partial x_2}(x)\psi = \cdots = \frac{\partial f}{\partial x_n}(x)\psi = 0 \right\}.$$

Any algebraic local cohomology class in $\mathcal{H}^n_{[O]}(\mathcal{O}_X)$ can be represented as a finite sum of the form (called: Čech representation) $\sum c_\lambda \left[\frac{1}{x^{\lambda+1}} \right] = \sum c_\lambda \left[\frac{1}{x_1^{\lambda_1+1} \cdots x_n^{\lambda_n+1}} \right]$ where $c_\lambda \in \mathbb{C}$ and $\lambda = (\lambda_1, \lambda_2, \ldots, \lambda_n) \in \mathbb{N}^n$. We can define a ordering of the term $\left[\frac{1}{x^\alpha} \right] = \left[\frac{1}{x_1^{\alpha_1}, \ldots, x_n^{\alpha_n}} \right]$ like a ordering of the (normal) term x^α. Let us fix a term order $\succ$. The details are in [3,5,6,7,8]. For an algebraic local cohomology class ψ of the form $\psi = c_\lambda \left[\frac{1}{x^\lambda} \right] + \sum_{\lambda \succ \lambda'} c_{\lambda'} \left[\frac{1}{x^\lambda} \right]$, $c_\lambda \neq 0$, we call $\left[\frac{1}{x^\lambda} \right]$ the **head term** and $\left[\frac{1}{x^{\lambda'}} \right]$, $\lambda \succ \lambda'$ the **lower terms**. We denote the head term of a cohomology class ψ by $ht(\psi)$. We define the set of terms of ψ as $T(\psi) = \{ \left[\frac{1}{x^\lambda} \right] | \psi = \sum_{\lambda \in \mathbb{N}^n} c_\lambda \left[\frac{1}{x^\lambda} \right], c_\lambda \neq 0, c_\lambda \in \mathbb{C} \}$. Moreover, for the set Ψ, $T(\Psi) = \bigcup_{\psi \in \Psi} T(\psi)$. Let Ψ be a set of algebraic local cohomology classes. We define the set of monomial elements of Ψ as $ML(\Psi)$, and the set of linear combination elements as $SL(\Psi_i)$, i.e., $\Psi = ML(\Psi) \cup SL(\Psi)$. We define the set of head terms of Ψ as $ht(\Psi)$ and the set of lower terms of Ψ as

$LL(\Psi) = \{[\frac{1}{x^{\lambda}}] \in T(\Psi)|[\frac{1}{x^{\lambda}}] \notin ht(\Psi)\}$. If we know a basis of H_f, then we can compute a normal form modulo J by using the basis.

Theorem 1 (Normal forms[6,8]). *Using the same notations as in the above argument, let Ψ be a basis of H_f. Suppose that $SL(\Psi)$ has an element whose form is*
$$\left[\frac{1}{x^{\tau+1}}\right] + \sum_{\tau+1 \succ \kappa+1} c_{(\tau+1,\kappa+1)} \left[\frac{1}{x^{\kappa+1}}\right] \text{ where } c_{(\tau+1,\kappa+1)} \in \mathbb{C} \text{ and } x^{\lambda+1} =$$
$(\lambda_1 + 1, \ldots, \lambda_n + 1) \in \mathbb{N}^n$. *Then, the following relations hold :*
if $[\frac{1}{x^{\lambda+1}}] \in LL(\Psi)$, then $x^{\lambda} \equiv \displaystyle\sum_{[\frac{1}{x^{\kappa+1}}]\in ht(\Psi)} c_{(\kappa+1,\lambda+1)} x^{\kappa} \bmod J$,

if $[\frac{1}{x^{\lambda+1}}] \notin LL(\Psi)$ and $[\frac{1}{x^{\lambda+1}}] \notin ht(\Psi)$, then $x^{\lambda} \equiv 0 \bmod J$, and

if $[\frac{1}{x^{\lambda+1}}] \in ht(\Psi)$, then $x^{\lambda} \equiv x^{\lambda} \bmod J$.

By Theorem 1, we can easily compute a normal form of a polynomial.

Example 3. Let us consider $f = x^3 + xy^5$. Then, a basis of H_f w.r.t. the weight lexicographic ordering such that $y \succ x$, is $\{[\frac{1}{xy}], [\frac{1}{x^2y}], [\frac{1}{xy^2}], [\frac{1}{x^2y^2}], [\frac{1}{xy^3}], [\frac{1}{x^2y^3}],$
$[\frac{1}{xy^4}], [\frac{1}{x^2y^4}], [\frac{1}{xy^5}], [\frac{1}{x^3y^4}] - 3[\frac{1}{xy^9}], [\frac{1}{x^3y^3}] - 3[\frac{1}{xy^8}], [\frac{1}{x^3y^2}] - 3[\frac{1}{xy^7}], [\frac{1}{x^3y}] - 3[\frac{1}{xy^6}]\}$. By Theorem 1, we have the following relations: $y^8 \equiv -3x^2y^3 \bmod J$, $y^7 \equiv -3x^2y^2 \bmod J$, $y^6 \equiv -3x^2y \bmod J$ and $y^5 \equiv -3x^2 \bmod J$.

In order to introduce our new algorithm for computing a basis of H_f, we require the following lemma which are from [3,8].

Lemma 1 ([8]). *Let Λ_{H_f} denote the set of exponents of head monomials in H_f and $\lambda = (\lambda_1, \ldots, \lambda_n) \in \mathbb{N}^n$. Let $\Lambda_{H_f}^{(\lambda)}$ denote a subset of $\Lambda_{H_f} : \Lambda_{H_f} = \{\lambda \in \mathbb{N}^n | \exists \psi \in H_f$ such that $ht(\psi) = [\frac{1}{x^{\lambda+1}}]\}$ and $\Lambda_H^{(\lambda)} = \{\lambda' \in \Lambda_{H_f} | \lambda' \prec \lambda\}$.*

 (C): "If $\lambda \in \Lambda_{H_f}$, then, for each $j = 1, 2, \ldots, n, (\lambda_1, \lambda_2, \ldots, \lambda_{j-1}, \lambda_j - 1, \lambda_{j+1}, \ldots, \lambda_n)$ is in $\Lambda_{H_f}^{(\lambda)}$, provided $\lambda_j \geq 1$."

Definition 1 (inverse orderings). *Let $\succ$ be a local or global term ordering. Then, the inverse ordering $\succ^{-1}$ of $\succ$ is defined by $x^{\alpha} \succ x^{\beta} \iff x^{\beta} \succ^{-1} x^{\alpha}$. Therefore, if $\succ$ is a global ordering ($\succ 1$), then $\succ^{-1}$ is the local ordering ($1 \succ^{-1}$). Conversely, if $\succ$ is a local ordering, then $\succ^{-1}$ is the global ordering.*

Our algorithm for change of ordering of standard bases, is essentially same as FGLM algorithm[1] in local ring. However, we do not need Mora's reduction, as we know Theorem 1.

Algorithm 1. [Standard Bases]

Input: $f \in \mathbb{C}[x_1, \ldots, x_n]$ a polynomial with an isolated singularity at the origin. $\succ$ is a local ordering.
Output: S: the reduced standard basis of $\langle \frac{\partial f}{\partial x_1}, \ldots, \frac{\partial f}{\partial x_n} \rangle$ w.r.t. $\succ$ in $\mathbb{C}[[x_1, \ldots, x_n]]$.
BEGIN
$\Psi \leftarrow$ Compute a basis of H_f by the algorithm [2] or [8]. (*1)

$T \leftarrow$ all terms of $\frac{\partial f}{\partial x_1}, \ldots, \frac{\partial f}{\partial x_n}$; $\quad C \leftarrow$ Compute the reduced Gröbner basis of T
$S \leftarrow \emptyset$; $\quad \Phi \leftarrow \emptyset$
while $C \neq \emptyset$ **do**
select the BIGGEST element p from C w.r.t. a global ordering $\succ^{-1}$
$C \leftarrow C \backslash \{p\}$
if $(p \equiv p' \bmod J)$ is linearly independent of the vectors in Φ **then**
$\Phi \leftarrow \Phi \cup \{p'\}$; $\quad M \leftarrow \{x_i p | 1 \leq i \leq n\}$
while $M \neq \emptyset$ **do**
select an element q from M; $M \leftarrow M \backslash \{q\}$
if q satisfies **(C) then** $\quad C \leftarrow C \cup \{q\}$
end-if
end-while
else if $p \equiv 0 \bmod J$ **then** $\quad S \leftarrow S \cup \{p\}$
else then from this dependence we get a new element g
$S \leftarrow S \cup \{g\}$
end-if
end-while
END

Let us remark that at (∗1), we need a basis $\Psi = \{\psi_1, \ldots, \psi_s\}$ which can be written $A \cdot T(\Psi) = \mathbf{0}$, to compute the reduced standard basis where A is a matrix in reduced row echelon form. The algorithms of [2,8] are able to output such matrices.

References

1. Faugère, J., Gianni, P., Lazard, D., Mora, T.: Efficient computation of zero-dimensional Gröbner bases by change of ordering. Journal of Symbolic Computation 16, 329–344 (1993)
2. Nabeshima, K., Tajima, S.: On efficient algorithms for computing parametric local cohomology classes associated with semi-quasihomogeneous singularities and standard bases. In: Proc. ISSAC 2014, pp. 104–111. ACM Press (2014)
3. Nakamura, Y., Tajima, S.: On weighted-degrees for algebraic local cohomologies associated with semiquasihomogeneous singularities. Advanced Studies in Pure Mathematics 46, 105–117 (2007)
4. Noro, M., Takeshima, T.: Risa/Asir- A computer algebra system. In: Proc. ISSAC 1992, pp. 387–396. ACM Press (1992)
5. Tajima, S., Nakamura, Y.: Algebraic local cohomology class attached to quasi-homogeneous isolated hypersurface singularities. Publications of the Research Institute for Mathematical Sciences 41, 1–10 (2005)
6. Tajima, S., Nakamura, Y.: Annihilating ideals for an algebraic local cohomology class. Journal of Symbolic Computation 44, 435–448 (2009)
7. Tajima, S., Nakamura, Y.: Algebraic local cohomology classes attached to unimodal singularities. Publications of the Research Institute for Mathematical Sciences 48, 21–43 (2012)
8. Tajima, S., Nakamura, Y., Nabeshima, K.: Standard bases and algebraic local cohomology for zero dimensional ideals. Advanced Studies in Pure Mathematics 56, 341–361 (2009)

Verification of Gröbner Basis Candidates

Masayuki Noro and Kazuhiro Yokoyama

Department of Mathematics, Rikkyo University, Japan
{noro,kazuhiro}@rikkyo.ac.jp

Abstract. We propose a modular method for verifying the correctness of a Gröbner basis candidate. For an inhomogeneous ideal I, we propose to check that a Gröbner basis candidate G is a subset of I by computing an exact generating relation for each g in G by the given generating set of I via a modular method. The whole procedure is implemented in Risa/Asir, which is an open source general computer algebra system. By applying this method we succeeded in verifying the correctness of a Gröbner basis candidate computed in Romanovski et al (2007). In their paper the candidate was computed by a black-box software system and it has been necessary to verify the candidate for ensuring the mathematical correctness of the paper.

Keywords: Gröbner basis, modular algorithm, verification.

1   Introduction

When it is difficult to compute a Gröbner basis of the ideal $I = \langle F \rangle$ generated by a given generating set $F = \{f_1, \ldots, f_m\}$ by the Buchberger algorithm or the F_4 algorithm over the rationals, it is often useful to apply modular methods such as Hensel lifting or Chinese remainder theorem. In most cases the obtained result G is merely a candidate of a Gröbner basis of I and we have to verify the correctness of the candidate. If it is proven that $G \subset I$ then the verification is easy by checking if G is a Gröbner basis of $\langle G \rangle$ and $F \subset \langle G \rangle$. Otherwise it becomes difficult in general. Arnold[3] showed that if I is homogeneous the verification is essentially reduced to check whether G is a Gröbner basis of $\langle G \rangle$. Idrees et al.[5] tried to extend this result to inhomogeneous ideals but their criterion is applicable to a limited case. Yokoyama [7] gives an improved criterion that can be applied to general inhomogeneous ideals but it is necessary to compute the Gröbner basis of $\langle F^h \rangle$ over a finite field, where F^h is the homogenization of the given generating set F. In many cases these methods work well. However there still exist cases where the Gröbner basis computation of $\langle F^h \rangle$ over a finite field is still hard to complete. Even in such cases, if we can compute the Gröbner bases of $\langle F \rangle$ over finite fields, then we will be able to get a probable Gröbner basis candidate G by Chinese remainder computation. For verifying the correctness of G, we propose to check the inclusion $G \subset I$ by computing an exact generating relation $g = h_1 f_1 + \cdots + h_m f_m$ for all $g \in G$ via modular computation. After computing the "coefficient polynomials" $h_1, \ldots, h_m$ over a finite field, we reduce

H. Hong and C. Yap (Eds.): ICMS 2014, LNCS 8592, pp. 419–424, 2014.

the number of terms in h_i's so that their coefficients are uniquely determined. Then we replace the coefficients by variables and we get a system of linear equations with respect to the variables. This system has at most one solution over the rationals and we can apply Hensel lifting to solve this system. If we can get the solution, then it directly gives the evidence of $g \in I$. The whole procedure is implemented in Risa/Asir. By applying this method we succeeded in verifying the correctness of a Gröbner basis candidate computed in Romanovski et al. [4]. In their paper the candidate was computed by a black-box software system and it has been necessary to verify the candidate for ensuring the mathematical correctness of the paper.

2   Verification of Gröbner Basis

Let p be a prime, $X = \{x_1, \ldots, x_n\}$ and ϕ_p the canonical projection from $\mathbb{Z}$ to $\mathbb{F}_p$. ϕ_p is naturally extended to $\mathbb{Z}_p^0 = \{a/b \mid p \nmid b\}$ and $\mathbb{Z}_p^0[X]$. $\mathrm{LC}(f)$ denotes the leading coefficient of f.

Definition 1. *Let p be a prime. $G \subset \mathbb{Z}_p^0[X]$ is a p-GB candidate for $F \subset \mathbb{Z}[X]$ if $\phi_p(\mathrm{LC}(g)) \neq 0$ for all $g \in G$ and $\phi_p(G)$ is a Gröbner basis of $\langle \phi_p(F) \rangle$.*

Let F be a subset of $\mathbb{Z}[X]$ and G a p-GB candidate for F. If $G \subset \langle F \rangle$ then the verification of G is easy:

G is a Gröbner basis of $\langle F \rangle \Leftrightarrow G$ is a Gröbner basis of $\langle G \rangle$ and $F \subset \langle G \rangle$.

If G is a Gröbner basis of $\langle G \rangle$ then it is easy to check $F \subset \langle G \rangle$ by the division algorithm, that is the normal form computation with respect G. However it is not easy to show $\langle G \rangle \subset \langle F \rangle$ in general. If F is homogeneous, Arnold showed that

G is a Gröbner basis of $\langle G \rangle \Rightarrow G$ is a Gröbner basis of $\langle F \rangle$.

We denote the homogenization of a polynomial f by f^h. For a set of polynomials F we set $F^h = \{f^h \mid f \in F\}$. For an inhomogeneous ideal $I = \langle F \rangle$, we have a criterion:

Theorem 1 ([7]). *Let $G \subset \mathbb{Z}_p^0[X]$ be a p-GB candidate for $F \subset \mathbb{Z}[X]$ with respect to a degree compatible order $\prec$ s.t. $I = \langle F \rangle \subset \langle G \rangle$ and G is a Gröbner basis of $\langle G \rangle$. Let k be a positive integer such that $\langle \phi_p(F)^h \rangle : t^k = \langle \phi_p(F)^h \rangle : t^\infty$, and $H \subset \mathbb{Q}[X]$ the reduced Gröbner basis of $F^h \cup \{t^k\}$ with respect to $\prec^h$, the homogenization of $\prec$. If $H \subset \mathbb{Z}_p^0[X]$ and $\phi_p(\mathrm{LC}(h)) \neq 0$ for all $h \in H$ then $I = \langle G \rangle$.*

We can cut off terms divisible by t^k during the Gröbner computation for $\langle F^h \cup \{t^k\} \rangle$. Thus it is expected that the Gröbner computation of $\langle F^h \cup \{t^k\} \rangle$ is easier than that of $\langle F^h \rangle$ if k is small. However, in order to compute k, we have to compute a Gröbner basis of $\langle \phi_p(F)^h \rangle$ and there are cases where the computation is much harder that that of $\langle \phi_p(F) \rangle$. A typical case is an ideal in Romanovski et al. [4]. In order to solve this case, we tried directly showing $g \in \langle F \rangle$ for $g \in G$ by constructing a generating relation for g by F.

3   Deciding Ideal Membership of Elements of a Candidate

Construction of a generating relation for g by $F = \{f_1, \ldots, f_m\}$, that is finding $h_1, \ldots, h_m$ s.t. $g = h_1 f_1 + \ldots + h_m f_m$, is not a new idea. Theoretically it can be constructed by gathering quotients obtained during remainder computations. But if we cannot complete an execution of the Buchberger algorithm, we cannot construct $h_1, \ldots, h_m$. In such a case it is useful to apply modular methods and there are several works on this direction [1–3]. We note that these works propose lifting methods for constructing a Gröbner basis candidate. In order to do this, the whole generating relations have to be lifted together and the time and space cost may be huge. In contrast to them, our method is to verify a Gröbner basis candidate. For this purpose, we only have to construct a system of linear equations per an element of the candidate from a generating relation over a finite field. Then we apply Hensel lifting for efficiently solving the system of linear equations.

3.1   A Modular Algorithm for Computing a Generating Relation

Algorithm 1 is a modular algorithm for computing a generating relation of a polynomial g by a set of polynomials F. T denotes the set of all monomials.

Algorithm 1. generating_relation(F, G, g)

Input : $F = \{f_1, \ldots, f_m\}, G \subset \mathbb{Z}[X]$ s.t. G is a p-GB candidate for F
         and $F \subset \langle G \rangle$, $g \in G$
Output : $h_1, \ldots, h_m \subset \mathbb{Q}[X]$ satisfying $g = h_1 f_1 + \cdots + h_m f_m$, or **failure**
$G_p = \phi_p(G) \leftarrow$ the reduced Gröbner basis of $\langle \phi_p(F) \rangle$
$(H_1, \ldots, H_m) \leftarrow H_i = \sum_j a_{ij} t_{ij}$ s.t. $\phi_p(g) = H_1 \phi_p(f_1) + \cdots + H_m \phi_p(f_m)$
         $(i = 1, \ldots, m, a_{ij} \in \mathbb{F}_p, t_{ij}$ are monomials$)$
$E \leftarrow (\sum_j c_{1j} t_{1j}) f_1 + \cdots + (\sum_j c_{1j} t_{mj}) f_m - g$
         $(c_{ij}$ are indeterminate coefficients$)$
Write E as $E = \displaystyle\sum_{i=1}^k e_i t_i$ $(t_i \in T, e_i$ is a linear form of $c_{ij})$
$Z \leftarrow$ the variables expressed by homogeneous linear forms of the free variables
         in the solution of $e_1 = \cdots = e_k = 0$ over $\mathbb{F}_p$.
$\bar{E} \leftarrow (\sum_{j, c_{1j} \notin Z} c_{1j} t_{ij}) f_1 + \cdots + (\sum_{j, c_{mj} \notin Z} c_{mj} t_{ij}) f_m - g$
Write $\bar{E}$ as $\bar{E} = \displaystyle\sum_{i=1}^k \bar{e}_i s_i$ $(s_i \in T, \bar{e}_i$ is a linear form of $c_{ij} \notin Z)$
if $\bar{e}_1 = \cdots = \bar{e}_l = 0$ has a solution $c_{ij} = b_{ij}$ over $\mathbb{Q}$ then
    return $(\displaystyle\sum_{j, c_{1j} \notin Z} b_{1j} t_{ij}, \ldots, \sum_{j, c_{mj} \notin Z} b_{mj} t_{ij})$
else
    return **failure**
end if

In Algorithm 1, we first compute the coefficient polynomials $H_1, \ldots, H_m \in \mathbb{F}_p[X]$ s.t. $\phi_p(g) = H_1\phi_p(f_1) + \cdots + \phi_p(f_m)$. Then the coefficients of H_i are replaced by indeterminate coefficients c_{ij}, and $\phi_p(g)$ and $\phi_p(f_1), \ldots, \phi_p(f_m)$ are replaced by g and $f_1, \ldots f_m$ respectively to obtain a system of linear equations over $\mathbb{Q}$ with respect to c_{ij}. We call it the initial system. If we solve the initial system over $\mathbb{F}_p$, then the solution contains free variables in general due to certain syzygies among $\phi_p(f_1), \ldots, \phi_p(f_m)$. By setting them to zero, the dependent variables represented by homogeneous linear forms with respect to the free variables are also set to zero. By setting all such variables to zero, we obtain a reduced system of linear equations. In general this system is over-determined and it is not ensured that it has a solution over $\mathbb{Q}$. Therefore Algorithm 1 is probabilistic in the sense that it may return **failure** even if $g \in \langle F \rangle$. However, if it successfully returns a set of polynomials, then it surely gives a generating relation of g by F over $\mathbb{Q}$.

The reason why we reduce the number of the variables are as follows.

1. The general solution of the initial system over $\mathbb{Q}$, if it exists, contains many free variables due to syzygies. Since we need only a single solution, it is inefficient to find the general solution.
2. By reducing the number of variables, the solution of the reduced system can be made unique. Then we can apply Hensel lifting for solving the reduced system over $\mathbb{Q}$.

3.2   Solving a Linear System by Hensel Lifting

The map ϕ_p is naturally extended to vectors or matrices over $\mathbb{Z}_p^0$. We suppose that the solution of $\phi_p(A)x = \phi_p(b)$ is unique for an $m \times n$ integer matrix $A = (a_{ij})$ $(m \geq n)$, an m-dimensional column vector $b = (b_i)$ and an n-dimensional unknown column vector $x = (x_j)$. Then the solution of $Ax = b$ is unique or it has no solution. Algorithm 2 is well known and it gives the solution of $Ax = b$ if it exists. In the algorithm $\texttt{inttorat(x,m)}$ tries to find a rational number $\frac{a}{b}$ s.t. $bx \equiv a \bmod m$, $|a|, |b| < \sqrt{\frac{m}{2}}$.

Algorithm 2. $\texttt{solve\_linear\_equation\_by\_hensel}(A, b)$

Input :   an $n \times m$ integer matrix s.t. $\text{rank}(A) = n$, an integer vector of size m
Output : the unique solution of $Ax = b$ or **failure**
$A' \leftarrow$ an $n \times n$ sub matrix of A s.t. $\det(\phi_p(A')) \neq 0$
$b' \leftarrow$ the sub-vector of b corresponding with the rows of A'
$(x, r, k) \leftarrow (0, b', 0)$
loop
    $y \leftarrow \texttt{inttorat}(x, p^k)$
    if $y \neq$ **failure** and $A'y = b'$
      if
        $Ay = b$ **then** return y
      else
        return **failure**
      end if

end if
$$s \leftarrow \phi_p(A')^{-1}\phi_p(r)$$
$$x \leftarrow x + p^k s$$
$$r \leftarrow \frac{r - A's}{p}$$
$$k \leftarrow k + 1$$
end loop

4 Implementation in Risa/Asir

We experimentally implemented Algorithm 1 in Risa/Asir. The detail of the implementation is as follows. We use the same notations in Algorithm 1.

1. Computation of a generating relation over a given finite field
   During computation of a Gröbner basis, at each step we keep the data which show how each intermediate basis element is represented by the previously computed ones. This is done by a built-in function nd_gr with options gentrace=1 and gensyz=1. A built-in function nd_btog computes a matrix which represents the Gröbner basis computed by nd_gr for the input polynomial set. The computation of this matrix over $\mathbb{Q}$ is hard in general and we apply nd_btog over a finite field for computing $(H_1, \ldots, H_m)$ in Algorithm 1.

2. Computation of $e_1, \ldots, e_k$
   In general each polynomial $h_i = \sum_j c_{ij}t_{ij}$ contains many terms but its coefficients are individual variables. Therefore each e_l can be computed as a sum of act where a is a coefficient of some $f \in F$, c is one of the variables c_{ij} and t is a monomial. In the current implementation, after computing $t_1, \ldots, t_k$ appearing in E, we apply a hash algorithm to efficiently find the monomial t_i s.t. $t = t_i$ for a particular act.

3. Computation of $\bar{e}_1, \ldots, \bar{e}_l$
   After computing the general solution of $e_1 = \cdots = e_k = 0$ over $\mathbb{F}_p$, a special solution is obtained by setting all free variables to zero. This procedure is done by an external program written in C. Then we repeat Step 2 for the reduced variables and we get $\bar{e}_1, \ldots, \bar{e}_l$.

4. Solving $\bar{e}_1 = \cdots = \bar{e}_l = 0$ over $\mathbb{Q}$
   This is done by Algorithm 2. By computing LU factorization of $\phi_p(A')$ over $\mathbb{F}_p$, $\phi_p(A')^{-1}\phi_p(r)$ is computed efficiently.

4.1 An Application

The motivation of this work was to verify a Gröbner basis candidate $G = \{g_1, \ldots, g_{69}\}$ given in Romanovski et al. [4]. G is a Gröbner basis candidate of an ideal $I = \langle F \rangle \subset \mathbb{Q}[a_{40}, a_{31}, a_{13}, a_{04}, b_{40}, b_{31}, b_{13}, b_{04}]$. F consists of inhomogeneous polynomials and $I \subset \langle G \rangle$. It is easy to compute a Gröbner basis of $\langle F \rangle$ over a finite field and G can be easily computed by CRT. However it is very hard at least in Risa/Asir to compute a Gröbner basis of $\langle F^h \rangle$ even over a finite field and it does not finish within a month. Some preliminary experiments show that a Gröbner basis of $\langle F \cup \{g_{60}, g_{61}\}\rangle$ can be computed rather easily by an ordinary method and it coincides with G, where

$$g_{60} = 349a_{31}a_{40} + 333b_{31}a_{40} - 333b_{04}a_{13} - 349b_{04}b_{13},$$
$$g_{61} = 555b_{31}a_{40} + 349a_{04}a_{13} - 555b_{04}a_{13} - 349b_{31}b_{40}.$$

Therefore if we can show $g_{60}, g_{61} \in \langle F \rangle$ then we can conclude that G is a Gröbner basis of I. We could compute generating relations for g_{60} and g_{61} by executing Algorithm 1 with Algorithm 2. The computation was done on Intel Xeon X5675 3.07GHz with 192GB of memory. The required memory was 32GB. We only show computational statics for g_{60} because those of g_{60} and g_{61} are almost the same. In the second table of Table 1, H_i, e_i, Z and $\bar{e}_i$ show the timing for the

Table 1. Computation of a generating relation of g_{60}

matrix size(initial/reduced)	182300× 440525 / 117585× 116556
computing time	9.7×10^4 sec
number of required steps in Hensel lifting	5030 steps
size of coefficients	about 20000digits

	H_i	e_i	Z	$\bar{e}_i$	LU	Hensel
time (sec)	4000	150	26000	50	3100	63000

corresponding objects in Algorithm 1. *LU* and *Hensel* show the timing for LU factorization of $\phi_p(A')$ and for the main loop in Algorithm 2, where `inttorat` was executed every 10 iterations. This result shows that the Hensel lifting is very successful because we could not solve the same system of linear equations within 10 months by a non-modular method.

References

1. Winkler, F.: A p-adic Approach to the Computation of Gröbner Bases. J. Symb. Comp. 6, 287–304 (1988)
2. Pauer, F.: On Lucky Ideals for Gröbner Basis Computations. J. Symb. Comp. 14, 471–482 (1992)
3. Arnold, E.: Modular algorithms for computing Gröbner bases. J. Symb. Comp. 35, 403–419 (2003)
4. Romanovski, V., Chen, X., Hu, Z.: Linearizability of linear systems perturbed by fifth degree homogeneous polynomials. J. Phys. A: Math. Theor. 40(22), 5905–5919 (2007)
5. Idrees, N., Pfister, G., Steidel, S.: Parallelization of modular algorithms. J. Symb. Comp. 46, 672–684 (2011)
6. Steidel, S.: Gröbner bases of symmetric ideals. J. Symb. Comp. 54, 72–86 (2013)
7. Yokoyama, K.: Usage of Modular Techniques for Efficient Computation of Ideal Operations - (Invited Talk). In: Gerdt, V.P., Koepf, W., Mayr, E.W., Vorozhtsov, E.V. (eds.) CASC 2012. LNCS, vol. 7442, pp. 361–362. Springer, Heidelberg (2012)

Cylindrical Algebraic Decomposition in the RegularChains Library

Changbo Chen[1] and Marc Moreno Maza[2]

[1] Chongqing Key Laboratory of Automated Reasoning and Cognition,
Chongqing Institute of Green and Intelligent Technology,
Chinese Academy of Sciences, China
changbo.chen@hotmail.com
http://www.orcca.on.ca/~cchen
[2] ORCCA, University of Western Ontario, Canada
moreno@csd.uwo.ca
http://www.csd.uwo.ca/~moreno

Abstract. Cylindrical algebraic decomposition (CAD) is a fundamental tool in computational real algebraic geometry and has been implemented in several software. While existing implementations are all based on Collins' projection-lifting scheme and its subsequent ameliorations, the implementation of CAD in the RegularChains library is based on triangular decomposition of polynomial systems and real root isolation of regular chains. The function in the RegularChains library for computing CAD is called CylindricalAlgebraicDecompose. In this paper, we illustrate by examples the functionality, the underlying theory and algorithm, as well the implementation techniques of CylindricalAlgebraicDecompose. An application of it is also provided.

Keywords: Cylindrical algebraic decomposition, triangular decomposition, RegularChains.

1 Introduction

Cylindrical Algebraic Decomposition (CAD) was introduced by Collins [6] for solving the real quantifier elimination (QE) problem. A CAD is a partition of the real space $\mathbb{R}^n$ into finitely many connected semi-algebraic subsets, called cells, such that any two cells are cylindrically arranged, that is the projection of them onto any low dimensional space are either disjoint or identical. Let F be a set of polynomials with real number coefficients in n variables. A CAD is called F-invariant if any polynomial of F is sign-invariant on any cell of the CAD. The rich properties of CAD make it become a fundamental tool in studying real algebraic geometry. Despite of the doubly exponential running time complexity in the worst case, the practical performance of CAD has been improved by many researchers [1]. Accompanying with these improvements, many software have been implemented to compute CADs, among which the best known are QEPCAD, Mathematica and Reduce.

H. Hong and C. Yap (Eds.): ICMS 2014, LNCS 8592, pp. 425–433, 2014.
© Springer-Verlag Berlin Heidelberg 2014

Most of the implementations for computing CADs are based on the original projection-lifting framework of Collins. In the projection phase, starting from an input set F of polynomials in n variables, one applies a pre-defined projection operator P to F and obtains a set $P(F)$ of polynomials in $n - 1$ variables. This process is recursively done for $P(F)$ until a set of univariate polynomials are computed. Let A be the set of all polynomials generated in such process. The projection phase guarantees that the zero sets of polynomials in A naturally defines a CAD of $\mathbb{R}^n$. The work of the lifting phase is to compute an explicit representation from such an implicitly defined CAD. In the base case, the real zeros of univariate polynomials in A are isolated, which divides the real line into disjoint open intervals. The real zeros and the intervals together form a CAD of $\mathbb{R}^1$. Assume a CAD of $\mathbb{R}^{n-1}$ is computed. For each cell C of it, one evaluates the polynomials of A in n variables at a sample point of the cell and obtains a set of univariate polynomials. Isolating the real roots of them allows one to deduce all the cells of the CAD of $\mathbb{R}^n$ whose projection are C.

In [4], a different method for computing CADs was proposed. It first produces a cylindrical decomposition of the complex space (CCD) through the computation of regular GCDs, and then refines the CCD into a CAD of the real space by isolating real roots of univariate polynomials with real algebraic number coefficients encoded by regular chains and isolating boxes. The efficiency of it was greatly improved in [2], where the computation of CCD is replaced by a new incremental algorithm. Both algorithms are based on triangular decomposition of polynomial systems and real root isolation of regular chains. For this reason, we call the CAD as computed in [4,3] RC-CAD. The algorithm of [4] was firstly implemented in the RegularChains library of MAPLE 14. The implementation was revised in MAPLE 16 and has remained the same in the subsequent versions of MAPLE. The algorithm of [3] was implemented in the RegularChains library, but not shipped with MAPLE. Any update of the implementation of both algorithms are now available through the RegularChains library (http://www.regularchains.org).

The purpose of this paper is to lift the veil of the implementation of RC-CAD. In the RegularChains library, the function for computing CCD and CAD are respectively CylindricalDecompose and CylindricalAlgebraicDecompose. In Section 2, we illustrate by examples how to use the two functions. In Section 3, we explain the underlying theory and algorithms of RC-CAD. The technical challenges for implementing the algorithms and our solutions are also discussed. Finally, in Section 4, we report on an application of our software.

2   Functionality

A cylindrical decomposition of the complex space, or complex cylindrical decomposition (CCD) is a partition of the complex space into cylindrically arranged constructible sets, each of which is the zero set of a regular system. Figure 1 shows a CCD represented in a piecewise format. Here the variable order is $x < y$. The "1's" in the formula are placeholders having no meanings. Such format can

```
> R := PolynomialRing([y, x]):
  F := [y^2-x]:
  CylindricalDecompose(F, R, output=piecewise);
```

$$\left\{ \begin{array}{ll} \left\{ \begin{array}{ll} 1 & y = 0 \\ 1 & \textit{otherwise} \end{array} \right. & x = 0 \\[3ex] \left\{ \begin{array}{ll} 1 & -y^2 + x = 0 \\ 1 & \textit{otherwise} \end{array} \right. & \textit{otherwise} \end{array} \right.$$

Fig. 1. Compute complex cylindrical decomposition by CylindricalDecompose

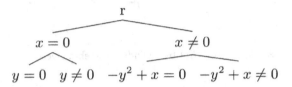

Fig. 2. A complex cylindrical tree

be interpreted as a tree shown in Figure 2, where each branch in the piecewise format corresponds to a path of the tree. The constraints on a path of the tree form a regular system. Such a CCD is sign-invariant w.r.t. $f := x^2 + y^2 - 1$, that is for a given path of the tree from the root to a leaf, either f vanishes at all points of the path or f vanishes at none of the points of the path.

An F-sign invariant CAD is depicted in Figure 3. The CylindricalAlgebraicDecompose command supports several different input and output formats. The input can be a list of polynomials, as shown in Figure 1, as well as a list of polynomial constraints, as shown in Figure 5. The format 'output'='cadcell' allows only true cells satisfying the input constraints are displayed. To get a sample point of a CAD cell, the function SamplePoints can be called. Here no cost occurs since sample points are computed along the computation of the CAD and are stored in the type cad_cell. A sample point is encoded by the type box, which is represented by a regular chain and an isolation cube. Such a representation allows one to easily test if the sign of a polynomial at the sample point by calling the function SignAtBox.

Due to the intrinsic doubly exponential complexity of CAD, it is not uncommon that the number of CAD cells is numerous. To get a compact output for the purpose of "solving" the input semi-algebraic system, the option 'output'='rootof' can be used. In this case, the solver will try to merge the adjacent CAD cells as much as possible in order to get a simple formula. See Figure 4 for an example.

```
> R := PolynomialRing([y, x]):
  F := [y^2-x]:
  CylindricalAlgebraicDecompose(F,R,output=piecewise);
```

$$
\left[\begin{array}{ll}
1 & x < 0 \\[4pt]
\left\{\begin{array}{ll}
1 & y < 0 \\
1 & y = 0 \\
1 & 0 < y
\end{array}\right. & x = 0 \\[12pt]
\left\{\begin{array}{ll}
1 & y < -\sqrt{x} \\
1 & y = -\sqrt{x} \\
1 & \text{And}\left(-\sqrt{x} < y, y < \sqrt{x}\right) \\
1 & y = \sqrt{x} \\
1 & \sqrt{x} < y
\end{array}\right. & 0 < x
\end{array}\right.
$$

Fig. 3. Compute CAD by CylindricalAlgebraicDecompose

```
> R := PolynomialRing([y, x]):
  CylindricalAlgebraicDecompose([x^2+y^2-1<=0],R,output=
  rootof);
```

$$
\left[\left[\text{And}(-1 \le x, x \le 1), \text{And}\left(-\sqrt{-x^2 + 1} \le y, y \le \sqrt{-x^2 + 1}\right)\right]\right]
$$

Fig. 4. Solve semi-algebraic systems by CylindricalAlgebraicDecompose

```
> R := PolynomialRing([y, x]):
  cad := CylindricalAlgebraicDecompose([x^2+y^2-1=0, x*y-1/2=0],R,output=cadcell);
  Display(cad, R);
  sp := map(SamplePoints, cad, R);
  Display(sp, R);
  s := SignAtBox(x^2+y^2-2, sp[1], R);
```

$$
cad := [\, cad\_cell, cad\_cell\,]
$$

$$
\left[\left[\begin{array}{l}
y = \dfrac{1}{2\,x} \\[8pt]
x = -\dfrac{1}{2}\sqrt{2}
\end{array}\right.
\;,\;
\left[\begin{array}{l}
y = \dfrac{1}{2\,x} \\[8pt]
x = \dfrac{1}{2}\sqrt{2}
\end{array}\right.\right]
$$

$$
sp := [\, box, box\,]
$$

$$
\left[\left[\begin{array}{l}
y = \left[-\dfrac{46341}{65536}, -\dfrac{1482907}{2097152}\right] \\[10pt]
x = \left[-\dfrac{46341}{65536}, -\dfrac{741455}{1048576}\right]
\end{array}\right.
\;,\;
\left[\begin{array}{l}
y = \left[\dfrac{185363}{262144}, \dfrac{741457}{1048576}\right] \\[10pt]
x = \left[\dfrac{741455}{1048576}, \dfrac{46341}{65536}\right]
\end{array}\right.\right]
$$

$$
s := -1
$$

Fig. 5. Compute CAD of a semi-algebraic system

3 Underlying Theory and Technical Contribution

In this section, we explain the algorithms and theory underlying RC-CAD as well as a universe tree data structure for implementing RC-CAD. There are two important ingredients in RC-CAD, namely a routine for computing CCD, and a routine for turning a CCD into a CAD.

Two algorithms have been proposed for computing CCD. The first one was proposed in [4]. It has two phases: InitialPartition and MakeCylindrical. Let F be a set of polynomials in n variables. InitialPartition partitions $\mathbb{C}^n$ into a family $\mathcal{C}$ of constructible sets, called cells, each of which is the zero set of a regular system. Moreover, for a given cell C and any polynomial $f \in F$, either f vanishes at all points of C or f vanishes at no points of C. The cells in the output of Initial-Partition are not necessarily cylindrically arranged. The cylindricity is achieved in a top-down fashion. The collection $\mathcal{C}$ of constructible sets is refined into a new family $\mathcal{D}$ of disjoint constructible sets, such that the projection of any two cells of $\mathcal{D}$ onto $\mathbb{C}^{n-1}$ are either identically equal or disjoint. By making a recursive call to MakeCylindrical on the projection of $\mathcal{D}$ on $\mathbb{C}^{n-1}$, one finally deduces a collection of cylindrically arranged cells. If the option 'method'='recursive' is enabled, CylindricalDecompose will use such an algorithm.

A second one was proposed in [3]. Let F be a set of m polynomials in n variables. The algorithm first builds an initial CCD $\mathcal{C}_0$, which consists of only one cell $\mathbb{C}^n$. It then pops a polynomial f_1 from F and refines $\mathcal{C}_0$ into $\mathcal{C}_1$ such that $\mathcal{C}_1$ is sign-invariant w.r.t. f_1. If F is not empty, a new polynomial f_2 is popped and $\mathcal{C}_1$ is refined w.r.t. f_2. This process is repeated until F gets empty. Making a CCD sign-invariant w.r.t. a polynomial is reduced to making every cell of the CCD sign-invariant w.r.t. a polynomial. This process has to preserve cylindricity, which is achieved by a refinement operation called IntersectPath in [3].

Let's illustrate this incremental algorithm by an example. Let $F := \{y^2 - x, x^2 + y^2 - 1\}$, where $f_1 = y^2 - x$, $f_2 = x^2 + y^2 - 1$. The evolution of the CCD tree during the computation is depicted by Figure 6. The first one is the initial tree. In the second tree, the node "any x" splits into two nodes to make the discriminant of f_1 w.r.t. y sign-invariant. In the third tree, the nodes "any y" split to make f_1 sign-invariant. Moreover, when $x = 0$, f_1 is replaced by its squarefree part modulo $x = 0$. In the fourth tree, the node $y \neq 0$ splits to make f_2 sign-invariant. In the fifth tree, the node $x \neq 0$ splits to make the resultant of f_1 and f_2 w.r.t. y sign-invariant. In the sixth tree, the node $x(x^2 + x - 1) \neq 0$ splits to make the discriminant of f_2 w.r.t. y sign-invariant. Finally the nodes $f_1 \neq 0$ splits to make f_2 sign-invariant.

The operation turning a CCD into a CAD is called MakeSemiAlgebraic. It is implemented in a recursive manner. For the base case $n = 1$, it collects all the polynomials in the equational nodes of the CCD tree, does univariate real root isolation, and picks sample points. Let $\mathcal{C}_{n-1}$ be a CAD of $\mathbb{R}^{n-1}$ derived from a CCD $\mathcal{T}$ of $\mathbb{C}^{n-1}$. Let C be a cell of $\mathcal{C}_{n-1}$ derived from a cell D of $\mathcal{T}$. Let s be a sample point of C. Let P be the set of polynomials appearing in the equational children of D. To compute the cells of a CAD of $\mathbb{R}^n$ whose projection onto $\mathbb{R}^{n-1}$ is C (these cells form a stack over C), one isolates the real

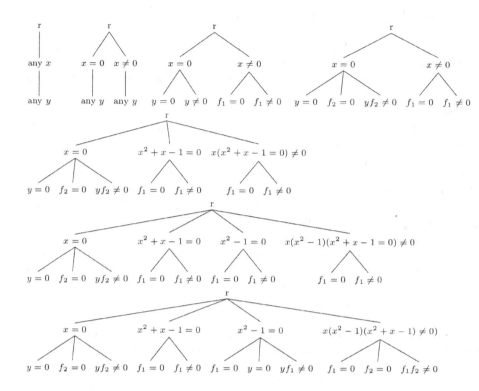

Fig. 6. The process of computing a CCD

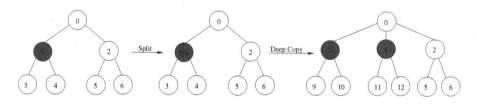

Fig. 7. The universe tree and the Split operation

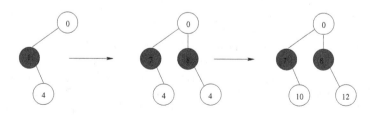

Fig. 8. A sub-tree evolves with the universe tree

roots of univariate polynomials $\tilde{p} := p([x_1, \ldots, x_{n-1}] = s, x_n)$, $p \in P$. Here the substitution is carried with interval arithmetic since the coordinate of s may be irrational real algebraic numbers. As a result, the coefficients of the univariate polynomials $\tilde{p}$ are approximated by intervals whose width can be reduced to arbitrarily small. If the width of the intervals are sufficiently small, the real roots of $\tilde{p}$ can be deduced from its sleeve polynomials, which are univariate polynomials with rational number coefficients, thanks to the fact that s is encoded by a regular chain and a box [8,9,5].

The data structure supporting the implementations of CCD and CAD is a *universe* tree [3]. It is a tree data structure equipped with a Split operation (7). The Split operation is frequently used in the incremental algorithm [3] for computed CCDs, where the nodes in a complex cylindrical tree are split to make new added or generated polynomials sign-invariant and maintain the tree cylindrical. This process is illustrated by Figure 6. As a result, the universe tree is always kept to be updated. Suppose now we'd like to do several operations on a sub-tree. In order to maintain data consistency, the sub-tree has to been updated according to the universe. Note that it is fine to only update the node to be immediately worked on. The update of the sub-tree is illustrated by Figure 8.

4   Application

In this section, we show the application of RC-CAD on solving two challenges [7].

```
> t0 := time():
  f_1 := x_1+x_1/(x_1^2+y_1^2)-(x_2+x_2/(x_2^2+y_2^2)):
  f_2 := y_1-y_1/(x_1^2+y_1^2)-(y_2-y_2/(x_2^2+y_2^2)):
  g_1 := numer(normal(f_1)): g_2 := numer(normal(f_2)):

  sys := &not( (x_1^2+y_1^2-1>0) &and (x_2^2+y_2^2-1>0) &and (g_1=0) &and (g_2=0)
  &implies ( (x_1=x_2) &and (y_1=y_2)  ) ):
  lsas := RegularChains:-TRDcadLogicFormulaToLsas(sys):
  nops(lsas); sys1 := lsas[1]: sys2 := lsas[2]:

  vars := SuggestVariableOrder(sys1): R := PolynomialRing(vars):
  out1 := CylindricalAlgebraicDecompose(sys1, R, output=rootof, precondition='TD',
  partial='true');

  vars := SuggestVariableOrder(sys2): R := PolynomialRing(vars):
  out2 := CylindricalAlgebraicDecompose(sys2, R, output=rootof, precondition='TD',
  partial='true');

  evalb( nops(out1)=0 and nops(out2)=0 ); t1 := time() - t0;
                                 2
                            out1 := [ ]
                            out2 := [ ]
                               true
                           t1 := 55.197
```

Fig. 9. Use CAD to solve Problem Joukowski-a

Challenge 1. *Demonstrate automatically the truth of Formula 1 over reals.*

$$\forall x_1 \forall x_2 \forall y_1 \forall y_2 \ (x_1^2 + y_1^2 > 1 \wedge x_2^2 + y_2^2 > 1 \wedge x_1 + \tfrac{x_1}{x_1^2+y_1^2} = x_2 + \tfrac{x_2}{x_2^2+y_2^2} \wedge$$
$$y_1 - \tfrac{y_1}{x_1^2+y_1^2} = y_2 - \tfrac{y_2}{x_2^2+y_2^2} \implies (x_1 = x_2 \wedge y_1 = y_2)) \tag{1}$$

Challenge 2. *Demonstrate automatically the truth of Formula 2 over reals.*

$$\forall x_1 \forall x_2 \forall y_1 \forall y_2 \ (y_1 > 0 \wedge y_2 > 0 \wedge x_1 + \tfrac{x_1}{x_1^2+y_1^2} = x_2 + \tfrac{x_2}{x_2^2+y_2^2} \wedge$$
$$y_1 - \tfrac{y_1}{x_1^2+y_1^2} = y_2 - \tfrac{y_2}{x_2^2+y_2^2} \implies (x_1 = x_2 \wedge y_1 = y_2)) \tag{2}$$

The first challenge is solved by RC-CAD within one minute on a laptop (Intel i7, 8Gb RAM, Ubuntu), as shown by Figure 9. The second challenge can be solved in a similar way in about 20 seconds, which is not shown here limited to space. Both answers are true. To achieve this, a universal quantifier elimination problem is converted to an existential one using the following equivalence:

$$\forall \mathbf{x}(A \implies (B \wedge C)) \text{ iff } \neg \exists \mathbf{x} \neg (A \implies (B \wedge C)) \text{ iff } \neg \exists \mathbf{x} \left((A \wedge \neg B) \vee (A \wedge \neg C)\right)$$

As a result, Formula 1 is true if and only if none of the two semi-algebraic systems $sys1$ and $sys2$ in Figure 9 has solutions. To solve a semi-algebraic system, the command CylindricalAlgebraicDecompose is called with three options. The option 'precondition'='TD' allows to precondition the input system by means of triangular decomposition. The option 'partial'='true' uses partial lifting.

Acknowledgements. This work was supported by the NSFC (11301524) and the CSTC (cstc2013jjys0002).

References

1. Caviness, B., Johnson, J. (eds.): Quantifier Elimination and Cylindrical Algebraic Decomposition. Texts and Monographs in Symbolic Computation. Springer (1998)
2. Chen, C., Moreno Maza, M.: Algorithms for computing triangular decomposition of polynomial systems. Journal of Symbolic Computation 47(6), 610–642 (2012)
3. Chen, C., Moreno Maza, M.: An incremental algorithm for computing cylindrical algebraic decompositions. In: Proc. ASCM 2012 (October 2012)
4. Chen, C., Moreno Maza, M., Xia, B., Yang, L.: Computing cylindrical algebraic decomposition via triangular decomposition. In: ISSAC 2009, pp. 95–102 (2009)
5. Cheng, J.S., Gao, X.S., Yap, C.K.: Complete numerical isolation of real zeros in zero-dimensional triangular systems. In: ISSAC, pp. 92–99 (2007)
6. Collins, G.E.: Quantifier elimination for real closed fields by cylindrical algebraic decomposition. In: Brakhage, H. (ed.) GI-Fachtagung 1975. LNCS, vol. 33, pp. 515–532. Springer, Heidelberg (1975)

7. Davenport, J.H., Bradford, R.J., England, M., Wilson, D.J.: Program verification in the presence of complex numbers, functions with branch cuts etc. In: SYNASC, pp. 83–88 (2012)
8. Lu, Z.Y., He, B., Luo, Y., Pan, L.: An algorithm of real root isolation for polynomial systems. In: Wang, D.M., Zhi, L. (eds.) Proceedings of Symbolic Numeric Computation 2005, pp. 94–107 (2005)
9. Xia, B., Zhang, T.: Real solution isolation using interval arithmetic. Comput. Math. Appl. 52(6-7), 853–860 (2006)

Hierarchical Comprehensive Triangular Decomposition*

Zhenghong Chen, Xiaoxian Tang, and Bican Xia

School of Mathematical Sciences, Peking University, China
{septemwnid,tangxiaoxian}@pku.edu.cn, xbc@math.pku.edu.cn

Abstract. The concept of *comprehensive triangular decomposition* (CTD) was first introduced by Chen *et al.* in their CASC'2007 paper and could be viewed as an analogue of comprehensive Gröbner systems for parametric polynomial systems. The first complete algorithm for computing CTD was also proposed in that paper and implemented in the `RegularChains` library in Maple. Following our previous work on generic regular decomposition for parametric polynomial systems, we introduce in this paper a so-called *hierarchical* strategy for computing CTDs. Roughly speaking, for a given parametric system, the parametric space is divided into several sub-spaces of different dimensions and we compute CTDs over those sub-spaces one by one. So, it is possible that, for some benchmarks, it is difficult to compute CTDs in reasonable time while this strategy can obtain some "partial" solutions over some parametric sub-spaces. The program based on this strategy has been tested on a number of benchmarks from the literature. Experimental results on these benchmarks with comparison to `RegularChains` are reported and may be valuable for developing more efficient triangularization tools.

Keywords: Comprehensive triangular decomposition, regular chain, hierarchical, generic regular decomposition, parametric polynomial system.

1 Introduction

Solving parametric polynomial system plays a key role in many areas such as automated geometry theorem deduction, stability analysis of dynamical systems, robotics and so on. For an arbitrary parametric system, in symbolic computation, solving this system is to convert equivalently the parametric system into new systems with special structures so that it is easier to analyze or solve the solutions to the new systems. There are two main kinds of symbolic methods to solve parametric systems, *i.e.*, the algorithms based on *Gröbner bases* [12, 14–17, 23] and those based on *triangular decompositions* [1, 2, 5, 9, 11, 13, 19, 20, 24–27].

The methods based on triangular decompositions have been studied by many researchers since Wu's work [24] on *characteristic sets*. A significant concept in the theories of triangular sets is *regular chain* (or *normal chain*) introduced

---

* The work was supported by National Science Foundation of China (Grants 11290141 and 11271034).

H. Hong and C. Yap (Eds.): ICMS 2014, LNCS 8592, pp. 434–441, 2014.

by Yang and Zhang [27] and Kalkbrener [11] independently. Gao and Chou proposed a method in [9] for identifying all parametric values for which a given system has solutions and giving the solutions by $p-chains$ without a partition of the parameter space. Wang generalized the concept of regular chain to *regular system* and gave an efficient algorithm for computing it [20–22]. It should be noticed that, due to their strong projection property, the regular systems or series may also be used as representations for parametric systems. Chen *et. al.* introduced the concept of *comprehensive triangular decomposition* (CTD) [5] to solve parametric systems, which could be viewed as an analogue of comprehensive Gröbner systems. Algorithm CTD for computing CTD was also proposed in [5].

There are several implementations based on the above triangularization methods, such as Epsilon [22], RegularChains [6] and wsolve [19].

Suppose $\mathbf{P} \subset \mathbb{Q}[U][X]$ is a parametric polynomial system where $X = (x_1, \ldots, x_n)$ are variables and $U = (u_1, \ldots, u_d)$ are parameters. The above mentioned algorithms all solve the system in $\mathbb{C}^{d+n}$ directly. That means all the unknowns (U and X) are viewed as variables and triangular decompositions are computed over $\mathbb{Q}$. It may happen that no triangular decompositions over $\mathbb{Q}$ can be obtained in a reasonable time for some systems while a triangular decomposition over $\mathbb{Q}[U]$ is much easier to be computed.

Based on this observation, we propose a strategy which computes CTDs for given parametric systems hierarchically and the CTDs are called *hierarchical comprehensive triangular decompositions* (HCTD). By "hierarchical" we mean that, roughly speaking, a generic regular decomposition is computed first over $\mathbb{Q}[U]$ and a parametric polynomial $B(U)$ is obtained at the same time such that the solutions to the original system in $\mathbb{C}^{d+n}$ can be expressed as the union of solutions to those regular systems in the decomposition provided that the parameter values satisfying $B(U) \neq 0$. Then, by applying similar procedure recursively, the solutions satisfying $B(U) = 0$ are obtained through adding $B(U) = 0$ to the system and treating some parameters as variables. We give an algorithm based on this hierarchical strategy which computes CTDs for given parametric systems. The algorithm has been implemented with Maple and tested on a number of benchmarks from the literature. Experimental results on these benchmarks with comparison to RegularChains are reported (see Tables 2) and may be valuable for developing more efficient triangularization tools. For some benchmarks, it is difficult to compute CTDs in reasonable time while our program can output "partial solutions" (see Table 4).

The rest part of this extended abstract is organized as follows. Section 2 introduces an algorithm, Algorithm HCTD, for computing CTDs hierarchically and an example is illustrated there. Section 3 compares the Algorithm HCTD and the Algrotihm CTD in [5] by experiments. Section 4 introduces another hierarchical strategy for computing CTD and the comparing experiments are also shown. Section 5 shows the benefit of the hierarchical strategy by experiments.

2   Algorithm HCTD

For the concepts and notations without definitions, please see [2, 6, 21].

Suppose $\mathbf{T}$ is a regular chain in $\mathbb{Q}[U][X]$ and $\mathbf{H} \subset \mathbb{Q}[U][X]$. $[\mathbf{T}, \mathbf{H}]$ is said to be a *regular system* [5] if $\mathrm{res}(H, \mathbf{T}) \neq 0$ for any $H \in \mathbf{H}$. For any $\mathbf{B} \subset \mathbb{Q}[U]$, $\mathrm{V}^U(\mathbf{B})$ denotes the set $\{(a_1, \ldots, a_d) \in \mathbb{C}^d | B(a_1, \ldots, a_d) = 0, \forall B \in \mathbf{B}\}$. For any $\mathbf{P} \subset \mathbb{C}[X]$, $\mathrm{V}(\mathbf{P})$ denotes the set $\{(b_1, \ldots, b_n) \in \mathbb{C}^n | P(b_1, \ldots, b_n) = 0, \forall P \in \mathbf{P}\}$. For any $\mathbf{P} \subset \mathbb{Q}[U][X]$, $\mathrm{V}(\mathbf{P})$ denotes the set $\{(a_1, \ldots, a_d, b_1, \ldots, b_n) \in \mathbb{C}^{d+n} | P(a_1, \ldots, a_d, b_1, \ldots, b_n) = 0, \forall P \in \mathbf{P}\}$. For $D \subset \mathbb{C}^{d+n}$, denote by $\Pi_U(D)$ the set $\{(a_1, \ldots, a_d) \in \mathbb{C}^d | (a_1, \ldots, a_d, b_1, \ldots, b_n) \in D\}$. Suppose $[\mathbf{T}, \mathbf{H}]$ is a regular system in $\mathbb{Q}[U][X]$. If $\mathbf{H} = \{H\}$, then $[\mathbf{T}, \mathbf{H}]$ is denoted by $[\mathbf{T}, H]$ for short.

Due to page limitation, we only present the specification of an algorithm for computing CTDs hierarchically.

Algorithm HCTD
Input: a finite set $\mathbf{P} \subset \mathbb{Q}[U][X]$, a non-negative integer m $(0 \leq m \leq d)$
output: finitely many 3-tuples $[\mathbf{A}_i, \mathbf{B}_i, \mathbb{T}_i]$, a polynomial B, where

> - $B \in \mathbb{Q}[u_{m+1}, \ldots, u_d]$, $\mathbf{A}_i, \mathbf{B}_i \subset \mathbb{Q}[U]$
> - $\mathbb{T}_i$ is a finite set of regular systems in $\mathbb{Q}[U][X]$

such that

> - $\cup_i \mathrm{V}^U(\mathbf{A}_i \backslash \mathbf{B}_i) = (\mathbb{C}^d \backslash \mathrm{V}^U(B)) \cap \Pi_U(\mathrm{V}(\mathbf{P}))$
> - for any i, j $(i \neq j)$, $\mathrm{V}^U(\mathbf{A}_i \backslash \mathbf{B}_i) \cap \mathrm{V}^U(\mathbf{A}_j \backslash \mathbf{B}_j) = \emptyset$
> - for any i, if $a \in \mathrm{V}^U(\mathbf{A}_i \backslash \mathbf{B}_i)$, then $[\mathbf{T}(a), \mathbf{H}(a)]$ is a regular system in $\mathbb{C}[X]$ for any $[\mathbf{T}, \mathbf{H}] \in \mathbb{T}_i$
> - for any i, if $a \in \mathrm{V}^U(\mathbf{A}_i \backslash \mathbf{B}_i)$, then
> $\mathrm{V}(\mathbf{P}(a)) = \cup_{[\mathbf{T}, \mathbf{H}] \in \mathbb{T}_i} \mathrm{V}(\mathbf{T}(a) \backslash \mathbf{H}(a))$.

The output of $\mathrm{HCTD}(\mathbf{P}, m)$ is called the m-HCTD of $\mathbf{P}$. Each $[\mathbf{A}_i, \mathbf{B}_i, \mathbb{T}_i]$ in the m-HCTD is called a **branch**. Each regular system in the set $\cup \mathbb{T}_i$ is called a **grape**. By Algorithm HCTD, for any $\mathbf{P}$, if $m = 0$, the output is the so-called *generic regular decomposition [8]* of $\mathbf{P}$; if $m = d$, the output is the *comprehensive triangular decomposition [5]* of $\mathbf{P}$. The Example 1 below shows how to get m-HCTD $(m = 0, \ldots, d)$.

Example 1. *Consider the parametric system*

$$\mathbf{P} = \begin{cases} 2x_2^2(x_2^2 + x_1^2) + (u_2^2 - 3u_1^2)x_2^2 - 2u_2 x_2^2(x_2 + x_1) + 2u_1^2 u_2(x_2 + x_1) \\ \quad - u_1^2 x_1^2 + u_1^2(u_1^2 - u_2^2), \\ 4x_2^3 + 4x_2(x_2^2 + x_1^2) - 2u_2 x_2^2 - 4u_2 x_2(x_2 + x_1) + 2(u_2^2 - 3u_1^2)x_2 + 2u_1^2 u_2, \\ 4x_1 x_2^2 - 2u_2 x_2^2 - 2u_1^2 x_1 + 2u_1^2 u_2. \end{cases}$$

where x_1, x_2 are variables and u_1, u_2 are parameters.

1. By the Algorithm RDU in [8], we compute a set $\mathbb{T}_1$ of regular systems and a polynomial $B_1(u_1, u_2)$ such that if $B_1(u_1, u_2) \neq 0$, then the solution set of $\mathbf{P} = 0$ is equal to the union of the solution sets of the regular systems in $\mathbb{T}_1$. Then we obtain the 0-HCTD of $\mathbf{P}$: $[\mathbf{A}_1, \mathbf{B}_1, \mathbb{T}_1]$.

2. Let $\mathbf{P}_1 = \mathbf{P} \cup \{B_1\}$. Regard $\{u_1, x_1, x_2\}$ as the new variable set. By the Algorithm RDU, we compute a set $\mathbb{S}_1$ of regular systems and a polynomial $B_2(u_2)$ such that if $B_1(u_1, u_2) = 0$ and $B_2(u_2) \neq 0$, then the solution set of $\mathbf{P} = 0$ is equal to the union of the solution sets of the regular systems in $\mathbb{S}_1$. For $\mathbb{S}_1$, applying the similar method as the Algorithm RegSer in [20] and the Algorithms Difference and CTD in [6], we obtain the 1-HCTD of $\mathbf{P}$: $[\mathbf{A}_1, \mathbf{B}_1, \mathbb{T}_1], \ldots, [\mathbf{A}_4, \mathbf{B}_4, \mathbb{T}_4]$.

3. Let $\mathbf{P}_2 = \mathbf{P}_1 \cup \{B_2\}$. Regard $\{u_2, u_1, x_1, x_2\}$ as the new variable set. By the Algorithm RDU, we compute a set of regular systems $\mathbb{S}_2$ and a polynomial $B_3 = 1$ such that if $B_1(u_1, u_2) = 0, B_2(u_2) = 0$ and $B_3 \neq 0$, then the solution set of $\mathbf{P} = 0$ is equal to the union of the solution sets of the regular systems in $\mathbb{S}_2$. For $\mathbb{S}_2$, applying the similar method as the Algorithms RegSer, Difference and CTD, we obtain the 2-HCTD of $\mathbf{P}$: $[\mathbf{A}_1, \mathbf{B}_1, \mathbb{T}_1], \ldots, [\mathbf{A}_6, \mathbf{B}_6, \mathbb{T}_6]$.

Table 1. $[\mathbf{A}_i, \mathbf{B}_i, \mathbb{T}_i]$

i	$\mathbf{A}_i$	$\mathbf{B}_i$	$\mathbb{T}_i$
1	$\emptyset$	$\{u_1 u_2(u_1^2 - u_2^2)\}$	$\{[\{-2x_1^2 + 3x_1 u_2 - u_2^2 + u_1^2, 2x_1 x_2 + u_1^2 - u_2 x_2\}, u_1]\}$
2	$\{u_1\}$	$\{u_2\}$	$\{[\{-2x_1 + u_2, u_2 - 2x_2\}, 1]\}$
3	$\{u_1 - u_2\}$	$\{u_2\}$	$\{[\{x_1, x_2 - u_2\}, 1], [\{2x_1 - 3u_2, x_2 + u_2\}, 1]\}$
4	$\{u_1 + u_2\}$	$\{u_2\}$	$\{[\{x_1, x_2 - u_2\}, 1], [\{2x_1 - 3u_2, 2x_2 + u_2\}, 1]\}$
5	$\{u_2\}$	$\{u_1\}$	$\{[\{2x_1^2 - u_1^2, 2x_2 - u_1^2\}, 1]\}$
6	$\{u_1, u_2\}$	$\{1\}$	$\{[\{x_2\}, 1], [\{x_1, x_2\}, 1], [\{2x_1^2 - u_1^2, 2x_2^2 - u_1^2\}, 1]\}$

3 Experiment of Comparison

We have implemented the Algorithm HCTD as a Maple function HCTD and tested a great many benchmarks from the references [5, 7, 12, 14]. Throughout this paper, all the computational results are obtained in Maple 17 using an Intel(R) Core(TM) i5 processor (3.20GHz CPU), 2.5 GB RAM and Windows 7 (32 bit). All the timings are given by seconds. The "timeout" mark means the time is greater than 1000 seconds. The Table 2 compares the functions HCTD (when $m = d$) and ComprehensiveTriangularize (CTD) in RegularChains.

In Table 2, the column "time" lists the timings of HCTD ($m = d$) and CTD; the column "branch" lists the numbers of branches output by HCTD and CTD; and the column "grape" lists the numbers of grapes output by HCTD and CTD. It is indicated by Table 2 that

- for the benchmarks 3–27, HCTD runs faster than CTD, especially, for the benchmark 27, CTD is timeout and HCTD completes the computation in time; for the benchmarks 28–40, CTD runs faster than HCTD, especially, for the benchmarks 38–40, HCTD is timeout and CTD solves the systems efficiently; for the benchmarks 41–49, both HCTD and CTD are timeout;
- for the benchmarks 14, 31, 32, 35 and 36, the number of branches output by HCTD is much bigger than that output by CTD;
- for the benchmarks 6, 10, 12, 29, 30, 32, 35 and 37, the number of grapes output by HCTD is much bigger than that output by CTD.

4   Different Hierarchical Strategy

To compute a m-HCTD for a given parametric system, as shown by Example 1, we first take $\{x_1, \ldots, x_n\}$ as variable set and then add one parameter into the variable set at each recursive step. A different hierarchical strategy may be that we add a prescribed number (say s) of parameters into the variable set at the first step and each recursive step.

The algorithm applying this different hierarchical strategy is called HCTDA and has been implemented as a function HCTDA. The comparing data of HCTD and HCTDA (for $s = 1$) is shown in Table 3. It is indicated by Table 3 that

- for the benchmarks 3–11, HCTD runs faster than HCTDA, especially, for the benchmarks 10–11, HCTDA is timeout and HCTD completes the computation in time; for the benchmarks 12–18, HCTDA runs faster than HCTD, especially, for the benchmarks 19–20, HCTD is timeout and HCTDA completes the computation in time;
- the difference of the numbers of branches (grapes) output by HCTD and HCTDA is not striking.

In fact, we can input different s when calling HCTDA. For many benchmarks in Table 2, the timings of different s are similar. There are some benchmarks on which the timings of HCTDA differ greatly for different s. Due to page limitation, we do not report the timings here.

5   Benefit of Hierarchical Strategy

We see that the benchmarks 41–49 in Table 2 are timeout when using both CTD and HCTD ($m = d$). In fact, for some polynomial systems from practical areas, the complexity of computing comprehensive triangular decomposition is way beyond current computing capabilities. However for these systems (especially the systems with many parameters), we may try to compute the m-HCTD for $m = 0, \ldots, d - 1$. In this way, although we cannot solve the system completely, we may still get partial solutions.

We have tried the timeout benchmarks 41-49 in Table 2. The experimental results are shown in Table 4, where the columns "$m = 0$", "$m = 1$", "$m = 2$", "$m = 3$" and "$m = 4$" denote the timings of calling Algorithm HCTD for $m = 0, 1, 2, 3, 4$; and the "error" mark means Maple returns an error message and stops computing. It is seen from the Table 4 that

- for all the benchmarks, we successfully get partial solutions;
- for most of the benchmarks, such as the benchmark 1 and benchmarks 3–7, we get results only when $m = 0$.

Table 2. Comparing HCTD and CTD

	benchmark	d	n	time		branch		grape	
				HCTD	CTD	HCTD	CTD	HCTD	CTD
1.	MontesS2	1	3	0.	0.	1	1	1	1
2.	MontesS4	2	2	0.	0.	1	1	1	1
3.	F8	4	4	0.437	1.014	18	14	14	9
4.	Hereman-2	1	7	0.093	0.468	2	2	10	6
5.	MontesS3	1	2	0.	0.031	3	2	2	2
6.	MontesS5	4	4	0.078	0.187	6	8	13	6
7.	MontesS6	2	2	0.015	0.047	4	3	5	4
8.	MontesS7	1	3	0.046	0.156	4	4	6	8
9.	MontesS8	2	2	0.	0.094	2	2	2	2
10.	MontesS12	2	6	0.593	7.925	5	5	61	27
11.	MontesS13	3	2	0.078	0.265	6	9	9	8
12.	MontesS14	1	4	0.452	4.353	6	3	28	12
13.	MontesS15	4	8	0.187	0.889	5	5	14	12
14.	MontesS16	3	12	1.198	1.825	37	8	11	7
15.	Bronstein	2	2	0.015	0.219	6	7	7	7
16.	AlkashiSinus	3	6	0.094	0.437	8	6	8	6
17.	Lanconelli	7	4	0.28	0.546	14	11	7	5
18.	zhou1	3	4	0.047	0.156	5	5	5	5
19.	zhou2	6	7	0.671	2.09	17	18	19	16
20.	zhou6	3	3	0.031	0.218	6	4	6	5
21.	SBCD13	1	3	0.015	0.094	2	2	9	6
22.	SBCD23	1	3	0.202	0.344	4	2	15	12
23.	F2	2	2	0.032	0.234	3	3	3	3
24.	F3	4	1	0.063	0.905	5	6	5	6
25.	F5	3	2	0.046	0.11	6	3	3	3
26.	F7	3	2	0.	0.016	2	2	2	2
27.	S2	4	1	44.544	timeout	150		92	
28.	MontesS9	3	3	0.693	0.468	21	13	16	13
29.	MontesS10	3	4	0.421	0.359	13	6	19	6
30.	MontesS11	3	3	0.858	0.655	12	16	20	10
31.	F4	4	2	11.637	0.375	20	3	3	3
32.	zhou5	4	5	5.616	2.902	51	19	97	22
33.	F6	4	1	0.296	0.14	13	3	11	3
34.	MontesS1	2	2	0.016	0.	4	2	3	3
35.	Hereman-8-8	3	5	96.439	10.468	108	9	161	14
36.	S3	4	3	2.618	1.436	35	13	17	11
37.	Maclane	3	7	5.242	4.009	17	9	155	27
38.	S1	3	2	timeout	4.04		10		10
39.	Neural	1	3	timeout	0.188		2		7
40.	Gerdt	3	4	timeout	0.842		4		6
41.	Lazard-ascm2001	3	4	timeout	timeout				
42.	Leykin-1	4	4	timeout	timeout				
43.	Cheaters-homotopy-easy	4	3	timeout	timeout				
44.	Cheaters-homotopy-hard	5	2	timeout	timeout				
45.	Lazard-ascm2001	3	4	timeout	timeout				
46.	MontesS18	2	3	timeout	timeout				
47.	Pavelle	4	4	timeout	timeout				
48.	p3p	5	2	timeout	timeout				
49.	z3	6	11	timeout	timeout				

Table 3. Comparing HCTD and HCTDA *(for s = 1)*

	benchmark	d	n	time HCTD	time HCTDA	branch HCTD	branch HCTDA	grape HCTD	grape HCTDA
1.	MontesS5	4	4	0.078	0.078	6	6	13	13
2.	zhou1	3	4	0.047	0.047	5	5	5	6
3.	MontesS9	3	3	0.693	0.796	21	21	16	27
4.	MontesS11	3	3	0.858	1.207	12	24	20	38
5.	MontesS12	2	6	0.593	0.671	5	5	61	60
6.	AlkashiSinus	3	6	0.094	0.109	8	10	8	10
7.	Bronstein	2	2	0.015	0.031	6	5	7	6
8.	MontesS7	2	2	0.046	0.266	4	6	4	6
9.	SBCD13	1	3	0.015	0.031	2	2	9	7
10.	F6	4	1	0.296	timeout	13		11	
11.	S2	4	1	44.544	timeout	150		92	
12.	Maclane	3	7	5.242	2.605	17	13	155	122
13.	SBCD23	1	3	0.202	0.109	4	2	15	13
14.	F4	4	2	11.637	1.653	20	26	3	3
15.	MontesS15	4	8	0.187	0.124	5	5	14	14
16.	F8	4	4	0.437	0.358	18	16	14	11
17.	MontesS16	3	12	1.198	0.951	37	21	11	8
18.	S3	4	3	2.618	1.81	35	29	17	15
19.	Neural	1	3	timeout	0.296		6		15
20.	Gerdt	3	4	timeout	288.352		4		11

Table 4. Timings of m-HCTD for different m

	benchmark	d	n	$m=0$	$m=1$	$m=2$	$m=3$	$m=4$
1.	Lazard-ascm2001	3	4	0.936	timeout			
2.	Leykin-1	4	4	0.203	20.436	timeout		
3.	Cheaters-homotopy-easy	4	3	3.681	timeout			
4.	Cheaters-homotopy-hard	5	2	39.640	timeout			
5.	Lazard-ascm2001	3	4	0.858	timeout			
6.	MontesS18	2	3	0.327	timeout			
7.	Pavelle	4	4	0.234	timeout			
8.	p3p	5	2	0.	0.	0.015	6.549	timeout
9.	z3	6	11	0.094	error			

References

1. Alvandi, P., Chen, C., Moreno Maza, M.: Computing the limit points of the quasi-component of a regular chain in dimension one. Computer Algebra in Scientific Computing, 30–45 (2013)
2. Aubry, P., Lazard, D., Moreno Maza, M.: On the theories of triangular sets. J. Symb. Comp. 28, 105–124 (1999)
3. Chen, C., Davenport, J., May, J.P., Moreno Maza, M., Xia, B., Xiao, R.: Triangular decomposition of semi-algebraic systems. In: Proc. ISSAC, pp. 187–194 (2010)
4. Chen, C., Davenport, J., Moreno Maza, M., Xia, B., Xiao, R.: Computing with semi-algebraic sets represented by triangular decomposition. In: Proc. ISSAC, pp. 75–82 (2011)

5. Chen, C., Golubitsky, O., Lemaire, F., Maza, M.M., Pan, W.: Comprehensive triangular decomposition. In: Ganzha, V.G., Mayr, E.W., Vorozhtsov, E.V. (eds.) CASC 2007. LNCS, vol. 4770, pp. 73–101. Springer, Heidelberg (2007)
6. Chen, C., Moreno Maza, M.: Algorithms for computing triangular decomposition of polynomial systems. J. Symb. Comp. 47 (6), 610–642 (2012)
7. Chou, S.-C.: Mechanical geometry theorem proving. Springer (1988)
8. Chen, Z., Tang, X., Xia, B.: Generic regular decompositions for parametric polynomial systems. Accepted by Journal of Systems Science and Complexity (2013), arXiv:1301.3991v1
9. Gao, X.-S., Chou, S.-C.: Solving parametric algebraic systems. In: Proc. ISSAC, pp. 335–341 (1992)
10. Gao, X.-S., Hou, X., Tang, J., Chen, H.: Complete solution classification for the perspective-three-point problem. IEEE Transactions on Pattern Analysis and Machine Intelligence 25(8), 930–943 (2003)
11. Kalkbrener, M.: A generalized euclidean algorithm for computing for computing triangular representationa of algebraic varieties. J. Symb. Comp. 15, 143–167 (1993)
12. Kapur, D., Sun, Y., Wang, D.: A new algorithm for computing comprehensive gröbner systems. In: Proc. ISSAC, pp. 25–28 (2010)
13. Moreno Maza, M.: On triangular decompositions of algebraic varieties. Technical Report TR 4/99, NAG Ltd., Oxford, UK (1999)
14. Montes, A., Recio, T.: Automatic discovery of geometry theorems using minimal canonical comprehensive Gröbner systems. In: Botana, F., Recio, T. (eds.) ADG 2006. LNCS (LNAI), vol. 4869, pp. 113–138. Springer, Heidelberg (2007)
15. Nabeshima, K.: A speed-up of the algorithm for computing comprehensive gröbner systems. In: Proc. ISSAC, pp. 299–306 (2007)
16. Suzuki, A., Sato, Y.: An alternative approach to comprehensive gröbner bases. In: Proc. ISSAC, pp. 255–261 (2002)
17. Suzuki, A., Sato, Y.: A simple algorithm to compute comprehensive gröbner bases. In: Proc. ISSAC, pp. 326–331 (2006)
18. Tang, X., Chen, Z., Xia, B.: Generic regular decompositions for generic zero-dimensional systems. Accepted by Science China: Information Sciences (2012), doi: 10.1007/s11432-013-5057-5
19. Wang, D.K.: Zero decomposition algorithms for system of polynomial equations. In: Computer Mathematics, pp. 67–70. World Scientific (2000)
20. Wang, D.M.: Computing triangular systems and regular systems. J. Symb. Comp. 30, 221–236 (2000)
21. Wang, D.M.: Elimination methods. Springer (2001)
22. Wang, D.M.: Elimination practice: software yools and applications. Imperial College Press (2004)
23. Weispfenning, V.: Comprehensive gröbner bases. J. Symb. Comp. 14, 1–29 (1992)
24. Wu, W.-T.: Basic principles of mechanical theorem proving in elementary geometries. Science in China Series A Mathematics, 507–516 (1977) (in Chinese)
25. Yang, L., Hou, X., Xia, B.: A complete algorithm for automated discovering of a class of inequality-type theorems. Science in China Series F Information Sciences 44(1), 33–49 (2001)
26. Yang, L., Xia, B.: Automatic inequality proving and discovering. Science Press (2008) (in Chinese)
27. Yang, L., Zhang, J.: Searching dependency between algebraic equations: An algorithm applied to automated reasoning. In: International Centre for Theoretical Physics, pp. 1–12 (1990)
28. Yang, L., Zhang, J., Hou, X.: Non-linear algebraic formulae and theorem automated proving. Shanghai Education Technology Publishers (1992) (in Chinese)

A Package for Parametric Matrix Computations

Robert M. Corless[1] and Steven E. Thornton[2]

[1] Western University, Canada
rcorless@uwo.ca
http://www.apmaths.uwo.ca/r̄corless/
[2] Western University, Canada
sthornt7@uwo.ca
steventhornton.ca

Abstract. Motivated by the problem of determining the Jordan and Weyr canonical forms of parametric matrices, we present a MAPLE package for doing symbolic linear algebra. The coefficients of our input matrices are multivariate rational functions, whose indeterminates are regarded as parameters and are subject to a system of polynomial equations and inequalities. Our proposed algorithms rely on the theory of regular chains and are implemented on top of the `RegularChains` library.

1 Introduction

This work is initially motivated by a desire to compute the Frobenius (rational) canonical forms of parametric matrices for applications in dynamical system theory. The Frobenius form can easily be extended to the Jordan and Weyr (see [15]) canonical forms. Additionally, the minimal polynomial is easily extracted form the Frobenius form.

Currently, computations on parametric matrices are considered difficult and costly because canonical forms such as the Frobenius, Jordan and Weyr forms are discontinuous; this requires special cases for completeness, and exhaustive analysis produces combinatorially many cases. Some papers considering special cases with parameters include [1] and [4]. There are a large number of methods for computing the Frobenius form of a constant matrix such as in [2], [11], [12], [13], [14] and [16]. We instead modify the algorithm of Storjohann from [17] and [18] for computations on parametric matrices, see Section 4. This algorithm requires the computation of the so-called *zig-zag form* before the Frobenius form can be computed. The zig-zag form itself is not directly useful for applications but provides a matrix from which the Frobenius form can easily be obtained. This will be discussed in detail in a forthcoming paper.

Determining the rank of a matrix is an even simpler problem than the computation of matrix canonical forms. Unfortunately, the computation for parametric matrices is a tedious process which, although of ongoing research interest by many groups, does not yet have a completely satisfactory solution implemented in any computer algebra system (CAS) that we are aware of. For instance, asking for the Jordan canonical form of a 5×5 integer matrix containing a single parameter fails in the computer algebra systems that we have tried.

H. Hong and C. Yap (Eds.): ICMS 2014, LNCS 8592, pp. 442–449, 2014.
© Springer-Verlag Berlin Heidelberg 2014

In Section 2, we present an algorithmic approach to extending methods of rank computation to parametric matrices with polynomial entries. External equality and inequality conditions on the parameters may be inherited in the problem being solved and will be considered in the computations. This proposed method is tailored to the problem of parametric matrix rank computation. That is, we avoid the usage of general tools for solving parametric polynomial systems, such as *comprehensive Gröbner bases* [19] *comprehensive triangular decomposition* [6], or *dynamic evaluation* [8,3]. In fact, we rely on the non-comprehensive triangular decomposition algorithms presented in [7] and [5] for the complex and real cases, respectively.

Our parametric computations of zig-zag forms and matrix ranks are implemented on top of the `RegularChains` library and are illustrated by screen shots of MAPLE sessions.

2   Preliminaries

Let $\mathbb{K}$ be either an algebraically closed field or a real closed field. Let $\alpha_1 < \cdots < \alpha_m$ be $m \geq 1$ ordered variables. We denote by $\mathbb{K}[\alpha] = \mathbb{K}[\alpha_1, \ldots, \alpha_m]$ the ring of polynomials in the variables $\alpha = \alpha_1, \ldots, \alpha_m$ with coefficients in $\mathbb{K}$. We denote by $\mathbb{K}(\alpha)$ the quotient field of $\mathbb{K}[\alpha]$, that is, the field of multivariate rational functions in α with coefficients in $\mathbb{K}$. If $\mathbb{K}$ is algebraically closed, we call a *constructible set* S of $\mathbb{K}[\alpha]$ the solution set of any polynomial system of the form

$$f_1(\alpha) = \cdots = f_a(\alpha) = 0, \ g(\alpha) \neq 0$$

where $f_1(\alpha), \ldots, f_a(\alpha), g(\alpha)$ belong to $\mathbb{K}[\alpha]$. If $f_1(\alpha), \ldots, f_a(\alpha)$ form a *regular chain* of $\mathbb{K}[\alpha]$ (see [7]) and if the polynomial $g(\alpha)$ is regular (i.e. neither zero nor a zero-divisor) modulo the saturated ideal of this regular chain, then the above system is called a *regular system* (see [6]). When this holds, we have $S \neq \varnothing$.

If $\mathbb{K}$ is a real closed field, we call a *semi-algebraic set* S of $\mathbb{K}[\alpha]$ the solution set of any polynomial system of the form

$$f_1(\alpha) = \cdots = f_a(\alpha) = 0, \ g(\alpha) \neq 0, \ p_1(\alpha) > 0, \ldots, p_b(\alpha) > 0, \ q_1(\alpha) \geq 0, \ldots, q_c(\alpha) \geq 0$$

where $f_1(\alpha), \ldots, f_a(\alpha), g(\alpha), p_1(\alpha), \ldots, p_b(\alpha), q_1(\alpha), \ldots, q_c(\alpha)$ are in $\mathbb{K}[\alpha]$. Under some technical assumptions (in particular assuming that $f_1(\alpha), \ldots, f_a(\alpha)$ is a regular chain and that each of the polynomial $p_1(\alpha), \ldots, p_b(\alpha)$ is regular modulo the saturated ideal of this regular chain, see [5]) then the above system is called a *regular semi-algebraic system*. When this holds, we have $S \neq \varnothing$.

Theorem 1 ([6,5]). *Assume $\mathbb{K}$ is algebraically closed (resp. real closed). Then, for every constructible set (resp. semi-algebraic set) $S \subseteq \mathbb{K}^m$ one can compute finitely many regular systems (resp. regular semi-algebraic system) $\Sigma_1, \ldots, \Sigma_\ell$ such that the union of their solution sets equals S; we call $\Sigma_1, \ldots, \Sigma_\ell$ a triangular decomposition of S. Moreover, for two constructible sets (resp. semi-algebraic sets) $S_1, S_2 \subset \mathbb{K}^m$, one can compute a triangular decomposition of their intersection $S_1 \cap S_2$, their union $S_1 \cup S_2$ and the set theoretical difference $S_1 \setminus S_2$.*

3 Parametric Rank

In this section, we discusses the computation of the rank of a matrix $A(\alpha) \in \mathbb{K}^{m \times n}(\alpha)$, where the parameters α are subject to a system S of polynomial constraints. The set S is defined by polynomials of $\mathbb{K}[\alpha]$ containing inequations implying that the denominator of every coefficient of $A(\alpha)$ is non-zero everywhere in this set. We define the polynomial system S' by adding to S the equations of $A(\alpha)X = 0$. The triangular decomposition of S' is denoted as $\mathcal{T}$.

The following procedure computes a decomposition of the zero set $\Sigma \subseteq \mathbb{K}^k$ of S into cells $C_0, C_1, \ldots, C_n$ such that for all $0 \leq r \leq n$ and all $\alpha \in C_i$ the rank of the specialized matrix $A(\alpha)$ is r. We use of the commands of the `RegularChains` library to state our algorithm. Assume first that $\mathbb{K}$ is algebraically closed.

Step 1: Let $\mathcal{T} := \texttt{Triangularize}(S', \mathbb{K}[\alpha_1 < \cdots < \alpha_k < x_1 < \cdots < x_n])$

Step 2: For $0 \leq r \leq n$, let C_r be the constructible set of $\mathbb{K}^k$ given by all regular systems $[T_j \cap \mathbb{K}[\alpha_1 < \cdots < \alpha_k], h_j]$ such that $[T_j, h_j] \in \mathcal{T}$ and the number of polynomials of T_j of positive degree in (at least) one of the variables $x_1 < \cdots < x_n$ is exactly $n - r$.

Step 3: For $r := n$ down to 1 do

$$C_r := \texttt{Difference}(C_r, C_{r-1} \cup \cdots \cup C_0)$$

Now, we state the algorithm for the case where $\mathbb{K}$ is real closed.

Step 1: Let $\mathcal{T} := \texttt{RealTriangularize}(S', \mathbb{K}[\alpha_1 < \cdots < \alpha_k < x_1 < \cdots < x_n])$

Step 2: For $0 \leq r \leq n$, let C_r be the semi-algebraic set of $\mathbb{K}^k$ given by all regular semi-algebraic systems $[T_j \cap \mathbb{K}[\alpha_1 < \cdots < \alpha_k], Q_j, \varnothing]$ such that $[T_j, Q_j, \varnothing] \in \mathcal{T}$ and the number of polynomials of T_j of positive degree in (at least) one of the variables $x_1 < \cdots < x_n$ is exactly $n - r$.

Step 3: For $r := n$ down to 1 do

$$C_r := \texttt{Difference}(C_r, C_{r-1} \cup \cdots \cup C_0)$$

Theorem 2. *Whether $\mathbb{K}$ is algebraically closed or real closed, the above procedure satisfies the claimed specification.*

We provide examples to which we have successfully applied the MAPLE implementation of the above procedure.

Example 1. Taking an example from [10] from control theory, we look for the conditions on the parameters such that the matrix is full rank. When it is full rank we know we have a controllable system.

$$E = \begin{pmatrix} 1 & 3 & 1 \\ 3 & 1 & 1 \\ 0 & 0 & 0 \end{pmatrix} \quad A_1 = \begin{pmatrix} 1 & 1 & 3 \\ 1 & 3 & 1 \\ 0 & 0 & 0 \end{pmatrix}, \quad A_2 = \begin{pmatrix} \lambda & 3\lambda & \lambda \\ 3\lambda + \mu & \lambda + \mu & \lambda + 3\mu \\ 0 & 0 & 0 \end{pmatrix}, \quad B = \begin{pmatrix} 1 \\ 0 \\ 1 \end{pmatrix}$$

```
> z3 := ZeroMatrix(3) : z1 := ZeroMatrix(3, 1) :
> E, B := Matrix([[1, 3, 1], [3, 1, 1, ], [0, 0, 0]]), Matrix([[0], [0], [1]]);
```

$$E, B := \begin{bmatrix} 1 & 3 & 1 \\ 3 & 1 & 1 \\ 0 & 0 & 0 \end{bmatrix} \begin{bmatrix} 0 \\ 0 \\ 1 \end{bmatrix}$$

```
> A1, A2 := Matrix([[1, 1, 3], [1, 3, 1], [0, 0, 0]]), Matrix([[λ, 3·λ, λ], [3
   ·λ + μ, λ + μ, λ + 3·μ], [0, 0, 0]]);
```

$$A1, A2 := \begin{bmatrix} 1 & 1 & 3 \\ 1 & 3 & 1 \\ 0 & 0 & 0 \end{bmatrix} \begin{bmatrix} \lambda & 3\lambda & \lambda \\ 3\lambda + \mu & \lambda + \mu & \lambda + 3\mu \\ 0 & 0 & 0 \end{bmatrix}$$

```
> A := Matrix([[-E, z3, z3, z3, B, z1, z1, z1, z1, z1], [-A1, -E, z3,
    z3, z1, B, z1, z1, z1, z1], [A2, A1, -E, z3, z1, z1, B, z1, z1,
    z1], [z3, A2, -A1, -E, z1, z1, z1, B, z1, z1], [z3, z3, A2, -A1,
    z1, z1, z1, z1, B, z1], [z3, z3, z3, A2, z1, z1, z1, z1, z1, B]])
    :
> rank, R := ParametricMatrixTools:-ComplexRank(A, [ ], [ ]) :
> seq(print( i, Display( rank[i], R)), i = 15 ..18);
```

$$15, \begin{cases} \lambda = 0 \\ \mu + 1 = 0 \end{cases}$$

$$16, \begin{cases} \lambda = 0 \\ \mu - 1 = 0 \end{cases} , \begin{cases} \lambda = 0 \\ \mu - 1 \neq 0 \\ \mu + 1 \neq 0 \end{cases}$$

$$17, \begin{cases} 2\mu - 1 = 0 \\ \lambda \neq 0 \\ 8\lambda + 1 \neq 0 \end{cases} , \begin{cases} 8\lambda + 1 = 0 \\ 16\mu - 8 = 0 \end{cases}$$

$$18, \begin{cases} \mu - 1 = 0 \\ \lambda \neq 0 \end{cases} , \begin{cases} \lambda \neq 0 \\ \mu - 1 \neq 0 \\ \mu + 1 \neq 0 \\ 2\mu - 1 \neq 0 \end{cases} , \begin{cases} \mu + 1 = 0 \\ \lambda \neq 0 \end{cases}$$

Fig. 1. Rank values and corresponding conditions on the parameters for Example 1

$$C = \begin{pmatrix} -E & 0 & 0 & 0 & B & 0 & 0 & 0 & 0 \\ -A_1 & -E & 0 & 0 & 0 & B & 0 & 0 & 0 \\ A_2 & -A_1 & -E & 0 & 0 & 0 & B & 0 & 0 \\ 0 & A_2 & -A_1 & -E & 0 & 0 & 0 & B & 0 \\ 0 & 0 & A_2 & -A_1 & 0 & 0 & 0 & 0 & B & 0 \\ 0 & 0 & 0 & A_2 & 0 & 0 & 0 & 0 & B \end{pmatrix}$$

As stated in [10], the matrix C only has full rank if $\lambda \neq 0$. We verify this by using our method to compute the set of all rank values. The case where C has full rank is displayed below; we also notice conditions on the value of μ for C to be full rank. Figure 1 gives the complete output with all possible rank values.

$$\left\{ \begin{cases} \mu - 1 = 0 \\ \lambda \neq 0 \end{cases} , \begin{cases} \mu + 1 = 0 \\ \lambda \neq 0 \end{cases} , \begin{cases} \lambda \neq 0 \\ \mu - 1 \neq 0 \\ \mu + 1 \neq 0 \\ 2\mu - 1 \neq 0 \end{cases} \right\}.$$

> `A := Matrix([[-1, 1, 1, 1, 1, 0, 1], [0, 0, 1, 0, 1, 0, 0],` $\left[0, \frac{1}{2}, 0, \frac{1}{2}, 0, 1,\right.$
> `0], [0, c·a, 0, a, 0, 0, 0], [0, 0, -c·a, 0, -a, 0, 1], [0, 0, 1, 0, 0, 0, 0],`
> `[1, 1, 0, 0, 0, 0, 0]]);`

$$A := \begin{bmatrix} -1 & 1 & 1 & 1 & 1 & 0 & 1 \\ 0 & 0 & 1 & 0 & 1 & 0 & 0 \\ 0 & \frac{1}{2} & 0 & \frac{1}{2} & 0 & 1 & 0 \\ 0 & ca & 0 & a & 0 & 0 & 0 \\ 0 & 0 & -ca & 0 & -a & 0 & 1 \\ 0 & 0 & 1 & 0 & 0 & 0 & 0 \\ 1 & 1 & 0 & 0 & 0 & 0 & 0 \end{bmatrix}$$

> `rank, R := ParametricMatrixTools:-RealRank(A, [ ],` $\left[a - \frac{1}{5}, \frac{6}{5} - a\right]$`, [c],`
> `[ ]):`
> `seq(print(i, map(Display, rank[i], R)), i = 6 .. 7);`

$6, \left\{\left\{\begin{array}{l} c - 2 = 0 \\ 5a < 6 \text{ and } 5a - 1 > 0 \end{array}\right., \left\{\begin{array}{l} 5a - 1 = 0 \\ c - 2 = 0 \end{array}\right., \left\{\begin{array}{l} 5a - 6 = 0 \\ c - 2 = 0 \end{array}\right.\right\}$

$7, \left\{\left\{\begin{array}{l} 5a - 1 = 0 \\ c > 0 \text{ and } c - 2 \neq 0 \end{array}\right., \left\{\begin{array}{l} 5a - 6 = 0 \\ c > 0 \text{ and } c - 2 \neq 0 \end{array}\right., 5a < 6 \text{ and } 5a - 1 > 0 \text{ and } c > 0\right.$

$\left. \text{and } c - 2 \neq 0\right\}$

Fig. 2. Rank values and corresponding conditions on the parameters for Example 2

Example 2. This is a modified version of the example in [9] where we introduce a new parameter c such that $c > 0$ and maintain the condition that $0.2 \leq a \leq 1.2$.

$$A = \begin{bmatrix} -1 & 1 & 1 & 1 & 1 & 0 & 1 \\ 0 & 0 & 1 & 0 & 1 & 0 & 0 \\ 0 & 1/2 & 0 & 1/2 & 0 & 1 & 0 \\ 0 & ca & 0 & a & 0 & 0 & 0 \\ 0 & 0 & -ca & 0 & -a & 0 & 1 \\ 0 & 0 & 1 & 0 & 0 & 0 & 0 \\ 1 & 1 & 0 & 0 & 0 & 0 & 0 \end{bmatrix}$$

We find that a rank of 6 or 7 is possible. The resulting conditions on a and c to have a rank of 6 are:

$$\begin{cases} c = 2 \\ \frac{1}{5} \leq a \leq \frac{6}{5} \end{cases}$$

and the conditions for rank 7 are

$$\begin{cases} c > 0 \\ c \neq 2 \\ \frac{1}{5} \leq a \leq \frac{6}{5} \end{cases}$$

while the commands executed are displayed in Figure 2.

4 Parametric Zig-zag Form

Parametric Polynomial. Let $f(x; \alpha)$ be a monic polynomial of degree r w.r.t. x. We write:

$$f(x; \alpha) = f_0(\alpha) + f_1(\alpha)x + \cdots + f_{r-1}(\alpha)x^{r-1} + x^r \qquad (1)$$

with coefficients $f_0(\alpha), \ldots, f_{r-1}(\alpha) \in \mathbb{K}(\alpha)$. The α-values are constrained to belong to a constructible (resp. semi-algebraic) set S such that the denominator of every coefficient $f_0(\alpha), \ldots, f_{r-1}(\alpha)$ is nonzero everywhere on S.

Zigzag Matrix. A parametric *Zigzag matrix* takes the form

$$\begin{bmatrix} C_{c_1(x;\alpha)} & B_1 & & & & & & \\ & C^T_{c_2(x;\alpha)} & & & & & & \\ & B_2 & C_{c_3(x;\alpha)} & B_3 & & & & \\ & & & C^T_{c_4(x;\alpha)} & & & & \\ & & & & \ddots & & & \\ & & & & & C^T_{c_{k-2}(x;\alpha)} & & \\ & & & & & B_{k-2} & C_{c_{k-1}(x;\alpha)} & B_{k-1} \\ & & & & & & & C^T_{c_k(x;\alpha)} \end{bmatrix}$$

for k even. Each polynomial $c_1(x; \alpha), \ldots, c_k(x; \alpha)$ takes the same form as Equation (1) and $C_{c_i(x;\alpha)}$ is the companion matrix of $c_i(x; \alpha)$. The blocks B_i have all entries zero except those in the upper left corner which are either 0 or 1; each block B_i has its size determined by its neighboring companion blocks. If there is an odd number of diagonal blocks we allow $\deg c_k = 0$ while $\deg c_i \geq 1$ holds for $1 \leq i < k$. This permits the kth diagonal block to have dimension zero and hence the block B_{k-1} above $C^T_{c_k(x;\alpha)}$ will also have dimension zero.

Theorem 3. *For every matrix $A(\alpha) \in \mathbb{K}^{n \times n}[\alpha]$, there exists a partition $(S_1, \ldots, S_N)$ of input constructible (resp. semi-algebraic) set S such that for each S_i, there exists a matrix $Z_i(\alpha) \sim A(\alpha)$ in Zigzag form where the denominators of the coefficients of the entries of $Z_i(\alpha)$ are all non-zero everywhere on S_i.*

We follow the same algorithm presented in Section 2 of [17]. Stages 1 and 3 must be modified for finding pivots vanishing nowhere on the underlying constructible (resp. semi-algebraic) set S. Once computation has split into two branches (one, called S_{neq}, where a pivot has been found and the set S has been replaced with S_{neq}, and another, called S_{eq}, where the pivot has not yet been found and the search for a pivot continues with S replaced by S_{eq}) the computations proceed in parallel (or by stack execution sequentially).

A sequential implementation has been written in MAPLE to compute the set of Zigzag forms similar to an input parametric matrix under algebraic or semi-algebraic constraints. The RegularChains library in MAPLE contains many useful procedures and sub-packages for performing polynomial computations with

parameters. See www.regularchains.org for details. The ConstructibleSet-
Tools and SemiAlgebraicSetTools sub-packages of RegularChains are useful for representing constructible sets and semi-algebraic sets respectively and,
performing set operations on them, as mentioned in Theorem 1. The General-
Construct procedure from the ConstructibleSetTools sub-package obtains
a triangular decomposition of an input system of polynomial equations and
inequations. Analogously, the RealTriangularize procedure computes a
triangular decomposition of a semi-algebraic set given by an input system of
polynomial equations, inequations and inequalities. The intersection and set theoretical difference computations needed are performed by the Intersection and
Difference commands of the ConstructibleSetTools and SemiAlgebraic-
SetTools sub-packages.

Example 1. Consider the 3×3 matrix with a single parameter α over the complex
numbers $\mathbb{C}$:

$$A(\alpha) = \begin{bmatrix} -1 & -\alpha - 1 & 0 \\ -1/2 & \alpha - 2 & 1/2 \\ -2 & 3\alpha + 1 & -1 \end{bmatrix}$$

with no input conditions on α. The Zigzag forms similar to this matrix are

$$Z_1(\alpha) = \begin{bmatrix} 0 & 0 & 4\alpha \\ 1 & 0 & 4(\alpha - 1) \\ 0 & 1 & \alpha - 4 \end{bmatrix} \quad \alpha + 3 \neq 0, \quad Z_2(\alpha) = \begin{bmatrix} 0 & -4 & 1 \\ 1 & -4 & 0 \\ \hline 0 & 0 & -3 \end{bmatrix} \quad \alpha + 3 = 0.$$

Clearly, $Z_1(\alpha)$ is already in Frobenius form whereas $Z_2(\alpha)$ requires additional
work to obtain the Frobenius form. This example turns out to have a continuous
Frobenius form in the parameter, hence the Frobenius form is $Z_1(\alpha)$ for all values
of α.

5   Future Implementation

Our aim with this software package is to be able to compute canonical forms
of parametric matrices. Specifically, we would like to be able to compute the
Frobenius (rational), Jordan and Weyr (see [15]) canonical forms of square matrices. Current research is being conducted on computing the Frobenius form
by computing a GCD free basis of the polynomials represented by the blocks
of the zig-zag form. This will later be extended into both the Jordan and Weyr
canonical forms.

References

1. Arnold, V.I.: On matrices depending on parameters. Russian Mathematical Surveys 26(2), 29–43 (1971)
2. Augot, D., Camion, P.: On the computation of minimal polynomials, cyclic vectors,
   and Frobenius forms. Linear Algebra and its Applications 260, 61–94 (1997)

3. Broadbery, P.A., Gómez-Díaz, T., Watt, S.M.: On the implementation of dynamic evaluation. In: Proc. of ISSAC, pp. 77–84 (1995)

4. Chen, G.: Computing the normal forms of matrices depending on parameters. In: Proceedings of ISSAC 1989, Portland, Oregon, pp. 244–249. ACM Press-Addison Wesley (1989)

5. Chen, C., Davenport, J.H., May, J.P., Moreno Maza, M., Xia, B., Xiao, R.: Triangular decomposition of semi-algebraic systems. J. Symb. Comput. 49, 3–26 (2013)

6. Chen, C., Golubitsky, O., Lemaire, F., Moreno Maza, M., Pan, W.: Comprehensive triangular decomposition. In: Ganzha, V.G., Mayr, E.W., Vorozhtsov, E.V. (eds.) CASC 2007. LNCS, vol. 4770, pp. 73–101. Springer, Heidelberg (2007)

7. Chen, C., Moreno Maza, M.: Algorithms for computing triangular decomposition of polynomial systems. J. Symb. Comput. 47(6), 610–642 (2012)

8. Della Dora, J., Dicrescenzo, C., Duval, D.: About a new method for computing in algebraic number fields. In: Caviness, B.F. (ed.) ISSAC 1985 and EUROCAL 1985. LNCS, vol. 204, pp. 289–290. Springer, Heidelberg (1985)

9. Dietz, S.G., Scherer, C.W., Huygen, W.: Linear parameter-varying controller synthesis using matrix sum-of-squares +relaxations. In: Brazilian Automation Conference, CBA (2006)

10. García-Planas, M.I., Clotet, J.: Analyzing the set of uncontrollable second order generalized linear systems. International Journal of Applied Mathematics and Informatics 1(2), 76–83 (2007)

11. Giesbrecht, M.: Fast algorithms for matrix normal forms. In: Proceedings of 33rd Annual Symposium on Foundations of Computer Science, pp. 121–130. IEEE (1992)

12. Giesbrecht, M.: Nearly optimal algorithms for canonical matrix forms. SIAM Journal on Computing 24(5), 948–969 (1995)

13. Kaltofen, E., Krishnamoorthy, M.S., Saunders, B.D.: Parallel algorithms for matrix normal forms. Linear Algebra and its Applications 136, 189–208 (1990)

14. Matthews, K.R.: A rational canonical form algorithm. Mathematica Bohemica 117(3), 315–324 (1992)

15. O'Meara, K., Clark, J., Vinsonhaler, C.: Advanced Topics in Linear Algebra: Weaving Matrix Problems Through the Weyr Form. Oxford University Press (2011)

16. Ozello, P.: Calcul exact des formes de Jordan et de Frobenius d'une matrice. PhD thesis, Université Joseph-Fourier-Grenoble I (1987)

17. Storjohann, A.: An $\mathcal{O}(n^3)$ algorithm for the Frobenius normal form. In: Proc. of ISSAC, pp. 101–105. ACM (1998)

18. Storjohann, A.: Algorithms for matrix canonical forms. PhD Thesis, Institut für Wissenschaftliches Rechnen, ETH-Zentrum, Zürich, Switzerland (2000)

19. Weispfenning, V.: Comprehensive Gröbner bases. J. Symb. Comput. 14(1), 1–30 (1992)

Choosing a Variable Ordering for Truth-Table Invariant Cylindrical Algebraic Decomposition by Incremental Triangular Decomposition

Matthew England, Russell Bradford, James H. Davenport, and David Wilson

University of Bath, UK
{M.England,R.J.Bradford,J.H.Davenport,D.J.Wilson}@bath.ac.uk

Abstract. Cylindrical algebraic decomposition (CAD) is a key tool for solving problems in real algebraic geometry and beyond. In recent years a new approach has been developed, where regular chains technology is used to first build a decomposition in complex space. We consider the latest variant of this which builds the complex decomposition incrementally by polynomial and produces CADs on whose cells a sequence of formulae are truth-invariant. Like all CAD algorithms the user must provide a variable ordering which can have a profound impact on the tractability of a problem. We evaluate existing heuristics to help with the choice for this algorithm, suggest improvements and then derive a new heuristic more closely aligned with the mechanics of the new algorithm.

1 Introduction

A *cylindrical algebraic decomposition* (CAD) is: a *decomposition* of $\mathbb{R}^n$, meaning a collection of cells which do not intersect and whose union is $\mathbb{R}^n$; *cylindrical*, meaning the projections of any pair of cells with respect to a given variable ordering are either equal or disjoint; and, *(semi)-algebraic*, meaning each cell can be described using a finite sequence of polynomial relations. The original CAD by Collins [1] was introduced as a tool for quantifier elimination over the reals. Since then CAD has also been applied to problems including epidemic modelling [9], parametric optimisation [18], theorem proving [22], motion planning [23] and reasoning with multi-valued functions and their branch cuts [14].

Traditionally, a CAD is built *sign-invariant* with respect to a set of polynomials such that each one has constant sign in each cell, meaning only one sample point per cell need be tested to determine behaviour. Collins' algorithm works in two phases. In the *projection* phase an operator is repeatedly applied to polynomials each time producing a set in one fewer variables. Then in the *lifting* phase CADs of real space are built incrementally by dimension according to the real roots of these polynomials. A full description is in [1] and [13] summarises improvements from the first 20 years ([4] references more recent developments).

In 2009 an approach to CAD was introduced which broke with the projection and lifting framework [12]. Instead, a *complex cylindrical decomposition* (CCD) of $\mathbb{C}^n$ is built using triangular decomposition by regular chains, and then real

H. Hong and C. Yap (Eds.): ICMS 2014, LNCS 8592, pp. 450–457, 2014.
© Springer-Verlag Berlin Heidelberg 2014

root isolation is applied to move to a CAD of $\mathbb{R}^n$. We can view the CCD as an enhanced projection since gcds are calculated as well as resultants. It means the second phase is less expensive than lifting since case distinction can avoid identifying unnecessary roots. We use PL-CAD for CADs built by projection and lifting and RC-CAD for CADs built with the new approach. The initial work was improved in [11] by introducing purpose-built algorithms to refine a CCD incrementally by constraint whilst maintaining cylindricity and recycling subresultant calculations. A modification of the incremental algorithm to work with relations instead of polynomials then allowed for simplification in the presence of *equational constraints* (ECs): equations whose satisfaction is logically implied by the input. The output was no longer sign-invariant for polynomials but *truth-invariant* for a formula (the conjunction of relations). Similar ideas had been developed for PL-CAD [21] but were difficult to generalise to multiple ECs.

In [2], a new variant of RC-CAD was presented. Here, instead of building a CAD for a set of polynomials or relations we build one for a sequence of quantifier free formulae (QFFs) such that each formula has constant truth value on each cell: a *truth-table invariant* CAD or TTICAD. It followed the development of TTICAD theory for PL-CAD (see [4], [5]) and combined it with the benefits of RC-CAD. The CCD is built using a tree structure incrementally refined by constraint. ECs are dealt with first, with branches refined for other constraints in a formula only if the EC is satisfied. Further, when there are multiple ECs in a formula branches can be removed when the constraints are not all satisfied. See [2] and [11] for full details. Building a TTICAD is often the best way to obtain a truth-invariant CAD for a single formula (if the formula has disjunctions then treating each conjunctive clause as a subformula allows simplification in the presence of any ECs) but is also the object required for applications like simplification of complex functions via branch cut analysis (see [3] [17]). The implementation of [2] in the RegularChains Library [24] (denoted RC-TTICAD) is our topic here.

All CAD algorithms require the user to specify an ordering on the variables. For PL-CAD this determines the order of projection and thus the sequence of Euclidean spaces considered en-route to $\mathbb{R}^n$. For RC-CAD if determines both the triangular decompositions performed and the refinement to $\mathbb{R}^n$. Depending on the application there may be a free or constrained choice. For example, in quantifier elimination we must order the variables as they are quantified but may change the ordering within quantifier blocks. Problems easy in one variable ordering can be infeasible in another, with [8] giving problems where one ordering leads to a cell count constant in the number of variables and another to one doubly exponential (irrespective of the algorithm used). Hence any choice must be made intelligently. We write $y \succ x$ if y is greater than x in an ordering (noting that PL-CAD eliminates variables from greatest to lowest in the ordering).

We start in Section 2 by evaluating (with respect to RC-TTICAD) existing heuristics for choosing the variable ordering. Then in Section 3 we suggest some extensions to improve their use before developing our own heuristic more closely aligned to RC-TTICAD. We give our conclusions in Section 4.

2 Evaluating Existing Heuristics

In what follows we assume f is a polynomial, v a variable and P the set of polynomials defining the input to RC-TTICAD. Let $\deg(f, v)$ be the degree of f in v, $\mathrm{tdeg}(f)$ the total degree of f and $\mathrm{lcoeff}(f, v)$ the leading coefficient of f when considered as a univariate polynomial in v. For a set let max be the maximum value, sum the sum of values and # the number of values. We start by considering two heuristics already in use for choosing the variable ordering in algorithms from the RegularChains Library [24].

Triangular: Start with the first criteria, breaking ties with successive ones.
1. Let $v^{[1]} = \max(\{\deg(f, v), \mid f \in P\})$. Then set $y \succ x$ if $y^{[1]} < x^{[1]}$.
2. Let $v^{[2]} = \max(\{\mathrm{tdeg}(\mathrm{lcoeff}(f, v)), \mid f \in P \text{ (containing } v)\})$.
   Then set $y \succ x$ if $y^{[2]} < x^{[2]}$.
3. Let $v^{[3]} = \mathrm{sum}(\{\deg(f, v), \mid f \in P\})$. Then set $y \succ x$ if $y^{[3]} < x^{[3]}$.

Brown: Start with the first criteria, breaking ties with successive ones.
1. Set $y \succ x$ if $y^{[1]} < x^{[1]}$ (as defined in the heuristic above).
2. Let $v^{[4]} = \max(\{\mathrm{tdeg}(t), \mid t \text{ is a monomial (containing } v) \text{ from a polynomial in } P\})$. Then set $y \succ x$ if $y^{[4]} < x^{[4]}$.
3. Let $v^{[5]} = \#(\{t, \mid t \text{ is a monomial (containing } v) \text{ from a polynomial in } P\})$. Then set $y \succ x$ if $y^{[5]} < x^{[5]}$.

These use only simple measures on the input. The first was implemented for [10] (although not detailed there) and is used for various algorithms in the REGU-LARCHAINS Library (being the default for SuggestVariableOrder). The second was first described in the CAD tutorial notes [7] and in [19] was shown to do well in choosing a variable ordering for QEPCAD (an implementation of PL-CAD).

The next two heuristics were developed for PL-CAD and work by running the projection phase for each possible variable ordering and picking an optimal ordering using a measure of the projection set. Our implementations use the projection polynomials generated by McCallum's operator [20] on P.

Sotd: Select the variable ordering with the lowest *sum of total degrees* for each of the monomials in each of the polynomials in the projection set.
Ndrr: Select the variable ordering with the lowest *number of distinct real roots* of the univariate projection polynomials.

Sotd was suggested in [15] where it was found to be a good heuristic for CAD in REDLOG (another implementation of PL-CAD). Ndrr was suggested in [6] as a means to identify differences occurring only in real space and thus missed by measures on degree. These heuristics are clearly more expensive but note that the lifting phase does the bulk of the work for PL-CAD, with the projection phase often trivial (and if not then the lifting phase is likely infeasible).

To evaluate the heuristics we generated 600 random examples, each with two QFFs themselves a conjunction of two constraints. There were 100 for each of six system types: **00, 10, 20, 11, 12, 22**. Each digit in these labels refers to the number of those constraints which are equalities (with the others strict inequalities). The polynomials defining the constraints were sparse and in three variables, generated using MAPLE's randpoly function. RC-TTICAD was applied to build CADs for the problems using each of the six possible variable orderings. A time out of 12 minutes a problem was used affecting only six examples (one with system type **20**, two with **10** and three with **00**). For the others, the cell count and computation time (in seconds) for each CAD was recorded.

Table 1 summarises this data, showing the average and median values for each system. As expected RC-TTICAD does better in the presence of ECs. We note the anomaly between system types **10** and **20**: it seems the savings from truncating branches where ECs are not simultaneously satisfied are wiped out by the costs of doing so. The savings would probably be restored if the QFFs contained further non-ECs requiring more processing per branch.

Next we note that the median cell counts and timings are considerably less than the mean average for every system type, indicating the presence of outliers. We provided a third piece of data: the median of the values for each problem when averaged over the six possible orderings. This will still avoid outlier problems but not outlier orderings. In every case this value is much closer to the mean average, indicating that most outlying data comes from bad orderings rather than bad problems, and thus highlighting the practical importance of the ordering.

We performed the following calculations for each problem and each heuristic:

1. Calculate the average cell count and timing for the problem from the six possible variable orderings.
2. Run and time each heuristic for choosing a variable ordering for the problem.
3. Record the cell count and timing of the heuristic's choice. If a heuristic chooses multiple orderings we take the first lexicographically.
4. Calculate the saving from using the heuristic's choice compared to the problem average, i.e. $(1) - (3)$ for cell counts and $(1) - (2) - (3)$ for timings.
5. Evaluate the savings as percentages of the problem average, i.e. $100(4)/(1)$.

Table 2 (the first four rows) shows averages of the values in (5) over problems of the same system type and the whole problem set. All four existing heuristics offer significant cell savings and so are making good selections of variable ordering. Although Sotd offers the highest cell savings overall, its higher costs means the Triangular heuristic is the most time efficient. The heuristics' costs decrease as

Table 1. The performance of RC-TTICAD over all variable orderings. Displayed are the mean and median values and the median of the values after averaging over orderings.

System	Cell Count			Computation Time		
	Mean	Median	Median of av.	Mean	Median	Median of av.
22	750.13	478	612.67	1.84	1.37	1.58
12	934.42	682	861.50	2.73	2.12	2.47
11	1355.45	839	1212.33	3.41	2.10	2.99
20	3271.51	2193	2918	8.90	6.02	7.92
10	2949.02	1528	2275	8.44	4.71	6.62
00	9838.76	4874	8566.67	34.46	17.05	29.88

a percentage of the CAD computation time for systems with fewer ECs and so Sotd can achieve a much higher saving for problems of type **00** than **22**. But there are other differences between systems not explained by running times, such as Brown generally saving more cells than Triangular but not for systems **20**.

3   Extensions, Improvements and a New Heuristic

Combining Measures. In [6] Ndrr was developed to help with problems where Sotd could not due to differences occurring in real space only. Hence a logical extension is to use their measures in tandem. We have used the same evaluation for heuristics SN (where Ndrr is used as a tie-break for Sotd) and NS (where Sotd is the tie-break) with results given in Table 2. In both cases the tie-breaker gives marginally higher cell savings than using the single heuristic, with NS giving the highest cell saving so far, but Brown remaining the most efficient for computation time. The costs of running these heuristics will be higher than using the single measure (at least for problems where the first measure tied) but these extra costs are usually less than the extra time savings obtained.

Greedy Heuristics. A *greedy* variant of Sotd was also suggested in [15] with the variable ordering decided alongside the projection phase. At each step the projection operator is evaluated with respect to all unallocated variables and the variable whose set has lowest sum of total degree of each monomial of each new polynomial is fixed in the ordering. We denote this GS in Table 2 where we see it offers less cell savings than full Sotd (though still competitive) but has lower costs and so gives more time savings. The cost of Sotd will increase alongside the number of admissible variable orderings and so for such problems the greedy variant may offer the only sensible approach. A greedy variant of Ndrr is not possible since that measure is on the univariate polynomials only.

Using Information from PL-TTICAD. The projection sets used so far are those for a sign-invariant PL-CAD, thus considering not the input constraints but the polynomials defining them. Since RC-TTICAD is building a TTICAD (smaller for all except systems **00**), a sensible extension is to use the projection phase from PL-TTICAD [5]. However, we cannot match the declared output structure exactly: PL-TTICAD uses (at most) one declared EC per QFF (with others treated the same as non-ECs). Hence, for QFFs with 2 ECs we will run the projection phase with the first of these declared (so for example, systems **20** are treated the same as **10**). We denote the heuristics applying Sotd and Ndrr with this projection set as S-TTI and N-TTI. From Table 2 we see they offer substantially more cell savings than their standard versions. They also achieve higher time savings: both due to the improved choices and lower running costs (since the TTICAD projection operator is a subset of the sign-invariant one). We can also run the greedy variant of Sotd with the PL-TTICAD projection phase (denoted GS-TTI in Table 2) which will lose some of the cell savings but increase the time savings.

Developing a New Heuristic. We now aim to develop a new heuristic, which considers more algebraic information than the input but is tailored to RC-TTICAD itself rather than a `PL-CAD` algorithm. The main saving offered by the regular chains approach is case distinction meaning that not all projection factors are considered universally. For example, the second coefficient in a polynomial is only considered when the first vanishes (and then only evaluated modulo that constraint). Consider a set of polynomials consisting of the following:

- the discriminants, leading coefficients and cross-resultants of the polynomials forming the first constraint in each QFF;
- if a QFF has no EC then also the (other) discriminants, leading coefficients and cross resultants of all polynomials defining constraints there;
- if a QFF has more than one EC then also the resultant of the polynomial defining the first with that of the second.

Here the resultants, discriminants and coefficients are taken with respect to the first variable in the ordering. These polynomials will all be sign-invariant in the output: see [2], [11] for the algorithm specifications and [16] for a fuller discussion and examples (from a study in the context of choosing the constraint ordering). This set does not contain all polynomials computed by `RC-TTICAD`, but those which are considered in their own right rather than modulo others.

We define a new heuristic to pick an orderings in two stages: First variables are ordered according to maximum degree of the polynomials forming the input (as with Triangular and Brown). Then ties are broke by calculating the set of polynomials above for each unallocated variable and ordering according to sum of degree (in that variable). This is denoted NewH in Table 2 and we see it achieves almost as many savings as S-TTI despite using fewer polynomials.

We could go further by including some more of the missing information. For example, we can use the degree of the omitted discriminants, resultants and

Table 2. Comparing the savings (as a percentage of the problem average) in cells (C) and net timings (NT) from various heuristics

Heuristic	22		12		11		20		10		00		All	
	C	NT	C	NT	C	NT	C	NT	C	NT	C	NT	C	NT
Triangular	32.6	33.9	33.9	34.0	40.9	41.3	47.9	46.8	47.7	47.2	56.0	58.8	43.0	43.6
Brown	37.6	39.1	39.3	39.8	45.9	47.1	45.0	44.3	51.6	50.9	61.9	64.5	46.8	47.5
Sotd	36.7	23.9	37.9	27.7	49.4	40.4	42.8	39.5	56.3	53.9	59.9	61.8	47.1	41.0
Ndrr	40.1	21.2	44.1	33.0	40.2	30.7	35.7	34.4	54.8	51.3	54.0	54.3	44.9	37.4
SN	37.0	24.3	37.2	27.4	49.2	40.4	42.5	39.6	56.0	53.5	60.4	62.5	47.0	41.1
NS	41.3	22.6	41.2	30.7	47.8	37.1	38.7	36.0	57.1	51.7	58.4	60.2	47.3	39.6
GS	35.0	32.7	33.7	32.5	49.5	46.5	39.8	38.9	52.3	52.1	52.5	55.9	43.8	43.3
S-TTI	42.7	40.4	46.4	43.2	55.0	49.1	48.4	48.1	61.2	60.2	59.9	61.7	52.2	50.3
N-TTI	48.5	37.1	46.8	40.5	48.6	42.3	47.8	46.9	59.0	55.3	54.0	54.3	50.7	46.0
GS-TTI	46.4	47.2	44.9	44.5	56.7	54.7	49.3	50.2	56.7	57.5	52.8	55.9	51.1	51.6
NewH	45.9	45.5	41.8	43.5	51.4	50.8	48.2	47.6	56.4	52.4	67.0	68.5	51.7	51.3
NewH-ext	46.2	45.9	42.2	43.3	51.6	51.4	49.3	49.5	55.9	52.0	67.0	68.5	52.0	51.7

leading coefficients as a third tie-break. This heuristic is denoted NewH-ext and the results of its evaluation are in the final row of Table 2. We see it achieves even higher cell savings (and the greatest time savings of any heuristic).

4   Conclusions

We have demonstrated that the variable ordering is important for RC-TTICAD and using any heuristic is advantageous. Simple measures on the input can be effective, but more cell savings can be obtained by using additional information. Existing heuristics obtained this from the projection phase of PL-CAD and we have suggested a new heuristic aligned to RC-TTICAD which identifies polynomials of most importance to the algorithm. It was sufficient for allocating two variables (and hence ordering three) as required by our problem set. Extending to problems with more variables is a topic of future work.

The heuristics performance varied with the system classes and so heuristics that changed along with this performed better. The precise relationships at work here are not always clear to see. Machine learning on the set of measures used by the heuristics may offer a meta-heuristic greater than the sum of its parts (as was found to be the case recently when choosing a variable ordering for QEPCAD [19]). Finally, we note that when using RC-TTICAD there are questions of problem formulation other that the variable ordering to use. As implied in Section 3, the order the constraints are presented affects the output. Advice on making this choice intelligently was recently derived in [16].

Acknowledgements. This work was supported by EPSRC grant EP/J003247/1. RC-TTICAD was developed by Changbo Chen, Marc Moreno Maza and the present authors.

References

1. Arnon, D., Collins, G.E., McCallum, S.: Cylindrical algebraic decomposition I: The basic algorithm. SIAM J. Comput. 13, 865–877 (1984)
2. Bradford, R., Chen, C., Davenport, J.H., England, M., Moreno Maza, M., Wilson, D.: Truth table invariant cylindrical algebraic decomposition by regular chains. To appear: Proc. CASC 2014 (2014), Preprint: http://opus.bath.ac.uk/38344/
3. Bradford, R., Davenport, J.H.: Towards better simplification of elementary functions. In: Proc. ISSAC 2002, pp. 16–22. ACM (2002)
4. Bradford, R., Davenport, J.H., England, M., McCallum, S., Wilson, D.: Cylindrical algebraic decompositions for boolean combinations. In: Proc. ISSAC 2013, pp. 125–132. ACM (2013)
5. Bradford, R., Davenport, J.H., England, M., McCallum, S., Wilson, D.: Truth table invariant cylindrical algebraic decomposition (submitted for publication, 2014), Preprint: http://opus.bath.ac.uk/38146/
6. Bradford, R., Davenport, J.H., England, M., Wilson, D.: Optimising problem formulation for cylindrical algebraic decomposition. In: Carette, J., Aspinall, D., Lange, C., Sojka, P., Windsteiger, W. (eds.) CICM 2013. LNCS, vol. 7961, pp. 19–34. Springer, Heidelberg (2013)

7. Brown, C.W.: Companion to the tutorial: Cylindrical algebraic decomposition. Presented at ISSAC 2004 (2004), http://www.usna.edu/Users/cs/wcbrown/research/ISSAC04/handout.pdf
8. Brown, C.W., Davenport, J.H.: The complexity of quantifier elimination and cylindrical algebraic decomposition. In: Proc. ISSAC 2007, pp. 54–60. ACM (2007)
9. Brown, C.W., El Kahoui, M., Novotni, D., Weber, A.: Algorithmic methods for investigating equilibria in epidemic modelling. J. Symbolic Computation 41, 1157–1173 (2006)
10. Chen, C., Davenport, J.H., Lemaire, F., Moreno Maza, M., Xia, B., Xiao, R., Xie, Y.: Computing the real solutions of polynomial systems with the REGULARCHAINS library in MAPLE. ACM C.C.A. 45(3/4), 166–168 (2011)
11. Chen, C., Moreno Maza, M.: An incremental algorithm for computing cylindrical algebraic decompositions. In: Proc. ASCM 2012. Preprint: http://arxiv.org/abs/1210.5543
12. Chen, C., Moreno Maza, M., Xia, B., Yang, L.: Computing cylindrical algebraic decomposition via triangular decomposition. In: Proc. ISSAC 2009, pp. 95–102. ACM (2009)
13. Collins, G.E.: Quantifier elimination by cylindrical algebraic decomposition – 20 years of progress. In: Quantifier Elimination and Cylindrical Algebraic Decomposition. Texts & Monographs in Symb. Com., pp. 8–23. Springer (1998)
14. Davenport, J.H., Bradford, R., England, M., Wilson, D.: Program verification in the presence of complex numbers, functions with branch cuts etc. In: Proc. 14th SYNASC 2012, pp. 83–88. IEEE (2012)
15. Dolzmann, A., Seidl, A., Sturm, T.: Efficient projection orders for CAD. In: Proc. ISSAC 2004, pp. 111–118. ACM (2004)
16. England, M., Bradford, R., Chen, C., Davenport, J.H., Moreno Maza, M., Wilson, D.: Problem formulation for truth-table invariant cylindrical algebraic decomposition by incremental triangular decomposition. In: Watt, S.M. (ed.) CICM 2014. LNCS (LNAI), vol. 8543, pp. 45–60. Springer, Heidelberg (2014), Preprint: http://opus.bath.ac.uk/39231/
17. England, M., Bradford, R., Davenport, J.H., Wilson, D.: Understanding branch cuts of expressions. In: Carette, J., Aspinall, D., Lange, C., Sojka, P., Windsteiger, W. (eds.) CICM 2013. LNCS, vol. 7961, pp. 136–151. Springer, Heidelberg (2013)
18. Fotiou, I.A., Parrilo, P.A., Morari, M.: Nonlinear parametric optimization using cylindrical algebraic decomposition. In: Proc. CDC-ECC 2005, pp. 3735–3740 (2005)
19. Huang, Z., England, M., Wilson, D., Davenport, J.H., Paulson, L.C., Bridge, J.: Applying machine learning to the problem of choosing a heuristic to select the variable ordering for cylindrical algebraic decomposition. In: Watt, S.M. (ed.) CICM 2014. LNCS (LNAI), vol. 8543, pp. 92–107. Springer, Heidelberg (2014), Preprint: http://opus.bath.ac.uk/39232/
20. McCallum, S.: An improved projection operation for cylindrical algebraic decomposition. In: Quantifier Elimination and Cylindrical Algebraic Decomposition. Texts & Monographs in Symb. Comp., pp. 242–268. Springer (1998)
21. McCallum, S.: On projection in CAD-based quantifier elimination with equational constraint. In: Proc. ISSAC 1999, pp. 145–149. ACM (1999)
22. Paulson, L.C.: Metitarski: Past and future. In: Beringer, L., Felty, A. (eds.) ITP 2012. LNCS, vol. 7406, pp. 1–10. Springer, Heidelberg (2012)
23. Wilson, D., Davenport, J.H., England, M., Bradford, R.: A "piano movers" problem reformulated. In: Proc. SYNASC 2013, pp. 53–60. IEEE (2014)
24. The REGULARCHAINS Library, http://www.regularchains.org

Using the Regular Chains Library to Build Cylindrical Algebraic Decompositions by Projecting and Lifting

Matthew England, David Wilson, Russell Bradford, and James H. Davenport

University of Bath, UK
{M.England,D.J.Wilson,R.J.Bradford,J.H.Davenport}@bath.ac.uk

Abstract. Cylindrical algebraic decomposition (CAD) is an important tool, both for quantifier elimination over the reals and a range of other applications. Traditionally, a CAD is built through a process of projection and lifting to move the problem within Euclidean spaces of changing dimension. Recently, an alternative approach which first decomposes complex space using triangular decomposition before refining to real space has been introduced and implemented within the REGULARCHAINS Library of MAPLE. We here describe a freely available package ProjectionCAD which utilises the routines within the REGULARCHAINS Library to build CADs by projection and lifting. We detail how the projection and lifting algorithms were modified to allow this, discuss the motivation and survey the functionality of the package.

1 Introduction

A *cylindrical algebraic decomposition* (CAD) is: a *decomposition* of $\mathbb{R}^n$, meaning a collection of cells which do not intersect and whose union is $\mathbb{R}^n$; *cylindrical*, meaning the projections of any pair of cells with respect to a given variable ordering are either equal or disjoint; and, *(semi)-algebraic*, meaning each cell can be described using a finite sequence of polynomial relations. CAD is best known for quantifier elimination over the reals, but has also found diverse applications such as motion planning [25] and reasoning with multi-valued functions [13].

The REGULARCHAINS Library [26] in MAPLE contains procedures to build CAD by first building a *complex cylindrical decomposition* (CCD) of $\mathbb{C}^n$ using triangular decomposition by regular chains, then refining to a CAD of $\mathbb{R}^n$. The core algorithm was developed in [11] with improvements detailed in [10] and [3].

These CAD algorithms are in contrast to the traditional approach of projection and lifting followed since Collins' original work [12]. Here, a *projection* phase repeatedly applies an operator to a set of polynomials (starting with those forming the input) each time producing another set in one fewer variables. Then the *lifting* phase builds CADs of $\mathbb{R}^i, i = 1 \ldots n$. $\mathbb{R}$ is decomposed into points and intervals corresponding to the real roots of the univariate polynomials. $\mathbb{R}^2$ is decomposed by repeating the process over each cell in $\mathbb{R}^1$ using the bivariate polynomials at a sample point. The output over each cell consists of *sections*

H. Hong and C. Yap (Eds.): ICMS 2014, LNCS 8592, pp. 458–465, 2014.
© Springer-Verlag Berlin Heidelberg 2014

(where a polynomial vanishes) and *sectors* (the regions between) which together form a *stack*. The union of these stacks gives the CAD of $\mathbb{R}^2$ and the process is repeated until a CAD of $\mathbb{R}^n$ is produced. Collins defined the projection operator so the CAD of $\mathbb{R}^n$ produced using sample points this way could be concluded *sign-invariant* for the input polynomials: each polynomial has constant sign on each cell. The key tool in the proof was showing polynomials to be *delineable* in a cell, meaning the zero set of individual polynomials are disjoint sections and the zero sets of different polynomials are identical or disjoint. For developments to Collins' algorithm see for example the introduction of [4].

We use PL-CAD for CADs built by projection and lifting and RC-CAD for CADs built via CCDs. We will discuss a freely available MAPLE package PROJECTION-CAD which builds PL-CADs by utilising routines developed for RC-CAD. We continue in Section 2 by describing the motivation for coupling these approaches before explaining the workings of the package in Section 3 and describing the current functionality in Section 4. Earlier versions of the package can be downloaded alongside [14] [15], with the latest version available from the authors. There are plans for its integration into the REGULARCHAINS Library [26] itself.

2   Motivation

PROJECTIONCAD uses routines in the REGULARCHAINS Library to build cells in the lifting phase. The advantages of utilising the routines are multiple:

- It avoids many costly algebraic number calculations by using efficient algorithms for triangular decomposition. When algebraic numbers are required (as sample points for lower dimension cells) they are represented as the unique root of a regular chain in a bounding box.
- It ensures ProjectionCAD will always use the best available sub-algorithms in MAPLE, such as the recently improved routines for real root isolation.
- It allows ProjectionCAD to match output formats with the RC-CAD algorithms. In particular, it allows for the use of the sophisticated *piecewise* interface [9] which highlights the tree-like structure of a CAD.

The PROJECTIONCAD package was developed to implement new theory for PL-CAD, most notably the work in [4], [5], [6] and [24]. More details of the functionality are given in Section 4. However, it has also allowed for easy comparison of PL-CAD and RC-CAD, leading to new developments for RC-CAD [3] [16]. A future aim is identification of problem classes suitable for one approach or the other.

3   CAD Construction in ProjectionCAD

The pseudo code in Algorithm 1 describes the framework which the main algorithms in PROJECTIONCAD follow. They apply to either polynomials or formulae. If the former then the CAD produced is sign-invariant for each polynomial. If the latter then each formula will have constant Boolean truth value on each cell and the CAD is said to be *truth table invariant* for the formulae: a TTI-CAD. The user may also have to supply additional information (such as which projection operator to use or which equational constraint to designate [21]). All algorithms require a specified variable ordering, which can have a significant affect on the tractability of using CAD [8]. We use ordered variables $x_1 \prec \ldots \prec x_n$ and say the *main variable* is the highest ordered variable present.

Algorithm 1. PL-CAD

Input : A variable ordering $x_1 \prec \ldots \prec x_n$ and F a sequence of polynomials (or quantifier-free formulae).

Output: A CAD of $\mathbb{R}^n$ sign-invariant for the polynomials (or truth invariant for the formulae) F; or FAIL if F is not well-oriented.

1 Run the projection phase using an appropriate projection operator on F.
2 **for** $i = 1, \ldots, n$ **do**
3 $\quad$ Assign to P_i the set of projection polynomials with main variable x_i.
4 Set C_1 to be a CAD of $\mathbb{R}^1$ formed by the decomposition of the real line according to the real roots of the polynomials in P_1.
5 **for** $i = 2, \ldots, n$ **do**
6 $\quad$ **for** each cell $c \in C_{i-1}$ **do**
7 $\quad\quad$ Check any necessary well-orientedness conditions.
8 $\quad\quad$ **if** *the input is not well oriented* **then**
9 $\quad\quad\quad$ **if** $dim(c) = 0$ **then**
10 $\quad\quad\quad\quad$ Assign to L a set containing the polynomials in P_i and any (non-constant) minimal delineating polynomials.
11 $\quad\quad\quad$ **else**
12 $\quad\quad\quad\quad$ return FAIL.
13 $\quad\quad$ **else**
14 $\quad\quad\quad$ Set $L := P_i$.
15 $\quad\quad$ Set $S_c := \texttt{GenerateStack}(c, L)$. // Apply Algorithm 2.
16 $\quad$ Set $C_i := \bigcup_c S_c$.
17 **return** C_n.

All algorithms in PROJECTIONCAD start with a projection phase (step 1) which uses a projection operator appropriate to the input to derive a set of projection polynomials. In steps 2–3 we sort these into sets P_i according to their main variables. The remainder of the algorithm defines the lifting phase. We start by decomposing $\mathbb{R}^1$ into cells according to the real roots of P_1 (step 4) and then repeatedly lift by generating stacks over cells until we have a CAD of $\mathbb{R}^n$. All cells are equipped with a *sample point* and a *cell index*. The index is an n-tuple of positive integers that corresponds to the location of the cell relative to the CAD. Cells are numbered in each stack during the lifting stage (from most negative to most positive), with sectors having odd numbers and sections having even numbers. Therefore the dimension of a given cell can be easily determined from its index: simply the number of odd indices in the n-tuple.

Before lifting over a cell we check any conditions necessary for the correctness of the theory being implemented (step 7). These are collectively refereed to as the input being *well-oriented* and require that projection polynomials are not *nullified* (meaning a polynomial with main variable x_i is not identically zero over a cell in $\mathbb{R}^{i-1}$). Which polynomials must be checked varies with the algorithm (see [20], [21], [4], [5] for details). If the conditions are not satisfied then an error message is returned in step 12, unless the cell in question is zero-dimensional when correctness can be restored by generating the stack with respect to minimal delineating polynomials (see [7]) as well as the projection polynomials in P_i (step 10). Note that input not well-oriented for one operator may be for another, and that Collins' operator is always successful (given sufficient resources).

Stacks are built by Algorithm 2 in step 15. These are collected together in step 16 to form a CAD of $\mathbb{R}^i$, with the final CAD of $\mathbb{R}^n$ returned. The correctness of Algorithm 1 follows from the correctness of Algorithm 2 and the correctness of the various PL-CAD theories implemented (see the citations in Section 4).

Algorithm 2. Stacks are generated following Algorithm 2. It finishes with a call to RegularChains:-GenerateStack, described in Section 5.2 of [11] (and implemented in MAPLE's RegularChains library). Algorithm 2 requires the input be projection polynomials: implying they satisfy the delineability conditions necessary for the cells produced when lifting to have the required invariance condition. The regular chains algorithm has stricter criteria, requiring in addition that the polynomials *separate above the cell*, meaning they are coprime and squarefree throughout. Hence Algorithm 2 must first pre-process to meet this condition.

In steps 1 and 2 we extract information from c. We identify those dimensions of the cell which are restricted to a point by consulting the cell index (indices with even integers) and collect together the equations defining these restrictions in steps $3-7$. There is no ambiguity in the ordering of the polynomials in E since a regular chain is defined by polynomials of different main variables [1]. If the cell is of full dimension then there is no need to process since the polynomials are delineable and taken from a squarefree basis. Otherwise, we process using Algorithms 3 and 4 in steps 10 and 11. The restriction is identified using a regular chain $\hat{rc}$ (step 9) together with the original bounding box. Note that $\hat{E}$ defines a single regular chain since the equations were extracted from one.

Algorithm 3. In order to make the polynomials coprime we use repeated calls to a triangular decomposition algorithm in step 3 (described in [19] and part of the RegularChains Library). Given lists of polynomials L_1 and L_2 and a regular chain, it returns a decomposition of the zeros of L_1 which are also also zeros of the regular chain but not zeros of L_2. We use $\hat{rc}$ for the regular chain, so we work on the restriction, and build up a list of coprime polynomials by ensuring existing ones (L_2) are not zeros in decompositions of the next one (L_1). Each time the decomposition is a list of either regular chains or *regular systems* (a regular chain and an inequality regular with respect to the chain [23]).

Algorithm 2. `GenerateStack`

Input : A cell c from a CAD of $\mathbb{R}^k$ and a set P of projection polynomials in $x_1 \ldots x_{k+1}$ (part of a squarefree basis).

Output: A set of cells $\mathcal{S}$ of $\mathbb{R}^{k+1}$ comprising a stack over c. The polynomials in P are sign-invariant on each cell of $\mathcal{S}$.

1 Set I and sp to be the cell index and sample point of c.
2 Set rc and bb to be the regular chain and bounding box encoding sp.
3 Set E to be the set of k polynomials whose zeros define rc, ordered by increasing main variable.
4 Set $\hat{E} := \{\,\}$.
5 **for** $i = 1, \ldots, k$ **do**
6 $\quad$ **if** *the i'th integer in I is even* **then**
7 $\quad\quad$ Add the ith polynomial in E to $\hat{E}$.

8 **if** $\hat{E} \neq \{\,\}$ **then**
9 $\quad$ Set $\hat{rc}$ to be the regular chain formed by $\hat{E}$.
10 $\quad$ $\hat{P} := \texttt{MakeCoprime}(P, \hat{rc}, c)$. // Apply Algorithm 3.
11 $\quad$ $\hat{P} := \texttt{MakeSquareFree}(P, \hat{rc}, c)$. // Apply Algorithm 4.

12 $\mathcal{S} := \texttt{RegularChains:-GenerateStack}(c, \hat{P}, k+1)$ // From [11].
13 **return** $\mathcal{S}$.

Algorithm 3. `MakeCoprime`

Input : A set of polynomials P, a regular chain $\hat{rc}$ and a cell c.

Output: A set of polynomials $\hat{P}$ which describe the same set of varieties, but which are coprime over c.

1 Set $\hat{P} = \{\,\}$.
2 **for** *polynomial* $p \in P$ **do**
3 $\quad$ $T := \texttt{Triangularize}([p], \hat{P}, \hat{rc})$. // From [19].
4 $\quad$ **for** *component* C *of* T **do**
5 $\quad\quad$ **if** $\mathrm{mvar}(C) \neq \mathrm{mvar}(p)$ **then**
6 $\quad\quad\quad$ next C.
7 $\quad\quad$ **if** C *has a zero compatible with the sample point of* c **then**
8 $\quad\quad\quad$ Add the polynomial in C with same main variable as p to $\hat{P}$.

9 **return** $\hat{P}$.

We consider each of these components in turn. If the main variable is lower then the solution is discarded. Otherwise we check if the component has a solution compatible with the sample point for the cell (as it may be a solution of $\hat{rc}$ other than one isolated by bb). This means isolating the real solutions (of the component excluding the top dimension) and refining their bounding boxes until they are either within bb or do not intersect at all. It is achieved using the

RealRootIsolate command in the RegularChains Library (see [2]). Finally if the component passes these tests then the polynomial in the main variable is extracted and added to the set returned from Algorithm 3 in step 8.

Algorithm 4. In order to make the polynomials squarefree we use repeated calls to an algorithm which does this modulo a regular chain ($\hat{rc}$: so that we are working on the restriction). It is an analogue of Musser's [22] with the gcd calculations performed modulo the regular chain as described in [18]. It assumes the polynomial is regular modulo the chain and so we first test for this. If not regular (the leading coefficient vanishes) then we consider the tail (polynomial minus the leading term) in step 5, if still in the main variable. The output of the factorization is either: rc and a list of polynomials forming a squarefree decomposition of p modulo rc; or a list of pairs of regular chains and squarefree decompositions where the regular chains are a decomposition of rc. In the latter case only one will be relevant for the root isolated by bb and we identify which using the RealRootIsolate command, similarly to Algorithm 3.

Algorithm 4. MakeSquareFree

Input : A set of polynomials P, a regular chain $\hat{rc}$ and a cell c.
Output: A set of polynomials $\hat{P}$ which describe the same set of varieties, but which are each squarefree over c.

1   Set $\hat{P} = \{\ \}$.
2   **while** P *is not empty* **do**
3   Remove a polynomial p from P.
4   **if** p *is not regular over* $\hat{rc}$ **then**
5   Set $\hat{p} = \text{tail}(p)$
6   **if** $\text{mvar}(\hat{p}) = \text{mvar}(p)$ **then**
7   Add $\hat{p}$ to P and continue from step 2.

8      $T := \text{SquarefreeFactorization}(p, \hat{rc})$.
9   Select $\mathcal{C}$ as the component in T compatible with the sample point of c.
10   Set $\hat{p}$ to be the product of polynomials in the decomposition in $\mathcal{C}$.
11   Add $\hat{p}$ to $\hat{P}$.

12 **return** $\hat{P}$.

---

4   Functionality of ProjectionCAD

We finish by listing some of the functionality of within PROJECTIONCAD, focusing on aspects not usually found in other CAD implementations:

- Sign-invariant CADs can be built using the Collins [12] or McCallum [20] projection operators.
- CADs can be built with the stronger property of *order-invariance* (where each polynomial has constant order of vanishing on each cell) [20].

- *Equational constraints* (ECs) are equations logically implied by the formula. They can be utilised via McCallum's reduced projection [21] and a more efficient lifting phased (detailed in Section 5 of [5]).
- TTICADs can be built for sequences of formulae, making use of ECs in each [4] [5]. TTICAD can be both a desired structure for applications [17] and an efficient way to build a truth-invariant CAD (allowing savings from ECs for conjunctive sub-formulae, not ECs of the whole formula).
- Minimal delineating polynomials [7] are built automatically, avoiding unnecessary failure declarations (which can occur in QEPCAD). See [14] for an example of this.
- User commands for stack generation and the construction of *induced CADs* (a CAD of $\mathbb{R}^i$, $i < n$ produced en route to a CAD of $\mathbb{R}^n$), allowing for easy experimentation with the theory.
- *Layered CADs* contain cells of only a certain dimension or higher. They can be produced (more efficiently than a full CAD) [24].
- *Variety CADs* contain only those cells that lie on the variety defined by an EC. They can be produced (more efficiently than a full CAD) [24].
- Layered and manifold TTICADs as well as layered-manifold CADs can be produced [24] (combining the savings from the different theories).
- Heuristics are available to help with choices such as variable ordering, EC designation and breaking up parent formulae for TTICAD [6].

Details can be found in the citations above and the technical reports [14], [15].

Acknowledgements. This work was supported by the EPSRC (grant number EP/J003247/1). We thank the developers of the REGULARCHAINS Library, especially Changbo Chen and Marc Moreno Maza, for access to their code and assistance working with it.

References

1. Aubry, P., Lazard, D., Moreno Maza, M.: On the theories of triangular sets. Journal of Symbolic Computation 28(1-2), 105–124 (1999)
2. Boulier, F., Chen, C., Lemaire, F., Moreno Maza, M.: Real root isolation of regular chains. In: Proc. ASCM 2009, pp. 15–29 (2009)
3. Bradford, R., Chen, C., Davenport, J.H., England, M., Moreno Maza, M., Wilson, D.: Truth table invariant cylindrical algebraic decomposition by regular chains. To appear: Proc. CASC 2014 (2014), Preprint: http://opus.bath.ac.uk/38344/
4. Bradford, R., Davenport, J.H., England, M., McCallum, S., Wilson, D.: Cylindrical algebraic decompositions for boolean combinations. In: Proc. ISSAC 2013, pp. 125–132. ACM (2013)
5. Bradford, R., Davenport, J.H., England, M., McCallum, S., Wilson, D.: Truth table invariant cylindrical algebraic decomposition (submitted 2014), Preprint: http://opus.bath.ac.uk/38146/
6. Bradford, R., Davenport, J.H., England, M., Wilson, D.: Optimising problem formulation for cylindrical algebraic decomposition. In: Carette, J., Aspinall, D., Lange, C., Sojka, P., Windsteiger, W. (eds.) CICM 2013. LNCS, vol. 7961, pp. 19–34. Springer, Heidelberg (2013)

7. Brown, C.W.: The McCallum projection, lifting, and order-invariance. Technical report, U.S. Naval Academy, Compt. Sci. Dept. (2005)
8. Brown, C.W., Davenport, J.H.: The complexity of quantifier elimination and cylindrical algebraic decomposition. In: Proc. ISSAC 2007, pp. 54–60. ACM (2007)
9. Chen, C., Davenport, J.H., May, J., Moreno Maza, M., Xia, B., Xiao, R., Xie, Y.: User interface design for geometrical decomposition algorithms in Maple. In: Proc. Mathematical User-Interface Workshop, 12 pages (2009)
10. Chen, C., Moreno Maza, M.: An incremental algorithm for computing cylindrical algebraic decompositions. In: Proc. ASCM 2012. Preprint: http://arxiv.org/abs/1210.5543
11. Chen, C., Moreno Maza, M., Xia, B., Yang, L.: Computing cylindrical algebraic decomposition via triangular decomposition. In: Proc. ISSAC 2009, pp. 95–102. ACM (2009)
12. Collins, G.E.: Quantifier elimination for real closed fields by cylindrical algebraic decomposition. In: Brakhage, H. (ed.) GI-Fachtagung 1975. LNCS, vol. 33, pp. 134–183. Springer, Heidelberg (1975)
13. Davenport, J.H., Bradford, R., England, M., Wilson, D.: Program verification in the presence of complex numbers, functions with branch cuts etc. In: Proc. SYNASC 2012, pp. 83–88. IEEE (2012)
14. England, M.: An implementation of CAD in Maple utilising McCallum projection. Technical report, Uni. of Bath, Dept. Comp. Sci., 2013-02 (2013), http://opus.bath.ac.uk/33180/
15. England, M.: An implementation of CAD in Maple utilising problem formulation, equational constraints and truth-table invariance. Technical report, Uni. of Bath, Dept. Comp. Sci., 2013-04 (2013), http://opus.bath.ac.uk/35636/
16. England, M., Bradford, R., Chen, C., Davenport, J.H., Moreno Maza, M., Wilson, D.: Problem formulation for truth-table invariant cylindrical algebraic decomposition by incremental triangular decomposition. In: Watt, S.M. (ed.) CICM 2014. LNCS (LNAI), vol. 8543, pp. 45–60. Springer, Heidelberg (2014)
17. England, M., Bradford, R., Davenport, J.H., Wilson, D.: Understanding branch cuts of expressions. In: Carette, J., Aspinall, D., Lange, C., Sojka, P., Windsteiger, W. (eds.) CICM 2013. LNCS, vol. 7961, pp. 136–151. Springer, Heidelberg (2013)
18. Li, X., Moreno Maza, M., Pan, W.: Computations modulo regular chains. In: Proc. ISSAC 2009, pp. 239–246. ACM (2009)
19. Moreno Maza, M.: On triangular decompositions of algebraic varieties. Presented at: Effective Methods in Algebraic Geometry, MEGA (2000)
20. McCallum, S.: An improved projection operation for cylindrical algebraic decomposition. In: Quantifier Elimination and Cylindrical Algebraic Decomposition. Texts & Monographs in Symbolic Computation, pp. 242–268. Springer (1998)
21. McCallum, S.: On projection in CAD-based quantifier elimination with equational constraint. In: Proc. ISSAC 1999, pp. 145–149. ACM (1999)
22. Musser, D.R.: Multivariate polynomial factorization. Journal of the ACM 22(2), 291–308 (1975)
23. Wang, D.: Computing triangular systems and regular systems. Journal of Symbolic Computation 30(2), 221–236 (2000)
24. Wilson, D., Bradford, R., Davenport, J.H., England, M.: Cylindrical algebraic sub-decompositions. To appear: Mathematics in Computer Science. Springer (2014)
25. Wilson, D., Davenport, J.H., England, M., Bradford, R.: A "piano movers" problem reformulated. In: Proc. SYNASC 2013. IEEE (2014)
26. The REGULARCHAINS Library, http://www.regularchains.org

An Improvement of Rosenfeld-Gröbner Algorithm

Amir Hashemi[1,2] and Zahra Touraji[1]

[1] Department of Mathematical Sciences,
Isfahan University of Technology Isfahan, 84156-83111, Iran
[2] School of Mathematics, Institute for Research in Fundamental Sciences (IPM),
Tehran, P.O. Box: 19395-5746, Iran
Amir.Hashemi@cc.iut.ac.ir, z.tooraji@math.iut.ac.ir
http://amirhashemi.iut.ac.ir/

Abstract. In their paper Boulier et al. (2009) described the Rosenfeld-Gröbner algorithm for computing a regular decomposition of a radical differential ideal generated by a set of polynomial differential equations, ordinary or with partial derivatives. In order to enhance the efficiency of this algorithm, they proposed their analog of Buchberger's criteria to avoid useless reductions to zero. For example, they showed that if p and q are two differential polynomials which are linear, homogeneous, in one differential indeterminate, with constant coefficients and with leaders θu and ϕu, respectively so that θ and ϕ are disjoint then the delta-polynomial of p and q reduces to zero w.r.t. the set $\{p, q\}$. In this paper we generalize this result showing that it remains true if p and q are products of differential polynomials which are linear, homogeneous, in the same differential indeterminate, with constant coefficients and θ and ϕ are disjoint where θu and ϕu are leaders of p and q, respectively. We have implemented the Rosenfeld-Gröbner algorithm and our refined version on the same platform in MAPLE and compare them via a set of benchmarks.

Keywords: Differential algebra, Rosenfeld-Gröbner, Buchberger first criterion.

1 Introduction

Differential algebra, founded by Ritt, is a very interesting subject to use methods from abstract algebra to study solutions of systems of polynomial nonlinear ordinary and partial differential equations (PDE's). One of the most important tools of differential algebra is the Ritt algorithm [14] which gives rise to algorithmic methods appear in this field. In this direction, several approaches and relative packages have been developed to employ these methods. Carra-Ferro and Ollivier developed the concept of *differential Gröbner bases* in [6,13], however the bases that they defined may be infinite. On the other hand, Mansfield [11] developed another concept of differential Gröbner bases. She has developed the MAPLE package diffgrob2 which is an effective and powerful package for simplifying

H. Hong and C. Yap (Eds.): ICMS 2014, LNCS 8592, pp. 466–471, 2014.

systems of linear or nonlinear PDEs. It can construct differential Gröbner bases for general linear systems and some nonlinear systems. Finally, Boulier et al. [4] presented *Rosenfeld-Gröbner* which takes as input a differential system and a ranking and outputs represents the radical of the differential ideal generated by the input system as a finite intersection of radical differential ideals presented by characteristic sets. The MAPLE package `DifferentialAlgebra` designed by Boulier et al. which computes such representation is now embedded in MAPLE. As an application of this theory in finding the Lie symmetries of differential equations we refer to [12,9]. For a comprehensive introduction of differential algebra and its computational aspects, we refer to Kolchin's book [10] and [15,8], respectively.

The main goal of this paper is to improve the Rosenfeld-Gröbner algorithm. In this direction, Boulier et al. [4] proposed their analog of Buchberger's criteria to avoid useless reductions to zero. They showed that if p and q are two differential polynomials which are linear, homogeneous, in one differential indeterminate, with constant coefficients and with leaders θu and ϕu, respectively so that θ and ϕ are disjoint then the delta-polynomial of p and q reduces to zero w.r.t. the set $\{p, q\}$. In this paper, we generalize this result showing that it remains true if p and q are products of differential polynomials which are linear, homogeneous, in the same differential indeterminate, with constant coefficients and θ and ϕ are disjoint where θu and ϕu are leaders of p and q, respectively. We have implemented the Rosenfeld-Gröbner algorithm and our refined version on the same platform in MAPLE and compare them via a set of benchmarks.

The paper is organized as follows. Section 2 contains an overview of the necessary background concerning the theory of differential algebras. In Section 3, we will state our main result and discuss the implementation issue of Rosenfeld-Gröbner algorithm equipped with the new analog of Buchberger first criterion.

2   Preliminaries

In this section, we present a brief overview of basic definitions and notations in differential algebra. For more details we refer the reader to [10,14].

Definition 1. *An operator* $\delta : R \to R$ *over the algebraic ring R is called a derivation operator, if for each $a, b \in R$ we have:*

$$\delta(a + b) = \delta(a) + \delta(b)$$
$$\delta(ab) = \delta(a)b + a\delta(b).$$

A differential ring is a pair (R, Δ) where R is a ring equipped with a collection $\Delta = \{\delta_1, \ldots, \delta_m\}$ of commuting derivations operators over it, satisfying:

$$\delta_i \delta_j a = \delta_j \delta_i a.$$

For simplicity, we suppress the dependence on Δ in the notation and denote a differential ring just by R. If $m = 1$, then R is called an ordinary differential *ring; otherwise it will be called* partially. *An algebraic ideal I of R is called a differential ideal when it is closed under the action of derivations of R, namely $\delta a \in I$ for each $\delta \in \Delta$ and $a \in I$.*

For example, the ring of polynomials $\mathbb{C}[x_1, \ldots, x_m]$ together with the set of operators $\partial/\partial x_1, \ldots, \partial/\partial x_m$ is a differential ring. Let R be a differential ring with $\Delta = \{\delta_1, \ldots, \delta_m\}$.

- We denote by Θ the free multiplicative commutative semigroup generated by the elements of Δ, namely

$$\Theta := \left\{ \delta_1^{t_1} \delta_2^{t_2} \ldots \delta_m^{t_m} \mid t_1, \ldots, t_m \in \mathbb{N} \right\}.$$

  Each element $\theta = \delta_1^{\alpha_1} \cdots \delta_m^{\alpha_m}$ of Θ is called a *derivation operator* of R and furthermore the sum $\operatorname{ord}(\theta) := \sum_{i=1}^{m} t_i$ is called the *order* of θ. Then θa is said to be a derivative of $a \in R$ of order $\operatorname{ord}(\theta)$.

- For an arbitrary subset S of R, we define $\Theta S := \{\theta s \mid s \in S, \theta \in \Theta\}$. It is the smallest subset of R containing S which is stable under derivation.

- An algebraic ideal of R is called a *differential ideal*, if it is closed under the derivation operators. We denote by (S) and $[S]$ respectively, the smallest algebraic and differential ideals of R containing S. In fact, $[S] = (\Theta S)$. This fact provides an algebraic approach to differential ideals which enables one to employ algebraic tools.

- For a field of characteristic zero K, a *differential polynomial ring*:

$$R := K\{u_1, \ldots, u_n\} := K[\Theta U]$$

  is the usual commutative polynomial ring generated by ΘU over K, where $U := \{u_1, \ldots, u_n\}$ is the set of *differential indeterminate*.

- For two certain derivatives θu and ϕu of a same differential indeterminate u, we denote by $\operatorname{lcd}(\theta u, \phi u)$ the *least common derivative* between θu and ϕu, easily seen to be:

$$\operatorname{lcd}(\theta u, \phi u) = \operatorname{lcm}(\theta, \phi) u.$$

In this paper we let K be a differential field of characteristic zero.

Definition 2. *Let $R = K\{U\}$ be a differential polynomial ring with the set of indeterminates $U = \{u_1, \ldots, u_n\}$. A ranking $>$ is an ordering over ΘU compatible with the derivation act over ΘU, i.e. for each derivation $\delta \in \Theta$ and for each $v, w \in \Theta U$ we have:*

- *$\delta v > v$,*
- *if $v > w$ then $\delta v > \delta w$.*

For each $\theta, \phi \in \Theta$ and $v, w \in U$, a ranking $>$ for which the statement $\operatorname{ord}(\theta) > \operatorname{ord}(\phi)$ implies that $\theta v > \phi w$ is called orderly. Simultaneously, if the assumption $v > w$ gives $\theta v > \phi w$, then $>$ is called elimination. Moreover, for a fixed ranking

$>$ over ΘU and for a differential polynomial $p \in R = K\{u_1, \dots, u_n\}$, the leader $\mathrm{ld}(p)$ of p is the highest derivative appearing in p with respect to $>$. If $\mathrm{ld}(p) = u$ and d is the degree of u in p then, the initial $\mathrm{in}(p) \in R$ is defined to be the coefficient of u^d in p. The separant s_p of p is the polynomial $\partial p / \partial u$. Finally, u^d is called the rank of p and we denote it by $\mathrm{rank}(p)$.

To analyze a given PDE system Σ, we use the Rosenfeld-Gröbner algorithm to decompose the radical of $[\Sigma]$, into some new PDE systems, presented by explicit generators. These generators have novel properties which leads to do a complete analysis of Σ. We can consider this algorithm as a differential analogues to Buchberger's algorithm. One of the main functions that we need in the Rosenfeld-Gröbner algorithm is Δ-polynomial which plays a similar role to S-polynomial in Buchberger algorithm.

Definition 3. Let us consider two differential polynomials p_1 and p_2 with $\mathrm{ld}(p_i) = \theta_i\, u_i$, $i = 1, 2$. Then, the Δ-polynomial of p_1 and p_2 is defined as

$$\Delta(p_1, p_2) = \begin{cases} s_{p_2} \frac{\theta_{1,2}}{\theta_1} p_1 - s_{p_1} \frac{\theta_{1,2}}{\theta_2} p_2 & u_1 = u_2, \\ 0 & u_1 \neq u_2, \end{cases}$$

where $\theta_{1,2} = \mathrm{lcd}(\theta_1, \theta_2)$.

The aim of calculating the Δ-polynomial of two differential polynomials is in fact to remove their leading derivatives to obtain (probably) a new leading derivative.

3 Statement of the Main Results

Buchberger's first criterion states that if the leading terms of two polynomials p and q are disjoint then the S-polynomial of p and q reduces to zero by $\{p, q\}$, see [1, Lemma 5.66]. The following differential analogues to this criterion was given in [4, Prop. 4].

Proposition 1. Let p and q be two differential polynomials which are linear, homogeneous, in one differential indeterminate and with constant coefficients. Further, suppose that we have $\mathrm{lcd}(\theta u, \phi u) = \theta \phi u$ where $\mathrm{ld}(p) = \theta u$ and $\mathrm{ld}(q) = \phi u$. Then the full differential remainder of $\Delta(p, q)$ w.r.t $\{p, q\}$ is zero.

Now we state the main result of this paper.

Theorem 1. Let p and q be two differential polynomials which are products of differential polynomials which are linear, homogeneous, in the same differential indeterminate and with constant coefficients. Furthermore, let θ and ϕ are disjoint where θu and ϕu are leaders of p and q. Then the full differential remainder of $\Delta(p, q)$ w.r.t $\{p, q\}$ is zero.

Proof. Due to space restriction, we only give the general idea of the proof for the case that p and q are products of two polynomials, and leave the detailed proof for the full version of paper. Let $p = f_1 f_2$ and $q = g_1 g_2$ where f_1, f_2, g_1, g_2 are differential polynomial which are linear, homogeneous and in one differential indeterminate with constant coefficients. Further, let $ld(f_1) = \theta u$ and $ld(g_1) = \phi u$. We prove that $\Delta(p, q) full-rem\{p, q\} = 0$. Suppose $ld(f_1) > ld(g_1), ld(f_1) > ld(f_2)$ and $ld(g_1) > ld(g_2)$. Then $s_p = i_p = f_2$ and $s_q = i_q = g_2$ and

$$
\begin{aligned}
\Delta(p, q) &= g_2 \phi p - f_2 \theta q \\
&= g_2 \sum_{\phi_i \phi_j = \phi} \phi_i f_1 \phi_j f_2 - f_2 \sum_{\theta_l \theta_k = \theta} \theta_l g_1 \theta_k g_2 \\
&= g_2 f_2 \phi f_1 - g_2 f_2 \theta g_2 + g_2 \sum_{\phi_i \phi_j = \phi, \mathrm{ord}(\phi_j) > 0} \phi_i f_1 \phi_j f_2 - f_2 \sum_{\theta_l \theta_k = \theta, \mathrm{ord}(\theta_k) > 0} \theta_l g_1 \theta_k g_2.
\end{aligned}
$$

Since f_1 and g_1 satisfy the conditions of Proposition 1, then $g_2 f_2 \phi f_1 - g_2 f_2 \theta g_2$ reduces to zero. Let us call r the rest of the above polynomial. To show that r reduces to zero, we consider two cases for the appearance of $ld(f_1)$ in r and we show that in both cases r reduces to zero.   □

We have implemented the MAPLE package `RosenfeldGrobner.mpl` containing a prototype implementation of the improved Rosenfeld-Gröbner algorithm which is available at the address `http://amirhashemi.iut.ac.ir/software.html`. In what follows we provide an example illustrating the efficiency of our algorithm.

```
> with(diffalg):
> R := differential_ring(derivations = [x, y], ranking = [u]):
> p := (u_{x,x,x,x} - u)^7 (u_{x,x} + u_x)^5:
> q := (u_{y,y,y} - u_{y,y})^4 (u_y + u)^5:
> r := delta_polynomial(p, q, R):
> a1, b1 := kernelopts(cputime, bytesused):
> differential_sprem(r, [p, q], R);
0
> a2, b2 := kernelopts(cputime, bytesused):
> a2-a1, b2-b1;
70.52, 598229337
```

These results show that the maple function `differential_sprem` takes 70.52 sec. to compute the full differential remainder of the polynomials p and q which satisfy the conditions of the above theorem.

Acknowledgements. The research of the first author was in part supported by a grant from IPM (No. 92550420).

References

1. Becker, T., Weispfenning, V.: Gröbner bases: a computational approach to commutative algebra. Springer (1993)
2. Buchberger, B.: An algorithm for finding the basis elements in the residue class ring modulo a zero dimensional polynomial ideal. German, English Translation: J. Symbolic Comput., Special Issue on Logic, Mathematics, and Computer Science: Interactions 41(3-4), 475–511 (2006)
3. Buchberger, B.: A criterion for detecting unnecessary reductions in the construction of Gröbner bases. In: Ng, K.W. (ed.) EUROSAM 1979 and ISSAC 1979. LNCS, vol. 72, pp. 3–21. Springer, Heidelberg (1979)
4. Boulier, F., Lazard, D., Ollivier, F., Petitot, M.: Computing representations for radicals of finitely generated differential ideals. AAECC 20, 73–121 (2009)
5. Boulier, F., Lazard, D., Ollivier, F., Petitot, M.: Representation for the radical of a finitely generated differential ideal. In: ISSAC 1995, pp. 158–166. ACM Press (1995)
6. Carra-Ferro, G.: Gröbner bases and differential ideals. In: Notes of AAECC 5, Menorca, Spain, pp. 129–140. Springer (1987)
7. Gallo, G.: Complexity issues in computational Algebra. Ph.D. thesis, Courant Institute of Mathematical Sciences, New York University (1992)
8. Hubert, E.: Notes on triangular sets and triangulation-decomposition algorithms II: Differential systems. In: Winkler, F., Langer, U. (eds.) SNSC 2001. LNCS, vol. 2630, pp. 40–87. Springer, Heidelberg (2003)
9. Hydon, P.E.: Symmetry methods for differential equations. Cambridge University Press, Cambridge (2000)
10. Kolchin, E.R.: Differential algebra and algebraic groups. Academic Press, New York (1973)
11. Mansfield, E.L.: Differential Gröbner bases. Ph.D. thesis, University of Sydney, Australia (1991)
12. Mansfield, E.L., Clarkson, P.A.: Applications of the differential algebra package diffgrob2 to classical symmetries of differential equations. J. Symb. Comp. 23, 517–533 (1997)
13. Ollivier, F.: Le problème de l'identifiabilité structurelle globale: approche théorique, méthodes effectives et bornes de complexité. Ph.D. thesis, École Polytechnique, Palaiseau, France (1990)
14. Ritt, J.F.: Differential Algebra. American Mathematical Society, New York (1948)
15. Sit, W.Y.: The Ritt-Kolchin theory for differential polynomials. In: Guo, L., Keigher, W.F., Cassidy, P.J., Sit, W.Y. (eds.) Differential Algebra and Related Topics, pp. 1–70. World Scientific Publishing Co. Inc. (2002)

Doing Algebraic Geometry
with the RegularChains Library

Parisa Alvandi[1], Changbo Chen[2], Steffen Marcus[3], Marc Moreno Maza[1],
Éric Schost[1], and Paul Vrbik[1]

[1] University of Western Ontario, Canada
{palvandi,moreno,eschost,pvrbik}@csd.uwo.ca
[2] Chongqing Institute of Green and Intelligent Technology, CAS, China
changbo.chen@hotmail.com
[3] The College of New Jersey, Ewing, USA
steffenmarcus@gmail.com

Abstract. Traditionally, Groebner bases and cylindrical algebraic decomposition are the fundamental tools of computational algebraic geometry. Recent progress in the theory of regular chains has exhibited efficient algorithms for doing local analysis on algebraic varieties. In this note, we present the implementation of these new ideas within the module `AlgebraicGeometryTools` of the `RegularChains` library. The functionalities of this new module include the computation of the (non-trivial) limit points of the quasi-component of a regular chain. This type of calculation has several applications like computing the Zarisky closure of a constructible set as well as computing tangent cones of space curves, thus providing an alternative to the standard approaches based on Groebner bases and standard bases, respectively. From there, we have derived an algorithm which, under genericity assumptions, computes the intersection multiplicity of a zero-dimensional variety at any of its points. This algorithm relies only on the manipulations of regular chains.

Keywords: Algebraic geometry, regular chains, local analysis.

1   Overview

Today, regular chains are at the core of algorithms computing triangular decomposition of polynomial systems and which are available in several software packages [7,9,10]. Moreover, those algorithms provide back-engines for computer algebra system front-end solvers, such as MAPLE's `solve` command.

One of the algorithmic strengths of the theory of regular chains is its *regularity test* procedure. Given a polynomial p and a regular chain R, both in a polynomial ring $\mathbf{k}[X_1, \ldots, X_n]$ over a field $\mathbf{k}$, this procedure computes regular chains $R_1, \ldots, R_e$ such that $R_1, \ldots, R_e$ is a decomposition of R in some technical sense[1] and for each $1 \leq i \leq e$ the polynomial p is either null or regular

---

[1] The radical of the saturated ideal sat(R) of R is equal to the intersection of the radicals of the saturated ideals of $R_1, \ldots, R_e$.

H. Hong and C. Yap (Eds.): ICMS 2014, LNCS 8592, pp. 472–479, 2014.

modulo the saturated ideal of R_i. In algebraic terms, this procedure decides whether the hypersurface $V(p)$ contains at least one irreducible component of the variety $V(\text{sat}(R))$. Thanks to the D5 Principle [4], this regularity test avoids factorization into irreducible polynomials and involves only polynomial GCD computations. This is a core routine in most operations on regular chains.

One of the technical difficulties of regular chain theory, however, is the fact that regular chains do not fit well in the "usual algebraic-geometric dictionary" (Chapter 4, [3]). Indeed, the "good" zero set encoded by a regular chain R is a constructible set $W(R)$, called the *quasi-component* of R, which does not correspond exactly to the "good" ideal encoded by R, namely sat(R), the *saturated ideal* of R. In fact, the affine variety defined by sat(R) equals $\overline{W(R)}$, that is, the Zariski closure of $W(R)$. This difficulty probably explains why the use of regular chains in computational algebraic geometry remains limited, despite their nice algorithmic properties such as the above mentioned regularity test.

In [1], three of the co-authors of this note have recently proposed a procedure for computing the *non-trivial limit points* of the quasi-component $W(R)$, that is, the set $\lim(W(R)) := \overline{W(R)} \setminus W(R)$ as a finite union of quasi-components of some other regular chains. This procedure, currently implemented in the case where sat(R) has dimension one, relies only on operations on regular chains, like the regularity test. As a byproduct, it becomes possible to compute $\overline{W(R)}$ without any Gröbner basis computations. We illustrate this feature in Section 2.

The regularity test for regular chains is a powerful tool which has been studied and applied within many situations including polynomial algebra [6], differential algebra [2,5], and computing with algebraic numbers [4], etc. Broadly speaking, it allows to extend an algorithm working over a field into an algorithm working over a direct product of fields. Or, to phrase it in another way, it allows to extend an algorithm working at point into an algorithm working at a group of points.

Following that strategy, three of the co-authors of this note have proposed, in another recent work [8], an extension of Fulton's algorithm for computing the intersection multiplicity of two plane curves at any of their intersection points. Indeed, as pointed out by Fulton in his *Intersection Theory*, the intersection multiplicity of two plane curves $V(f)$ and $V(g)$ satisfy a series of seven properties which *uniquely* define $I(p; f, g)$ at each point $p \in V(f, g)$. Moreover, the proof of this remarkable fact is constructive, which leads to an algorithm, that we call Fulton's algorithm. Unfortunately, this algorithm implicitly assumes that the coordinates of the point p are rational numbers.

Another limitation of Fulton's algorithm is that it does not generalize to n polynomials $f_1, \ldots, f_n$ in n variables $x_1, \ldots, x_n$. In [8], two extensions of Fulton's algorithm are proposed. First, thanks to the regularity test for regular chains, the construction is adapted such that it can work correctly at any point of $V(f, g)$, rational or not. Secondly, thanks to the above mentioned procedure for computing the limit points of a quasi-component, an algorithmic criterion is proposed for reducing the case of n variables to the bivariate one. These algorithmic tools are now implemented in the module `AlgebraicGeometryTools` and are illustrated in Sections 3 and 4.

2   Computation of Limit Points

For a regular chain [2] $R \subset \mathbf{k}[X_1, \ldots, X_n]$, recall that the *quasi-component* of R
is $W(R) := V(R) \setminus V(h_R)$, that is, the common zeros of R that do not cancel
the product h_R of the initials of R. As mentioned in Section 1, computing the
non-trivial limit points, that is, the set $\lim(W(R)) := \overline{W(R)} \setminus W(R)$ has many
applications. The algorithm proposed in [1] relies on the Puiseux series solutions
of the quasi-component $W(R)$.

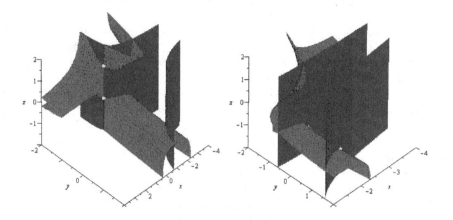

Fig. 1. Limit points of one-dimensional regular chain

```
> with(AlgebraicGeometryTools):
> R := PolynomialRing([x, y, t]);
> F := [t*y^2 + y + 1, (t + 2)*t*x^2 + (y +1)* (x + 1)];
> C := Chain(F, Empty(R), R);
> lm := LimitPoints(C, R, false, true);
> Display(lm, R);
```

$$R := polynomial\_ring$$

$$F := \left[t\,y^2 + y + 1, (t+2)\,t\,x^2 + (y+1)\,(x+1) \right]$$

$$C := regular\_chain$$

$$lm := [\,regular\_chain, regular\_chain, regular\_chain, regular\_chain\,]$$

$$\left\| \left[\begin{cases} x + 1 = 0 \\ y + \dfrac{1}{2} = 0 \\ t + 2 = 0 \end{cases} , \begin{cases} x + 1 = 0 \\ y - 1 = 0 \\ t + 2 = 0 \end{cases} , \begin{cases} x + \dfrac{1}{2} = 0 \\ y + 1 = 0 \\ t = 0 \end{cases} , \begin{cases} x - 1 = 0 \\ y + 1 = 0 \\ t = 0 \end{cases} \right] \right\|$$

Fig. 2. Limit point computation with `AlgebraicGeometryTools`

---

[2] We refer to [1] for the basic concepts and properties of regular chain theory.

The MAPLE session displayed on Figure 2 illustrates a limit point computation with $R = \{ty^2+y+1, (t+2)tx^2+(y+1)(x+1)\}$ for the variable ordering $t < y < x$. Thus we have $h_R = t(t - 2)$. As shown in Figure 1, there are four limit points - see the yellow dots - which are returned by the command LimitPoints in the form of regular chains. Other formats are possible; they are controlled by the last two arguments of the LimitPoints command.

3   Intersection Multiplicity in the Plane

Let us now turn to intersection multiplicity computation. We start with the bivariate case, illustrating it with the two plane curves depicted on Figure 3. The command TriangularizeWithMultiplicity, see Figure 4, computes the five intersection points of these curves and, for each of them, returns the corresponding intersection multiplicity. Observe that four of these five points are

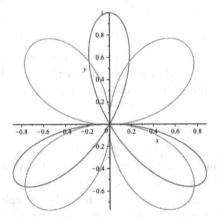

Fig. 3. Two plane curves with an intersection multiplicity of 14 at the origin

```
> R := PolynomialRing([x,y]);
> F := [(x^2+y^2)^2+3*x^2 * y-y^3, (x^2+y^2)^3-4 *x^2 *y^2];
> dec := TriangularizeWithMultiplicity(F, R);
> Display(dec,R);
>
```

$$R := polynomial\_ring$$

$$F := \left[\left(x^2 + y^2\right)^2 + 3\,x^2\,y - y^3, \left(x^2 + y^2\right)^3 - 4\,x^2\,y^2\right]$$

$$dec := [[1, regular\_chain], [14, regular\_chain]]$$

$$\left[\left[\left[1, \begin{cases} (64\,y + 80)\,x^2 + 20\,y - 15 = 0 \\ 16\,y^2 - 5 = 0 \\ 64\,y + 80 \neq 0 \end{cases}\right], \left[14, \begin{cases} x = 0 \\ y = 0 \end{cases}\right]\right]\right]$$

Fig. 4. Intersection multiplicity computation with AlgebraicGeometryTools

described by a single regular chain; moreover, at each of those, the intersection multiplicity is 1. In fact, the computation of their intersection multiplicity is performed as a single computation since there is no need to write explicitly the coordinates of each of these points.

4   Intersection Multiplicity in Higher Dimension and Computation of Tangent Cones

Fulton's algorithm does not apply directly to n polynomials $f_1, \ldots, f_n$ in n variables $x_1, \ldots, x_n$. However, the criterion proved in [8] allows to reduce the computation of the intersection multiplicity of $f_1, \ldots, f_n$ at a point p to an intersection multiplicity calculation in a lower dimension space. We recall this criterion. Assume that $h_n = V(f_n)$ is non-singular at p. Let v_n be tangent hyperplane of h_n at p. Assume that h_n meets each component (through p) of the curve $\mathcal{C} = V(f_1, \ldots, f_{n-1})$ transversally (that is, the tangent cone $TC_p(\mathcal{C})$ intersects v_n only at the point p). Let $g \in k[x_1, \ldots, x_n]$ be the degree 1 polynomial defining v_n. Then, we have

$$I(p; f_1, \ldots, f_n) = I(p; f_1, \ldots, f_{n-1}, g).$$

Assume[3] that the coefficient of x_n in g is non-zero, thus $g = x_n - g'$, where $g' \in k[x_1, \ldots, x_{n-1}]$. Hence, we can rewrite the ideal $\langle f_1, \ldots, f_{n-1}, g \rangle$ as $\langle g_1, \ldots, g_{n-1}, g \rangle$ where g_i is obtained from f_i by substituting x_n with g'. Then, we have

$$I_n(p; f_1, \ldots, f_n) = I_{n-1}(p|_{x_1, \ldots, x_{n-1}}; g_1, \ldots, g_{n-1}).$$

In the example from Figure 5, the tangent hyperplane $y = 0$ of $V(y - z^3)$ at the origin and each component (through the origin) of the curve $\mathcal{C} := V(x, x + y^2 - z^2) = V(x, (y-z)(y+z))$ meet transversally. Therefore, we have $I_3((0, 0, 0); x, x + y^2 - z^2, y - z^3) = I_2((0, 0); x, x - z^2) = 2$, as shown by the calculation on Figure 6.

Verifying this transversality condition requires the computation of tangent cones of and tangent planes. The module `AlgebraicGeometryTools` provides commands `TangentCone` (of space curves) and `TangentPlane` for that purpose. Each tangent cone is computed as a limit of secants and reduces to compute limit points of quasi-components of one-dimensional regular chains.

In Figure 8, the transversality condition does not hold. This is detected using the `TangentCone` and `TangentPlane` commands. Another strategy is then attempted, we call it *cylindrification*. We explain its principle under simplifying assumptions. Assume that among $f_1, \ldots, f_n$ one polynomial, say f_n has degree one in x_n and assume that its coefficient in x_n is invertible in the local ring at p. Then, one replaces $f_1, \ldots, f_{n-1}$ by $g_1, \ldots, g_{n-1}$ where g_i is the pseudo-remainder of f_i by f_n w.r.t x_n. It is not hard to see that $I(p; f_1, \ldots, f_n) = I(p; g_1, \ldots, g_{n-1}, f_n)$ holds. Moreover, the transversality condition clearly holds for $g_1, \ldots, g_{n-1}, f_n$. Therefore, we are back in the above case where we could reduce computations from n to $n-1$ variables. Returning to the example of Figure 8, cylindrification replaces the three surfaces on the left of Figure 7 with the surfaces on the right.

---

[3] One can always reduce to this case by means of a linear change of coordinates.

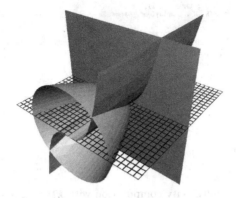

Fig. 5. Three surfaces: case where reduction 3D to 2D is straightforward

```
> R    := PolynomialRing( [x,y,z] ):
> F    := [x, x-y^2-z^2,y-z^3];
> dec := Triangularize(F, R); Display(dec, R);
>
```

$$F := \left[x,\ -y^2 - z^2 + x,\ -z^3 + y \right]$$
$$dec := [\ regular\_chain,\ regular\_chain\ ]$$

$$\left[\left[\begin{array}{c} x = 0 \\ y - z^3 = 0 \\ z^4 + 1 = 0 \end{array} \right.,\ \left\{ \begin{array}{c} x = 0 \\ y = 0 \\ z = 0 \end{array} \right. \right]$$

```
> IsTransverse(dec[1], F[3], F[1..2], R); IsTransverse(dec[2], F[3], F[1..2], R);
```
$$true$$
$$true$$

```
> dec := TriangularizeWithMultiplicity(F, R); Display(dec, R);
```
$$dec := [\ [\ 1,\ regular\_chain\ ],\ [\ 2,\ regular\_chain\ ]\ ]$$

$$\left[\left[1,\ \left[\begin{array}{c} x = 0 \\ y - z^3 = 0 \\ z^4 + 1 = 0 \end{array} \right. \right],\ \left[2,\ \left\{ \begin{array}{c} x = 0 \\ y = 0 \\ z = 0 \end{array} \right. \right] \right]$$

Fig. 6. Intersection multiplicity computation with `AlgebraicGeometryTools`

Fig. 7. Three surfaces: case where reduction 3D to 2D requires cylindrification

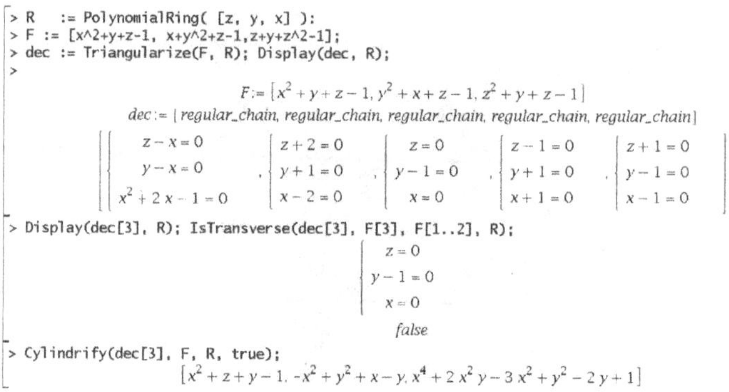

```
> R    := PolynomialRing( [z, y, x] ):
> F := [x^2+y+z-1, x+y^2+z-1,z+y+z^2-1];
> dec := Triangularize(F, R); Display(dec, R);
>
```

$$F := \left[x^2 + y + z - 1, y^2 + x + z - 1, z^2 + y + z - 1 \right]$$

$$dec := [\ regular\_chain,\ regular\_chain,\ regular\_chain,\ regular\_chain,\ regular\_chain\ ]$$

$$\left[\left|\begin{matrix} z - x = 0 \\ y - x = 0 \\ x^2 + 2x - 1 = 0 \end{matrix}\right|, \left|\begin{matrix} z + 2 = 0 \\ y + 1 = 0 \\ x - 2 = 0 \end{matrix}\right|, \left|\begin{matrix} z = 0 \\ y - 1 = 0 \\ x = 0 \end{matrix}\right|, \left|\begin{matrix} z - 1 = 0 \\ y + 1 = 0 \\ x + 1 = 0 \end{matrix}\right|, \left|\begin{matrix} z + 1 = 0 \\ y - 1 = 0 \\ x - 1 = 0 \end{matrix}\right|\right]$$

```
> Display(dec[3], R); IsTransverse(dec[3], F[3], F[1..2], R);
```

$$\left|\begin{matrix} z = 0 \\ y - 1 = 0 \\ x = 0 \end{matrix}\right.$$

$$false$$

```
> Cylindrify(dec[3], F, R, true);
```

$$\left[x^2 + z + y - 1, -x^2 + y^2 + x - y, x^4 + 2x^2 y - 3x^2 + y^2 - 2y + 1 \right]$$

Fig. 8. Intersection multiplicity computation with `AlgebraicGeometryTools`

5   Concluding Remarks

The new module `AlgebraicGeometryTools` of the `RegularChains` library performs various operations on one-dimensional objects (computation of limit points of constructible sets, tangent cones of space curves) as well as intersection multiplicity computation for zero-dimensional varieties. All these operations rely only on regular chain techniques, that is, no calculations of Gröbner bases or standard bases are performed.

Extending limit point computations to higher dimension is work in progress. The cylindrification strategy (for reducing intersection multiplicity calculation from n to $n-1$ variables) may fail in some situations; improving this situation is also work in progress.

Benchmarks reported in the PhD dissertation of the last author shows that the command `TriangularizeWithMultiplicity` is computationally efficient in the sense that it can process almost all examples which can be processed by the `Triangularize`[4]

The `RegularChains` library is available at www.regularchains.org.

Acknowledgments. This work was supported by the NSFC (11301524) and the CSTC (cstc2013jjys0002).

References

1. Alvandi, P., Chen, C., Moreno Maza, M.: Computing the limit points of the quasi-component of a regular chain in dimension one. In: Gerdt, V.P., Koepf, W., Mayr, E.W., Vorozhtsov, E.V. (eds.) CASC 2013. LNCS, vol. 8136, pp. 30–45. Springer, Heidelberg (2013)

---

[4] The `Triangularize` command decomposes a polynomial system into regular chains but discards any multiplicity information.

2. Boulier, F., Lazard, D., Ollivier, F., Petitot, M.: Computing representations for radicals of finitely generated differential ideals. Appl. Algebra Eng. Commun. Comput. 20(1), 73–121 (2009)

3. Cox, D., Little, J., O'Shea, D.: Ideals, Varieties, and Algorithms, 2nd edn. Springer (1996)

4. Della Dora, J., Dicrescenzo, C., Duval, D.: About a new method for computing in algebraic number fields. In: Buchberger, B. (ed.) ISSAC 1985 and EUROCAL 1985. LNCS, vol. 203, pp. 289–290. Springer, Heidelberg (1985)

5. Hubert, E.: Factorization-free decomposition algorithms in differential algebra. J. Symb. Comput. 29(4-5), 641–662 (2000)

6. Kalkbrener, M.: Algorithmic properties of polynomial rings. J. Symb. Comput. 26(5), 525–581 (1998)

7. Lemaire, F., Moreno Maza, M., Xie, Y.: The RegularChains library. In: Maple 10, Maplesoft, Canada (2005), http://www.regularchains.org

8. Marcus, S., Moreno Maza, M., Vrbik, P.: On Fulton's algorithm for computing intersection multiplicities. In: Gerdt, V.P., Koepf, W., Mayr, E.W., Vorozhtsov, E.V. (eds.) CASC 2012. LNCS, vol. 7442, pp. 198–211. Springer, Heidelberg (2012)

9. Wang, D.K.: The Wsolve package, http://www.mmrc.iss.ac.cn/~dwang/wsolve.html

10. Wang, D.M.: Epsilon 0.618, http://www-calfor.lip6.fr/~wang/epsilon

On Multivariate Birkhoff Rational Interpolation

Peng Xia[1], Bao-Xin Shang[2], and Na Lei[3,⋆]

[1] School of Mathematics, Liaoning University, Shenyang 110036, P.R. China
`xpxiapengxp7@gmail.com`
[2] School of Mathematics, Jilin University, Changchun 130012, P.R. China
`shbxin@gmail.com`
[3] School of Mathematics, Key Lab. of Symbolic Computation and Knowledge
Engineering (Ministry of Education), Jilin University, Changchun 130012, P.R. China
`leina@jlu.edu.cn`

Abstract. Multivariate Birkhoff rational interpolation is the most general algebraic interpolation scheme. There is few literature on this problem due to the complex structure of the rational function and the non-continuity of the orders of the derivative interpolating conditions. In this paper, by adding the lacking derivative conditions and setting the artificial interpolating values as undetermined parameters, we propose a parametric linearization method to convert the problem of finding a multivariate Birkhoff rational interpolation function into solving a parametric polynomial system in which the coefficients in the numerator and denominator are the unknowns. We use the parametric triangular decomposition to solve the system and prove the solution provides a Birkhoff rational interpolation function as long as there exist proper parameters such that the denominator does not vanish at each interpolating point. The algorithm is implemented in Maple 15.

Keywords: Birkhoff rational interpolation, triangular decomposition, parametric polynomial system.

1 Introduction

Birkhoff interpolation is one of the most general and important polynomial interpolation problems. There exists a rich literature on the existence, uniqueness and representations of the problem, such as [1–5]. In recent years, many scholars applied the Birkhoff interpolation in numerically solving boundary value problems [6] and initial-value problems [7].

Rational functions sometimes are superior to polynomials with the same interpolation data because they can achieve more accurate approximations with the same amount of computation [8]. In addition, rational interpolation has a natural way in interpolating poles whereas polynomial interpolation does not [9]. In this paper, we study the Birkhoff rational interpolation problem. The interpolation scheme consists of two components.

⋆ Corresponding author.

H. Hong and C. Yap (Eds.): ICMS 2014, LNCS 8592, pp. 480–483, 2014.

a) A set of nodes Z, $Z = \{Y_i\}_{i=1}^m = \{(y_{i,1}, \ldots, y_{i,n})\}_{i=1}^m$, where $Y_i \in K^n$, K is a field.

b) The derivative conditions S_i at each node Y_i, $i = 1, \ldots, m$, where S_i is a subsets of $\mathbb{N}_0^n$. Some S_i $(i = 1, \ldots, m)$ may not be lower sets.

The multivariate Birkhoff rational interpolation problem is to find a rational function $r(X) = \dfrac{p(X)}{q(X)}$ satisfying

$$L_{i,\alpha} = D^\alpha r(Y_i) = \frac{\partial^{\alpha_1 + \cdots + \alpha_n}}{\partial x_1^{\alpha_1} \cdots \partial x_n^{\alpha_n}} r(Y_i) = c_{i,\alpha}, \forall \alpha \in S_i, \tag{1}$$

where $p(X) \in \mathcal{P}_{T_1} = \left\{ p \mid p(X) = p(x_1, \ldots, x_n) = \sum_{\alpha_i \in T_1} a_i x_1^{\alpha_1} \cdots x_n^{\alpha_n} \right\}$, $q(X) \in \mathcal{Q}_{T_2} = \left\{ q \mid q(X) = q(x_1, \ldots, x_n) = \sum_{\beta_i \in T_2} b_i x_1^{\beta_1} \cdots x_n^{\beta_n} \right\}$, a_i, $b_i \in K, T_1, T_2$ are subsets of $\mathbb{N}_0^n$, $c_{i,\alpha} \in K$ are given constants, $L_{i,\alpha}$ are functionals related to the corresponding interpolation conditions.

The key character of multivariate Birkhoff rational interpolation is that the orders of the derivative conditions at some nodes are non-continuous. Without the non-continuity, i.e. all the S_i's are lower sets $(i = 1, \ldots, m)$, the problem degenerates into an Hermit rational interpolation. Salzer [10] proposed a linearization technique to deal with univariate Hermit rational interpolation. Whereas there is no similar results for multivariate Hermit case, not even for Birkhoff rational interpolation. In this paper, we propose a parametric linearization method to solve the problem of multivariate Birkhoff rational interpolation and implement the algorithm in Maple 15. The function is described in section 2 and one example is shown in section 3. In section 4 the main idea of the algorithm is introduced.

2   Functionality

We created a function BirkhoffRationalInterp(Y,F) in Maple to implement the multivariate Birkhoff rational interpolation algorithm.

Calling sequence
BirkhoffRationalInterp(Y,F).

Parameters
Y–list of points. Each point is represented as a row vector.

F–list of matrices. The i-th matrix is determined by the interpolation conditions corresponding to the i-th point Y_i. The number of the rows of the i-th matrix equals to the number of the interpolation conditions according to the i-th point. Each row of the i-th matrix $[\alpha_1, \ldots, \alpha_n, c_{i,\alpha}]$ denotes a interpolation condition $D^\alpha r(Y_i) = c_{i,\alpha}$ where $\alpha = (\alpha_1, \ldots, \alpha_n)$.

Description
The BirkhoffRationalInterp command constructs the multivariate Birkhoff rational interpolation functions in a field K (default is $\mathbb{R}$). The output of this command is a list of the rational functions.

The packages "Groebner" and "RegularChains" are required.

3   Example

Given a interpolation problem (see Table 1). Let $Y := [[0,0],[0,1],[1,0],[1,1]]$;

Table 1. Interpolation problem

Y_i	(0,0)	(0,1)	(1,0)	(1,1)
S_i	$\{(0,0),(0,1),(1,1)\}$;	$\{(0,0),(1,0),(1,1)\}$;	$\{(0,0),(1,1)\}$;	$\{(0,0),(1,0),(0,1)\}$
$c_{i,\alpha}$	$\{\ 6\ ,\ 5\ ,\ 0\ \}$;	$\{\ 7\ ,\ 2\ ,\ -2\ \}$;	$\{\ 6\ ,\text{-}5/2\ \}$;	$\{20/3,\text{-}7/9\ ,16/9\}$

$F_1 := \text{Matrix}([[0,0,6],[0,1,5],[1,1,0]])$, $F_2 := \text{Matrix}([[0,0,7],[1,0,2],[1,1,-2]])$,
$F_3 := \text{Matrix}([[0,0,6],[1,1,-\frac{5}{2}]])$, $F_4 := \text{Matrix}([[0,0,\frac{20}{3}],[1,0,\frac{16}{9}],[0,1,-\frac{7}{9}]])$.

The output of the command BirkhoffRationalInterp(Y,$[F_1, F_2, F_3, F_4]$) is a list $[r_1(x,y), r_2(x,y)]$, where

$$r_1(x,y) = \frac{6 - 44.217y + 233.040x + 77.917y^2 - 221.333xy - 108.216x^2}{1 - 8.203y + 35.048x + 12.874y^2 - 34.997xy - 14.244x^2},$$

$$r_2(x,y) = \frac{6 - 37.464y + 2887.787x - 196.995y^2 - 261.344xy - 2552.415x^2}{1 - 7.077y + 430.953x + -26.560y^2 - 46.423xy - 375.057x^2}.$$

4   Underlying Theory

The main idea of the algorithm is as follows.

We firstly generalize the univariate linearization method proposed by Salzer in [10] to multivariate Hermite case. Let $L(\alpha) = \{\beta \in \mathbb{N}_0^n : \beta_i < \alpha_i, \ i = 1,\ldots,n\}$.

Lemma 1. *If $q(Y_i) \neq 0$ $(i = 1,\ldots,m)$, the Hermite rational interpolation systems*

$$D^\alpha(p/q)(Y_i) = f_{i,\alpha}, \ i = 1,\ldots,m, \ \boldsymbol{\alpha} \in A_i \tag{2}$$

is equivalent to the system

$$D^\alpha p(Y_i) = \sum_{\sigma \in L(\alpha)} f_{i,\sigma} D^{\alpha-\sigma} q(Y_i), \ i = 1,\ldots,m, \ \boldsymbol{\alpha} \in A_i, \tag{3}$$

where A_i, $i = 1,\ldots,m$, are lower sets, $f_{i,\sigma}$, $\boldsymbol{\sigma} \in L(\alpha)$, $i = 1,\ldots,m$, are the given derivative values.

Secondly, for a given Birkhoff interpolation problem, we add the lacking derivative conditions and setting the artificial interpolating values as undetermined parameters, then we obtain a parametric Hermit rational interpolation problem.

Let $\widetilde{S}_i = S_i$. For each $\alpha \in \widetilde{S}_i$, if $\exists \beta \in L(\alpha)$ and $\beta \notin \widetilde{S}_i$, then we add β to $\widetilde{S}_i$, and set $c_{i,\beta}$ as an undetermined parameter. Finally, a parametric Hermit rational system is derived.

$$D^\alpha(p/q) = c_{i,\alpha}, \ \forall \alpha \in \widetilde{S}_i, \ i = 1, \ldots, m, \tag{4}$$

where $c_{i,\alpha}$, is a given constant if $\alpha \in S_i$, an undetermined parameter otherwise.

Now the original problem reduces to solving a parametric polynomial system. We use the parametric triangular decomposition to solve the system and the following theorem proves the solution provides a Birkhoff rational interpolation function as long as there exist proper parameters such that the denominator does not vanish at each interpolating point.

Theorem 1. *If p/q is a solution of* (1), *then there exist some parameters $c_{i,\beta}$ such that p, q satisfy*

$$D^\alpha p(Y_i) = \sum_{\sigma \in L(\alpha)} c_{i,\sigma} D^{\alpha-\sigma} q(Y_i), \ i = 1, \ldots, m, \ \alpha \in \widetilde{S}_i. \tag{5}$$

Conversely, if p, $q \in K[X]$ is a solution of (5), *and q satisfies $q(Y_i) \neq 0$, $i = 1, \ldots, n$, then p/q satisfies* (1).

Acknowledgements. The authors would like to thank Dr. Changbo Chen for his valuable suggestions on solving the parametric polynomial system.

References

1. Turán, P.: On some open problems of approximation theory. Journal of Approximation Theory 29, 23–85 (1980)
2. Shi, Y.G.: Theory of Birkhoff Interpolation. Nova Science, New York (2003)
3. Palacios, F., Rubi, P., Diaz, J.L.: Order regularity of two-node Birkhoff interpolation with lacunary polynomials. Applied Mathematics Letters 22(3), 386–389 (2009)
4. Chai, J.J., Lei, N., Li, Y., Xia, P.: The proper interpolation space for multivariate Birkhoff interpolation. Journal of Computational and Applied Mathematics 235, 3207–3214 (2011)
5. Kowalski, K.: Reelle Hermite-Birkhoff-Interpolation. Results in Mathematics 62, 405–414 (2012)
6. Costabile, F.A., Longo, E.: A Birkhoff interpolation problem and application. Calcolo 47, 49–63 (2010)
7. Dehghana, M., Aryanmehra, S., Eslahchi, M.R.: A technique for the numerical solution of initial-value problems based on a class of Birkhoff-type interpolation method. Journal of Computational and Applied Mathematics 244, 125–139 (2013)
8. Meinguet, J.: On the solvability of the Cauchy interpolation problem. In: Talbot, A. (ed.) Approximation Theory, pp. 137–163. Academic Press, New York (1970)
9. Lei, N., Liu, T., Zhang, S., Feng, G.: Some problems on multivariate rational interpolation. Journal of Information and Computaional Science 3(3), 453–461 (2006)
10. Salzer, H.E.: Note on osculatory rational interpolation. Mathematics of Computation 16, 486–491 (1962)

Computing Moore-Penrose Inverses
of Ore Polynomial Matrices

Yang Zhang

Department of Mathematics, University of Manitoba,
Winnipeg, MB R3T 2N2, Canada
yang.zhang@umanitoba.ca

Abstract. In this paper we define and discuss the generalized inverse and Moore-Penrose inverse for Ore polynomial matrices. Based on GCD computations and Leverrier-Faddeeva method, some fast algorithms for computing these inverses are constructed, and the corresponding Maple package including quaternion case is developed.

Keywords: Moore-Penrose inverse, Ore polynomial matrices, Quaternion.

1 Introduction

Ore polynomial matrices are matrices over Ore algebras (including differential operators and difference operators). It has a long research history, at least dated back to Jacobson's seminal work in 1940s. In past ten years, Ore matrices have attracted more and more people in computer algebra area, and many important properties of Ore matrices have been discussed by using symbolic computation methods, for example, various fast algorithms for computing Hermite forms and Smith forms, fraction-free algorithms for computing Popov forms (see, for example, [2] and [8]).

It is well-known that the generalized inverse of matrices over commutative rings (or fields), especially the Moore-Penrose inverse, play important roles in matrix theory and have applications in many areas: solving matrix equations, statistics, engineering, etc. This motivates us to consider the generalized inverse of Ore polynomial matrices.

First we define the generalized inverse and the Moore-Penrose inverse for Ore polynomial matrices, and prove some basic properties including uniqueness. Unlike the commutative case, the generalized inverse for a given Ore polynomial matrix may not exist. We use blocked matrices and greatest common right (left) divisor computations to give some sufficient and necessary conditions for Ore polynomial matrices to have the generalized inverses and the Moore-Penrose inverses. Moreover when these inverses exist, we construct algorithms to compute them. In quaternion case, we define generalized characteristic polynomials and give an analogy version of Leverrier-Faddeeva algorithm.

All our algorithms are implemented in the symbolic programming language Maple, and tested on examples. Our aim is to develop a Maple package for computing the generalized inverses and the Moore-Penrose inverse of Ore polynomial

H. Hong and C. Yap (Eds.): ICMS 2014, LNCS 8592, pp. 484–491, 2014.
© Springer-Verlag Berlin Heidelberg 2014

matrices, in particular, for quaternion case. To our best knowledge, it is the first Maple package in this direction.

2   Definitions, Properties and Algorithms

Let D be a division ring (or called skew field) and $\sigma : D \to D$ be an automorphism of D. A σ-derivation $\delta : D \to D$ is a mapping satisfying: for any $a, b \in D$,

$$\delta(a + b) = \delta(a) + \delta(b), \quad \delta(ab) = \sigma(a)\delta(b) + \delta(a)b.$$

The Ore polynomial ring $R = D[x; \sigma, \delta]$ is defined as the set of usual polynomials in $D[x]$ under the usual addition, but with multiplication defined by

$$xa = \sigma(a)x + \delta(a), \quad \text{for any } a \in D.$$

The Ore polynomial matrices $R^{m \times n}$ will be the set of all $m \times n$ matrices with Ore polynomial entries. More properties can be found in, for example, Jacobson[12] and Lam[14].

To consider Moore-Penrose inverses, we assume that D has an involution "$*$", that is, "$*$" is an anti-automorphism of order 1 or 2 on D. It is easy to check that "$*$" can be extended to $D[x; \sigma, \delta]$ as an involution as follows. For any $f = \sum_{i=0}^{n} a_i x^i \in R = D[x; \sigma, \delta]$, we define

$$f^* = \sum_{i=0}^{n} a_i^* x^i.$$

Furthermore, "$*$" can be extended to Ore polynomial matrices $R^{m \times n}$ in a common way. Hence for any $A \in R^{m \times n}$ we can define the involution of transpose A^T of A as: $A^T = (A_{ij}^*)^T \in R^{n \times m}$.

Next we give the definition of Moore-Penrose inverses, and refer the reader to [3] and [21] for details in commutative case.

Definition 1. *A matrix $A^\dagger \in R^{n \times m}$ is called a Moore-Penrose inverse of $A \in R^{m \times n}$ if $A^\dagger$ satisfies:*

$$(i) \ AA^\dagger A = A \quad (ii) \ A^\dagger AA^\dagger = A^\dagger \quad (iii) \ (AA^\dagger)^T = AA^\dagger \quad (iv) \ (A^\dagger A)^T = A^\dagger A.$$

People are also interested in the matrices which only satisfy some of the above equations. If a matrix $A^{\{1\}}$ satisfies (i), then we say that $A^{\{1\}}$ is a $\{1\}$-inverse of A or a generalized inverse of A. Similarly, $\{1, 2\}$-inverse (or more generally $\{i, j, k\}$-inverse) can be defined.

Recall that $A \in D^{m \times m}$ is unitary if $AA^T = A^T A = I_m$. One can prove the following elementary properties that will be often used throughout this paper.

Theorem 1. *Let $A \in R^{m \times n}$ and $B \in R^{n \times l}$. Then*

(i) $(AB)^T = B^T A^T$ and $AA^T = (AA^T)^T$.

(ii) If A has a Moore-Penrose inverse $A^\dagger$, then $(A^T)^\dagger = (A^\dagger)^T$, $A^\dagger (A^\dagger)^T A^T = A^\dagger = A^T (A^\dagger)^T A^\dagger$ and $A^\dagger AA^T = A^T = A^T AA^\dagger$.

(iii) If A has a Moore-Penrose inverse $A^\dagger$, then $A^\dagger$ is unique.

(iv) Let A have the Moore-Penrose inverse $A^\dagger$. If $U \in D^{m \times m}$ is a unitary matrix, then $(UA)^\dagger = A^\dagger U^T$.

Rao condition is a common assumption in commutative case. Now we can define it over division ring D with an involution " $*$ ": for any $a_1, \ldots, a_s \in R$,

$$a_1 = a_1 \cdot a_1^* + \cdots + a_s \cdot a_s^* \quad \text{implies} \quad a_2 = \cdots = a_s = 0.$$

Clearly, Ore polynomial ring $R = D[x; \sigma, \delta]$ also satisfies Rao condition, in particular, when $D = \mathbb{C}(x)$ rational function field over complex number field $\mathbb{C}$ or $D = \mathbb{H}(x)$ quotient skew field over quaternion polynomial ring $\mathbb{H}[x]$.

Throughout this paper, we assume that D is a division ring with an involution " $*$ " which satisfies Rao condition, and $R = D[x; \sigma, \delta]$ is an Ore polynomial ring over D.

Note that we require that $A^\dagger$ (and other generalized inverses) must be in $R^{n \times m}$, not in matrices over its quotient skew field $Q(R)$. Thus unlike the matrices over fields, the Moore-Penrose inverse for some Ore polynomial matrices might not exist.

In this paper, we first discuss the existence of the Moore-Penrose inverses including using Jacobson normal forms. Some interesting results in commutative case (see [3] and [22]) can be extended to Ore polynomial matrices. Here we just list two interesting results.

Theorem 2. *If $E \in R^{m \times m}$ is a symmetric projection, that is, $E = E^2 = E^T$, then $E \in D^{m \times m}$.*

Theorem 3. *Let $A \in R^{m \times n}$. Then A has the Moore-Penrose inverse $A^\dagger$ if and only if $A = U \begin{bmatrix} A_1 & A_2 \\ 0 & 0 \end{bmatrix}$ with $U \in D^{m \times m}$ unitary and $A_1 A_1^T + A_2 A_2^T$ a unit in $R^{r \times r}$ with $r \leq \min\{m, n\}$. Moreover,*

$$A^\dagger = \begin{bmatrix} A_1^T (A_1 A_1^* + A_2 A_2^T)^{-1} & 0 \\ A_2^T (A_1 A_1^T + A_2 A_2^T)^{-1} & 0 \end{bmatrix} U^T.$$

Next we outline how to use computing greatest common right (and left) divisors (gcrd) methods to find generalized inverses. First, from Section 3.7 of Jacobson[12], we know that for $a, b \in R$, not both zero, we can compute the

GCRD $g = \text{gcrd}(a, b)$, and $u, v \in R$ such that $ua + vb = g$, and $s, t \in R$ such that $sa = -tb = \text{lclm}(a, b)$. Furthermore we have

$$U = \begin{bmatrix} u & v \\ s & t \end{bmatrix} \in R^{2\times 2} \quad \text{such that} \quad U \begin{bmatrix} a \\ b \end{bmatrix} = \begin{bmatrix} g \\ 0 \end{bmatrix}.$$

Similarly, using greatest common left divisors and least common right multiplies, we can find a $V \in R^{2\times 2}$ such that $\begin{bmatrix} a & b \end{bmatrix} V = \begin{bmatrix} d & 0 \end{bmatrix}$. For general matrices, using the above fact, we can prove the following theorem by induction:

Theorem 4. *Given $A = (a_{ij}) \in R^{m\times n}$, let $r, c \in R$ be the gcrd of the entries on the first column and the gcld of the entries on the first row of A, respectively, that is, $r = \text{gcrd}(a_{11}, \ldots, a_{m1})$ and $c = \text{gcld}(a_{11}, a_{12}, \ldots, a_{n1})$. Then there exist unimodular matrices $U \in R^{m\times m}$, $V \in R^{n\times n}$ such that*

$$UA = \begin{bmatrix} r & * \\ \mathbf{0} & * \end{bmatrix}, \quad AV = \begin{bmatrix} c & \mathbf{0} \\ * & * \end{bmatrix}.$$

Note that R is noetherian, a right inverse of $A \in R^{n\times n}$ is also a left inverse of A. The following theorem provides a recursive method to compute {1}-inverse.

Theorem 5. *Suppose that $0 \neq a \in R, b = (b_1, \ldots, b_n) \in R^{1\times n}$ and $A \in R^{m\times n}$.*

(i) If $\begin{bmatrix} a & b \\ \mathbf{0} & A \end{bmatrix} \in R^{(m+1)\times(n+1)}$ has a {1}-inverse, then $\text{gcld}(a, b_1, \ldots, b_n) = 1$.

(ii) If $\begin{bmatrix} a & \mathbf{0} \\ \mathbf{0} & A \end{bmatrix} \in R^{(m+1)\times(n+1)}$ has a {1}-inverse, then $a \in D$ and A has a {1}-inverse.

Now we give an algorithm to compute {1}-inverse. The advantages of our algorithm is that we could use some well-known fast algorithms for computing gcrd, gcld, lclm and lcdm (see, for example, [4] and [15]).

Algorithm: Computing {1}-inverse

Input: $A \in R^{m\times n}$.

Output: {1}-inverse $A^{\{1\}}$ of A or no such {1}-inverse exists.

1. Compute a unimodular $U \in R^{m\times m}$ such that

$$UA = \begin{bmatrix} r & * \\ \mathbf{0} & * \end{bmatrix}.$$

2. Compute the gcld of the first row of UA. If gcld$\neq 1$, return "no such {1}-invere exists". Otherwise goto next step.
3. Compute a unimodular $V \in R^{n\times n}$ such that

$$UAV = \begin{bmatrix} 1 & \mathbf{0} \\ * & * \end{bmatrix}.$$

4. Do row transformations, find a unimodular $U_1 \in R^{m \times m}$ such that

$$U_1 U A V = \begin{bmatrix} 1 & \mathbf{0} \\ \mathbf{0} & A_1 \end{bmatrix}.$$

5. Recursively apply Step 1 for A_1, and so on.
6. Finally we have two unimodular matrices U_0, V_0 such that

$$U_0 A V_0 = A_0,$$

where $(A_0)_{ii} = 1$ or 0, $i = 1, \ldots, \min\{m, n\}$ and other entries equal zero.
7. return $A^{\{1\}} = V_0 A_0^T U_0$.

In fact, in Step 6, we could rearrange the rows and columns of A_0 such that $(A_0)_{ii} = 1$, $i = 1, \ldots, r \leq \min\{n, m\}$, and other entries are equal to zero. Moreover, we can prove that r is equal to the rank of A.

Using above algorithm, we can compute other kinds of inverses. For example, to compute $\{1, 2\}$-inverse of A, we first use above algorithm to find U_0, V_0 such that

$$U_0 A V_0 = \begin{bmatrix} I & \mathbf{0} \\ \mathbf{0} & \mathbf{0} \end{bmatrix}.$$

Then any $\{1, 2\}$-inverse of A is of the form $V_0 \begin{bmatrix} I & S \\ T & TS \end{bmatrix} U_0$, where S, T are arbitrary matrices with appropriate sizes.

As one of most important subclasses of Ore polynomials over division rings, we are interested in quaternion (skew) polynomial rings. The algebra $\mathbb{H}$ of real quaternion was discovered by Sir Rowan Hamilton in 1843, which is a four-dimensional non-commutative algebra over real number field $\mathbb{R}$ with canonical basis $1, \mathbf{i}, \mathbf{j}, \mathbf{k}$ satisfying the conditions:

$$\mathbf{i}^2 = \mathbf{j}^2 = \mathbf{k}^2 = \mathbf{ijk} = -1,$$

that implies

$$\mathbf{ij} = -\mathbf{ji} = \mathbf{k}, \ \mathbf{jk} = -\mathbf{kj} = \mathbf{i}, \ \text{and} \ \mathbf{ki} = -\mathbf{ik} = \mathbf{j}.$$

Elements in $\mathbb{H}$ can be written as a unique way $\alpha = a + b\mathbf{i} + c\mathbf{j} + d\mathbf{k}$, where a, b, c and d are real numbers, that is, $\mathbb{H} = \{a + b\mathbf{i} + c\mathbf{j} + d\mathbf{k} \mid a, b, c, d \in \mathbb{R}\}$. The conjugate of α is defined as $\bar{\alpha} = a - b\mathbf{i} - c\mathbf{j} - d\mathbf{k}$.

Since $\mathbb{H}$ is a skew field, there are several forms of quaternion polynomials depending on the positions of coefficients. In our case, we will use one which put the coefficients on the left side of a variable x, which it is also called regular quaternion polynomials in [5] or quaternion simple polynomials in [17]. Furthermore, some Ore polynomials over $\mathbb{H}$ can be defined in a natural way, for example, $\mathbb{H}[x; -]$ and $\mathbb{H}(t)[x; \frac{\partial}{\partial t}]$. Some properties of quaternion (skew) polynomials and matrices over them have been discussed with many applications in control theory and physics (see, for example, [18] and [14]).

Based on the special properties of quaternion, we give another method to compute the Moore-Penrose inverse in quaternion case. First we define and discuss generalized characteristic polynomial for quaternion polynomials, and then consider the Leverrier-Faddeeva algorithm (see, [1], [6] and [7]) in quaternion case. Finally we explore the interpolation for quaternion polynomials and quaternion polynomial matrices and construct a fast algorithm. The detailed results will be included in a full paper after the conference.

3   Implementation

Our Maple package includes two parts: general Ore polynomial matrices and quaternion polynomial matrices.

In Part I, dedicated to general Ore polynomial matrices, all commands are compatible with OreTools and OreAlgebras in Maple 17. We use the same commands to set up Ore polynomials over fields and do basic computations including gcrd, gcld, lclm and lcdm. Here are a few key commands in our package:

- $OreMat(A, m, n)$: set up an $m \times n$ Ore matrix.
- $Rgcrd(A, 1, j)$: compute $\mathrm{gcrd}((A)_{11}, (A)_{j1})$ and make row transformation such that $(A)_{11} = \mathrm{gcrd}((A)_{11}, (A)_{j1})$ and $(A)_{j1} = 0$.
- $Cgcld(A, 1, j)$: compute $\mathrm{gcld}((A)_{11}, (A)_{1j})$ and make column transformation such that $(A)_{11} = \mathrm{gcld}((A)_{11}, (A)_{1j})$ and $(A)_{1j} = 0$.
- Iinverse(A) returns {1}-inverse of A.
- MPinverse(A) returns the Moore-Penrose inverse of A.

Note that our methods work for Ore polynomials over division rings, in particular, over quaternions. As we know that there are no quaternion package in Maple 17. Although there is a quaternion package available on Maple Help website, it only includes some basic operations. No quaternion polynomials and matrices are included. This motivates us to develop a Maple package for quaternion polynomials and matrices.

In Part II, dedicated to quaternion polynomial matrices, we develop this package from the beginning to keep consistence, that is, set up Maple commands for quaternion operations first, which include norm, conjugate, similar, etc.

Secondly we set up Maple commands for quaternion polynomial operations. Simple quaternion polynomials and quaternion skew polynomials with conjugate " $-$ " are pre-defined. People can use the $SetQuaternionRing$ command to define various quaternion (skew) polynomials.

We prove that the Extended Euclidean Algorithm also works for quaternion polynomials and use it to compute gcrd, lclm, etc. Some commands are as follows:

- $qGCRD(f, g)$ returns the monic GCRD of quaternion polynomials of f and g.
- $qExtendedGCRD(f, g, A, 'u', 'v')$ returns the monic GCRD of quaternion polynomials of f and g, and two pairs $\{u_1, v_1\}$ and $\{u_2, v_2\}$ such that

$$u_1 f + u_2 g = \mathrm{GCRD}(f, g), \quad v_1 f + v_2 g = 0.$$

where $v_1 f$ is the LCLM of f and g. The parameter A presents a quaternion polynomial defined by the $SetQuaternionRing$ function.

Regarding matrices over quaternion (skew) polynomials, we develop basic matrix operations and three kinds of row (column) transformations. Combing Part I and Part II, we can compute the generalized inverses for quaternion (skew) polynomial matrices.

In the final part of our package, we implement the Leverrier-Faddeeva algorithm for quaternion polynomials and use Lagrange interpolation in quaternion to construct a fast algorithm. Here are some key commands:

- *LFMPinverse*(*A*) returns the Moore-Penrose inverse of quaternion polynomial matrix *A* by using Leverrier-Faddeeva algorithm.
- *Linterpolation*($[a_1, .., a_n], [b_1, .., b_n]$) returns a quaternion polynomial *f* such that $f(a_i) = b_i$.
- *LLFMPinverse*(*A*) returns the Moore-Penrose inverse of quaternion polynomial matrix *A* by using Lagrange interpolation and Leverrier-Faddeeva algorithm.

Acknowledgment. This research was supported by the grants from the National Sciences and Engineering Research Council (NSERC) of Canada and URGP from the University of Manitoba.

References

1. Barnett, S.: Leverrier's algorithm: a new proof and extensions. SIAM J. Matrix Anal. Appl. 10, 551–556 (1989)
2. Beckermann, B., Cheng, H., Labahn, G.: Fraction-free Row Reduction of Matrices of Ore Polynomials. Journal of Symbolic Computation 41(5), 513–543 (2006)
3. Ben-Israel, A., Greville, T.N.E.: Generalized Inverses: Theory and Applications, 2nd edn. Springer, New York (2003)
4. Bostan, A., Chyzak, F., Li, Z., Salvy, B.: Fast computation of common left multiples of linear ordinary differential operators. In: Proceedings of the 2012 International Symposium on Symbolic and Algebraic Computation, pp. 99–106. ACM Press (2012)
5. Damiano, A., Gentili, G., Sreuppa, D.: Computations in the ring of quaternionic polynomials. J. of Symbolic Computation 45, 38–45 (2010)
6. Decell, H.P.: An application of the Cayley-Hamilton theorem to generalized matrix inversion. SIAM Rev. 7, 526–528 (1965)
7. Faddeeva, V.N.: Computational methods of linear algebra. Dover books on advanced mathematics. Dover Publications (1959)
8. Giesbrecht, M., Sub Kim, M.: Computation of the Hermite form of a Matrix of Ore Polynomials. Journal of Algebra 376, 341–362 (2013)
9. Gordon, B., Motzkin, T.S.: On the zeros of polynomials over division rings. Trans. Amer. Math. Soc. 116, 218–226 (1965)
10. Horn, R.A., Johnson, C.R.: Matrix analysis. Cambridge University Press (1990)
11. Huang, L., Zhang, Y.: Maple package for quaternion matrices, The University of Manitoba (2013)
12. Jacobson, N.: The Theory of Rings. Amer. Math. Soc., New York (1943)
13. Jones, J., Karampetakis, N.P., Pugh, A.C.: The computation and application of the generalized inverse via Maple. J. Symbolic Comput. 25, 99–124 (1998)

14. Lam, T.Y.: A first course in noncommutative rings. Graduate Texts in Mathematics, vol. 131. Springer (2001)
15. Liand, Z., Nemes, I.: A modular algorithm for computing greatest common right divisors of Ore polynomials. In: Proceedings of the 1997 International Symposium on Symbolic and Algebraic Computation, pp. 282–289. ACM Press (1997)
16. Niven, I.: The roots of a quaternion. Amer. Math. Monthly 49, 386–388 (1942)
17. Opfer, G.: Polynomials and Vandermonde matrices over the field of quaternions. Electronic Transactions on Numerical Analysis 36, 9–16 (2009)
18. Pereira, R., Rocha, P.: On the determinant of quaternionic polynomial matrices and its application to system stability. Math. Meth. Appl. Sci. 31, 99–122 (2008)
19. Pereira, R., Rocha, P., Vettori, P.: Algebraic tools for the study of quaternionic behavioral systems. Linear Algebra and its Applications 400, 121–140 (2005)
20. Pereira, R., Vettori, P.: Stability of Quaternionic linear systems. IEEE Transactions on Automatic Control 51, 518–523 (2006)
21. Penrose, R.: A generalized inverse for matrices. Proc. Cambridge Philos. Soc. 51, 406–413 (1955)
22. Puystjens, R.: Moore-Penrose inverses for matrices over some Noetherian rings. J. Pure Appl. Algebra 31(1-3), 191–198 (1984)
23. Zhang, F.: Quaternions and matrices of quaternions. Linear Algebra and its Applications 251, 21–57 (1997)
24. Ziegler, M.: Quasi-optimal arithmetic for quaternion polynomials. In: Ibaraki, T., Katoh, N., Ono, H. (eds.) ISAAC 2003. LNCS, vol. 2906, pp. 705–715. Springer, Heidelberg (2003)

Software Using the GRÖBNER COVER for Geometrical Loci Computation and Classification

Miguel A. Abánades[1], Francisco Botana[2], Antonio Montes[3], and Tomás Recio[4]

[1] Universidad Complutense de Madrid, Spain
abanades@ajz.ucm.es
[2] Universidad de Vigo, Spain
fbotana@uvigo.es
http://webs.uvigo.es/fbotana/
[3] Universitat Politècnica de Catalunya, Spain
antonio.montes@upc.edu
http://www-ma2.upc.edu/montes/
[4] Universidad de Cantabria, Spain
tomas.recio@unican.es
http://www.recio.tk

Abstract. We describe here a properly recent application of the GRÖBNER COVER algorithm (GC) providing an algebraic support to Dynamic Geometry computations of geometrical loci. It provides a complete algebraic solution of locus computation as well as a suitable taxonomy allowing to distinguish the nature of the different components. We included a new algorithm LOCUS into the Singular grobcov.lib library for this purpose. A web prototype has been implemented using it in Geogebra.

Keywords: Locus, Taxonomy, Dynamical Geometry, Groebner Cover.

1 Introduction

[1] One of the defining characteristics of Dynamical Geometry (DG) is obtaining geometrical loci problems. Neverthless, the existing DG software are not able to give a satisfactory answer. This is the case for the first standard DG systems developed in the late 80's (such as Cabri and The Geometer's Sketchpad), as well as for more recent ones, such as GeoGebra or Java Geometry Expert. In DG systems, it is often the case that locus determination is purely graphical, producing an output that is not robust enough and not reusable by the given software. Moreover, extraneous objects are frequently appended to the true locus as side products of the locus determination process.

---

[1] Authors partially supported by the Spanish "Ministerio de Economía y Competitividad" and by the European Regional Development Fund (ERDF), under the Project MTM2011–25816–C02–02. The third author was supported by projects Gen. Cat. DGR 2009SGR1040 and MICINN MTM2009-07242.

H. Hong and C. Yap (Eds.): ICMS 2014, LNCS 8592, pp. 492–499, 2014.

Using the GRÖBNER COVER we are able to give an exact algebraic result, allowing moreover to give a taxonomy of the different locus components. In section 2 we give a summary of the GRÖBNER COVER. In section 3 we give the locus taxonomy obtained using it, and describe the new LOCUS algorithm. Finally, in section 4 we show characteristic examples justifying the taxonomy and showing its functionality. A web prototype [3] has been implemented using the new algorithm in Geogebra.

2   The GRÖBNER COVER

The GRÖBNER COVER algorithm for discussing parametric polynomial ideals gives a canonical description, classifying the solutions by their characteristics (number of solutions, dimension, etc.). This is used here for defining a taxonomy of geometrical loci and to implement it in the new algorithm LOCUS. It is included in the Singular grobcov.lib library allowing its use by DG software.

The GRÖBNER COVER provides the analog of the *reduced Gröbner basis* of an ideal for parametric ideals. Its existence was proved by Wibmer's Theorem [6], and the method and algorithms were developed in [5]. Montes implemented in Singular the grobcov.lib library [7], whose actual version incorporates Kapur-Sun-Wang algorithm [2] for computing the initial Gröbner System used in GROB-COV algorithm, as described in [4], and recently also the LOCUS algorithm described here. A more detailed description can be seen in [1].

Let $\mathbf{y} = y_1, \ldots, y_n$ be the set of variables and $\mathbf{u} = u_1, \ldots, u_m$ the set of parameters. Given a generating set $F = \{f_1, \cdots, f_s\} \subset \mathbb{Q}[\mathbf{u}][\mathbf{y}]$ of the parametric ideal $I = \langle F \rangle$ and a monomial order $\succ_{\mathbf{y}}$ in the variables, the GROBCOV algorithm determines

- the unique *canonical partition* of the parameter space $\mathbb{C}^m$ into locally closed sets (*segments*) with associated *generalized reduced Gröbner basis*:

$$GC = \{(S_1, B_1, \mathrm{lpp}_1), \ldots, (S_r, B_r, \mathrm{lpp}_r)\}.$$

- The segments S_i are disjoint locally closed subsets of $\mathbb{C}^m$ and $\oplus_i S_i = \mathbb{C}^m$.
- The basis B_i of a segment S_i has *fixed set of leading power products* (lpp), who ensures that the type of solutions is the same over all points of the segment, and is the *generalized reduced Gröbner basis* of $\langle F \rangle$ over the segment S_i.
- The lpp's are included in the output, even if they can be seen on the basis, to characterize the segments and facilitate the applications.
- Moreover, if the ideal is homogeneous, the lpp's are characteristic of the segment, no other segment having the same lpp's.

The generalized reduced Gröbner basis B_i of a segment S_i is formed by a set of monic I-regular functions over S_i. An I-regular function, representing an element of the basis, allows a full-representation in terms of a set of polynomials that specialize for every point $\mathbf{u}_0$ of the segment, either to the corresponding element of the reduced Gröbner basis of the specialized ideal $I_{\mathbf{u}_0}$ after normalization, or to zero. It also allows a generic representation given by a single polynomial that specializes well on an open subset of the segment and to zero on the remaining points of it. Usually it is sufficient with the generic representation, and we can, if needed, compute the full representation from it using the EXTEND algorithm.

The segments S_i are given in canonical P-representation, given by a set of prime ideals of the form

$$\text{Prep}(S) = \{\{\mathfrak{p}_i, \{\mathfrak{p}_{ij} : 1 \leq j \leq r_i\}\} : 1 \leq i \leq s\}$$

representing the set:

$$S = \bigcup_{i=1}^{s} \left(\mathbb{V}(\mathfrak{p}_i) \setminus \bigcup_{j=1}^{r_i} \mathbb{V}(\mathfrak{p}_{ij}) \right).$$

Each $\mathbb{V}(\mathfrak{p}_i) \setminus \bigcup_{j=1}^{r_i} \mathbb{V}(\mathfrak{p}_{ij})$ is a *component* of the segment, and its representative $\{\mathfrak{p}_i, \{\mathfrak{p}_{ij} : 1 \leq j \leq r_i\}\}$, by abuse of language, is also denoted a component when there is no ambiguity. $\mathfrak{p}_i$ is called the *top* of the component, and $\{\mathfrak{p}_{ij} : 1 \leq j \leq r_i\}$ the *holes*.

3   Locus Taxonomy

A geometric locus is a set of points satisfying some conditions. Locus computation is an important issue in Dynamic Geometry, where the term locus generally refers to loci of the following kind: determine the trajectory determined by the different positions of a point P (tracer), corresponding to the different positions of a second point M (mover) along the path where it is constrained by the construction. Nevertheless, the actual existing DG software do not provide correct algebraic solutions.

Using the GRÖBNER COVER we are able to give a precise algebraic answer. We shall consider more general locus problems in the plane with a tracer point $P(u_1, u_2)$, whose coordinates $\mathbf{u} = (u_1, u_2)$ are considered as parameters and the remaining coordinates, distances, etc. $\mathbf{y} = y_1, \ldots, y_n$ of the construction as variables. The locus problem will give rise to a system $F \subset \mathbb{Q}[\mathbf{u}][\mathbf{y}]$. The locus determination consists now in obtaining the conditions over the parameters $\mathbf{u}$ for which there are solutions for the $\mathbf{y}$.

In the GRÖBNER COVER the values of the parameters and variables are considered over $\mathbb{C}$. Thus we can provide only locus solutions over the complex $\mathbb{C}$, whereas DG is interested in the real projection, who is not always obvious. Moreover, we consider only problems that can be be formulated exactly in terms of equations with coefficients in $\mathbb{Q}$. We also restrict the study to plane loci problems, even if it can be generalized to higher dimensional spaces. Let

$$\mathbb{V}(F) = \{(\mathbf{u}, \mathbf{y}) \subset \mathbb{C}^{2+n} : \forall f \in F, f(\mathbf{u}, \mathbf{y}) = 0\}$$

be the set of solutions of F. Denote π_1 and π_2 the projections onto the parameter and variable spaces:

$$\pi_1 : \mathbb{C}^{2+n} \longrightarrow \mathbb{C}^2 \qquad\qquad \pi_2 : \mathbb{C}^{2+n} \longrightarrow \mathbb{C}^n$$
$$(\mathbf{u}, \mathbf{y}) \mapsto \mathbf{u} \qquad\qquad\qquad (\mathbf{u}, \mathbf{y}) \mapsto \mathbf{y}$$

Table 1. Locus algorithm

$L \leftarrow \mathbf{Locus}(G)$
Input: $G = \{(S_i, B_i, \mathrm{lpp}_i) : i \leq i \leq s\}$ the Gröbner cover of an ideal in $\mathbb{Q}[\mathbf{u}][\mathbf{y}]$,
   where $S_i = \cup_j C_{ij}$ and $C_{ij} = \{(\mathfrak{p}_{ij}, \{\mathfrak{p}_{ijk} : 1 \leq k \leq r_{ij}\}) : 1 \leq j \leq r_i\}$.
Output: L, the components of the P-representation of the locus
   $L = \{\{\mathfrak{q}_i, \{\mathfrak{q}_{ij} : 1 \leq j \leq s_i\}, \mathrm{type}_i\} : 1 \leq i \leq s\}$
begin
  $C_1 =$ Select the segments of G with $\dim(\mathrm{lpp}_i) = 0$ # normal-segments
  $C_1 =$ Specialize the basis on every component of C_1 and mark the component
    'Normal' if the basis continues to depend on the $\mathbf{u}$'s and 'Special' if not
  $C_2 =$ Select all the components of the segments of G with $\dim(\mathrm{lpp}_i) > 0$
    # non-normal segments
  $L_1 = \mathbf{LCUnion}(C_1)$,   marking the components of L_1 as
    'Normal' or 'Special' inheriting the character of the full
  $L_2 = \mathbf{LCUnion}(C_2)$;
  Mark the components of L_2 of $\dim(C) = 0$ and $\dim(C) > 0$ respectively as
    'Accumulation' and 'Degenerate' components
  $L = L_1 \cup L_2$
end

The taxonomy that we give is motivated by the interpretation of the solutions in a lot of loci problems of different kind (see [1]):

Generic Locus L. associated to the parametric polynomial system $F(\mathbf{u}, \mathbf{y})$ is the set $L = \pi_1(\mathbb{V}(F)) \subset \mathbb{C}^2$,(i.e. the set of values of the coordinates of the tracer for which there exist solutions).

Taxonomy
- **Normal locus:** are the points $\mathbf{u} \in \mathbb{C}^2$ of the locus L for which $\dim(\pi_2(\mathbb{V}(F) \cap \pi_1^{-1}(\mathbf{u}))) = 0$ (i.e. the set of points in the parameter space that correspond to a single (or a finite number of) positions of the variables).
  - **Normal components:** A component C_s of the normal locus is *normal* if $\dim(C_s) = \dim(\pi_2(\mathbb{V}(F) \cap \pi_1^{-1}(C_s)))$ (i.e. the components of the normal locus whose different points are generated by different points of the variables).
  - **Special components:** A component C_s of the normal locus is *special* if $\dim(C_s) > 0$ and $\dim(\pi_2(\mathbb{V}(F) \cap \pi_1^{-1}(C_s))) = 0$ (i.e. the components of the normal locus of dimension 1 that are generated by a single (or a finite number of) points of the variables).
- **Non-normal locus:** are the points $\mathbf{u} \in \mathbb{C}^2$ of the locus L for which $\dim(\pi_2(\mathbb{V}(F) \cap \pi_1^{-1}(\mathbf{u}))) > 0$ (i.e. the set of points in the parameter space that correspond to infinite positions of the variables).
  - **Degenerate components:** are the components C_d of the non-normal locus with $\dim(C_d) > 0$ (i.e. the components of the non-normal locus of dimension 1).

- **Accumulation components (points):** are the components C_a of the non-normal locus with $\dim(C_a) = 0$ (i.e. the zero dimensional components of the non-normal locus).

Problems in section 4 are chosen to justify the taxonomy. The geometric relevance of this algebraic classification of the different components of a locus is open to interpretation by the user. Dynamic Geometry systems could present the collection of different parts (with the corresponding typology) of the computed locus, letting to the user the decision of which pieces are to be considered or discarded in a particular context.

Based on our experience (see section 4), we tend to discard the degenerate components as geometrically irrelevant, as they usually correspond to degenerate instances of a construction, such as two coincident vertices in a triangle. However, we consider the accumulation points as forming part of the (geometric) locus, since they represent special points that are determined by infinitely many values of the variables. The special components are generally also discarded, as they are generated by a single position of the mover, but sometimes they can be useful.

We designed the LOCUS algorithm (Table 1) that takes the output of the GROBCOV and classifies the appropriated components of the segments following the defined taxonomy. We show now some examples justifying our taxonomy and its functionality.

4   Applications and Functionality

4.1   Pascal's Limaçon

The following Problem is considered here to justify the definitions of "Normal" and "Special" components of the "Normal Locus". Consider the following locus problem. Let O be a fixed point on a circle c of radius r, and l be a line passing

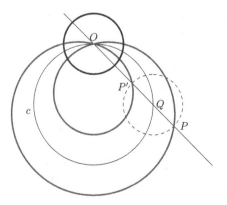

Fig. 1. Pascal limaçon

through $O(0, r)$ and $Q(y_1, y_2)$, a general point on c. Let $P(u_1, u_2)$ be a point on l such that $distance(P, Q) = k$, where k is a constant. The limaçon of Pascal is the locus set traced by P as Q moves along c, as shown on Figure 1. Setting $r = 2$ and $k = 1$, the ideal (set of equations) determining the locus is:

$$F = \langle y_1^2 + y_2^2 - 4, \; (u_1 - y_1)^2 + (u_2 - y_2)^2 - 1, \; (2 - y_2)u_1 + y_1(u_2 - 2) \rangle$$

Computing the solution using grobcov and locus algorithms in Singular we do the following:

Input:

```
LIB "grobcov.lib";
ring R=(0,u1,u2),(y1,y2),lp;
short=0;
ideal F= y1^2+y2^2-4, (u1-y1)^2+(u2-y2)^2-1, (2-y2)*u1+y1*(u2-2);
def L=locus(grobcov(F));
"locus(grobcov(F))="; L;
```

Output:

```
locus(grobcov(F))=
[1]:
   [1]:
      _[1]=(u1^4+2*u1^2*u2^2-9*u1^2+u2^4-9*u2^2+4*u2+12)
   [2]:
      [1]:
         _[1]=1
   [3]:
      Normal
[2]:
   [1]:
      _[1]=(u1^2+u2^2-4*u2+3)
   [2]:
      [1]:
         _[1]=1
   [3]:
      Special
```

LOCUS algorithm determines and characterizes two components: a "Normal" component that is the Pascal limaçon's concoid, and a "Special" circle generated by the single mover point O, for which the construction is degenerate. Usually these Special components are to be discarded by Dynamical Geometry software, and the algorithm returns it as "Special" letting to the user the decision about its consideration.

4.2   Offset of a Circle

Now we want to justify the definition of "Accumulation" and "Degenerate" components of a locus considering the locus of the offset of a circle of radius 1 at distance 1. The given circle has as equation $g : y_1^2 + y_2^2 - 1$. The family of circles who generate the envelope is $F = (u_1 - y_1)^2 + (u_2 - y_2)^2 - 1$, where (u_1, u_2) is some point of the envelope. To compute the envelope we have to add the equation $\dfrac{\partial F}{\partial y_1}\dfrac{\partial g}{\partial y_2} - \dfrac{\partial F}{\partial y_2}\dfrac{\partial g}{\partial y_1}$, ensuring that the envelope is tangent in each point to a curve of the family. We have to consider thus the following ideal:

$$H = \langle y_1^2 + y_2^2 - 1, (u_1 - y_1)^2 + (u_2 - y_2)^2 - 1, y_1 u_2 - y_2 u_1 \rangle.$$

The standard method will eliminate (y_1, y_2) to obtain the envelope. But we can consider (y_1, y_2) as the mover and take (u_1, u_2) as the tracer using the ring R=(0,u1,u2),(y1,y2),lp; and the command locus(grobcov(H)).

Doing so we also obtain two components: The "Normal" component consisting of the circle $u_1^2 + u_2^2 - 4$, plus the "Accumulation" point component (u_1, u_2) at the center generated by all the circles. The accumulation points are to be considered as part of the offset by Dynamical Geometry software,.

4.3   Improvements: Detecting *Bad* Mover Positions

Locus algorithm as described in Table 1 is incomplete, and must be improved. It assumes that the generic segment of the Gröbner Cover has basis $\{1\}$, as we do not expect a locus that contents the whole plane except some curves. Nevertheless, in the next example the generic segment does not have basis $\{1\}$. The reason is that there is a point of the mover for which the construction is degenerate, and would give rise to a bidimensional solution. It is necessary to eliminate this point of the mover to obtain the correct expected locus.

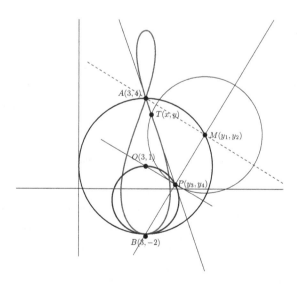

Fig. 2. Locus described by T (and P) as M runs along its circle

We programmed our algorithm to be able to detect and eliminate such bad mover point positions producing degenerate solutions in the locus.

Consider the following locus construction (see Figure 2). The mover $M(y_1, y_2)$ runs over the circle with center at $O(3, 1)$ and radius OA, where $A = (3, 4)$. We consider the line parallel to the line AM passing through O and the line perpendicular to it passing through the point $B = (3, -2)$. Both lines intersect at point $P(y_3, y_4)$. Consider the line AP and the circle with center M and radius MP. We consider this intersection as the tracer point(s): $T(x, y)$ and $T' = P$.

The polynomial system describing the problem is the ideal F given by

$$F = \langle (y_1 - 3)^2 + (y_2 - 1)^2 - 9, \ (4 - y_2)(y_3 - 3) + (y_1 - 3)(y_4 - 1),$$
$$(y_1 - 3)(x_1 - y_1) - (4 - y_2)(y_4 - y_2),$$
$$(y_4 - 4)x - (y_3 - 3)y + 4y_3 - 3y_4,$$
$$(x - y_1)^2 + (y - y_2)^2 - (y_1 - y_3)^2 - (y_2 - y_4)^2 \rangle$$

For the computation we use the ring `ing R=(0,x,y),(x1,x2,y1,y2),1p;`. With this improvement we obtain the proper two irreducible "Normal" components

$$\mathbb{V}(x^2 - 6x + y^2 + y + 7),$$
$$\mathbb{V}(x^4 - 12x^3 + 2x^2y^2 - 13x^2y + 236x^2 - 12xy^2 + 78xy - 1200x + y^4$$
$$-13y^3 + 60y^2 - 85y + 1495).$$

The output produced by the algorithm includes a message informing of the removal of *bad* mover positions (point $A(3, 4)$ in our case).

Geometrically, the problem is that when the mover is on the point ($y_1 = 3, y_2 = 4$), the line AM is not defined. The algorithm eliminates the segments of the Gröbner Cover containing this point that is detected on the generic segment.

This locus is example 9 in our prototype [3]. By clicking the *Find locus* button, we obtain the locus description shown on Figure 2.

References

1. Abanades, M., Botana, F., Montes, A., Recio, T.: An Algebraic Taxonomy for Locus Computation in Dynamic Geometry. Computer Aided Design (to appear, 2014)
2. Kapur, D., Sun, Y., Wang, D.: A new algorithm for computing comprehensive Gröbner systems. In: Proceedings of ISSAC 2010, pp. 29–36. ACM Press, New York (2010)
3. Botana, F.: Locus Prototype (2012), http://webs.uvigo.es/fbotana/LocusGC/
4. Montes, A.: Using Kapur-Sun-Wang algorithm for the Gröbner Cover. In: Sendra, R., Villarino, C. (eds.) Proceedings of EACA 2012, pp. 135–138. Universidad de Alcalá de Henares (2012)
5. Montes, A., Wibmer, M.: Gröbner bases for polynomial systems with parameters. Jour. Symb. Comp. 45, 1391–1425 (2010)
6. Wibmer, M.: Gröbner bases for families of affine or projective schemes. Jour. Symb. Comp. 42, 803–834 (2007)
7. Singular, http://www.singular.uni-kl.de/ (last accessed February 2014)

Using MAPLE's RegularChains Library to Automatically Classify Plane Geometric Loci

Francisco Botana[1] and Tomás Recio[2]

[1] Universidad de Vigo, Spain
fbotana@uvigo.es
http://webs.uvigo.es/fbotana/
[2] Universidad de Cantabria, Spain
tomas.recio@unican.es
http://www.recio.tk

Abstract. We report a preliminary discussion on the usability of the RegularChains library of MAPLE for the automatic computation of plane geometric loci and envelopes in graphical interactive environments. We describe a simple implementation of a recently proposed taxonomy of algebraic loci, and its extension to envelopes of families of curves is also discussed. Furthermore, we sketch how currently unsolvable problems in interactive environments can be approached by using the RegularChains library.

Keywords: parametrical systems solving, constructible sets, dynamic geometry.

1 Introduction

A Dynamic Geometry System (DGS) is a computer program that allows an accurate on–screen drawing of geometric diagrams and their interactive manipulation by mouse dragging or similar device. A key issue of these systems is their ability to display the trajectory of a point that depends on another one bound to a linear path, that is, a geometric locus. Traditionaly, DGS strategy to display loci consists of sampling the linear path and, for each sample, plotting the corresponding position of the locus point. Some ad–hoc heuristics are then applied to join contiguous points, ending with a visually continuous locus.

A locus can be seen as the projection on the space of its coordinates of the surface defined by the problem constraints, and Gröbner based elimination was proposed as a technique to find algebraic knowledge about loci [1]. Nevertheless, since Zariski closures are obtained as result, loci equations can include spurious points, and complete 1–dimensional objects can even appear due to construction degeneracies. A finer analysis of loci problems can be done through a recently proposed taxonomy [2], and it has been implemented using the GröbnerCover algorithm [3]. Here we describe an alternative implementation of the taxonomy using MAPLE's RegularChains library. The results seem to be competitive when dealing with the above class of loci. Despite license problems can emerge when

H. Hong and C. Yap (Eds.): ICMS 2014, LNCS 8592, pp. 500–503, 2014.
© Springer-Verlag Berlin Heidelberg 2014

linking MAPLE with widespread free DGSs as GeoGebra, the special attention given by RegularChains to constructible sets justifies considering this library. Furthermore, other subpackages will allow extending dynamic geometry issues to real geometry.

2   Loci as Parametric Problems

We consider a locus problem where all constraints can be described as polynomials. Let (x, y) the coordinates of a generic locus point, and $x_1, \ldots, x_n$ the remaining variables. So, the problem can be seen as a parametric polynomial system $F \subset \mathbb{Q}[x, y][x_1, \ldots, x_n]$, where x, y are parameters and $x_1, \ldots, x_n$ variables. See [4] for an precise setting description and a full account on the taxonomy. Roughly speaking, the taxonomy classifies locus points as

- normal: if for these parameter values the system has finite number of solutions;
- non-normal: if the system has infinite solutions for these parameter values.

Thus, in order to perform this classification we can count the number of solutions of F in RegularChains by using the command ComplexRootClassification. For the sake of illustration, we consider the limaçon of Pascal, an algebraic curve obtained as a locus as follows. Let $Q(x_1, x_2)$ be a point on the circle centered at the origin and with radius 2, $O(0, 2)$ and $P(x, y)$ a point at distance 1 from Q and lying on the line $\overline{OQ}$. As Q glides on the circle, P describes the limaçon. The polynomial system is

$$F = \{x_1^2 + x_2^2 - 4, (y-2)x_1 - x_2 x + 2x, x_1^2 + x_2^2 + (-2x)x_1 + (-2y)x_2 + x^2 + y^2 - 1\},$$

and the MAPLE code for classifying the locus points,

```
with(RegularChains); with(ConstructibleSetTools):
with(ParametricSystemTools); R := PolynomialRing([x1, x2, x, y]):
F := [x1^2+x2^2-4,(y-2)*x1-x2*x+2*x,..
..x1^2+x2^2-2*x*x1-2*y*x2+x^2+y^2-1]:
crc := ComplexRootClassification(F, 2, R):
map(x -> [Info(x[1],R),x[2]],crc);
```

returns three constructible sets where the system has exactly one complex solution,

$$[[x^4 + (2y^2 - 9)x^2 + y^4 - 9y^2 + 4y + 12], [y - 2, x^2 + y^2 - 4y + 4, 2y - 3]],$$

$$[[4x^2 - 15, 2y - 3], [1]], [[x^2 + y^2 - 4y + 3], [2y - 3]],$$

and two other constructible sets determining two system solutions

$$[[x, y - 2], [1]], [[4x^2 - 3, 2y - 3], [1]].$$

Therefore, all locus points are normal. In order to further analyze these points (recall the taxonomy [4]), we count the number of solutions of the original system plus each constructible set, once inverted the roles of parameters and variables. If we get infinite solutions for a constructible set, we declare it as a special locus part. In the case we are dealing with, that happens for the third constructible set, $[[x^2 + y^2 - 4y + 3], [2y - 3]]$:

```
R := PolynomialRing([x, y, x1, x2]):
F := [x1^2+x2^2-4,(y-2)*x1-x2*x+2*x,..
..x1^2+x2^2-2*x*x1-2*y*x2+x^2+y^2-1,x^2+y^2-4*y+3]:
H := [2*y-3]:
crc := ComplexRootClassification(F, H, 2, R):
map(x -> [Info(x[1],R),x[2]],crc);
```

returns

$$[[[[x_1, x_2 - 2], [1]], \infty]],$$

meaning that the circle centered at O must be labeled as a special locus part. Note that, although algebraically pertinent, this circle comes from a degeneracy in the construction, since it stems from the coincidence of points O and Q.

3   Envelopes as Loci

Envelope computation can be also seen as solving a parametric system. As in loci, elimination can include spurious factors. Thus, applying the taxonomy can drive to improvements when automatically computing envelopes in a DGS. Envelope points are classified as normal or non–normal points as in loci. Consider, for instance, the envelope of horizontal lines through a point on the unit circle. Using the above MAPLE commands we obtain

```
R := PolynomialRing([x, y, x1, x2]):
F := [x1^2+x2^2-1, y-x2, x1]:
crc := ComplexRootClassification(F, 2, R):
map(x -> [Info(x[1],R),x[2]],crc);
```

$$[[[[y + 1], [1]], [[y - 1], [1]], 1]],$$

that is, the envelope is $y = \pm 1$, where all points are normal. Nevertheless, studying the number of solutions of the system, interchanging roles of parameter and variables, we get for the first constructible set

```
R := PolynomialRing([x, y, x1, x2]):
F := [x1^2+x2^2-1, y-x2, x1,y+1]:
H := [1]:
crc := ComplexRootClassification(F, H, 2, R):
map(x -> [Info(x[1],R),x[2]],crc);
```

$$[[[[x_1, x_2 + 1], [1]], \infty]].$$

Thus, strictly following the taxonomy, the line $y = -1$ should be declared as special, while it is an ordinary part of the envelope. On the contrary, the envelope of lines passing through $A(0, 1)$ and a point gliding along the unit circle exactly consists of A, whereas MAPLE finds a constructible set where all points are normal, $[[y - 1], [x]]$, and another one with a non–normal point, $[[x, y - 1], [1]]$, that is, the point A. Note that the first constructible set should be removed from the envelope result (it comes from a construction degeneration) while the taxonomy would label it as special. Currently, we do not know how to automatically distinguish the two cases.

4   Further Work

Many dynamic geometry constructions can only be expressed in a essential semi-algebraic way. Each time a segment is used in a locus construction, for example, it is highly probable that the above computations include extra parts, since traditionally DGSs using these methods replace the segment by the whole line. Also, envelopes are defined in $\mathbb{R}$, while we work in $\mathbb{C}$. The subpackage SemiAlge-braicSetTools will allow to extend the class of dynamic geometry problems able to be automatically solved. As a simple illustration, consider two circles each one with a point gliding along it. The locus of their midpoint is a 2–dimensional part of the plane, currently only descriptible in any DGS by displaying a more or less accurate screen. There are techniques and algorithms capable of giving more precise answers. Our future work will study such problems in the context of dynamic geometry.

Acknowledgment. Both authors were partially supported by the Spanish Ministerio de Economía y Competitividad and by the European Regional Development Fund (ERDF), under the Project MTM2011–25816–C02–02.

References

1. Botana, F., Valcarce, J.: A Software Tool for the Investigation of Plane. Loci. Math. Comput. Simul. 61, 139–152 (2003)
2. Abánades, M., Botana, F., Montes, A., Recio, T.: An Algebraic Taxonomy for Loci Computation in Dynamic Geometry. Comput. Aided Des. (submitted, 2014)
3. Montes, A., Wibmer, M.: Gröbner Bases for Polynomial Systems with Parameters. J. Symb. Comput. 45, 1391–1425 (2010)
4. Abánades, M., Botana, F., Montes, A., Recio, T.: Software Using the Gröbner Cover for Geometrical Loci Computation and Classification. In: Hong, H., Yap, C. (eds.) ICMS 2014. LNCS, vol. 8592, Springer, Heidelberg (2014)

Solving Parametric Polynomial Systems
by RealComprehensiveTriangularize

Changbo Chen[1] and Marc Moreno Maza[2]

[1] Chongqing Key Laboratory of Automated Reasoning and Cognition,
Chongqing Institute of Green and Intelligent Technology,
Chinese Academy of Sciences, China
changbo.chen@hotmail.com
http://www.orcca.on.ca/~cchen
[2] ORCCA, University of Western Ontario, Canada
moreno@csd.uwo.ca
www.csd.uwo.ca/~moreno

Abstract. In the authors' previous work, the concept of comprehensive triangular decomposition of parametric semi-algebraic systems (RCTD for short) was introduced. For a given parametric semi-algebraic system, say S, an RCTD partitions the parametric space into disjoint semi-algebraic sets, above each of which the real solutions of S are described by a finite family of triangular systems. Such a decomposition permits to easily count the number of distinct real solutions depending on different parameter values as well as to conveniently describe the real solutions as continuous functions of the parameters. In this paper, we present the implementation of RCTD in the **RegularChains** library, namely the RealComprehensiveTriangularize command. The use of RCTD is illustrated by the stability analysis of several biological systems.

Keywords: Parametric polynomial system, real comprehensive triangular decomposition, **RegularChains**.

1 Introduction

Parametric polynomial systems arise naturally in many applications. For this reason, the computer algebra community has contributed many techniques to deal with such systems, by means of various tools such as cylindrical algebraic decomposition [9], quantifier elimination [2,1], comprehensive Gröbner bases [14], discriminant varieties [11] or border polynomials [15], as well as comprehensive triangular decomposition (CTD).

The concept of comprehensive triangular decomposition was introduced in [5] in order to study the specialization properties of regular chains. As a byproduct, one can determine for which parameter values, a parametric polynomial system has complex solutions, or, in general, compute the projection of a constructible set, or do quantifier elimination over an algebraically closed field. In [3], different variants of CTDs were proposed, including disjoint squarefree CTD and real

H. Hong and C. Yap (Eds.): ICMS 2014, LNCS 8592, pp. 504–511, 2014.
© Springer-Verlag Berlin Heidelberg 2014

CTD (RCTD). The former is used to count and describe distinct complex solutions depending on parameters, whereas the latter is used to count and describe distinct real solutions depending on parameters.

Informally speaking, an RCTD of a parametric semi-algebraic system S is a partition of the parametric space into disjoint semi-algebraic sets together with a family of triangular systems, such that above each cell in the partition, the real solutions of S are disjoint (if finitely many) and are described by a sub-family of the triangular systems. RCTD is implemented as RealComprehensiveTriangularize in the RegularChains library.

In Section 2, we illustrate the functionality of RCTD through the analysis of the stability of a simple dynamical system. In Section 3, we present some more advanced applications of RCTD. In Sections 4 and 5, we briefly explain the theory behind RCTD and list the related RegularChains commands.

2   Functionality

In the field of biology, a very important problem is to study the stability of the equilibria (or steady states) of a biological system. For a biological system, say BS, modeled by a system of autonomous differential equations, say

$$DS : \frac{d\mathbf{x}(t)}{dt} = F(\mathbf{u}, \mathbf{x}),$$

the equilibria of BS (or DS) are defined as the real zeros of $F(\mathbf{u}, \mathbf{x})$. Here $\mathbf{u}$ denotes parameters.

Assuming that F is a vector of rational functions in $\mathbb{Q}[\mathbf{u}, \mathbf{x}]$, the study of the equilibria of BS can often be reduced to solving a parametric semi-algebraic system. Taking also into consideration the fact that certain degenerated behaviors have no practical interest, in [6,3], the authors introduced the concept of *real comprehensive triangular decomposition* (RCTD). Broadly speaking, for a parametric semi-algebraic system S, a RCTD is given by

(*a*) a partition of the whole parameter space such that above each cell
    (*i*) either the corresponding constructible system induced by S has infinitely
         many complex solutions,
    (*ii*) or S has no real solutions,
    (*iii*) or S has finitely many real solutions which are continuous functions of
         the parameters and with disjoint graphs.
(*b*) in Case (*iii*), a description of the solutions of S as functions of the parameters
    by means of triangular systems.

RealComprehensiveTriangularize is the RegularChains command computing RCTDs. We apply it to the stability analysis of the dynamical system below, an instance of the *multiple switch model* proposed in [8] by Cinquin and Demongeot:

$$\frac{dx}{dt} = -x + \frac{s}{1 + y^2}$$

$$\frac{dy}{dt} = -y + \frac{s}{1 + x^2},$$

where x, y denote concentrations of two proteins and s denotes the strength of unprocessed protein expression. One wishes to determine the values of the parameter s for which this system is bistable, that is, those parameter values for which the system has two asymptotically stable equilibria. Figure 1 shows how to set up the problem while Figure 2 illustrates the stability analysis conducted through RCTD computation. On this latter, one can see that the biological system is bistable if and only if $s > 2$ holds. Moreover, in this case, the concentrations x and y are expressed as functions of s. The following plots illustrate the bi-stability.

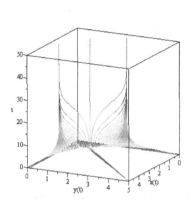
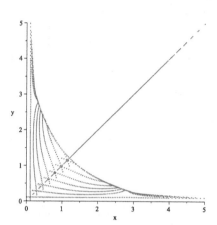

Let us explain now the usage of RealComprehensiveTriangularize as shown in Figure 2. The function takes three arguments: (1) a semi-algebraic system, (2) the number of its parameters and (3) a polynomial ring for which the list of its variables are sorted in descending order. For this example, the second argument "1" means that the last variable, that is s, is the only parameter. The output consists of (i) two squarefree semi-algebraic systems in a triangular shape and, (ii) three semi-algebraic sets forming the partition of the parameter space. In this output, the list of indices following a semi-algebraic set C specifies the triangular systems describing the solutions above C. If this list of indices is empty, then the input parametric system has no real solutions above C.

After an RCTD is computed, the different numbers of real solutions (depending on the parameter values) are stored in a data-structure. To know the parameter values corresponding to a particular number of real solutions, one can call RealComprehensiveTriangularize with the computed RCTD as input, as well as the polynomial ring and the prescribed number of real solutions.

The real solutions are encoded by squarefree semi-algebraic systems. As we can see on Figure 2, such system consists of a set of equations in a triangular shape and a set of positive inequalities. Moreover, the (real) zero set of any squarefree semi-algebraic system is non-empty and, its dimension is the same as the number of parameters.

The biological system is described by the following system of differential equations. Its right hand side encodes the equilibria:

```
> ode := {diff(x(t),t) = -x(t)+s/(1+y(t)^2), diff(y(t),t)=-y(t)+s/(1+x(t)^2)}:
  F := [-x+s/(1+y^2), -y+s/(1+x^2)]:
```

The following two Hurwitz determinants determine the stability of the hyperbolic equilibria:

```
> D1 := -(diff(F[1],x)+diff(F[2],y)): #D1 is 2
  D2 := diff(F[1],x)*diff(F[2],y)-diff(F[1],y)*diff(F[2],x):
```

The semi-algebraic system below encodes the asymtotically stable hyperbolic equilibria:

```
> P := [numer(normal(F[1]))=0, numer(normal(F[2]))=0, x>0, y>0, s>0, numer(D2)>0];
```
$$P := \left[-y^2 x - x + s = 0, \ -y x^2 - y + s = 0, \ 0 < x, \ 0 < y, \ 0 < s, \ 0 < 1 + 2 x^2 + x^4 + 2 y^2 + 4 y^2 x^2 + 2 y^2 x^4 + y^4 \right.$$
$$\left. + 2 y^4 x^2 + y^4 x^4 - 4 y x s^2 \right]$$

Fig. 1. Study of the stability of equilibria of a biological system: problem set-up

3   Application

In this section, we show how RealComprehensiveTriangularize is used to analyze the stability of a real life biological system from [13], which was also solved by other parametric solving tools in [12].

The system models the antagonistic interactions between cyclin-dependent kinases and the anaphase promoting complex. It is described by the following dynamical system.

$$\frac{dx}{dt} = k_1 - (k_2' + k_2'' y)x$$
$$\frac{dy}{dt} = \frac{(k_3' + k_3'' A)(1-y)}{J_3 + 1 - y} - \frac{k_4 m x y}{J_4 + y},$$

where x and y denotes the concentrations of cyclin B/Cdk dimers and active Cdh1/APC complexes, $k_1, k_2', k_2'', k_3', k_3'', k_4$ are rate constants, J_3, J_4 are Michaelis constants, and m is a real parameter representing cell "mass". Figure 3 shows how to compute conditions on m such that the system is bistable by RealComprehensiveTriangularize. The whole computational process is similar to that of Section 2. Isolating the real roots of the univariate polynomial in the output, we obtain three real roots: $0.1097139798, 0.5273193027, 1.132028425$. So the system is bistable if and only if $0.1097139798 < m < 0.5273193027$ or $m > 1.132028425$.

4   Related Notions and Commands

The notion of a *regular chain*, introduced independently in [10] and [16], is closely related to that of a triangular decomposition of a polynomial system. Broadly speaking, a *triangular decomposition*[1] of a polynomial system S is a set of simpler (in a precise sense) polynomial systems $S_1, \ldots, S_e$ such that

---

[1] http://en.wikipedia.org/wiki/Triangular_decomposition

$$p \text{ is a solution of } S \Leftrightarrow \exists i : p \text{ is a solution of } S_i. \tag{1}$$

When the purpose is to describe all the solutions of S, whether their coordinates are real numbers or not, (in which case S is said to be *algebraic*) those simpler systems are required to be regular chains[2]. If the coefficients of S are real numbers and if only the real solutions are required (in which case S is said to be *semi-algebraic*), then those real solutions can be obtained by a triangular decomposition into so-called *regular semi-algebraic systems*, a notion introduced in [4]. In both cases, each of these simpler systems has a triangular shape and remarkable properties, which justifies the terminology. We refer to [7] for a formal presentation on the concepts of a regular chain and a triangular decomposition of a polynomial system

Compute a real comprhensive triangular decomposition of P w.r.t. the parameter s:

```
> R := PolynomialRing([y, x, s]): ctd := RealComprehensiveTriangularize(P, 1, R);
ctd := [[[1, squarefree_semi_algebraic_system], [2, squarefree_semi_algebraic_system]], [[semi_algebraic_set,
    []], [semi_algebraic_set, [1]], [semi_algebraic_set, [2]]]]
```

Derive the values of *s* such that P has 2 positive real solutions, that is the biological system is bistable:

```
> ctd2 := RealComprehensiveTriangularize(ctd, R, 2);Display(ctd2[2][1][1],R); Display(ctd2[1][1]
   [2],R);
```

$$ctd2 := [[[1, \textit{squarefree\_semi\_algebraic\_system}]], [[\textit{semi\_algebraic\_set}, [1]]]]$$

$$[2 < s]$$

$$x\,y - 1 = 0$$

$$x^2 - s\,x + 1 = 0$$

$$y > 0$$

$$x > 0$$

$$8\,x\,s^3 - 6\,x\,s^5 - 4\,s^2 + 5\,s^4 - s^6 + x\,s^7 > 0$$

Fig. 2. Study of the stability of equilibria of biological system: solution with RealComprehensiveTriangularize

Consider a multivariate semi-algebraic system

$$f_1 = 0, \ldots, f_m = 0, h_1 \neq 0, \ldots, h_t \neq 0, p_1 > 0, \ldots, p_s > 0, \tag{2}$$

with $f_1, \ldots, f_m, h_1, \ldots, h_t, p_1, \ldots, p_s \in \mathbb{Q}[u_1, \ldots, u_d, y_1, \ldots, y_\ell]$. where $u_1, \ldots, u_d$ are regarded as parameters and $y_1, \ldots, y_\ell$ are regarded as unknowns. Various ways of solving System (2) are available and implemented in the **RegularChains** library.

One may want to express $y_1, \ldots, y_\ell$ as functions of $u_1, \ldots, u_d$. This is essentially done by the command **RealComprehensiveTriangularize** which was presented in the previous sections.

---

[2] More generally, a triangular decomposition into regular chains of a polynomial system S with coefficients in an arbitrary field $\mathbb{K}$ describes the solutions of S whose coordinates are in the algebraic closure of $\mathbb{K}$.

```
> F := [k1-(k21+k22*y)*x, (k31+k32*A)*(1-y)/(J3+1-y)-k4*m*x*y/(J4+y)]:
  F := normal(subs({k1=1/25, k21=1/25, k22=1, k31=1, k32=10, k4=35, J3=1/25, J4=
  1/25,A=0}, F));
```

$$F := \left[\frac{1}{25} - xy - \frac{1}{25}x, \ -\frac{25\left(875\,xy^2\,m - 910\,xy\,m - 25\,y^2 + 24\,y + 1\right)}{(25\,y - 26)\,(25\,y + 1)} \right]$$

```
> D1 := numer(normal(-(diff(F[1],x)+diff(F[2],y)))):
  D2 := numer(normal(diff(F[1],x)*diff(F[2],y)-diff(F[1],y)*diff(F[2],x))):
```

```
> P := [numer(F[1])=0, numer(F[2])=0, x>0, y>0, m>0, 25*y-26<>0, D1>0, D2>0]:
```

```
> R := PolynomialRing([x, y, m]):
  ctd := RealComprehensiveTriangularize(P, 1, R, 2);
```

$$ctd := [\, [\, [\, 1, \ squarefree\_semi\_algebraic\_system \,]], [\, [\, semi\_algebraic\_set, \ [\,1\,]\,]]]$$

```
> out := Info(ctd[2][1][1], R);
```

$$out := \Big[\,\big[\,\text{And}\big(\,RootOf\big(28983500\,\_Z^3 - 51273600\,\_Z^2 + 22577975\,\_Z - 1898208, \ index = real_1\big)$$

$$< m, \ m < RootOf\big(28983500\,\_Z^3 - 51273600\,\_Z^2 + 22577975\,\_Z - 1898208, \ index = real_2\big)\big)\,\big],$$

$$\big[\,RootOf\big(28983500\,\_Z^3 - 51273600\,\_Z^2 + 22577975\,\_Z - 1898208, \ index = real_3\big) < m\,\big]\,\Big]$$

Fig. 3. Determine when a biological system is bistable

Alternatively, one may want to simply determine the $(u_1, \ldots, u_d)$-values for which there exists at least one $(y_1, \ldots, y_\ell)$-value satisfying System (2). This question is often refered as *existential quantifier elimination* and can be answered by RealComprehensiveTriangularize as well. However, the commands QuantifierElimination and Projection are more specialized answers to this question. In some circumstances, they may run faster than RealComprehensive-Triangularize or provide more compact answers.

Of practical interest is a variant of the previous question, which is specified as follows. Given a non-negative integer range $[k_1, k_2]$, with $k_1 \leq k_2$, determine the $(u_1, \ldots, u_d)$-values for which there exist at least k_1 and at most k_2 $(y_1, \ldots, y_\ell)$-values satisfying System (2). This question is often refered as *real root classification* [15] and can be answered by the RealRootClassification command of the RegularChains library.

Finally, we observe that, in practice, parametric semi-algebraic systems are often given by linear polynomials, that is, multivariate polynomials of total degree 1. For this case, the LinearSolve command of the RegularChains library implements a variant of the Fourier-Motzkin Algorithm.

5   Theoretical Concepts Underlying RCTDs

This section gathers the mathematical definitions and properties underlying the notion of RCTD. Recall that $\mathbb{K}$ is a field; we denote by $\mathbf{K}$ its algebraic closure. A (squarefree) regular chain T of $\mathbb{K}[\mathbf{u}, \mathbf{y}]$ *specializes well* at a point $u \in \mathbf{K}^d$ if $T(u)$ is a (squarefree) regular chain of $\mathbf{K}[\mathbf{y}]$ and $\text{init}(T)(u) \neq 0$ holds, where $\text{init}(T)$ denotes the product of the initials of T.

For instance, the regular chain $T = \begin{cases} (s+1)z \\ (x+1)y+s \\ x^2+x+s \end{cases}$ with $s < x < y < z$

does not specialize well at $s = 0$ or $s = -1$. Indeed, we have:

$$T(0) = \begin{cases} z \\ (x+1)y \\ (x+1)x \end{cases} \quad T(1) = \begin{cases} 0 \\ (x+1)y-1 \\ x^2+x-1 \end{cases}.$$

Definition 1. *Let $F \subset \mathbb{K}[\mathbf{u}, \mathbf{y}]$. A CTD of the zero set $V(F)$ of F is given by:*

- *a finite partition $\mathcal{C}$ of the parameter space into constructible sets,*
- *above each $C \in \mathcal{C}$, there is a set of regular chains $\mathcal{T}_C$ such that:*
  - *each regular chain $T \in \mathcal{T}_C$ specializes well at any $u \in C$ and*
  - *for any $u \in C$, we have $V(F(u)) = \bigcup_{T \in \mathcal{T}_C} W(T(u))$.*

where $W(T(u))$ denotes the zeros of $T(u)$ which do not cancel $\mathrm{init}(T)(u)$.

Example 1. *A CTD of $F := \{x^2(1+y) - s, y^2(1+x) - s\}$ is as follows:*

1. $s \neq 0 \longrightarrow \{T_1, T_2\}$
2. $s = 0 \longrightarrow \{T_2, T_3\}$

where

$$T_1 = \begin{cases} x^2y+x^2-s \\ x^3+x^2-s \end{cases} \quad T_2 = \begin{cases} (x+1)y+x \\ x^2-sx-s \end{cases} \quad T_3 = \begin{cases} y+1 \\ x+1 \\ s \end{cases}.$$

Definition 2. *Let $F \subset \mathbb{K}[\mathbf{u}, \mathbf{y}]$. A DSCTD of $V(F)$ is given by :*

- *a finite partition $\mathcal{C}$ of the parameter space,*
- *each cell $C \in \mathcal{C}$ is associated with a set of squarefree regular chains $\mathcal{T}_C$ such that:*
  - *each squarefree regular chain $T \in \mathcal{T}_C$ specializes well at any $u \in C$ and*
  - *for any $u \in C$, $V(F(u)) = \biguplus_{T \in \mathcal{T}_C} W(T(u))$, where $\uplus$ denotes disjoint union.*

Example 2. *A DSCTD of $F := \{x^2(1+y) - s, y^2(1+x) - s\}$ is as follows (where T_1, T_2, T_3 are as above):*

1. $s \neq 0$, $s \neq 4/27$ and $s \neq -4 \longrightarrow \{T_1, T_2\}$
2. $s = -4 \longrightarrow \{T_1\}$
3. $s = 0 \longrightarrow \{T_3, T_4\}$
4. $s = 4/27 \longrightarrow \{T_2, T_5, T_6\}$

$$T_4 = \begin{cases} y \\ x \\ s \end{cases} \quad T_5 = \begin{cases} 3y-1 \\ 3x-1 \\ 27s-4 \end{cases} \quad T_6 = \begin{cases} 3y+2 \\ 3x+2 \\ 27s-4 \end{cases}.$$

We conclude by stating a sketch of the algorithm computing an RCTD. Let $S \subset \mathbb{Q}[\mathbf{u}][\mathbf{y}]$ be a parametric semi-algebraic system. For simplicity, we assume that S consists of equations only.

(1) Compute a DSCTD $(\mathcal{C}, (\mathcal{T}_C, C \in \mathcal{C}))$ of S.
(2) Refine each constructible set cell $C \in \mathcal{C}$ into connected semi-algebraic sets by cylindrical algebraic decomposition.
(3) For each connected cell C above which S has finitely many complex solutions: compute the number of real solutions of $T \in \mathcal{T}_C$ at a sample point u of C and remove those Ts which have no real solutions at u.

Acknowledgments. This work was supported by the NSFC (11301524) and the CSTC (cstc2013jjys0002).

References

1. Basu, S., Pollack, R., Roy, M.-F.: Algorithms in real algebraic geometry. Algorithms and Computations in Mathematics, vol. 10. Springer (2006)
2. Caviness, B., Johnson, J. (eds.): Quantifier Elimination and Cylindical Algebraic Decomposition. Texts and Mongraphs in Symbolic Computation. Springer (1998)
3. Chen, C.: Solving Polynomial Systems via Triangular Decomposition. PhD thesis, University of Western Ontario (2011)
4. Chen, C., Davenport, J.H., May, J., Maza, M.M., Xia, B., Xiao, R.: Triangular decomposition of semi-algebraic systems. J. Symb. Comp. 49, 3–26 (2013)
5. Chen, C., Golubitsky, O., Lemaire, F., Moreno Maza, M., Pan, W.: Comprehensive triangular decomposition. In: Ganzha, V.G., Mayr, E.W., Vorozhtsov, E.V. (eds.) CASC 2007. LNCS, vol. 4770, pp. 73–101. Springer, Heidelberg (2007)
6. Chen, C., Maza, M.M.: Semi-algebraic description of the equilibria of dynamical systems. In: Gerdt, V.P., Koepf, W., Mayr, E.W., Vorozhtsov, E.V. (eds.) CASC 2011. LNCS, vol. 6885, pp. 101–125. Springer, Heidelberg (2011)
7. Chen, C., Moreno Maza, M.: Algorithms for computing triangular decomposition of polynomial systems. J. Symb. Comput. 47(6), 610–642 (2012)
8. Cinquin, O., Demongeot, J.: Positive and negative feedback: striking a balance between necessary antagonists (2002)
9. Collins, G.E.: Quantifier elimination for real closed fields by cylindrical algebraic decomposition. In: Brakhage, H. (ed.) GI-Fachtagung 1975. LNCS, vol. 33, pp. 515–532. Springer, Heidelberg (1975)
10. Kalkbrener, M.: Three contributions to elimination theory. PhD thesis, Johannes Kepler University, Linz (1991)
11. Lazard, D., Rouillier, F.: Solving parametric polynomial systems. J. Symb. Comput. 42(6), 636–667 (2007)
12. Niu, W., Wang, D.M.: Algebraic approaches to stability analysis of biological systems. Mathematics in Computer Science 1, 507–539 (2008)
13. Tyson, J., Novak, B.: Regulation of the eukaryotic cell cycle: Molecular antagonism, hysteresis, and irreversible transitions. Journal of Theoretical Biology 210(2), 249–263 (2001)
14. Weispfenning, V.: Comprehensive Gröbner bases. J. Symb. Comp. 14, 1–29 (1992)
15. Yang, L., Hou, X.R., Xia, B.: A complete algorithm for automated discovering of a class of inequality-type theorems. Science in China, Series F 44(6), 33–49 (2001)
16. Yang, L., Zhang, J.: Searching dependency between algebraic equations: an algorithm applied to automated reasoning. Technical Report IC/89/263, International Atomic Energy Agency, Miramare, Trieste, Italy (1991)

QE Software Based on Comprehensive Gröbner Systems

Ryoya Fukasaku

Graduate School of Science, Tokyo University of Science, Japan
1414704@ed.tus.ac.jp

Abstract. We introduce two quantifier elimination softwares, one is in the domain of an algebraically closed field and another is of a real closed field. Both softwares are based on the computations of comprehensive Gröbner systems.

Keywords: Quantifier Elimination, Comprehensive Gröbner System.

1 Introduction

In recent years several drastic improvements have been achieved for the computation of a comprehensive Gröbner system(CGS) [5,6,12,9,7,8,3,4,10]. We now have satisfactorily practical algorithms to compute CGS's. In particular, the algorithm introduced in [10] produces a concise CGS very often. In this paper, we introduce our two quantifier elimination(QE) implementations based on the computation of CGS's, one is in the domain of an algebraically closed field(ACF) and the other is of a real closed field(RCF). We simply call them ACF-QE and RCF-QE in this paper.

It is rather straightforward to construct a ACF-QE algorithm using CGS computations. As long as a corresponding CGS computation terminates, we can immediately obtain an equivalent quantifier free formula. If we use the CGS computation algorithm of [10], the obtained formula has a simple form in most cases. In case the corresponding CGS computation does not terminate in a realistic length of time, we have to abandon this approach if we use a CGS computation as a black box tool. Looking at the recent CGS algorithms in deep, we have found that we can replace a component of the CGS computation by another computation. This observation leads us to a new ACF-QE algorithm which consists of hybrid computations of CGS's and parametric unary GCD's. Our implementation on the computer algebra system Risa/Asir[1] is superior to other existing ACF-QE implementations in most cases.

We have also implemented RCF-QE algorithm on Risa/Asir. Our algorithm is essentially based on the RCF-QE algorithm of Weispfenning introduced in the early 1990's[14]. Since his algorithm is based on comprehensive Gröbner bases and there existed no efficient algorithm to compute comprehensive Gröbner systems at that time, there have been very few implementation to date. Though our work on RFC-QE is still ongoing, our implementation is superior to other existing RFC-QE implementations in many cases. Our implementation only deals

H. Hong and C. Yap (Eds.): ICMS 2014, LNCS 8592, pp. 512–517, 2014.

with the following basic formulas (1) for ACF-QE and (2) for RCF-QE:

(1) $\exists \bar{X}(f_1(\bar{Y}, \bar{X}) = 0 \wedge \cdots \wedge f_s(\bar{Y}, \bar{X}) = 0 \wedge g_1(\bar{Y}, \bar{X}) \neq 0 \wedge \cdots \wedge g_t(\bar{Y}, \bar{X}) \neq 0)$

(2) $\exists \bar{X}(f_1(\bar{Y}, \bar{X}) = 0 \wedge \cdots \wedge f_s(\bar{Y}, \bar{X}) = 0 \wedge g_1(\bar{Y}, \bar{X}) \neq 0 \wedge \cdots \wedge g_t(\bar{Y}, \bar{X}) \neq 0$
$\wedge h_1(\bar{Y}, \bar{X}) > 0 \wedge \cdots \wedge h_u(\bar{Y}, \bar{X}) > 0)$

where $\bar{X}$ and $\bar{Y}$ denote variables and f_i, g_j, h_k are polynomials in $\mathbb{Q}[\bar{X}, \bar{Y}]$.

In section 2, we give a rough sketch of our new ACF-QE algorithm together with a quick review of underlying theory of CGS. In section 3, we review the important result on real root counting introduced in [11] which is a base for the RCF-QE algorithm of [14]. The novelty of our work is just employing recent CGS computation algorithm instead of CGB(comprehensive Gröbner basis)[13] computation, nevertheless our implementation achieves remarkable success in some examples. In section 4, we report on some experimental results.

2   ACF-QE

In this section, we use the following notations.

K denotes a field and C its algebraic closure. $K[\bar{Y}, \bar{X}]$ denotes a polynomial ring with variables $\bar{Y} = Y_1, \ldots, Y_m$ and $\bar{X} = X_1, \ldots, X_n$. σ denotes a homomorphism from $K[\bar{Y}]$ to C, i.e. a specialization of $\bar{Y}$ with elements $c_1, \ldots, c_m$ of C, it is also naturally extended to a homomorphism from $K[\bar{Y}, \bar{X}]$ to $C[\bar{X}]$. $T(\bar{X})$ denotes the set of terms consisting of $\bar{X}$. An admissible term order on $T(\bar{Y}, \bar{X})$ such that each X_i is greater than any term in $T(\bar{Y})$ is denotes by $\bar{X} \gg \bar{Y}$. We fix an admissible term order $>$ on $T(\bar{X})$, $LM(h)$, $LT(h)$ and $LC(h)$ denotes the leading monomial, the leading term and the leading coefficient respectively of $h \in K[\bar{Y}, \bar{X}]$ w.r.t. $>$ regarding $K[\bar{Y}, \bar{X}]$ as a polynomial ring $(K[\bar{Y}])[\bar{X}]$ over the coefficient ring $K[\bar{Y}]$. Note that $LM(h) = LC(h)LT(h)$. For an ideal I of a polynomial ring over K, its variety in C is denoted by $\mathbb{V}_C(I)$.

We begin with the following result concerning stability of Gröbner basis, which is an easy consequence of Theorem 3.1 of [2] as observed in [3,4].

Theorem 1. Let I be an ideal of $K[\bar{Y}, \bar{X}]$ and G be its Gröbner basis w.r.t. $>$ regarding $K[\bar{Y}, \bar{X}]$ as a polynomial ring $(K[\bar{Y}])[\bar{X}]$. Let $G = \{g_1, \ldots, g_s, \ldots, g_t\}$ be indexed so that the following properties (i) and (ii) hold :

(i) $G \cap K[\bar{Y}] = \{g_{s+1}, \ldots, g_t\}$

(ii) $\sigma(g_{s+1}) = \sigma(g_{s+2}) = \cdots = \sigma(g_t) = 0$

Let $\{LT(g_{n_1}), \ldots, LT(g_{n_l})\}$ be the minimal subset of $\{LT(g_1), \ldots, LT(g_s)\}$ concerning the order of divisibility, that is each term of $\{LT(g_1), \ldots, LT(g_s)\}$ is divisible by some term of $\{LT(g_{n_1}), \ldots, LT(g_{n_l})\}$ and any term of $\{LT(g_{n_1})$ $, \ldots, LT(g_{n_l})\}$ is not divisible each other.

If $\sigma(LM(g_{n_1})) \neq 0, \ldots, \sigma(LM(g_{n_l})) \neq 0$, then $G' = \{\sigma(g_{n_1}), \ldots, \sigma(g_{n_l})\}$ is a Gröbner basis of $\langle \sigma(I) \rangle$ w.r.t. $>$ regardless whether $\sigma(LM(g_i)) = 0$ or not for each $i \in \{1, \ldots, s\} - \{n_1, \ldots, n_l\}$.

Definition 1. A finite set $\{\mathcal{P}_1, \ldots, \mathcal{P}_s\}$ of subsets of $\overline{K}^m$ is called an *algebraic partition* of $\overline{K}^m$ if it satisfies the following:

(i) $\cup_{i=1}^s \mathcal{P}_i = \overline{K}^m$
(ii) $\mathcal{P}_i \cap \mathcal{P}_j = \emptyset$ for each distinct i, j.
(iii) Each $\mathcal{P}_i$ is a basic constructible set, that is $\mathcal{P}_i = \mathbb{V}(\langle P_i \rangle) - \mathbb{V}(\langle Q_i \rangle)$ for some finite subsets P_i, Q_i of $K[\bar{Y}]$.

Definition 2. Let $>$ be an admissible term order on $T(\bar{X})$. For a finite subset F of $K[\bar{Y}, \bar{X}]$, a finite set $\mathcal{G} = \{(\mathcal{P}_1, G_1), \ldots, (\mathcal{P}_s, G_s)\}$ which satisfies the following properties (i)-(iii) is called a *CGS(comprehensive Gröbner system)* of F with parameters $\bar{Y}$ and main variables $\bar{X}$ w.r.t. $>$.

(i) Each G_i is a finite subset of $K[\bar{Y}, \bar{X}]$.
(ii) $\{\mathcal{P}_1, \ldots, \mathcal{P}_s\}$ is an algebraic partition of $\overline{K}^m$.
(iii) For each $\bar{c} \in \mathcal{P}_s$, $G_i(\bar{c}, \bar{X}) = \{g(\bar{c}, \bar{X}) : g \in G_i\}$ is a Gröbner basis of the ideal $\langle F(\bar{c}, \bar{X}) \rangle$ in $\overline{K}[\bar{X}]$ w.r.t. $>$.

In addition, if each $G_i(\bar{c}, \bar{X})$ is a reduced(minimal) Gröbner basis, $\mathcal{G}$ is said to be reduced(minimal). (We do not require the polynomials to be monic.) Each $\mathcal{P}_i$ is called a *segment* of $\mathcal{G}$.

We can eliminate all quantifiers from the basic formula (1) of the introduction by computing only one CGS.

Algorithm
Input: A basic formula in a form of (1)
Output: An equivalent quantifier free formula
Let $\bar{Z} = Z_1, \ldots, Z_t$ be new variables. Compute a minimal CGS $\mathcal{G} = \{(G_1, P_1, Q_1), \ldots, (G_r, P_r, Q_r)\}$ of $\{f_1(\bar{Y}, \bar{X}), \ldots, f_s(\bar{Y}, \bar{X}), g_1(\bar{Y}, \bar{X})Z_1 - 1, \ldots, g_t(\bar{Y}, \bar{X})Z_t - 1\}$ with parameters $\bar{Y}$ and main variables $\bar{X}, \bar{Z}$. We order G_i's, so that each $G_1, \ldots, G_k$ contains at least one polynomial including some main variable, and each $G_{k+1}, \ldots, G_r$ contains only polynomials of parameters. When $k = r$, the output is true, otherwise the output formula is given by $\phi_1 \vee \cdots \vee \phi_k \vee \theta_{k+1} \vee \cdots \vee \theta_r$, where each ϕ_i and θ_j is given as follows. Let $P_i = \{p_1(\bar{Y}), \ldots, p_a(\bar{Y})\}, Q_i = \{q_1(\bar{Y}), \ldots, q_b(\bar{Y})\}$, then $\phi_i \equiv p_1(\bar{Y}) = 0 \wedge \cdots \wedge p_a(\bar{Y}) = 0 \wedge (q_1(\bar{Y}) \neq 0 \vee \cdots \vee q_b(\bar{Y}) \neq 0)$. For $j = k+1, \ldots, r$, let $P_j = \{p_1(\bar{Y}), \ldots, p_a(\bar{Y})\}, Q_j = \{q_1(\bar{Y}), \ldots, q_b(\bar{Y})\}$ and $G_j = \{h_1(\bar{Y}), \ldots, h_c(\bar{Y})\}$, then $\theta_j \equiv p_1(\bar{Y}) = 0 \wedge \cdots \wedge p_a(\bar{Y}) = 0 \wedge (q_1(\bar{Y}) \neq 0 \vee \cdots \vee q_b(\bar{Y}) \neq 0) \wedge h_1(\bar{Y}) = 0 \wedge \cdots \wedge h_c(\bar{Y}) = 0$.

This algorithm deeply depends on the computation of the minimal CGS. If it does not terminate, what we can do is just waiting. In order to handle a hard case where the corresponding CGS computation does not terminate in a realistic length of time, we introduce a new algorithm which consists of hybrid computations of CGS's and parametric unary GCD's.

Each of the algorithms of CGS introduced in [9,3,4,10] is a modification of Suzuki-Sato's CGS algorithm [12]. In those algorithms, we incrementally divide parametric spaces, and proceed a Gröbner basis computation for each space in parallel. According to our experiments, when the CGS computation does not terminate in a realistic length of time, in many cases there are only a few Gröbner bases computations which do not terminate. For a quantifier elimination, we do not actually need a CGS. For a divided parametric space, if the Gröbner bases computation does not terminate, we can quit it and consider the original formula with the additional condition used for the divided parametric space. In the CGS algorithm of [10], the divided parametric space is given in a form of $\mathbb{V}(P) - \mathbb{V}(Q)$ for finite subsets P and Q of $K[\bar{Y}]$. In this parametric space, the original formula is equivalent to the following form:

$$\exists \bar{X}(f_1(\bar{Y}, \bar{X}) = 0 \wedge \cdots \wedge f_s(\bar{Y}, \bar{X}) = 0 \wedge g_1(\bar{Y}, \bar{X}) \neq 0 \wedge \cdots \wedge g_t(\bar{Y}, \bar{X}) \neq 0$$
$$\wedge p_1(\bar{Y}) = 0 \wedge \cdots \wedge p_a(\bar{Y}) = 0)$$

where $P = \{p_1(\bar{Y}) \ldots, p_a(\bar{Y})\}$. In our new algorithm, we proceed ACF-QE algorithm only by using the computation of parametric unary GCD's. Since we have new extra conditions $p_1(\bar{Y}) = 0 \wedge \cdots \wedge p_a(\bar{Y}) = 0$, there is a much better chance that the computation terminates than the computation for the original formula. This rather simple idea leads us to a drastic improvement as described in the introduction.

3 RCF-QE

The following theorem introduced in [11] is a base for the RCF-QE algorithm of [14].

Theorem 2. *Let K denote a field, R a real closed field such that $K \subset R$. Let I be a zero-dimensional ideal of a polynomial ring $K[\bar{X}]$. $\mathbf{V}_R(I)$ denotes an affine variety of I in R. Considering $K[\bar{X}]/I$ as a K linear space, for $f \in K[\bar{X}]$ let m_f denotes a linear map $K[\bar{X}]/I \to K[\bar{X}]/I$ defined by $m_f([h]) = [f][h] = [fh]$. Let B be a symmetric bilinear form $K[\bar{X}]/I \times K[\bar{X}]/I \to K$ defined by $B([f], [g]) = Tr(m_f \cdot m_g)$ and B_h be a symmetric bilinear form $K[\bar{X}]/I \times K[\bar{X}]/I \to K$ defined by $B_h(f, g) = B(hf, g)$ for $h \in K[\bar{X}]$. Let Q_h be the quadratic form of coefficients matrix B_h and $\sigma(Q_h)$ be its signature. Then we have the following property.*

$$\sigma(Q_h) = \#\{\bar{x} \in \mathbf{V}_R(I) : h(\bar{x}) > 0\} - \#\{\bar{x} \in \mathbf{V}_R(I) : h(\bar{x}) < 0\}$$

The RCF-QE algorithm of [14] employs comprehensive Gröbner bases. For the application to RCF-QE we need parametric monomial reductions. Though the concept of comprehensive Gröbner bases is mathematically interesting itself, in the computations of parametric monomial reductions, we need useless computations of polynomials of parameters which vanish in a segment. For avoiding such unnecessary computations, the use of minimal CGS's is suitable. We have implemented RCF-QE algorithm on Risa/Asir using also the CGS computation algorithm of [10] . Though our algorithm is rather naive, our implementation is superior to other existing RFC-QE implementations in many cases.

4 Experimental Data

We give some computation data of two examples, one is for ACF-QE and another for RCF-QE. Each computation is done on a standard laptop computer with a CPU Intel Core i5 2GHZ and Memory 4GB.

ACF-QE
$$\exists x \exists y \exists z \exists w (ax^5 + by^5 + abz^5 + w^5 - 1 = 0 \land xy + azw - 1 = 0 \land abx^5 + y^5 - 1 = 0 \land ax + by \neq 0)$$

Computation is done by 6 programs. In the following tables CGS-QE and Hybrid-QE denote our implementation on Risa/Asir. CGS-QE is the implementation of a naive algorithm based CGS computation, Hybrid-QE is our improved implementation. Resolve and Reduce denote ACF-QE packages based on GCD computations of parametric unary polynomials implemented in Mathematica 9, Projection denotes a ACF-QE package based on regular chain computations implemented in Maple 17. Table 1 contains the computation time(seconds). Table 2 contains the number of segments which can be considered as a measure how the output formula is complicated.

Table 1. Computation time

Hybrid-QE	CGS-QE	Resolve	Reduce	Projection
4sec.	> 1hour	> 1hour	> 1hour	860sec.

Table 2. Expression length

Hybrid-QE	CGS-QE	Resolve	Reduce	Projection
10	-	-	-	38

RCF-QE
$$\exists x \exists y \exists z (xy + axz + yz = 1 \land xyz + xz + xy + b = 0 \land xz + yz - az - x - y = 1)$$

Our implementation on Risa/Asir terminates within 2 seconds, whereas any of the other existing packages for RCF-QE do not terminate within one hour.

References

1. Risa/Asir, http://www.math.kobe-u.ac.jp/Asir/asir.html
2. Kalkbrener, M.: On the Stability of Gröbner Bases Under Specializations. J. Symbolic Computation 24(1), 51–58 (1997)
3. Kapur, D., Sun, Y., Wang, D.: A New Algorithm for Computing Comprehensive Gröbner Systems. In: Proceedings of International Symposium on Symbolic and Algebraic Computation, pp. 29–36. ACM-Press (2010)

4. Kurata, Y.: Improving Suzuki-Sato's CGS Algorithm by Using Stability of Gröbner Bases and Basic Manipulations for Efficient Implementation. Communications of JSSAC 1, 39–66 (2011)
5. Montes, A.: A new algorithm for discussing Gröbner bases with parameters. Journal of Symbolic Computation 33(2), 183–208 (2002)
6. Manubens, M., Montes, A.: Improving DISPGB algorithm using the discriminant ideal. Journal of Symbolic Computation 41, 1245–1263 (2006)
7. Manubens, M., Montes, A.: Minimal Canonical Comprehensive Gröbner System. Journal of Symbolic Computation 44, 463–478 (2009)
8. Montes, A., Wibmer, M.: Gröbner Bases for Polynomial Systems with parameters. Journal of Symbolic Computation 45, 1391–1425 (2010)
9. Nabeshima, K.: A Speed-Up of the Algorithm for Computing Comprehensive Gröbner Systems. In: Proceedings of International Symposium on Symbolic and Algebraic Computation, pp. 299–306. ACM-Press (2007)
10. Nabeshima, K.: Stability Conditions of Monomial Bases and Comprehensive Gröbner Systems. In: Gerdt, V.P., Koepf, W., Mayr, E.W., Vorozhtsov, E.V. (eds.) CASC 2012. LNCS, vol. 7442, pp. 248–259. Springer, Heidelberg (2012)
11. Pedersen, P., Roy, M.F., Szpirglas, A.: F., and Szpirglas,A.: Counting real zeroes in the multivariate case. In: Eysette, F., Galigo, A. (eds.) Proceedings of the MEGA 92. Computational Algebraic Geometry. Progress in Mathematics, vol. 109, pp. 203–224. Birkhäuser, Boston (1993)
12. Suzuki, A., Sato, Y.: A Simple Algorithm to Compute Comprehensive Gröbner Bases Using Gröbner Bases. In: International Symposium on Symbolic and Algebraic Computation, pp. 326–331. ACM-Press (2006)
13. Weispfenning, V.: Comprehensive Gröbner Bases. Journal of Symbolic Computation 14(1), 1–29 (1992)
14. Weispfenning, V.: A New Approach to Quantifier Elimination for Real Algebra. In: Caviness, B.F., Johnson, J.R. (eds.) Quantifier Elimination and Cylindrical Algebraic Decomposition. Texts and monographs in symbolic computation, pp. 376–392. Springer (1998)

SyNRAC: A Toolbox for Solving Real Algebraic Constraints

Hidenao Iwane[1,2], Hitoshi Yanami[2], and Hirokazu Anai[2,3,1]

[1] National Institute of Informatics
2-1-2 Hitotsubashi, Chiyoda-ku, Tokyo 101-8430, Japan
[2] Social Innovation Laboratories, Fujitsu Laboratories Ltd.
4-1-1 Kamikodanaka, Nakahara-ku, Kawasaki 211-8588, Japan
{iwane,yanami,anai}@jp.fujitsu.com
[3] Institute of Mathematics for Industry, Kyushu University
744 Motooka, Nishi-ku, Fukuoka 819-0395, Japan

Abstract. We introduce various aspects of the design and the implementation of a symbolic/symbolic-numeric computation toolbox, called SyNRAC. SyNRAC is a package of commands written in the Maple language and the C language. This package indeed provides an environment for dealing with first-order formulas over the reals.

Keywords: quantifier elimination.

1 Introduction

Many mathematical and engineering problems can be naturally translated to formulas consisting of polynomial equations and inequalities, quantifiers and logical operators. Such formulas construct sentences in the first-order theory of real closed fields (RCFs) and are called first-order formulas (FOFs). *Quantifier elimination* (QE) is an algorithm for computing an equivalent quantifier-free formula for a given FOF. In the 1930's, A. Tarski [13] showed that an RCF allows QE and gave the first QE procedure for RCFs. For example, the formula $\exists x(x^2 + bx + c = 0)$ can be reduced to a quantifier-free formula $b^2 - 4c \geq 0$ by QE. If all variables are quantified, QE decides whether the given formula is true or false (this is a decision problem).

QE over the reals is a very powerful concept for solving problems containing real algebraic constraints. One of the big advantages of QE is that it can deal with parametric and non-convex constraints. For example, QE can exactly prove real theorems, perform geometric reasoning, solve polynomial optimization problems, transportation problems, scheduling problems, mechanical engineering, and stability analysis, and so on. Practically efficient software systems for QE have been developed on several computer algebra systems, such as QEP-CAD [3], REDLOG [6], Mathematica [12], and SyNRAC[18,10,8].

We started the development of SyNRAC in 2002, first appeared in a literature in 2003, with a focus on the implementation of purely symbolic QE. The solver

H. Hong and C. Yap (Eds.): ICMS 2014, LNCS 8592, pp. 518–522, 2014.

Fig. 1. Logo of SyNRAC

Table 1. Notations of quantifiers and logical operators in SyNRAC

quantifiers/operators	notation in SyNRAC
$\exists x f$	Ex([x], f)
$\forall x f$	All([x], f)
$f \wedge g$	And(f, g)
$f \vee g$	Or(f, g)
$\neg f$	Not(f)
$f \rightarrow g$	Impl(f, g)
$f \leftarrow g$	Repl(f, g)
$f \leftrightarrow g$	Equiv(f, g)

to be addressed includes several symbolic-numeric QE algorithms that have recently been implemented to improve the efficiency of its symbolic counterparts without loss of exactness. SyNRAC stands for a $\underline{\text{S}}$ymbolic-$\underline{\text{N}}$umeric toolbox for $\underline{\text{R}}$eal $\underline{\text{A}}$lgebraic $\underline{\text{C}}$onstraints. SyNRAC can be freely downloaded from

> http://jp.fujitsu.com/group/labs/en/techinfo/freeware/synrac/.

2 Commands in SyNRAC

In this section we show some examples to illustrate how to use commands in SyNRAC. The SyNRAC library provides the four commands:

- qe which does QE over the reals,
- synsimpl which simplifies a given FOF [16,5], and
- RegionPlot and RegionPlot3d which draw the feasible regions of the input quantifier-free formula with two and three variables, respectively.

We first need to load the SyNRAC library by the with command in Maple.

```
> with(SyNRAC):
```

The qe command takes only one argument which is an FOF. The next example shows how to use the command to solve the problems $\exists x(x^2 + bx + c = 0)$ and $\forall x(ax^2 + bx + c > 0)$. The notations of the quantifier symbols and the logical operators are shown in Table 1.

```
> qe(Ex([x], x^2 + b*x + c = 0));
-b^2+4*c <= 0
> qe(All([x], a*x^2+b*x+c>0));
Or(And(-a < 0,-4*a*c+b^2 < 0),And(-a = 0,-b = 0,-c < 0))
```

Next, we consider the following FOF:

$$\exists x_1 \exists x_2 \, (y_1 = x_1^2 + x_2^2 \wedge y_2 = 5 + x_2^2 - x_1 \wedge -5 \leq x_1 \leq 5 \wedge -5 \leq x_2 \leq 5).$$

Using the qe command we obtain the following quantifier-free formula.

```
> F := qe(Ex([x1,x2], And(y1=x1^2+x2^2,y2=x2^2-x1+5,
                          -5<=x1,x1<=5,-5<=x2,x2<=5)));
F := Or(And(y2^2-y1-10*y2 <= -25,-4*y1+4*y2 <= 21,y1-y2 <= 25,-y2^2+y1
+60*y2-925 = 0,2*y2 <= 61),And(y2^2-y1-10*y2 <= -25,-4*y1+4*y2 <= 21,
y1-y2 <= 25,-y2^2+y1+60*y2 <= 925,2*y2 < 61),And(-2*y2 < -11,-4*y1+4*
y2 <= 21,y1-y2 <= 25,-y2^2+y1+60*y2-925 = 0,2*y2 <= 61),And(-2*y2 <
-11,-4*y1+4*y2 <= 21,y1-y2 <= 25,-y2^2+y1+60*y2 <= 925,2*y2 < 61),And(
y2^2-y1-60*y2 <= -925,-4*y1+4*y2 <= 21,-y2^2+y1+10*y2-25 = 0,-2*y2 <=
-11,y1-y2 <= 15),And(2*y2 < 61,-4*y1+4*y2 <= 21,-y2^2+y1+10*y2-25 = 0,
-2*y2 <= -11,y1-y2 <= 15),And(y2^2-y1-60*y2 <= -925,-4*y1+4*y2 <= 21,
-y2^2+y1+10*y2 <= 25,-2*y2 < -11,y1-y2 <= 15),And(2*y2 < 61,-4*y1+4*y2
<= 21,-y2^2+y1+10*y2 <= 25,-2*y2 < -11,y1-y2 <= 15))
```

Although the qe command may return a redundant formula, we can obtain a simpler equivalent formula by the synsimpl command.

```
> synsimpl(F);
And(-4*y1+4*y2 <= 21,Or(And(y1-y2 <= 25,y2^2-y1-10*y2 <= -25),And(y2
<= 26,-2*y2 < -11,y1-y2 <= 15),And(y2 <= 25,-y2 <= -10,y1-y2 <= 25),
And(y2 <= 26,-y2 <= -25,-y2^2+y1+60*y2 <= 925),And(y1-y2 <= 15,y2^2-
y1-60*y2 <= -925),And(-y2 <= -26,2*y2 < 61,y1-y2 <= 15)))
```

By RegionPlot, we can draw feasible regions (see Fig. 2). In the following example the fourth argument 'gridrefine' of RegionPlot has the same meaning of an option of the Maple built-in command plots[implicitplot]. Please see the help of plots[implicitplot] for more details. We can also use the other options of plots[implicitplot].

```
> with(plots):
> display(RegionPlot(F, y1=0..30, y2=0..10, gridrefine=5));
```

3   QE Algorithms

The focus of the implemented QE algorithms is on practically effective QE for certain industrial/engineering problems [17,9,11]. In this section we show the algorithms utilized in the qe command.

3.1   Special QE by the Sturm-Habicht Sequence

The Sturm-Habicht sequence associated to a polynomial in $\mathbb{R}[x]$ tells us the number of real zeros in the given intervals. In 1993, by using this L. González-Vega [7] proposed a new QE algorithm in the form $\forall x(f(x) > 0)$ where $f(x) \in \mathbb{R}[x]$.

In 2000, H. Anai and S. Hara [1] proposed a special QE algorithm for a similar type of formulas $\forall x(x \geq 0 \rightarrow f(x) > 0)$ and called this condition the *sign definite condition* (SDC). They also showed that many important design specifications such as H_∞ norm constraints, stability margins etc, which are frequently used as indices of the robustness, reduce to SDCs [2].

We have developed improved algorithms that construct more concise formulas by logical formula manipulation [8].

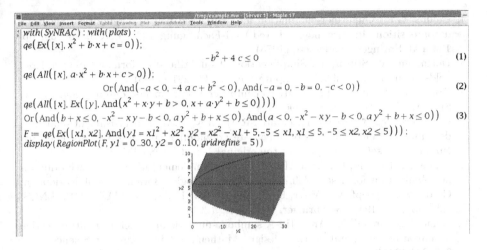

Fig. 2. Examples of SyNRAC

3.2 Special QE by Virtual Substitution

V. Weispfenning [14,15] gave specialized QE algorithms that are applicable to the formulas with a low degree with respect to the quantified variables. These algorithms have been developed and applied to various problems. We have implemented the algorithms for the linear and quadratic cases.

3.3 QE by Cylindrical Algebraic Decomposition

In 1975, G. E. Collins [4] discovered a new general QE algorithm based on *cylindrical algebraic decomposition* (CAD) that was far more efficient than any previous approach. However, QE based on CAD is not considered to be practical on computers, since CAD usually consists of many purely symbolic computations and has high computational complexity. We have implemented a symbolic-numeric CAD which avoids heavy symbolic computation by interval arithmetic techniques without loss of exactness [10].

References

1. Anai, H., Hara, S.: Fixed-structure robust controller synthesis based on sign definite condition by a special quantifier elimination. In: Proceedings of American Control Conference, vol. 2, pp. 1312–1316 (2000)
2. Anai, H., Hara, S.: A parameter space approach to fixed-order robust controller synthesis by quantifier elimination. International Journal of Control 79(11), 1321–1330 (2006)
3. Brown, C.W.: QEPCAD B: A program for computing with semi-algebraic sets using CADs. SIGSAM Bulletin 37, 97–108 (2003)

4. Collins, G.E.: Quantifier elimination for real closed fields by cylindrical algebraic decomposition. In: Brakhage, H. (ed.) GI-Fachtagung 1975. LNCS, vol. 33, pp. 134–183. Springer, Heidelberg (1975)
5. Dolzmann, A., Sturm, T.: Simplification of quantifier-free formulas over ordered fields. Journal of Symbolic Computation 24, 209–231 (1995)
6. Dolzmann, A., Sturm, T.: REDLOG computer algebra meets computer logic. ACM SIGSAM Bulletin 31, 2–9 (1996)
7. González-Vega, L., Recio, T., Lombardi, H., Roy, M.-F.: Sturm-Habicht sequences determinants and real roots of univariate polynomials, pp. 300–316. Springer, Vienna (1998), softcover reprint of the original 1st ed. 1998 edition
8. Iwane, H., Higuchi, H., Anai, H.: An effective implementation of a special quantifier elimination for a sign definite condition by logical formula simplification. In: Gerdt, V.P., Koepf, W., Mayr, E.W., Vorozhtsov, E.V. (eds.) CASC 2013. LNCS, vol. 8136, pp. 194–208. Springer, Heidelberg (2013)
9. Iwane, H., Yanami, H., Anai, H.: A symbolic-numeric approach to multi-objective optimization in manufacturing design. Mathematics in Computer Science 5(3), 315–334 (2011)
10. Iwane, H., Yanami, H., Anai, H., Yokoyama, K.: An effective implementation of symbolic–numeric cylindrical algebraic decomposition for quantifier elimination. Theoretical Computer Science 479, 43–69 (2013)
11. Matsui, Y., Iwane, H., Anai, H.: Two controller design procedures using SDP and QE for a power supply unit. In: Development of Computer Algebra Research and Collaboration with Industry. COE Lecture Note, vol. 49, pp. 43–51 (2013)
12. Strzeboński, A.W.: Cylindrical algebraic decomposition using validated numerics. Journal of Symbolic Computation 41(9), 1021–1038 (2006)
13. Tarski, A.: A decision method for elementary algebra and geometry, 2nd edn. University of California Press (1952)
14. Weispfenning, V.: The complexity of linear problems in fields. Journal of Symbolic Computation 5, 3–27 (1988)
15. Weispfenning, V.: Quantifier elimination for real algebra - the quadratic case and beyond. Applicable Algebra in Engineering, Communication and Computing 8, 85–101 (1993)
16. Wilson, D.J., Bradford, R.J., Davenport, J.H.: Speeding up cylindrical algebraic decomposition by Gröbner bases. CoRR, abs/1205.6285 (2012)
17. Yanami, H.: Multi-objective design based on symbolic computation and its application to hard disk slider design. Journal of Math-for-Industry 1, 149–156 (2009)
18. Yanami, H., Anai, H.: The Maple package SyNRAC and its application to robust control design. Future Generation Computer Systems 23(5), 721–726 (2007)

An Algorithm for Computing Tjurina Stratifications of μ-Constant Deformations by Using Local Cohomology Classes with Parameters

Katsusuke Nabeshima[1] and Shinichi Tajima[2]

[1] Institute of Socio-Arts and Sciences, The University of Tokushima,
1-1 Minamijosanjima, Tokushima, Japan
nabeshima@tokushima-u.ac.jp
[2] Graduate School of Pure and Applied Sciences, University of Tsukuba,
1-1-1 Tennoudai, Tsukuba, Japan
tajima@math.tsukuba.ac.jp

Abstract. Algebraic local cohomology classes attached to semi-quasiho-mogeneous hypersurface isolated singularities are considered. A new algorithm, that utilizes local cohomology classes with parameter, is proposed to compute Tjurina stratifications associated with μ-constant deformations of weighted homogeneous isolated singularities. The proposed algorithm has been implemented in a computer algebra system. Usage of the implementation is also described.

Keywords: μ-constant deformations, local cohomology, singularities.

1 Introduction

We introduce an algorithm for computing Tjurina stratifications of μ-constant deformations, and its implementation.

Let $f \in \mathbb{C}[x_1, \ldots, x_n]$ be a semi-quasihomogeneous polynomial (with parameters). Assume that the weighted homogeneous part of f has an isolated singularity at the origin. Then, f can be regarded as a μ-constant deformation. In this paper, we consider Tjurina numbers of μ-constant deformations by using local cohomology classes with parameters.

Let H_{J_f} be a set of local cohomology classes, which is annihilated by the Jacobi ideal $J_f = \langle \frac{\partial f}{\partial x_1}, \ldots, \frac{\partial f}{\partial x_n} \rangle$. Then, H_{J_f} is a vector space and the dimension of the vector space H_{J_f} is equal to the Milnor number of the isolated singularity. Let H_{T_f} be a set of local cohomology classes, which is annihilated by the ideal $\langle f, \frac{\partial f}{\partial x_1}, \ldots, \frac{\partial f}{\partial x_n} \rangle$. Then, H_{T_f} is a vector space and the dimension of the vector space H_{T_f} is equal to the Tjurina number of the isolated singularity [9]. By utilizing the relation between H_{J_f} and H_{T_f}, we are able to analyze the properties of the μ-constant deformation. We construct in particular, by using local cohomology classes with parameters, an algorithm for computing Tjurina stratifications of μ-constant deformations.

H. Hong and C. Yap (Eds.): ICMS 2014, LNCS 8592, pp. 523–530, 2014.
© Springer-Verlag Berlin Heidelberg 2014

The existing algorithm due to B. Martin and G. Pfister[3] is based on the deformation theory and utilizes the Kodaira-Spencer maps. In contrast, the proposed algorithm, which has been implemented in the computer algebra system Risa/Asir [7], is free from the Kodaira-Spencer maps.

2 Preliminaries

Here, we briefly recall the notion of semi-quasihomogeneity, Milnor number, Tjurina number and the notion of algebraic local cohomology supported at a point. For details, we refer the reader to [2,5,6,10,11,12,13]. The set of natural numbers $\mathbb{N}$ includes zero. $\mathbb{C}$ is the field of complex numbers.

Let us fix a **weight** vector $\mathbf{w} = (w_1, w_2, \ldots, w_n)$ in $\mathbb{N}^n$ for a fixed coordinate system $x = (x_1, x_2, \ldots, x_n)$. We define a **weighted degree** of the term $x^\alpha = x_1^{\alpha_1} x_2^{\alpha_2} \cdots x_n^{\alpha_n}$, with respect to $\mathbf{w}$ by $|x^\alpha|_{\mathbf{w}} = \sum_{i=1}^n w_i \alpha_i$.

Definition 1 ([1]). A nonzero polynomial f in $\mathbb{C}[x]$ is **quasihomogeneous of type** $(d; \mathbf{w})$ if all terms of f have the same weighted degree d with respect to $\mathbf{w}$, i.e., $f = \sum_{|x^\alpha|_{\mathbf{w}} = d} c_\alpha x^\alpha$ where $c_\alpha \in \mathbb{C}$. We define a weighted degree of f by $\deg_{\mathbf{w}}(f) = \max\{|x^\alpha|_{\mathbf{w}} | x^\alpha \text{ is a term of } f\}$.

We define $\mathrm{ord}_{\mathbf{w}}(f) = \min\{|x^\alpha|_{\mathbf{w}} \ |x^\alpha \text{ is a term of } f\}$ ($\mathrm{ord}_{\mathbf{w}}(0) = -1$). The polynomial f is called **semi-quasihomogeneous of type** $(d; \mathbf{w})$ if f is of the form $f = f_0 + g$ where f_0 is a quasihomogeneous polynomials of type $(d; \mathbf{w})$ with an isolated singularity at the origin, $f = f_0$ or $\mathrm{ord}_{\mathbf{w}}(f - f_0) > d$. The polynomial f_0 is called **quasihomogeneous part** of f and a term of g is called **upper monomial**.

Let X be an open neighborhood of the origin O of the n-dimensional complex space $\mathbb{C}^n$ with coordinates $x = (x_1, x_2, \ldots, x_n)$ and let $\mathcal{O}_X$ be the sheaf on X of holomorphic functions. Let f be a holomorphic function defined on X with an isolated singularity at the origin O and let J_f denote the Jacobi ideal $\left\langle \frac{\partial f}{\partial x_1}, \frac{\partial f}{\partial x_2}, \ldots, \frac{\partial f}{\partial x_n} \right\rangle$ in $\mathcal{O}_{X,O}$ generated by the partial derivatives $\frac{\partial f}{\partial x_1}, \frac{\partial f}{\partial x_2}, \ldots, \frac{\partial f}{\partial x_n}$, where $\mathcal{O}_{X,O}$ is the stalk at O of the sheaf $\mathcal{O}_X$. Let $\langle f, J_f \rangle$ denote the ideal $\left\langle f, \frac{\partial f}{\partial x_1}, \frac{\partial f}{\partial x_2}, \ldots, \frac{\partial f}{\partial x_n} \right\rangle$ in $\mathcal{O}_{X,O}$ generated by f and $\frac{\partial f}{\partial x_1}, \frac{\partial f}{\partial x_2}, \ldots, \frac{\partial f}{\partial x_n}$. The **Milnor** number of the singularity, denoted μ, is given by

$$\mu = \dim_{\mathbb{C}}(\mathcal{O}_{X,O}/J_f).$$

The **Tjurina** number of the singularity, denoted τ, is given by

$$\tau = \dim_{\mathbb{C}}(\mathcal{O}_{X,O}/\langle f, J_f \rangle).$$

It is known that the Milnor number μ is a topological invariant and the Tjurina number τ is an analytic invariant of the singularity. In general, the Tjurina number τ is less than or equal to the Milnor number μ ($\tau \leq \mu$). If $f \in \mathbb{C}[x_1, \ldots, x_n]$ is a quasihomogeneous polynomial, then the Milnor number μ is equal to the Tjurina number τ. Conversely, if the Milnor number μ is equal to the Tjurina number τ, then there exists a holomorphic coordinate transformation such that f is represented as a quasihomogeneous polynomial ([8]). According to this fact, the numerical invariant $\mu - \tau$ can be regarded as a complex analytic invariant that measures non quasihomogeneity of a hypersurface isolated singularity.

Next we quickly review algebraic local cohomology. Let $\mathcal{H}^n_{[O]}(\mathcal{O}_X)$ denote the set of algebraic local cohomology classes, defined by

$$\mathcal{H}^n_{[O]}(\mathcal{O}_X) = \lim_{k \to \infty} Ext^n_{\mathcal{O}_X}(\mathcal{O}_X/\langle x_1, x_2, \ldots, x_n \rangle^k, \mathcal{O}_X),$$

where $\langle x_1, x_2, \ldots, x_n \rangle$ is the maximal ideal generated by $x_1, x_2, \ldots, x_n$. Consider the pair $(X, X - O)$ and its relative Čech covering. Then, any section of $\mathcal{H}^n_{[O]}(\mathcal{O}_X)$ can be represented as an element of relative Čech cohomology. Any algebraic local cohomology class in $\mathcal{H}^n_{[O]}(\mathcal{O}_X)$ can be represented as a finite sum of the form (called: Čech representation) $\sum c_\lambda \left[\frac{1}{x^{\lambda+1}} \right] = \sum c_\lambda \left[\frac{1}{x_1^{\lambda_1+1} x_2^{\lambda_2+1} \cdots x_n^{\lambda_n+1}} \right]$ where $c_\lambda \in \mathbb{C}$ and $\lambda = (\lambda_1, \lambda_2, \ldots, \lambda_n) \in \mathbb{N}^n$. Note that the multiplication is defined as

$$x^\alpha \left[\frac{1}{x^{\lambda+1}} \right] = \begin{cases} \left[\dfrac{1}{x^{\lambda+1-\alpha}} \right], & \lambda_i \geq \alpha_i, i = 1, \ldots, n, \\ \\ 0, & otherwise, \end{cases}$$

where $\alpha = (\alpha_1, \ldots, \alpha_n) \in \mathbb{N}^n$ and $\lambda + 1 - \alpha = (\lambda_1 + 1 - \alpha_1, \ldots, \lambda_n + 1 - \alpha_n)$.

We introduce a vector space H_{J_f} defined to be the set of algebraic local cohomology classes in $\mathcal{H}^n_{[O]}(\mathcal{O}_X)$ that are annihilated by the Jacobi ideal J_f

$$H_{J_f} := \left\{ \psi \in \mathcal{H}^n_{[O]}(\mathcal{O}_X) \,\middle|\, \frac{\partial f}{\partial x_1}(x)\psi = \frac{\partial f}{\partial x_2}(x)\psi = \cdots = \frac{\partial f}{\partial x_n}(x)\psi = 0 \right\}.$$

We also introduce a vector space H_{T_f} defined to be the set of algebraic local cohomology classes in $\mathcal{H}^n_{[O]}(\mathcal{O}_X)$ that are annihilated by the ideal $\langle f, J_f \rangle$

$$H_{T_f} := \left\{ \psi \in \mathcal{H}^n_{[O]}(\mathcal{O}_X) \,\middle|\, f(x)\psi = \frac{\partial f}{\partial x_1}(x)\psi = \cdots = \frac{\partial f}{\partial x_n}(x)\psi = 0 \right\}.$$

It is known that H_{J_f} and H_{T_f} are finite dimensional vector spaces. Moreover, $\dim_{\mathbb{C}}(H_{J_f}) = \mu$ (Milnor number) and $\dim_{\mathbb{C}}(H_T) = \tau$ (Tjurina number). In [4,13], algorithms for computing a basis of H_{J_f} (and H_{T_f}) are introduced.

We represent an algebraic local cohomology class $\sum c_\lambda \left[\frac{1}{x^{\lambda+1}} \right]$ as a polynomial in n variables $\sum c_\lambda x^\lambda$ (called: **polynomial representation**) to manipulate algebraic local cohomology classes efficiently (see [4,13]). Hereafter, we adapt "polynomial representation" to represent an algebraic local cohomology class.

3   Basic Facts

Here, we describe some basic facts which are utilized for constructing our new algorithm. Let $\varphi : H_{J_f} \to H_{J_f}$ be a map defined by $\varphi(\eta) = f\eta$. Since the kernel of the map φ, $Ker(\varphi) = \{\eta \in H_{J_f} \mid f\eta = 0\}$ is equal to the vector space H_{T_f}, we have the following results ([9]).

Proposition 1. The sequence $0 \longrightarrow H_{T_f} \longrightarrow H_{J_f} \longrightarrow \varphi(H_{J_f}) \longrightarrow 0$ of vector spaces defined by φ is exact where $\varphi(H_{J_f}) = \{f\eta | \eta \in H_{J_f}\}$.

Corollary 1. $\dim_{\mathbb{C}}(H_{T_f}) - \dim_{\mathbb{C}}(H_{J_f}) + \dim_{\mathbb{C}}(\varphi(H_{J_f})) = 0$.

Proposition 2. The annihilator ideal, in the local ring $\mathcal{O}_{X,O}$, of $\varphi(H_{J_f})$ is the ideal quotient $\{h \in \mathcal{O}_{X,O} \mid fh \in J_f\}$:

$$Ann_{\mathcal{O}_{X,O}}(\varphi(H_{J_f})) = \{h \in \mathcal{O}_{X,O} \mid fh \in J_f\}.$$

Proposition 1 says that $\varphi(H_{J_f})$ can be regarded as a vector space representing the non-quasihomogeneity of the singularities. Corollary 1 says that the dimension of the vector space $\varphi(H_{J_f})$ coincides with $\mu - \tau$. Proposition 1 tells us that the ideal quotient $\{h \in \mathcal{O}_{X,O} \mid fh \in J_f\}$ can be explicitly determined by computing the annihilator in $\mathcal{O}_{X,O}$ of $\varphi(H_{J_f})$.

As an algorithm introduced in [4] can compute algebraic local cohomology classes with parameters in H_{J_f} associated with a μ-constant deformation f of an weighted homogeneous isolated singularity, we are able to analyze the parameter dependency of Tjurina numbers by computing the space $\varphi(H_{J_f})$.

Before describing the main algorithm, we see in the next section the implementation of [4] for computing local cohomology classes with parameters.

4   Local Cohomology with Parameters

In [4], we have introduced an algorithm for computing parametric local cohomology classes associated with the Jacobi ideals of semi-quasihomogeneous hypersurface isolated singularities. There are two main advantages of the algorithm. First, it generates a nice decomposition, in the context of deformation theory, of the parameter space according to a structure of local cohomology and second, it efficiently computes parametric local cohomology. This algorithm has been implemented in a computer algebra system Risa/Asir[1]. Here, we quickly see how the implementation works.

Let $(s, t) = (s_1, s_2, \ldots, s_{m_1}, t_1, t_2, \ldots, t_{m_2})$ be parameters in $\mathbb{C}^m$ with $m = m_1 + m_2$. In the outputs of the algorithm, a list of two list $[[p_1, \ldots, p_k], [q_1, \ldots, q_l]]$ is used to represent algebraically constructible set (stratum) $\mathbb{V}(p_1, \ldots, p_k) \backslash \mathbb{V}(q_1, \ldots, q_l)$ in $\mathbb{C}^m$, where $p_1, \ldots, p_k, q_1, \ldots, q_l \in \mathbb{C}[s, t]$ and $\mathbb{V}(p_1, \ldots, p_k) = \{\bar{a} \in \mathbb{C}^m | p_1(\bar{a}) = \cdots = p_k(\bar{a}) = 0\}$. The list represents conditions of parameters.

Example 1. A polynomial $f = x^3 + y^9 + sx^2y^3 + ty^{10} \in (\mathbb{C}[s,t])[x,y]$ is semi-quasihomogeneous of type $(9; (3, 1))$ where x, y are variables and s, t are parameters. Then, $x^3 + y^9 + tx^2y^3$ is the quasihomogeneous part of f. Set $J_f = \langle \frac{\partial f}{\partial x}, \frac{\partial f}{\partial y} \rangle$. Our implementation outputs a basis of H_{J_f} as follows:

```
[676] para_qhcoho(x^3+y^9+s*x^2*y^3,t*y^10,[3,1],[s,t],[x,y],1);
non-zero dim. are [[[4*s^3+27],[1]]]
```

---

[1] Risa/Asir([7]) is an open source general computer algebra system. http://www.math.kobe-u.ac.jp/Asir/asir.html

```
No. of segments is  3
[[s],[1]]
[[y^7*x,y^6*x,y^5*x,y^4*x,y^3*x,y^2*x,y*x,x,y^7,y^6,y^5,y^4,y^3,y^
2,y,1],[]]


[[0],[4*t*s^4+27*t*s]]
[[y^2*x,y*x,x,y^7,y^6,y^5,y^4,y^3,y^2,y,1],[(s^2*y*x^3-3/2*s*y^4*x
^2+9/4*y^7*x+1/2*s^2*y^10-5/9*t*s^2*y^9+50/81*t^2*s^2*y^8)/(s^2),(
s^2*x^3-3/2*s*y^3*x^2+9/4*y^6*x+1/2*s^2*y^9-5/9*t*s^2*y^8)/(s^2),(
s*y^2*x^2-3/2*y^5*x-1/3*s^2*y^8)/(s),(s*y*x^2-3/2*y^4*x)/(s),(s*x^
2-3/2*y^3*x)/(s)]]


[[t],[4*s^4+27*s,t]]
[[y^2*x,y*x,x,y^7,y^6,y^5,y^4,y^3,y^2,y,1],[(s^2*y*x^3-3/2*s*y^4*x
^2+9/4*y^7*x+1/2*s^2*y^10)/(s^2),(s^2*x^3-3/2*s*y^3*x^2+9/4*y^6*x+
1/2*s^2*y^9)/(s^2),(s*y^2*x^2-3/2*y^5*x-1/3*s^2*y^8)/(s),(s*y*x^2-
3/2*y^4*x)/(s),(s*x^2-3/2*y^3*x)/(s)]]
```

The meaning of the output is the following.

- if parameters s, t belong to $\mathbb{V}(4s^3 + 27)$, the quasihomogeneous part of f has non isolated singularities,
- if parameters s, t belong to $\mathbb{V}(s)$, then a basis of H_{J_f} is $\{y^7x, y^6x, y^5x, y^4x, y^3x, y^2x, yx, x, y^7, y^6, y^5, y^4, y^3, y^2, y, 1\}$,
- if parameters s, t belong to $\mathbb{C}^2 \backslash \mathbb{V}(4ts^4 + 27ts)$, then a basis of H_{J_f} is $\{y^2x, yx, x, y^7, y^6, y^5, y^4, y^3, y^2, y, 1, yx^3 - \frac{3}{2s}y^4x^2 + \frac{9}{4s^2}y^7x + \frac{1}{2}y^{10} - \frac{5}{9}ty^9 + \frac{50}{81}t^2y^8, x^3 - \frac{3}{2s}y^3x^2 + \frac{9}{4s^2}y^6x + \frac{1}{2}y^9 - \frac{5}{9}ty^8, y^2x^2 - \frac{3}{2s}y^5x - \frac{1}{3}sy^8, yx^2 - \frac{3}{2s}y^4x, x^2 - \frac{3}{2s}y^3x\}$, and
- if parameters s, t belong to $\mathbb{V}(t) \backslash \mathbb{V}(4s^4 + 27s, t)$, then a basis of H_{J_f} is $\{y^2x, yx, x, y^7, y^6, y^5, y^4, y^3, y^2, y, 1, yx^3 - \frac{3}{2s}y^4x^2 + \frac{9}{4}y^7x + \frac{1}{2}y^{10}, x^3 - \frac{3}{2s}y^3x^2 + \frac{9}{4s^2}y^6x + \frac{1}{2}y^9, y^2x^2 - \frac{3}{2s}y^5x - \frac{1}{3}sy^8, yx^2 - \frac{3}{2s}y^4x, x^2 - \frac{3}{2s}y^3x\}$.

Now, as we have seen local cohomology classes with parameters, can be computed by the implementation[4].

5   The Main Algorithm and Examples

Corollary 1 together with the algorithm [4] allows us to design an algorithm to compute Tjurina Stratifications.

Here, we give the resulting algorithm for computing Tjurina stratifications.

Algorithm 1. [Tjurina Stratification]
Input: $f \in (\mathbb{C}[s_1, \ldots, s_{m_1}, t_1, \ldots, t_{m_2}])[x_1, \ldots, x_n]$ a semi-quasihomogeneous polynomial with parameters $s_1, \ldots, s_{m_1}, t_1, \ldots, t_{m_2}$.
Output: a Tjurina stratification of μ-constant deformations.

Step 1: Compute a basis of H_{J_f} by the algorithm [4].

Step 2: Compute a basis Q of $\varphi(H_{J_f})$ for each stratum where φ is from section 3. (This is possible to compute. For example, Gaussian elimination method can be extendable to handle the parametric cases.)

Step 3: For each stratum, compute $\mu - \sharp(Q)$ where μ is the Milnor number and $\sharp(Q)$ is the cardinality of the set Q. Then, $\mu - \sharp(Q)$ is the Tjurina number in the stratum.

Algorithm 1 has been implemented in the computer algebra system Risa/Asir. There exists a command "tjurinast" in the implementation, which outputs the Tjurina number and a basis of $\varphi(H_{J_f})$ for each stratum. We see some examples.

Example 2. A polynomial $f = x^3 + sx^2y^4 + y^{12} + ty^{13} + uy^{14} \in (\mathbb{C}[s,t,u])[x,y]$ is semi-quasihomogeneous of type $(12; (4,1))$ where x,y are variables and s,t,u are parameters. Then, $x^3 + sx^2y^4 + y^{12}$ is the quasihomogeneous part of f. Note that the quasihomogeneous part of f contains a parameter s. Set $J_f = \langle \frac{\partial f}{\partial x}, \frac{\partial f}{\partial y} \rangle$. Our implementation outputs a Tjurina stratification of f as follows:

```
tjurinast(x^3+s*x^2*y^4+y^12,t*y^13+u*y^14,[4,1],[s,t,u],[x,y]);
zero dim.   [[[4*s^3+27],[1]]]
-----------

[[s],[1]]
basis
[]
22-0=22
-----------

[[0],[4*t*s^4+27*t*s]]
basis
[1,(t*y-13/12*t^2+2*u)/(t)]
22-2=20
-----------

[[t],[4*u*s^4+27*u*s,t]]
basis
[1]
22-1=21
-----------

[[t,u],[4*s^4+27*s,t]]
basis
[]
22-0=22
```

The meaning of the output is the following.

- if parameters s,t,u belong to $\mathbb{V}(4s^3 + 27)$, then the quasihomogeneous part of f has non isolated singularities,
- if parameters s,t,u belong to $\mathbb{V}(s)$, then $\varphi(H_{J_f}) = \{0\}$ and the Tjurina number of the isolated singularity is 22,
- if parameters s,t,u belong to $\mathbb{C}^3 \backslash \mathbb{V}(4ts^4 + 27ts)$, then a basis of $\varphi(H_{J_f})$ is $\{1, y - \frac{13}{12}t + \frac{2u}{t}\}$ and the Tjurina number of the isolated singularity is 20,

- if parameters s, t, u belong to $\mathbb{V}(t) \backslash \mathbb{V}(4us^4 + 27us, t)$, then a basis of $\varphi(H_{J_f})$ is $\{1\}$ and the Tjurina number of the isolated singularity is 21, and
- if parameters s, t, u belong to $\mathbb{V}(t, u) \backslash \mathbb{V}(4s^4 + 27s, t)$, then $\varphi(H_{J_f}) = \{0\}$ and the Tjurina number of the isolated singularity is 22.

There exists a command "tjurinast1" in the implementation, which outputs only a Tjurina stratification of μ-constant deformations.

Example 3. A polynomial $f = x^5 + y^{11} + txy^9 + ux^2y^7 \in (\mathbb{C}[t, u])[x, y]$ is semi-quasihomogeneous of type $(55; (11, 5))$ where x, y are variables and t, u are parameters. Then, $x^5 + y^{11}$ is the quasihomogeneous part of f. Our implementation outputs a Tjurina stratification of f as follows:

```
[3562] tjurinast1(x^5+y^11,t*x*y^9+u*x^2*y^7,[11,5],[t,u],[x,y]);
zero dim.   []
-----------
[[0],[-9*u*t^5+58*u^2*t^3-88*u^3*t]]
40-6=34
-----------
[[t^2-4*u],[t,u]]
40-5=35
-----------
[[u],[t,u]]
40-5=35
-----------
[[9*t^2-22*u],[t,u]]
40-6=34
-----------
[[t],[t,u]]
40-6=34
-----------
[[t,u],[1]]
40-0=40
```

The meaning of the output is the following.
· The polynomial f always has an isolated singularity at the origin.

conditions of parameters	Tjurina numbers
$\mathbb{C}^2 \backslash \mathbb{V}(-9ut^5 + 58u^2t^3 - 88u^3t)$	34
$\mathbb{V}(t^2 - 4u) \backslash \mathbb{V}(t, u)$	35
$\mathbb{V}(u) \backslash \mathbb{V}(t, u)$	35
$\mathbb{V}(9t^2 - 22u) \backslash \mathbb{V}(t, u)$	34
$\mathbb{V}(t) \backslash \mathbb{V}(t, u)$	34
$\mathbb{V}(t, u)$	40

As we described in Example 2, the proposed algorithm can handle the case where even the quasihomogeneous part contains parameters. An algorithm of Martin and Pfister[3] can not handle such case. As the basic structures of Tjurina stratifications are determined by the structures of parametric local cohomology classes, the proposed algorithm outputs a nice decomposition of the parameter spaces. This is one of the advantages. Another advantage is that the computation consists of only linear algebra computation. Therefore, the proposed algorithm is more efficient than an existing algorithm, in computational complexity.

References

1. Arnold, V.I.: Normal forms of functions in neighbourhoods of degenerate critical points. Russian Math. Survey 29, 10–50 (1974)
2. Grothendieck, A.: Local Cohomology, notes by R. Hartshorne. Lecture Notes in Math., vol. 41. Springer (1967)
3. Martin, B., Pfister, G.: The kernel of the Kodaira-Spencer map of the versal μ-constant deformation of an irreducible plane curve with C^*-action. Journal of Symbolic Computation 7, 527–531 (1989)
4. Nabeshima, K., Tajima, S.: On efficient algorithms for computing parametric local cohomology classes associated with semi-quasihomogeneous singularities and standard bases. In: Proc. ISSAC 2014, pp. 351–358. ACM-Press (2014)
5. Nakamura, Y., Tajima, S.: On weighted-degrees for algebraic local cohomologies associated with semi-quasihomogeneous singularities. Advanced Studies in Pure Mathematics 46, 105–117 (2007)
6. Nakamura, Y., Tajima, S.: Algebraic local cohomologies and local b-functions Attached to semi-quasihomogeneous singularities with $L(f) = 2$. IRMA Lectures in Mathematics and Theoretical Physics 20, 103–116 (2012)
7. Noro, M., Takeshima, T.: Risa/Asir- A computer algebra system. In: Proc. ISSAC 1992, pp. 387–396. ACM-Press (1992)
8. Saito, K.: Quasihomogeneous isolated Singularitäten von Hyperflächen. Invent. Math. 14, 123–142 (1971)
9. Tajima, S.: Parametric local cohomology classes and Tjurina stratifications for μ-constant deformations of quasi-homogeneous singularities. In: Topics on Real and Complex Singularities, pp. 189–200. World Scientific (2014)
10. Tajima, S., Nakamura, Y.: Algebraic local cohomology class attached to quasi-homogeneous isolated hypersurface singularities. Publications of the Research Institute for Mathematical Sciences 41, 1–10 (2005)
11. Tajima, S., Nakamura, Y.: Annihilating ideals for an algebraic local cohomology class. Journal of Symbolic Computation 44, 435–448 (2009)
12. Tajima, S., Nakamura, Y.: Algebraic local cohomology classes attached to unimodal singularities. Publications of the Research Institute for Mathematical Sciences 48, 21–43 (2012)
13. Tajima, S., Nakamura, Y., Nabeshima, K.: Standard bases and algebraic local cohomology for zero dimensional ideals. Advanced Studies in Pure Mathematics 56, 341–361 (2009)

An Implementation Method of Boolean Gröbner Bases and Comprehensive Boolean Gröbner Bases on General Computer Algebra Systems

Akira Nagai and Shutaro Inoue

Tokyo University of Science, 1-3, Kagurazaka, Shinjuku-ku, Tokyo, Japan
1414703@ed.tus.ac.jp, sinoue@rs.kagu.tus.ac.jp

Abstract. We study an implementation method to compute Boolean Gröbner bases introduced in our previous work [15] in more detail. We extend our method for computing comprehensive Boolean Gröbner bases with a technique introduced in [10]. Our work has been implemented on the computer algebra system Risa/Asir. It enables us to do our recent work of a non-trivial application of Boolean Gröbner bases.

Keywords: Boolean Gröbner Bases, Boolean rings.

1 Introduction

A commutative ring $\mathbf{B}$ such that $\forall a \in \mathbf{B}\ a^2 = a$ is called a *Boolean ring*. A residue class ring $\mathbf{B}/I$ of the polynomial ring $\mathbf{B}[X_1, \ldots, X_n]$ by the ideal $I = \langle X_1^2 + X_1, \ldots, X_n^2 + X_n \rangle$ denoted by $\mathbf{B}(X_1, \ldots, X_n)$ also becomes a Boolean ring which is called a *Boolean polynomial ring*. A Gröbner basis in a Boolean polynomial ring is called a *Boolean Gröbner basis*. The notion of Boolean Gröbner basis was first introduced in [5],[6] together with an algorithm using special monomial reductions. The first implementation of Boolean Gröbner bases is written by the programming language Prolog([7]), two years later an improved version was also written by the parallel logic programming language KLIC([8]). Since we have to prepare a special data structure for representing Boolean polynomials and their special monomial reductions, it was extremely hard to implement the computation of Boolean Gröbner bases on a general computer algebra system. In [15], we introduced a computation method of Boolean Gröbner bases which can be easily implemented on most computer algebra system with a facility to compute Gröbner bases in polynomial rings over the Galois field $\mathbb{GF}_2$. Our method is also applicable for the computation of comprehensive Boolean Gröbner bases. In this short paper, we give a more detailed description of our method together with an improved version of our method for the computation of comprehensive Boolean Gröbner bases. Our work has been implemented on the computer algebra system Risa/Asir([14]). It enables us to do our recent work of a non-trivial application of Boolean Gröbner bases. The reader is referred to [13] for detailed descriptions of Boolean Gröbner bases.

H. Hong and C. Yap (Eds.): ICMS 2014, LNCS 8592, pp. 531–536, 2014.
© Springer-Verlag Berlin Heidelberg 2014

2   Computation of Boolean Gröbner Bases

Throughout of the paper we use **B** to represent some Boolean ring. $\bar{X}$ denotes some variables $X_1, \ldots, X_n$.

Definition 1. *Let St denote an enumerable set of strings (of some computer language). $\mathcal{P}_{FC}(St)$ denotes the set of all finite or co-finite subsets of St, i.e. $\mathcal{P}_{FC}(St) = \{S \subset St | S \text{ is finite or } St \setminus S \text{ is finite}\}$.*

In the rest of the paper we consider $\mathcal{P}_{FC}(St)$ as a Boolean ring and we concentrate on this Boolean ring for the computation of Boolean Gröbner bases. Note that an atomic element in this Boolean ring is nothing but a singleton of a string.

Let $f_1(\bar{X}), \ldots, f_l(\bar{X}) \in \mathcal{P}_{FC}(St)(\bar{X})$. When we compute a Boolean Gröbner basis of the ideal $\langle f_1(\bar{X}), \ldots, f_l(\bar{X}) \rangle$, we do not need to use whole $\mathcal{P}_{FC}(St)$ (which is infinite). Let $\{s_1, s_2, \ldots, s_k\}$ be the set of all string that is contained in a coefficient of some f_i and $e_1 = \{s_1\}, e_2 = \{s_2\}, \ldots, e_k = \{s_k\}$. Then, the finite Boolean subring **B** generate by $e_1, \ldots, e_k$ is enough to work. By Stone's representation theorem **B** is isomorphic to some direct product of $\mathbb{GF}_2$, more precisely it is isomorphic to $\mathbb{GF}_2^{k+1}$. Note that we have an extra atomic element $1 + e_1 + e_2 + \cdots + e_k$ of **B**.

Definition 2. *For any element $b \in \mathbf{B}$, we can express b in the form $b = b_1 e_1 + b_2 e_2 + \cdots + b_k e_k + b_{k+1}(1 + e_1 + e_2 + \cdots + e_k)$, where $b_i \in \{0, 1\}, 1 \leq i \leq k+1$. Let θ be an isomorphism from **B** to $\mathbb{GF}_2^{k+1}$ defined by $\theta(b) = (b_1, \ldots, b_{k+1})$. For each $i = 1, \ldots, k+1$, a projection π_i is an epimorphism from **B** to $\mathbb{GF}_2$ defined by $\pi_i(b) = \theta(b)_i$(the i-th component of $\theta(b)$). We also define a monomorphism π_i^{-1} from $\mathbb{GF}_2$ to **B** by $\pi_i^{-1}(0) = 0$ and $\pi_i^{-1}(1) = e_i$ for each $i = 1, \ldots, k$ and $\pi_{k+1}^{-1}(1) = 1 + e_1 + \cdots + e_k$. θ, π_i and π_i^{-1} are naturally extended to an isomorphisms from $\mathbf{B}(\bar{X})$ to $\mathbb{GF}^{k+1}(\bar{X})$, an epimorphism from $\mathbf{B}(\bar{X})$ to $\mathbb{GF}(\bar{X})$ and a monomorphism from $\mathbb{GF}(\bar{X})$ to $\mathbf{B}(\bar{X})$ respectively.*

We can easily prove the following lemma.

Lemma 3. *For each $f \in \mathbf{B}(\bar{X})$ $f = \pi_1^{-1}(\pi_1(f)) + \cdots + \pi_{k+1}^{-1}(\pi_{k+1}(f))$.*

Since $\mathbb{GF}_2$ is a field, a Boolean Gröbner basis in the Boolean polynomial ring $\mathbb{GF}_2(\bar{X})$ can be computed by a usual Buchberger algorithm and we can compute it in most computer algebra system which has a facility to compute Gröbner bases in a polynomial ring over $\mathbb{GF}_2$.

For a finite set of polynomials $F = \{f_1, \ldots, f_l\} \subset \mathcal{P}_{FC}(St)(\bar{X})$, let $e_1, \ldots, e_k$ and **B** be as above. We abuse the notations π_i and π_i^{-1} for a set, i.e. $\pi_i(P) = \{\pi_i(f) | f \in P\}$ and $\pi_i^{-1}(Q) = \{\pi_i^{-1}(f) | f \in Q\}$.

Now we are ready to describe our implementation method.

Algorithm: Boolean GB
input: F a finite subset of $\mathcal{P}_{FC}(St)(\bar{X})$ and a term order $>$ on $T(\bar{X})$
output: G a reduced Boolean Gröbner basis of $\langle F \rangle$ w.r.t. $>$

For each $i = 1, \ldots, k + 1$ compute the reduced Boolean Gröbner basis G_i of the ideal $\langle \pi_i(F) \rangle$ in $\mathbb{GF}_2(\bar{X})$.
Set $G = \cup_{i=1}^{k+1} \pi_i^{-1}(G_i)$.
In order to get a stratified Boolean Gröbner basis, we further need the following manipulation.

Algorithm: Stratification
input: G a reduced Boolean Gröbner basis in $\mathcal{P}_{FC}(St)(\bar{X})$
output: G' a stratified Boolean Gröbner basis
Let $\{t_1, \ldots, t_s\}$ be the set of all leading terms(i.e. initials) of some polynomial in G. For each $i = 1 \ldots, t_s$ $g_i = \sum_{LT(g) = t_i, g \in G} g$. Set $G' = \{g_1, \ldots, g_s\}$.

Example 1. *Compute the stratified Boolean Gröbner basis of $F = \{(\{s_1, s_2\} + 1)(XY + X + Y), \{s_1\}(X + 1), XY\}$ in $\mathcal{P}_{FC}(St)(X, Y)$ w.r.t. the lex term order such that $X > Y$.*
We apply projection map for F.
$\pi_1(F) = \{0, X + 1, XY\}$, $\pi_2(F) = \{0, 0, XY\}$, $\pi_3(F) = \{XY + X + Y, 0, XY\}$
Compute the reduced Boolean Gröbner basis G_i of $\pi_i(F)$ for each i.
$G_1 = \{X, Y\}$, $G_2 = \{XY\}$, $G_3 = \{X + 1, Y\}$.
Compute $\pi_1^{-1}(G_1) = \{\{s_1\}X, \{s_1\}Y\}$, $\pi_2^{-1}(G_2) = \{\{s_2\}XY\}$ and $\pi_3^{-1}(G_3) = \{(1 + \{s_1, s_2\})X + (1 + \{s_1, s_2\}), (1 + \{s_1, s_2\})Y\}$.
We obtain a reduced Boolean Gröbner basis
$G = \{\{s_1\}X, \{s_1\}Y, \{s_2\}XY, (1 + \{s_1, s_2\})X + (1 + \{s_1, s_2\}), (1 + \{s_1, s_2\})Y\}$.
Applying the stratification to G we finally produce the stratified Boolean Gröbner basis $G' = \{\{s_2\}XY, (\{s_2\} + 1)X + \{s_1\}, (\{s_2\} + 1)Y\}$.

3 Comprehensive Boolean Gröbner Bases

In this section we concentrate on the computation of comprehensive Boolean Gröbner bases. Unlike the construction of comprehensive Gröbner bases (or comprehensive Gröbner systems) in polynomial rings over fields, the construction of comprehensive Boolean Gröbner bases is extremely easy.

Theorem 4. *Let F be a finite set of Boolean polynomials in $\mathbf{B}(\bar{A}, \bar{X})$ with variables $\bar{A}$ and $\bar{X}$. Regarding $\mathbf{B}(\bar{A}, \bar{X})$ as a Boolean polynomial rings $\mathbf{B}(\bar{A})(\bar{X})$ with variables $\bar{X}$ over the coefficient Boolean ring $\mathbf{B}(\bar{A})$, compute a(stratified) Boolean Gröbner basis of F in this Boolean polynomial ring. Then it becomes a(stratified) comprehensive Boolean Gröbner basis of F with parameters $\bar{A}$.*

The free software [8] contains an implementation to compute comprehensive Boolean Gröbner bases based on this theorem. Unfortunately, when we have many parameters this implementation is not practical. We can not apply our method presented in the previous section either, since the number of the atomic elements of $\mathbf{B}(\bar{A})$ is equal to p^q where p is the number of the atomic element of $\mathbf{B}$ and q is the number of variables in $\bar{A}$.

The following theorem introduced in [11] enables us to have a more practical algorithm.

Theorem 5. *Let F be a finite set of Boolean polynomials in $\mathbf{B}(\bar{A}, \bar{X})$ with variables $\bar{A}$ and $\bar{X}$. Give a term order $>$ of $T(\bar{X})$, compute a Boolean Gröbner basis G of F in $\mathbf{B}(\bar{A}, \bar{X})$ w.r.t. a term order of $T(\bar{A}, \bar{X})$ which extends $>$ so that $\bar{X}$ is lexicographically greater than $\bar{A}$. Then it becomes a comprehensive Boolean Gröbner basis of F with parameters $\bar{A}$ on the variety of $\mathbf{B}(\bar{A}) \cap \langle F \rangle$.*

Note that the obtained comprehensive Boolean Gröbner basis is not stratified nor even reduced after a specialization in general.

Of course we can apply our method to compute a Boolean Gröbner basis G for the computation of a comprehensive Boolean Gröbner basis by this theorem. This observation was given in our previous paper [15].

When the computation of the Boolean Gröbner basis G does not terminate in a realistic length of time, we have to give up. In [10] we gave a computation method to compute an elimination ideal $\mathbf{B}(\bar{A}) \cap \langle F \rangle$ without computing a Boolean Gröbner basis of the whole ideal $\langle F \rangle$. This technique leads us to have an alternative method to compute comprehensive Boolean Gröbner bases based on our method.

Algorithm: Comprehensive Boolean GB
input: A finite set F of $\mathcal{P}_{FC}(St)(\bar{A}, \bar{X})$ and a term order $>$
output: G a comprehensive stratified Boolean Gröbner basis w.r.t. $>$
on the variety of $\langle F \rangle \cap \mathcal{P}_{FC}(St)(\bar{A})$
Compute the elimination ideal $\langle F \rangle \cap \mathcal{P}_{FC}(St)(\bar{A})$.
Compute atomic elements $e_1, \ldots, e_k$ of the Boolean ring
$$\mathcal{P}_{FC}(St)(\bar{A})/\langle F \rangle \cap \mathcal{P}_{FC}(St)(\bar{A}).$$
For each $i = 1, \ldots, k$ compute the reduced Boolean Gröbner basis G_i of the ideal $\langle \pi_i(F) \rangle$ in $\mathbb{GF}_2(\bar{X})$. Set $G = \mathbf{Stratification}(\cup_{i=1}^{k} \pi_i^{-1}(G_i))$.

Unless the elimination ideal $\langle F \rangle \cap \mathcal{P}_{FC}(St)(\bar{A})$ is rich enough, this algorithm is not practical. For example, when $\langle F \rangle \cap \mathcal{P}_{FC}(St)(\bar{A})$ is a trivial ideal $\{0\}$ the ideal $\mathcal{P}_{FC}(St)(\bar{A})/\langle F \rangle \cap \mathcal{P}_{FC}(St)(\bar{A})$ is nothing but the whole ideal $\mathcal{P}_{FC}(St)(\bar{A})$. In such a case the number k would be extremely big as mentioned right after the theorem 4.

4   Conclusion and Remarks

We can compute a Boolean Gröbner basis by the computation of Gröbner bases of $\mathbb{GF}_2$ with a proper term order. Look at the Example 1. Using extra variables S_1 and S_2 to represent singletons $\{s_1\}$ and $\{s_2\}$, F is translated into the set of polynomials $\{(S1 + S2 + 1)(XY + Y + X), S1(X + 1), XY, S1S2, S1 + S2 + 1\}$ in $\mathbb{GF}_2(X, Y, S1, S2)$. The reduced Boolean Gröbner basis of $\langle (S1 + S2 + 1)(XY + Y + X), S1(X + 1), XY, S1S2, S1 + S2 + 1 \rangle$ in $\mathbb{GF}_2(X, Y, S1, S2)$ w.r.t. the lex term order such $X > Y > S2 > S1$ has the form:

$$\{XY, S2X + X + S1, S1X + S1, S2Y + Y, S1Y, S2S1\}.$$

By Theorem 5, it is a comprehensive Boolean Gröbner basis of $\{(S1 + S2 + 1)(XY + Y + X), S1(X + 1), XY, S1S2, S1 + S2 + 1\}$ in $\mathbb{GF}_2(X, Y)$ with parameters $S1, S2$. Obvious it is also a comprehensive Boolean Gröbner basis in $P_{FC}(St)(X, Y)$. Hence, $\{XY, \{s2\}X + X + \{s1\}, \{s1\}X + \{s1\}, \{s2\}Y + Y, \{s1\}Y\}$ is a Boolean Gröbner basis of $F = \{((\{s_1, s_2\} + 1)(XY + X + Y), \{s_1\}(X + 1), XY\}$ in $P_{FC}(St)(X, Y)$ w.r.t. the lex term order such that $X > Y$.

When we have many strings, this computation will be extremely slow comparing it with our method. The obtained Boolean Gröbner basis is not even reduced in general.

We conclude this paper with the following remark. There are many sophisticated implementations to compute Boolean Gröbner bases such as [1,2,3]. Though any of them deals with only Boolean polynomials over $\mathbb{GF}_2$, we can implement our method on those systems.

References

1. Decker, W., Greuel, G.-M., Pfister, G., Schönemann, H.: Singular 3-1-2 - A computer algebra system for polynomial computations (2010),
   http://www.singular.uni-kl.de/
2. Bosma, W., Cannon, J., Playoust, C.: The Magma algebra system. I. The user language. J. Symbolic Comput. 24(3-4), 235–265 (1997),
   http://magma.maths.usyd.edu.au/magma/
3. Brickenstein, M., Dreyer, A.: A framework for Gröbner-basis computations with Boolean polynomials. J. Symbolic Comput. 44(9), 1326–1345 (2009),
   http://polybori.sourceforge.net/ (PolyBoRi Polynomials over Boolean Rings)
4. Noro, M., et al.: A Computer Algebra System Risa/Asir (2009),
   http://www.math.kobe-u.ac.jp/Asir/asir.html
5. Sakai, K., Sato, Y.: Boolean Gröbner bases. ICOT Technical Memorandum 488 (1988)
6. Sakai, K., Sato, Y., Menju, S.: Boolean Gröbner bases (revised). ICOT Technical Report 613 (1991)
7. Sato, Y.: Set Constraint Solvers (Prolog Version) (1996), http://www.jipdec.or.jp/archives/icot/ARCHIVE/Museum/FUNDING/funding-95-E.html; Weispfenning, V.: Gröbner Bases in polynomial ideals over commutative regular rings. In: Davenport, J.H. (ed.) ISSAC 1987 and EUROCAL 1987. LNCS, vol. 378, pp. 336–347. Springer, Heidelberg (1989)
8. Sato, Y.: Set Constraint Solvers (KLIC Version) (1998), http://www.jipdec.or.jp/archives/icot/ARCHIVE/Museum/FUNDING/funding-98-E.html
9. Sato, Y., Inoue, S.: On the Construction of Comprehensive Boolean Gröbner Bases. In: Proceedings of the 7th Asian Symposium on Computer Mathematics (ASCM 2005), pp. 145–148 (2005)
10. Sato, Y., Nagai, A., Inoue, S.: On the computation of elimination ideals of boolean polynomial rings. In: Kapur, D. (ed.) ASCM 2007. LNCS (LNAI), vol. 5081, pp. 334–348. Springer, Heidelberg (2008)
11. Inoue, S.: On the Computation of Comprehensive Boolean Gröbner Bases. In: Gerdt, V.P., Mayr, E.W., Vorozhtsov, E.V. (eds.) CASC 2009. LNCS, vol. 5743, pp. 130–141. Springer, Heidelberg (2009)

12. Rudeanu, S.: Boolean functions and equations. North-Holland Publishing Co., American Elsevier Publishing Co., Inc., Amsterdam, New York (1974)
13. Sato, Y., et al.: Boolean Gröbner bases. J. Symbolic Comput. 46, 622–632 (2011)
14. Inoue, S.: BGSet Boolean Gröebner bases for Sets (2009), http://www.mi.kagu.tus.ac.jp/~inoue/BGSet/
15. Inoue, S., Nagai, A.: On the implementation of Boolean Gröbner bases. In: Proceedings of the Joint Conference of ASCM 2009 and MACIS 2009. COE Lect. Note, vol. 22, pp. 58–62. Kyushu Univ. Fac. Math. (2009)

A Method to Determine if Two Parametric Polynomial Systems Are Equal

Jie Zhou and Dingkang Wang

KLMM, Academy of Mathematics and Systems Science, CAS, Beijing, China
jiezhou@amss.ac.cn, dwang@mmrc.iss.ac.cn

Abstract. The comprehensive Gröbner systems of parametric polynomial ideal were first introduced by Volker Weispfenning. Since then, many improvements have been made to improve these algorithms to make them useful for different applications. In contract to reduced Groebner bases, which is uniquely determined by the polynomial ideal and the term ordering, however, comprehensive Groebner systems do not have such a good property. Different algorithm may give different results even for a same parametric polynomial ideal. In order to treat this issue, we give a decision method to determine whether two comprehensive Groebner systems are equal. The polynomial ideal membership problem has been solved for the non-parametric case by the classical Groebner bases method, but there is little progress on this problem for the parametric case until now. An algorithm is given for solving this problem through computing comprehensive Groebner systems. What's more, for two parametric polynomial ideals and a constraint over the parameters defined by a constructible set, an algorithm will be given to decide whether one ideal contains the other under the constraint.

Keywords: Constructible Set, Quasi-algebraic set, Gröbner Bases, Comprehensive Gröbner System.

1 Introduction

The comprehensive Gröbner systems of parametric polynomial ideal were introduced by Volker Weispfenning in 1992 [12]. Many engineering problems are parameterized and have to be repeatedly solved for different values of parameters. The comprehensive Gröbner systems can give the structure of solution space(finitely many, infinitely many, or the dimension of the solutions), which is similar to the properties of the Gröbner bases.

Let k be a field, $k[U][X]$ be the polynomial ring with the parameters $U = \{u_1, \ldots, u_m\}$ and the variables $X = \{x_1, \ldots, x_n\}$, where U and X are disjoint. K is an algebraically closed field of k, F be a subset of $k[U][X]$. A specialization σ is the homomorphism from $k[U][X]$ to $K[X]$. The comprehensive Gröbner systems for F is a finite set $\mathcal{G} = \{(A_1, G_1), \ldots, (A_l, G_l)\}$, which satisfy $\sigma_{\bar{a}}(G_i)$ is a Gröbner basis for the ideal $\langle \sigma_{\bar{a}}(F) \rangle$ in $K[X]$ for any $\bar{a} \in A_i$ and $i = 1, \ldots, l$.

Many algorithms have been provided for computing the comprehensive Gröbner systems, including CGB (V.Weispfenning, 1992)[12], CCGB(V.Weispfenning,

H. Hong and C. Yap (Eds.): ICMS 2014, LNCS 8592, pp. 537–544, 2014.
© Springer-Verlag Berlin Heidelberg 2014

2003)[13], ACGB(Y.Sato and A.Suzuki, 2003)[9], SACGB(Y.Sato and A.Suzuki, 2006)[10], HSGB(González-Vega et al., 2005)[2] and BUILDTREE (A.Montes, 2002)[5]. A speed-up of the algorithm was given by Nabeshima [7]. A newest version for computing CSG was provided by Kapur, Sun and Wang by removing redundant segments [3]. There is an related concept of Gröbner cover introduced by Montes and Wibmer in 2010 [6]. In contract to reduced Gröbner bases, which is uniquely determined by the polynomial ideal and the term ordering, however, the comprehensive Gröbner systems do not have such a good property. Different algorithm may output different results. Weispfenning[13], Manubens and Montes[4], Wibmer[14] have done some researches about the canonical Gröbner system. In this paper, we compare two *CGS* from another aspect.

In ISSAC'09, Suzuki and Sato[8] gave a method to compute the inverses in residue class rings of parametric polynomial ideals. For a given parametric ideal $I \subset k[U][X]$, and a polynomial $f \in k[U][X]$, they first compute a *CGS* $\mathcal{G}$ of $I + \langle fy - 1 \rangle$. For any branch $(A, G) \in \mathcal{G}$. if there is a polynomial which can be expressed as $y - h$, where $h \in k[U][X]$, then f is invertible in $K[X]/(I : f^\infty)$ under the constraint A. In order to judge whether f is invertible in $K[X]/I$, it still need to decide whether I and $I : f^\infty$ are equal under the constraint A. This is the motivation of the paper.

The ideal membership problem of non-parametric case has been totally solved in the past[1]. But there is little research about the problem of parametric case until now. This paper can solve this problem through computing CGS of the parametric polynomial ideal. In the paper, we also give a method to decide whether two comprehensive Gröbner systems are equal. As a consequence, for two parametric polynomial ideals and a constructible set, the method can judge whether one of ideals is contained in the other one under the constraint of the constructible set.

This paper is organized as follow. Section 2 gives some preliminaries about the constructible set and the quasi-algebraic set. In section 3, the method of solving the ideal membership problem is presented. The inclusion and equivalence relation about two parametric ideal are also given in this section. Finally, some conclusions are given in section 4.

2   Notations and Preliminary

2.1   Notations

Let k be a field, K be the algebraic closure of k, R be a polynomial ring $k[U]$ in parameters $U = \{u_1, \ldots, u_m\}$, and $R[X]$ be a polynomial ring over R in variables $X = \{x_1, \ldots, x_n\}$ where X and U are disjoint. Let $PP(X)$ be the sets of power products of X, and $\prec$ be an admissible monomial ordering on $PP(X)$. As before, for a polynomial $f \in R[X] = k[U][X]$, the leading power product, leading coefficient and leading monomial of f w.r.t. the ordering $\prec$ are denoted by $\mathrm{lpp}(f)$, $\mathrm{lc}(f)$ and $\mathrm{lm}(f)$ respectively. Note that $\mathrm{lc}(f) \in k[U]$ and $\mathrm{lm}(f) = \mathrm{lc}(f)\mathrm{lpp}(f)$.

For arbitrary $\bar{a} \in K^m$, a **specialization** of R induced by $\bar{a}$ is a homomorphism $\sigma_{\bar{a}} : R \longrightarrow K$. That is, for $\bar{a} \in K^m$, the induced specialization $\sigma_{\bar{a}}$ is defined as follows:

$$\sigma_{\bar{a}} : f \longrightarrow f(\bar{a}),$$

where $f \in R$. Every specialization $\sigma_{\bar{a}} : R \longrightarrow K$ extends canonically to a specialization $\sigma_{\bar{a}} : R[X] \longrightarrow K[X]$ by applying $\sigma_{\bar{a}}$ coefficient-wise. For a subset F of $k[U][X]$, $\sigma_{\bar{a}}(F) = \{\sigma_{\bar{a}}(f) \mid f \in F\}$.

Definition 1 (Member). *Let F be an subsets of parametric polynomial ring $k[U][X]$, $f \in k[U][X]$. We say f is a* **member** *of the ideal generated by E, if for any $\bar{a}$ in K^m, $\sigma_{\bar{a}}(f)$ is a member of the ideal generated by $\sigma_{\bar{a}}(E)$ in $K[X]$.*

Definition 2 (Contain). *Let E, F be two subsets of the parametric polynomial ring $k[U][X]$. We say E* **contains** *F, if the ideal generated by $\sigma_{\bar{a}}(E)$ contains the ideal generated by $\langle \sigma_{\bar{a}}(F) \rangle$ in $K[X]$ for any $\bar{a} \in K^m$. If E contains F, and F contains E, we say E and F are* **equal**.

For any $\bar{a} \in K^m$, if every element in the Gröbner bases of $\langle \sigma_{\bar{a}}(F) \rangle$ is contained in the ideal $\langle \sigma_{\bar{a}}(E) \rangle$, it is obvious that E contains F. For different $\bar{a} \in K^m$, the Gröbner bases of $\langle \sigma_{\bar{a}}(F) \rangle$ may be different, so we need to study the structure of the Gröbner bases of $\langle \sigma_{\bar{a}}(F) \rangle$ with respect to the parametric space K^m. Before that, we introduce the notations about quasi-algebraic set and the constructible set.

For any subset $E = \{e_1, \ldots, e_s\}$ of $k[U]$, the set of common zeros in K^m of E is a Zariski closed set, denoted by $\mathbb{V}(E)$. For a single polynomial h in R, we denote the complement of $\mathbb{V}(h)$ in K^m by $\mathbb{V}(h)^c$, which is a basic Zariski open set. A **quasi-algebraic set** is the intersection of a Zariski closed set with a basic Zariski open set, and a **constructible set** is a finite union of quasi-algebraic set [11]. We denote $\mathbb{V}(E) \cap \mathbb{V}(h)^c$ by $\mathbb{V}(E) \backslash \mathbb{V}(h)$.

In this paper, we only consider the constructible set has a form $\mathbb{V}(E) \backslash \mathbb{V}(N)$, where $E = \{e_1, \ldots, e_s\}$ and $N = \{n_1, \ldots, n_t\}$ are subsets of R. It is obvious that $\mathbb{V}(E) \backslash \mathbb{V}(N) = \cup_{i=1}^{t}(\mathbb{V}(E) \backslash \mathbb{V}(n_i))$. We say a constructible set $\mathbb{V}(E) \backslash \mathbb{V}(N)$ is **consistent** if it is not empty.

Now we can describe the structure of the Gröbner bases of a parametric ideal. For a parametric polynomial system $F \subset R[X]$, a comprehensive Gröbner system of F is defined below.

Definition 3 (CGS). *Let F be a subset of $R[X]$, $A_1, \ldots, A_l$ be algebraical constructible subsets of K^m, $G_1, \ldots, G_l$ be subsets of $R[X]$, and S be a subset of K^m such that $S \subset A_1 \cup \cdots \cup A_l$. A finite set $\mathcal{G} = \{(A_1, G_1), \ldots, (A_l, G_l)\}$ is called a* **comprehensive Gröbner system** *on S for F, if $\sigma_{\bar{a}}(G_i)$ is a Gröbner basis for the ideal $\langle \sigma_{\bar{a}}(F) \rangle$ in $K[X]$ for any $\bar{a} \in A_i$ and $i = 1, \ldots, l$. Each (A_i, G_i) is called a branch of $\mathcal{G}$. Particularly, if $S = K^m$, then $\mathcal{G}$ is called a comprehensive Gröbner system for F.*

In above, the constructible set A_i can be expressed as $A_i = \mathbb{V}(E_i) \backslash \mathbb{V}(N_i)$, where E_i, N_i are subsets of $k[U]$.

Notes that, for many algorithm of computing CGS, such as the algorithm given in [3,7], the output of these algorithm has the following property: for any branch (A, G) of a CGS, for each $g \in G$, $\sigma_{\bar{a}}(\mathrm{lc}(g)) \neq 0$ for any $\bar{a} \in A$. So in the paper, we always assume the CGS has the above property.

2.2 Some Preliminaries

Given two polynomial f, g in $k[U][X]$, and a term ordering " $\prec$ ". If there is a term $c_\alpha X^\alpha$ in f with the coefficient $c_\alpha \neq 0$, and X^α is a multiple of $\mathrm{lpp}(g)$, we say f can be reduced by g, and $r = \mathrm{lc}(g)f - c_\alpha X^\gamma g$ is the remainder of f reduced by g through one step reduction, where $X^\gamma = \frac{X^\alpha}{\mathrm{lpp}(g)}$. Continuing reduce r by g until no term of the remainder is a multiple of $\mathrm{lpp}(g)$, assume the remainder is r_0, we say the r_0 is the **remainder** of f reduced by g.

For a subset F of $k[U][X]$, and a polynomial $g \in k[U][X]$, we can define the reduction of g by F as the following lemma. The pseudo division algorithm in $k[U][X]$ is similar to the division algorithm in $k[X]$, more details about the division algorithm can refer to the book [1].

Lemma 1. *Let $F = \{f_1, \ldots, f_s\}$ be a subset of $k[U][X]$, and " $\prec$ " be a term ordering. Then every $g \in k[U][X]$ can be represented as:*

$$\prod_{i=1}^{s} \mathrm{lc}(f_i)^{\delta_i} g = p_1 f_1 + \cdots + p_s f_s + r,$$

for some elements $h_1, \ldots, h_s, r$ in $k[U][X]$, nonnegative integers $\delta_1, \ldots, \delta_s$, such that:

i.) $p_i = 0$ or $\mathrm{lpp}(p_i f_i) \preceq \mathrm{lpp}(g)$,
ii.) $r = 0$ or no term of r is a multiple of any $\mathrm{lpp}(f_i)$, $i = 1, \ldots, s$.

At the end of this part, we review some properties about the constructible set and the quasi-algebraic set.

For a constructible set $A = \mathbb{V}(E) \setminus \mathbb{V}(N)$, where $E, N = \{n_1, \ldots, n_l\}$ are subsets of $k[U]$. We are only interested in those constructible sets which are consistent. Since $A = \mathbb{V}(E) \setminus \mathbb{V}(N) = \bigcup_{i=1}^{l}(\mathbb{V}(E) \setminus \mathbb{V}(n_i))$, we only need to know whether the quasi-algebraic set $\mathbb{V}(E) \setminus \mathbb{V}(n_i)$ is empty, for $i = 1, \ldots, l$.

Lemma 2. *Let $A = \mathbb{V}(E) \setminus \mathbb{V}(h)$ be a quasi-algebraic set, where E is a subset of $k[U]$ and h is a polynomial in $k[U]$. Then A is consistent if and only if h is not in the radical ideal generated by E in $k[U]$.*

In the following lemma, we show any finite intersection of quasi-algebraic set is a quasi-algebraic set.

Lemma 3 ([11]). *Let $A_1 = \mathbb{V}(E_1) \setminus \mathbb{V}(h_1)$, $A_2 = \mathbb{V}(E_2) \setminus \mathbb{V}(h_2)$ be two quasi-algebraic sets, where E_1, E_2 are subsets of $k[U]$, and h_1, h_2 are polynomials in $k[U]$. Then $A_1 \cap A_2$ is also a quasi-algebraic set, and $A_1 \cap A_2 = \mathbb{V}(E_1, E_2) \setminus \mathbb{V}(h_1 h_2)$.*

Given a quasi-algebraic set $A = \mathbb{V}(E)\backslash\mathbb{V}(h) \subset K^m$, and a parametric polynomial $f = c_1 X^{\alpha_1} + \cdots + c_t X^{\alpha_t} \in k[U][X]$, $c_i \in k[U]$ for $i = 1,\ldots,t$. If for any $\bar{a} \in A$, the specialization $\sigma_{\bar{a}}(f) = 0$, then A must be a subset of the common zeros of the coefficients, i.e. $A \subset \mathbb{V}(c_1,\ldots,c_t)$. Let $\overline{A}$ be the Zariski closure of A,

$$\mathbb{V}(\langle E \rangle : h^\infty) = \overline{\mathbb{V}(\langle E \rangle)\backslash\mathbb{V}(h)} = \overline{A} \subset \overline{\mathbb{V}(c_1,\ldots,c_t)} = \mathbb{V}(c_1,\ldots,c_t),$$

so c_i is in the radical ideal generated by the saturated ideal $\langle E \rangle : h^\infty$ in $k[U]$ for $i = 1,\ldots,t$. On the other hand, we can regard the polynomials in E and h as polynomial in $k[U,X]$, then f is in the radical ideal generated by the saturated ideal $\langle E \rangle : h^\infty$ in $k[U,X]$.

Lemma 4. *Given a quasi-algebraic set $A = \mathbb{V}(E)\backslash\mathbb{V}(h) \subset K^m$, and a parametric polynomial $f \in k[U][X]$. If for any $\bar{a} \in A$, the specialization $\sigma_{\bar{a}}(f) = 0$, then $\langle E, fhv - 1 \rangle = \langle 1 \rangle = k[v,U,X]$, where v is an auxiliary variable different from X an U.*

3   The Computations about Two Parametric Ideals

In this section, first we give the method to judge whether a parametric polynomial is a member of a parametric polynomial ideal. Then we give the method to determine the inclusion and equivalence relationship about two parametric polynomial ideals. Several examples will be given for illustrating our methods.

3.1   The Membership Problem of Parametric Ideal

Given a subset F of parametric polynomial ring $k[U][X]$, and a parametric polynomial f in $k[U][X]$. If f is a member of the ideal generated by F in $k[U,X]$, it is obvious for any $\bar{a} \in K^m$, $\sigma_{\bar{a}}(f)$ is a member of the ideal generated by $\sigma_{\bar{a}}(F)$. But there are some situations, f is not a member of the ideal generated by F in $k[U,X]$, f is still a member of ideal generated by F. For example, $F = \{a^3b^2x^2 - y^2, ab^2x^2 - b^2xy^2\}$, $f = abx^2 - b^3y^6$, it is easy to check f is not in the ideal generated by F in $k[a,b,x,y]$. But we will see, for any $(a,b) \in \mathbb{C}^2$, $\sigma_{\bar{a}}(f)$ is a member of the ideal generated by $\sigma_{\bar{a}}(F)$, so f is a member of the ideal generated by F.

In order to check whether a parametric polynomial is a member of a parametric ideal, we have following theorem.

Theorem 4. *Let F be a subset of parametric polynomial ring $k[U][X]$, and f be a parametric polynomial in $k[U][X]$. Assume $\mathcal{G} = \{(A_1, G_1),\ldots,(A_l, G_l)\}$ be a comprehensive Gröbner system of F w.r.t. a term ordering " $\prec$ ". For any branch (A, G) in $\mathcal{G}$, r is the remainder of f reduced by G. If for any $\bar{a} \in A$, $\sigma_{\bar{a}}(r) = 0$, then f is a member of the parametric ideal generated by F.*

We continue the above example to illustrate the result given in Theorem 1.

Example 1. Let $F = \{a^3b^2x^2 - y^2, ab^2x^2 - b^2xy^2\}$ be a subset of $\mathbb{Q}[a, b][x, y]$, and $f = abx^2 - b^3y^6$ in $\mathbb{Q}[a, b][x, y]$. Check whether f is a member of the ideal generated by F.

First, a *CGS* $\mathcal{G}$ of F w.r.t. a lexicographic ordering $x \succ y$ is computed.

$$\mathcal{G} = \{(A_1, G_1), (A_2, G_2), (A_3, G_3)\},$$

where $A_1 = \mathbb{V}(\emptyset) \backslash \mathbb{V}(ab), G_1 = \{ab^2y^4 - y^2, xy^2 - b^2y^6, ab^2x^2 - b^2xy^2\}; A_2 = \mathbb{V}(a) \backslash \mathbb{V}(1), G_2 = \{y^2\}; A_3 = \mathbb{V}(b) \backslash \mathbb{V}(a), G_3 = \{y^2\}$.

For branch (A_1, G_1), f is reduced by G_1 to 0. For branch (A_2, G_2), f is reduced by G_2 to abx^2. Since under the constraint A_2, $a = 0$, so for any $\bar{a} \in A_2$, $\sigma_{\bar{a}}(abx^2) = 0$. For branch (A_3, G_3), f is reduced by G_3 to abx^2. Since under the constraint A_3, $b = 0$, so for any $\bar{a} \in A_3$, $\sigma_{\bar{a}}(abx^2) = 0$. By the Theorem 1, f is a member of parametric ideal generated by F.

3.2   The Equivalence Relationship about Two Parametric Ideal

In this part, we give the method to determine whether a parametric polynomial ideal E contains another parametric polynomial ideal F.

Let $\mathcal{G}_1 = \{(A_1, G_1), \ldots, (A_l, G_l)\}$ be a *CGS* of E w.r.t. a term ordering " $\prec_1$ ", and $\mathcal{G}_2 = \{(B_1, H_1), \ldots, (B_r, H_r)\}$ be a *CGS* of F w.r.t. a term ordering " $\prec_2$ ". For any branch $(A, G) \in \mathcal{G}_1$, if $\sigma_{\bar{a}}(F)$ is contained in $\langle \sigma_{\bar{a}}(E) \rangle$, it is obvious E contains F. We have the following theorem.

Theorem 5. *Let* $\mathcal{G}_1, \mathcal{G}_2$ *be as above. For any branch* $(A, G) \in \mathcal{G}_1$ *and* $(B, H) \in \mathcal{G}_2$, *assume* $G = \{g_1, \ldots, g_s\}, H = \{h_1, \ldots, h_t\}$, *and* r_i *be the remainder of* h_i *reduced by* G *for* $i = 1, \ldots, t$. *If for any* $\bar{a} \in A \cap B$, $\sigma_{\bar{a}}(r_i) = 0$, *then* E *contains* F, *where* $i = 1, \ldots, t$.

Remark 1. If we only need to know whether F is contained in E under some constructible set A, we only need to compute a *CGS* of F on A, then use the Theorem 2.

If E contains F and F contains E, E and F are equal. We have the following consequence of Theorem 2.

Corollary 6. *Let* $\mathcal{G}_1, \mathcal{G}_2$ *be as above. For any branch* $(A, G) \in \mathcal{G}_1$ *and* $(B, H) \in \mathcal{G}_2$, *assume* $G = \{g_1, \ldots, g_s\}, H = \{h_1, \ldots, h_t\}$, r_i *be the remainder of* h_i *reduced by* G *w.r.t. "* $\prec_1$ *" , and* q_j *be the remainder of* g_j *reduced by* H *w.r.t. "* $\prec_2$ *" . If for any* $\bar{a} \in A \cap B$, $\sigma_{\bar{a}}(r_i) = 0$ *and* $\sigma_{\bar{a}}(q_j) = 0$, *then* E *and* F *are equal, where* $i = 1, \ldots, t, j = 1, \ldots s$.

In the ISSAC'09, Suzuki and Sato give a method to compute the inverse in residue class rings of parametric polynomial ideals. Given a parametric polynomial ideal $I \subset k[U][X]$, and $f \in k[U][X]$, they first compute a *CGS* of $\langle I + fy - 1 \rangle$ w.r.t. a block order "$y >> X >> U$" in $k[U][X, y]$. By their method, it only can decide whether f is invertible in $K[X]/(I : f^\infty)$ in every branch directly.

In order to judge whether f is invertible in $K[X]/I$, it still need to compare whether I and $I : f^\infty$ are equal in every branch. The following is the example come from [8].

Example 2. Let $I = \{ax_3^2 + 2x_2x_3 + bx_1^2x_3 + (-b+d)x_1x_3 - dx_3 + ax_2^2 + abx_1^2x_2 + adx_1x_2 + x_1^2 + cx_1 + e, ax_2 + ax_1x_2 + x_1x_3\}$ be a set of parametric polynomials, $f = x_1 + ax_2 + x_3$ be a parametric polynomial in $\mathbb{Q}[U][X]$, where $U = \{a, b, c, d, e\}$ are parameters and $X = \{x_1, x_2, x_3\}$ are variables. We need to check under what specialization σ from $\mathbb{Q}[U]$ to $\mathbb{C}$, $\sigma(f)$ is invertible in $\mathbb{C}[X]/\langle \sigma(I) \rangle$, where $\mathbb{C}$ is the complex field.

Suzuki and Sato computes a CGS of $I + \langle fy - 1 \rangle$ in $\mathbb{Q}[a, b, c, d, e][x_1, x_2, x_3, y]$ w.r.t. a lexicographic term ordering $y \succ x_1 \succ x_2 \succ x_3$. There are six branches where f is invertible in $\mathbb{C}[X]/(I : f^\infty)$. We only choose two of them to study whether I and $I : f^\infty$ are equal in these branch.

Branch 1: (A_1, G_1)
$A_1 = \mathbb{V}(a) \backslash \mathbb{V}(e(c+d)(b-c-d+2))$,
$G_1 = \{(2x_2 - d)x_3 + x_1^2 + cx_1 + e, x_1x_3, (-2x_2 + d)x_3^2 - ex_3, (4x_2^2 + (-2c - 4d)x_2 + dc + d^2)x3 - ex_1 - e^2y - ec\};$

Branch 6: (A_6, G_6)
$A_6 = \mathbb{V}(e, c+d, b+2, a-1) \backslash \mathbb{V}(d)$,
$G_6 = \{-x_3^2 + (-2x_2 + d)x_3 - x_2^2 + (d+1)x_2, -x_3 - x_2 - x_1 + d, -dy + 1\}$.
    In the branch (A_1, G_1), $G_1' = G_1 \cap \mathbb{Q}[U][X] = \{(2x_2 - d)x_3 + x_1^2 + cx_1 + e, x_1x_3, (-2x_2 + d)x_3^2 - ex_3\}$ is the Gröbner basis of $I : f^\infty$ under the constraint of A_1. We first compute a CGS $\mathcal{G}_1$ of I under the constraint A_1,

$$\mathcal{G}_1 = \{\mathbb{V}(a) \backslash \mathbb{V}(e(c+d)(b-c-d+2)), \{(2x_2-d)x_3+x_1^2+cx_1+e, x_1x_3, (-2x_2+d)x_3^2-ex_3\}\}.$$

It is obvious I and $I : f^\infty$ are equal under the constraint A_1, so $\sigma_{\bar{a}}(f)$ is invertible in $\mathbb{C}[X]/\langle \sigma_{\bar{a}}(I) \rangle$ for $\bar{a} \in A_1$.
    In the branch (A_6, G_6), $G_6' = G_6 \cap \mathbb{Q}[U][X] = \{-x_3^2 + (-2x_2 + d)x_3 - x_2^2 + (d+1)x_2, -x_3 - x_2 - x_1 + d\} = \{g_1, g_2\}$ is the Gröbner basis of $I : f^\infty$ under the constraint of A_6. We first compute a CGS $\mathcal{G}_2$ of I under the constraint A_6:
$\mathcal{G}_2 = \{\mathbb{V}(e, c+d, b+2, a-1) \backslash \mathbb{V}(d), \{x_2^4 + 4x_2^3x_3 - dx_2^3 - 2x_2^3 + 6x_2^2x_3^2 - 3dx_2^2x_3 - 4x_2^2x_3 + dx_2^2 + x_2^2 + 4x_2x_3^3 - 3dx_2x_3^2 - 2x_2x_3^2 + dx_2x_3 + x_3^4 - dx_3^3, x_1x_3 + x_3^2 + 3x_2^2x_3 - dx_3^3 - 2x_2^2 + 3x_2x_3^2 - 2dx_2x_3 - 2x_2x_3 + dx_2 + x_2 + x_3^2 - dx_3^3, x1x_2 + x_1x_3 + x_2, x_1^2 - dx_1 + x_2^2 + 2x_2x_3 - dx_2 + x_3^2 - dx_3\}\} = \{A_6, \{h_1, h_2, h_3, h_4\}\}$.
It is obvious there is no term of $g_1 = -x_3^2 + (-2x_2 + d)x_3 - x_2^2 + (d+1)x_2$, or $g_2 = -x_3 - x_2 - x_1 + d \in G_6'$ can be reduced by $\{h_1, h_2, h_3, h_4\}$, so the remainder of g_1, g_2 reduced by $\{h_1, h_2, h_3, h_4\}$ are $r_1 = g_1, r_2 = g_2$ respectively. For $\bar{a} = (1, -2, 1, -1, 0) \in A_6$, $\sigma_{\bar{a}}(r_1) = -x_3^2 + (-2x_2 - 1)x_3 - x_2^2 \neq 0$. So I and $I : f^\infty$ are not equal under the constraint A_6. That is, $\sigma_{\bar{a}}(f)$ is invertible in $\mathbb{C}[X]/\langle \sigma_{\bar{a}}(I : f^\infty) \rangle$ but not invertible in $\mathbb{C}[X]/\langle \sigma_{\bar{a}}(I) \rangle$ for $\bar{a} \in A_6$.

4   Conclusions

In this paper, we give the method to solve the membership problem about parametric polynomial ideals, and determine whether two parametric polynomial ideal are equal.

Given a parametric polynomial f and an ideal F in $k[U][X]$, before computing a CGS of F, we can first compute a Gröbner bases G of F in $k[U, X]$ and the remainder of f reduced by G. If the remainder is zero, then f is obvious the member of F. Otherwise, we use the theorem 4 to decide whether f is a member of F. Similarly, for two subset I, J of $k[U][X]$, if the reduced Gröbner bases of I and J in $k[U, X]$ are same, I and J must equal.

The authors are grateful to Professor Y. Sato for his helpful suggestions.

Acknowledgements. This work was supported by National Key Basic Research Project of China (Grant No. 2011CB302400), National Natural Science Foundation of China (Grant No. 11371356, 60970152).

References

1. Cox, D., Little, J., O'Shea, D.: Ideals, Varieties, and Algorithms, 3rd edn. Springer, New York (2007)
2. Gonzalez–Vega, L., Traverso, C., Zanoni, A.: Hilbert stratification and parametric Gröbner bases. In: Ganzha, V.G., Mayr, E.W., Vorozhtsov, E.V. (eds.) CASC 2005. LNCS, vol. 3718, pp. 220–235. Springer, Heidelberg (2005)
3. Kapur, D., Sun, Y., Wang, D.K.: A new algorithm for computing comprehensive Gröbner systems. In: Proc. ISSAC 2010, pp. 29–36. ACM Press, New York (2010)
4. Manubens, M., Montes, A.: Minimal canonical comprehensive Gröbner systems. Journal of Symbolic Computation 44(5), 463–478 (2009)
5. Montes, A.: A new algorithm for discussing Gröbner bases with parameters. Journal of Symbolic Computation 33(2), 183–208 (2002)
6. Montes, A., Wibmer, M.: Gröbner bases for polynomial systems with parameters. Journal of Symbolic Computation 45(12), 1391–1425 (2010)
7. Nabeshima, K.: A speed-up of the algorithm for computing comprehensive Gröbner systems. In: Proc. ISSAC 2010, pp. 299–306. ACM Press, New York (2007)
8. Sato, Y., Suzuki, A.: Computation of inverses in residue class rings of parametric polynomial ideal. In: Proc. ISSAC 2009, pp. 311–316. ACM Press, New York (2009)
9. Suzuki, A., Sato, Y.: An alternative approach to comprehensive Gröbner bases. Journal of Symbolic Computation 36(3), 649–667 (2003)
10. Suzuki, A., Sato, Y.: A simple algorithm to compute comprehensive Gröbner bases using Gröbner bases. In: Proc. ISSAC 2006, pp. 326–331. ACM Press, New York (2006)
11. Sit, W.: Computations on Quasi-Algebraic Sets (2001)
12. Weispfenning, V.: Comprehensive Gröbner bases. Journal of Symbolic Computation 14(1), 1–29 (1992)
13. Weispfenning, V.: Canonical comprehensive Gröbner bases. Journal of Symbolic Computation 36(3), 669–683 (2003)
14. Wibmer, M.: Gröbner bases for families of affine or projective schemes. Journal of Symbolic Computation 42(8), 803–834 (2007)

An Implementation Method of a CAS with a Handwriting Interface on Tablet Devices

Mitsushi Fujimoto

Department of Mathematics, Fukuoka University of Education, Japan
fujimoto@fue.ac.jp
http://www.fue.ac.jp/~fujimoto/

Abstract. Most of computer algebra systems were designed to have a command line interface. However, the interaction with CAS engine through terminal on tablets is too inefficient. A user have to switch the software keyboard over and over again to complete the input operation. We need a GUI for computer algebra system on tablets. This article describes an implementation method of a handwriting CAS app.

Keywords: tablet devices, CAS, GUI, handwriting recognition.

1 Introduction

The author developed AsirPad [1], a computer algebra system on Linux PDA Zaurus. It has a handwriting interface for mathematical expressions and can communicate with a CAS engine Risa/Asir [2] through the OpenXM protocol [3]. It was used to present a lecture on RSA cryptography at a junior high school [4]. Ordinary calculator is not available because encryption and decryption in RSA use division of large numbers. The students learned how to encrypt/decrypt their messages through calculations by AsirPad. They could input mathematical expressions and calculate without any special training. This experience encouraged us to explore the possibility of mobile devices in math classroom.

In 2010, we started a project to develop a Math e-Learning system Mathellan for pen-based mobile devices [5]. We are planning to rebuild AsirPad for a client of Mathellan. However, the mainstream of the current mobile devices is shifting from PDA to smartphones or tablet devices. Therefore, we needed a new development environment. AsirPad consists of two main components: a CAS engine and a handwriting interface. We used a cross-build environment by QEMU and chroot to make an executable binary of Risa/Asir for the Android platform. Furthermore, we adopted a cross-platform application framework Qt to build the handwriting interface of AsirPad. Qt can be used to build applications for various operating systems. We can develop GUI for various mobile devices with the same source code by Qt.

2 Use of CAS from Tablet Devices

There are four methods to access to computer algebra systems from tablet devices. The first method is to use a native application including a computer

H. Hong and C. Yap (Eds.): ICMS 2014, LNCS 8592, pp. 545–548, 2014.
© Springer-Verlag Berlin Heidelberg 2014

algebra engine on tablets (Method 1). The second method is to access to a computer algebra system on the other machine through an application on tablets (Method 2). The third method is to access to a computer algebra system on the other machine through a web browser on tablets (Method 3). The last method is to use a work sheet including a computer algebra kernel (Method 4). The following is a table for computer algebra systems on tablets we investigated.

Table 1. CASs on tablet devices

Method	App name	CAS engine	OS	Paid
	MathStudio	original	Android/iOS	✓
	Mathomatic	Mathomatic	Android/iOS	✓
	PariDroid	PARI	Android	
	Maxima on Android	Maxima	Android	
Method 1	JavaYacas	Yacas	Android	
(Native CAS App)	Yacas for iPhone	Yacas	iOS	✓
	iCAS	Reduce	iOS	✓
	PocketCAS	Giac/Xcas	iOS	✓
	Pi Cubed	original	iOS	✓
	Python Math	SymPy	iOS	
Method 2	WolframAlpha	Mathematica	Android/iOS	✓
(CAS through App)	SageMath	GAP, Maxima, etc.	Android/iOS	
	WolframAlpha	Mathematica	N/A	
Method 3	Sage	GAP, Maxima, etc.	N/A	
(CAS through Web)	FriCAS	Axiom	N/A	
	Omega	Maxima	N/A	
Method 4	Maple Player	Maple	iOS	
(CAS work sheet)	Wolfram CDF Player	Mathematica	iOS (not yet)	

3 Implementation of a CAS Engine for Tablet Devices

A CAS engine / GUI / communication mechanism with CAS engine / internal form for mathematical expressions are needed to realize a handwriting CAS app. In this section, we concentrate on building a CAS engine by Method 1. Android is a UNIX-based operating system. Thus, UNIX-based computer algebra systems are suitable for Android platform.

Most of computer algebra systems need some external libraries, e.g., garbage collector or arbitrary-precision arithmetic library. Some of them are not presumed to be built by cross-build environments. In such a case, self-build environment is the best way to avoid troubles with building. Risa/Asir needs a garbage collector Boehm-GC. We made a binary of Risa/Asir for Zaurus as the CAS engine of AsirPad by this way since a self-build environment for Zaurus was available. Unfortunately, this way is unavailable for Android. Thus, we need a cross-build environment to get an executable binary for Android. We recommend a cross-build environment constructed by QEMU and chroot to avoid troublesome issues.

3.1 Cross-Build Environment by QEMU and Chroot

By the following commands, Debian/ARM root filesystem is created in a Debian Linux machine.

1. Execute debootstrap
   ```
   $ sudo debootstrap --foreign --arch armel squeeze armel_squeeze
   http://ftp.jp.debian.org/debian/
   ```
2. Copy a statically-linked version of QEMU to rootfs:
   ```
   $ sudo cp /usr/bin/qemu-arm-static armel_squeeze/usr/bin/
   ```
3. Execute debootstrap again:
   ```
   $ sudo chroot armel_squeeze /debootstrap/debootstrap
   --second-stage
   ```

3.2 Cross-Build of a CAS Engine

QEMU is a machine emulator, and a developer can login to Debian/ARM environment by chroot: `$ sudo chroot ./armel_squeeze`. We can build source codes in this environment as if we are in a self-build environment. It means that we do not need to modify source codes for cross-build.

Now, we shall compile the Risa/Asir's source code for Android platform by this cross-build environment.

```
# cd /home/devel
# export CFLAGS="-O2 -Wall -D ANDROID -fsigned-char -static"
# cd asir2000
# ./configure
# make
# make install
# make install-lib
```

In case of this method, it is important to use statically-linked libraries by -static option for GCC because Android OS does not have GLIBC, a standard C library.

4 Creation of GUI

Console based computer algebra systems are useful as CAS engines. However, the interaction with CAS engine through terminal on tablets is too inefficient. We need a GUI for computer algebra system on tablets.

The GUI of 'Maxima on Android' [6] by Y. Honda was developed using WebView, a GUI framework of Android. This GUI uses MathJax to display mathematical expressions, and can output computational results beautifully. WebView is a core class in the WebKit. This method can be used to develop applications for iOS having WebKit as a standard framework. We think that GUI should be developed by the method not depending on platform like this. On the other hand, the GUI of AsirPad was developed using Qt. Qt is a cross-platform application framework, and can be used to build applications for various operating systems: Windows, MacOS X, Linux, Android and iOS.

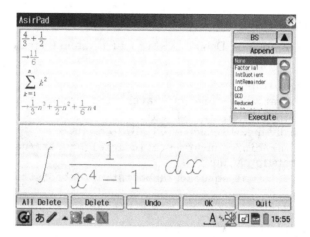

Fig. 1. GUI of AsirPad created by Qt

References

1. Fujimoto, M., Suzuki, M.: AsirPad - A Computer Algebra System with a Pen-based Interface on PDA. In: Proceedings of the Seventh Asian Symposium on Computer Mathematics, Korea Institute for Advanced Study, pp. 259–262 (2005), Demo movie, http://www.inftyproject.org/demo/AsirPad_Demo.zip
2. Noro, M., et al.: A computer algebra system Risa/Asir, http://www.math.kobe-u.ac.jp/Asir/asir.html
3. Maekawa, M., Noro, M., Takayama, N., Tamura, Y., Ohara, K.: The Design and Implementation of OpenXM-RFC 100 and 101. In: Computer Mathematics (Proceedings of the Fifth Asian Symposium on Computer Mathematics), pp. 102–111. World Scientific (2001)
4. Fujimoto, M., Suzuki, M., Kanahori, T.: On a classroom experiment using PDA and handwriting interface (in Japanese). IPSJ Symposium Series 2006(8), 331–338 (2006)
5. Fujimoto, M., Watt, S.M.: An Interface for Math e-Learning on Pen-Based Mobile Devices. In: Proceedings of the Workshop on Mathematical User-Interfaces (2010), http://www.activemath.org/workshops/MathUI/10/proc/FujimotoWatt.html
6. Honda, Y.: Maxima on Android (2012), https://sites.google.com/site/maximaonandroid/

New Way of Explanation of the Stochastic Interpretation of Wave Functions and Its Teaching Materials Using KETpic

Kenji Fukazawa

National Institute of Technology, Kure, Japan
fukazawa@kure-nct.ac.jp
http://www.kure-nct.ac.jp/global/index.html

Abstract. Many students face problems when studying Quantum Mechanics to grasp the stochastic interpretation of the wave functions. In this talk, we show the special solutions of Schrödinger equations, and describe our approach to explain the stochastic interpretation, which is based on these special solutions. We also present some examples of the special solutions, and the teaching materials we developed concerning such solutions, which have been obtained by using a special software called KETpic.

Keywords: stochastic interpretation, wave functions, KETpic.

1 Introduction

In Quantum Mechanics, the interpretation of wave functions is a subject that students find traditionally problems to grasp, because one cannot introduce the stochastic interpretation logically. Historically speaking, Born could not find any other reasonable interpretation, and had no choice but to insist the stochastic one. In many textbooks of Quantum Mechanics[1], they explain that Nature has the stochastic character, and that is the reason of the stochastic interpretation of wave functions. Many students fail to accept such an explanation. In this talk, we show the special solutions of Schrödinger equations, and give our way to explain the stochastic interpretation, which is based on the special solutions. We also present some examples of the special solutions, and the teaching materials concerning the solutions by the use of KETpic[2].

KETpic is a tool to insert mathematical figures in LATEX documents, and is implemented as an add-on of many popular Computer Algebra Systems (CAS) such as Scilab, Mathematica, and many others.

The structure of this paper is as follows: in Sect. 2, we explain the special solutions in the simple case and our approach to explain the interpretation of wave functions. In Sect. 3, we present some examples of the special solutions, and the related teaching materials by using KETpic.

H. Hong and C. Yap (Eds.): ICMS 2014, LNCS 8592, pp. 549–553, 2014.

2 Special Solutions for Schrödinger Equations and the Interpretation of Wave Functions

We have found the special solutions for Schrödinger equations, whose spatial width at arbitrary time is equal to the wave length multiplied by a positive integer or half-integer. Note that the ordinary plane wave solutions are wide infinitely in spatial directions. The simplest case is that of a free particle with mass m, and one can easily verify that the plane wave solution with one wave length wide

$$\Psi = N \left[\theta(\omega t - \boldsymbol{k} \cdot \boldsymbol{r} + \pi) - \theta(\omega t - \boldsymbol{k} \cdot \boldsymbol{r} - \pi) \right] e^{-i(\omega t - \boldsymbol{k} \cdot \boldsymbol{r})} \tag{1}$$

satisfies the Schrödinger equations

$$i\hbar \frac{\partial \Psi}{\partial t} = -\frac{\hbar^2}{2m} \boldsymbol{\nabla}^2 \Psi \tag{2}$$

where N denotes the normalization constant and θ denotes the unit step function.

When one considers the meaning of wave functions, one should note that the squared amplitude of a wave is proportional to its energy in classical physics. If this applies to wave functions, the squared absolute value of wave function $|\Psi(t, r)|^2$ is proportional to its mass density, according to Special relativity. Hence, you can define the position of the center of mass of a wave function, in the same manner as the definition of that of a rigid body:

$$\boldsymbol{r} = \frac{\iiint_D \boldsymbol{r}' \, |\Psi(t, \boldsymbol{r}')|^2 \, d\boldsymbol{r}'}{\iiint_D |\Psi(t, \boldsymbol{r}')|^2 \, d\boldsymbol{r}'} \tag{3}$$

where D denotes the region of the wave function. If one accepts the assumption that this position is equal to the position of a particle, one has the method to calculate the position of the particle from a wave function at arbitrary time.

Note that the definition of the center of mass of the body with the infinite length loses its meaning, because of its infinitude. Thus one finds that in the case of the ordinary plane wave solutions, the center of mass of a wave function cannot be defined, and one cannot adapt the above asuumption to the wave functions.

Other physical quantities of the particle can be calculated in the same manner. One finds that the obtained formulas are equivalent to the formulas based on the ordinary stochastic interpretation. In this way, we think that students are able to accept the stochastic interpretation of wave functions smoothly.

3 Examples of the Special Solutions and Teaching Materials

We consider some concrete examples of the special solutions and present the teaching materials in relation to such solutions.

As a first example, we consider a popular problem in many textbooks and show how to solve it in a straightforward way. We consider the particle between two walls in one dimension, and shows the dynamic solutions of this problem. The potential in this case is:

$$U(x) = \begin{cases} 0 & (0 < x < L) \\ +\infty & (others) \end{cases} \tag{4}$$

where L is the distance of two walls. The Schrödinger equation reads

$$i\hbar \frac{\partial \Psi}{\partial t} = -\frac{\hbar^2}{2m} \frac{\partial^2 \Psi}{\partial x^2} \tag{5}$$

between the walls, and has the dynamic solutions

$$\Psi(t, x) = \begin{cases} \psi_1(t, x) - \psi_1(t, 2L - x) & (0 \leqq t < T) \\ \psi_1(t, 2L + x) - \psi_1(t, 2L - x) & (T \leqq t < 2T) \end{cases} \tag{6}$$

where

$$\psi_1(t, x) = A\left[\theta(\omega t - kx + \pi) - \theta(\omega t - kx - \pi)\right] e^{-i(\omega t - kx)} \tag{7}$$

with an amplitude A, and T is the time to move from one wall to another: $T = L/v = kL/\omega$. Note that the periodic boundary condition restricts the wave length as $\lambda/2 = L/n$ (n: positive integer). Note also that Eq. (6) satisfies the boundary conditions

$$\Psi(t, x) = 0 \qquad \text{at} \quad x = 0, \; L. \tag{8}$$

The solution (6) is the dynamic solution, which corresponds to the classical solution of a moving particle between two walls, and has the discrete energy level

$$E_n = \frac{\pi^2 \hbar^2}{2mL^2} n^2. \tag{9}$$

Note that this solution is not usually stated in ordinary textbooks. The solution stated in them is the stationary wave solution only, which has the same energy level (9), and does not corresponds to the classical solution.

A sequence of figures reported in the teaching materials we developed concerning the dynamic solution (6) is shown from Fig. 1 to Fig. 5. In this sample, we pick up the only imaginary part of the wave function, because we cannot draw both parts simultaneously. We also draw only the half period of its imaginary part for simplicity. In these figures, the wave functions are drawn at several times, where the parameters are set as $L = 10, \lambda = L, v = 1, A = 1$, so that $k = 2\pi/\lambda = \pi/5, \omega = kv = \pi/5, T = L/v = 10$. Furthermore, we append a dot and a dashed line in each figure, which show the position of the center of mass of the imaginary part of the wave function.

We have used KETpic to generate this sample for our teaching materials. The reason to use KETpic is that this software works as a plug-in of a popular CAS of our choice. Briefly, KETpic is a library of macros to generate LaTeX source codes

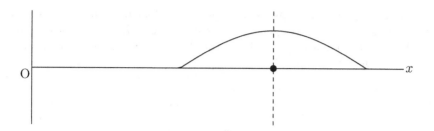

Fig. 1. Wave function at t = 4 and its center of mass

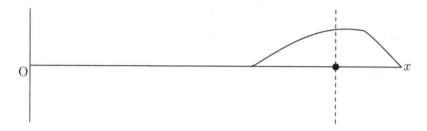

Fig. 2. Wave function at t = 6 and its center of mass

Fig. 3. Wave function at t = 8 and its center of mass

for high-quality scientific artwork. Such macros can be implemented in different Computer Algebra Systems (CAS), thus, yielding different versions (plug-ins) of the program. Depending on the CAS they are based on, these plug-ins typically run in a quite different way, but this process is usually transparent to end-users. Once KETpic is loaded, users are simply requested to execute commands in the CAS of their choice in order to plot graphs and other mathematical data. KETpic commands generate additional source code and files, which are subsequently compiled in LATEX in the usual manner. As a result, accurate graphical figures can be obtained.

In this example, the drawn function is well known, so that one can use other tools, such as Tikz or Asymptote. But drawing the dot at the position of the center of mass is not so easy with these tools, because one must calculate the position by integrating the squared wave function, with dividing the interval of

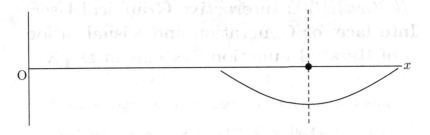

Fig. 4. Wave function at t = 10 and its center of mass

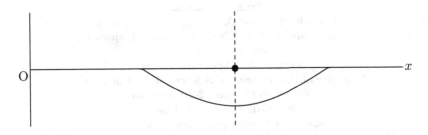

Fig. 5. Wave function at t = 10 and its center of mass

integration. Utilizing KETpic, one can write a scripting code and accomplish this task very easily by taking advantage of the programming features it provides. This flexibility is the primary benefit of KETpic, and that is the main reason to explain why we choose it for producing the figures of our teaching material.

As other examples, we consider two-dimensional versions of the first example, the case surrounded by four walls and the case surrounded by a circular wall, which will be presented in our talk.

References

1. For example, Schiff, L.: Quantum Mechanics, 3rd edn. McGraw-Hill, New York (1968)
2. Kaneko, M., et al.: A simple method of the TEXsurface drawing suitable for teaching materials with the aid of CAS. In: Bubak, M., van Albada, G.D., Dongarra, J., Sloot, P.M.A. (eds.) ICCS 2008, Part II. LNCS, vol. 5102, pp. 35–45. Springer, Heidelberg (2008)

IFSGen4LaTeX: Interactive Graphical User Interface for Generation and Visualization of Iterated Function Systems in LaTeX

Akemi Gálvez[1], Kiyoshi Kitahara[2], and Masataka Kaneko[3]

[1] Dpt. of Applied Mathematics & Comp. Sci., Universidad de Cantabria
E.T.S.I. Caminos, Canales y Puertos, Avda. de los Castros, s/n
E-39005, Santander, Spain
galveza@unican.es
[2] Division of Liberal Arts, Kogakuin University
Shinjuku Campus, 1-24-2 Nishi-Shinjuku, Shinjuku-ku
163-8677, Tokyo, Japan
kitahara@cc.kogakuin.ac.jp
[3] Faculty of Pharmaceutical Sciences, Toho University
Narashino Campus, 2-2-1 Miyama
274-8510, Funabashi, Japan
masataka.kaneko@phar.toho-u.ac.jp

Abstract. This paper presents a new interactive, user-friendly graphical user interface (GUI) for generation and visualization of IFS. The program, called *IFSGen4LaTeX*, is particularly designed for proper LaTeX editing in a WYSIWYG mode. During a working session, IFS are created interactively from scratch and visualized by using the *IFSGen4LaTeX* GUI; simultaneously, its engine generates source code that, once inserted in LaTeX and compiled, displays the same fractal images in LaTeX. This process leads to substantial savings in CPU time and memory storage, since the resulting source code is astonishingly small, but still of excellent visual quality because the number of iterations is decided by the user.

Keywords: Mathematical editing, fractal images, iterated function systems, graphical user interface, LaTeX.

1 Introduction

Fractals are among the most exciting and intriguing mathematical objects ever discovered. In addition to their impressive visual beauty, they have many remarkable mathematical properties [1, 3, 22]. The development of fractal geometry provides rigorous concepts and techniques for the mathematical analysis of irregular processes [20]. For example, the fractal dimension of strange attractors associated with chaotic dynamical systems gives very valuable information about the irregular evolution observed in such systems [5–7, 10, 11, 15, 17–19].

Owing to these reasons, fractals are receiving great interest from the mathematical community. In fact, several computer tools for generating fractal images

H. Hong and C. Yap (Eds.): ICMS 2014, LNCS 8592, pp. 554–561, 2014.

have been developed during the last few decades, see e.g. [8, 12, 14]. However, their manipulation and visualization for mathematical editing purposes is still challenging, since the typical approach of embedding them as figures is not very flexible and do not always perform properly in terms of computer storage and visual quality. An illustrative example of this problem is given by LaTeX, the standard computer editor for high-quality typesetting of scientific documents. As good as it is for automating most aspects of typesetting and publishing, including numbering, cross-referencing, tables, figures, page layout, table of contents and handling of bibliographic entries, it only provides a basic and very limited set of graphical capabilities to yield drawings. The only feasible alternative is to use a graphical editor to generate our artwork and then invoke the resulting image file from LaTeX. Typically, this option produces a large collection of heavy images files, thus requiring huge storing capacity and preventing users from their transferring on the web. Further, incompatibilities may arise as image files should comply with a limited set of prescribed formats.

These problems are exacerbated for irregular objects such as fractals, which are not suited for vectorization, leading to large bitmap files. A particular type of fractals are the *Iterated Function Systems* (IFS), defined by a finite number of affine transformations (rotations, translations, and scalings), and therefore represented by a relatively small set of input data. Instead of storing the bitmap information of the image, the IFS image is generated from successive iterations of affine functions, with the great advantage that the final image becomes resolution independent. In this paper we claim that the code to generate such images by using IFS can readily be embedded into a LaTeX source file and compiled in the usual way. The resulting text files are astonishingly small when compared with their image file counterparts but still offer high-resolution quality, as the resolution only depends on the number of iterations, which can be performed in real-time. As a result, this process leads to much higher compression ratios than other conventional formats such as JPEG, GIF, PNG, EPS, and the like.

A clear drawback is that some expertise is required to perform this process. To overcome this limitation, in this paper we present a new interactive, user-friendly graphical user interface (GUI) for generation and visualization of IFS. The program, called *IFSGen4LaTeX*, is particularly designed for proper LaTeX editing in a WYSIWYG (what you see is what you get) mode. During the working session, IFS are created interactively from scratch and visualized by using the *IFSGen4LaTeX* GUI; simultaneously, its engine generates source code that, once inserted in LaTeX and compiled, generates the same fractal images in LaTeX. The resulting source files are astonishingly small, and still of very high quality, as the number of iterations is decided by the user. This process leads to substantial savings in CPU time and memory storage with excellent visual quality.

The structure of this talk is as follows: Section 2 describes the main concepts and definitions about IFS. Section 3 reports technical details of our program. Three examples of its application to generate IFS images and its associated source code to be embedded into LaTeX are given in Section 4. The paper closes with the main conclusions and some future work.

2 Mathematical Background

Several fractal modeling and generation methods have been developed during the last decades. Among them, the IFS models popularized by Barnsley in the 80s, are particularly interesting due to their mathematical simplicity [1]. IFS have been applied in many disciplines, such as artificial landscape modeling in computer graphics [3, 22]. In short, an IFS consists of a collection of functions that map a region onto smaller parts. Iteration of these mappings results in convergence to a self-similar fractal that is called the *attractor* of the IFS.

2.1 Basic Concepts and Definitions

Since we are interested in 2D images, we restrict our attention to contractive affine transformations on the Euclidean (complete) metric space $(\mathbb{R}^2, d_2)$, while the contractive affine transformations w_i are of the form:

$$\begin{bmatrix} x^* \\ y^* \end{bmatrix} = w_i \begin{bmatrix} x \\ y \end{bmatrix} = \begin{bmatrix} a_i & b_i \\ c_i & d_i \end{bmatrix} \cdot \begin{bmatrix} x \\ y \end{bmatrix} + \begin{bmatrix} e_i \\ f_i \end{bmatrix} \tag{1}$$

or, equivalently, $\mathbf{w}_i(\mathbf{x}) = \mathbf{A}_i.\mathbf{x} + \mathbf{b}_i$ where $\mathbf{b}_i$ is a translation vector and $\mathbf{A}_i$ is a 2×2 matrix with eigenvalues λ_1, λ_2, such that $|\lambda_i| < 1$. In fact, $|det(\mathbf{A}_i)| < 1$ meaning that w_i shrinks distances between points. In other words, the fractal is made up of the union of several copies of itself, where each copy is transformed by a function w_i. Such a function is mathematically a 2D affine transformation, so the IFS is defined by a finite number of affine transformations (rotations, translations, and scalings), and therefore represented by a relatively small set of input data. From now on, an IFS will be represented by a list of the pairs $(\mathbf{A}, \mathbf{b})$. Let us now define a transformation, T, in the compact subsets of X, $\mathcal{H}(X)$, by: $T(A) = \bigcup_{i=1}^{n} w_i(A)$. If all the w_i are contractions, T is also a contraction in $\mathcal{H}(X)$ with the induced Hausdorff metric. Then, T has a unique fixed point, $|\mathcal{W}|$, called the *attractor of the IFS*. Considering a set of probabilities $p_1, \ldots, p_n \in (0, 1)$, with $\sum_{i=1}^{n} p_i = 1$, $|\mathcal{W}|$ supports several measures in a natural way. We refer to $\{X; w_1, \ldots, w_n; p_1, \ldots, p_n\}$ as an *IFS with Probabilities* (IFSP). Given a set $\{p_1, \ldots, p_n\}$, there exists an unique Borel regular measure $\nu \in \mathcal{M}(X)$, called the *invariant measure of the IFSP*, such that $\nu(S) = \sum_{i=1}^{n} p_i \nu(w_i^{-1}(S))$, $S \in \mathcal{B}(X)$, where $\mathcal{B}(X)$ denotes the Borel subsets of X. Using the Hutchinson metric on $\mathcal{M}(X)$, it is possible to show that the Markov operator $M : \mathcal{M}(X) \longrightarrow \mathcal{M}(X)$ associated to the IFSP and defined by: $(M\nu)(S) = \sum_{i=1}^{n} p_i \nu(w_i^{-1}(S))$, where S is a Borel subset of X is a contraction with a unique fixed point, $\nu \in \mathcal{M}(X)$ [16]. Furthermore, $support(\nu) = |\mathcal{W}|$. Thus, given an arbitrary initial measure $\nu_0 \in \mathcal{M}(X)$ the sequence $\{\nu_k\}_{k=0,1,2,\ldots}$ constructed as $\nu_{k+1} = M(\nu_k)$ converges to the invariant measure of the IFSP. Also, a similar iterative deterministic scheme can be derived to obtain $|\mathcal{W}|$.

2.2   Fractal Image Rendering: The Chaos Game

In addition to this deterministic algorithm, there exists a more efficient method for the generation of the attractor of an IFS, known as the *probabilistic algorithm* (also referred to as the *chaos game*). This algorithm follows from the result $\{x_k\}_{k>0} = |\mathcal{W}|$ provided that $x_0 \in |\mathcal{W}|$, where: $x_k = w_i(x_{k-1})$ with probability $p_i > 0$ (see, for instance, [2]).

The fractal image is determined only by the set of contractive mappings; the set of probabilities gives the efficiency of the rendering process. Thus, one of the main problems of the chaos game algorithm is to find the optimal set of probabilities to render the fractal attractor associated with an IFS at a given resolution with the minimum number of iterations. Several heuristic methods for this problem have been proposed in the literature, but none of them solves the general case [3, 4, 9]. The most standard was suggested by Barnsley [1] and has been widely used in the literature. For each of the mappings, this method (called *Barnsley's algorithm*) selects a probability value that is proportional to the area of the figure associated with the mapping. Taking into account that the area filled by a linear mapping w_i is proportional to its contractive factor, s_i, this algorithm proposes the following set of probabilities: $p_i = s_i/S$, ($i = 1, \ldots, n$), with $S = \sum_{j=1}^{n} s_j$. However, this choice of the probabilities is far from being the most efficient in some situations. Another algorithm, proposed in 1996 and known as *multifractal algorithm*, provides a method for obtaining the most efficient choice for the probabilities in the case of non-overlapping IFS models [13]. Intuitively, a non-overlapping IFS is one whose self-similar parts do not overlap each other. The basic idea of the proposed method consists of using a multifractal analysis to characterize the performance of the different sets of probabilities in the rendering process [12, 13]. As a result, the standard choice for the probabilities described above do not correspond to the best choice, given by a multifractal measure with the smallest strength of singularities, under the condition: $log(p_i) = Dlog(w_i) \Leftrightarrow p_i = w_i^D$, where D is the unique real number satisfying $\sum_{i=1}^{n} w_i^D = 1$ [13, 14]. Then, the most efficient choice corresponds to: $p_i = s_i^D$, ($i = 1, \ldots, n$), where D denotes the *similarity dimension*. If the contractive mappings are non-overlapping, the only real number satisfying this constraint is the fractal dimension of the fractal. Otherwise, the number obtained will be an approximation of such dimension.

3   The Program

The program *IFSGen4LATEX* consists basically of two major components:

1. a *computational library (toolbox)*: it contains a collection of functions and routines implemented to perform the numerical and graphical tasks.
2. a *graphical user interface (GUI)*: although this toolbox is enough to meet all our computation needs, potential end-users might be challenged for using it properly unless they are really proficient on *Matlab*'s syntax and functionalities and the toolbox routines. To overcome these limitations, we created a

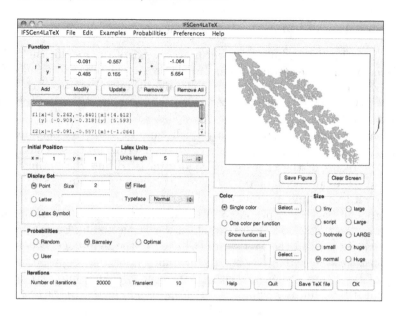

Fig. 1. Screenshot of the *IFSGen4LaTeX* program for the leaf IFS

GUI providing recognizable visual cues to help the user navigate efficiently through information. *Matlab* provides a mechanism to generate GUIs by using the so-called *guide* (*GUI development environment*). In our program we do an effective use of powerful interface tools (e.g., drop-down menus for choice lists, radio buttons for single choice from multiple options, text boxes for displaying messages, list boxes and edit boxes for input/output user interaction, push buttons, etc.) designed according to the type of values being used. Although its implementation requires - for complex interfaces - a high level of expertise, it allows end-users to deal with the toolbox with a minimal knowledge and input, thus facilitating its use and dissemination.

Regarding the implementation, our program has been implemented in *Matlab* v2013a [21] by using the native *Matlab* programming language on a 2.6 GHz. Intel Core i7 processor with 8 GB of RAM. It supports major computer operating systems (Microsoft Windows, Linux, and Mac OS X).

4   Illustrative Examples

Figures 1 to 3 show three screenshots of our graphical user interface and its application to three different examples. In all cases, the user has to input the IFS code of the contractive functions (top-left area) following the pattern in Eq. (1), an initial position, the unit length used for conversion into LaTeX code, the number of iterations (bottom-left), and the graphical options (bottom-right area) such as the color (single color or one color per function) and size (according

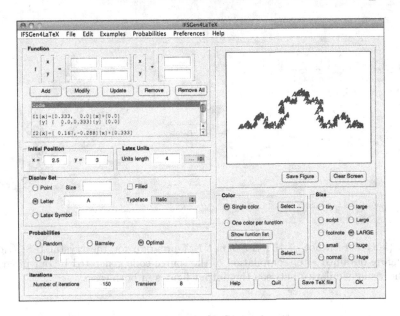

Fig. 2. Screenshot of the *IFSGen4LaTeX* program for the curve of Koch IFS

to LaTeX syntax). Note that the list of functions can be modified by removing all or some previous functions and adding new ones by using a set of five push buttons. The list of functions along with their IFS codes appear in the text box below such buttons. Our program also allows the user to choose between the deterministic or the probabilistic method. In the latter case, the probabilities can be automatically computed by our program by using either Barnsley's or the optimal methods. Additionally, the user has also the option to select any set of probabilities of his/her choice. Finally, we can also select different elements for the graphical representation: points, letters, and LaTeX symbols (our three examples have actually been chosen to represent these three cases, respectively). Those elements have different options, such as point size and filling (on/off), the typeface of the letter (normal, bold, italic, etc.), and others. The program also accepts a sequence of LaTeX or any other usual expression commonly accepted in the $...$ environment. The fractal images are displayed in the top-right area. The user can also save the output figure into supported graphical formats (including JPG, PNG, GIF, EPS, and others) and in *Matlab*'s native FIG format.

First example in Fig. 1 shows the leaf IFS, obtained with 2 contractive functions and 20000 iterations with a filled `point` primitive of point size 2 by using a single green color and the Barnsley method for the probabilities. Second example in Fig. 2 shows the curve of Koch IFS, obtained with 4 contractive functions and 150 iterations with the A letter primitive in blue color, font style *italic*, font size \LARGE, and the optimal set of probabilities. Finally, third example in Figure 3(left) shows the crystal IFS generated by 400 iterations from the initial LaTeX symbol \heartsuit, with red color, and font size \footnotesize. In all cases, simultaneously to displaying the fractal on the graphical area of the working

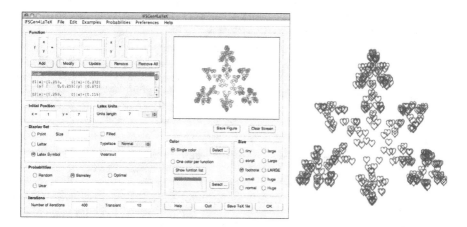

Fig. 3. (left) Screenshot of the *IFSGen4LaTeX* program for the crystal IFS; (right) LaTeX figure obtained after compilation of the source code generated by our program. Note the visual similarity between the output figures in both cases.

window of our program, *IFSGen4LaTeX* also generates source code that can be subsequently inserted into a LaTeX file and compiled to generate the same graphical image in LaTeX automatically. For instance, Fig. 3(right) shows the similar figure to that on the left as it appears in LaTeX after compilation. The resulting file size is incredibly small, requiring only 398 LaTeX words in text format to generate the image. Of course, similar results can be obtained with the previous examples, which are omitted here for the sake of limitations of space.

5   Conclusions and Future Work

This paper introduces a new program, called *IFSGen4LaTeX*, for generation and graphical visualization of IFS fractals in LaTeX. In our setup the IFS are created interactively by using a graphical interface; simultaneously, it generates source code that, once inserted in LaTeX and compiled, displays the same fractal images in LaTeX. This process leads to substantial savings in CPU time and memory storage, while preserving an excellent visual quality. Future work includes the extension of this program to other types of fractals and other relevant mathematical objects. This research has been supported by the Computer Science National Program of the Spanish Ministry of Economy and Competitiveness, Project Ref. TIN2012-30768, University of Cantabria (Santander, Spain), Kogakuin University (Tokyo, Japan), and Toho University (Funabashi, Japan).

References

1. Barnsley, M.F.: Fractals Everywhere, 2nd edn. Academic Press (1993)
2. Elton, J.H.: An ergodic theorem for iterated maps. Ergodic Theory Dynam. Syst. 7, 481–488 (1987)

3. Falconer, K.: Fractal Geometry: Mathematical Foundations and Applications. Wiley (1990)
4. Forte, B., Vrscay, E.R.: Solving the inverse problem for measures using iterated function systems: a new approach. Adv. Appl. Prob. 27, 800–820 (1995)
5. Gálvez, A.: Numerical-symbolic *Matlab* program for the analysis of three-dimensional chaotic systems. In: Shi, Y., van Albada, G.D., Dongarra, J., Sloot, P.M.A. (eds.) ICCS 2007, Part II. LNCS, vol. 4488, pp. 211–218. Springer, Heidelberg (2007)
6. Gálvez, A., Iglesias, A.: Symbolic/numeric analysis of chaotic synchronization with a CAS. Future Generation Computer Systems 25(5), 727–733 (2007)
7. Gálvez, A.: Matlab Toolbox and GUI for analyzing one-dimensional chaotic maps, pp. 321–330. IEEE Computer Society Press, Los Alamitos (2008)
8. Gálvez, A.: IFS Matlab Generator: a computer tool for displaying IFS fractals, pp. 132–142. IEEE Computer Society Press, Los Alamitos (2009)
9. Graf, S.: Barnsley's scheme for the fractal encoding of images. Journal of Complexity 8, 72–78 (1992)
10. Gutiérrez, J.M., Iglesias, A., Rodríguez, M.A.: Logistic map driven by correlated noise. In: Garrido, P.L., Marro, J. (eds.) Second Granada Lectures in Computational Physics, pp. 358–364. World Scientific, Singapore (1993)
11. Gutiérrez, J.M., Iglesias, A., Rodríguez, M.A.: Logistic map driven by dichotomous noise. Physical Review E 48(4), 2507–2513 (1993)
12. Gutiérrez, J.M., Iglesias, A., Rodríguez, M.A., Rodríguez, V.J.: Fractal image generation with Iterated Function Systems. In: Keranen, V., Mitic, P. (eds.) Mathematics With Vision, Proceedings of the First International Symposium of Mathematica, pp. 175–182. Computational Mechanics Publications (1995)
13. Gutiérrez, J.M., Iglesias, A., Rodríguez, M.A.: A multifractal analysis of IFSP invariant measures with application to fractal image generation. Fractals 4(1), 17–27 (1996)
14. Gutiérrez, J.M., Iglesias, A., Rodríguez, M.A., Rodríguez, V.J.: Efficient rendering of fractal images. The Mathematica Journal 7(1), 7–14 (1997)
15. Gutiérrez, J.M., Iglesias, A.: A Mathematica package for the analysis and control of chaos in nonlinear systems. Computers in Physics 12(6), 608–619 (1998)
16. Hutchinson, J.: Fractals and self-similarity. Indiana Univ. Math. Jour. 30, 713–747 (1981)
17. Iglesias, A., Gálvez, A.: Analyzing the synchronization of chaotic dynamical systems with Mathematica: Part I. In: Gervasi, O., Gavrilova, M.L., Kumar, V., Laganá, A., Lee, H.P., Mun, Y., Taniar, D., Tan, C.J.K. (eds.) ICCSA 2005. LNCS, vol. 3482, pp. 472–481. Springer, Heidelberg (2005)
18. Iglesias, A., Gálvez, A.: Analyzing the synchronization of chaotic dynamical systems with Mathematica: Part II. In: Gervasi, O., Gavrilova, M.L., Kumar, V., Laganá, A., Lee, H.P., Mun, Y., Taniar, D., Tan, C.J.K. (eds.) ICCSA 2005. LNCS, vol. 3482, pp. 482–491. Springer, Heidelberg (2005)
19. Iglesias, A., Gálvez, A.: Revisiting some control schemes for chaotic synchronization with Mathematica. In: Sunderam, V.S., van Albada, G.D., Sloot, P.M.A., Dongarra, J. (eds.) ICCS 2005. LNCS, vol. 3516, pp. 651–658. Springer, Heidelberg (2005)
20. Mandelbrot, B.B.: The Fractal Geometry of Nature. W. H. Freeman and Co. (1982)
21. The Mathworks Inc.: Using Matlab. Natick, MA (1999)
22. Peitgen, H.O., Jurgens, H., Saupe, D.: Chaos and Fractals. New Frontiers of Science. Springer (1993)

GNU TeXMACS
towards a Scientific Office Suite

Massimiliano Gubinelli[1],
Joris van der Hoeven[2], François Poulain[2], and Denis Raux[2]

[1] Université Paris Dauphine, France
m.gubinelli@gmail.com
[2] École polytechnique, France
vdhoeven@lix.polytechnique.fr

Abstract. GNU TeXMACS is a free mathematical text editor, which can also be used as an interface for several computer algebra systems and other mathematical software, such as Scilab, GNU R, etc. Its primary aim is to offer an alternative to LaTeX, which achieves a similar typesetting quality, but also provides a user friendly WYSIWYG interface. This user friendliness makes TeXMACS suitable for a broader audience, such as high school education.

1   Introduction

The GNU TeXMACS project aims to provide a free, polyvalent and user-friendly scientific office suite, which can easily be interfaced with a wide range of external mathematical software. The system can be downloaded from www.texmacs.org. It should be noticed that TeXMACS has been developed from scratch in C++ and SCHEME. In particular, the software does not rely on TeX or LaTeX.

With respect to standard office suites such as Microsoft Word or Open Office, we offer better support for mathematical typesetting, formula editing, and other features useful for scientists. With respect to TeX/LaTeX [4],[5] and its various front-ends, TeXMACS has the advantage of being completely *wysiwyg* (what you see is what you get). Indeed, the development of our system was initially motivated by the following reasons:

- An editor should allow the author to concentrate on *what* is written and not on *how* it is written. In particular, editors should be as *wysiwyg* as possible.
- With the advent of a wide variety of mathematical software, it should be possible to make documents more active. One might wish to incorporate computer algebra sessions and spreadsheets, for instance.
- Scientific editors should become more integrated, taking example on office suits for non-scientific users. For instance, they should provide tools for drawing technical pictures, making presentations from a laptop, annotating texts, etc.

In this paper, we will present a quick survey of the traditional features of TeXMACS and continue with more recent improvements and new features. The next major release of TeXMACS 2.1 is planned for later this year.

H. Hong and C. Yap (Eds.): ICMS 2014, LNCS 8592, pp. 562–569, 2014.

The core of T_EX_{MACS} consists of a free wysiwyg scientific text editor, which includes a mathematical formula editor, the possibility to write structured texts, and to extend the editor using personal style files or customizations of the user interface. Advanced typesetting algorithms are used, which allow for the creation of high quality documents.

Gradually, more and more features have been added to the software, thereby moving towards our goal to provide a fully fledged scientific office suite. T_EX_{MACS} currently offers an editor for graphical pictures, a presentation mode, a rudimentary spreadsheet facility, integrated version control, etc. In addition, T_EX_{MACS} has been interfaced to many external mathematical computation systems. These interfaces can be used either in shell like sessions, inside spreadsheets, or on the fly inside regular text.

Our main objectives for the next major version T_EX_{MACS} 2.1 are to increase the portability of the software and to further improve the user experience. For these reasons, we completely redesigned the graphical user interface (see Figure 1), which is now based on Qt instead of Xwindows. Recent versions of T_EX_{MACS} are available under Linux, MacOS and Windows. We took special care at following standard user interface conventions for each of these operating systems (regarding keyboard shortcuts, for instance). The existing L^AT_EX converters were also greatly improved and we now provide a native converter to Pdf. We finally improved the font support and the internationalization of T_EX_{MACS}.

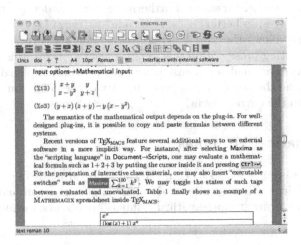

Fig. 1. The new graphical user interface of T_EX_{MACS} under MacOS

2   T_EX_{MACS} as a Structured Text Editor

The backbone of T_EX_{MACS} is a wysiwyg structured text editor. The user interface is redundant by design, so as to make the software suitable for users with diverse backgrounds. For instance, in order to create a new section, the user has the following options:

1. Use the `Insert`→`Section`→`Section` menu item.
2. In the second icon bar, click on the ▓ icon, followed by a click on `Section`.
3. As in LaTeX, type `\SECTION` followed by ↵.
4. Use the keyboard shortcut `Alter-1`.

In a similar way as in LaTeX, authors are invited to concentrate on intent rather than presentation. Nevertheless, most tags have a sufficiently distinctive presentation for making the structure apparent from the mere rendering of the document.

All documents are internally represented and manipulated as trees. The structure of the documentation is made more visible to the user by putting non intrusive boxes around all tags which contain the cursor. The innermost tag is called the current focus and is highlighted using a special color. Various editing operations allow the user to directly operate on the structure of the document. For instance, if the current focus is a section title, then there are actions for changing it into a subsection title, to jump to the next and the previous section, to toggle the numbering, or to get contextual help on the section tag.

An analogue of the "LaTeX source code" is available in TeX$_{\mathrm{MACS}}$ using `Document`→`Source`→`Edit source tree`. However, from our standpoint, there is no real concept of the "source code". In reality, documents are trees, which can be rendered in different ways so as to make certain tags more or less explicit. In particular, the presentation of the "source code" can be customized using `Document`→`Source`→`Preferences`. Furthermore, we consider $\sqrt{x}$ to be just as good (and arguably even better) a "source code" as `\sqrt{x}`.

TeX$_{\mathrm{MACS}}$ also allows the user to create new style sheets or to modify the definitions of existing presentation macros. In recent versions, this kind of customizations have been made even easier: we both provide simplified widgets for editing macros and the possibility to jump directly to the definition of the macro corresponding to the current focus.

3   Mathematical Formulas

Special care has been taken so as to make the input of mathematical formulas particularly efficient. First of all, we designed a special input method which allows users to enter most mathematical symbols are obtained using a small set of basic rules:

- Characters which are naturally obtained as "superpositions" or "concatenations" of symbols on your keyboard are entered in a straightforward way. For instance, `->` yields $\rightarrow$, `<=` yields $\leqslant$, `+-` yields $\pm$ and `<<` yields $\ll$.
- The "variant" key ↹ may be used in order to obtain variants of a given symbol or keyboard shortcut. For instance, `<=↹` yields $\leq$, `<=↹↹` yields $\Leftarrow$, `<↹` yields $\prec$ and `<|↹` yields $\lhd$. All Greek letters can be obtained as variants of the Roman ones: `A↹` yields α, `L↹` yields λ and so on. Sometimes, additional variants are available: `B↹↹` yields $\flat$ and `E↹↹` yields the mathematical constant e.

– The / and @ keys are used for obtaining negations and symbols inside other symbols. For instance, `<=/` yields $\nleq$ and `@+` yields $\oplus$. More elaborated examples are `@-+` and `<-=-/-` , which yield $\boxplus$ resp. $\nleqq$

Efficient shortcuts are also available for most mathematical constructs: `F` starts a fraction, `Alter-S` a square root, `_` a subscript, `^` a superscript, etc. Being faithful to the principle of redundancy, these actions can also be performed through the menus, the icons, or *via* LaTeX equivalents.

Inside a mathematical formula, the cursor keys allow you to move around in a graphically intuitive way. In particular, when done with a particular subformula, it usually suffices to press the right arrow key $\rightarrow$ in order to return to the main formula.

Wysiwyg editors are especially interesting for more complex formulas. For instance, it is easy to insert new rows and columns inside a matrix, or to copy and paste submatrices (and not only rows).

Another particular feature of TeX_MACS is that formulas carry more semantics than in LaTeX, when entered appropriately. For instance, "invisible" multiplication (as in x y) should be entered explicitly using `*`, whereas function application (as in sin x) should be entered using `Space`.

Recent versions of TeX_MACS integrate a parser for mathematical formulas and a syntax checker. When activating "semantic editing" from the icon menu, the focus box indicates the arguments of mathematical operators and its color changes to red whenever a formula is syntactically incorrect. The mathematical formula parser is based on a fixed grammar which works on a wide variety of mathematical texts. Nevertheless, the user can explicitly modify binding forces when needed and define specific notations using the standard macro mechanism. We refer to [2] for more details.

4 Interfaces with External Software

TeX_MACS has been interfaced with many external systems [1]. In particular, we have interfaces for the computer algebra systems AXIOM, MACAULAY2, MAPLE, MATHEMGIX, MATHEMATICA, MAXIMA, PARI, REDUCE, SAGE, etc. We also have interfaces for other mathematical software, such as OCTAVE, R, SCILAB, etc.

The traditional way to use "plug-ins" is through "shell sessions":

```
Maxima 5.9.0 http://maxima.sourceforge.net
[...]
(%i1) diff (x^x^x^x, x);
```

$$(\%o1) \quad x^{x^{x^x}} \left(x^{x^x} \log (x) \left(x^x \log (x) (\log (x) + 1) + x^{x-1} \right) + x^{x^x - 1} \right)$$

```
(%i2) integrate (%o1, x);
```

$$(\%o2) \quad e^{e^{e^{x \log (x)} \log (x)} \log (x)}$$

TeX_MACS also supports two-dimensional input, through toggling of `Focus→Input options→Mathematical input`:

$$(\%\text{i}3) \quad \begin{vmatrix} x+y & y \\ x-y^2 & y+z \end{vmatrix}$$

$$(\%\text{o}3) \quad (y+x)(z+y) - y(x-y^2)$$

The semantics of the mathematical output depends on the plug-in. For welldesigned plug-ins, it is possible to copy and paste formulas between different systems.

Recent versions of $\mathrm{T\!E\!X}_{\mathrm{MACS}}$ feature several additional ways to use external software in a more implicit way. For instance, after selecting `Maxima` as the "scripting language" in `Document→ Scripts`, one may evaluate a mathematical formula such as $1+2+3$ by putting the cursor inside it and pressing `Ctrl-↵`. For the preparation of interactive class material, one may also insert "executable switches" such as `Maxima` $\sum_{k=1}^{100} k^2$. We may toggle the states of such tags between evaluated and unevaluated. Table 1 finally shows an example of a MATHEMAGIX spreadsheet inside $\mathrm{T\!E\!X}_{\mathrm{MACS}}$.

Table 1. Computation of successive derivatives in a spreadsheet

x^x
$(\log(x)+1)x^x$
$\left(\log(x)^2 + \frac{1}{x} + 2\log(x) + 1\right)x^x$
$\left(\log(x)^3 + 3\log(x)^2 + 3\left(\frac{1}{x}+1\right)\log(x) + \frac{3}{x} - \frac{1}{x^2} + 1\right)x^x$
$\left(\log(x)^4 + 4\log(x)^3 + 6\left(\frac{1}{x}+1\right)\log(x)^2 + \left(\frac{12}{x} - \frac{4}{x^2} + 4\right)\log(x) + 1 + \frac{6}{x} - \frac{1}{x^2} + \frac{2}{x^3}\right)x^x$
$\left(\log(x)^5 + 5\log(x)^4 + 10\left(\frac{1}{x}+1\right)\log(x)^3 + \left(\frac{30}{x} - \frac{10}{x^2} + 10\right)\log(x)^2 + \left(5 + \frac{30}{x} - \frac{5}{x^2} + \frac{10}{x^3}\right)\log(x) + 1 + \frac{10}{x} + \frac{5}{x^2} - \frac{6}{x^4}\right)x^x$

To the left:

x^x
=derive($a1,x$)
=derive($a2,x$)
=derive($a3,x$)
=derive($a4,x$)
=derive($a5,x$)

5   Towards a Scientific Office Suite

We have seen that $\mathrm{T\!E\!X}_{\mathrm{MACS}}$ integrates a structured text editor, a formula editor, a spreadsheet facility and many interfaces to external programs. Let us describe a few other tools that are available nowadays inside our system, which make $\mathrm{T\!E\!X}_{\mathrm{MACS}}$ a fairly complete scientific office suite.

5.1   Presentation Mode

A wysiwyg editor such as $\mathrm{T\!E\!X}_{\mathrm{MACS}}$ is particularly useful for preparing laptop presentations. For this, it suffices to select `beamer` as the document style in the `Document→Style` menu. In addition, several standard themes can be used. Presentations are organized as successions of "screens".

Special markup is provided for showing and hiding content in specified orders. For instance, the "unroll" tag allows item lists to be unrolled progressively. There is also support for general "overlays", where the user has full control over the order in which content appears and disappears on specified ranges of overlays.

The `Insert`→`Animation` menu allows for the insertion of animated content. For the moment, only simple animations are implemented, but more elaborate graphical effects and artwork are planned for future versions, as well as support for embedded videos.

5.2 Technical Pictures

Existing pictures can be embedded inside a document using `Insert`→`Image`→ `Insert image` or `Insert`→`Image`→`Link image`. In addition, T_EX_{MACS} includes a native editor for drawing simple technical pictures. One advantage of this integrated drawing tool is that it is easy to include mathematical formulas or other T_EX_{MACS} markup inside the picture. One may use `Insert`→`Image`→`Draw image` to start a new drawing and `Insert`→`Image`→`Draw over selection` to draw a picture on top of the current selection (typically an external picture, or a mathematical formula). Currently, T_EX_{MACS} implements the most basic primitives for drawing vector graphics. We have plans to extend the T_EX_{MACS} macro mechanism to graphics and to offer full support for the SVG standard.

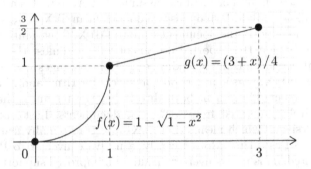

Fig. 2. Toy example of a technical picture created with the drawing tool

5.3 Version Control

T_EX_{MACS} comes with an efficient tool for computing and visualizing "structured differences" between two versions of a document. The default way to visualize changes is to show the old and new versions side by side in different colors. Authors may quickly go trough the changes introduced by a coauthor and select which versions they prefer. T_EX_{MACS} also integrates support for external versioning software. For the moment, we only provide an interface for Svn, but it would be easy to add interfaces for other systems.

6 Compatibility with Other Formats

Unfortunately, T_EX_{MACS} is not yet as wide-spread as L^AT_EX. For the sake of backward compatibility, T_EX_{MACS} provides high quality converters from and to L^AT_EX. However, these converters cannot be perfect for several reasons.

The main reason is that LATEX is not a format, like Html, but rather a programming language. In particular, the only program which parses all LATEX files correctly is LATEX itself. Since TEX_{MACS} is *not* a LATEX front-end, it follows that we can only ensure correct conversions for a (quite large) sublanguage of LATEX.

The other main reason is that TEX_{MACS} has a more powerful typesetting engine than LATEX and that it provides several extensions (like a graphical editor or animations) which are not available in LATEX. Therefore, a conversion to LATEX may downgrade your document, both in typesetting quality and in structure. For instance, TEX_{MACS} pictures are exported as postscript images, so their structure is lost.

Nevertheless, during recent years, we have invested a lot of energy in making the LATEX converters as good as possible, in both directions. In particular, we support the most frequently used LATEX styles. The behaviour of the converters can be fine-tuned for specific needs via the user preferences. For instance, should macros be expanded or not during conversions? Are preambles allowed to contain additional macro definitions? Etc.

An interesting recent addition is the possibility to use the converters in a "conservative" fashion [3]. For instance, assume that Alice writes a document in LATEX and sends it to Bob. Bob opens the document in TEX_{MACS}, makes a few modifications and exports the document back to LATEX. Conservative converters have the property that the exported document will be almost the same as the Alice's original version: ideally speaking, only Bob's changes will really be exported. Besides LATEX, reasonably good converters for Html and MathML have also been implemented, again in both directions. For instance, the TEX_{MACS} website is entirely generated from TEX_{MACS} documents. Last but not least, TEX_{MACS} is wysiwyg, which means that TEX_{MACS} traditionally features a lossless converter to Postscript. More recently, a native converter to Pdf has also been implemented. This new converter includes an improved support for images, fonts and certain types of graphics. In particular, the quality of Pdf documents with oriental languages is much better nowadays.

7   Customization

TEX_{MACS} can be customized in many ways. Besides the possibility to write your own style files for the presentation of documents, it is also possible to customize the behaviour of the editor, or to write new plug-ins for external software. We will briefly give a few examples below. Questions can be asked on our mailing lists:

    http://www.texmacs.org/tmweb/home/ml.en.html

7.1   User-Defined Macros

The user may define new typesetting constructs or customize the rendering of the standard styles using a special macro language. For instance, one may define a macro cd for square commutative diagrams using

$$\langle\text{assign}|cd|\langle\text{macro}|A|B|C|D|\begin{matrix} A & \to & B \\ \downarrow & & \downarrow \\ C & \to & D \end{matrix}\rangle\rangle$$

This macro may then be used by typing `\CD↵` as in

$$\begin{matrix} A \oplus B & \to & C \\ \downarrow & & \downarrow \\ D & \to & E \otimes F \end{matrix}$$

For more information, we refer to Help→Manual→Writing your own style files and Help→Reference guide.

7.2   Customizing the Interface and Scheme Extensions

Following the example of GNU EMACS, the user interface and most of the editing functions of TEXMACS are written in Scheme, a high level "extension language". This makes it possible for the user to customize the behaviour of TEXMACS and write extensions to the editor. Simple customizations can be put in the file my-init-texmacs.scm of the .TeXmacs/progs subdirectory of your home directory. For instance, assume that this file contains the following code:

```
(kbd-map
  (:mode in-text?)
  ("T h ." (make 'theorem))
  ("D e f ." (make 'definition)))
```

Then the keyboard shortcuts ⇧TH. and ⇧DEF. can be used inside text mode in order to insert a theorem resp. a definition. In a similar way, you may customize the menus, or add more complex extensions to the editor. For more details, we refer to Help→Scheme extensions.

References

1. Grozin, A.G.: TeXmacs interfaces to Maxima, MuPAD and Reduce. In: Gerdt, V.P. (ed.) Proc. Int. Workshop Computer Algebra and its Application to Physics, number 11-2001-279 in JINR E5, Dubna, p. 149 (June 2001), Arxiv cs.SC/0107036
2. van der Hoeven, J.: Towards semantic mathematical editing. Technical report, HAL, submitted to JSC (2011), http://hal.archives-ouvertes.fr/hal-00569351
3. van der Hoeven, J., Poulain, F.: Conservative conversion between LaTeX and TeXmacs. Technical report, HAL (2014), http://hal.archives-ouvertes.fr/hal-00952926
4. Knuth, D.E.: The TEXbook. Addison Wesley (1984)
5. Lamport, L.: LATEX, a document preparation system. Addison Wesley (1994)

Computer Software Program for Representation and Visualization of Free-Form Curves through Bio-inspired Optimization Techniques

Andrés Iglesias[1,2] and Akemi Gálvez[2]

[1] Department of Information Science, Toho University
Narashino Campus, Faculty of Sciences
2-2-1 Miyama, 274-8510, Funabashi, Japan
[2] Dpt. of Applied Mathematics & Comp. Sci., Universidad de Cantabria
E.T.S.I. Caminos, Canales y Puertos, Avda. de los Castros, s/n
E-39005, Santander, Spain
{iglesias,galveza}@unican.es
http://personales.unican.es/iglesias

Abstract. Free-form parametric curves are becoming increasingly popular in many theoretical and applied domains because of their ability to model a wide variety of complex shapes. In real-world applications those shapes are usually given in terms of data points, for which a fitting curve is to be obtained. Unfortunately, this is a very difficult task for classical optimization techniques. Recently, it has been shown that bio-inspired optimization techniques can be successfully applied to overcome this limitation. This paper introduces a new interactive, user-friendly computer software program for the representation and visualization of free-form parametric curves from sets of data points. Given a cloud of data points as initial input, the user is prompted to a graphical interface where he/she can choose the bio-inspired technique of his/her preference, set up the control parameters interactively, and obtain the mathematical representation and graphical visualization of the underlying shape. The paper discusses the main features of this software. An illustrative example of its application is also briefly reported.

Keywords: computer software, mathematical representation, scientific visualization, free-form curves, bio-inspired optimization.

1   Introduction

Free-form parametric curves are becoming increasingly popular in many theoretical and applied domains. They are a key tool, for instance, to obtain an accurate approximating curve to sets of data points in theoretical fields such as numerical analysis, data fitting, approximation theory, and geometric modeling and processing [1,2,3,4,5,6]. They are also widely used in industrial and applied domains such as computer-aided design and manufacturing (CAD/CAM), computer-numerically-controlled milling and machining, automotive, aerospace

H. Hong and C. Yap (Eds.): ICMS 2014, LNCS 8592, pp. 570–577, 2014.

and ship hull building industries, computer graphics and animation, entertainment industries (video-games, computer movies), and many others.

The immense popularity of free-form curves can be mostly attributed to their ability to model a wide variety of complex shapes. In real-world applications those shapes are usually given in terms of data points acquired by using technologies such as 3D laser scanners, and other digitizing devices. The resulting cloud of data points is then approximated by using mathematical entities such as curves and surfaces, leading to a least-squares minimization problem [6,7,8,9,10]. Unfortunately, this problem is very difficult to deal with, as it usually requires to solve a highly nonlinear continuous optimization problem. This problem is also multivariate, as it typically involves a large number of unknown variables for massive sets of data points, a case that happens very often in real-world examples. It is also overdetermined, because we expect to obtain the approximating curve with many fewer parameters that the number of data points. Finally, the problem is known to be multimodal; that is, the least-squares objective function can exhibit several (global and/or local) good solutions.

A number of different techniques have been developed during the last decades to tackle this issue (see [3,11,12,13] for a comprehensive introduction to the field). However, the problem has proved to be more elusive than it appeared at first sight, and the scientific community is still looking for new mathematical and computational methods to solve it. Among the myriad of methods proposed in the field, those based on bio-inspired optimization techniques are receiving increasing attention during the last few years, owing to their ability to perform well under very unfavorable conditions, such as multimodal, multivariate, nonlinear optimization problems, noisy data points, little knowledge about the problem to be solved, and many others [14]. As a result, a number of bio-inspired optimization methods have been applied to the problem of data fitting through free-form parametric curves and surfaces [15,16,17,18,19,20,21,22,23,24,25,26,27]. Unfortunately, such methods are hard to use and require a certain level of expertise in order to apply them efficiently. Consequently, there is an increasing demand of computer solutions for user-friendly manipulation of these methods. The computer software program presented in this paper is aimed at filling this gap.

The structure of this paper is as follows: our program and its main features (implementation, architecture, installation, and workflow) are discussed in Section 2. Then, an illustrative example about the application of this software is briefly reported in Section 3. The paper closes with the main conclusions and our plans for future work.

2   The Program

In this paper we introduce a new interactive, user-friendly computer software program called *BioFit* (version 1.0) for the representation and visualization of free-form parametric curves from sets of data points by using different bio-inspired optimization techniques. Given an input data consisting of a cloud of data points (acquired through a 3D laser scanner or other digitizing devices),

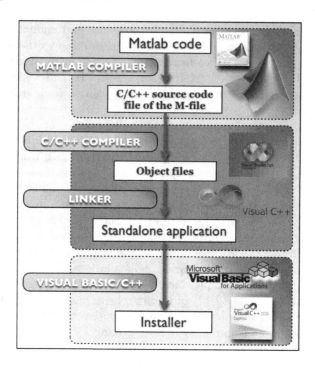

Fig. 1. Workflow for generation of C/C++-based standalone *BioFit* application and the application installer from the original *Matlab* source code.

our program allows the user to choose the bio-inspired technique of his/her preference, set up the control parameters interactively through a user graphical interface, and obtain the mathematical representation and graphical visualization of the underlying shape. This section describes its main features in detail.

2.1   Implementation

Our program has been originally developed by the authors in *Matlab* v2013a by using a a 2.6 GHz. Intel Core i7 processor with 8 GB of RAM. In our development, we took advantage of a very useful functionality of Matlab: the possibility of compile and link its different modules and windows along with their associated libraries and underlying code to generate standalone applications. Figure 1 shows the different steps of this process. The top-down arrows indicate the compilation flow while the dotted curved rectangles enclose the different programming environments required for each step indicated within. Roughly, this process can be summarized as follows: firstly, the chosen module is compiled by using the *Matlab* compiler. The result is a set of multiple C or C++ source code modules that are actually versions in C/C++ of the initial M-files implemented in *Matlab*. These files are subsequently compiled in a C/C++ environment to

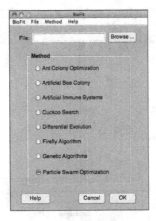

Fig. 2. Main window for choice of the bio-inspired method for curve fitting

generate the object files. Then, those files are linked with the C++ graphics library, M-file library, Built-In library, API library and ANSI C/C++ library files. The final result is an executable program running on our computer.

We remark here the excellent integration of all these tools to generate optimized code that can easily be invoked from different programming environments for several operating systems, providing both great portability and optimal communication between all modules. Complementary, we also constructed an installer for our program by using both Visual C++ and Visual Basic.

Regarding the operating system, our program supports many platforms, including Microsoft Windows (XP, Vista, 7, and 8), Linux and Mac OS X. Figures in this paper correspond to the Mac OS X v10.8 platform version.

2.2 Architecture of the Program

Our program consists of two major components:

- a *computational kernel*, comprised of a set of libraries for several bio-inspired optimization techniques. This kernel allows the user to compute an accurate mathematical representation of the shape in terms of free-form curves belonging to a parametric family of his/her choice. The kernel has a modular structure, in the sense that each library is associated with a particular bio-inspired optimization technique, and libraries operate in a fully independent way. As a consequence, modification of a particular library does not affect the performance of any other. In addition, the kernel is also extensible, as new libraries for other bio-inspired methods can be added as add-ons at any time. Consequently, the system can be continuously enhanced with new features and functionalities. All features in this paper correspond to the version 1.0 of this software.

- a powerful *graphical user interface* (GUI). The main goal of this user interface is to allow unexperienced users to be able to apply different bio-inspired

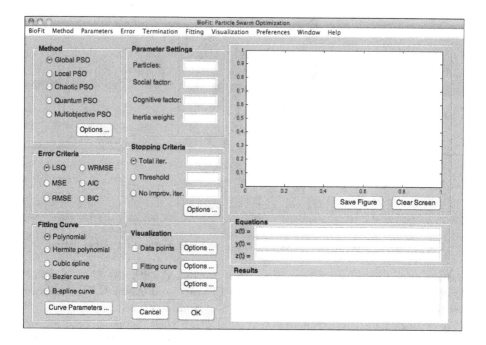

Fig. 3. Screenshot of the window for the particle swarm optimization method

techniques in a very friendly way, without really being proficient in the subject. Figures 2 to 4 show three screenshots of our graphical interface (see Sections 2.4 and 3 for details).

2.3   Installation

The program has been created as an standalone application working on major computer operating systems (Microsoft Windows, Linux, and Mac OS X). It comes with an installer so that it can readily be installed in your computer by double clicking on the installer icon and then following the traditional *Next* > *Next* > ... installation sequence.

2.4   Workflow

Once installed, the user simply has to double click on the application icon to reach the graphical interface shown in Figure 2. In that window, the user introduces the initial input (a file containing the collection of data points to be fitted) by using a button to select the file and its corresponding file path. Then, the user selects the bio-inspired optimization method of his/her choice from a list of radio buttons, one for each method included in the application. So far, the following methods (or families of methods) are included in our program:

– ant colony optimization

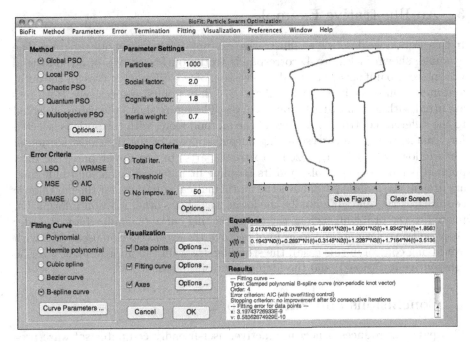

Fig. 4. Example of application of the particle swarm optimization method to an industrial workpiece (a paint spray gun model).

- artificial bee colony
- artificial immune systems
- cuckoo search
- differential evolution
- firefly algorithm
- genetic algorithms
- particle swarm optimization

Clicking on the OK button the user is prompted to a new window where he/she can set up the control parameters interactively, and obtain the mathematical representation and graphical visualization of the underlying shape. The configuration and options available in that window depend on the particular technique chosen by the user. Figure 3 shows the window for the case of the particle swarm optimization method. As the reader can see, the window contains a number of interactive tools and controls (menus, radio buttons, check buttons, push buttons, editable text boxes, and so on) to define even the more subtle details. Note also that our user-oriented design minimizes the time required for the user to get accustomed to the program and all its functionalities.

3   An Illustrative Example

In this section we describe an illustrative example of the use of our program. The example, shown in Figure 4, corresponds to the outline curves of an industrial workpiece: a paint spray gun model consisting of two different curves for the outer and inner boundary lines, with 542 and 276 data points, respectively. Both sets are fitted with fourth-order B-spline curves. The figure shows the values assigned to the different control parameters of the method along with the options selected for visualization. The corresponding output appears in the rightmost section of the window, which is organized into three parts. In the upper part, the data points (as red cross symbols) and its B-spline fitting curve (as a solid blue line) are drawn; the middle part shows the mathematical equations for the parametric coordinates $x(t)$ and $y(t)$; finally, the lower part shows other relevant information related to the problem, such as the type of fitting curve (a clamped fourth-order B-spline curve with non-periodic knot vector), the error criterion used (AIC with overfitting control in this case), the stopping criterion (no improvement after 50 consecutive iterations), and the fitting error obtained.

4   Conclusions and Future Work

This paper introduces a new interactive, user-friendly computer software program for the representation and visualization of free-form parametric curves from sets of data points. The paper discusses the main features of this software. An illustrative example of its application is also briefly reported. Future work includes the extension of this work to other bio-inspired techniques such as bacterial foraging, bat algorithm, and many others. We are also interested to apply this program to real-world problems. This research has been supported by the Computer Science National Program of the Spanish Ministry of Economy and Competitiveness, Project Ref. TIN2012-30768, Toho University (Funabashi, Japan), and University of Cantabria (Santander, Spain).

References

1. Barnhill, R.E.: Geometric Processing for Design and Manufacturing. SIAM, Philadelphia (1992)
2. Castillo, E., Iglesias, A.: Some characterizations of families of surfaces using functional equations. ACM Transactions on Graphics 16(3), 296–318 (1997)
3. Dierckx, P.: Curve and Surface Fitting with Splines. Oxford University Press, Oxford (1993)
4. Iglesias, A., Echevarría, G., Gálvez, A.: Functional networks for B-spline surface reconstruction. Future Generation Computer Systems 20(8), 1337–1353 (2004)
5. Iglesias, A., Gálvez, A.: A new artificial intelligence paradigm for computer-aided geometric design. In: Campbell, J., Roanes-Lozano, E. (eds.) AISC 2000. LNCS (LNAI), vol. 1930, pp. 200–213. Springer, Heidelberg (2001)
6. Ma, W.Y., Kruth, J.P.: Parameterization of randomly measured points for least squares fitting of B-spline curves and surfaces. Computer Aided Design 27(9), 663–675 (1995)

7. Echevarría, G., Iglesias, A., Gálvez, A.: Extending neural networks for B-spline surface reconstruction. In: Sloot, P.M.A., Tan, C.J.K., Dongarra, J., Hoekstra, A.G. (eds.) ICCS 2002, Part II. LNCS, vol. 2330, pp. 305–314. Springer, Heidelberg (2002)
8. Wang, W.P., Pottmann, H., Liu, Y.: Fitting B-spline curves to point clouds by curvature-based squared distance minimization. ACM Transactions on Graphics 25(2), 214–238 (2006)
9. Yang, H.P., Wang, W.P., Sun, J.G.: Control point adjustment for B-spline curve approximation. Computer-Aided Design 36, 639–652 (2004)
10. Gálvez, A., Iglesias, A., Cobo, A., Puig-Pey, J., Espinola, J.: Bézier curve and surface fitting of 3D point clouds through genetic algorithms, functional networks and least-squares approximation. In: Gervasi, O., Gavrilova, M.L. (eds.) ICCSA 2007, Part II. LNCS, vol. 4706, pp. 680–693. Springer, Heidelberg (2007)
11. Piegl, L., Tiller, W.: The NURBS Book. Springer, Heidelberg (1997)
12. Varady, T., Martin, R.R., Cox, J.: Reverse engineering of geometric models - an introduction. Computer-Aided Design 29(4), 255–268 (1997)
13. Varady, T., Martin, R.: Reverse engineering. In: Farin, G., Hoschek, J., Kim, M. (eds.) Handbook of Computer Aided Geometric Design. Elsevier Science (2002)
14. Engelbretch, A.P.: Fundamentals of Computational Swarm Intelligence. John Wiley and Sons, Chichester (2005)
15. Gálvez, A., Iglesias, A.: Efficient particle swarm optimization approach for data fitting with free knot B-splines. Computer-Aided Design 43(12), 1683–1692 (2011)
16. Hoffmann, M.: Numerical control of Kohonen neural network for scattered data approximation. Numerical Algorithms 39, 175–186 (2005)
17. Jing, L., Sun, L.: Fitting B-spline curves by least squares support vector machines. In: Proc. of the 2nd. Int. Conf. on Neural Networks & Brain, Beijing, China, pp. 905–909. IEEE Press, Beijing (2005)
18. Gálvez, A., Iglesias, A., Puig-Pey, J.: Iterative two-step genetic-algorithm method for efficient polynomial B-spline surface reconstruction. Information Sciences 182(1), 56–76 (2012)
19. Gálvez, A., Iglesias, A.: Particle swarm optimization for non-uniform rational B-spline surface reconstruction from clouds of 3D data points. Information Sciences 192(1), 174–192 (2012)
20. Sarfraz, M., Raza, S.A.: Capturing outline of fonts using genetic algorithms and splines. In: Proc. of Fifth International Conference on Information Visualization, IV 2001, pp. 738–743. IEEE Computer Society Press (2001)
21. Gálvez, A., Iglesias, A.: Firefly algorithm for polynomial Bézier surface parameterization. Journal of Applied Mathematics, Article ID 237984, 9 pages (2013)
22. Gálvez, A., Iglesias, A.: A new iterative mutually-coupled hybrid GA-PSO approach for curve fitting in manufacturing. Applied Soft Computing 13(3), 1491–1504 (2013)
23. Yoshimoto, F., Harada, T., Yoshimoto, Y.: Data fitting with a spline using a real-coded algorithm. Computer Aided Design 35, 751–760 (2003)
24. Ulker, E., Arslan, A.: Automatic knot adjustment using an artificial immune system for B-spline curve approximation. Information Sciences 179, 1483–1494 (2009)
25. Gálvez, A., Iglesias, A.: Firefly algorithm for explicit B-spline curve fitting to data points. Mathematical Problems in Engineering, Article ID 528215, 12 pages (2013)
26. Gálvez, A., Iglesias, A.: From nonlinear optimization to convex optimization through firefly algorithm and indirect approach with applications to CAD/CAM. The Scientific World Journal, Article ID 283919, 10 pages (2013)
27. Zhao, X., Zhang, C., Yang, B., Li, P.: Adaptive knot adjustment using a GMM-based continuous optimization algorithm in B-spline curve approximation. Computer Aided Design 43, 598–604 (2011)

On Some Attempts to Verify the Effect of Using High-Quality Graphics in Mathematics Education

Kiyoshi Kitahara[1], Tadashi Takahashi[2], and Masataka Kaneko[3]

[1] Kogakuin University, Japan
kitahara@cc.kogakuin.ac.jp
[2] Konan University, Japan
takahasi@konan-u.ac.jp
[3] Toho University, Japan
masataka.kaneko@phar.toho-u.ac.jp

Abstract. In this paper, we will show some of our attempts to verify the effect of using high-quality graphics in collegiate mathematics education through two types of experiments. In the first experiment, we gave a lesson on the law of logarithms usually done without using graphics. We used teaching materials containing graphics to give students some hints. To prepare the graphics, we utilized an extension of TeX capabilities for flexible page layout. Then we estimated the effects of the lesson through a statistical approach. In the second experiment, we detected the change of students' brain activity by making behavioral observation and neuroimaging simultaneously. For this lesson, we chose the comparison of the degree of growth between two functions as the theme, and prepared some graphs for them. To generate these graphs, we utilized the programmability of the computer algebra system for automatically changing the scale. We showed them to three students and observed their responses. Simultaneously we monitored their brain activities through EEG (ElectroEncephaloGram) measurements. We observed that the judgment of these students changed when they saw a triggering figure, and some change in the trend of the EEG signal was observed at that time. From the results of these experiments, it is indicated that using effective figures in materials might have a great influence on learners' reasoning processes.

Keywords: graphics, TeX, computer algebra system, brain activity.

1 Introduction

According to the results of our questionnaire survey, one major opinion of collegiate mathematics teachers in Japan is that there is no necessity to use high-quality graphics in education[1]. However, from our experiences, graphics use seems to play a crucial role in some classroom situations. Therefore, we usually try to use various graphics in teaching materials edited by the popular TeX tool.

H. Hong and C. Yap (Eds.): ICMS 2014, LNCS 8592, pp. 578–585, 2014.

Moreover, we have been trying to verify the effect of using graphics by comparing students' responses in cases when high-quality graphics are used and in cases when they are not used[2].

To generate high-quality graphics, the computer algebra system (CAS) should be the most preferred tool due to its computing and programming capabilities. However it is not always easy to handle the resulting graphical images in the documents edited by TEX. For instance, some elaborations are needed to put the generated image in a suitable position and to flexibly arrange the layout of other components in a harmonious way. Though there exist some TEX graphic systems such as PStricks[3] and TikZ[4], their computing capabilities are fairly limited. As a handy tool for both generation of high-quality graphics with CAS and easy arrangement of the components in TEX documents, we have been using KETpic. KETpic is a macro package designed to generate TEX-readable code for CAS creating graphical output. Its package and related documentations can be freely downloaded at the website: **http://ketpic.com**. The procedure to generate graphics with KETpic is summarized in Figure 1. Applicable CASs are Mathematica, Maple, Matlab, Scilab and R. Here we mainly use the Scilab version. To make a flexible page layout in step III possible, the TEX macro package named "ketlayer" has also been prepared[5].

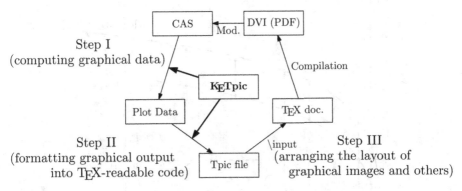

Figure 1. KETpic cycle

The aim of this presentation is to show some of our new attempts to verify the effect of using high-quality graphics in collegiate mathematics education. From the two kinds of experiments conducted to verify the effectiveness of using graphics in mathematics education, we can claim that these methodologies have a great possibility to become objective ways to verify the effect of various mathematical software.

2 Estimating the Effect Statistically

In this section, we will show the method and result of conducting a lesson which aims to teach the law of logarithmic functions:

$$\log_a xy = \log_a x + \log_a y$$

This topic is usually taught through deduction based on the definition of logarithm without using graphics. We planned some alternative approach to explain this law, using material with graphs to help students visualize the law. The subjects are 56 students at a university in Japan who have already studied this law through the deductive method. Our strategy is to compare the precise graphs of functions $y = \log_a x$ and $y = \log_a(cx)$ in two ways. As an example, we gave students the graphs in Figure 2 contained in printed teaching materials.

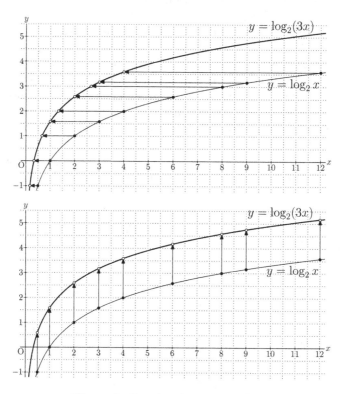

Figure 2. Graphs given to students

By observing the first figure, students can understand that the graph of $y = \log_2(3x)$ can be obtained by reducing the graph of $y = \log_2 x$ by $\frac{1}{3}$ in the x direction. Then, by comparing to the second figure, students can observe that the graph of $y = \log_2(3x)$ coincides with the result of parallel translation in the y direction. The key fact of our strategy is that the distance of this parallel translation is equal to $\log_2 3$. It is not so easy for most students to understand this fact only by observing Figure 2. Therefore, to support students' understanding, we showed each of the graphs in Figure 3 at one time on a PC screen and projector. In these graphs, the rays are parallel translated until their initial points are located on the x axis. Especially, the last graph (6) indicates that the ray at the position $x = 1$ should be given most attention. Since this explanation is also applicable to the case of $y = \log_a(cx)$ for any positive value of c, students will be able to grasp the mechanism of this law.

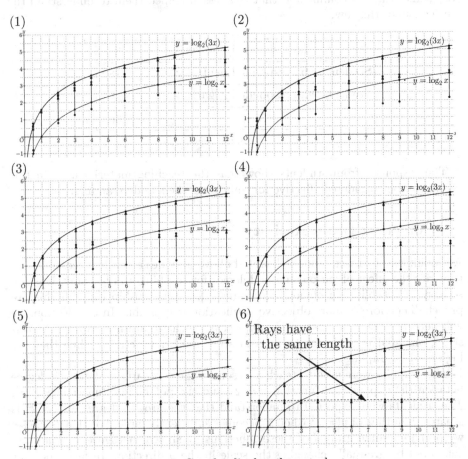

Figure 3. Graphs displayed to students

To show these graphs, it is essential that they are located in exactly the same position on each page. Moreover, it is desirable that comments and other graphical components be put on the preferred position as shown in (6). Using the ketlayer environment, we can readily meet these requirements. In fact, we only used the following simple commands in Step III of KETpic cycle (Figure 1):

```
\begin{layer}{130}{0}
  \putnotes{60}{35}{{\input{fig6.tex}}}
    %specifying the preferred position for putting the graph
  \arrowline{62}{85}{40}{145}
  \putnotee{15}{40}{Rays have}
  \putnotee{17}{44}{the same length}
\end{layer}
```

After the lesson, we asked students whether this explanation is easier to understand compared to the deductive one or not. The result is shown in Table 1. Also some students commented that this lesson helped them to understand the mechanism of this law.

Students who responded positively	36
Students who responded neutrally	18
Students who responded negatively	2
Total	56

Table 1. Students' responses

The responses from students show that the teaching material used in this lesson can be considered effective for improving their understanding.

3 Detecting Brain Activity

Although the statistical approach stated above provides an easily accessible way to verify the effect of various methodologies for teaching and learning, it is not so easy to establish an obvious cause-and-effect relationship through this approach. Therefore, a more objective verification is desirable. In this section, we will demonstrate a neuroscientific approach which we are now trying. At this stage, our goal is to detect changes in students' brain activity while they see high-quality graphics.

The graphics were used to teach a lesson on the comparison of growth degree between the exponential function $y = 2^x$ and the polynomial function $y = x^4$. The students were expected to understand that the growth degree of $y = 2^x$ is greater than that of $y = x^4$ when x becomes sufficiently large. For that purpose, we prepared some graphs of these functions as shown in Figure 4. They are generated by gradually changing the scale in the y direction. In fact, the unit lengths of x and y axes are the same in graph (1), and those in the y direction of graphs (2) (3) (4) and (5) are reduced to the ratios $\frac{1}{10}$, $\frac{1}{100}$, $\frac{1}{1000}$ and $\frac{1}{10000}$ respectively. To generate these graphs, we used the following simple commands of the Scilab version for KETpic.

We showed these graphs one by one to three students. The students were asked which function they thought increased more rapidly. We observed their behavior and monitored their brain activities through EEG measurement. We used the EEG devise "Polymate V" and attached two electrodes to the positions displayed in Figure 5. These are positions F3 and F4 in the international 10-20 system.

```
function Out=Functions(Ra,YM)
  Setscaling(1/Ra);
  Setwindow([0,15.5],[-0.5,YM]);
  G1=Plotdata("2^x","x");
  G2=Plotdata("x^4","x");
  Out=list(G1,G2)
endfunction

for K=[10,100,1000,10000];
  G=Functions(K,10.1*K);
  Openfile("fig"+string(K)+".tex");
  Beginpicture('1cm');
    Drwline(G);
  Endpicture(1);
  Closefile();
end
```

Commands in Step I
of KₑTpic cycle (Figure 1)

Commands in Step II
of KₑTpic cycle (Figure 1)

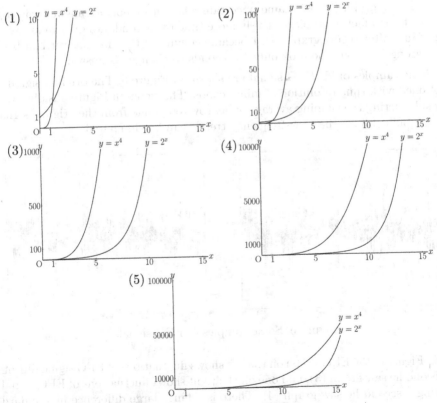

Figure 4. Graphs used for the EEG measurement

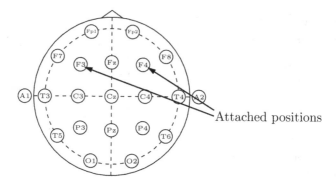

Figure 5. The positions where electrodes are attached

In the lesson, the students first answered that $y = x^4$ grew more rapidly. But, when they saw triggering figure (4) or (5), they changed their answers. Moreover, in the case of one student, some trend change of EEG signal was observed at that time. In the rest of this section, we will look at this case in more detail.

This student changed his answer after he saw graph (4). In the interview after the experiment, he stated as follows:

1. "Seeing graph (3), I became uncertain whether or not my answer was correct." (In fact, he took relatively more time (45 seconds) to give an answer.)
2. "Just after seeing graph (4), I became convinced that my answer had been wrong." (In fact, he took only 15 seconds to change his answer.)

Some samples of his EEG signal are shown in Figure 6. The original signal is obtained with time resolution of milliseconds. The signal in Figure 6 is obtained by subtracting the running average of nearby 100 points from the original signal at the point F3, so that low-frequency trends can be eliminated.

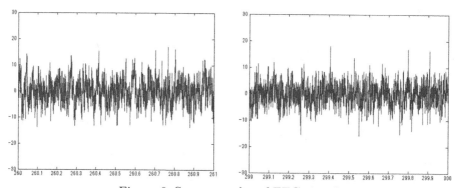

Figure 6. Some samples of EEG signal

In Figure 6, the EEG graph on the left shows fluctuations of EEG signal during a second he saw graph (3), and one on the right shows fluctuations of EEG signal during a second he saw graph (4). There is a fairly large difference in standard deviation between these two cases. In fact, its realized value is $4.4000[\mu V]$ for the left one, and 3.8661[μV] for the right one. This result might indicate that some change in brain activity was induced by his seeing triggering figure (4).

It is not easy to interpret this difference in EEG signal, since the relationship between specific brain activity and its influence on electric potential over the scalp has not been fully understood. There are at least two possibilities to interpret the above mentioned difference:

1. Recognizing the difference between graph (3) and (4), the student repeated the reasoning process for comparing growth degree. The difference was caused by the occurrence of this new process.
2. According to the student's statement, there is a difference in his emotional state between these two cases. The difference was caused by this transition from uncertainty to certainty.

Since both of these new developments in reasoning and emotion can be regarded as the result of using figure (4), it is reasonable to claim that the effect of using high-quality graphics can be verified through this experiment.

4   Conclusions and Future Study

The results of this study show that using effective graphics can greatly influence students' reasoning. The verification method used here should be applicable to various educational resources. In future, we must clarify the following points through similar experiments:

1. Among the many topics in collegiate mathematics education, for which ones are graphics effective?
2. How can we evaluate (or compare) the effect of various graphics through the neuroscientific approach?
3. Among the many methodologies to generate graphics, which ones are suitable for each theme?

Acknowledgments. This work is partially supported by Grant-in-Aid for Scientific Research (C) 24501075. Professor Masahiro Nakagawa at Nagaoka University of Technology provided the use of Polymate V. Also Professor Ikusaburou Kurimoto at Kisarazu National College of Technology and Mr. Masashi Ikeda helped to execute this experiment and analyze the results. The authors are grateful for their assistance.

References

1. Kaneko, M., Takato, S.: The effective use of LATEX drawing in linear algebra. The Electronic Journal of Mathematics and Technology 5(2), 129–148 (2011)
2. Kaneko, M., et al.: A scheme for demonstrating and improving the effect of CAS use in mathematics education. In: Proc. ICCSA 2013, pp. 62–71 (2013)
3. Graphic system for TEX, http://sourceforge.net/projects/pgf/
4. TEX users group home page, http://tug.org/PSTricks/main.cgi/
5. Kaneko, M., Takato, S.: A CAS macro package as TEX graphical command generator and its applications. In: Proc. ICCSA 2011, pp. 72–81 (2011)

Math Web Search Interfaces
and the Generation Gap of Mathematicians

Andrea Kohlhase[1,2]

[1] Jacobs University, Germany
[2] Forschungsinformationszentrum Karlsruhe, Germany
a.kohlhase@jacobs-university.de

Abstract. New technologies and interfaces are changing the way users engage with technology, mathematicians are no exception. In a previous study we found some interesting attitudes/practices of professional mathematicians with respect to search interfaces, that sets them apart from other web searchers. In a nutshell, this study explores whether and if so, how math search interfaces are distinctly perceived by younger and older mathematicians and we offer first design implications.

Keywords: math search interfaces, repertory grid analysis, generation gap.

1 Introduction

In [Koh14] we presented ten behavioral patterns of mathematicians with respect to math search interfaces. Note that in this paper, mathematicians are people with a research interest in mathematics. We were able to show that mathematicians and non-mathematicians do approach math search very differently.

For instance, mathematicians strategically use the search engine "Google" for *finding* specific objects of interest, i.e., specifically looking for identifiable information. This comprises that mathematicians not only know previous to the search what exactly they are looking for and thus, anticipating the exact search result, but also that they know how to formalize the search query. In contrast, Google is best-known for its *browsing* quality, i.e., getting an impression of what data are available (e.g., for an overview or for inspiration) and possibly refining the search as a consequence. Very simply put, mathematicians would look for a definition of "Cauchy sequence" instead of looking for information about "Cauchy sequence".

But there were noticeable differences: some expected the Google experience in all text-based math search interfaces, others restricted it purely to Google itself. It was conjectured that the generation gap in the math community could account for this. This would have the particular consequence that math search interfaces for the new generation of mathematicians have to be different as the usability criteria in-between generations are not shared.

To verify that there indeed is a "generation gap" between older and younger mathematicians, we looked closer into the available data from the study done in [Koh14].

H. Hong and C. Yap (Eds.): ICMS 2014, LNCS 8592, pp. 586–593, 2014.

2   The Study

We used repertory grid interviews ("**RGI**" ;see for example [HW00, Jan03, Kel03]) as main methodology to elicit evaluation schemes ("**constructs**") with respect to selected math search interfaces ("**mathUI**") and to understand how mathematicians classify those mathUIs. The main advantage of the method is its semi-empirical nature. On the one hand, it allows to get deep insights into the topic at hand through deconstruction and intense discussion. On the other hand, the grids produced in such RGI sessions can be analyzed with a General Procrustes Analysis to obtain statistically significant correlations between the elicited constructs or the given mathUIs.

We decided to use 12 RGIs from the set of interviews conducted in [Koh14]. All interviewees were mathematicians, but 6 we assigned to the 'older' group "**mathPROFs**" and 6 to the 'younger' one "**mathSTUDs**". Our criterion for assigning a group did not rest on the age of the interviewees, but on their social-ization time within the math community. We considered a subject 'old' when he or she were longer than five years in a leading position in the community, e.g. as a professor. The underlying reason consisted in our interest in differences in mathematical practices, which depend more on community status than on life age. Nevertheless, the distinct criteria coincided in most cases.

3   The Generation Gap for Mathematicians

In this study we are using the same analysis tools as in [Koh14], therefore we like to ask the reader to look for details of the method there. Basically, we obtained a total of 67 evaluation schemes by 12 participants for the following 17 mathUIs:

zbMathNew zbMath.org: a mathematical abstracting and reviewing service

zbMathOld : a former version of zbMathNew

MathSciNet ams.org/mathscinet: a mathematical abstracting and reviewing service

Google-Scholar scholar.google.com: search for scholarly literature on the Web

Google google.com: search on the Web

TIB tib.uni-hannover.de: a university online catalogue

vifamath vifamath.de: a virtual library of mathematics

arXiv arxiv.org: a (mathematical) open e-print archive

ResearchGate researchgate.net: a scientific network platform

mathoverflow mathoverflow.net: a mathematical answer and question plat-form

MSC-Map map.mathweb.org: an interactive map for mathematics based on the math subject classification (MSC)

arxiv-Catchup arxiv.org/catchup: catching up with newest mathematical up-loads to arXiv

FormulaSearch zbmath.org/formulae: searching for formulae in the zbMathNew database

myLibrary : a physical library as math search interface

`myOffice` : the personal office as math search interface
`myColleagues` : personal colleagues as math search interface
`Bibliography` : a bibliography as math search interface

The General Procrustes analysis on the elicited data was done with Idiogrid [Gri02], the biplots and cluster dendrograms were generated with OpenRep-Grid [Ope].

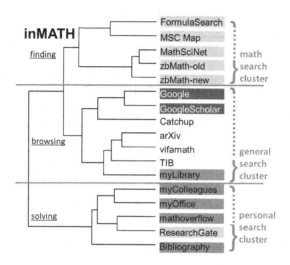

Fig. 1. Element Clusters for all Mathematicians

Recall that **dendrograms** are a visual representation of correlation data. Two elements, e.g., in Fig. 1, are closely correlated, if their scores on the RGI elements are similar. The distance to the next upper level of two elements/groups of elements *indicates* this relative closeness.

In Fig. 1 we see the result of the cluster analysis of elements of all mathematicians, the so-called "**inMATH**" group, as analyzed in [Koh14]. To be able to better compare the distinct elements, the colors in all dendrograms are adapted to the ones used in the left-hand side of Fig. 2 for the results of the 'purest' mathematical group, i.e., the MATHPROFs group.

We can derive directly that `MathSciNet`, `zbMathOld` and `zbMathNew` are correlated very closely and that the innovative interfaces `FormulaSearch` and `MSC-Map` are perceived by the mathematicians as potential mathematical search tools, thus, they are close to the former ones. All of them are considered specifically fit math search tools, so we called this the "math search cluster". In another main cluster, the "personal search cluster", all the mathUIs are comprised that offer very fine-grained, personal math search interfaces. The third main "general search cluster" contains more general search tools, here, particularly `Google` resp. `Google-Scholar` were marked by many interviewees.

To understand what the differences in the MATHPROFs versus the MATHSTUDs group are, we split up the group and obtain the two separate dendrograms in Fig. 2.

The 'Pure' Math Cluster It is obvious that both groups agree on the yellow cluster within the math search cluster: `MathSciNet`, `zbMathOld` and `zbMathNew` do have similar ratings across the generation boundary. Thus, they are considered undisputedly as interfaces for math search of similar quality. Note that we have to check for the corresponding evaluation schemes to decide whether this means a fitting, good quality or not.

The "Innovative" and "Standard Search" Clusters. The closeness of the innovative interfaces `FormulaSearch` and `MSC-Map` respectively the standard web search interfaces `Google` and `Google-Scholar` is unchallenged as well. This is surprising as such as the innovative interfaces are very different interfaces, as one, e.g., accepts only LaTeX input, the other is a map service with zoom in and out facilities. The standard web search interfaces cluster, even though one is responsible for search on the entire net, whereas the other only considers scholarly objects as search data, thus, it is interesting per se that they cluster.

Nevertheless, their clustering has changed. For the MATHPROFs group the innovative, but math specific interfaces are still closer to the 'pure', yellow math-UIs than all others, but they are themselves considered a little bit closer to the standard search cluster.

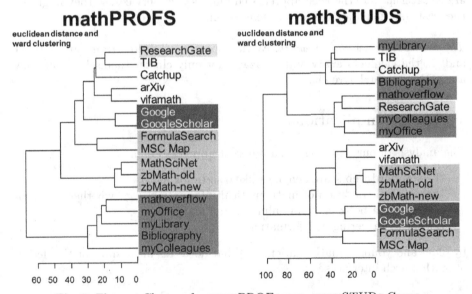

Fig. 2. Element Clusters for MATHPROFs resp. MATHSTUDs Group

The Personal Search Cluster. Another immediate observation consists in the differences between the MATHPROFs group and the MATHSTUDs group with respect to the red personal search cluster.

The former perceives them as most different from all the other mathUIs (see left-hand side of Fig. 2). Note that there was also a strong personal search cluster in the INMATH group (Fig. 1), but `ResearchGate` was replaced by `myLibrary`.

The mathUI `ResearchGate` Here, the connection with real people seems to be the critical difference. Even though the 'older' mathematicians appreciated the personal flavor of a scientific network like `ResearchGate`, they did not appreciate the network itself as being valuable for their search behavior. The library, on the other side, (even though only seldom used) was associated with a librarian, which filtered the available information for them in a very directed way.

In contrast, the MATHSTUDs group experiences people and networks as personal support. They also visit only infrequently the real library, but they do associate a rather inconvenient, even inaccessible pool of information with it. Note that all the physical mathUIs (and such that have a physical component like people) are in one cluster on the right-hand side of Fig. 2, all the virtual ones in the other.

The mathUIs `arxiv-Catchup` *and* `TIB`. The exception is the `arxiv-Catchup` service. We suspect that the service of "catching up with new information" itself is based on physical world metaphors and doesn't fit to the information strategy of the 'younger' generation. Note that for the MATHPROFs group `arxiv-Catchup` and `TIB` are placed similarly, but we believe that other reasons underly this fact. In the interviews it became clear that the three mathUIs `ResearchGate`, `TIB` and `arxiv-Catchup` were the least appreciated math search services, so they might be the least functional for the 'older' generation.

The mathUIs `arXiv` *and* `vifamath`. With respect to all dendrograms in Figures 1 and 2, the elements `arXiv` and `vifamath` not only cluster themselves but they are in the same relative cluster.

4   Design Implications

The analysis of our data suggest a different appreciation of

- the standard web search engines like `Google` for math,
- the physical condition of mathematical information, e.g., whether it is represented in a book or in an online portal, and
- the social gathering of information

by older and younger mathematicians. What does this mean now for the design of math search interfaces?

4.1   The Google Search Design Factor

One best practice of a well-known set of usability guidelines [NM90] states that *"any extra unit of information in a dialogue competes with the relevant units of information"* and therefore a minimalist design is to be preferred (see also [Nie99]). For a search interface Google's layout is a prototypical example. But Google has achieved more than good design, it turned itself into the "**Google search design factor**", consisting according to JARVIS in [Jar09, 391ff.] of

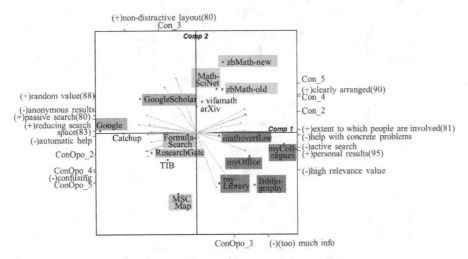

Fig. 3. Biplot for MATHPROFs Group

1. Aesthetics of simplicity turns into an ethics of simplicity. Simplicity stands for successfully elaborated complexity.
2. Google rewards openness.
3. The effects of the long tail.

For the mathematicians in general part (1) of the Google factor was not relevant at all (see Patterns 6 and 7 in [Koh14]). They evaluated Google according to its capability of finding (Pattern 3) as fit for math, but not for its aesthetics. In Fig. 3 we can see how the mathUIs are distributed according to their structure coefficients in the plane spanned by the first principal components in the MATHPROFs group. Only the more relevant constructs are shown. Note the rather negative connotations close to Google. In contrast, in Fig. 4, you can see the ones for the MATH-STUDs group. Here, Google and co. are clearly more positively connotated, we thus suspect that they do care for the Google search design factor.

Older mathematicians were and still are astonished about what the Google search engine can accomplish. They might associate the plainness of the main page with simplicity. This is not the case for the younger mathematicians. They care for the Google factor, but do not depend on it. Minimalistic design does not necessarily mean non-complex design, but rather a clear and focused structure. For the main page of Google this has been done, but it doesn't mean that a one-line search box is wanted everywhere. It is important though that the reduction is transparent (see Pattern 9 in [Koh14]), that is, that mathematicians can retrace the simplifications. Especially older mathematicians handle the tools at hand very deliberately. So, one part of the Google factor can be translated to

> **Implication 1:** *"The interface complexity has to be reduced as much as possible, but at the same time the reductions need to be as transparent as possible."*

The MATHSTUDs group do stress the accessibility to the original doc and the currentness of information when using, for example, Google. Google aims for

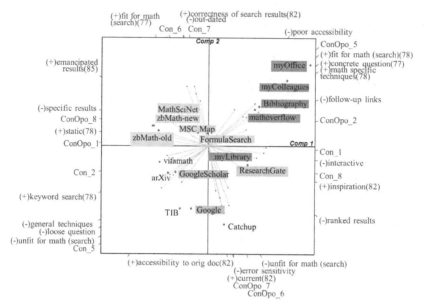

Fig. 4. Biplot for MATHSTUDs Group

completeness of data at any point in time by the means of rewarding openness (2.) of the Google factor). Therefore, we suggest that mathUIs should evaluate openness higher than completeness. More precisely:

> **Implication 2:** *"To draw younger mathematicians in, math search interfaces should strive for and reward up-to-dateness and accessibility."*

The third part (3) of the Google factor refers to Google's capacity to find even the rarest items. This is extremely important for mathematicians as Pattern 3 in [Koh14] indicated. In Fig. 3 as well as Fig. 4 the constructs indicate that this search mode is important for MATHPROFs (e.g., "anonymous results" as evaluation of Google) as well as MATHSTUDs (e.g., "specific results" for the 'pure' math cluster).

4.2   The Library Design Factor

We already argued above that physicality of information objects as offered by libraries, for instance, doesn't seem to be especially attractive to younger mathematicians any longer. Moreover, even for the MATHPROFs group the mere existence in the real world, its embodiment, doesn't count as much as its functionality. Moreover, mathematicians seem to believe that there is nothing lost, when turning to virtual facilities. So, we conclude very simply (but for different reasons in each generation):

> **Implication 3:** *"Math search doesn't have to take physical forms of information objects into account. Mathematical information transition does not depend on its embodiment."*

4.3   The ResearchGate Design Factor

In HCI it is said that Social Media become more and more important for social interaction among people. In [Koh14] Pattern 4 verified that mathematicians make use of social interaction in their math workflows as mathematical practices. Therefore, it is not astounding that younger mathematicians turn towards social media and try to initiate exactly those kind of social interactions as mathematical practices that are common among the traditional math community. Some mathUIs like `mathoverflow` have already succeeded to even convince older mathematicians, others like `ResearchGate` still have to win them over. As of now networking is not included in mathematical practices, so a new task for math search interfaces include:

> **Implication 4:** *"Math search interfaces need to understand networking and its use for mathematical practices to integrate them into their services."*

5   Conclusion

We have presented a study concerned with the generation gap of mathematicians with respect to math search interfaces. We found differences and suggested first design implications for future math search interfaces. Note that some of the implications can be generalized to math interfaces.

References

[Gri02] Grice, J.W.: Idiogrid: Software for the management and analysis of repertory grids. Behavior Research Methods, Instruments, & Computers 34, 338–341 (2002)

[HW00] Hassenzahl, M., Wessler, R.: Capturing design space from a user perspective: The repertory grid technique revisited. International Journal of Human-Computer Interaction 12, 441–459 (2000)

[Jan03] Jankowicz, D.: The Easy Guide to Repertory Grids. Wiley (2003)

[Jar09] Jarvis, J.: What Would Google Do? HarperCollins Publishers Limited (2009)

[Kel03] Kelly, G.: A Brief Introduction to Personal Construct Theory. In: International Handbook of Personal Construct Technology, pp. 3–20. John Wiley & Sons (2003)

[Koh14] Kohlhase, A.: Search interfaces for mathematicians. In: Watt, S.M., Davenport, J.H., Sexton, A.P., Sojka, P., Urban, J. (eds.) CICM 2014. LNCS (LNAI), vol. 8543, pp. 153–168. Springer, Heidelberg (2014)

[Nie99] Nielsen, J.: Designing Web Usability: The Practice of Simplicity. New Riders Press (1999)

[NM90] Nielsen, J., Molich, R.: Heuristic evaluation of user interfaces. In: Proceedings of the SIGCHI Conference on Human Factors in Computing Systems, CHI 1990, pp. 249–256. ACM, New York (1990)

[Ope] `openrepgrid.org`

Practice with Computer Algebra Systems in Mathematics Education and Teacher Training Courses

Hideyo Makishita

Shibaura Institute of Technology, Japan
`hideyo@shibaura-it.ac.jp`

Abstract. PISA survey research revealed that Japanese high school students have difficulty using mathematics. Responding to those survey results and aiming at improving this situation, the Japanese Ministry of Education, Culture, Sports, Science and Technology created a new subject designated as "Application of Mathematics" in which mathematics is developed with close involvement in culture. The design of the subject "Application of Mathematics" is based on two fundamental pillars: "human activity and mathematics", and "mathematical considerations for social life". In this talk, the author wants to add a new perspective: "Application of Mathematics to Mathematics". Accordingly, the talk will present some examples from WASAN problems of this concept using Computer Algebra Systems (CAS), such as Mathematica, Maple, Maxima, Scilab and R, and TeX documents used as classroom materials through KₑTpic. The talk encompasses the author's practice in the teacher-training course and high school classes with CAS. This approach is particularly effective for students in the field of geometry.

Keywords: Application of Mathematics, Scilab, KₑTpic, WASAN.

1   Introduction

The PISA survey identified that Japanese high school students have difficulty related to their attitudes about and capabilities for application of mathematics. In fact, the author, as a teacher of mathematics, feels that although students are good at solving mathematical problems listed in a textbook, many students are puzzled when faced with a scene to discuss or with daily situations that must be resolved mathematically. This state of affairs, which is regarded as attributable to the fact that they get bewildered at how to apply mathematics to daily problems, has been researched. In fact, educational materials dealing with actual scenarios that can be resolved by mathematics are rarely used in conventional classes, which might explain some of the bewildered feeling of many students.

Given that background, the Ministry of Education, Culture, Sports, Science and Technology newly created a mathematical subject called "Application of Mathematics" that includes "mathematics and human activities" and "mathematical consideration for social life" as a mathematical subject for high schools.

H. Hong and C. Yap (Eds.): ICMS 2014, LNCS 8592, pp. 594–600, 2014.

As described in this paper, the author insists that for cultivation of student's capability to use mathematics subjectively, mathematics should be visible all the time. To achieve this, it is considered that teachers should teach students and develop educational materials with emphasis on "Application of Mathematics to Mathematics".

Then, such research becomes important for development of educational materials with the perspective that mathematics is practically applicable to mathematics. That point is demonstrated at lesson studies to be used in designing a future curriculum. At the same time, teachers who can practice and teach application of mathematics should be cultivated to an urgent degree.

As a concrete example of the author's idea of "Application of Mathematics to Mathematics", roles of mathematics as a language for describing an event using formula manipulation software Scilab and K$_E$Tpic are presented in this paper. This perspective is derived from the perspective of drawing figures. Section 2 presents and explanation of how to draw an inner center and an inscribed circle of a triangle is shown. Section 3 shows how to draw a third circle mathematically in the problem of WASAN using the relation between two circumscribed circles and common tangent.

2   Language for Describing Events: Expression by CAS

Bisectors of three inner angles of a triangle meet at one point. This is designated as the inner center, which is located an equal distance from three sides; a circle contacting the three sides can be drawn around the inner center. Such a circle is designated as an inscribed circle.

University students of the teacher-training course for aspiring mathematics teachers of high schools were once requested to find an inner center of a triangle. Using a ruler and a compass, they were able to draw one accurately based on the definition. According to the author, this is a category of solving an ordinary mathematical problem.

They were then asked to draw an inner center using other mathematical structures without using a ruler or a compass on the assumption of using CAS. Many did not know at all how to do it. Although equations of a straight line and a circle were taught in high school in the course unit "Graphics and equations", with instruction in the functions of the ruler and compass, their understanding of structures through mathematical concepts might have been insufficient.

The benefits of "Application of Mathematics to Mathematics" are acquired by replacing an issue by another mathematical structure as performed using the ruler and compass. CAS is suited to experiences of an actual sensation of roles of mathematics as a language for describing phenomena. Therefore, it might be said that it is also suited for acquisition of the concept and attitude of "Application of Mathematics to Mathematics", which the author proposes.

2.1   Drawing Inner Center of Triangle

To draw an inner center of a triangle using CAS and without using a ruler and a compass, the goal here is how to create a bisector of the inner angle.

Here, based on the contents of mathematics that are studied up until high school, drawing an inner center of a triangle using two methods from elementary geometry and vectors are proposed.

2.2   Bisecting Angle Using Elementary Geometry (Diagonal Line of Rhombus)

As properties of graphics, we studied the properties of a rhombus as follows:

Using two diagonal lines of the rhombus, it is divided into four congruent right triangles. From this, as Fig.2, if a rhombus OGJH is created around ∠GOH to include vertex O, then the bisector of ∠GOH is generated as its diagonal line.

2.3   Bisecting Angle Using Elementary Geometry

Students have already studied the following theorem:

Theorem 1. *As Fig.1, in triangle OAB, the intersection point of bisector of ∠AOB and side AB is designated as D, |AD| : |BD| = |OA| : |OB| is established. The opposite is also true. The opposite of the theorem used here is also true.*

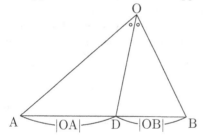

Fig. 1.

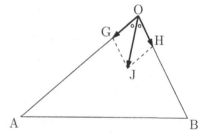

Fig. 2.

This famous problem is frequently selected as a practice exercise presented in high school textbooks in Japan.

When the lengths of three sides of triangle OAB are given, the intersection point D of side AB can be obtained using the opposite of this proposition and bisector OD of ∠AOB is obtained. Similarly, a bisector is obtained for the other two angles. An intersection point of bisectors of three angles is sought as the inner center.

2.4   Method Using Unit Vector Sum

Theorem 2. *As Fig.2, a bisector of an angle can be expressed using the sum of two nonparallel unit vectors. That is to say, for two vectors $\overrightarrow{OA}$ and $\overrightarrow{OB}$, the sum of each vector $\dfrac{\overrightarrow{OA}}{|\overrightarrow{OA}|}$ and $\dfrac{\overrightarrow{OB}}{|\overrightarrow{OB}|}$, i.e., $\dfrac{\overrightarrow{OA}}{|\overrightarrow{OA}|} + \dfrac{\overrightarrow{OB}}{|\overrightarrow{OB}|}$ becomes a bisector of ∠AOB.*

2.5 Method Using Position Vector

Theorem 3. *If the vertex of triangle ABC is expressed by position vectors* $A(\vec{a})$, $B(\vec{b})$, *and* $C(\vec{c})$, *then the position vector of the center of gravity* $G(\vec{g})$ *is expressed by the following well-known expression.*

$$\vec{g} = \frac{\vec{a} + \vec{b} + \vec{c}}{3}$$

This expression appears in the mathematics textbook of Japanese high schools. This theorem is famous for every student.

2.6 Position Vector of Inner Center

In Japan, students learn a circumcenter, an orthocenter and excenters in addition to the inner center, center of gravity, constituting a total of five centers of triangle. Here, the inner center position is described.

Theorem 4. *If the vertex of triangle ABC is expressed by position vectors* $A(\vec{a})$, $B(\vec{b})$, *and* $C(\vec{c})$, *then the position vector of inner center* $I(\vec{i})$ *is expressed as follows: However,* $BC = a$, $CA = b$, $AB = c$.

$$\vec{i} = \frac{a\vec{a} + b\vec{b} + c\vec{c}}{a + b + c}$$

2.7 How to Find the the Radius of an Inscribed Circle of Triangle

If the area of triangle ABC is S, the lengths of sides are a,b,c and the radius of an inscribed circle of triangle ABC measures r, the relationship shown on the follow is satisfied. where $2s = a + b + c$.

$$S = sr$$

There is a formula for areas of triangle which depends on only lengths of sides. (Heron's formula) :

$$S = \sqrt{s(s - a)(s - b)(s - c)}$$

We can compute the the radius of an inscribed circle of triangle by 2 formulas.

$$r = \frac{S}{s} = \sqrt{\frac{(s - a)(s - b)(s - c)}{s}}$$

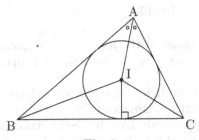

Fig. 3.

2.8   How to Use the Scilab and KₑTpic in Finding the Inner Center and the Radius of an Inscribed Circle of Triangle

Scilab includes many of mathematical functions. It has a high level programming language 2-D and 3-D graphical functions.

Scilab prepares graphics functions to visualize, annotate and export data and many ways to create and customize various types of plots and charts. KₑTpic has been developed as a plug-in based on CAS. Mathematics teachers create a figure using KₑTpic program in CAS along with mathematical drawing procedures.

In this section, the author shows how to compute the inner center of triangle and the radius of an inscribed circle of triangle by Scilab. Here is an algorithm to find the inner center and the radius of an inscribed circle of triangle.

Scilab	mathematical meaning
function Out=fn(A,B,C);	← A,B,C express the position vector.
a=norm(B-C);	← $a = \|\overrightarrow{CB}\|$
b=norm(C-A);	← $b = \|\overrightarrow{AC}\|$
c=norm(A-B);	← $c = \|\overrightarrow{BA}\|$
I=(a*A+b*B+c*C)/(a+b+c);	← Point I is expressed as the inner center.
s=(a+b+c)/2;	
S=sqrt(s*(s-a)*(s-b)*(s-c));	← Heron's Formula
Out=S/s;	← $\dfrac{S}{s}$ means the radius of an inscribed circle.
endfunction	

3   Using CAS in WASAN

Japanese old Mathematics, known as WASAN, is well known to have evolved uniquely, especially during the Edo period (1603–1868). It includes numerous problems related to geometry and beautiful figures.The term Sangaku as Fig.4 refers to Ema (votive tablets) on which mathematical problems were written and which were dedicated to shrines and temples. It is often said that the custom of offering Sangaku began in 1660's.

3.1   Translation of the Problem

Question: As the Fig.5 shows, if the medium circle diameter is 9 sun and the small circle diameter is 4 sun, what is the large circle diameter?
Answer: 36 sun.
Explanation (formula): First divide the diameter of the medium circle by that of the small circle and take the square root of that number. Then, subtract 1 from that number and square the result. Last divide the diameter of the medium circle by that number, one can find the diameter of the large circle; then simplified.

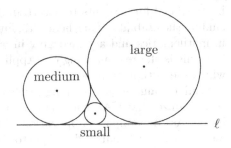

Fig. 4.   Fig. 5.

3.2 Explanation

Many problems related to the construction of a figure are solvable using the loci of points. When one wants to find the position of a point, one can draw two loci which respectively satisfy the two given conditions, and find where they intersect. Such a method is called the intersection of loci. This Sangaku introduces how to describe the circle O_3 by the intersection of loci with Scilab and K$_E$Tpic. The following shows its process. The Fig.6 shows it also, letting the line ℓ be the common tangent line.

(1) First, describe the circle O_1, and describe the parabola C_1 at O_1 as a focal point.

(2) Secondly, describe the circle O_2, and describe the parabola C_2 at O_2 as a focal point.

(3) The parabola C_1 meets the parabola C_2 at O_3. Finally, one can describe circle O_3 at an intersection point O_3.

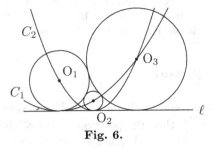

Fig. 6.

This section is cited from my paper as follows [1], but those figures are drawn by Scilab and K$_E$Tpic except Fig.4.

4   Conclusion and Future Work

With CAS, if the respective diameters of a large circle, medium circle, and a small circle are known, as is true also of the Sangaku shown in Section 3, then a drawing can be produced easily by designating a center and radius.

Drawing the Sangaku described in Section 3 is performed such that the third circle is drawn mathematically using the relation between two circumscribed circles and common tangent, with characteristics of a parabola. This problem is used in the lesson of quadratic curve for a Japanese high school. CAS is good at drawing such graphics. As described above, the use of CAS such as Scilab and K$_E$Tpic for mathematical drawing is effective to encourage students to experience the application of mathematics to mathematical problems, which will also be beneficial for students to ascertain the meaning of learning mathematics including the usability of mathematics. Furthermore, it will attract student interest in mathematics and act favorably in emotional aspects of learning mathematics. This is the real meaning of "Application of Mathematics to Mathematics", which the author proposes.

Students aiming at becoming teachers of mathematics in the future were able to recognize that CAS such as Scilab and KETpic including use of LATEX are useful. As an example of application of mathematics to mathematics, the author intends to apply Scilab and KETpic to the drawing of figures used in WASAN and to Sangaku in the future, with eventual disclosure of the results to the public.

Acknowledgements. This research is supported in part by Grant-in-Aid for Scientific Research (C) 26350198. The author would like to thank his colleagues for their valuable comments and input. In addition, the author has received significant advice from KETpic research members.

References

1. Makishita, H.: Solving Problems from Sangaku with Technology: For Good Mathematics in Education. The Bulletin of the Graduate School of Education of Waseda University, Separate 19(1), pp. 275–290 (2011)
2. Satoh, K., Makishita, H., Itoh, H.: Sangaku Dojoh, Kenseisha (2002)

Appendix

If the vertex of triangle ABC is expressed by position vectors $A(\overrightarrow{a})$, $B(\overrightarrow{b})$, and $C(\overrightarrow{c})$, then the position vector of the center of circumcenter O, orthocenter H, excenters I_A, I_B, and I_C are expressed by the following well-known expression.

(1) Circumcenter : $O\left(\dfrac{\sin 2A\,\overrightarrow{a} + \sin 2B\,\overrightarrow{b} + \sin 2C\,\overrightarrow{c}}{\sin 2A + \sin 2B + \sin 2C}\right)$

(2) Orthocenter : $H\left(\dfrac{\tan A\,\overrightarrow{a} + \tan B\,\overrightarrow{b} + \tan C\,\overrightarrow{c}}{\tan A + \tan B + \tan C}\right)$

(3) Excenters : $I_A\left(\dfrac{-a\,\overrightarrow{a} + b\,\overrightarrow{b} + c\,\overrightarrow{c}}{-a+b+c}\right)$, $I_B\left(\dfrac{a\,\overrightarrow{a} - b\,\overrightarrow{b} + c\,\overrightarrow{c}}{a-b+c}\right)$

$I_C\left(\dfrac{a\,\overrightarrow{a} + b\,\overrightarrow{b} - c\,\overrightarrow{c}}{a+b-c}\right)$

Development of Visual Aid Materials
in Teaching the Bivariate Normal Distributions

Toshifumi Nomachi[1], Toshihiko Koshiba[2], and Shunji Ouchi[3]

[1] Yuge National College of Maritime Technology, Japan
nomati@gen.yuge.ac.jp
[2] Anan National College of Technology, Japan
hyotanzima20@gmail.com
[3] Simonoseki City University, Japan
ouchi@shimonoseki-cu.ac.jp

Abstract. We have evaluated the orthant probabilities (i.e., all components are positive) and the upper probabilities of bivariate normal distributions by using Scilab software. The calculated values are presented here in a tabular form. They may be used as a teaching aid material in statistics courses at colleges. We have explained the orthant and upper probabilities using 3D diagrams as a visual aid teaching material in the present work. Moreover, we have evaluated probabilities for more generalized domains by using Scilab software.

Keywords: Scilab software, orthant probability, 2-dimensional normal distribution, Gaussian function, teaching statistical material.

1 Introduction

It is well known that the random errors often have normal distributions which are also called by Gaussian distributions or bell shaped distributions. But it is very difficult to calculate the probabilities of normal distributions by definite integration of Gaussian functions. So we use the numerical values of upper probabilities of standard normal distribution which are obtained by using some computer software. In our previous work (2013) we [1] have investigated the table of the probabilities of the standard normal distribution by using Scilab software. However, it is desirable to treat two random variables which have two-dimensional probability distribution in the case of two factors in the collegiate study. Toda and Ono [2] (1978) introduced the algorism of computing the upper probabilities of two-dimensional standard normal distribution. Furthermore Miwa et.al [3] (2003) evaluated orthant probabilities of multi-dimensional normal distribution. Here we have evaluated the orthant probabilities of bivariate standard normal distribution, i.e. all components are positive, by using Scilab software. We present and explain the tables of upper probabilities of two-dimensional standard normal distribution by applying Scilab software. Especially in the present paper we explain these probabilities with making 3-dimentional diagrams as visual teaching collegiate materials. The visual technique is based on LaTeX, using Scilab software.

H. Hong and C. Yap (Eds.): ICMS 2014, LNCS 8592, pp. 601–606, 2014.

Besides that in this work, we have evaluated the probabilities for more generalized domains of the two-dimensional standard normal distribution by using Scilab software, without previous tables.

2 Orthant Probabilities

The orthant probability is the probability that all random variables of the probability distribution are positive. In this paper, we treat only the bivariate normal distributions. Bivariate probability distributions may involve with both two-dimension and one-dimension. In the present work we assume a bivariate standard normal random vector (U, V) has a mean vector $(0, 0)$ and a correlation coefficient ρ. That is the orthant probability is $P(U > 0, V > 0 \,;\rho)$.

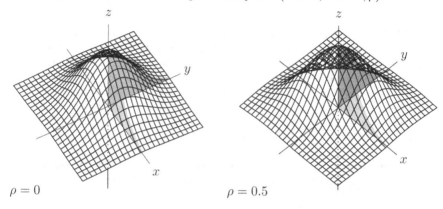

$\rho = 0$ $\qquad\qquad\qquad\qquad\qquad$ $\rho = 0.5$

Fig. 1. Two-dimensional Standard Normal Distribution with $\rho = 0$ and 0.5

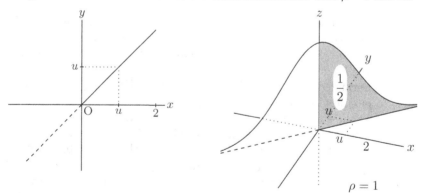

$\rho = 1$

Fig. 2. Area of $(y = x, x > 0)$ and Normal Ortham Probabirity for $\rho = 1$

It is well known that $P(U > 0,\ V > 0 \,;\rho) = \dfrac{1}{2}(\dfrac{1}{2} + \dfrac{1}{\pi}\sin^{-1}\rho)$.

We have dealt with the following three cases.

For Case 1, $\rho = 1$ and $\{U > 0\}$. Then $\{V > 0\}$ with probability 1. Therefore, as seen in Fig. 2, $P(U > 0,\ V > 0 \,; 1) = P(U > 0) = P(V > 0) = \dfrac{1}{2}$.

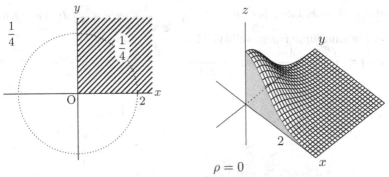

Fig. 3. Area of $(U > 0, V > 0)$ and Two-dimensional Normal Ortham Probabirity for $\rho = 0$

For Case 2, $\rho = 0$ then U and V are independent mutually. Then as seen in Fig. 3, $P(U > 0,\ V > 0;\ 0) = P(U < 0) \cdot P(V < 0) = \dfrac{1}{4}.$

For Case 3, we assume $0 < \rho < 1$ without loss of generality. We have evaluated these values by using Scilab software. The computational method is described in the following section and the results from the calculation using the software are given in tabular form in the appendix.

Especially, it is easily seen that for $\rho = 0.5$ the corresponding probability is 0.3333 that matches the results up to fourth-order.

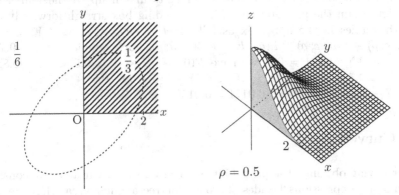

Fig. 4. Area of $(U > 0, V > 0)$ and Two-dimensional Normal Ortham Probabirity for $\rho = 0.5$

3   Upper Probabilities

This section describes the method we used to compute the upper probabilities of the probability distribution i.e., $P(U > x,\ V > y\ ; \rho)$ for any x and y.

Without loss of generality may assume x and y are non-negative. In the case of $x = 0$ and $y = 0$, the upper probability is an orthant probability. The algorism of computing the upper probabilities of two-dimensional standard normal distribution is well known. Let $L(x, y\ ; \rho) = P(U > x,\ V > y\ ; \rho)$. The standard normal upper probability is $Q(h) = 1 - \int_{-\infty}^{h} \frac{1}{\sqrt{2\pi}} \exp(-\frac{z^2}{2}) \, dz$, i.e., $Q(h)$ is the standard normal upper probability. Then

$$L(x, \, y; \rho) = \int_{0}^{\rho} \frac{1}{2\pi\sqrt{1-t^2}} \exp\left[-\frac{1}{2(1-t^2)}(x^2 - 2xyt + y^2)\right] \, dt + Q(x)Q(y).$$

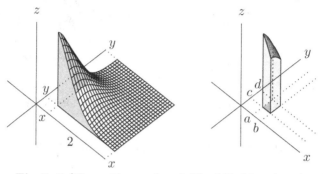

Fig. 5. P $(U > x, V > y; \rho)$ and $P(a \leqq U \leqq b, c \leqq V \leqq d; \rho)$

We have calculated these integrations by using cdfnor (Cumulated Distribution Functions) in Scilab software. The calculated values of the upper probabilities are given in the appendix in tabular form. In the case of $\rho = 0.5$ and $x = 0$ and $y = 0$, the upper probability is an orthant probability whose value is equal to 0.3333. This shows us the results match up to the fourth order. We can obtain the probabilities of any bounding box straightforwardly by using the tables in the appendix as follows: $P(a < U < b, \, c < V < d; \rho) = L(a, \, c; \rho) - L(b, \, c; \rho) - L(a, \, d; \rho) + L(b, \, d; \rho)$. For example, for $a = 0.7$, $b = 1.4$, $c = 0.5$, and $d = 0.8$, we have $P(0.7 < U < 1.4, \, 0.5 < V < 0.8; \, \rho = 0.5) = L(0.7, 0.5; \, 0.5) - L(1.4, 0.5; \, 0.5) - L(0.7, 0.8; \, 0.5) + L(1.4, 0.8; \, 0.5) = 0.0747 - 0.0513 - 0.0249 + 0.0171 = 0.0156$.

4   Curved Area

After having obtained the probabilities for any rectangular area, we considered polynomial expressions besides the ones for rectangular area. Here we evaluated the probability for $\{y > x^2\}$ by using the function int2d(X, Y, f) in Scilab software.

First we divided the inside area into four triangles (A1, A2, A3, A4) as shown in the Fig. 6. A calculation gives a result equal to 0.2298967. It is of interest to validate the accuracy of the calculation. Without loss of generality, we may restrict the area only to positive components. In other words, it is possible to evaluate the probability for an area of $\{y > x^2\}$ and $x > 0$. So next we divided the corresponding area over with three triangles B1, B2, B3. The value of integration for B3 is 0.0000168, which is very small one. The value of integration corresponding to B1, B2 and B3 is 0.192226. However the value of integration corresponding to A1 and A2 is 0.147533.

The values of integration for A1 is 0.1236266 and for B1 is (0.1356191), respectively. The values of integration for A2 is 0.0239064, and corresponding to B2 and B3 is (0.0355147), respectively. The differences are 0.0119925 and 0.0116079.

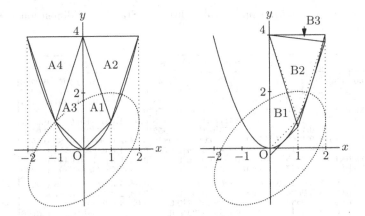

Fig. 6. Area of $\{Y > X^2\}$ and $\{Y > X^2, X > 0, Y > 0\}$ for $\rho = 0.5$

5   Conclusion

In the present work, we have examined a feasible way to explicitly introduce college students the bivariate normal orthant probabilities. First by using Scilab software, we have evaluated the normal probabilities for rectangular domains with fourth-order or in other words the calculation results match up to the fourth digits after the decimal point. Then by applying the results, bell shaped 3D-figures like Mt. Fuji were formed. For general domains we have obtained the numerical integration values of Gaussian function with second-order by using the function (int2d) in Scilab software. As further work, we have to develop in order to improve the calculation accuracy and to automatically form the bell shaped 3D-figures.

Acknowledgments. This work is subsidized by Nagaoka University of Technology, Toyohashi University of Technology, Institute of National Colleges of Technology. We would like to thank Professor Setsuo Takato for his helpful suggestions.

References

1. Takato, S., Koshiba, T., Nomachi, T.: Evaluation for the tool and the element of the staistical materials (in Japanese), RIMS Kokyuroku (2013)
2. Toda, H., Ono, R.: Algolizm for the computing of two-dimensional normal distribution function (in Japanese). Japanese Journal of Aplied Statistics (1978)
3. Miwa, T., Hayter, A.J., Kuriki, S.: The evaluation of general non-centred orthant probabilities. Journal of the Royal Statistical Society (2003)

Appendix

Table 1. $L(u, v; \rho = 0.5)$: Upper Probabilities of Standard 2-dimensional Normal Distribution

$u\backslash v$	0.0	0.1	0.2	0.3	0.4	0.5	0.6	0.7	0.8	0.9
0.0	0.3333	0.2782	0.2197	0.1917	0.1723	0.1543	0.1371	0.1210	0.1059	0.0920
0.1	0.2782	0.2491	0.2033	0.1766	0.1586	0.1420	0.1262	0.1113	0.0975	0.0847
0.2	0.2197	0.2033	0.1806	0.1612	0.1450	0.1298	0.1154	0.1018	0.0891	0.0774
0.3	0.1917	0.1766	0.1612	0.1461	0.1317	0.1179	0.1048	0.0925	0.0809	0.0703
0.4	0.1723	0.1586	0.1450	0.1317	0.1187	0.1063	0.0945	0.0834	0.0730	0.0634
0.5	0.1543	0.1420	0.1298	0.1179	0.1063	0.0952	0.0846	0.0747	0.0654	0.0568
0.6	0.1371	0.1262	0.1154	0.1048	0.0945	0.0846	0.0752	0.0664	0.0581	0.0505
0.7	0.1210	0.1113	0.1018	0.0925	0.0834	0.0747	0.0664	0.0585	0.0513	0.0445
0.8	0.1059	0.0975	0.0891	0.0809	0.0730	0.0654	0.0581	0.0513	0.0449	0.0390
0.9	0.0920	0.0847	0.0774	0.0703	0.0634	0.0568	0.0505	0.0445	0.0390	0.0339
1.0	0.0793	0.0730	0.0668	0.0606	0.0547	0.0490	0.0435	0.0384	0.0336	0.0292
1.1	0.0678	0.0624	0.0571	0.0518	0.0467	0.0419	0.0372	0.0328	0.0287	0.0250
1.2	0.0575	0.0530	0.0484	0.0440	0.0397	0.0355	0.0316	0.0278	0.0244	0.0212
1.3	0.0484	0.0445	0.0407	0.0370	0.0334	0.0299	0.0265	0.0234	0.0205	0.0178
1.4	0.0404	0.0372	0.0340	0.0309	0.0278	0.0249	0.0221	0.0195	0.0171	0.0149
1.5	0.0334	0.0307	0.0281	0.0255	0.0230	0.0206	0.0183	0.0162	0.0142	0.0123
1.6	0.0274	0.0252	0.0231	0.0209	0.0189	0.0169	0.0150	0.0133	0.0116	0.0101
1.7	0.0223	0.0205	0.0188	0.0170	0.0154	0.0138	0.0122	0.0108	0.0094	0.0082
1.8	0.0180	0.0165	0.0151	0.0137	0.0124	0.0111	0.0099	0.0087	0.0076	0.0066
1.9	0.0144	0.0132	0.0121	0.0110	0.0099	0.0089	0.0079	0.0069	0.0061	0.0053

Creating Interactive Graphics
for Mathematics Education Utilizing KETpic

Shunji Ouchi[1], Yoshifumi Maeda[2], Kiyoshi Kitahara[3] and Naoki Hamaguchi[4]

[1] Shimonoseki City University, Japan
ouchi@shimonoseki-cu.ac.jp
[2] Nagano National College of Technology, Japan
maeda@nagano-nct.ac.jp
[3] Kogakuin University, Japan
kitahara@cc.kogakuin.ac.jp
[4] Nagano National College of Technology, Japan
hama@ge.nagano-nct.ac.jp

Abstract. In teaching mathematics, there are instances when we need to graphically present mathematical concepts and solid figures to clarify students' understanding of them. For the last few years, we have been creating graphics that illustrate these various concepts dynamically through careful utilization of KETpic. Examples we will look at include an interactive graphic developed to clearly illustrate the line of intersection of two solid surfaces. In this case, we can easily show the cross-section of the intersection following the cut. A second example is of an interactive graphic produced in order to dynamically present the correspondence relation between the $z-$plane and $w-$plane in a complex function $w = f(z)$. Here, by using the navigation buttons embedded in the graphic, we can demonstrate how the regions on the $w-$plane change in relation to the $z-$plane. Other graphics we have produced will also be introduced.

Keywords: computer algebra systems, Scilab, interactive graphics, mathematics education, hyperlink, TEX, KETpic

1 Introduction

Our project team has been developing effective teaching materials for mathematics (including statistics) education, predominately at the early grades of tertiary education, through careful utilization of KETpic.

Briefly, KETpic consists of a library of macros to generate LaTeX source codes for high-quality scientific artwork. Such macros can be implemented in different Computer Algebra Systems (CAS) such as Scilab, Maple and R, thus providing different plug-ins for the program. How the plug-ins run may vary based on the specific CAS, but as this process is transparent to the end-user, it should minimize the time required to learn the program. After loading KETpic in the CAS, users simply need to execute commands following system requests in order to plot graphs and other mathematical data. CAS-embedded KETpic commands

H. Hong and C. Yap (Eds.): ICMS 2014, LNCS 8592, pp. 607–613, 2014.
© Springer-Verlag Berlin Heidelberg 2014

generate additional LATEX source code and files, this generated output can then easily be compiled in LATEX. As a result, precise and visually compelling graphical figures can be obtained either on a PC display or as printed output [2], [3].

For the last few years, we have been creating graphics that illustrate mathematical concepts and solid figures dynamically for mathematics education, here referred to as *interactive graphics* [1]. We will introduce interactive graphics we have created to date and outline programs for creating the graphics.

2 How to Create Interactive Graphics

The idea of creating interactive graphics is based on that of producing a flip book. We produce a PDF file of multiple slides including graphics using a CAS-based KETpic plug-in and advance those slides by using software capable of viewing PDF files.

ketslide and *ketlayer* are KETpic style files which can easily produce PDF slides incorporating high-quality graphics. *ketlayer* enables us to precisely embed graphics and symbols in LATEX documents in the exact position we wish them to be included. We introduce two examples of interactive graphics. The aim in creating the first example is to better illustrate the fact that the function $z = x^2 - y^2$ has a saddle point at the origin. Initially we create multiple (in this case 244) graphs showing the intersection of the surface $z = x^2 - y^2$ and a plane parallel to the coordinate plane. Next we make a PDF file of multiple slides incorporating these graphs. Figure 1 to 3 show a sample of the graphs.

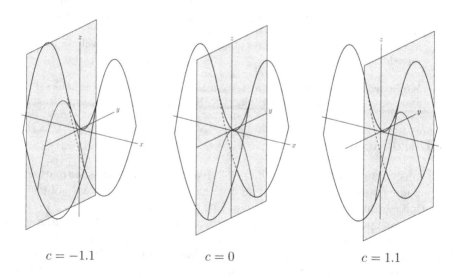

$c = -1.1$ $c = 0$ $c = 1.1$

Fig. 1. Intersection of surface $z = x^2 - y^2$ and plane $x = c$

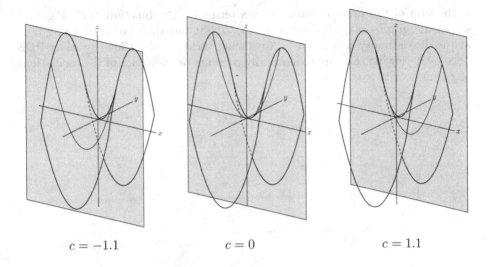

Fig. 2. Intersection of surface $z = x^2 - y^2$ and plane $y = c$

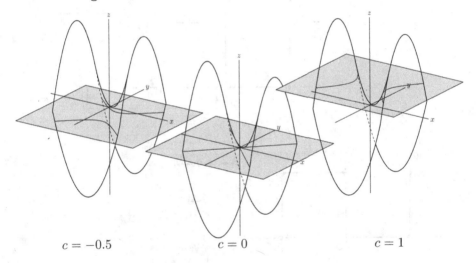

Fig. 3. Intersection of surface $z = x^2 - y^2$ and plane $z = c$

The second example was designed to help the learner better understand the Gibbs phenomenon. The function f defined by

$$f(x) = \begin{cases} 0 & (-1 \le x < 0) \\ 1 & (0 \le x < 1) \end{cases}, \quad f(x+2) = f(x)$$

is discontinuous at $x = k$ for all integer values of k.

$$f_N(x) = \frac{1}{2} + \sum_{n=1}^{N} \frac{1}{n\pi} \left(1 - (-1)^n\right) \sin n\pi x$$

is the sum of the first N Fourier series terms for the function $f(x)$. Figure 4 shows the graphs of $f_N(x)$ for $N = 10, 55, 100$ and their enlarged views near at the discontinuity point $x = 1$. Advancing these graphs' slides by using PDF document viewing software dynamically presents the behavior of the oscillations of $f_N(x)$ near to $x = 1$.

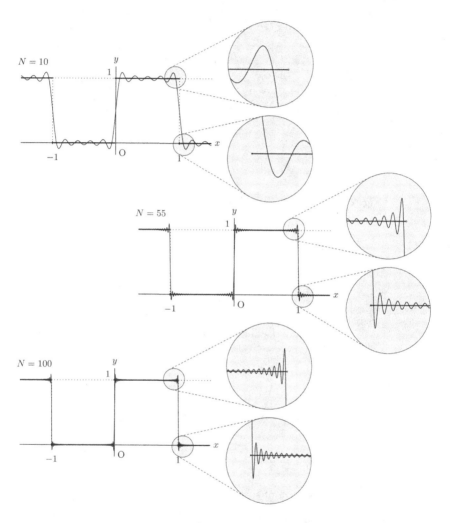

Fig. 4. Displaying Gibbs phenomenon

3   Outline of the Program to Create Interactive Graphics

In this section, we outline the main part of the program for creating interactive graphics utilizing the Scilab-based KₑTpic plug-in. The program is composed of two parts:

(1) produces TₑX files for graphs.

```
Ketinit();                          //initialize global variables
Fname='filename';
cd('c:/work/figs/');        //folder containing stored graph data
 //here, insert program for initialization
                          //define size of graph,number of graphs,etc
 for k=1:L                                  //L is number of graphs
     Openfile(Fname+'_'+string(k)+'.tex');
     Program for producing the k-th graph
     Closefile();
 end
```

(2) making multiple slides including the graphs produced in part (1).

```
cd('c:/work/pages/');       //folder containing stored slide data
Openfile(Fname+'.tex');
Texcom('\newslide{\bf\color{NavyBlue}\title});
                                            //create new slide
 for k=1:L
     if k<>1 then
           Texcom('\sameslide')
     end

     Texcom('\begin{layer}{110}{0}');
     Texcom('\putnotese{\Xichi}{\Yichi}
           {\input{\zu/'+Fname+'_'+string(k)+'.tex}}');
               //\Xichi and \Yichi are parameters to determine
                    where to locate graph on slide
     Texcom('\end{layer}');
     Texcom('\bun');              //\bun is description about graph
 end
 Closefile();
```

The arguments that are passed to the program above are specified in the following LATEX document.

```
\newcommand{\honbun}[3]{             //display description on slide
   \begin{layer}{100}{0}
     \putnotese{#1}{#2}{
       \begin{minipage}{35zw}{\large #3}
       \end{minipage}}
   \end{layer}}

\begin{document}
\def\dai{title name}
\def\bun{\honbun{x-coord}{y-coord} //location of description below
   { //here, insert description about the contents of the slide
   }}
```

```
\def\Xichi{x-coordinate of place to put graph}
\def\Yichi{y-coordinate of place to put graph}
\input{page/slide file name}
\end{document}
```

4 Interactive Graphics with Embedded Hyperlinks

Finally, we introduce interactive graphics enhanced by using the TeX macro package hyperref. The example given here is a graphic which dynamically presents the correspondence relation between the $z-$plane and $w-$plane in a complex function $w = f(z)$. Figure 5 is a sample of the graphs to be incorporated in the interactive graphic.

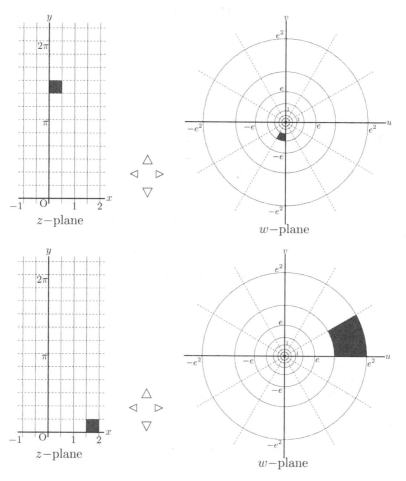

Fig. 5. Correspondence relation between z – plane and w – planein function $w = e^2$

All rectangular regions on the z−plane are hyperlinked to the corresponding regions (the images making up the graphic) on the w−plane and the navigation buttons, shown as "△", "▽", "▷" and "◁" in figure 5. This allows us to explore how the corresponding regions on the w−plane change for a rectangular grid representing the z−plane by clicking the buttons.

5   Conclusion

The interactive graphics outlined here are relatively simple to create, and once produced can easily be utilized in the classroom. We plan to produce further examples in the future.

Acknowledgements. This work was partially supported by KAKENHI (Grant-in-Aid for Scientific Research C, Number 24501075) from the Japan Society for the Promotion of Science.

References

1. Maeda, Y., Takato, S.: The possibility and usefulness of KₑTpic (in Japanese). RIMS Kokyuroku 1865, 72–78 (November 2013)
2. Kaneko, M., Takato, S.: The Effective Use of LaTeX Drawing in Linear Algebra-Utilization of Graphics Drawn with KₑTpic. The Electronic Journal of Mathematics and Technology 5(2), 129–148 (2011)
3. Ouchi, S., Takato, S.: High-Quality Statistical Plots in LaTeX for Mathematics Education Using an R-based KₑTpic Plug-In. In: Proceeding of the 15th ATCM Conference-Kuala Lumpur, 265–275 (December 2010)

A Tablet-Compatible Web-Interface for Mathematical Collaboration

Marco Pollanen[1], Jeff Hooper[2], Bruce Cater[3], and Sohee Kang[4]

[1] Trent University, Canada
marcopollanen@trentu.ca
http://euclid.trentu.ca/math/marco/
[2] Acadia University, Canada
jeff.hooper@acadiau.ca
http://math.acadiau.ca/hooper/
[3] Trent University, Canada
bcater@trentu.ca
http://www.trentu.ca/economics/staff_cater.php
[4] University of Toronto Scarborough, Canada
soheekang@utsc.utoronto.ca
http://www.utsc.utoronto.ca/cms/sohee-kang

Abstract. Mathematical novices – including students in introductory mathematics and statistics service courses – increasingly need to engage in online mathematical collaboration. Using currently-available interfaces for their mobile and touch-enabled devices, however, this group faces difficulties, for those interfaces are text-based and not directly suitable for mathematical communication and collaboration.

To address the deficiency of digital input methods and interfaces for mathematics, we introduce a cross-platform synchronous communication interface for mathematical collaboration. The interface is designed to be intuitive for multiple user groups ranging from novices to experts. We demonstrate that it is possible to create a Web-based communication interface that simultaneously incorporates TeX-, palette- and pen-based input methods, and that is compatible with both touch-enabled tablet and traditional keyboard-mouse user interface principles. The design principles we introduce may be valuable for the design of other mathematical user interfaces on touch-enabled devices, such as with Computer Algebra System interaction.

Keywords: Mathematical Collaboration, Mathematical User Interfaces, Formula Input

1 Introduction

Communication technologies are now used to great effect in post-secondary education, increasing, for example, outside-the-classroom student-teacher contact. This type of interaction correlates positively with key educational indicators, including academic performance, student retention and student satisfaction [8].

H. Hong and C. Yap (Eds.): ICMS 2014, LNCS 8592, pp. 614–620, 2014.

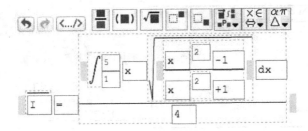

Fig. 1. BrEdiMa: A Structure-based Editor for Mathematical Input [9]

The use of technology for the communication of mathematical ideas can be a particularly effective pedagogical tool [4]. Its potential, however, has yet to be fully realized. This may be due to two factors: most internet communication technologies are text-based and both the text and non-text based mathematics input methods that have been developed have been designed mostly for experts, rather than for the increasingly novice student user-base.

2 Barriers to Communication

Communicating mathematics online using text-based technologies is problematic for two reasons. There are hundreds of commonly used mathematical symbols, many of which have no commonly accepted textual equivalent and must therefore be described. The inherently two-dimensional structure of mathematical notation requires spatial relationships between symbols; such relationships are difficult to communicate in inline text. Consequently, the standards that exist for the text-based entry of mathematics suffer from having a steep learning curve and very low human readability. For instance,

$$\lim\_{x\to\infty}\frac{\sqrt{8+x} - 3x^{1/3}}{x^2 -3x +2}$$

and

$$\int\_0^2 r \sqrt{ 5 - \sqrt{ 4-r^2}} dr$$

are LATEX representations for the two first-year university calculus expressions

$$\lim_{x\to\infty} \frac{\sqrt{8+x} - 3x^{1/3}}{x^2 - 3x + 2} \quad\text{and}\quad \int_0^2 r\sqrt{5 - \sqrt{4 - r^2}}dr \ .$$

Neither LATEX string is intuitive for novices, and small errors in either can seriously affect the mathematical meaning.

The main alternative to inline text-based input is structure-based direct-manipulation editors, such as those found in Microsoft Word or in BrEdiMa (Figure 1).

In such an editor, the user inserts individual symbols and mathematical structures from palettes of symbols. As with inline text editors, however, structure-based direct-manipulation editors suffer from severe usability problems [12]. In

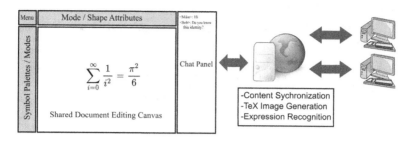

Fig. 2. Schematic of iCE UI Layout and Communication

[14] it is argued that structure-based editors usually force a user to write a formula in a different manner than they would on paper. For example, consider the expression $\sqrt{x}/y$. The default behavior of a structure-based editor forces the user to input the fraction first, followed by the square root symbol, and then the x and y. Intuitively, however, somebody writing this expression with pen and paper would probably write the square root of x first followed by the fraction bar and then the y. So in essence, the user of a structure based editor has to use an unintuitive order to input the expression. This requires the user to have the ability to mentally parse the desired mathematical expression and reorder for input, which can be difficult for students and other novice users. The user must adapt to the technology as it is non-intuitive and has not been designed with casual users in mind.

While digital pen-based input methods would allow users to write mathematics as they would on paper, their use introduces new problems. For instance, robust handwriting recognition algorithms for mathematics are still in their infancy, and pen-input typically requires special hardware that most students do not possess.

In this paper we outline the development of a real-time multi-modal web-based open-source mathematics collaboration interface that works on all commonly available computing devices, ranging from computers to tablets to smartphones, and is intuitive for first-time users who are mathematical novices.

3   Interface Choice

The goal of iCE (interface for Collaborative Equations) is to be a hybrid environment that allows users familiar with any mathematical editor input model (e.g. palette/structure based, TeX, or pen-based) to communicate and collaborate mathematically as quickly and effortlessly as possible. It is structured as a shared SVG document simultaneously being edited by multiple users in a collaborative whiteboard model with a chat-pane on the side to emulate verbal conversation. This model best replicates the usual in-person mathematical collaboration model where a mathematical conversation is usually assisted by facilitating technology, such as a piece of paper or chalkboard [3].

iCE allows for unconstrained input, permitting users to enter mathematical symbols in any order. This approach has been shown to allow faster input, has a minimal learning curve ([3] and [15]) and allows users to enter mathematics as they usually do on a chalkboard and thus minimizes destructive interference and cognitive load [11].

4   Browser-Based Interfaces

With the advent of Web 2.0 and cloud computing, the web browser is increasingly becoming the standard interface to access full-featured applications. In particular, Google Docs and Office 365 have become mainstream browser-based environments for collaborative documents. Thus, in developing a cross-platform mathematics collaboration environment, it is natural for it to be browser-based as well.

The goal of iCE is to create a communication interface that can incorporate collaboration into a variety of web applications through any major browser (Internet Explorer, Chrome, Firefox, Safari) across multiple platforms (PC, Mac, Linux, iPad, iPhone, and Android Smartphones) without installing any software or plugins.

Web-based applications can be difficult to develop due to the application being embedded in an existing browser interface; the browser framework leads to many UI restrictions, limited protocol support, and sandbox restrictions. A cross-browser application must also deal with many API inconsistencies. When both personal computers and mobile devices are to be supported, the application's design must take into account a number of major differences between client instances: keyboard/mouse versus touch UI, screen size, and input accuracy. In addition, mobile devices typically have limited computational power and so CPU intensive operations must be delegated to web services.

Browser-based applications for communication of or editing of mathematical content face additional difficulties, as in-browser layout and display of mathematical expressions is problematic, and even more so if the content is made to be interactive: copied, scaled, and manipulated. While Mathematical Markup Language (MathML) was intended to be the standard for mathematics on the internet, it is thus far not fully supported [1]. For these reasons, iCE has been designed around a front-end of Javascript/SVG (based on SVG-Edit [16]) with calls to several web services to keep it lightweight.

5   User-Interface Interactions

iCE implements an unconstrained direct editor equation model based on a diagram editor. Users are free to place elements anywhere on a whiteboard canvas. However, in this case the elements are not restricted to geometric objects, but include resizable mathematical symbols. This approach is consistent with both mouse and touch-based UI interactions and elements may be manipulated as follows:

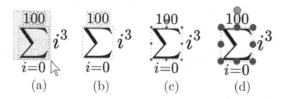

Fig. 3. Object manipulation in iCE: (a) selecting a group of symbols; (b) the symbol group after selection; (c) resizing widgets on a PC; (d) resizing widgets on an iPad.

Moving a Symbol: Individual symbols on the canvas may easily be dragged to any location.

Group Selection: By pressing down (via mouse or finger) on an empty spot of the canvas and dragging out a rectangular outline, all symbols falling inside the rectangle are selected. They will then be treated as a group until unselected by clicking or tapping on a blank section of the canvas. The group of symbols in a selection may, collectively, be dragged to a new location, have their attributes changed, or be deleted.

Resizing: By clicking or tapping on a symbol, resizing widgets appear and that individual symbol may be resized by dragging these widgets. Due to the limited precision of the finger on a touch-based device, the size of these widgets will vary depending on the device. See Figure 3.

In any equation input system the keyboard plays an important role. Since hardware keyboards have fixed keys this makes the symbol mapping complicated. On the other hand, touch-based devices rely on virtual keyboards. While custom virtual keyboards can include mathematics symbols, virtual keyboards on touch devices generally occupy a large percentage of the screen and have problems with users accurately selecting the correct symbol.

Keyboard input on iCE takes one of three forms: text-mode, symbol-mode, and TeX mode. Consistent with a diagram editor, *text-mode* may be selected and a corresponding text-box may be placed anywhere on the screen. Within each text-box, the text may use different fonts with attributes appropriate for labeling a diagram, such as size, style, color, etc. Synchronization across participants occurs only upon completion of the entire text.

On the other hand, the symbol mode incorporates a persistent on-screen cursor that can input text or symbols using shortcuts from the keyboard while it is in almost any mode. The cursor may be placed anywhere on the canvas by cursor keys or by clicking (tapping) on an empty space. Unlike text-mode, symbol-mode is synchronized across participants in real time.

To accommodate the large number of possible math symbols, many symbols are mapped to the same key. By pressing a key in rapid succession a number of different symbols are cycled through (for example, aside from its usual use, the keyboard key 'A' cycles through 'A', alpha (α), for all ($\forall$), logical 'and' ($\wedge$), aleph ($\aleph$), and the angle symbol ($\angle$). When the mapping is not obvious, symbols may also be inserted by selecting them from a palette.

The final keyboard mode is a variation of the text-box mode. Users enter TeX code into a text-box and on completion the TeX code is compiled and turned into symbolic content through a web service call. This symbolic content is then redistributed to all users.

6 Web Services

iCE is a collaborative interface and thus makes use of several network components. In its implementation, it relies on a server built using node.js [10] for communication synchronization. To keep iCE lightweight, additional web services can be utilized for CPU-intensive tasks. Currently, TeX code is converted to SVG content through the use of MathJax [6] as a web service. In addition, selections of symbols may also be converted to TeX through a web service call to the spatial recognition algorithm XPRESS [14].

7 Further Enhancements

Other web services we plan to implement to make the interface truly multi-modal:

Handwriting Recognition: iCE allows for pen input (if available on the platform). However, this is maintained as a digital ink layer and is not processed further. There are a number of handwriting recognition approaches (e.g., [17]) for mathematics that could be explored in the future. Much like the current call to XPRESS for TeX rendering, this intensive task is best left on the server and not implemented in client-side Javascript.

Voice Recognition: The HTML5 API [5] allows for Javascript to access a browser's audio stream which may allow iCE input to be paired with mathematical voice recognition [2].

CAS Assistance: It may be possible to make communication faster and more effortless by allowing calls to a server-side Computer Algebra System (CAS) to assist with calculations. This would involve an additional parsing component, since this requires using content markup as opposed to the use of presentation-oriented markup like TeX.

8 Conclusion / Discussion

Communication technology in mathematics lags far behind its use in other academic disciplines. In this paper we introduce a multi-modal mathematics collaboration interface that is designed to be fast and intuitive for both novice and expert users. We hope that this new interface approach will lead to improvements in the design of future interfaces for mathematical input that will have a positive impact on mathematical education and collaboration.

References

1. Cervone, D.: MathJax: a platform for mathematics on the Web. Notices of the AMS 59(2), 312–316 (2012)
2. Fateman, R.: How can we speak math? (2013), http://www.eecs.berkeley.edu/~fateman/papers/speakmath.pdf (Unpublished manuscript)
3. Gozli, D.G., Pollanen, M., Reynolds, M.: The characteristics of writing environments for mathematics: Behavioral consequences and implications for software design and usability. In: Carette, J., Dixon, L., Coen, C.S., Watt, S.M. (eds.) Calculemus/MKM 2009. LNCS (LNAI), vol. 5625, pp. 310–324. Springer, Heidelberg (2009)
4. Hooper, J., Pollanen, M., Teismann, H.: Effective online office hours in the mathematical sciences. Journal of Online Learning and Teaching 2(3) (2006)
5. HTML 5.1 API, http://www.w3.org/TR/html51/
6. MathJax homepage, http://www.mathjax.org
7. Miner, R.: The importance of MathML to mathematics communication. Notices of the AMS 52(5), 532–538 (2005)
8. Nadler, M.K., Nadler, L.B.: Out-of-class communication between faculty and students: A faculty perspective. Communication Studies 51(2), 176–188 (2000)
9. Nakano, Y., Murao, H.: BrEdiMa: yet another Web-browser tool for editing mathematical expressions. In: Proceedings of Math UI 2006 (2006)
10. Node.js homepage: http://nodejs.org/
11. Oviatt, S.L., Arthur, A.M., Brock, Y., Cohen, J.: Expressive pen-based interfaces for math education. In: CSCL 2007, pp. 573–582 (2007)
12. Padovani, L., Solmi, R.: An investigation on the dynamics of direct-manipulation editors for mathematics. In: Asperti, A., Bancerek, G., Trybulec, A. (eds.) MKM 2004. LNCS, vol. 3119, pp. 302–316. Springer, Heidelberg (2004)
13. Pollanen, M.: Interactive Web-based mathematics communication. Journal of Online Mathematics and its Applications 6(4) (2006)
14. Pollanen, M., Wisniewski, T., Yu, X.: Xpress: a novice interface for the real-time communication of mathematical expressions. In: Proceedings of MathUI 2007 (2007) (online)
15. Pollanen, M., Reynolds, M.: A model for effective real-time entry of mathematical expressions. Research, Reflections and Innovations in Integrating ICT in Education, Formatex, pp. 320–324 (2009)
16. SVG-Edit homepage: https://code.google.com/p/svg-edit/
17. Tapia, E., Rojas, R.: Recognition of on-line handwritten mathematical formulas in the E-Chalk System. In: Proceedings of the Seventh International Conference on Document Analysis and Recognition, pp. 980–984 (2003)

Development and Evaluation of a Web-Based Drill System to Master Basic Math Formulae Using a New Interactive Math Input Method

Shizuka Shirai[1] and Tetsuo Fukui[2]

[1] Graduate School of Human Environmental Sciences, Mukogawa Women's University, Japan
shizukas@acm.org,
[2] Mukogawa Women's University, Japan
fukui@mukogawa-u.ac.jp
http://www.mukogawa-u.ac.jp/~hi/fukui/

Abstract. We present a web-based drill system named DIGITAL-WORK which assists learners in mastering some basic formulae using a new interactive math input method. This method enables users to format any mathematical expression in WYSIWYG by converting from colloquial style strings in fuzzy mathematical notation. The purpose of this study is to investigate whether students can smoothly learn basic math formulae with our drill system. In this paper, we report the results from a field survey that was conducted in an actual remedial math class of 20 junior high school students. The results of our survey showed that 85% of them found this system to be more fun than learning on paper.

Keywords: math e-assessment systems, mathematics interfaces.

1 Introduction

Online testing is an important function of e-assessment systems. Available types of tests on most e-assessment systems are binary or multiple choice, matching types, or (numerical) completion types. However, in the field of math and science education, it is preferable that learners be able to respond to mathematical questions directly with mathematical expressions. In recent years, a few e-assessment systems have enabled responses with mathematical expressions by using a computer algebra system (CAS) [8,9]. These systems have been used for instruction to students as drills and as homework at many universities.

As of 2014, there are two ways for learners to respond with mathematical expressions on these e-assessment systems, by text-based interfaces or by template-based interfaces. Text-based interface for math input is very familiar in that it operates with a keyboard. However, learners must input an answer according to CAS command syntax such as Maple and Maxima. Thus, novice learners of mathematics must learn not only the mathematics but also the CAS command syntax which is unrelated to mathematics. Furthermore, when learners input a

H. Hong and C. Yap (Eds.): ICMS 2014, LNCS 8592, pp. 621–628, 2014.

CAS command string with this interface, it is difficult for them to imagine the two-dimensional desired mathematical expressions (e.g., super and subscripts, fractions). Therefore, such a situation is not educationally ideal.

Template-based interface for math input has the advantage that learners are able to operate clearly in WYSIWYG by using GUI math template icons (e.g., mathematical symbol icons). However, this interface has disadvantages in that learners must use both keyboard and mouse in turns and look for a mathematical operator or special symbol from a large number of templates. As a result, this interface often strains learners when they repeatedly have to enter their answers with mathematical expressions such as when doing drill work.

The troublesome math input operations may not be a problem in the case of advanced mathematics when a learner spends a long time thinking about a particular mathematical question in comparison with the time it takes to input the answer. However, in the case of mental arithmetic such as math drills to practice basic formulae, the complicated math input operation may disturb the learning process. Furthermore, our preliminary examination reveals that many students feel the procedure of submitting an answer by shifting from keyboard to mouse and clicking the submission icon to be troublesome on an e-assessment system. Therefore, there are the following two problems with using the current e-assessment systems for drill work instruction to practice basic formulae.

1. It is troublesome to input mathematical expressions as an answer using the current math input methods.
2. In a cycle of repetitive drill work, the procedure of submitting an answer becomes a bottleneck (In other words, it stops the learning temporarily).

To improve these problems, we have adopted an interactive math input method in terms of conversion from fuzzy mathematical strings in WYSIWYG and developed an original assessment system, named DIGITAL-WORK that specialized in drill work to assist learners in mastering some basic formulae. An interactive math input method has been proposed by Fukui in 2012 [3]. This method implemented a math input interface, named MathTOUCH [10] which enables users to format any mathematical expression by inputting a colloquial style string in fuzzy mathematical notation and by selecting the desired candidate shown by the system. Therefore, we expect that this interface is user-friendly for novice mathematics learners because the user will be able to easily input a mathematical formula with only the keyboard, without having to learn a new language or syntax [4,5].

The purpose of this study is to investigate whether students can smoothly learn some basic mathematics formulae with our drill system. In this paper, we report the results from a field survey that was conducted in a ninth grade remedial math class.

This paper is organized as follows. Section 2 describes the outlines of this input method and MathTOUCH. Section 3 introduces the whole of our drill system (DIGITAL-WORK). Section 4 describes the results of our investigation regarding learner's reaction to our drill system. A final section concludes.

2 The Interactive Math Input Method [3]-[7]

2.1 Math Conversion Procedure from a Fuzzy Mathematical String

Math conversion procedure from a fuzzy mathematical string is a procedure similar to conversion of kana spelling (phonetic representation) into kanji (Chinese characters) when we input a Japanese sentence on a computer. The math conversion procedure enables mathematical expression input on a computer according to the procedures outlined in steps 1-3 below.

(Step 1) First, the system accepts input of the character string written by a user in fuzzy mathematical notation.

(Step 2) The system predictively calculates candidates for the two dimensional mathematical expression which the user wishes to appear and shows them in WYSIWYG.

(Step 3) Next this system accepts an adoption operation for each math element which the user wishes for from the list of corresponding candidates; and finally when all the elements are chosen, the formatting expression process is complete.

Here, the procedure is described with the fuzzy mathematical notation for a mathematical expression as follows:

> **Fuzzy mathematical notation for a mathematical expression**
> Set the fuzzy key letters (or words) corresponding to the elements of a mathematical expression linearly in order of colloquial (or reading) style, without considering two-dimensional placement and delimiter.

In other words, a fuzzy key letter (or word) consists of the ASCII code(s) corresponding to the initial or the clipped form (like LATEX-form) of the objective mathematical symbol. Therefore, one fuzzy key often supports many mathematical symbols. For example, when a user wants to input α^2, the fuzzy string is denoted by "a2" where "a" stands for the "alpha" symbol and it is unnecessary to include a power sign (like the caret letter (^)). In the case of $\frac{1}{\alpha^2+2}$, the fuzzy string is denoted by "1/a2+2" where it is not necessary to surround the denominator (which is generally the operand of an operator) in parentheses since such parentheses are never printed. Table 1 presents other examples in fuzzy mathematical notation.

Thanks to fuzzy mathematical notation, the user will be able to input almost any mathematical expression with only a keyboard, without learning a new language or syntax. Therefore, this method poses less operating trouble than a text-based method using a CAS command.

2.2 Characteristics of the Math Input Interface with Fuzzy Mathematical Notation

The math input interface implementing the above math conversion method with the fuzzy mathematical notation is called MathTOUCH. Figure 1 presents the

Table 1. Examples of mathematical expression using fuzzy string notation

Mathematical category	Examples of math expression	Strings of fuzzy mathematical notation
fraction	$\frac{2}{3}$	2/3
polynomial	$5x^2 + 2$	5x2+2
equation	$(x - \frac{1}{2})^2 = x^2 - x + \frac{1}{4}$	(x-1/2)2=x2-x+1/4
square root	$\sqrt{3}$	root3

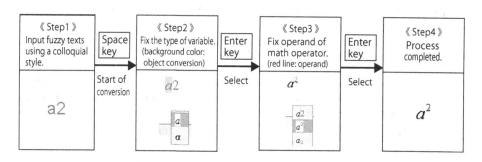

Fig. 1. Conversion procedure from fuzzy mathematical text using MathTOUCH

procedure for inputting a^2 with MathTOUCH. In Step 1, type in the fuzzy string as "a2" on the keyboard. Then hit the space key to start the math formatting process with MathTOUCH. Perform the conversion procedure for all elements in order from left to right in the fuzzy mathematical string. In Step 2, a list of the conversion candidates, in this case a and α, is shown directly under the first target letter "a". The current (marked) candidate is switched by the space or arrow key and adopted by using the enter key. In Step 3, if a highlighted element is adopted, then the current target moves on to next element to the right. In this case, the target is an unexpressed operator - a power sign between a and 2 in a^2. Then the list of candidates is presented $a2$, a^2, a_2 and so on. When all the elements in the mathematical expression are fixed by sequential adoption operations as in Step2, the math formatting process is complete (Step4).

Characteristics of MathTOUCH are described as follows:

- It enables users to input the desired mathematical expressions in WYSIWYG with only a keyboard, without learning a complex syntax.
- It allows users to input any expression dealt with in the widespread categories of mathematics from a junior high school level to a university level.
- Math input performance with MathTOUCH is better than with a template-based interface [4].
- Its input performance improves by the learning function of the fuzzy key dictionary data after users have used MathTOUCH repeatedly [5,6].

3   A Web-Based Drill System

3.1   Development Policies

As stated in section 1, the goal of this system is to support novices in basic mathematics on computers by reducing user stress. Specifically, we assume the following three policies for a math e-assessment system in developing this.

1. The type of learners which the system is designed to support are novices in elementary mathematics. They will be junior high or high school students.
2. We wish to minimize the trouble with inputting operations with mathematical expressions.
3. The specialized web-based drill system will enable students to master basic formulae by repetitively entering answers for problems they have solved in their heads.

As a result, we have adopted MathTOUCH as the math input interface for our system in order to enable doing repetitive drill work smoothly with only the keyboard.

3.2   System Configurations and Features

This system has been developed as a web application system using Tomcat with MySQL. Hence, it will be able to be used by most computers across the Internet. In other words, this drill system will enable us to provide services not only to schools but also to homes.

Additionally, DIGITAL-WORK has the following three features that would be necessary for math drills.

Learning history function Learners will be able to check their own learning history consisting of learning date, content, score, and question numbers where they got wrong answers.

Learning function specialized in the case of wrong answers Learners will be able to relearn the specialized set of questions which they previously got wrong answers to.

Randomized question generator function This system will make questions randomly for learners from the prepared question bank.

3.3   The Learning Cycle on this Drill System

First, learners look at a question displayed in the question and assessment area (Figure 2(1)). They are also able to read some hints in the same area if necessary. Next, they input an answer as a mathematical expression using MathTOUCH which is displayed in the response area (Figure 2(2)). After that, if the learners submit their answer by pushing the enter key on the keyboard, the feedback of their assessment is shown in the question and assessment area immediately. Pushing the enter key again, the learning cycle proceeds to the next question. In this way, a cycle of repeating drill work with this system is achieved using only the keyboard by virtue of MathTOUCH math input interface as tailored to our e-assessment system.

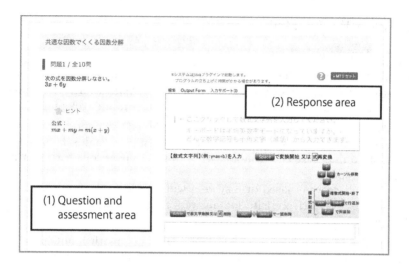

Fig. 2. Screenshot of DIGITAL-WORK

4 Experimental Evaluation

4.1 Design and Procedure

To evaluate our system, we conducted a field survey with junior high school students in a summer remedial class in math at Mukogawa Women's University Junior & Senior High School to measure their subjective satisfaction with DIGITAL-WORK. The class was carried out over two days (each lesson was 50 minutes), and twenty students were taking this class. On the first day, we introduced DIGITAL-WORK to the students and showed them how to use it, especially how to input their answer for a math drill question with MathTOUCH. On the second day, the students worked drills for basic math formulae which were taught in their junior high school class in the first semester. Therefore, the math drill questions were prepared based on the authorized textbook *Mathematics in Junior High School 3* [2] as in Table 2. After the drill work, we gave them a questionnaire regarding subjective satisfaction about DIGITAL-WORK and MathTOUCH using a 5-point rating scale from 1(strongly disagree) to 5(strongly agree). Our questionnaire consisted of ten questions (See the second column of Table 3).

4.2 Results and Discussion

An overview of the results of the questionnaire is given in Table 3. Regarding DIGITAL-WORK, the mean overall score from students was 4.12. This result represents a high evaluation because the mean value was significantly higher than the average score on a 1-5 rating scale for subjective satisfaction (3.6) [1].

Table 2. Example of Math drill questions

Example of questions	Suggested answer
Expand the following expression: $-3(2x + 3)$	$-6x - 9$
Expand the following expression: $(x + 1)(x + 4)$	$x^2 + 5x + 4$
Expand the following expression: $(x - 5)^2$	$x^2 - 10x + 25$
Expand the following expression: $(x + 1)(x - 1)$	$x^2 - 1$

Table 3. Result of the questionnaire regarding subjective satisfaction (N=20)

System	Contents of the questionnaire	Mean	SD
	I could master the basic formulae.	4.05	1.00
	The function repeating missed questions was useful.	4.35	1.04
DIGITAL	Being able check the result of drills was helpful.	4.40	1.05
-WORK	I like the design.	3.55	1.05
	Learning the basic formulae on DIGITAL-WORK is more fun than learning on paper.	4.25	1.02
	It was easy to master the use of MathTOUCH. (Learnability)	3.80	1.47
Math	Mathematical expressions could be inputted smoothly. (Efficiency)	3.65	1.23
TOUCH	I remember how to use MathTOUCH even 1 day later. (Memorability)	3.90	1.07
	It was easy for me to correct mis-entered operations. (Error)	4.00	1.03
Comprehensive Evaluation	Would you like to learn the basic formulae on DIGITAL-WORK again in class?	4.05	1.32
	Overall mean questionnaire rating for DIGITAL-WORK	4.12	1.06
	Overall mean questionnaire rating for MathTOUCH	3.84	1.20

The results showed that 85% of students found DIGITAL-WORK to be more fun than learning basic formulae on paper. Most of the mean scores to the five questions regarding DIGITAL-WORK are over 4.0, except the satisfaction with the screen design where the result was 3.55. Hence, it is necessary to improve the screen design of our system because it is an important factor for learning motivation. With respect to MathTOUCH, all of the mean scores to the four questions asked of students were above 3.6, with the overall mean result at 3.84. However, satisfaction regarding efficiency was 3.65 because an error happened in the system when the students struck the enter key repeatedly. We need to improve the system so that it is stable under any user operations.

The total results of DIGITAL-WORK and MathTOUCH indicated that students were able to work drills themselves smoothly. The result of the comprehensive evaluation shows that 70% of students answered that they would like to learn basic math drills again with our system in the future. In fact, many of the free responses to this questionnaire from the students showed that they felt our system to be more fun than learning on paper because they were able to work drills smoothly.

5   Conclusion and Future Work

We have developed a web-based drill system to master basic math formulae using an interactive math input method with fuzzy mathematical notation. We have also investigated whether students can smoothly learn the same basic math formulae with our drill system in a field survey which was conducted in an actual remedial math class.

The results showed that 85% of students found DIGITAL-WORK to be more fun than learning basic formulae on paper and that 70% of students answered that they would like to learn basic math drills again using our system. Therefore, this has shown that our drill system was accepted by novice learners of mathematics in the case of mental arithmetic such as repeated drills.

Finally, we expect that our web-based drill system will assist many students in learning basic mathematics. The most important avenues for future research are to improve the screen design in our system and to prepare a sufficient amount of content for math drill questions for junior and high school mathematics.

References

1. Nielsen, J.: Usability Engineering. Tokyo Denki University Press, Tokyo (2002)
2. Okabe, T., et al.: Mathematics in Junior High School 3. Suuken Shuppan, Tokyo (2012)
3. Fukui, T.: An Intelligent Method of Interactive User Interface for Digitalized Mathematical Expressions (in Japanese). In: RIMS Kokyuroku, vol. 1780, pp. 160–171 (2012)
4. Fukui, T.: Evaluation of an Intelligent Method of Interactive User Interface for Digitalized Mathematical Expressions (in Japanese). In: Sushiki-shori, vol. 18, no.2, pp. 47–50 (2012)
5. Fukui, T.: The performance of interactive user interface for digitalized mathematical expressions using an intelligent formatting from linear strings (in Japanese). In: RIMS Kokyuroku, vol. 1785, pp. 32–44 (2012)
6. Fukui, T.: A Design of the Editing Function of an Interactive Interface for Equation Edit Using Linear String Formatting From the Key Dictionary by Machine Learning (in Japanese). In: Proceedings of the 74th National Convention of IPSJ, vol. 2F-1, pp. 4.1–4.2 (2012)
7. Fukui, T.: An Intelligent User Interface Technology for Easy Formatting of Digitalized Mathematical Expressions – a Mathematical Expression Editor on Web-Browser (in Japanese). In: Interaction 2013 IPSJ Symposium Series, vol. 2013(1), 2EXB-50, pp. 537–540 (2013)
8. STACK, http://stack.bham.ac.uk/
9. Maple, T.A.: http://maplesoft.com/products/mapleta/
10. MathTOUCH, http://math.mukogawa-u.ac.jp/

Generating Data of Mathematical Figures
for 3D Printers with KETpic
and Educational Impact of the Printed Models

Setsuo Takato[1], Naoki Hamaguchi[2], and Haiduke Sarafian[3]

[1] Toho University, Japan
takato@phar.toho-u.ac.jp
[2] Nagano National College of Technology, Japan
hama@nagano-nct.ac.jp
[3] Pennsylvania State University, USA
has2@psu.edu

Abstract. KETpic has the capability to create graphic objects which are used in LaTeX documents. These features alone make KETpic an ideal portable code language for printing mathematical materials for in-class use or textbooks for publication, and are available at no cost. The latest version of KETpic supports the making of easy-to-use 3D graphs. Recently new commands for generating data in obj format have been introduced in KETpic. The data is easily converted to stl format, making prints of 3D models possible. As a result, a LaTeX document with a figure and a 3D model are obtained simultaneously. It is the view of the authors that combining printed materials and 3D models is the preferred mode of approach in math education.

Keywords: LaTeX, 3D printer, Teaching Materials.

1 Introduction

Teachers at the collegiate level often need various 3D figures in their math classes. For example, in calculus classes, teachers can better explain the meanings of double integral and repeated integral using Fig. 1 and Fig. 2

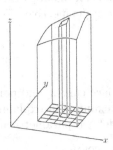

Fig. 1.

Fig. 2.

H. Hong and C. Yap (Eds.): ICMS 2014, LNCS 8592, pp. 629–634, 2014.

Additionally, figures such as Fig. 3 and Fig. 4 are used in linear algebra and vector analysis classes.

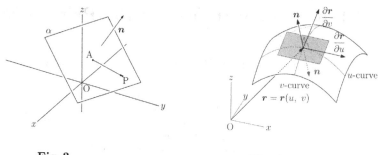

Fig. 3.   **Fig. 4.**

It is hard for ordinary teachers to draw these figures precisely on the black board, and moreover, students may have great difficulty in copying them down in their notebooks.

KETpic[1,2] is a macro package which is based on CASs, such as Mathematica, Maple, Scilab, R, and generates a sequence of graphical codes for LaTeX. KETpic also includes functions for making tables in flexible formatting, creating LaTeX macros with graphical elements, and placing components freely on LaTeX sheets. Therefore, suitable mathematical figures for use in printed materials and math textbooks are easily generated with LaTeX and KETpic together. The plotting and manipulation of three dimensional figures are supported as well. The plotting is basically constructed with skeleton, silhouette, and boundary lines.

Recently, new commands for generating data in obj format have been implemented in KETpic. The data can be immediately converted to stl format, with which 3D printers can make 3D models. It should be noted that the definition of a mathematical figure is the same as that for making printed material with LaTeX plotting. Only the path for generating plotting data has to be changed. For example, "G=Spacecurve(...)" defines the graph of a parametric function. Subsequently, "Drwline(Projpara(G))" is used when putting it into a LaTeX document, whereas "Objcurve(G)" is used when generating obj data. As a result, a LaTeX document with the figure and the 3D model are obtained simultaneously. In the following sections, we describe some commands for making 3D models, and compare printed materials and 3D models from the perspective of educational impact. In this article, Scilab will be used.

2   Commands for 3D Data

KETpic can generate LaTeX figures of surfaces, space curves, and polyhedra.

We take the graph of $z = \cos \sqrt{x^2 + y^2}$ as an example in explaining how to make a figure of a surface. This function can be defined parametrically as,

$$x = v \cos u, \ y = v \sin u, \ z = \cos v. \tag{1}$$

We follow the process below.

(1) Define the function:

```
Fd=list("p","x=V*cos(U)","y=V*sin(U)","z=cos(V)",
                    "U=[0,2*%pi]","V=[0,4*%pi]","n");
```

(2) Find the boundaries and carry out hidden line elimination:

```
G1=Sfbdparadata(Fd);
```

(3) Generate the output file:

```
Openfile(Filename.tex);
    Drwline(Projpara(G1));
Closefile();
```

(4) Input the file into a LATEX document:

```
\input{filename.tex};
```

From this, Fig. 5a is obtained. Also we can add details such as wireframes.

Fig. 5a.   **Fig. 5b.**

Meanwhile, the procedures for making 3D models are as follows:
(1) The definition of the function is the same as above.
(2) The above-mentioned step (2) is unnecessary.
(3) Make the obj-formatted data file:

```
Openobj(Filename.obj);
    Objsurf(Fd,1);
Closeobj();
```

Here, the last parameter (1 or −1) is for setting the outside of the surface.
(4) Use 3D viewing software such as

Meshlab[3], Blender[4], or i3dViewer[5] for iPad.

to display it on computer or tablet screens.

Fig. 5 is a screenshot in Meshlab.
(5) Use the command "Objthicksurf" at (3) to solidify the surface.

```
Openobj(Filename.obj);
    Objthicksurf(Fd,0.02,-0.02,1,"n+");
Closeobj();
```

Converting obj format to stl format, we obtain data for 3D printers.

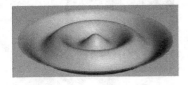

Fig. 6.   **Fig. 7.**

Fig. 8 is a photo of the printed-out 3D model.

The process for space curves is similar to that of surfaces.

(1) Define and make the space curve data:

```
Sc=Spacecurve("[cos(t),sin(t),0.1*t)","t=[0,4*%pi]");
```

(2) Use "Drwline" to make the LaTeX file.

```
Drwline(Projpara(Sc));
```

Use "Objcurve" to make the obj-formatted file.

```
Objcurve(Sc);
```

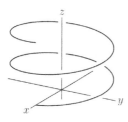

Fig. 8.

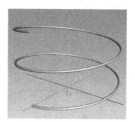

Fig. 9.

3   Comparison of Impact

KETpic has now grown into a tool that gives teachers the ability to make materials with mathematical 3D figures that can be used in different ways. They may be distributed as printed materials, 3D data may be sent to students' tablets, and actual 3D models may be shown. In this section we compare these.

Needless to say, real 3D models contain much more information, which sometimes hinders students from understanding the nature of the point being explained. Moreover, it might be hard to add the necessary expressions, symbols, lines, and notes, etc.

For example, to find the volume of the intersection of two cylinders mutually crossing at a right-angle is a typical problem in calculus classes. Teachers often want to show the shape to students. Fig. 11 is the 3D model, which cannot be said to be better than the figure in the printed materials in Fig. 10. Students will fail to grasp the shape of the inside body with only Fig. 11.

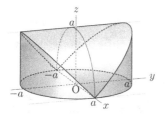

Fig. 10.

Fig. 11.

It should be noted here that examples of the 3D figures may be shown on tablets. In Fig. 12a and Fig. 12b, the outside parts have been separated so that the inside body may be seen.

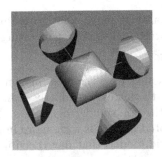

Fig. 12a.   **Fig. 12b.**

In the case of more complicated 3D shapes, on the other hand, a 3D model is probably more effective. For example, the function defined by

$$z = \frac{x^2 - y^2}{x^2 + y^2} \tag{2}$$

is discontinuous at $(x, y) = (0, 0)$. Accordingly, the graph of the function has a complex shape near the origin. Fig. 13 and Fig. 14, respectively, are the figure for printed materials or presentations and the 3D model.

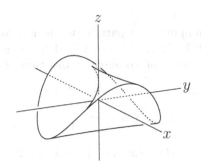

Fig. 13.   **Fig. 14.**

4   Conclusions and Future Work

KETpic has implemented functionality to generate data for 3D printers, which enables teachers to make materials with 3D models in various ways:

1. To distribute printed materials
2. To present slides on the screen
3. To have students manipulate figures on their tablets
4. To show or let students touch solid models

5. To use some combination of the above-mentioned

Though we can choose other tools such as Blender to generate data for 3D printers, KETpic has the advantages of

1. the ability to make complicated mathematical figures.
2. similarity to LATEX in generating figures, which is effective when used together.

Solid models have the most information about the real objects, and they may impact math classes accordingly. However, they sometimes hinder students from grasping the point under discussion. There are definitely more than a few cases when printed materials have more impact in raising students' insight.

The solid models that have been used until now have practical limitations as teaching materials.

1. It can be said that it is impossible to distribute them to all students considering the high cost of making them.
2. Shown from the front of the classroom, they may be too small for all to see.
3. Transparent models are not so good.
4. The weight makes it hard to make models such as space curves.

Nevertheless, teachers may experiment with combinations of materials, which should prompt them to investigate how to make effective use of solid models in conjunction with other materials.

Acknowledgments. This research is supported in part by Grant-in-Aid for Scientific Research (C) 24501075, 40259825. It is also subsidized by Nagaoka University of Technology, Toyohashi University of Technology, Institute of National Colleges of Technology.

References

1. Sekiguchi, M., Yamashita, S., Takato, S.: Development of a Maple Macro Package Suitable for Drawing Fine TEX-Pictures. In: Iglesias, A., Takayama, N. (eds.) ICMS 2006. LNCS, vol. 4151, pp. 24–34. Springer, Heidelberg (2006)
2. Takato, S., Galvez, A., Iglesias, A.: Use of ImplicitPlot in Drawing Surfaces Embedded into LaTeX Document. In: Proceedings of ICCSA 2009, Part II, pp. 115–122 (2009)
3. Meshlab, http://meshlab.sourceforge.net/
4. Blender, http://www.blender.org/
5. i3dViewer, http://www.allmastersoftware.com/i3dViewer/i3dViewer.htm

A Touch-Based Mathematical Expression Editor

Wei Su[1], Paul S. Wang[2], and Lian Li[1]

[1] School of Information Science & Engineering, Lanzhou University, China
suwei@lzu.edu.cn
http://wme.lzu.edu.cn/suwei
[2] Department of Computer Science, Kent State University, USA

Abstract. MathEdit is an interactive tool for creating and editing mathematical expressions on the Web. It is an open-source program implemented in standard XHTML and JavaScript to run in regular browsers. The tool supports both WYSIWYG editing and command-line editing operations. Recently, a touch version of MathEdit is under development. In MathEdit touch version, we design a virtual mathematical keyboard for users to enter mathematical symbols and expression templates conveniently. The navigation method of highlighting current node instead of blinking cursor is used in MathEdit. Users can select a sub-expression via a virtual key or touch move operation.

1 Introduction

The *Web-based Mathematics Education* (WME) project [9] at the *Institute for Computational Mathematics* (ICM/Kent) was started in the mid 1990's to build an innovative on-Web mathematics education environment for middle school teachers and students. As part of WME, work also began on an interactive visual editor for mathematical expressions that runs in standard Web browsers and works with standard mathematics representations such as MathML [19], infix, and LaTeX.

In this direction, Lanzhou University and Kent State University jointly have developed MathEdit [13,18], an open-source tool running in standard browsers for entering and editing mathematical expressions for the Web. MathEdit can be accessed on http://www.mathedit.org. MathEdit allows users to create and edit mathematical expressions with a convenient and intuitive graphical user interface (GUI) as well as an efficient command-line environment with character-string input. Using well-defined API functions, MathEdit can also be embedded in the interactive Web application systems by authors to create mathematical expressions.

Through the GUI and character-string input box, user actions, mouse clicks, and keyboard input, are treated as commands. Commands invoke JavaScript functions that operate on HTML and MathML DOM trees to support editing and visual navigation of mathematical expressions. MathML Presentation and Content codes are basic to the internal operations of MathEdit. But MathEdit also provides format conversion that can convert the format for the expressions

H. Hong and C. Yap (Eds.): ICMS 2014, LNCS 8592, pp. 635–640, 2014.

among MathML, OpenMath [15], LaTeX, and infix. Each edit operation basically adjusts the DOM tree of MathML markup kept internally for the mathematical expression being constructed or edited. The effect of each editing operation is reflected in the visual display immediately.

MathEdit provides different user input modes, convenient API for the host webpage. Preference settings and customizations at the user and program levels make it possible to use MathEdit for different purposes and at different levels of mathematics.

2   Comparing MathEdit with Other Systems

In the past decade, many companies and research institutes have developed mathematical expression editors [1,5,6,8,10,11,16,17,20]. Table 1 compare serval state-of-the-art editors with MathEdit in direct-manipulation input, linear-format input, output format, editing mode.

Firstly, let us focus on input method and output format. Most of the editors can support visual direct-manipulation: they provide well-defined template, visual navigation by mouse and arrow cursor, and shortcut key. But MathEdit,

Table 1. Compare the other editors with MathEdit

Item		MathType	WebEQ	MathEX	Amaya	Lyx	TEXmacs	ASCIIMathML	MathEdit
Direct-manipulation	Template	√	√	√	√	√			√
	Visual Navigation	√	√	√	√	√			√
	Shortcut Key	√	√	√	√		√		√
Linear-format Input	Infix							√	√
	MML Content		√	√					√
	MML Presentation		√			√		√	√
	OpenMath								√
	LaTeX	√	√*				√	√	√
Output Format	Infix							√	√
	MML Content		√	√	√				√
	MML Presentation	√	√	√	√			√	√
	OpenMath			√					√
	LaTeX	√	√*				√	√	√
	Picture	√	√					√	√
Content-based Editing					√				√
Presentation-based Editing		√	√			√	√	√	√
Web-based			√	√				√	√

MathEX [12], and WebEQ [17] support more flexible customizable templates. Though most people prefer to use templates and a mouse when first learning an application, in the long run it is often more convenient to use keyboard shortcuts for common operations. All the editors use either MathML Presentation, or LaTeX, or both as their main mathematical representation format. For linear-format input, LaTeX are most popular input format which is supported in MathEdit, MathType [4] (MathType Version 6 begin to support LaTeX input), TeXmacs [14], ASCIIMathML [2] and WebEQ (WebEQ use WebTeX which syntax and commands are similar to the mathematics mode part of LaTeX). The table also shows us supporting for directly editing MathML Presentation are also an important feature for the modern mathematical expression editor. Only MathEdit and MathEX can support OpenMath which is a semantic markup language. MathEdit is only one which supports combination of infix and direct-manipulation input editing method.

For the editing mode, the other editors usually aim either to capture the meaning or to describe the visual appearance. The Amaya [1], LyX [7], TeXmacs and MathType using MathML Presentation, LaTeX, or native formats to store expressions are suitable for describing the expression appearance. Most Computer Algebra Systems, such as Maple and Maxima, and a few independent editors, such as MathEX, use infix and MathML Content to capture the meaning of expressions. MathEdit is different which satisfies both the need for visual display and the need for expression processing.

Otherwise in the Web-based editors, MathEX and WebEQ use Java Applet to embed into other Web application. While MathEdit, developed by JavaScript, is the only visual editor in the table 1 which can be seamless fused into other Web application such as Web-based text editor.

3   Touch-Based Operation for Mathematical Expression Editing

The rapidly-developing world of multi-touch tabletop and surface computing is opening up new possibilities for editing mathematical expressions. Recently, a touch version of MathEdit is under development. Figure 1 shows the user interface of MathEdit touch version. Instead of a physical keyboard and mouse, the user of multi-touch device interacts directly with a virtual keyboard and touch-sensitive screen. In MathEdit touch version, we design a virtual mathematical keyboard for users to enter mathematical symbols and expression templates conveniently. Figure 2 shows us the screenshot of virtual keyboard. The virtual keyboard consists of digits, English letters, Greek letters, special letters, mathematical operators, transcendental functions and expression templates. The literature [3] analyzes 20,000 mathematical documents from the mathematical arXiv server from 2000-2004, the period corresponding to the new mathematical subject classification. It quantify empirically the use of common expressions in the mathematical literature and gives us a statistical result of usage frequency of mathematical notations. The layout of each key on the virtual mathematical keyboard of MathEdit touch version is based on the utilization rate of mathematical symbols and templates in [3]. Such a layout of keys could make users enter the expressions more easily and quickly.

Fig. 1. The User Interface of MathEdit

Another important user interface aspect of a WYSIWYG editor is navigating to the precise point within the expression where an editing operation is to take place. Windowing environments have taught users to experience computers with one hand, focusing on a single point. However the size of human fingers and the lack of sensing precision make precise touch screen interactions difficult. In MathEdit, editing operations, such as insert, edit, delete, covert, and replace, are relative to the current node, which is highlighted visually and visual navigation refers to moving the current expression to different positions in the expression being edited. The current node could be either a single mathematical notation or a sub-expression. A single mathematical notation can be selected by a finger touch operation. There are two methods for selecting a sub-expression in MathEdit. A sub-expression could be selected by press a virtual key (See Figure 3). As shown in Figure 4, another method of selecting a sub-expression is to move your finger from current selected area to your desired area. The minimum sub-expression which covers all the touched symbols by moving finger will be selected. For current node, there are three different operation statuses: entry before, entry after, and replace the current node. The operation status can be switched by touching on space area of work canvas after a sub-expression is selected. The highlighted current node will show different colors for different operation status.

The first screen of virtual mathematics keyboard   The first screen of virtual mathematics keyboard

The first screen of virtual mathematics keyboard

Fig. 2. The virtual keyboard of MathEdit

Fig. 3. Selecting a sub-expression by pressing a virtual key

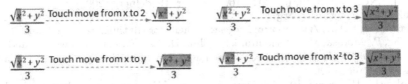

Fig. 4. Selecting a sub-expression by touch move

4   Conclusion

In the paper, MathEdit, a Web-based open source mathematical expression editor is introduced. The design and implementation of MathEdit are presented together with a comparison with several other mathematical expression editors. A new feature of MathEdit, editing mathematical expression on multi-touch device, is addressed in the paper. The screens of touch mobile devices are generally small, which limits the size of displayed pages and the space that can be devoted to user operation interface such as menu, toolbar. In MathEdit, we integrate the general virtual touch keyboard with mathematical symbols and templates and design a virtual mathematical keyboard. The virtual mathematical keyboard can save the display space and also improve the expression input efficiency. The navigation method of highlighting current node instead of blinking cursor is used in MathEdit. Users can select a sub-expression via a virtual key or touch move operation. In a certain degree, it overcomes the difficulty of precise interactions on a touch screen with human fingers.

Acknowledgments. This work is supported by Natural Science Foundation of China (61003139, 60903102), Innovation Found For Technology Based Firms(12C26216206998) of China, Fundamental Research Funds for Central Universities (lzujbky-2013-39, lzujbky-2013-188, lzujbky-2013-187)the MOE-Intel Joint Research Fund (MOE-INTEL-11-03). Any opinions, findings, and conclusions or recommendations expressed in this material are those of the authors and do not necessarily reflect the views of the funding agencies.

References

1. Amaya Homepage, http://www.w3.org/Amaya/
2. ASCIIMathML Homepage, http://www1.chapman.edu/~jipsen/mathml/asciimath.html
3. So, C.M., Watt, S.M.: Determining Empirical Characteristics of Mathematical Expression Use. In: Kohlhase, M. (ed.) MKM 2005. LNCS (LNAI), vol. 3863, pp. 361–375. Springer, Heidelberg (2006)
4. Document of MathType, http://www.dessci.com/en/products/mathtype
5. Alabi, K.: Generation, Documentation and presentation of mathematical equations and symbolic scientific expressions using pure HTML and CSS. In: Proceedings of the 16th International Conference on World Wide Web, Banff, Alberta, Canada, May 08-12 (2007)
6. Padovani, L., Solmi, R.: An Investigation on the Dynamics of Direct-Manipulation Editors for Mathematics. In: Asperti, A., Bancerek, G., Trybulec, A. (eds.) MKM 2004. LNCS, vol. 3119, pp. 302–316. Springer, Heidelberg (2004)
7. LyX, http://www.lyx.org/
8. Mathmled, http://www.newmexico.mackichan.com/MathML/mathmled.htm
9. Wang, P., Mikusa, M., Al-shomrani, S., Chiu, D., Lai, X., Zou, X.: Features and Advantages of WME: a Web-based Mathematics Education System. In: IEEE Southeast Conference (2005)
10. Librrecht, P., Jednoralski, D.: Drag-and-drop of Formula from a Browser. In: Proceeding of Mathematical User-Interfaces Workshop 2006, St Anne's Manor, Workingham, United Kingdom (August 2006)
11. Dooley, S.S.: Editing Mathematical Content and Presentation Markup in Interactive Mathematical Documents. In: Proceedings of ISSAC (2002)
12. Dooley, S.S.: MathEX: A Direct-Manipulation Structural Editor for Compound XML Documents. In: Proceeding of Mathematical User-Interfaces Workshop 2007, chloss Hagenberg, Linz, Austria (June 2007)
13. Wei, S., Wang, P.S., Lian, L.: An On-line MathML Editing Tool for Web Applications. In: Proceeding of International Multi-Symposiums on Computer and Computational Sciences 2007 (IMSCCS 2007), The University of Iowa, Iowa City, Iowa, USA (August 2007)
14. TeXmacs, http://www.texmacs.org/
15. The OpenMath Standard 2.0 Draft, http://www.openmath.org/
16. The W3C MathML software list, http://www.w3.org/Math/Software/
17. WebEQ Documentation, http://www.dessci.com/en/products/webeq
18. Su, W., Wang, P.S., Li, L., Li, G., Zhao, Y.: MathEdit, A Browser-based Visual Mathematics Expression Editor. In: Proceedings of ATCM 2006, Hong Kong, China (2006)
19. W3C Math, http://www.w3.org/Math
20. Doleh, Y., Wang, P.S.: A System Independent User Interface for an Integrated Scientific Computing Environment. In: Proceedings of the ISSAC 1990, pp. 88–95. Addison-Wesley (August 1990) ISBN 0-201-54892-5

Establishment of KETpic Programming Styles for Drawing

Satoshi Yamashita[1], Yoshifumi Maeda[2], Hisashi Usui[3], Kiyoshi Kitahara[4], Hideyo Makishita[5], and Kazushi Ahara[6]

[1] Kisarazu National College of Technology, Japan
yamasita@kisarazu.ac.jp
[2] Nagano National College of Technology, Japan
maeda@nagano-nct.ac.jp
[3] Gunma National College of Technology, Japan
usui@nat.gunma-ct.ac.jp
[4] Kogakuin University, Japan
kitahara@cc.kogakuin.ac.jp
[5] Shibaura Institute of Technology, Japan
hideyo@shibaura-it.ac.jp
[6] Meiji University, Japan
kazuaha63@hotmail.co.jp

Abstract. When collegiate mathematics teachers make their original teaching materials, they often use TeX and CAS in order to insert figures and tables into the materials. TeX and CAS have their own programming languages, respectively. Programs must be written in a good style so that other people can read them. KETpic is a plug-in based on CAS to enable teachers to create figures as they like. KETpic has a programming language for drawing but its programming style is not yet established as a good programming style. In this paper we propose the requirements for a good KETpic programming style.

1 Introduction

When collegiate mathematics teachers make their students understand a new mathematical concept, the teachers often hand out their original teaching materials to the students. The materials need to include accurate and impressive figures and tables which urge the students to understand the concept. In order to create accurate figures, teachers often use Computer Algebra System (in short, CAS), such as Mathematica, Maple, Maxima, Matlab, Scilab and R, and they often use TeX to make the manuscript of the materials. The figures created by CAS are changed into the graphics files formatted into EPS or PDF, and are inserted in a TeX document. It is difficult for teachers to insert satisfactory accessories, such as characters, expressions, ticks and scales, in graphics files. Since 2006 we have developed KETpic as a plug-in based on CAS to enable teachers to create figures as they like. Because the figures created by KETpic are accurate and impressive line drawing, students can understand the concept, writing necessary information in the figures of their teaching material. We think that these

H. Hong and C. Yap (Eds.): ICMS 2014, LNCS 8592, pp. 641–646, 2014.

figures are suitable for mathematics class materials. In Section 2 we explain the usage and the characteristic of KETpic. We are going to clarify the necessary functions for creating figures in mathematics class materials by questionnaire and interview to collegiate mathematics teachers.

To produce a figure using KETpic, teachers create a KETpic program in CAS along with mathematical drawing procedures. During this creation they must recognize the global image of the figure clearly and concentrate their energy on qualitative improvement of the figure. This thinking activity is called ???symbolic thinking???. In order to perform symbolic thinking, they have to master the programming manner of writing for drawing. By doing so, other teachers can also use the KETpic program to create their original figures. To realize the above, both the TeX document of the class materials and the KETpic program for drawing must be written in code which every teacher can read. TeX programming styles by Knuth, Larrabee and Roberts[1] and the programming styles for programming languages by Kernighan and Plauger[2] are famous as good programming styles. However, a good programming style for drawing is not yet known. It is necessary to establish KETpic programming style as a good programming style for drawing. We investigated many KETpic programs for drawing which were written by collegiate mathematics teachers. In Section 3 we introduce the manner of this investigation and show requirements for a good programming style for drawing.

Moreover, based on KETpic programming styles for drawing, we are going to start the portal site for supporting creation of mathematics class materials with figures. In Section 4 we show future works on KETpic programming styles.

2   The Characteristic of KETpic

KETpic has been developed as a plug-in based on CAS in order to create figures inserted into a TeX document. We explain the usage of KETpic by the case where OS is Microsoft Windows and CAS is Scilab. You set the folder for work as "c:/work" and you can use KETpic as follows:

1. Start a CAS (Scilab) and load KETpic (ketpicsciL5) into the CAS. Initialize KETpic and define a figure file name "Figure tex". The first chracter of a KETpic command is capital.

   ```
   1. cd("c:/work");
   2. Ketlib=lib("ketpicsciL5");
   3. Ketinit();
   4. Fname=Figure tex;
   ```

2. Make plotting data of a figure using KETpic commands as follows:

   ```
   6. Setangle(60,30);
   7. Fd=list("z=sin(2*sqrt(abs(x^2+y^2)))","x=R*cos(T)",...
   8.         "y=R*sin(T)","R=[0,4]","T=[0,2*%pi]","e");
   9. S=Sfbdparadata(Fd);
   10. Ax=Xyzax3data("x=[0,5]","y=[0,5]","z=[0,4]");
   ```

```
11. Axo=Crvsfparadata(Ax,S,Fd);
12. Axi=CrvsfHiddenData();
13.
14. Setwindow([-5,5],[-2.5,4]);
15. Ps=Skeletonparadata(S,list(Axo));
16. Paxo=Projpara(Axo);
17. Paxi=Projpara(Axi);
18. Windisp(Ps,Paxo,"c")
```

A space figure is drawn as a figure projected on the plane defined by line 6. On line 6 the normal vector of the plane projected space figures is defined by $(\sin 60° \cos 30°,\ \sin 60° \sin 30°,\ \cos 60°)$. From line 7 to line 12 you can make plotting data of a surface and the axis of coordinates in space. From line 7 to line 8 a surface $z = \sin(2\sqrt{x^2 + y^2})$ is defined and on line 9 you can get plotting data of the ridgeline of the surface by the projection defined by line 6. On line 10 the axis of coordinates are defined. On line 11 the portions of the axis which are sticking out of the surface are defined and on line 12 the portions of axis which hide in the surface are defined. From line 15 to line 17 you can obtain plotting data of the figures projected these figures on the plane defined by line 6. In the usual projection, the projection of the ridgeline of the surface S and the projection of the axis Axo may overlap in one point. In order to express depth perception at the point, the projection of the ridgeline of the surface S removes the far portion from the axis Axo.

3. Produce tpic specials codes, which are the TeX codes for drawing, of figures using KETpic commands and write out them into a figure file "Figure tex" as follows:

```
20. Openfile(Fname,"0.5cm");
21. Drwline(Ps,Paxo);
22. Dottedline(Paxi);
23. Xyzaxparaname(Ax);
24. Expr([1,3.5],"se","z=\sin(2\sqrt{x^2+y^2}));
25. Closefile("0");
```

On line 20 "0.5cm" means the unit length of the output display. On line 25 the argument "0" means that the axis of coordinates which the plane projected space figures has are not displayed ("1" means that they are displayed).

You should just insert the figure file "Figure tex" into a TeX document by not using a TeX command \includegraphics but using \input. We have developed "ketlayer.sty" as a TeX style file to arrange a figure file in the position which you want to put. You can use the layer environment defined by "ketlayer.sty" as follows:

```
1. \documentclass{article}
2. \usepackage{amsmath,amssymb}
3. \usepackage{ketlayer}
4.
```

```
5. \begin{document}
6. \begin{layer}{60}{40}
7. \putnotese{0}{0}{\input{Figure tex}}
8. \end{layer}
9. \end{document}
```

On line 3 you can use the style file "ketlayer.sty". From line 6 to line 8 you can put coordinates into the output display. On line 7 you can place a figure file on the southeast direction on the basis of coordinates (0, 0) as is shown in Figure 1. If you change the 2nd argument on line 6 into 0 from 40, you can erase coordinates from the output display (see Figure 2).

As is shown in Figure 2, the figure produced by using KETpic has the following characteristics:

- The figure is drawn by the simple line like comics.
- The figure has accurate length and position.
- In order to express space figures in the plane projected them, the skelton method and the ridgeline method are used in the figure.

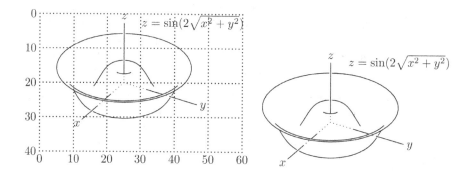

Fig. 1. The output display   **Fig. 2.** A surface in space

3   The Investigation of KETpic Programs

In 2012 we investigated KETpic programs which drew the ellipse. The makers of these KETpic programs were 8 mathematics teachers of program beginners, 12 mathematics teachers of KETpic users and 4 students of KETpic beginners. As a result, the following requirements for a good programming style for drawing became clear[3]:

- The maker has to know the basic knowledge of programming.
- The maker has to know the basic knowledge of mathematics.
- The maker has to know the basic knowledge of KETpic commands.

Since 2013 we have investigated 104 KETpic programs which were written by two collegiate mathematics teachers; 51 programs were written by a program beginner and 53 programs were written by a programmer. As a result, the following 9 requirements for a good programming style for drawing became clear:

1. The maker has to arrange a command in a suitable position.
2. The maker has to attach the suitable name for a variable or plotting data.
3. The maker has to use the calculation function of CAS.
4. The maker has to use KETpic commands appropriately.
5. The maker has to divide a program into readable blocks.
6. The maker has to use a reference point in order to arrange a character and an expression in a suitable position.
7. The maker has to use list structure appropriately.
8. The maker has to use for syntax appropriately.
9. The maker has to define local variables.

Results of the program investigation became as in Table 1. The beginner was bad on all requirements other than the 3rd requirement. The programmer did not use the calculation function of CAS but used the approximate value calculated by hand calculation. We think that, when he made figures by using KETpic, he did not notice that he could use the calculation function of CAS. The requirements which the programmer was bad exceeding 50% were the 4th "KETpic commands" and the 9th "Local variable". It shows that the programmer would not master KETpic and not try to tune finely in order to create the optimal figures.

Table 1. Results of the program investigation

requirements	beginner		programmer	
	bad/total	rate (%)	bad/total	rate (%)
1. Arrange commands	26/51	51.0	9/53	17.0
2. Suitable names	31/51	64.7	20/53	37.7
3. Calculation of CAS	14/39	38.9	14/27	51.9
4. KETpic commnads	37/51	72.6	29/53	54.7
5. Readable blocks	19/51	37.3	0/53	0.0
6. Reference points	32/51	62.8	11/50	22.0
7. List structure	1/1	100.0	1/2	50.0
8. for syntax	1/1	100.0	0/2	0.0
9. Local variables	35/47	74.5	17/32	53.1

4 Future Works

We found 9 requirements for a good programming style for drawing. In Section 2 we showed an ideal KETpic program based on a good programming style for drawing. This KETpic program is satisfied with 9 requirements and it is proved that other persons can read it. We will continue investigating a program further and will check whether 9 requirements are enough. Finally, future works are as follows:

- We are going to clarify the necessary functions for creating figures in mathematics class materials by questionnaire and interview to collegiate mathematics teachers.
- We are going to start the portal site for supporting creation of mathematics class materials with figures.

References

1. Knuth, D.E., Larrabee, T., Roberts, P.M.: Mathematical writing. The Mathematical Association of America (1989)
2. Kernighan, B.W., Plauger, P.J.: The elements of programming style, 2nd edn. Bell Telephone Laboratories, Incorporated (1978)
3. Yamashita, S., Usui, H., Kitahara, K., Takato, S.: The Elements of Programming Style for Making Class Materials with Figures. In: The 13th ICCSA, pp. 73–80 (2013)

Integration of Libnormaliz
in CoCoALib and CoCoA 5

John Abbott[1], Anna Maria Bigatti[1], and Christof Söger[2]

[1] Univ. degli Studi di Genova, Italy
{abbott,bigatti}@dima.unige.it
http://www.dima.unige.it/~abbott
http://www.dima.unige.it/~bigatti
[2] Univ. Osnabrück, Germany
csoeger@uos.de
http://www.math.uos.de/normaliz

Abstract. *libnormaliz* is a C++ library for computations with rational cones and affine monoids and *CoCoALib/CoCoA-5* offers a general environment for computations in Commutative Algebra. For mutual benefit we have developed a simple and fast interface between the two software libraries.

We present how this integration was designed, and then describe in detail the Normaliz functions we have made available in CoCoALib (and also in CoCoA-5).

1 Introduction

libnormaliz [4] is a C++ library for computations with rational cones and affine monoids; **Normaliz** is a simple, stand-alone system built on top of libnormaliz. **CoCoALib** [1] is a C++ library designed to offer an easy-to-use, general environment for efficient computations in Commutative Algebra; **CoCoA-5** [3] is an interactive system built on top of CoCoALib.

For mutual benefit we have developed a simple and fast interface between the two libraries; thus, for instance, in the presence of libnormaliz, CoCoALib acquires new data-structures and operations giving almost direct access to libnormaliz capabilities. These extensions also appear in CoCoA-5, so libnormaliz operations become accessible when working on rings and monoid algebras in CoCoA-5; in this way CoCoA-5 becomes a convenient, sophisticated front-end for libnormaliz.

There are several different approaches for software interoperation: the system Normaliz communicates with the systems Singular and Macaulay2 via files and using their respective user-interface languages. A more flexible approach has been adopted by Sage [8]: it is built on top of many existing open-source packages and accesses their combined power through a common, Python-based language. Yet another possibility is typified by CoCoAServer (of which an advanced prototype already existed in the CoCoA distribution): it follows the client-server model, and

H. Hong and C. Yap (Eds.): ICMS 2014, LNCS 8592, pp. 647–653, 2014.
© Springer-Verlag Berlin Heidelberg 2014

offers symbolic computation services over a socket and using an OpenMath[1]-like language.

For combining Normaliz and CoCoALib we opted for direct integration by adding the relevant interface code in C++. This approach makes the communication costs much lower, and also facilitates a "symmetric/symbiotic integration": for instance the newly developed extension to Normaliz, called **nmzIntegrate** (see Sec. 5), uses features from CoCoALib. An object oriented design allows for efficient management of transported values and the corresponding representation transformations.

In Sections 2 and 3 we present how this integration was designed and how the same simple design techniques can be used for integrating other external libraries into CoCoALib (and also subsequently into CoCoA-5); indeed we have applied the same structure for integrating the software libraries Frobby and GSL (in part).

Then in Section 4 we describe in detail the Normaliz functions we have made available in CoCoALib (and also in CoCoA-5). We offer specific ring theoretic functions applicable to monomial subalgebras, monomial ideals and binomial ideals in polynomial rings. Additionally, we provide direct access to libnormaliz via a general function NmzComputation which faithfully reflects the internal structure of the libnormaliz design.

2   Integrating an External Library in CoCoALib

Using the case of libnormaliz as a guide example, here we show the general guidelines for integrating an (optional) external library into CoCoALib.

The code for the communication between the two libraries is all enclosed in the file ExternalLibs-Normaliz.C (and its corresponding .H file). It contains the definition of the (CoCoALib) class **cone**, which embodies the libnormaliz class, the functions performing the data conversions between the two libraries, and the functions and constructors actually available to the CoCoALib user.

Other libraries are integrated in CoCoALib with a similar structure through the files ExternalLibs-Frobby and ExternalLibs-GSL (see [2] in this volume).

The only mathematical library which is necessary for the compilation of CoCoALib is GMP [6]: it plays a fundamental rôle in CoCoA, Normaliz, and in a lot of other mathematical software. Apart from GMP no external library is linked by default. A CoCoALib user who wants to access libnormaliz functions must download and compile libnormaliz (which also requires the BOOST library), and then configure and compile CoCoALib simply by giving the two commands

```
./configure --with-libnormaliz=<path-to-libnormaliz.a>
make
```

then the code contained in ExternalLibs-Normaliz.C will actually be compiled, together with its dedicated test suites (in the CoCoALib test directory

---

[1] OpenMath [7] is a standard for representing the semantics of mathematical objects, so they can be reliably communicated between programs.

`src/tests/`) and its explanatory examples (in the CoCoALib example directory `examples/`).

Currently we are working on a simple design to generate a meaningful compilation error message in the case some CoCoALib/C++ program calls a Normaliz function when libnormaliz is not present (and similarly for any other optional external library).

3 Porting a CoCoALib Function to CoCoA-5

One of the main goals in the design of the new interpreter for CoCoA-5, the interactive system based on CoCoALib, was that it should be easy to expose a CoCoALib function to a user of the interactive CoCoA-5 system.

The old CoCoA-4 interpreter made this operation quite cumbersome, consequently very many functions offered by CoCoA-4 were actually implemented using the interpreted CoCoA-4 language, and resided in CoCoA "packages". Some packages still persist in CoCoA-5, but their functions are gradually being translated into C++ with the double benefit of making them faster and making them available in CoCoALib; so CoCoALib users will have access to all the mathematical capabilities of CoCoA.

An obvious pleasant consequence is that it is just as easy to make any function defined in `ExternalLibs-Normaliz` available in CoCoA-5. In this way CoCoA-5 becomes a convenient, sophisticated front-end for libnormaliz.

Also at this level we are working on a simple design to generate a meaningful CoCoA-5 error message in the case a CoCoA-5 user calls a Normaliz function when CoCoA-5 has been compiled without libnormaliz being linked in (and similarly for any other optional external library).

3.1 Some Implementation Specifics

The code related to the CoCoA-5 built-in functions from libnormaliz is implemented in `BuiltInFunctions-Normaliz.C`.

There are three ways to make a built-in function in CoCoA-5:

One-Liner Built-in Function
This quick approach is applicable when the CoCoA-5 function name is the same as the CoCoALib function name, and the arguments are the same (with simple types). For instance, the function `NmzHilbertBasis` expects a matrix and outputs a matrix. Mathematically the input is a set of vectors, but by requiring them to be in a matrix we have an automatic guarantee that all the vectors have the same length, and that their coordinates are "compatible". In this specific case the coordinates are all integers, but for simplicity we use CoCoA-5's default ring $\mathbb{Q}$ for matrix entries. So the input (and output) is a matrix over $\mathbb{Q}$, which is a CoCoA-5 object directly represented by a CoCoALib object. So the code to port this to CoCoA-5 is just one line

```
DECLARE_COCOALIBFORC5_FUNCTION1(NmzHilbertBasis, MatrixValue)
```

This is a very handy macro saying in just one line that the CoCoA-5 function `NmzHilbertBasis` has 1 argument of type `MatrixValue` (the intepreter wrapper for the CoCoALib type `matrix`). The output type is automatically determined and wrapped up and sent to CoCoA-5 (in this case again a `matrix` which is wrapped into a `MatrixValue`).

Standard Builtin Function

This method is for functions with a fixed number of arguments, but whose types are not fixed or where the CoCoALib function has a different name. For instance the function `NmzNormalToricRing` expects as input a list of power-products, but lists in CoCoA-5 can have elements of any type (even mixed types), so they are far more flexible than the corresponding C++ counterparts. The implementation is

```
DECLARE_STD_BUILTIN_FUNCTION(NmzNormalToricRing, 1)
{ vector<RingElem> v=runtimeEnv->evalArgsAsRingElemList(ARG(0));
  return Value::from(NmzNormalToricRing_forC5(v));
}
END_STD_BUILTIN_FUNCTION
```

meaning that the CoCoA-5 function `NmzNormalToricRing` has 1 argument which is expected to be a CoCoA-5 "list of ring elements". In case no error is thrown (and automatically caught, generating a CoCoA-5 error message), this list is stored into a C++ `vector<RingElem>` to be used by CoCoALib. The return value type is again automatically determined and wrapped up for CoCoA-5 by the handy function `Value::from` (which in this case a converts the CoCoALib output `vector<RingElem>` into a `ListValue` for CoCoA-5).

Functions with a Variable Number of Arguments

The function `NmzComputation` takes either one or two arguments, and this requires a little more hand-work. It intimately reflects the design of libnormaliz: to render its structure the input and output in CoCoA-5 is via a `record` (see Sec. 4.4) and again this aspect does require some *ad hoc* code.

4   Normaliz Functions in CoCoA

In this section we describe in detail the Normaliz functions we have made available in CoCoALib and also in CoCoA-5. We offer (almost) the complete functionality of libnormaliz, and additionally, specific ring theoretic functions applied to monomial subalgebras and monomial ideals in polynomial rings.

4.1   Datatypes in CoCoALib

The fundamental cone object of libnormaliz has as template parameter the integer type used in the computations. In CoCoALib we implemented a "double cone" object which contains two libnormaliz cones: one cone with template parameter `long` and one for the arbitrary precision type `mpz_class` from the GMP library. For many computations machine integer precision is sufficient; and if it suffices, it is much faster than using `mpz_class`; we observed speed differences of a factor 5 or more.

For this reason the CoCoALib "double cone" tries first to execute the requested computation with `long`. If libnormaliz reports an arithmetic overflow it switches to the `mpz_class` cone and sets a flag indicating `long` arithmetic is not sufficient in this case. The user of CoCoALib does not have to worry about overflow problems, it is handled completely automatically by the "double cone".

4.2 Functions in CoCoALib

The cone object is accompanied by several functions which return a single property of the cone, *e.g.* `HilbertBasis(c)` for a previously created cone `c`. These functions trigger a computation if necessary, otherwise they just return the already computed cached result. The computed answer is stored inside the libnormaliz cone, which also handles what has to be computed and what is already known. The cone also has a `myCompute` method which allows the user to initiate the computation of multiple properties at once, for example to compute the Hilbert basis and the Hilbert series, this combined computation is faster than two separate computations since they have a large common part.

Input and output is realized with CoCoALib data types, *e.g.* a list of generators as `std::vector< std::vector<BigInt> >`. The Hilbert series is returned as the specialized CoCoALib class `HPSeries` and the Hilbert polynomial as a CoCoALib polynomial (`RingElem`). This allows easy application of other CoCoALib functions.

Additionally to the full functionality of libnormaliz via the flexible member function `myCompute`, we offer more direct specialized functions to work with monomial subalgebras and monomial ideals. They work directly with CoCoALib data-structures, instead of a `cone`, and are themselves used by the corresponding CoCoA-5 functions described in the next subsection. They communicate via the CoCoALib class `PPVector` which represents a list of power-products in a "power-product monoid" (`PPMonoid`), monomials in a given polynomial ring.

4.3 Datatypes in CoCoA-5

CoCoA-5 has fewer datatypes than CoCoALib, but they are very flexible. For the input and output of the Normaliz functions we use

`MATRIX` for representing a list of vectors of integers

`LIST` for representing a list of monomials

`RECORD` for representing the full flexibility of a libnormaliz/CoCoALib cone.

4.4   Functions in CoCoA-5

In CoCoA-5 we have functions to work with toric rings. A toric ring S is a monomial subalgebra of a polynomial ring generated by a set of monomials. The functions NmzNormalToricRing and NmzIntClosureToricRing compute the normalization and integral closure of the toric ring generated by the given monomials. NmzIntClosureMonIdeal takes a list of monomials which generate a monomial ideal I and computes its integral closure. In all cases the result is returned as a list of monomials representing the monomial subalgebra or monomial ideal, respectively.

Additionally, we provide direct access to libnormaliz via the general function NmzComputation which faithfully reflects the internal structure of the libnormaliz design.

We show the usage of NmzComputation via an example. First we create a record, which is a CoCoA-5 construct to store labeled data. In the following example the record C represents the cone and has only a single entry, the generators for which we want to consider the normalization. In general we can combine multiple fields like inequalities and equations. The possible input combinations are exactly the same as for input files of Normaliz itself.

```
M := mat([[0,0,0,1],[2,0,0,1],[0,3,0,1],[0,0,5,1],[-1,-1,0,1]]);
C := record[ normalization := M];
```

Now we use NmzComputation with the record C as argument, to perform the most complete computation libnormaliz can do. The result is a record again, containing all interesting data that was computed.

```
indent(NmzComputation(C));
```

If we are just interested in special properties we can use NmzComputation with a list of properties that we want to get computed as second argument. To see a special field use a command like

```
NmzComputation(C,["HilbertBasis"]).HilbertBasis;
```

to get just the Hilbert basis. For simplicity this common special task can also be done by calling

```
NmzHilbertBasis(M);
```

which will give you the same information as the command above.

5   Using CoCoALib in NmzIntegrate

So far we have described how Normaliz functions have extended the capabilities of CoCoALib/CoCoA-5. Here we look at the converse: the newly developed extension to Normaliz, called NmzIntegrate, which uses features from CoCoALib.

NmzIntegrate is a program to compute the *generalized Ehrhart function*

$$E(f, k) = \sum_{x \in M, \deg x = k} f(x),$$

of a normal affine monoid M with grading deg and with respect to the polynomial f. This is done via the *generalized Ehrhart series*, which is the ordinary generating function

$$E_f(t) = \sum_{k=0}^{\infty} E(f, k)t^k.$$

Normaliz itself handles the special case $f = 1$, where the points in each degree are simply counted. See [5] for more background on this topic.

In the generalized Ehrhart series computation we need to handle multivariate polynomials. NmzIntegrate uses CoCoALib for representing and operating on multivariate polynomial operations, most importantly multiplications and (linear) substitutions. These operations are the most time consuming part in NmzIntegrate and therefore it profits directly from their sophisticated handling in CoCoALib. Without this "help" from CoCoALib, it would have been much harder to develop NmzIntegrate (and it is unlikely that the result would be so refined).

References

1. Abbott, J., Bigatti, A.M.: CoCoALib: a C++ library for doing Computations in Commutative Algebra, http://cocoa.dima.unige.it/cocoalib
2. Abbott, J., Bigatti, A.M.: What is new in coCoA? In: Hong, H., Yap, C. (eds.) ICMS 2014. LNCS, vol. 8592, pp. 352–358. Springer, Heidelberg (2014)
3. Abbott, J., Bigatti, A.M., Lagorio, G.: CoCoA-5: a system for doing Computations in Commutative Algebra, http://cocoa.dima.unige.it/cocoalib
4. Bruns, W., Ichim, B., Römer, T., Söger, C.: Normaliz, http://www.mathematik.uni-osnabrueck.de/normaliz
5. Bruns, W., Söger, C.: The computation of generalized Ehrhart series in Normaliz. J. Symb. Comput. (to appear), Preprint, arXiv:1211.5178
6. Granlund, T.: The GMP development team: GNU MP: The GNU Multiple Precision Arithmetic Library https://gmplib.org
7. OpenMath Society, http://www.openmath.org
8. Stein, W.A., et al.: Sage Mathematics Software, http://www.sagemath.org

Elements of Design for Containers and Solutions in the **LinBox** Library[*][**]

Extended Abstract

Brice Boyer[1], Jean-Guillaume Dumas[2], Pascal Giorgi[3],
Clément Pernet[4], and B. David Saunders[5]

[1] Department of Mathematics, North Carolina State University, USA
`bbboyer@ncsu.edu`
[2] Laboratoire J. Kuntzmann, Université de Grenoble, France
`Jean-Guillaume.Dumas@imag.fr`
[3] LIRMM, CNRS, Université Montpellier 2, France
`pascal.giorgi@lirmm.fr`
[4] Laboratoire LIG, Université de Grenoble et INRIA, France
`clement.pernet@imag.fr`
[5] University of Delaware, Computer and Information Science Department, USA
`saunders@udel.edu`

Abstract. We describe in this paper new design techniques used in the C++ exact linear algebra library LinBox, intended to make the library safer and easier to use, while keeping it generic and efficient. First, we review the new simplified structure for containers, based on our *founding scope allocation* model. We explain design choices and their impact on coding: unification of our matrix classes, clearer model for matrices and submatrices, *etc.* Then we present a variation of the *strategy* design pattern that is comprised of a controller–plugin system: the controller (solution) chooses among plug-ins (algorithms) that always call back the controllers for subtasks. We give examples using the solution `mul`. Finally we present a benchmark architecture that serves two purposes: Providing the user with easier ways to produce graphs; Creating a framework for automatically tuning the library and supporting regression testing.

Keywords: LinBox, design pattern, algorithms and containers, benchmarking, matrix multiplication algorithms, exact linear algebra.

1 Introduction

This article follows several papers and memoirs concerning LinBox[1] (*cf.* [2, 7, 8, 13, 19]) and builds upon them. LinBox is a C++ template library for fast and

---

[*] This material is based on work supported in part by the National Science Foundation under Grant CCF-1115772 (Kaltofen) and Grant CCF-1018063 (Saunders).

[**] This material is based on work supported in part by the Agence Nationale pour la Recherche under Grant ANR-11-BS02-013 HPAC (Dumas, Giorgi, Pernet).

[1] See `http://www.linalg.org`

H. Hong and C. Yap (Eds.): ICMS 2014, LNCS 8592, pp. 654–662, 2014.
© Springer-Verlag Berlin Heidelberg 2014

exact linear algebra, designed with generality and efficiency in mind. The LinBox library is under constant evolution, driven by new problems and algorithms, by new computing paradigms, new compilers and architectures. This poses many challenges: we are incrementally updating the *design* of the library towards a 2.0 release. The evolution is also motivated by developing a high-performance mathematical library available for researchers and engineers that is easy to use and help produce quality reliable results and quality research papers.

Let us start from a basic consideration: we show in the Table 1 the increase in the "lines of code" size[2] of LinBox and its coevolved dependencies Givaro and Fflas–Ffpack[3].

Table 1. Evolution of the number of lines of code in LinBox

LinBox	$1.0.0^{\dagger\ddagger}$	$1.1.0^{\dagger\ddagger}$	$1.1.6^{\ddagger}$	$1.1.7^{\ddagger}$	1.2.0	1.2.2	1.3.0	1.4.0
loc (×1000)	77.3	85.8	93.5	103	108	109	112	135
Fflas–Ffpack	n/a	n/a	n/a	1.3.3	1.4.0	1.4.3	1.5.0	1.8.0
loc	—	—	—	11.6	23.9	25.2	25.5	32.1
Givaro	n/a	n/a	3.2.16	3.3.3	3.4.3	3.5.0	3.6.0	3.8.0
loc	—	—	30.8	33.6	39.4	41.1	41.4	42.8
total	77.3	85.8	124	137	171	175	179	210

This increase affects the library in several ways. First, it demands a stricter development model, and we are going to list some techniques we used. For instance, we have transformed Fflas–Ffpack (*cf.* [10]) into a new standalone header library, resulting in more visibility for the Fflas–Ffpack project and also in better structure and maintainability of the library. A larger template library is harder to manage. There is more difficulty to trace, debug, and write new code. Techniques employed for easier development include reducing compile times, enforcing stricter warnings and checks, supporting more compilers and architectures, simplifying and automating version number changes, automating memory leak checks, and setting up buildbots to check the code frequently.

This size increase also requires more efforts to make the library user friendly. For instance, we have: Developed scripts that install automatically the latest stable/development versions of the trio, resolving version dependencies; Eased the discovery of Blas/Lapack libraries; Simplified and sped up the checking process, covering more of the library; Updated the documentation and distinguished user and developer oriented docs; Added comprehensive benchmarking tools.

Developing generic high performance libraries is difficult. We can find a large literature on coding standards and software design references in (*cf.* [1, 11, 15, 17, 18]), and draw from many internet sources and experience acquired by/from free software projects. We describe advances in the design of LinBox in the next three sections. We will first describe the new *container* framework in Section 2, then, in

---

[2] Using sloccount, available at http://sourceforge.net/projects/sloccount/

[3] Symbol † when Givaro is included and † when contains Fflas–Ffpack.

Section 3, the improved *matrix multiplication* algorithms made by contributing special purpose matrix multiplication plugins, and, finally, we present the new *benchmark/optimization* architecture (Section 4).

2   Containers Architecture

LinBox is mainly conceived around the RAII (Resource Acquisition Is Initialization, see [17]) concept with reentrant function. We also follow the founding scope allocation model (or *mother model*) of [8] which ensures that the memory used by objects is allocated in the constructor and freed only at its destruction. The management of the memory allocated by an object is exclusively reserved to it.

LinBox uses a variety of container types (representations) for matrix and vectors over fields and rings. The fragmentation of the containers into various matrix and blackbox types has been addressed and simplified. The many different matrix and vector types with different interfaces has been reduced into only two containers: `Matrix` and `Vector`.

2.1   General Interface for Matrices

First, in order to allow operations on its elements, a container is parameterized by a field object (Listing 1.1), not the field's element type. This is simpler and more general. Indeed, the field element type can be inferred from a `value_type` type definition within the field type. Then, the storage type is given by a second template parameter that can use defaults, *e.g.* dense Blas matrices (stride and leading dimension or increment), or some sparse format.

```
template< class _Field, class _Storage = denseDefault >
class Vector ;
```

Listing: Matrix or Vector classes in LinBox.

In the founding scope allocation model, we must distinguish containers that own (responsible for dynamically allocated memory) and containers that share memory of another. `SubMatrix` and `SubVector` types share the memory; `Matrix` and `Vector` own it. All matrix containers share the common `BlackBox` interface described in the next paragraphs, it accommodates both owner and sharer container types, and defines the minimal methods required for a template `BlackBox` matrix type: *Input/Output*. Our matrix containers all read and write from Matrix Market format[4] which is well established in the numerical linear algebra community and facilitates sharing matrices with other software tools. The MatrixMarket header comment provides space for metadata about the provenance of a matrix and our interest in it. However, because of our many entry domains

---

[4] See `http://math.nist.gov/MatrixMarket/`.

and matrix representations, extensions are necessary to the MatrixMarket format. For instance, the header comment records the modulus and irreducible polynomial defining the representation of a matrix over $GF(p^e)$. We can further adapt the header to suit our needs, for instance create new file formats that save space (*e.g.* CSR fashion saves roughly a third space over COO, *cf.* Harwell-Boeing format). Structured matrices (Toeplitz, Vandermonde, *etc.*) can have file representations specified.

Apply method. This is essential in the BlackBox interface (Sections 2.2 and 3).

Rebind/Conversions. In addition to the rebind mechanism (convert from one field to the other), we add conversion mechanisms between formats, for instance all sparse matrix formats can convert to/from CSR format: this 'star' mechanism can simplify the code (to the expense of memory usage) and may speed it up when some central formats are well tuned for some task.

This is a common minimal interface to all our matrix containers that can be used by all algorithms. This interface provides the basic *external* functionality of a matrix as a "linear mapping" (black box). This interface is shared by: *dense* containers (Blas-like,...); *permutation* containers (compressed Lapack or cycle representation); *sparse* containers (based on common formats or on STL containers such as map, deque,...); *structured* containers (Diagonal, Hankel, Butterfly,...); *compound* containers (Compose, Submatrix,...). Additional functions of a container can be added, and flagged with a trait, for example those that support internal changes as for Gaussian elimination.

2.2   The apply Method

The apply method (left or right) is arguably the most important feature in the matrix interface and the LinBox library. It performs what a linear application is defined for: apply to a vector (and by extension a block of vectors, *i.e.* a matrix).

We propose the new interface (Listing 1.2), where _In and _Out are vector or matrices, and Side is Tag::Right or Tag::Left, whether the operation $y \leftarrow A^\top x$ or $y \leftarrow Ax$ is performed. We also generalize to the operation $y \leftarrow \alpha Ax + \beta y$.

```
template< class _In, class _Out >
_Out& apply(_Out &y, const _In& x, enum Side) ;
```

Listing: Apply methods.

This method is fundamental as it is the building block of the BlackBox algorithms (for instance block-Wiedemann) and as the matrix multiplication, main operation in linear algebra, needs to be extremely efficient (Section 3). The implementation of the apply method can be left to a mul solution, which can include a helper/method argument if the apply parameters are specialized enough.

3 Improving **LinBox** Matrix Multiplication

We propose a *design pattern* (the closest pattern to our knowledge is the *strategy* one, see [6, Fig 2.]) in Section 3.1 and we show a variety of new algorithms where it is used in the **mul** solution (Section 3.2).

3.1 Plugin Structure

We propose in Figure 1 a generalization of the *strategy design pattern* of [6, Fig 2.], where distinct algorithms (modules) can solve the same problem and are combined, recursively, by a controller. The main advantage of our pattern is that the modules always call the controller of a function so that the best version will be chosen at each level. An analogy can be drawn with dynamic systems — once the controller sends a correction to the system, it receives back a new measure that allows for a new correction.

For instance, we can write (Figure 2) the standard cascade algorithms (see [10]) in that model. Cascade algorithms are used to combine several algorithms that are switched using thresholds, ensuring better efficiency than that of any of the algorithms individually. This method allows for the reuse of modules and ensures efficiency. It is then possible to adapt to the architecture, the available modules, the resources. The only limitation is that the choice of module

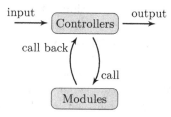

Fig. 1. Controller/Module design pattern

must be fast. On top of this design, we have Method objects that allow caller selection of preferred algorithms, shortcutting the strategy selection.

This infrastructure supports modular code. For instance, Fflas–Ffpack has seen major modularization (addition, scaling, reduction,...) Not only does it enable code to be hardly longer than the corresponding pseudocode listings,

Algorithm 1. `Algo`: controller	**Algorithm 2.** `Algo`: recursive module

Input: A and B, dense, with resp. dimensions $n \times k$ and $k \times n$.
 Input: H Helper
 Output: $C = A \times B$
 if $\min(m,k,n) < H.\texttt{threshold}()$
 then
   | `Algo(C,A,B,BaseCase())` ;
 else
   | `Algo(C,A,B,RecursiveCase())`
 end
	Input: A, B, C as in controller.
	 Input: H, `RecursiveCase` Helper
	 Output: $C = A \times B$
	 Cut A,B,C in S_i, T_i
	 ...
	 $P_i = \texttt{Algo}(S_i, T_i, H)$
	 ...

Fig. 2. Conception of a recursive controlled algorithm

[5], (compared to $\approx 2.5\times$ on some routines before) but it also automatically brings performance, because we can separately improve individual modules and immediately have the benefit throughout the whole library.

3.2 New Algorithms for the mul Solution

New algorithms and techniques improve on matrix multiplication in several ways: reducing memory consumption, reducing runtime, using graphics capabilities, generalizing the Blas to integer routines.

Reduced memory. The routine fgemm in Fflas uses by default the classic schedules for the multiplication and the product with accumulation (*cf.* [5]), but we also implement the low memory routines therein. The new algorithms are competitive and can reach sizes that were limiting. One difficulty consists in using the memory contained in a submatrix of the original matrix, that one cannot free or reallocate.

Using Bini's approximate formula. In [3], we use Bini's approximate matrix multiplication formula to derive a new algorithms that is more efficient that the Strassen–Winograd implementation in fgemm by $\approx 5 - 10\%$ on sizes 1500–3000. This is a cascade of Bini's algorithm and Strassen–Winograd algorithm and/or the nave algorithm (using Blas). The idea is to analyze precisely the error term in the approximate formula and make it vanish.

Integer Blas. In order to provide fast matrix multiplication with multiprecision integers, we rely on multimodular approach through the Chinese remainder theorem. Our approach is to reduce as much as possible to fgemm. Despite, the existence of fast multimodular reduction (resp. reconstruction) algorithm [12], the nave quadratic approach can be reduced to fgemm which makes it more efficient into practice. Note that providing optimized fast multimodular reduction remains challenging. This code is directly integrated into Fflas.

Polynomial Matrix Multiplication over small prime fields. The situation is similar to integer matrices since one can use evaluation/interpolation techniques through DFT transforms. However, the optimized Fast Fourier Transform of [16] makes fast evaluation (resp. interpolation) competitive into practice. We thus rely on this scheme together with fgemm for pointwise matrix multiplications. One can find some benchmark of our code in [14].

Sparse Matrix–Vector Multiplication. For sparse matrices a main issue is that the notion of *sparsity* is too general *vs.* the specificity of real world sparse matrices: the algorithms have to adapt to the shape of the sparse matrices. There is a huge literature from numerical linear algebra on SpMV (Sparse Matrix Vector multiplication) and on sparse matrix formats, some of which are becoming standard (COO, CSR, BCSR, SKY,...). In [4] we developed some techniques to improve the SpMV operation in LinBox. Ideas include the separation of the ± 1 for removing multiplications, splitting in a sum (HYB for hybrid format) of sparse matrix whose formats are independent and using specific routines. For instance, on $\mathbf{Z}/p\mathbf{Z}$ with word size p, one can split the matrix ensuring no reduction is needed in the dot product and call Sparse Blas (from Intel MKL or Nvidia cuBLAS for instance) on each matrix. One tradeoff is as usual between available memory, time spent on optimizing *vs.* time spent on apply, and all the more so because we allow the concurrent storage of the transpose in an optimized fashion, usually yielding huge speedups. This can be decided by *ad hoc.* optimizers.

Work on parallelizations using OpenCL, OpenMP or XKaapi for dense or sparse matrix multiplication include [4, 9, 20].

4 Benchmarking for Automated Tuning and Regression Testing

Benchmarking was introduced in LinBox for several reasons. First, It gives the user a convenient way to produce quality graphs with the help of a graphing library like gnuplot[5] and provides the LinBox website with automatically updated tables and graphs. Second, it can be used for regression testing. Finally, it will be used for selecting default methods and setting thresholds in installation time autotuning.

4.1 Performance Evaluation and Automated Regression Testing

Our plotting mechanism is based on two structures: PlotStyle and PlotData. The PlotGraph structure uses the style and data to manage the output. We allow plotting in standard image formats, html and LaTeX tables, but also in raw csv or xml for file exchange, data comparisons and extrapolation. This mechanism can also automatically create benchmarks in LinBox feature matrix (this is a table that describes what solutions we support, on which the fields).

Saving graphs in raw format can also enable automatic regression testing on the buildbots that already checked our code. For some specifically determined matrices (of various shapes and sizes and over several fields), we can accumulate the timings for key solutions such as (rank, det, mul,...) over time. At each new release, when the documentation is updated, we can check any regression on these base cases and automatically update the regression plots.

4.2 Automated Tuning and Method Selection

Some of the code in LinBox is already automatically tuned (such as thresholds in fgemm), but we improve on it.

Instead of searching for a threshold using fast dichotomous techniques, for instance, we propose to interpolate curves and find the intersection. Using least squares fitting, we may even tolerate outliers (but this is time consuming).

---

[5] http://www.gnuplot.info/.

Automatically tuning a library is not only about thresholds, it may also involve method/algorithm selection. Our strategy is the following: a given algorithm is tuned for each `Helper` (method) it has. Then the solution (that uses these algorithms) is tuned for selecting the best methods. At each stage, defaults are given, but can be overridden by the optimizer. The areas where a method is better are extrapolated from the benchmark curves.

References

1. Alexandrescu, A.: Modern C++ design: generic programming and design patterns applied. C++ in-depth series. Addison-Wesley (2001)
2. Boyer, B.: Multiplication matricielle efficace et conception logicielle pour la bibliothéque de calcul exact LinBox. PhD thesis, Université de Grenoble (June 2012)
3. Boyer, B., Dumas, J.-G.: Matrix multiplication over word-size prime fields using Bini's approximate formula (submitted, May 2014),
   http://hal.archives-ouvertes.fr/hal-00987812
4. Boyer, B., Dumas, J.-G., Giorgi, P.: Exact sparse matrix-vector multiplication on GPU's and multicore architectures. In: Proceedings of the 4th International Workshop on Parallel and Symbolic Computation, PASCO 2010, pp. 80–88. ACM, New York (2010)
5. Boyer, B., Dumas, J.-G., Pernet, C., Zhou, W.: Memory efficient scheduling of Strassen-Winograd's matrix multiplication algorithm. In: Proceedings of the 2009 International Symposium on Symbolic and Algebraic Computation, ISSAC 2009, pp. 55–62. ACM, New York (2009)
6. Cung, V.-D., Danjean, V., Dumas, J.-G., Gautier, T., Huard, G., Raffin, B., Rapine, C., Roch, J.-L., Trystram, D.: Adaptive and hybrid algorithms: classification and illustration on triangular system solving. In: Dumas, J.-G. (ed.) Proceedings of Transgressive Computing 2006, Granada, España (April 2006)
7. Dumas, J.-G., Gautier, T., Giesbrecht, M., Giorgi, P., Hovinen, B., Kaltofen, E., Saunders, B.D., Turner, W.J., Villard, G.: LinBox: A generic library for exact linear algebra. In: Proceedings of the 2002 International Congress of Mathematical Software, Beijing China, World Scientific Pub. (August 2002)
8. Dumas, J.-G., Gautier, T., Pernet, C., Saunders, B.D.: LINBOX founding scope allocation, parallel building blocks, and separate compilation. In: Fukuda, K., van der Hoeven, J., Joswig, M., Takayama, N. (eds.) ICMS 2010. LNCS, vol. 6327, pp. 77–83. Springer, Heidelberg (2010)
9. Dumas, J.-G., Gautier, T., Pernet, C., Sultan, Z.: Parallel computation of echelon forms. In: Proceedings of the 20th International Conference on Parallel Processing, Euro-Par 2014, Porto, Portugal. LNCS, vol. 8632 (August 2014) (to appear)
10. Dumas, J.-G., Giorgi, P., Pernet, C.: Dense linear algebra over word-size prime fields: the Fflas and Ffpack packages. ACM Trans. Math. Softw. 35(3), 1–42 (2008)
11. Gamma, E.: Design Patterns: Elements of Reusable Object-Oriented Software. Addison-Wesley Professional Computing Series. Addison-Wesley (1995)
12. von zur Gathen, J., Gerhard, J.: Modern Computer Algebra. Cambridge University Press, New York (1999)
13. Giorgi, P.: Arithmètique et algorithmique en algèbre linéaire exacte pour la bibliothèque LinBox. PhD thesis, École normale supérieure de Lyon (December 2004)
14. Giorgi, P., Lebreton, R.: Online order basis and its impact on block Wiedemann algorithm. In: Proceedings of the 2014 International Symposium on Symbolic and Algebraic Computation, ISSAC 2014. ACM (to appear, 2014)

15. Gregor, D., Järvi, J., Kulkarni, M., Lumsdaine, A., Musser, D., Schupp, S.: Generic programming and high-performance libraries. International Journal of Parallel Programming 33, 145–164 (2005), 10.1007/s10766-005-3580-8
16. Harvey, D.: Faster arithmetic for number-theoretic transforms. Journal of Symbolic Compututations 60, 113–119 (2014)
17. Stroustrup, B.: The design and evolution of C++. Programming languages/C++. Addison-Wesley (1994)
18. Sutter, H., Alexandrescu, A.: C++ Coding Standards: 101 Rules, Guidelines, and Best Practices. The C++ In-Depth Series. Addison-Wesley (2005)
19. Turner, W.J.: Blackbox linear algebra with the library. PhD thesis, North Carolina State University (May 2002)
20. Wezowicz, M., Saunders, B.D., Taufer, M.: Dealing with performance/portability and performance/accuracy trade-offs in heterogeneous computing systems: a case study with matrix multiplication modulo primes. In: Proc. SPIE, vol. 8403, pp. 08–08–10 (2012)

Recent Developments in Normaliz

Winfried Bruns and Christof Söger

University of Osnabrück, Germany
{wbruns,csoeger}@uos.de
http://www.home.uni-osnabrueck.de/wbruns/
http://www.math.uni-osnabrueck.de/normaliz/

Abstract. The software Normaliz implements algorithms for rational cones and affine monoids. In this note we present recent developments. They include the support for (unbounded) polyhedra and semi-open cones. Furthermore, we report on improved algorithms and parallelization, which allow us to compute significantly larger examples.

Keywords: Hilbert basis, Hilbert series, rational cone, polyhedron.

1 Introduction

Normaliz [2] is a software for computations with rational cones and affine monoids. It pursues two main computational goals: finding the Hilbert basis, a minimal generating system of the monoid of lattice points of a cone; and counting elements degree-wise in a generating function, the Hilbert series. For the mathematical background we refer the reader to [1].

Normaliz (present public version 2.11) is written in C++ (using Boost and GMP/MPIR), parallelized with OpenMP, and runs under Linux, MacOs and MS Windows. It bases on its C++ library libnormaliz which offers the full functionality of Normaliz. There are file based interfaces for Singular, Macaulay 2 and Sage, and C++ level interfaces for CoCoA, polymake, Regina and GAP (in progress). There is also the GUI interface jNormaliz.

Normaliz has found applications in commutative algebra, toric geometry, combinatorics, integer programming, invariant theory, elimination theory, mathematical logic, algebraic topology and even theoretical physics.

2 Hilbert Bases and Hilbert Series

We will first describe the main functionality of Normaliz. The basic objects that constitute the input of Normaliz are a finitely generated rational cone C in $\mathbb{R}^d$ together with a sublattice L of $\mathbb{Z}^d$.

Definition 1. *A (rational) polyhedron P is the intersection of finitely many (rational) halfspaces. If it is bounded, then it is called a polytope. If all the halfspaces are linear, then P is a cone.*

H. Hong and C. Yap (Eds.): ICMS 2014, LNCS 8592, pp. 663–668, 2014.

The dimension *of P is the dimension of the smallest affine subspace* aff(P) *containing P.*

An *affine monoid is a finitely generated submonoid of* $\mathbb{Z}^d$ *for some d.*

By the theorem of Minkowski-Weyl, $C \subset \mathbb{R}^d$ is a (rational) cone if and only if there exist finitely many (rational) vectors $x_1, \ldots, x_n$ such that

$$C = \{a_1 x_1 + \cdots + a_n x_n : a_1, \ldots, a_n \in \mathbb{R}_+\}.$$

For Normaliz, cones C and lattices L can either be specified by generators $x_1, \ldots, x_n \in \mathbb{Z}^d$ or by constraints, i.e., homogeneous systems of diophantine linear inequalities, equations and congruences. Normaliz also offers to define an affine monoid as the quotient of $\mathbb{Z}_+^n$ modulo the intersection with a sublattice of $\mathbb{Z}^n$. From version 2.11 on, Normaliz can handle rational polyhedra. This recent extension is described in Section 3.

In the following we will assume that C is pointed, i.e. $x, -x \in C \Rightarrow x = 0$. By Gordan's lemma the monoid $M = C \cap L$ is finitely generated. This affine monoid has a (unique) minimal generating system called the *Hilbert basis* Hilb(M), see Figure 1 for an example. The computation of the Hilbert basis is the first main tasks of Normaliz.

One application is the computation of the *normalization* of an affine monoid N; this explains the name Normaliz. The normalization is the intersection of the cone generated by M with the sublattice gp(M) generated by M. One calls M *normal*, if it coincides with its normalization.

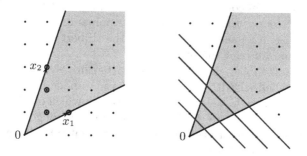

Fig. 1. A cone with the Hilbert basis (circled points) and grading

The second main task is to compute the Hilbert (or Ehrhart) series of a graded monoid. A *grading* of a monoid M is simply a homomorphism deg : $M \to \mathbb{Z}^g$ where $\mathbb{Z}^g$ contains the degrees. The *Hilbert series* of M with respect to the grading is the formal Laurent series

$$H(t) = \sum_{u \in \mathbb{Z}^g} \#\{x \in M : \deg x = u\} t_1^{u_1} \cdots t_g^{u_g} = \sum_{x \in M} t^{\deg x},$$

provided all sets $\{x \in M : \deg x = u\}$ are finite. At the moment, Normaliz can only handle the case $g = 1$, and therefore we restrict ourselves to this case.

We assume in the following that $\deg x > 0$ for all nonzero $x \in M$ and that there exists an $x \in \mathrm{gp}(M)$ such that $\deg x = 1$. (Normaliz always rescales the grading accordingly.)

The basic fact about $H(t)$ in the $\mathbb{Z}$-graded case is that it is the Laurent expansion of a rational function at the origin:

Theorem 1 (Hilbert, Serre; Ehrhart). *Suppose that M is a normal affine monoid. Then*

$$H(t) = \frac{R(t)}{(1 - t^e)^r}, \qquad R(t) \in \mathbb{Z}[t],$$

where r is the rank of M and e is the least common multiple of the degrees of the extreme integral generators of $\mathrm{cone}(M)$. As a rational function, $H(t)$ has negative degree.

Usually one can find denominators for $H(t)$ of much lower degree than that in the theorem, and Normaliz tries to give a more economical presentation of $H(t)$ as a quotient of two polynomials. One should note that it is not clear what the most natural presentation of $H(t)$ is in general (when $e > 1$).

A rational cone C and a grading together define the rational polytope $Q = C \cap A_1$ where $A_1 = \{x : \deg x = 1\}$. In this sense the Hilbert series is nothing but the Ehrhart series of Q.

Note that the coefficients of the Hilbert series are computed by a quasi-polynomial. Its leading coefficient is the suitably normed volume of Q.

3   Polyhedra and Inhomogeneous Systems

A main addition to the functionality of Normaliz is the direct support for (un-bounded) polyhedra. For computations it is useful to homogenize coordinates by embedding $\mathbb{R}^d$ as a hyperplane in $\mathbb{R}^{d+1}$, namely via

$$\kappa : \mathbb{R}^d \to \mathbb{R}^{d+1}, \qquad \kappa(x) = (x, 1).$$

If P is a (rational) polyhedron, then the closure of the union of the rays from 0 through the points of $\kappa(P)$ is a (rational) cone $C(P)$, called the *cone over P*. The intersection $C(P) \cap (\mathbb{R}^d \times \{0\})$ can be identified with the *recession* (or *tail*) *cone*

$$\mathrm{rec}(P) = \{x \in \mathbb{R}^d : y + x \in P \text{ for all } y \in P\}.$$

It is the cone of unbounded directions in P. The recession cone is pointed if and only if P has a vertex. The theorem of Minkowski-Weyl can then be generalized as follows:

Theorem 2 (Motzkin). *The following are equivalent for $P \subset \mathbb{R}^d$, $P \neq \emptyset$:*

1. *P is a (rational) polyhedron;*
2. *$P = Q + C$ where Q is a (rational) polytope and C is a (rational) cone.*

If P has a vertex, then the smallest choice for Q is the convex hull of its vertices, and $C = \mathrm{rec}(P)$ is uniquely determined.

Clearly, P is a polytope if and only if $\mathrm{rec}(P) = \{0\}$. Normaliz computes the recession cone and the polytope Q if P is defined by constraints. Conversely it finds the constraints if the vertices of Q and the generators of C are specified.

Suppose that P is given by a system

$$Ax \geq b, \qquad A \in \mathbb{R}^{m \times d}, \ b \in \mathbb{R}^m,$$

of linear inequalities (equations are replaced by two inequalities). Then $C(P)$ is defined by the *homogenized system*

$$Ax - x_{d+1}b \geq 0,$$

whereas the $\mathrm{rec}(P)$ is given by the *associated homogeneous system* $Ax \geq 0$. The solution set of the associated homogeneous system is always called the recession cone of the system, even if P is empty.

Via the concept of dehomogenization, Normaliz allows for a more general approach. The *dehomogenization* is a linear form δ on $\mathbb{R}^{d+1}$. For a cone $\widetilde{C}$ in $\mathbb{R}^{d+1}$ and a dehomogenization δ, Normaliz computes the polyhedron $P = \{x \in \widetilde{C} : \delta(x) = 1\}$ and the recession cone $C = \{x \in \widetilde{C} : \delta(x) = 0\}$. In particular, this allows other choices of the homogenizing coordinate.

Let $P \subset \mathbb{R}^d$ be a rational polyhedron and $L \subset \mathbb{Z}^d$ be an *affine sublattice*, i.e., a subset $w + L_0$ where $w \in \mathbb{Z}^d$ and $L_0 \subset \mathbb{Z}^d$ is a sublattice. In order to investigate (and compute) $P \cap L$ one again uses homogenization: P is extended to $C(P)$ and L is extended to $\mathcal{L} = L_0 + \mathbb{Z}(w, 1)$. Then one computes $C(P) \cap \mathcal{L}$. Via this "bridge" one obtains the following inhomogeneous version of Gordan's lemma:

Theorem 3. *Let P be a rational polyhedron with vertices and $L = w + L_0$ an affine lattice as above. Set $\mathrm{rec}_L(P) = \mathrm{rec}(P) \cap L_0$. Then there exist $x_1, \ldots, x_m \in P \cap L$ such that*

$$P \cap L = \{(x_1 + \mathrm{rec}_L(P)) \cap \cdots \cap (x_m + \mathrm{rec}_L(P))\}.$$

If the union is irredundant, then $x_1, \ldots, x_m$ are uniquely determined.

The Hilbert basis of $\mathrm{rec}_L(P)$ is given by $\{x : (x, 0) \in \mathrm{Hilb}(C(P) \cap \mathcal{L})\}$ and the minimal system of generators can also be read off the Hilbert basis of $C(P) \cap \mathcal{L}$: it is given by those x for which $(x, 1)$ belongs to $\mathrm{Hilb}(C(P) \cap \mathcal{L})$. Normaliz computes the Hilbert basis of $C(P) \cap L$ only at "levels" 0 and 1.

We call $M = \mathrm{rec}_L(P)$ the *recession monoid* of P with respect to L (or L_0). It is justified to say that $P \cap L$ a *module* over $\mathrm{rec}_L(P)$. In the light of the theorem, it is a finitely generated module with a unique minimal system of generators.

After the introduction of coefficients from a field K, $\mathrm{rec}_L(P)$ is turned into an affine monoid algebra, and $N = P \cap L$ into a finitely generated torsionfree module over it. As such it has a well-defined *module rank* $\mathrm{mrank}(N)$, which is computed by Normaliz via the following combinatorial description: Let $x_1, \ldots, x_m$ be a system of generators of N as above; then $\mathrm{mrank}(N)$ is the cardinality of the set of residue classes of $x_1, \ldots, x_m$ modulo $\mathrm{rec}_L(P)$.

Clearly, to model $P \cap L$ we need linear diophantine systems of inequalities, equations and congruences which now will be inhomogeneous in general. Conversely, the set of solutions of such a system is of type $P \cap L$.

If $\mathbb{Z}^d$ is endowed with a grading whose restriction to M satisfies our conditions, then the Hilbert series

$$H_N(t) = \sum_{x \in N} t^{\deg x}$$

is well-defined, and the qualitative statement above about rationality remains valid. However, the degree may now be ≥ 0. Again, one has an associated quasipolynomial with constant leading coefficient given by

$$q_{r-1} = \mathrm{mrank}(N) \frac{\mathrm{vol}(Q)}{(r-1)!}, \qquad Q = \mathrm{rec}(P) \cap A_1.$$

The *multiplicity* of N is $\mathrm{mrank}(N) \, \mathrm{vol}(Q)$.

4   Further Extensions

Normaliz now can compute the Hilbert function of a semiopen cone. Such a semiopen cone is given by $C' = C \setminus \mathcal{F}$, where C is a cone and $\mathcal{F}$ is a union of faces (not necessarily facets) of C. Typical applications come from mixed systems of homogeneous inequalities and strict inequalities. This situation could also be modeled by inhomogeneous constraints, but if only few faces are excluded it is beneficial to compute in the original cone and just exclude $\mathcal{F}$.

Additionally, we implemented two new methods of computing the lattice points of a rational polytope. One is a specialization of the so-called dual mode Hilbert basis computation to this case. The other one approximates the rational polytope by a lattice polytope.

The extension NmzIntegrate (introduced in 2.9) counts lattice points with a polynomial weight to compute the *generalized Ehrhart series*, see [4].

5   Algorithmic Improvements

Most of the algorithms in Normaliz base on a *triangulation* of the cone, i.e. a subdivision into *simplicial cones*. Simplicial cones are generated by linearly independent vectors and therefore they are much easier to handle than general cones. The improvements focus on handling large triangulations.

A triangulation is a non-disjoint decomposition of the cone, the simplicial cones intersect in lower dimensional cones. Especially for Hilbert series computations an exact (disjoint) decomposition is needed. Since version 2.7 a principle described by Köppe and Verdoolaege in [5] is used to gain it from the triangulation Γ. It allows the independent handling of the simplicial cones in Γ and thus is superior over the old method, where the simplicial cones had to be compared with each other. This exact decomposition of the cone is then used to obtain a disjoint decomposition of the monoid $M = C \cap L$ of the form

$$M = \bigcup_{\sigma \in \Gamma} \bigcup_{y} (y + M_\sigma),$$

where y runs over a special finite subset of $\sigma \cap L$ and the M_σ are free monoids. Such a disjoint union is called *Stanley decomposition*, named after R. Stanley who proved its existence in 1982.

The *pyramid decomposition* is a newly developed method to compute huge triangulations. It splits the cone in smaller pieces, the pyramids, and handles them completely independent of each other. The result is an algorithm following the "divide-and-conquerer" principle. It gives formidable improvements for larger examples, both in computation time and memory usage, and enables Normaliz to handle triangulations with more than 10^{11} simplicial cones. We refer the reader to [3] for an exact description.

For Hilbert basis computations of combinatorial examples we had introduced a *partial triangulation* in version 2.5. It has now been tuned to check the normality of even larger monoids. For example, the exact decomposition is used to avoid duplicate points in the intersections of the simplicial cones. It reduces computation time and memory requirements, together with intermediate reductions; see [7] for more details.

Together with the parallelization of the algorithms, these improvements enable us to compute significantly larger examples. One interesting class are the cut monoids of graphs for which Sturmfels and Sullivant conjecture normality if the graph is free of K_5-minors (K_5 is the complete graph on 5 vertices). With the partial triangulation implementation of Normaliz 2.5 we were able to validate the conjecture for all graphs up to 8 vertices. The recent version could verify the conjecture for all graphs up to 10 vertices (see [7]), using a result of Ohsugi [6]. The biggest of these examples produced a partial triangulation with more than $15 \cdot 10^9$ simplicial cones, almost $7 \cdot 10^8$ candidates for the Hilbert basis, and took 30 hours with 20 threads on our compute server.

References

1. Bruns, W., Gubeladze, J.: Polytopes, rings and K-theory. Springer (2009)
2. Bruns, W., Ichim, B., Römer, T., Söger, C.: Normaliz. Algorithms for rational cones and affine monoids, http://www.math.uos.de/normaliz
3. Bruns, W., Ichim, B., Söger, C.: The power of pyramid decomposition in normaliz, preprint, arXiv:1206.1916
4. Bruns, W., Söger, C.: The computation of generalized Ehrhart series in Normaliz. J. Symb. Comput. (to appear), Preprint, arXiv:1211.5178
5. Köppe, M., Verdoolaege, S.: Computing parametric rational generating functions with a primal Barvinok algorithm. Electr. J. Comb. 15, R16, 1–19 (2008)
6. Ohsugi, H.: Normality of cut polytopes of graphs is a minor closed property. Discrete Math. 310, 1160–1166 (2010)
7. Söger, C.: Parallel Algorithms for Rational Cones and Affine Monoids. Dissertation thesis (2014), urn:nbn:de:gbv:700-2014042212422

The Basic Polynomial Algebra Subprograms

Changbo Chen[1], Svyatoslav Covanov[2], Farnam Mansouri[2],
Marc Moreno Maza[2], Ning Xie[2], and Yuzhen Xie[2]

[1] Chongqing Key Laboratory of Automated Reasoning and Cognition,
Chongqing Institute of Green and Intelligent Technology,
Chinese Academy of Sciences, China
changbo.chen@hotmail.com
[2] University of Western Ontario, Canada
{moreno,covanov,fmansou3,nxie6,yxie}@csd.uwo.ca

Abstract. The Basic Polynomial Algebra Subprograms (BPAS) provides arithmetic operations (multiplication, division, root isolation, etc.) for univariate and multivariate polynomials over prime fields or with integer coefficients. The code is mainly written in CilkPlus [10] targeting multicore processors. The current distribution focuses on dense polynomials and the sparse case is work in progress. A strong emphasis is put on adaptive algorithms as the library aims at supporting a wide variety of situations in terms of problem sizes and available computing resources. One of the purposes of the BPAS project is to take advantage of hardware accelerators in the development of polynomial systems solvers. The BPAS library is publicly available in source at www.bpaslib.org.

Keywords: Polynomial arithmetic, parallel processing, multi-core processors, Fast Fourier Transforms (FFTs).

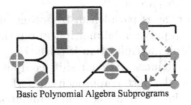

Basic Polynomial Algebra Subprograms

1 Design and Specification

Inspired by the Basic Linear Algebra Subprograms (BLAS), BPAS functionalities are organized into three levels. At Level 1, one finds basic arithmetic operations that are specific to a polynomial representation or specific to a coefficient ring. Examples of Level-1 operations are multi-dimensional FFTs/TFTs and univariate real root isolation. At Level 2, arithmetic operations are implemented for all types of coefficients rings that BPAS supports (prime fields, ring of integers, field of rational numbers). Level 3 gathers advanced arithmetic operations taking as input a zero-dimensional regular chain, e.g. normal form of a polynomial, multivariate real root isolation.

H. Hong and C. Yap (Eds.): ICMS 2014, LNCS 8592, pp. 669–676, 2014.
© Springer-Verlag Berlin Heidelberg 2014

Level 1 functions are highly optimized in terms of data locality and parallelism. In particular, the underlying algorithms are nearly optimal in terms of cache complexity [5]. This is the case, for instance, for our modular multi-dimensional FFTs/TFTs [14], modular dense polynomial arithmetic [15] and Taylor shift [3] algorithms.

At Level 2, the user can choose between algorithms that either minimizes work (at the possible expense of decreasing parallelism) or maximizes parallelism (at the possible expense of increasing work). For instance, five different integer polynomial multiplication algorithms are available, namely: Schönhage-Strassen, 8-way Toom-Cook, 4-way Toom-Cook, divide-and-conquer plain multiplication and the two-convolution method [2].

- The first one has optimal work (i.e. algebraic complexity) but is purely serial due to the difficulties of parallelizing 1D FFTs on multicore processors.
- The next three algorithms are parallelized but their parallelism is static, that is, independent of the input data size; these algorithms are practically efficient when both the input data size and the number of available cores are small, see [12] for details.
- The fifth algorithm relies on modular 2D FFTs which are computed by means of the row-column scheme; this algorithm delivers high scalability and can fully utilize the hardware on fat multicore nodes.

Another example of Level 2 functionality is parallel Taylor shift computation for which four different algorithms are available: the two plain algorithms presented in [3], Algorithm (E) of [7] and an optimized version of Algorithm (F) of [7].

- The first two are highly effective when both the input data size and the number of available cores are small.
- The third algorithm creates parallelism by means of a divide-and-conquer procedure and relies on polynomial multiplication; this approach is effective when 8-way Toom-Cook multiplication is selected.
- The fourth algorithm reduces a Taylor shift computation to a single polynomial multiplication; this latter approach outperforms the other three, as soon as the two-convolution multiplication dominates its counterparts, that is, when either input data size and the number of available cores become large.

This variety of parallel solutions leads, at Level 3, to adaptive algorithms which select appropriate Level 2 functions depending on available resources (number of cores, input data size). An example is parallel real root isolation. Many procedures for this purpose are based on a *subdivision scheme*. However, on many examples, this scheme exposes only a limited amount of opportunities for concurrent execution, see [3]. It is, therefore, essential to extract as much as parallelism from the underlying routines, such as Taylor shift computations.

2   User Interface

Inspired by computer algebra systems like AXIOM [9] and Magma [1], the BPAS library makes use of type constructors so as to provide genericity. For instance SparseUnivariatePolynomial (SUP) can be instantiated over any BPAS ring. On the other hand, for efficiency consideration, certain polynomial type constructors, like DistributedDenseMultivariateModularPolynomia (DDMMP), are only available over finite fields in order to ensure that the data encoding a DDMMP polynomial consists only of consecutive memory cells.

For the same efficiency consideration, the most frequently used polynomial rings, like DenseUnivariateIntegerPolynomial (DUZP) and DenseUnivariateRationalNumberPolynomial (DUQP) are primitive types. Consequently, DUZP and SUP<Integer> implement the same functionalities; however the implementation of the former is further optimized.

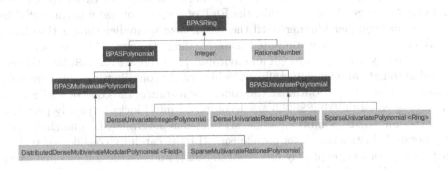

Fig. 1. A snapshot of BPAS algebraic data structures

Figure 1 shows a subset of BPAS's tree of algebraic data structures. Dark and blue boxes correspond respectively to abstract and concrete classes. BPAS counts many other classes for instance Intervals and RegularChains.

Figure 2 first shows how two dense univariate polynomials are read from a file and how their product is computed. Then, on the same code fragment, a (zero-dimensional) regular chain is read from a file and its real roots are isolated.

3   Implementation Techniques

Modular FFTs are at the core of asymptotically fast algorithms for dense polynomial arithmetic operations. A substantial body of code of the BPAS library is, therefore, devoted to the computation of one-dimensional and multi-dimensional FFTs over finite fields. In the current release, the characteristic of those fields is of machine word size while larger characteristics are work in progress.

The techniques used for the multi-dimensional FFTs are described in [14,15] while those for one-dimensional FFTs are inspired by the design of the FFTW [4].

```
#include <bpas.h>

int main(int argc, char *argv[]) {
        /* Univariate Integer Polynomial Multiplication */
        DUZP a(128), b(128);
        a.read("a_input.dat"); b.read("b_input.dat");
        DUZP c = a * b;

        /* Real Root Isolation */
        mpq_class width(1, 20);
        RegularChains rcs;
        rcs.read("rcs_input.dat");
        Intervals boxes = realRootIsolation(rcs, width);

        return 0;
}
```

Fig. 2. A snapshot of BPAS code

BPAS one-dimensional FFTs code is optimized in terms of cache complexity and register usage. To achieve this, the FFT of a vector of size n is computed in a divide-and-conquer manner until the vector size is smaller than a threshold, at which point FFTs are computed using a tiling strategy. This threshold can be specified by the user through an environment variable HTHRESHOLD or determined automatically when installing the library. At compile time, this threshold is used to generate and optimize the code. For instance, the code of all FFTs of size less or equal to HTHRESHOLD are decomposed into blocks (typically performing FFTs on 8 or 16 points) for which straight-line program (SLP) machine code is generated. Instruction level parallelism (ILP) is carefully considered: vectorized instructions are explicitly used (SSE2, SSE4) and instruction pipeline usage is highly optimized. Other environment variables are available for the user to control different parameters in the code generation.

Table 1. One-dimensional modular FFTs: Modpn vs BPAS

Size	Modpn	BPAS	Speedup
16777216	6.232	1.391	4.48
33554432	12.987	2.957	4.392
67108864	26.783	6.266	4.274
134217728	55.329	13.235	4.181
268435456	113.8	27.901	4.079

Table 1 compares running times (in sec. on Intel Xeon 5600) of one-dimensional modular FFTs computed by the Modpn library [11] and BPAS, both using serial C code in this case. The first column of Table 1 gives the size of the input vector; coefficients are in a prime field whose characteristic is a 57-bit prime.

Modular FFTs support the implementation of several algorithms performing dense polynomial arithmetic. As an example, we consider parallel multiplication of dense polynomials with integer coefficients by means of the *two-convolution method* [2] and which is illustrated on Figure 3. Given two univariate polynomials $a(y)$, $b(y)$ with integer coefficients, their product $c(y)$ is computed as follows.

(S1) Convert $a(y)$, $b(y)$ to bivariate integer polynomials $A(x, y)$, $B(x, y)$ s.t. $a(y) = A(\beta, y)$ and $b(y) = B(\beta, y)$ hold at $\beta = 2^M$, $K = \deg(A, x) = \deg(B, x)$, where M is essentially the maximum bit size of a coefficient in a and b.

(S2) Consider $C^+(x, y) \equiv A(x, y) \, B(x, y) \bmod \langle x^K + 1 \rangle$ and $C^-(x, y) \equiv A(x, y) \, B(x, y) \bmod \langle x^K - 1 \rangle$. Compute $C^+(x, y)$ and $C^-(x, y)$ modulo machine-word primes so as to use modular 2D FFTs.

(S3) Consider $C(x, y) = \frac{C^+(x,y)}{2} (x^K - 1) + \frac{C^-(x,y)}{2} (x^K + 1)$ and evaluate $C(x, y)$ at $x = \beta$, which finally gives $c(y) = a(y) \, b(y)$.

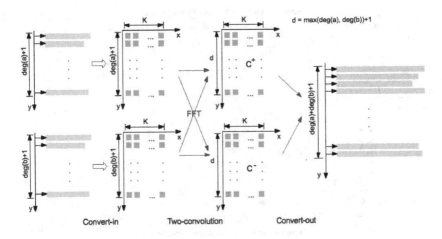

Fig. 3. Multiplication scheme for dense univariate integer polynomials

The conversions from the univariate polynomials $a(y)$, $b(y)$ to the bivariate polynomials $A(x, y)$, $B(x, y)$ in Step (S1) as well as the conversions from the bivariate polynomials $C^+(x, y)$ and $C^-(x, y)$ in Step (S3) require only additions and shift operations on machine words. Moreover, the polynomials $C^+(x, y)$ and $C^-(x, y)$ are reconstructed from their modular images (in practice two modular images are sufficient) within Step (S3). Consequently, the data produced by 2D FFT computations is converted in a *single pass* into the final result $c(y)$. Similarly the bivariate polynomials $A(x, y)$, $B(x, y)$ are obtained from $a(y)$, $b(y)$ (here again by means of additions and shift operations on machine words) in a single pass. Since BPAS' 2D FFT computations are optimal in terms of cache complexity [15], the whole multiplication procedure is optimal for that same complexity measure. Last, but not least, BPAS' 2D FFTs are computed by the row-column scheme which provides lots of parallelism with limited overheads on multicore architectures. As a result, our multiplication code, based on this two-convolution method scales well on multicores as illustrated hereafter.

4   Experimental Evaluation

As mentioned above, one of the main purposes of the BPAS library is to take advantage of hardware accelerators and support the implementation of polynomial system solvers. With this goal, polynomial multiplication plays a central role. Moreover, both sparse and dense representations are important. Indeed, input polynomial systems are often sparse while many algebraic transactions, like substitution, tend to densify data. Parallel sparse polynomial arithmetic has been studied by Gastineau and Laskar in [6] and by Monagan and Pearce in [13].

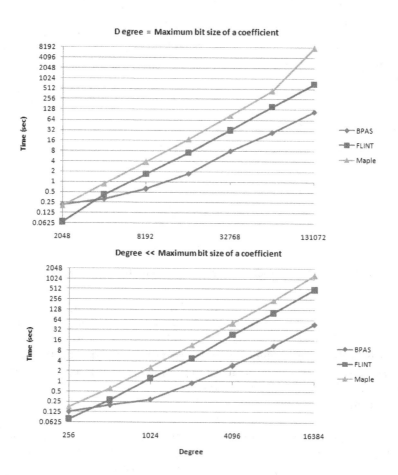

Fig. 4. Dense integer polynomial multiplication: BPAS vs FLINT vs Maple

Up to our knowledge, BPAS is the first publicly available library for parallel dense integer polynomial arithmetic. For this reason, we compare BPAS' parallel dense polynomial multiplication against state-of-the-art counterpart implementation in FLINT 2.4.3 and Maple 18. On Figure 4, the input of each test case is a pair of polynomials of degree d where each coefficient has bit size N. Two plots are provided: one for which d = N holds and one for which d is much smaller than N.

The BPAS library is implemented with the multi-threaded language CilkPlus [10] and we compiled our code with the CilkPlus branch of GCC[1]. Our experimental results were obtained on an 48-core AMD Opteron 6168, running at 900Mhz with 256 GB of RAM and 512KB of L2 cache.

Table 2 shows that the work overhead (measured by Cilkview, the performance analysis tool of CilkPlus) of the BPAS method w.r.t. to a method based on Schönhage & Strassen algorithm (KS) is only around 2 (see Column 3), whereas BPAS provides large amount of parallelism (see Column 2).

Table 2. Cilkview analysis of BPAS and KS (* shows the number of instructions)

Size	Work(KS)*	Work(BPAS)*	Span(BPAS)*	$\frac{\text{Work(BPAS)}}{\text{Span(BPAS)}}$	$\frac{\text{Work(BPAS)}}{\text{Work(KS)}}$
2048	795,549,545	1,364,160,088	41,143,119	33.16	1.715
4096	4,302,927,423	5,663,423,709	96,032,325	58.97	1.316
8192	16,782,031,611	23,827,123,688	292,735,521	81.39	1.420
16384	63,573,232,166	100,688,072,711	1,017,726,160	98.93	1.584
32768	269,887,534,779	425,149,529,176	3,804,178,563	111.76	1.575

5   Application

Turning to parallel univariate real root isolation, we have integrated our parallel integer polynomial multiplication into the algorithm proposed in [3]. To this end, we perform the Taylor Shift operation, that is, the map $f(x) \longmapsto f(x+1)$, by means of Algorithm (E) in [7], which reduces calculations to integer polynomial multiplication in large degrees and to using algorithm of [3] in small degrees. In Table 3, we call BPAS this adaptive algorithm combining FFT-based arithmetic (via Algorithm (E)) and plain arithmetic (via [3]).

Table 3. Univariate real root isolation running time for four examples

	Size	BPAS	CMY [3]	realroot	#Roots
Cnd	32768	18.141	125.902	816.134	1
	65536	66.436	664.438	7,526.428	1
Chebycheff	2048	608.738	594.82	1,378.444	2047
	4096	8,194.06	10,014	35,880.069	4095
Laguerre	2048	1,336.14	1,324.33	3,706.749	2047
	4096	20,727.9	23,605.7	91,668.577	4095
Wilkinson	2048	630.481	614.94	1,031.36	2047
	4096	9,359.25	10,733.3	26,496.979	4095

We run these two parallel real root algorithms, BPAS and CMY [3], which are both implemented in CilkPlus, against Maple 18 serial realroot command, which implements a state-of-the-art algorithm. Table 3 shows the running times

---

[1] http://gcc.gnu.org/svn/gcc/branches/cilkplus/

(in sec.) of four well-known test problems, including `Cnd`, `Chebycheff`, `Laguerre` and `Wilkinson`. Moreover, for each test problem, the degree of the input polynomial varies in a range. The results reported in Table 3 show that integrating parallel integer polynomial multiplication into our real root isolation code has substantially improved the performance of the latter.

Acknowledgments. This work was supported by the NSFC (11301524) and the CSTC (cstc2013jjys0002).

References

1. Bosma, W., Cannon, J., Playoust, C.: The Magma algebra system. I. The user language. J. Symbolic Comput. 24(3-4), 235–265 (1997)
2. Chen, C., Mansouri, F., Moreno Maza, M., Xie, N., Xie, Y.: Parallel Multiplication of Dense Polynomials with Integer Coefficient. Technical report, The University of Western Ontario (2013)
3. Chen, C., Moreno Maza, M., Xie, Y.: Cache complexity and multicore implementation for univariate real root isolation. J. of Physics: Conf. Series 341 (2011)
4. Frigo, M., Johnson, S.G.: The design and implementation of FFTW3 93(2), 216–231 (2005)
5. Frigo, M., Leiserson, C.E., Prokop, H., Ramachandran, S.: Cache-oblivious algorithms. ACM Transactions on Algorithms 8(1), 4 (2012)
6. Gastineau, M., Laskar, J.: Highly scalable multiplication for distributed sparse multivariate polynomials on many-core systems. In: Gerdt, V.P., Koepf, W., Mayr, E.W., Vorozhtsov, E.V. (eds.) CASC 2013. LNCS, vol. 8136, pp. 100–115. Springer, Heidelberg (2013)
7. von zur Gathen, J., Gerhard, J.: Fast algorithms for taylor shifts and certain difference equations. In: ISSAC, pp. 40–47 (1997)
8. Hart, W., Johansson, F., Pancratz, S.: FLINT: Fast Library for Number Theory. V. 2.4.3, http://flintlib.org
9. Jenks, R.D., Sutor, R.S.: AXIOM, The Scientific Computation System. Springer (1992)
10. Leiserson, C.E.: The Cilk++ concurrency platform. The Journal of Supercomputing 51(3), 244–257 (2010)
11. Li, X., Moreno Maza, M., Rasheed, R., Schost, É.: The modpn library: Bringing fast polynomial arithmetic into maple. J. Symb. Comput. 46(7), 841–858 (2011)
12. Mansouri, F.: On the parallelization of integer polynomial multiplication. Master's thesis, The University of Western Ontario, London, ON, Canada (2014), http://www.csd.uwo.ca/~moreno/Publications/farnam-thesis.pdf
13. Monagan, M.B., Pearce, R.: Parallel sparse polynomial multiplication using heaps. In: ISSAC, pp. 263–270. ACM (2009)
14. Moreno Maza, M., Xie, Y.: FFT-based dense polynomial arithmetic on multi-cores. In: Mewhort, D.J.K., Cann, N.M., Slater, G.W., Naughton, T.J. (eds.) HPCS 2009. LNCS, vol. 5976, pp. 378–399. Springer, Heidelberg (2010)
15. Moreno Maza, M., Xie, Y.: Balanced dense polynomial multiplication on multicores. Int. J. Found. Comput. Sci. 22(5), 1035–1055 (2011)
16. Schönhage, A., Strassen, V.: Schnelle multiplikation großer zahlen. Computing 7(3-4), 281–292 (1971)

Function Interval Arithmetic*

Jan Duracz[1], Amin Farjudian[2], Michal Konečný[3], and Walid Taha[4]

[1] jan@duracz.net
http://duracz.net/jan
[2] School of Computer Science, University of Nottingham Ningbo, China
Amin.Farjudian@nottingham.edu.cn
www.cs.nott.ac.uk/~avf
[3] School of Engineering and Applied Science, Aston University, Birmingham, UK
M.Konecny@aston.ac.uk
www-users.aston.ac.uk/~konecnym
[4] Halmstadt University, Sweden & Rice University, Houston, Texas, USA
Walid.Taha@hh.se
http://bit.ly/WT-EMG

Abstract. We propose an arithmetic of *function intervals* as a basis for convenient rigorous numerical computation. Function intervals can be used as mathematical objects in their own right or as *enclosures* of functions over the reals. We present two areas of application of function interval arithmetic and associated software that implements the arithmetic: (1) Validated ordinary differential equation solving using the AERN library and within the Acumen hybrid system modeling tool. (2) Numerical theorem proving using the PolyPaver prover.

Keywords: Validated Numeric Computation, ODEs, Theorem Proving.

1 Background

In validated numerical computation, all values are computed together with rigorous upper bounds on their errors or uncertainty. Applications of this approach range from pure mathematics to the development of safety-critical systems.

Using more conventional methods, one can achieve such level of reliability by a combination of: (1) Approximate numerical computation based on floating-point numbers; (2) a specification of error bounds and a formal proof that the implementation of the algorithm stays within the bounds. While the conventional approach is appropriate in many applications, a formally proved numerical analysis can easily become too complex. Validated numerical computation often offers a viable alternative in such cases.

Keeping the derived error bounds relatively small is essential. We are interested in methods that support reducing the bounds arbitrarily close to their theoretical limits.

* This work was supported by EPSRC grant EP/C01037X/1, Altran Technologies SA, the US NSF CPS award 1136099 and Swedish KK-Foundation CERES Centre.

H. Hong and C. Yap (Eds.): ICMS 2014, LNCS 8592, pp. 677–684, 2014.
© Springer-Verlag Berlin Heidelberg 2014

Interval Computation. It is impossible to represent all real numbers or functions in finite space. Computation over these objects is realizable only through computation over their valid finite representations. For the purposes of validated computation, the most widely used approach is to represent a real number by an interval with floating-point endpoints that enclose the real number, bounding it from below and from above.

A driving force behind the development of interval computation has been a number of notable applications to rigorous differential equation solving, global optimization, and theorem proving. In the past couple of decades these application areas have also been combined in the construction of rigorous algorithms for reachability analysis of dynamical systems. A famous example of such advances is Tucker's proof that the Lorenz attractor is a strange attractor [Tuc02]. Interval computation has gained more recognition since the formation of the IEEE Interval Standard Working Group P1788 in 2008.

Interval computations use and produce enclosures of numerical quantities and functions. In their simplest form, such enclosures are just real intervals, most often used to bound single real values. Cartesian products of real intervals are interval vectors, or boxes. A box can be used to approximate a single real vector, a geometric entity, or a part of a function graph. Our focus is the approximation of functions, using enclosures that are more refined than boxes.

Function intervals are intervals whose endpoints are functions and can serve as enclosures of functions in the same way that real intervals can serve as enclosures of numbers. More formally, a function interval is a pair

$$[f(x_1, \ldots, x_n), g(x_1, \ldots, x_n)]$$

where f and g belong to the set $D \to \mathbb{R}$ of real-valued functions on a box domain $D \subset \mathbb{R}^n$. A function interval $[f, g]$ is typically used to enclose a single continuous function h in $D \to \mathbb{R}$—i. e., $f \leq h \leq g$—and its *accuracy* increases as its width —i. e., $\|g - f\|_\infty$—decreases.

Established special cases of function intervals include Berz and Makino's Taylor Models (TMs) [MB02] and affine forms [CS93]. Arithmetics of TMs and affine forms have been used with considerable success to bound rounding errors in floating-point computations [DK14a] and to approximate the solutions of differential equations in beam physics simulations [MB09]. Our implementation of function intervals is more general than TMs and affine forms.

2   Functionality

The AERN library[1] formally defines *functional interval arithmetic* as an *algebraic structure* over function intervals and provides an implementation based on polynomials. The algebraic structure includes the following operations:

– *constructors*: constant functions $0, \pi, [0, 1]$, projections, such as $\lambda(x, y).x$

---

[1] AERN is freely available from `https://github.com/michalkonecny/aern`

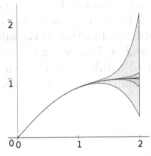

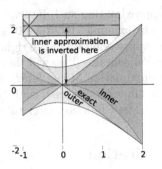

Fig. 1. Enclosures of erf(x) computed by AERN with various effort settings

Fig. 2. The product $[-1,1]\cdot x$, its outer and inner approximations by AERN

- *pointwise operations*: the field operations $+, -, *, /$, common elementary functions such as e^x, $\sqrt{x}$, minimum, maximum and absolute value
- *analytic operations*: integration, such as $\int_0^x f(\xi, y)\, d\xi$
- *domain-changing operations*: evaluation (*e. g.*, $f(1)$, $f([0,1])$), composition (*e. g.*, $f(g(1,x),x)$), adding a variable (*e. g.*, $f(x) \mapsto f(x,y)$), restricting the domain of a variable (*e. g.*, $f(x,y)|_{x\in[0,1]}$)

Most of these operations can be computed only approximately, producing enclosures of the exact results. Each approximate operation has an optional parameter that gives the user control over the trade-off between computation effort and accuracy. For example, consider the task of approximating the error function

$$\mathrm{erf}(x) = \frac{2}{\sqrt{\pi}} \int_0^x e^{-t^2}\, dt$$

over the domain $x \in [0,2]$. In our implementation of this function using AERN[2], the following parameters are available to control the approximation effort:

- The precision of polynomial coefficients and constants such as π
- An upper bound on the polynomial degree
- The Taylor degree for approximating e^x

Fig. 1 shows the enclosures of the error function computed by AERN for several settings of the above effort parameters. The only parameter that changes is the upper bound on the polynomial degree and it ranges from 5 to 25.

Defaults are available for these effort parameters. There is also an iRRAM-style [Mül01] adaptive mode, in which the user specifies the desired accuracy of the result and the library adaptively increases effort parameters until the accuracy is reached.

AERN provides facilities to approximate not only real functions but also real interval functions, such as the product

$$[-1, 1] \cdot x = [\min(-x, x), \max(-x, x)] . \tag{1}$$

---

[2] The code is available at https://github.com/michalkonecny/aern/blob/master/aern-poly-plot-gtk/demos/erf.hs

When approximating such intervals, it is often not sufficient to provide an enclosure. For example, to prove $0 \in [-\varepsilon, \varepsilon] + [-1, 1] \cdot x$, it is useful to compute an *inner* approximation of the right-hand-side interval function. If the inner approximation contains 0, so will the exact function. Fig. 2 shows an outer and an inner approximation of the function (1) computed within AERN[3]. Near $x = 0$ the plotted inner enclosure is slightly inverted. Adding $[-\varepsilon, \varepsilon]$ turns the enclosure into a consistent function interval containing 0.

3   Applications

We demonstrate the practical utility of function interval arithmetic via three concrete applications in computational mathematics.

Solving ODE IVPs. The arithmetic was used to compute enclosures for solutions of ordinary differential equation initial value problems (ODE IVPs) by means of a direct implementation of the interval Picard operator of Edalat and Pattinson [EP07]. Moreover, function interval arithmetic provides a conceptually simple way of extending Edalat and Pattinson's work to the case of uncertain initial conditions. Fig. 3 shows a parametric plot of an enclosure produced by AERN for the following Lorenz IVP with an uncertain initial value:

$$\begin{cases} y_1' = 10(y_2 - y_1) \qquad y_2' = y_1(28 - y_3) - y_2 \qquad y_3' = y_1 y_2 - 8y_3/3 \\ y(0) \in (15 \pm 0.01, 15 \pm 0.01, 36 \pm 0.01) \end{cases} \qquad (2)$$

Note that the rectangular initial value uncertainty results in a non-rectangular intermediate value uncertainty as time progresses. The short elongated shapes visible in Fig. 3 are enclosures of the uncertainty sets for sample time points.

Enclosing Zeno Behavior. A restricted version of function interval arithmetic and our ODE solving method are included in the Acumen[4] tool for modeling and rigorous simulation of hybrid dynamical systems. In combination with a novel method for event processing [KTD+13], the Acumen tool can compute a tight enclosure of a trajectory that contains infinitely many events due to so-called Zeno behavior. An example of such enclosure is shown in Fig. 4.

Theorem Proving. The arithmetic has been used to automatically prove theorems that take the form of inclusions of non-linear interval expressions, such as:

$$1 - e^{x^2} \left(\frac{0.3480242}{1 + 0.47047x} - \frac{0.0958798}{(1 + 0.47047x)^2} + \frac{0.7478556}{(1 + 0.47047x)^3} \right) \in \frac{2}{\sqrt{\pi}} \int_0^x e^{-t^2}\, dt \pm 0.00005 \qquad (3)$$

---

[3] The code is available at https://github.com/michalkonecny/aern/blob/master/aern-poly-plot-gtk/demos/thickprod.hs
[4] Freely available from www.acumen-language.org

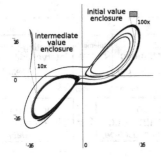
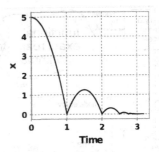

Fig. 3. An AERN enclosure of all solutions of the Lorenz IVP (2)

Fig. 4. An Acumen enclosure of a trajectory with Zeno behavior at time 3

A version of AERN has been embedded in our numerical theorem prover Poly-Paver[5]. PolyPaver proves such inclusions by constructing an outer approximation of the contained function and an inner approximation of the containing interval function and showing that the inclusion holds for the approximations. (Fig. 2 gives an example of an inner approximation.) PolyPaver has been successfully applied to proving correctness theorems for tight accuracy properties of floating-point programs [DK14b]. For example, we proved that a Riemann integrator produces a value close to the exact integral.

4   Underlying Theory

In this section we first give an overview of the main types available in the AERN library, and then explain why these types feature a generalized notion of interval. This is followed by a description of the role of abstract types in specifying and checking the reliability of our implementation. Finally, we highlight some aspects of Domain Theory, which inspires and models the essence of our approach to approximating real numbers, intervals and functions.

Types. The main types provided by the AERN library and their relationships within and outside AERN are outlined in Fig. 5. Specifically, there are two abstract types, one defining an algebraic structure of approximations to the real numbers and intervals and the other one defining a structure of approximations to continuous real functions and function intervals. The former is implemented by a floating-point interval arithmetic and the latter is implemented by a polynomial interval arithmetic.

Generalized Intervals. The two algebraic structures are closely linked to the continuous lattice of generalized real intervals. A generalized interval (sometimes called directed or modal interval) [Kau80] is a pair $[c, d]$ with no requirement

---

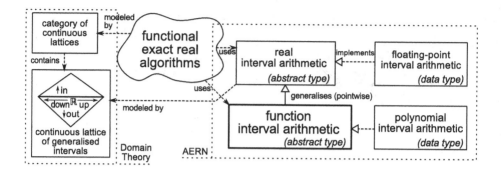

Fig. 5. Main abstract and concrete types in AERN

that $c \leq d$. If $c > d$, the interval has no canonical set interpretation. The lattice is ordered by the *refinement* relation, denoted $\sqsubseteq$ and defined as follows:

$$[a, b] \sqsubseteq [c, d] \iff (a \leq c \text{ and } d \leq b)$$

If $c \leq d$, then the relation $\sqsubseteq$ can be intuitively interpreted as "$[a, b]$ contains less information than $[c, d]$". The intuition is that the more points an approximating set has, the less it is saying about the location of the approximated object. We consider *inconsistent* intervals, *i. e.*, intervals $[c, d]$ where $c \not\leq d$ because an inner approximation of a normal interval can lead to the bounds of the interval crossing. This happens, for example, near point 0 in Fig. 2 and also when approximating π when proving inclusion (3).

Properties. The abstract types are formalized in the Haskell programming language using its type class feature. A type class is somewhat similar to an interface in object-oriented languages. It facilitates the specification of operations and their signatures. Moreover, AERN also specifies a comprehensive list of algebraic laws that each implementation of the type class should satisfy. For example, one of these laws is the *commutativity of addition*, modified as follows to suit approximate operations:

$$(x \mathbin{\langle +\rangle} y) \sqsubseteq (y \mathbin{\rangle + \langle} x)$$

where $\langle + \rangle$ and $\rangle + \langle$ produce outer and inner approximations of the exact sum, respectively. AERN randomly generates thousands of tests for dozens of such properties using the Haskell QuickCheck library, giving the implementation a very thorough check.

Domain Theory. One of the strongest features of interval-based frameworks is their solid theoretical foundation. In particular, we rely on the rich theory of *continuous lattices* and *continuous partial orders* [GHK+03], which possess useful order-theoretic, algebraic, and topological properties.

The types in Fig 5 implement an algebraic structure on the set of generalized real and function intervals. The operations are all isotonic with respect to the refinement relation $\sqsubseteq$, which facilitates reasoning about soundness and convergence of the resulting algorithm implementations.

One of the most important properties of continuous lattices is that they form a Cartesian-closed category. This means that the function spaces in the category are also continuous lattices and we can use the rich mathematical theory of such lattices over the function spaces as well.

For example, take the space of continuous generalized interval functions. To approximate elements of this space in software, a countable basis is required. The set $\mathcal{B}_{\mathrm{box}}$ of "box approximations"—i. e., piece-wise constant interval functions that have finitely many "steps" and rational coordinates—is a suitable basis for this space.

Nonetheless, from a practical point of view, the basis $\mathcal{B}_{\mathrm{box}}$ is far from ideal. For example, to enclose a linear function $y(x) = a_0 + a_1 x$ with accuracy $\leq 2^{-n}$, one generally needs $O(2^n)$ boxes, while a more succinct representation could be devised, such as the pair (a_0, a_1). A similar argument can be made using a quadratic function approximated by affine intervals except that one needs $O(\sqrt{2^n})$ intervals to get accuracy $\leq 2^{-n}$. The trade-off between space and accuracy improves with increasing polynomial degree.

The above observation suggests that polynomial approximations provide a more practical basis for enclosing functions than $\mathcal{B}_{\mathrm{box}}$. This view is also supported by the result that, using a specific polynomial approximation, the ODE IVP $y'(x) = f(x), y(0) = 0$ can be solved in polynomial time if f is a polynomial-time computable real function [MM93].

5   Technical Contribution

A number of challenges in developing AERN have been related to polynomial approximation of various operations. Perhaps surprisingly, the most complex operation to implement was multiplication. A multiplication of generalized intervals was first introduced by Warmus [War56], but the $\sqsubseteq$-isotonic version that we have adapted was given by Kaucher [Kau80]. The operation is defined in 16 cases, distinguished by the signs of the endpoints of the two interval operands. The challenge is that when the operands are function intervals, the sign of their endpoints may be changing at different places over the domain of their variables. Thus an arbitrary subset of the 16 cases can arise for one pair of functions. This problem is solved in AERN by merging the formulas for the results in all cases that cannot be ruled out using pointwise min and max. Two of Kaucher's 16 original cases also contain min and max in the formulas.

The challenge of implementing pointwise approximate min and max for polynomials has been addressed by a combination of Bernstein approximation and domain translation. The degree of Bernstein approximation used for min and max is one of the effort parameters for any expression that includes multiplication. The enclosures of the product in Fig. 2 have been computed with the use of pointwise min and max.

Almost all polynomial operations have the potential of exceeding the maximum degree and maximum term size limits. To approximate a polynomial by another with lower degree or fewer terms, some terms are carefully eliminated. The version of AERN used in PolyPaver uses the Chebyshev basis to reduce the loss of accuracy due to degree reductions. We plan to port this feature to the main AERN library.

Another challenge was implementing random generation of floating-point intervals and polynomial intervals required for randomized testing of algebraic properties. The generation in AERN produces a distribution of intervals that contains singletons, consistent non-singletons and anti-consistent non-singletons with equal probability. Moreover, special values such as 0 and 1 are generated with a relatively high probability.

References

[CS93] Comba, J.L.D., Stolfi, J.: Affine arithmetic and its applications to computer graphics. Presented at SIBGRAPI 1993, Recife, PE, Brazil (October 1993)

[DK14a] Darulova, E., Kuncak, V.: Sound compilation of reals. SIGPLAN Not. 49(1), 235–248 (2014)

[DK14b] Duracz, J., Konečný, M.: Polynomial function intervals for floating-point software verification. Annals of Mathematics and Artificial Intelligence 70, 351–398 (2014)

[EP07] Edalat, A., Pattinson, D.: A domain-theoretic account of Picard's theorem. LMS Journal of Computation and Mathematics 10, 83–118 (2007)

[GHK+03] Gierz, G., Hofmann, K.H., Keimel, K., Lawson, J.D., Mislove, M.W., Scott, D.S.: Continuous Lattices and Domains. Encycloedia of Mathematics and its Applications, vol. 93. Cambridge University Press (2003)

[Kau80] Kaucher, E.: Interval analysis in the extended interval space IR. Computing Suppl. 2, 33–49 (1980)

[KTD+13] Konečný, M., Taha, W., Duracz, J., Duracz, A., Ames, A.: Enclosing the behavior of a hybrid system up to and beyond a Zeno point, pp. 120–125. IEEE (2013)

[MB02] Makino, K., Berz, M.: New applications of Taylor model methods. In: Automatic Differentiation of Algorithms: From Simulation to Optimization, ch. 43, pp. 359–364. Springer (2002)

[MB09] Makino, K., Berz, M.: Rigorous integration of flows and odes using taylor models. In: Proceedings of the 2009 Conference on Symbolic Numeric Computation, SNC 2009, pp. 79–84. ACM, New York (2009)

[MM93] Müller, T.N., Moiske, B.: Solving initial value problems in polynomial time. In: Proc. 22 JAIIO - PANEL 1993, Part 2, pp. 283–293 (1993)

[Mül01] Müller, N.T.: The iRRAM: Exact arithmetic in C++. In: Blank, J., Brattka, V., Hertling, P. (eds.) CCA 2000. LNCS, vol. 2064, pp. 222–252. Springer, Heidelberg (2001)

[Tuc02] Tucker, W.: A rigorous ODE solver and Smale's 14th problem. In: Foundations of Computational Mathematics, pp. 53–117 (2002)

[War56] Warmus, M.: Calculus of approximations. Bull. Acad. Polon. Sci. Cl. III IV(5), 253–259 (1956)

Generating Optimized Sparse Matrix Vector Product over Finite Fields

Pascal Giorgi and Bastien Vialla

LIRMM, CNRS, Université Montpellier 2
{pascal.giorgi,bastien.vialla}@lirmm.fr

Abstract. Sparse Matrix Vector multiplication (SpMV) is one of the most important operation for exact sparse linear algebra. A lot of research has been done by the numerical community to provide efficient sparse matrix formats. However, when computing over finite fields, one need to deal with multi-precision values and more complex operations. In order to provide highly efficient SpMV kernel over finite field, we propose a code generation tool that uses heuristics to automatically choose the underlying matrix representation and the corresponding arithmetic.

Keywords: sparse linear algebra, finite fields, SpMV.

1 Introduction

Modern sparse linear algebra is fundamentally relying on iterative approaches such as Wiedemann or Lanczos. The main idea is to replace the direct manipulation of a sparse matrix with its Krylov subspace. In such approach, the cost is therefore dominated by the computation of the Krylov subspace, which is done by successive matrix-vector products. Let $A \in \mathcal{F}^{n \times n}$ be a sparse matrix with $O(n \log^{o(1)} n)$ non zero entries where $\mathcal{F}$ is a finite field. The matrix-vector product $y = Ax$ where $y, x \in \mathcal{F}^n$ costs $O(n \log^{o(1)} n)$ operations in $\mathcal{F}$. We call this operation SpMV in the rest of this paper. SpMV is a particular operation in the linear algebra framework, since it requires as much memory accesses as arithmetic operations. Basically, one entry in the matrix contributes to the SpMV computation only once. Therefore, on modern processor where memory hierarchy has a larger impact then arithmetical operations, data access is a major challenge to reach good performances. This challenge has been widely studied by the numerical community, and led to many different matrix storage for floating point numbers.

Over finite fields the situation is slightly different. Basically, the underlying arithmetic is more complex. Indeed, modern processor does not provide efficient support for modular operations. Furthermore, finite fields can be large and then requiring multiple precision arithmetic. Our main concern is weather the numerical formats are still satisfying when computing over finite fields and which arithmetic strategy is the most suited to the particularity of SpMV. This question has

---

[1] This work has been supported by the Agence Nationale pour la Recherche under Grants ANR-11-BS02-013 HPAC, ANR-12-BS02-001 CATREL.

H. Hong and C. Yap (Eds.): ICMS 2014, LNCS 8592, pp. 685–690, 2014.

been already addressed in many papers, as in [3,6], but no general optimization approach has been designed. In this paper, we propose a general framework which incorporates most of the optimization techniques for SpMV over finite field. We provide a software tool, available at www.lirmm.fr/~vialla/spmv.html, that emphasizes our approach for prime fields at any precision.

2   General Optimization Approach

It is well know that SpMV performance is limited by memory accesses. Indeed, the irregularity access of the $x[i]$'s during SpMV does not allow the processor to prefetch data to the cache memory in advance. In order to minimize cache misses, one need to minimize the memory footprint of the matrix while preserving a cache aware structure. To further speed-up SpMV over finite field, one need to minimize the number of clock cycle per arithmetic operation. This could be done by minimizing the modular reductions or taking care of particular entries in the matrix, i.e. ones and minus ones. One can evaluate *a priori* the impact of these optimizations by using the roofline model of [9].

Preprocessing the matrix in advance is a key tool to detect the most suited optimization. This can be done at runtime, as in the OSKI library [8]. Our proposed approach is to do this at compile time through two steps: 1) preprocessing the matrix to provide an optimization profile.; 2) generating an optimized SpMV with this profile.

3   Optimized SpMV Generator

The workflow of our generator is given in Figure 1. It receives a matrix A and a prime number p such that the SpMV with A is performed over $\mathbb{F}_p$. Depending on the prime p and some characteristics of A, such as the number of ± 1 or the dimension, the generator choose the best suited matrix format and an arithmetic strategy. Then, it generates an optimization profile that can be used to compile a SpMV implementation for A over $\mathbb{F}_p$.

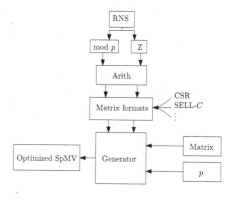

Fig. 1. Generator workflow

3.1   Matrix Formats

A lot of matrix format have been proposed by the numerical community for SpMV: e.g. CSR, COO, BlockCSR[8], Compress Sparse Block [4]. In general, matrices over finite fields do not have any structural properties than can be used to improve performances. Therefore, we choose to focus on the CSR and

SELL-C format [7], and some adaptation avoiding the storage of ± 1. However, our approach is generic and more format can be added if necessary.

The CSR format compress data according to the row indices. It needs three arrays: `val[]` for the matrix entries; `idx[]` for the column indices; `ptr[]` for the number of non-zero entries per row. The SELL-C format is a variant of CSR which is designed to incorporate SIMD operations. It sorts the rows according to their sparsity and split the matrix by chunk of size C. Each chunk is padded with zeros such that each row in a same chunck has exactly the same sparsity. The parameter C is a chosen to be a multiple of the SIMD unit's width. In order to minimize the memory footprint, our generator adapt the data types of every arrays, e.g. 4 bytes for `idx[]` when column dimension is $< 65\,536$. It also choose a data type related to $\|A\|_\infty$ rather than p for `val[]`.

3.2   Delayed Modular Reduction

As demonstrated in [5], performing modular reductions only when necessary leads to better performances. Hence, the computation is relaxed over the integer and needs that no overflow occurs.

Depending on the finite field, our generator will compute *a priori* the maximum value of k such that $k(p-1)\|A\|_\infty$ does not overflow. Note that knowing k at compile time will allow the compiler to perform loop unrolling. As delayed modular reduction is fundamentally tied with the underlying data type, our approach is to use the best suited one to reduce the number of modular reductions. Nevertheless, some compromises must be done between the cost of the standard operations $(+, \times)$ *vs* the number of reductions.

3.3   Hybrid

In most applications over finite fields, many matrix entries are ± 1. Thus, one can avoid superfluous multiplication within SpMV and further reduce the memory footprint of the matrix. This approach have been developed in [3] using a splitting of the matrix A in 3 matrices: A_1 storing only 1's , A_{-1} storing -1's and finally, A_λ store the rest of the entries. SpMV is then computed independently for each matrices and the results are sum up, i.e. $y = Ax = A_1 x + A_{-1} x + A_\lambda x$.

The drawback of this method is to amplify the number of cache misses arising during the reading of the vector x. Indeed, most of the matrices have a spatial locality in their row entries which is useful to avoid cache misses.

Our proposed hybrid approach is to keep this spatial locality such that SpMV still can be performed row by row. CSR format is well designed for this approach. Indeed, for each row we can store in `idx[]`, the column indices of the entries different ± 1, then the indices of ones and the minus ones. We can do exactly the same for `ptr[]`, and `val[]` only stores the entries different from ± 1. We call this hybrid format CSRHyb. Note this approach, cannot be directly applied to SELL-C since the zero padding may introduce too much memory overhead. To circumvent this, one must store entries different from ± 1 in SELL-C format and the ± 1 in CSR format, but these two formats must be interleaved by row.

4   Benchmarks

Our benchmarks have been done on matrices arising in mathematical applications. The Table 1 gives the characteristic of such matrices (available at http://hpac.imag.fr). In this table, nnz is the number of non zero entries, nnz_{row} is the average number of non zero entries in a row, and k_{max} is the maximum number of non zero entries in a row.

Table 1. List of matrices arising in mathematical applications

Name	Dimensions	nnz	nnz_{row}	k_{max}	± 1	$\neq \pm 1$	Problems
cis.mk8-8.b5	564 480 × 376 320	3.3M	6	6	3.3M	0	(A)
GL7d17	1 548 650 × 955 128	25M	16	69	25M	382K	(B)
GL7d19	1 911 130 × 1 955 309	37M	19	121	36M	491K	(B)
GL7d22	349 443 × 822 922	8.2M	23	403	7M	307K	(B)
M06-D9	1 274 688 × 1 395 840	9.2M	7	10	9.2M	0	(E)
rel9	5 921 785 × 274 669	23M	3	4	23M	19K	(C)
relat9	9 746 231 × 549 336	38M	3	4	23M	29K	(C)
wheel_ 601	902 103 × 723 605	2.1M	2	602	38M	29K	(F)
ffs619	653 365 × 653 365	65M	100	413	60M	5M	(D)
ffs809	3 602 667 × 3 602 667	360M	100	452	335M	25M	(D)

(A) Simplicial complexes from homology; (B) Differentials of the Voronoï complex of the perfect forms (C) Relations;(D) Function field sieve ; (E) Homology of the moduli space of smooth algebraic curves $M_{g,n}$; (F) Combinatorial optimization problems.

We used g++ 4.8.2 and an Intel bi-Xeon E5-2620, 16GB of RAM for our benchmarks. We performed a comparison with the best SpMV available at this time in the LinBox library[1] (rev 4901) based on the work of [3].

4.1   Prime Field $\mathbb{F}_p$ with Small p

In this section we consider the case where $(p-1)^2$ fits the mantissa of a double floating point number, e.g. 53 bits. In this case, the modular reduction is costly compare to the standard operations. However, it does not worth it to extend the precision beyond 53 bits to avoid most of the reductions. Our strategy is then to use **double** and to find the largest k such that $k||A||_\infty (p-1) < 2^{53}$ and perform reduction at least every k entries in a row. If the matrix does not have too many ± 1, the SELL-C format will be chosen to better exploit the SIMD vectorization, otherwise the CSRHyb format will be preferred.

The Figure 2 gives the relative performances of our optimized SpMV against the one of LinBox for the prime field $\mathbb{F}_{1048583}$. One can see that our code is always faster than the CSR implementation of LinBox, up to a speed-up of 2.2.

---

[1] www.linalg.org

Indeed, this can be explain by the fact that most matrices have a many ±1 entries and that the CSR of LinBox is not handling such particularity. The implementation of the hybrid format from [3] is not yet fully operational in LinBox and we did not get the chance yet to compare to it. However, following the speed-up of the hybrid format *vs* the CRS one given in [3, Figure 3], which is less than 1.5, we are confident in the performance of our optimized SpMV.

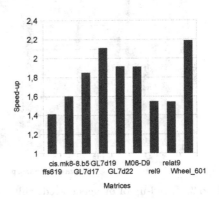

Fig. 2. Speed-up of our generated SpMV against LinBox over $\mathbb{F}_{1048583}$

4.2 Prime Field $\mathbb{F}_p$ with p Multiple Precision

Our motivations come primarily from the computation of discrete logarithms over finite fields [1]. We focus only on matrices which have small entries compare to the prime p, e.g. less than a machine word, since it is mostly the case in mathematical applications.

In order to compute with multiple precision integers, one can use the well known GMP library[2] which is the fastest one for each single arithmetic operations. However, when dealing with vectors of small integers, e.g. ≈ 1024 bits, the GMP representation through pointer makes it difficult to exploit cache locality. In such a case, one should prefer to use a fixed precision representation through a residue number system [2], called RNS for short. Such approach provides intrinsic data locality and parallelism which are good for SpMV and its SIMD vectorization. The difficulty is then transferred to the reduction modulo p that cannot be done in the RNS basis. However, one can use the explicit chinese remainder theorem [2,6] to provide a reduction that can use SIMD instructions. Furthermore, one can use matrix multiplication to perform modular reduction of a vector of RNS values and then better exploit data locality and SIMD.

In order to minimize the memory footprint of the matrix, we propose to store the matrix entries in `double` and convert them to RNS on the fly. The m_i's are chosen so that $(m_i - 1)^2$ fits in 53-bits to allow floating point SIMD in the RNS arithmetic. Larger m_i's could be chosen to reduce the RNS basis but this would induce a more complex SIMD vectorization which makes it harder to reach sustainable performances. In our multiple precision SpMV, the m_i's are chosen so that $k_{max}(m_i - 1)^2 < 2^{53}$ and $M = \prod_i m_i > (p-1) \cdot k_{max} \cdot ||A||_\infty$. This both ensures that the modular reduction mod m_i and mod p can be done only once per row. The matrix format is CSRHyb since the SIMD vectorization is done over the arithmetic rather then being on SpMV operations.

---

[2] https://gmplib.org/

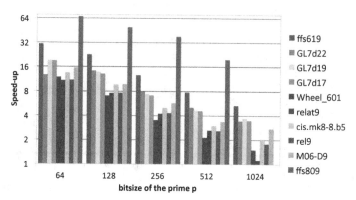

Fig. 3. Speed-up of our generated multiple precision SpMV over LinBox

The Figure 3 gives the relative performances of our optimized SpMV against the one of LinBox for prime fields of bitsize, 64, 128, 256, 512, and 1024. One can see that our code is always faster than LinBox one, up to a speed-up of 67. Indeed, this can be explain by the fact that most matrices have many ± 1 entries. But, mainly because LinBox stores matrix entries as GMP integers while our SpMV stores them as `double`. This has two consequences on LinBox SpMV performances. First, matrix entries are not store contiguously and then many cache misses are done. Secondly, LinBox use the arithmetic of GMP which is not using any SIMD vectorizations for mixed precision arithmetic.

References

1. Barbulescu, R., Bouvier, C., Detrey, J., Gaudry, P., Jeljeli, H., Thomé, E., Videau, M., Zimmermann, P.: Discrete logarithm in gF(2^{809}) with FFS. In: Krawczyk, H. (ed.) PKC 2014. LNCS, vol. 8383, pp. 221–238. Springer, Heidelberg (2014)
2. Bernstein, D.J.: Multidigit modular multiplication with the explicit Chinese remainder theorem (1995), http://cr.yp.to/papers/mmecrt.pdf
3. Boyer, B., Dumas, J.-G., Giorgi, P.: Exact sparse matrix-vector multiplication on gpu's and multicore architectures. In: Proc. of the 4th International Workshop on Parallel and Symbolic Computation, pp. 80–88 (July 2010)
4. Buluç, A., Williams, S., Oliker, L., Demmel, J.: Reduced-bandwidth multithreaded algorithms for sparse matrix-vector multiplication. In: Proc. of the 2011 IEEE International Parallel & Distributed Processing Symposium, pp. 721–733 (2011)
5. Dumas, J.-G., Giorgi, P., Pernet, C.: Dense Linear Algebra over Finite Fields: the FFLAS and FFPACK package. ACM Trans. Math. Soft. 35, 19:1–19:42 (2008)
6. Jeljeli, H.: Accelerating iterative spmv for discrete logarithm problem using gpus. hal-00734975 (2013), http://hal.inria.fr/hal-00734975/en/
7. Kreutzer, M., Hager, G., Wellein, G., Fehske, H., Bishop, A.-R.: A unified sparse matrix data format for modern processors with wide simd units. arXiv:1307.6209 (2013), http://arxiv.org/abs/1307.6209
8. Vuduc, R., Demmel, J.W., Yelick, K.A.: OSKI: A library of automatically tuned sparse matrix kernels. In: Proc. of SciDAC 2005, Journal of Physics: Conference Series, pp. 521–530 (June 2005)
9. Williams, S., Waterman, A., Patterson, D.: Roofline: An insightful visual performance model for multicore architectures. Commun. ACM 52(4), 65–76 (2009)

swMATH – An Information Service
for Mathematical Software

Gert-Martin Greuel[1] and Wolfram Sperber[2]

[1] Department of Mathematics, Centre for Computer Algebra,
University of Kaiserslautern, Postbox 3049, 67653 Kaiserslautern
greuel@mathematik.uni-kl.de
[2] FIZ Karlsruhe/Zentralblatt MATH, Franklinstr. 11, 10587 Berlin, Germany
wolfram@zentralblatt-math.org

Abstract. swMATH is a novel information service for mathematical
software. It offers open access to a comprehensive database with information on mathematical software and provides a systematic collection of
references and linking to software-relevant mathematical publications.

1   Introduction

Efficient knowledge services are the bridge for the dissemination and acceptance
of research results, hidden knowledge is a barrier for use. In view of the growing
number of publications the mathematics community became aware of the relevance of *information about publications* long time ago. The 'Jahrbuch über die
Fortschritte der Mathematik' (JFM) [1], founded in 1868, was the first mathematical information service. The JFM collected reviews of the mathematical
literature, and printed and disseminated them annually in special volumes. The
'Zentralblatt für Mathematik und ihre Grenzgebiete' [2] (and later the Mathematical Reviews) took over and speeded up this process until nowadays the
electronic databases zbMATH and MathSciNet [3] have replaced Zentralblatt
resp. Mathematical Reviews, still providing short reviews of published papers.
They are important and widely used tools for searching, browsing and filtering
information on mathematical publications, authors and journals. MathSciNet
and zbMATH together provide and develop the Mathematics Subject Classification (MSC) scheme for structuring the mathematical publications by their
topic.

But in the 21th century, the situation of mathematical knowledge is more complex and goes much beyond publication of scientific articles and books. 'Mathematics inside' this is valid today for all key technologies, not only in the sciences.
And mathematical software is the nexus. When David Hilbert says "The tool
implementing the mediation between theory and practice, between thought and
observation is mathematics", we can continue to say that mathematical software
is the connecting bridge between mathematical theories and concrete applications.

Thus, mathematical software has become an emerging domain of mathematical knowledge. However, in contrast to mathematical literature, no complete

H. Hong and C. Yap (Eds.): ICMS 2014, LNCS 8592, pp. 691–701, 2014.

database providing metadata or short reviews of mathematical software exists. In fact, developing an efficient information service for mathematical software is a challenging task. The existing portals or linked lists for mathematical software are not sufficient for several reasons: they do not aim to cover all existing software, either they provide a platform just for developers, or they are restricted to special mathematical topics, e.g., the portal for optimisation software Plato [4]. Information about software is often sparse and insufficient for the user. Moreover, in many cases these platforms are outdated after some time.

In this article we report on the new database swMATH [5] on mathematical software and describe the principles behind this novel kind of information service. A core feature of our approach is to systematically connect mathematical software with publications that either describe or use the software.

2   Mathematical Software

By mathematical software we understand computer programs that implement mathematical objects and relations and which are used to analyse, solve, simulate or model a mathematical problem. It usually relies on mathematical publications that provide methods and algorithms describing a concrete path to solve a mathematical problem. Software represents, often quite deep, mathematical knowledge in an active way, including automated mathematical reasoning and conclusions.

Therefore, mathematical software has its own characteristics that differs substantially from mathematical literature:

- Mathematical software has often experimental character.
- It is focused more on modelling, simulating and solving mathematical problems rather than on structural description of mathematical concepts and theories.
- Mathematical software is of dynamic nature and a 'living' object under permanent development by improvement and extension. It can 'die' and lose its usefulness, in contrast to mathematical publications which remain valid for ever.
- Implementation of mathematical software depends on the environmental features: hardware, operating system, programming language, interfaces, other software.
- Hence, mathematical software cannot be combined in a free way as this is valid for theoretical results. Therefore, the software must provide appropriate interfaces.
- Quality of mathematical software can't be simple evaluated as 'correct' or 'non-correct', it also depends on a lot of other factors such as performance, ease of use, operating system and programming language, and special features. The evaluation of the quality of software is a difficult and complex problem.
- Granularity of software varies from big general purpose systems or libraries to specialised and small packages.

- Typically, software modules are developed by groups and not by individuals such that the authorship is often not transparent or unknown.
- Some portals for mathematical software are well established in the Web. Each of these portals uses its own description scheme. No widely accepted standards for metadata and content analysis of mathematical software exist until now.

These characteristics show already that it is impossible to use similar methods for providing information about mathematical software as for mathematical literature. Moreover, some general remarks show the difficulties to find information about mathematical software.

- Direct information about software is often sparse.
- Many software packages do not have a homepage or a reasonable documentation.
- The information about a software is, in the same way as the software itself, under permanent change.
- The dependency of software from the technical environment leads to problems with evaluation, long-term availability, and reproducibility of results.
- Software can have internal structures, e.g., subroutines, function calls, etc. These can be important, but up to now, software information services don't use or don't have such internal information.

Major challenges that arise from this analysis are: to develop methods and tools for a (machine-based) content analysis of software and to develop a model for a standardised description of software as a basis for retrieval.

3   Content Analysis and Retrieval in the Digital Age

Models for a unified description and content analysis of digital objects were (and are) widely discussed, e.g., the Dublin Core Metadata Initiative (DCMI) [6]. The DCMI Initiative bases on the century-old methods and experiences of librarians in the organisation of traditional libraries. The Dublin Core (DC) metadata scheme for an object consists of an identifier and a simple set of elements such as creator, title, subject, date etc. (for more information see [7]). It can be easily extended and adapted to other information objects corresponding to the requirements of specific user communities. Thus, more specific information about a document going beyond the document level of library catalogs can be given.

Moreover, Semantic Web technologies, e.g., the Resource Description Framework (RDF), provide methods and tools to encode the information in machine-understandable way. Also description models for software are under development, like the SoftWare Ontology Project (SWOP) [8]. But up to now, a widely accepted model for the description of software is missing.

Therefore, we had to develop an own description model for the database swMATH. The following list seems to cover most *aspects of interesting information about a specific software package:*

- bibliographic and content information about the software: title, author(s), mathematical subjects, keywords, classification, documentation, homepage
- technical parameters of the software: hardware specification, operating system, programming language
- versioning of the software
- licensing and usability conditions of the software
- dependencies from and to other software
- references to related publications
- references to similar software
- applications for which the software is used
- special features of the software
- acceptance and use of the software

A database as an information service for mathematical software in general should fulfil the following *requirements*:

- *'weak' completeness* (an information service of mathematical software should cover the currently most important mathematical software)
- *structured* (a well-organized and structured collection of mathematical software)
- *usefulness* (relevant information about objects of type 'mathematical software')
- *flexibility* (the description scheme has to be extensible)
- *maintainability* (it should be mainly maintained automatically)

The needs and expectations of the user community to information services and retrieval functionalities are permanently increasing. Users need efficient retrieval tools for the growing flood of publications. Therefore, content analysis of objects must be permanently improved and extended. In the case of mathematical software, we face the special problem that the desired information is often hard to find or even not available. But even if the information is somewhere available, it must be found, collected and put into a uniform format.

Our approach, described below, tried to meet the above requirements and the expectations of the user community. However, sometimes these expectations are unrealistic because the users are either unaware on the inherent difficulties, or because they are used to the powerful and sophisticated tools offered by the big search engines like Google with almost unlimited financial and personal resources. Nevertheless we believe that the database swMATH provides a service with information on mathematical software which is otherwise, including the big search engines, hard to find.

4   A Publication-Based Approach

The aim of our initiative was not to create a repository of mathematical software but to collect information on mathematical software with useful searching, browsing and filtering functionalities.

Our idea was, to systematically analyse the links between a software and all available publications that describe, use, or cite the software. Of course, the problem arises, how to get access to the relevant mathematical publications. A worldwide repository of all mathematical publications does not exist. Only a small percentage of the publications is Open Access and freely digitally available, but scattered on various home pages or preprint servers in all possible kinds of formats. Already this shows that an Internet search, trying to continuously find and use the publications relevant to mathematical software is hopeless.

Therefore, we chose an approach based on the reviewing and abstracting database zbMATH for mathematical publications. zbMATH lists basically all peer-reviewed mathematical publications since 1868. Of course, for mathematical software only the last, say, twenty five years are relevant. Although zbMATH does not contain the full texts of the publications, it contains in many cases enough information on a publication that allows to identify relevant software. A reference to a software in the bibliographic data means that the software is essential for the publication. We use the software references in publications indexed in zbMATH in two ways: for detecting software and also for describing the software. From the publications, reviews or abstracts we use the classification code, key words and phrases and also (partially) information about authors, as information about the software.

4.1   Advantage of the Publication-Based Approach

This approach has several advantages. Today, more and more mathematical knowledge results from the usage of mathematical software. Mathematical theory and software are combined to solve or analyse mathematical objects, e.g., numerical methods for partial differential equations, symbolic methods for structural knowledge about algebraic or geometric objects, statistical methods for random data, but also formal methods for theorem proving. This is reflected in the database zbMATH through the permanently increasing number of references in publications to software. Moreover, as the database zbMATH is continuously growing and always up to date, the provided data referring to software are also continuously growing and up to date.

Another advantage of our approach is, that the required peer-reviewing for publications indexed in zbMATH is an (indirect) quality filter for the used software. The number of publications referencing a software is also an indicator for the acceptance and usefulness of a software. Direct methods for evaluating the quality of a software are under development, but they are difficult. For example, the guidelines for peer reviewing of contributions to the journal 'Mathematical Programming Computation' [9] require also the evaluation of the software' [10] but these are currently yet in an experimental stage. In any case, direct full and fair evaluation is almost impossible for just one large software system, and it is completely out of range to do this for several thousand software packages.

swMATH is mainly targeted to full software packages or software libraries (in the following called modules), but not to single routines, as, e.g., in the netlib [11]. But if a mathematical software is referenced in zbMATH, the software is also listed in swMATH.

Large software systems like 'Maple', 'Mathematica', 'Matlab' or 'R' are not a monolith but have an internal structure, like special libraries for different mathematical disciplines and tasks. Such sub-libraries are also listed in zbMATH but up to now without explicitly declaring hierarchical structures.

The general approach which finally led to the existing swMATH service consists of several steps:

1. Using mathematical publications for identifying mathematical software.
2. Extracting and analysing the existing information in publications about the software.
3. Searching for further information about the software (e.g., the homepage of the software).

4.2   Detecting Software

Software names are heterogeneous, many software packages have artificial names as 'BayesTree' or 'r2d2lri': such artificial tokens are easy to identify and they are identified in the first step. But software names referring to famous persons like 'KANT' or 'EDISON' or to mathematical terms like 'SINGULAR' or 'ellipse' or to not mathematic-specifics terms like 'race' or 'Sage' are ambiguous and its meaning is context-sensitive. Hence, some heuristic methods had to be developed for a term-based search by characteristic tokens for software such as 'package' or 'module' in the zbMATH data. We searched especially in the fields 'reviews or abstracts', 'title', 'keywords' and increasingly also in the field 'references' of a publication. The resulting candidates had then to be checked manually.

Up to now, more than 6,600 mathematical software packages could be identified. The list of identified software is continuously updated following the daily update of zbMATH. Typically, the number of software modules is increasing but sometimes the number decreases by removing falsely identified software or duplicates.

4.3   Software References in Publications

As said above, software references in publications are a central method of our approach. Heuristic methods search daily for references to software in the zb-MATH data. At the moment, more than 68.000 publications containing software references are identified. Note that publications with references to software have different roles, depending of the type of the publication.

4.4   Standard Publications and User Publications

If a software itself is the subject of a publication we call it 'standard publication' (for this software). If a software is just used and applied in a publication, we call the article a 'user publication'. The type of a publication (standard publication or user publication) is relevant for the metadata extraction.

4.5   Content Metadata

Standard Publications: The bibliographic information of a standard publication is nearly optimal for a high-quality description of the content of a software. Review or abstracts and key phrases of standard publications can often be used directly for the description of the referenced software. Also, the authors of standard publications are useful, at least as contact persons. The MSC code of the publication is not optimal for the classification of the software but it is helpful to characterise the mathematical subject. So, also the MSC codes are collected.

User Publications: User publications provide valuable additional information about a software, especially links to theoretical foundations and links to application areas. In other words, user publications provide important information for embedding the software in a broader context. Key phrases and MSC codes of the user publications are also integrated into the content analysis.

4.6   Homepages

Homepages of a software are an important source for more information about the software. They often cover detailed technical parameters, license and usability conditions and available versions (e.g. actual and out-to-dated versions or versions depending on license conditions). But content, structure and form of the information on the homepages are very heterogeneous. If a software is identified, we know at least the name of the software. The name, also in combination with other terms (especially the term 'software'), is the starting point to search and identify the homepage (if it exists) in the Web. Unfortunately, the heterogeneity makes an automatic analysis difficult. Up to now, the information from homepages is only partially used in zbMATH.

4.7   Remarks

1. *Support*: swMATH is planned as a community-driven open-access service. It was developed in a joint project by Oberwolfach, FIZ Karlsruhe and in cooperation with the Berlin Research Center of Mathematics MATHEON, ZIB Berlin, WIAS Berlin and the Felix-Klein Center for Mathematics in Kaiserslautern from 2012 to 2014. The project was funded by the Leibniz-Association for a period of three years. After the end of project, the service is continued by FIZ Karlsruhe.
2. *zbMATH Database*: All data in zbMATH are stored and maintained in a relational database. No indexed item will ever be removed.
3. *Persistent Identifiers*: Each software item in swMATH has a unique identifier. This could be used for referencing.
4. *Out-of-Date Software and Actuality*: The homepages of the software are checked periodically. Principally, out-of-date software can be removed from the swMATH Web site.

5. *Benefits*: Anyone interested in information about a specific software or in finding a software which helps to solve a problem benefits in an obvious way. But also software developers benefit, in particular form the list of publications using their software. They might be unaware of some of them and the context of the usage gives them interesting feedback. Moreover, the list is an obvious and independent indication of the usefulness of the software and helps the developers to get their work acknowledged. It is well known that software developers in mathematics do often not receive the credits they deserve. It is hoped that the visibility in swMATH may help to improve the situation.

6. *Social Networking*: A living service like swMATH needs the support of the mathematical community. We invite the community to help us to maintain and improve the swMATH service by information about missing software or to send us corrections and additional information about a software. Software developers and providers of specialised portals for mathematical software are specially invited to help to improve swMATH. As an example, the portal Oberwolfach References to Mathematical Software (ORMS) is linked to swMATH. ORMS provides manually-curated high-quality information of selected software modules and provides additional information going beyond swMATH. On the other hand, swMATH extends ORMS by information about related publications, which is not contained in ORMS.

7. *zbMATH-Links to Mathematical Software*: The publication items in zbMATH referencing mathematical software are linked with the corresponding items in swMATH. In this way swMATH enhances the information fo articles in zbMATH with reference to a software in a substantial way.

8. *Further Data Bases*: Most swMATH items result from the analysis of the zbMATH database. Besides zbMATH also the databases ioport [12] and MathEduc [13] are used.

5   The Web Site of swMATH

It was the intention of swMATH to provide a general, simple, and easy-to-use information service about mathematical software.

The swMATH database has a clear structure: the homepage with the basic searching functionality. A search leads to a result list with short information. Clicking on a software on the list leads to the Web page with detailed information of the software. In the following, the swMATH Web site is discussed in detail.

5.1   swMATH Homepage

The swMATH homepage is a search interface with a search field for simple search (full-text search about all information in swMATH) and buttons for advanced search and browsing. Further buttons lead to useful information of the service, especially for feedback, help, and links to the swMATH providers and further mathematical information services.

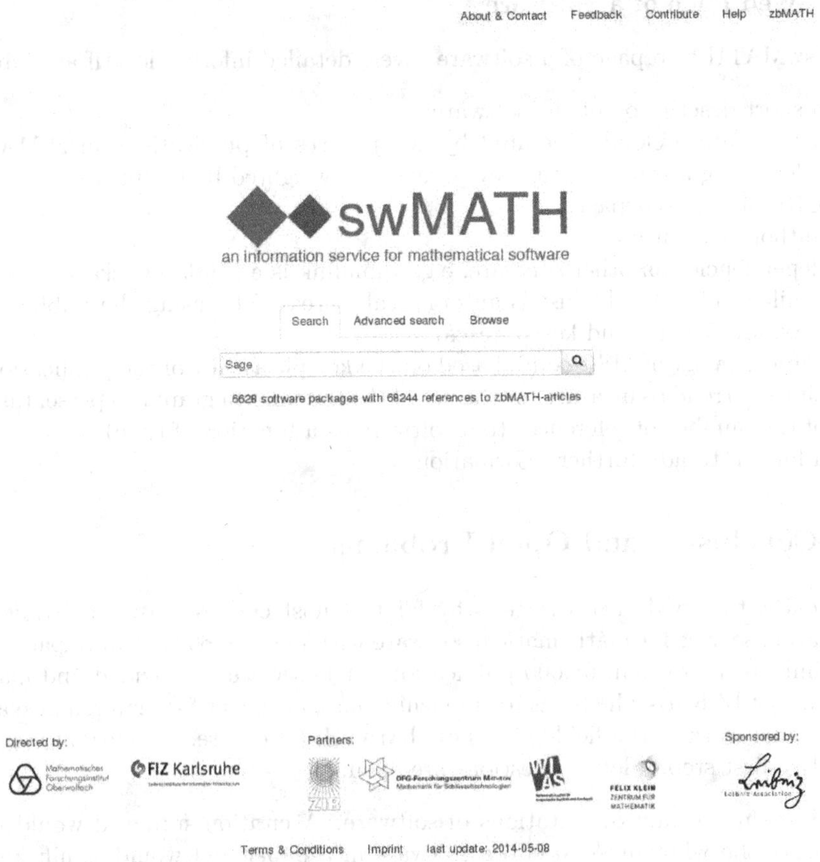

About & Contact   Feedback   Contribute   Help   zbMATH

◆◆ swMATH

an information service for mathematical software

Search   Advanced search   Browse

Sage

8628 software packages with 68,244 references to zbMATH-articles

Directed by: Partners: Sponsored by:

◆ FIZ Karlsruhe

Terms & Conditions   Imprint   last update: 2014-05-08

Fig. 1. The homepage of swMATH, searching for 'Sage'

5.2   Search Result: The Result List

As usually, a list of software packages related to " Sage" will be presented, containing the identified software and short description of it. In more detail, the results cover

- the name of the software
- the number of papers referencing the software
- a (persistent) swMATH identifier of the form sw????? , where ? stands for a digit
- a snippet of the description text of the software

Search terms are emphasized in the result list written in 'bold'.

Ranking: The list can be sorted both alphabetically or by relevance. The relevance ranking is based on different factors, especially on the number of publications referencing a software and key phrases. Hence, large software packages will be usually listed on the top of the list.

5.3 Web Page of a Software

The swMATH web page of a software covers detailed information (if available):

- a short description of the software
- a key phrase cloud generated by key phrases of publications in zbMATH referencing a software; the key phrases are weighted by frequencies
- URL of the homepage
- authors / creators
- dependencies on other software, e.g., Simulink is a Matlab package
- similar software (this list is automatically created by using the publication lists, MSC codes and key phrases)
- subject (a list of MSC codes based on the key phrase list of the publications)
- citation trend as measure for the use of the software (a graphical presentation of the number of references to a software as a function of time)
- a button to add further information

6 Conclusion and Open Problems

Currently, the swMATH service is by far the most comprehensive existing information service for mathematical software with about 6.600 software packages and links to more than 68.000 publications. The service is provided and maintained by FIZ Karlsruhe by using a machine-based concept for maintenance and by manual work of the field-editors of zbMATH. Of course, the current service is only a first step, a lot of questions are open:

- How can we improve citations of software? A citation standard would improve the willingness to cite a software in a paper and would significantly improve the results of our publication-based approach.
- How can we get more relevant data of the software, e.g., license conditions?
- Which features should be used as quality parameters?
- What is the long-term information of software?
- How should we present applications of a software?
- What is with other types of mathematical software, e.g., benchmarks, test-data collections, mathematical models, visualizations, etc.?
- ...

Also, the heuristic concepts used in swMATH need further development. Nevertheless, we regard swMATH as a useful service to the mathematical community to get information about mathematical software and also as a contribution to strengthen the area of mathematical software by developing a suitable information infrastructure.

With this paper we like to initiate a discussion about the concept, methods, and usability of the swMATH service. You are invited to use the swMATH service and send us your feedback and comments. Enjoy!

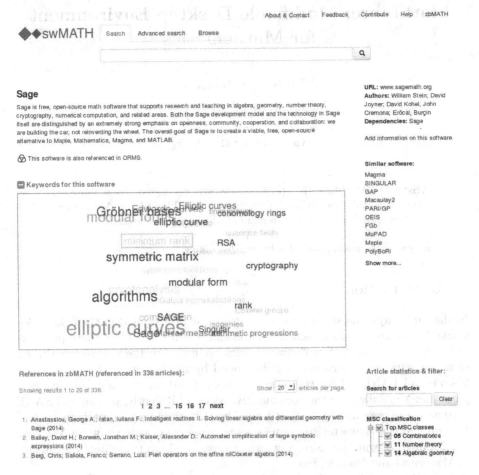

Fig. 2. A cutout of the Web page of the software 'Sage'

References

1. The database JFM, http://www.emis.de/MATH/JFM/
2. The database zbMATH, http://www.zentralblatt-math.org/zbmath/
3. The database MathSciNet, http://www.ams.org/mathscinet/
4. The Plato portal for optimisation software, http://plato.la.asu.edu/bench.html
5. The database swMATH, http://www.swmath.org
6. The Dublin Core Metadata Initiative, http://dublincore.org/
7. The DCMI terms, http://dublincore.org/documents/dcmi-terms/
8. The SoftWare Ontology Project, http://softwareontology.wordpress.com/
9. The journal Mathematical Programming Computation, http://mpc.zib.de
10. Reviewing in Mathematics Computation, http://mpc.zib.de/index.php/MPC/about/editorialPolicies#peerReviewProcess
11. The software repository netlib, http://www.netlib.org/
12. The database for computer science, http://www.zentralblatt-math.org/ioport/
13. The database for mathematical education,
    http://www.zentralblatt-math.org/matheduc/

MathLibre: Modifiable Desktop Environment for Mathematics

Tatsuyoshi Hamada

Fukuoka University, Japan
hamada@holst.sm.fukuoka-u.ac.jp
http://www.mathlibre.org/

Abstract. MathLibre is a project to archive open source mathematical software and documents and offer them with a Live Linux. Anyone can build modified and localized version of MathLibre, very easily.

Keywords: mathematical software, live linux.

1 Introduction

In the early days, the symbolic computation was investigated in the field of Artificial Intelligence and Physics. Macsyma and Reduce were the famous products of the first generation of computer algebra systems. For the last 20 years, a lot of mathematical research systems developed by mathematicians. For example, KANT and PARI/GP for the number theory, GAP for the finite groups, Singular and Macaulay2 for the commutative algebra. Recently, we can find SAGE system, it is a very famous project leaded by the community of mathematicians. We can find the characteristic properties that many systems are published with open source software licenses. Now, we can use Maxi-ma (the direct descendant of Macsyma) and Reduce, freely.

In mathematical investigations, these professional systems has become more important, but installing them are bothersome for many people. We want to introduce these systems for our colleagues and students without non-essential troubles.

2 MathLibre Is a Bootable Linux

MathLibre[1] is a project to archive open source mathematical software and documents. It's a direct descendant of *KNOPPIX/Math* (cf. [1]). These are collaborative works with OpenXM (cf. [2]). KNOPPIX/Math project began in February 2003. After that, every spring, we distributed 1000 pieces of CD/DVD of KNOPPIX/Math in the annual meeting of Mathematical Society of Japan.

In ASCM2005 in Seoul, we introduced KNOPPIX/Math Korean edition. We changed the project name in March 2012, and distributed MathLibre at ICME-12 in Seoul and AMC2013 in Busan. The current product is MathLibre 2014.

---

[1] http://www.mathlibre.org/

H. Hong and C. Yap (Eds.): ICMS 2014, LNCS 8592, pp. 702–705, 2014.

It's supporting the virtual machine, USB bootable stick and hard disk installation. KNOPPIX is a Live Linux, developed by Klaus Knopper in Germany, KNOPPIX is a branch of a distribution, Debian GNU/Linux. Debian always has some releases in active maintaining: "stable", "testing", "unstable", and the alpha testing version: "experimental". KNOPPIX is a mixed distribution of these four releases, so it's little bit complicated of customizing and upgrading. So, we changed the base to another live distribution, it's *Live Systems Project*[2]. Live Systems Project was used for the official Debian Live images. It's developed by Daniel Baumann and Live Systems Team.

We created the easy mechanism of language setting, so we can freely distribute the localized system for your colleagues and students. In the last years, we were making the localized version of MathLibre, and introduced these systems in some congresses held in China, Hungary, Japan, Korea and Taiwan. We have experiences of distributing over 10000 pieces of DVDs in these years. We will describe the practices of MathLibre and how to modify this system.

3   Practices of Distributing Mathematical Software

We participated ICM2006 Madrid in Spain and ICM2010 Hyderabad in India. We had exhibition booths of "mathsoftware.org" in these congresses, and we will have an exhibition booth in ICM2014 Seoul. In India, we prepared 1000 pieces of DVDs for participants, but, we had finished to distribute all DVDs in the first three days. In India, Linux system is popular, so KNOPPIX/Math was very attractive for them. When we stayed in Hyderabad, we had new colleagues in India. Now, we have a collaborative work with MTTS. MTTS:Mathematics Training and Talent Search programme is a national level four weeks intensive summer training programmed in mathematics and has been running since 1993 in India.

4   How to Use MathLibre

MathLibre is a bootable Live Linux, after successfully making and rebooting the MathLibre DVD, we can find the boot menu (Fig. 1). Press the Enter key and, after a display of the boot-sequence messages, the desktop environment of MathLibre will be displayed (Fig. 2). In some cases, it will reboot Windows; if this is the case, the BIOS settings need to be reconfigured. When a PC is rebooted, the message "BIOS Setup" is briefly displayed. After pressing the correct function key, usually <F2> or <F8>, we can find the "Boot" menu in the BIOS configuration. By changing the order of booting, we can boot from a DVD or a USB storage device.

MathLibre includes over 100 mathematical software packages, such as CoCoA, GAP, GeoGebra, gfan, KSEG, Macaulay2, Maxima, Octave, OpenXM, PARI/GP, Polymake, R, Risa/Asir, Singular, SAGE, and others. Select the

---

[2] http://live.debian.net/

Math menu from the start menu at the bottom left-hand side of the screen, as you would do for Windows Start. Alternatively, double click the "Math Software" icon; there is a collection of start-up icons for mathematical software and a "MathLibre Start" button, which leads to an HTML file that contains short introductions and links to the developers of the various software packages.

We can use DVD as an install media for Debian GNU/Linux. When you boot MathLibre, you can find menus of "Live" and "Install". Selecting the menu "Install", you can install a Linux system with many mathematical systems with using whole area of your hard disk drive. If you want to use MathLibre with Windows or Mac OS simultaneously, we recommend to use a virtual machine, for example, Oracle VirtualBox, VMware Player for Windows or VMware Fusion for Mac OS. MathLibre includes Windows Application InfraRecorder and VirtualBox in DVD. Both of them are open source software. InfraRecorder is developed by Christian Kindahl, and VirtualBox was created by innotek GmbH, purchased in 2008 by Sun Microsystems, and now developed by Oracle cooperation. We can create a disk copy with InfraRecorder to use the boot image with VirtualBox. In DVD you can find the setting file for VirtualBox in vbox directory. You can easily make your environment for daily use.

Another method of using MathLibre, we can create bootable USB stick. We made a short experimental script "mkusbmath", when you boot with MathLibre and input the command with sudo, you will get bootable USB stick with persistent home directory. In any cases, you can install additional software packages on it, because it's using *aufs*, the union file system.

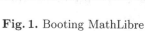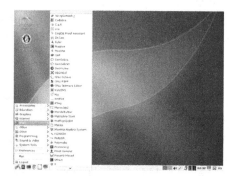

Fig. 1. Booting MathLibre   **Fig. 2.** MathLibre Desktop

5   How to Customize MathLibre

For customizing, we need Debian GNU/Linux for building MathLibre, of course, you can use MathLibre as an alternative. For building, you need "live-build" and "git" package. Making a clone repository from GitHub, and make it, you will build an ISO hybrid image for DVD and USB stick.

```
% git clone https://github.com/knxm/mathlibre
% cd mathlibre
% make
```

Live Systems Project is well developed and maintained project, but it's not supporting easy localization, we created some sample settings. If you want to localized version of MathLibre, for example you would like to build Korean language edition, please input the following:

```
% make lang=ko
```

You can find more configuration files in the directory "`lang`". If you can't find the setting file your want, you can make new one based with other setting files.

The list of packages are in "`config/package-lists`". You can add or remove package name in the files `*.list.chroot`. If you want to some additional software, but it's not included in the Debian official packages, the easiest way is to use "`config/includes.chroot`". For example, SAGE is a very huge system, we are putting **sage** directory in `config/includes.chroot/usr/local`. If you would like to know the details of customizing, please refer [3].

6 Conclusion

We can find similar projects, *Sage Debian Live*[3], it's using the same base system, but the main target is USB stick only. And *lmonade*[4] is a light-weight meta distribution which does not need administration mode. It's building with Gentoo Prefix. And *Mathemagix*[5] has a plan to provide a new high level language, it's an ambitious project.

MathLibre is supporting various environments, Live DVD and USB, virtual machine, hard disk installation, and some language settings. Considering the portability and customizability, MathLibre is meaningful for researchers, educators and students. Especially, even if it does not have the Internet connections, anyone can explore the world of mathematics and mathematical software systems.

References

1. Hamada, T., Suzaki, K., Iijima, K., Shikoda, A.: KNOPPIX/Math: Portable and Distributable Collection of Mathematical Software and Free Documents. In: Iglesias, A., Takayama, N. (eds.) ICMS 2006. LNCS, vol. 4151, pp. 385–390. Springer, Heidelberg (2006)
2. Noro, M., Ohara, K., Takayama, N.: OpenXM: Open message eXchange for Mathematics, http://www.math.sci.kobe-u.ac.jp/OpenXM/
3. Debian Live Manual, http://live.debian.net/manual/stable/

---

[3] http://sagedebianlive.metelu.net/

[4] http://www.lmona.de/

[5] http://www.mathemagix.org/

Software Packages
for Holonomic Gradient Method

Tamio Koyama[1], Hiromasa Nakayama[2], Katsuyoshi Ohara[3], Tomonari Sei[4],
and Nobuki Takayama[5]

[1] Department of Mathematical Informatics, Graduate School of Information Science
and Technology, University of Tokyo, JSPS Research Fellow, Japan
tkoyama@stat.t.u-tokyo.ac.jp
http://www.stat.t.u-tokyo.ac.jp/~tkoyama/
[2] Department of Mathematics, Tokai University, Japan
nakayama@tokai-u.jp
http://sm.u-tokai.ac.jp/~nakayama/
[3] Faculty of Mathematics and Physics, Kanazawa University, Japan
ohara@air.s.kanazawa-u.ac.jp
http://air.s.kanazawa-u.ac.jp/~ohara/
[4] Department of Mathematics, Keio University , Japan
sei@math.keio.ac.jp
http://www.math.keio.ac.jp/~sei/
[5] Department of Mathematics, Kobe University, Japan
takayama@math.kobe-u.ac.jp
http://www.math.kobe-u.ac.jp/~taka/

Abstract. We present software packages for the holonomic gradient
method (HGM). These packages compute normalizing constants and the
probabilities of some regions. While many algorithms which compute integrals over high-dimensional regions utilize the Monte-Carlo method,
our HGM utilizes algorithms for solving ordinary differential equations
such as the Runge-Kutta-Fehlberg method. As a result, our HGM can
evaluate many integrals with a high degree of accuracy and moderate
computational time. The source code of our packages is distributed on
our web page [12].

Keywords: holonomic gradient method, normalizing constant, region
probability, Bingham prior, R project.

1 Introduction

The numerical evaluation of the normalizing constant for a given statistical distribution is a fundamental problem in statistics. For example, the normalizing
constant of the Gaussian distribution is expressed in terms of a rational expression of a parameter of the distribution named the standard deviation. However,
normalizing constants of many interesting statistical distributions do not have
such closed expressions.

H. Hong and C. Yap (Eds.): ICMS 2014, LNCS 8592, pp. 706–712, 2014.

The *holonomic gradient method*, HGM in short, is a general method to evaluate normalizing constant numerically for several parameters in the framework of Zeilberger's holonomic systems approach [11]. In fact, broad classes of normalizing constants are holonomic functions with respect to parameters. Then, such normalizing constants satisfy holonomic systems of linear partial differential equations.

The HGM consists of three steps for a given normalizing constant. (1) Finding a holonomic system satisfied by the normalizing constant. We may use computational or theoretical methods to find it. Gröbner basis and related methods are used. (2) Finding an initial value vector for the holonomic system. This is equivalent to evaluating the normalizing constant and its derivatives at a point. This step is usually performed by a series expansion. (3) Solving the holonomic system numerically. We utilize several methods in numerical analysis such as the Runge-Kutta method of solving ordinary differential equations and solvers of systems of linear equations.

The HGM was proposed in 2011 by a group of people including us [6] and has given several new results. For example, the orthant probability is the normalizing constant of the multivariate normal distribution restricted to the first orthant. The HGM can evaluate it in a high accuracy up to the 20 dimensional case when the mean vector is near the origin. In the 20 dimensional case, we numerically solve an ordinary differential equation of rank $2^{20} = 20,148,576$.

We have developed software packages for the HGM. Packages based on computer algebra systems help us to solve steps (1) and (2). We have implemented the step (3) for the Fisher-Bingham distribution, the Bingham distribution, the orthant probability, the Fisher distribution on SO(3), some of A-distributions, and the distribution function of the largest root of a Wishart matrix in the language C and/or in the system for statistics R [7]. An implementation for the polyhedral probability is a project in progress. We find an interesting interplay with systems for polytopes in the project. Further references and current implementations are listed in [12].

This paper is dedicated to Kenta Nishiyama in our memory.

2   Distributions and Algorithms

We give a brief discussion on the Bingham distribution and the orthant probability in view of the HGM in this section. As to the Fisher distributions on $SO(3)$, the largest roots of Wishart matrices, Fisher-Bingham distributions, and A-distributions, we refer to papers in [12].

2.1   Bingham Distribution

The Bingham distribution is a probability distribution on the (p-1)-dimensional sphere defined as

$$\frac{1}{Z(\Sigma)} \exp(x^\top \Sigma^{-1} x) \mu(dx) \quad (x \in S^{p-1})$$

where Σ is a p times p positive definite matrix, $\mu(dx)$ is the uniform measure on the sphere, and $Z(\Sigma)$ is the normalizing constant ($Z(\Sigma)$ is also denoted by $c(\Sigma)$ in literatures). We denote by $x^\top$ the transpose of x. We can assume without loss of generality that the matrix Σ^{-1} is a diagonal matrix such as $\mathrm{diag}(\theta_1, \ldots, \theta_{p-1}, 0)^\top$.

In [9], the HGM for this normalizing constant is discussed, and an explicit form of a Pfaffian equation associated with the normalizing constant has been given. The size of the matrix in the Pfaffian equation is p. In the current implementation, we evaluate the initial value for the HGM by a series expansion. The complexity to evaluate it is proportional to the number of the terms in the truncated series. Hence, the complexity is $O(p^N)$. Here, we denote by N the degree of the truncated series. Thus, the computational complexity of the HGM for this problem is estimated as

Theorem 1. *The complexity of the series expansion method and the HGM for the normalizing constant of the Bingham distribution on the $(p-1)$-dimensional sphere is bounded by*

$$O(p^N) + O(p^2) \times \text{(steps of the Runge-Kutta method)}.$$

Note that the holonomic system for the normalizing constant of the Bingham distribution and it's holonomic rank are not determined rigorously. There might exist a smaller system than that in [9]. Thus, the complexity in the above theorem is the upper bound of the complexity. We conjecture that the above complexity gives the lower bound of the complexity of the HGM for the Bingham distributions.

We provide a package of the HGM for the system for statistics R [7]. The function `hgm.ncBingham(th, ...)` in our R package `hgm` performs the HGM for Bingham distributions with the deSolve package. The initial value for the HGM is computed by the power series expansion. This function also computes derivatives of the normalizing constant of the Bingham distribution at any specified point. The variable `th` is a $(p-1)$-dimensional vector which specifies the first $(p-1)$ components of the parameter vector of the Bingham distribution on the $(p-1)$-dimensional sphere. The p-th parameter is assumed to be zero.

For $\Sigma^{-1} = \mathrm{diag}(1,3,5,0)$, we can obtain the normalizing constant as

```
hgm.ncBingham(c(1,3,5))
```

after loading the package with the command `library('hgm')`

2.2   The Orthant Probability

The orthant probability is the probability with which the random vector, which is normally distributed with the mean vector μ and the covariance matrix Σ, falls in the first orthant, and it can be written as

$$\int_0^\infty \cdots \int_0^\infty \frac{1}{(2\pi)^{d/2}|\Sigma|^{1/2}} \exp\left(-\frac{1}{2}(x-\mu)^\top \Sigma^{-1}(x-\mu)\right) dx_1 \ldots dx_d.$$

where we denote by d the dimension.

In [3], the HGM for the orthant probability is discussed, and an explicit form of a holonomic system and a Pfaffian equation associated with the probability is given. The holonomic rank of the system, which equals to the size of the Pfaffian equation, is 2^d. The initial value of the HGM for the orthant probability can be given exactly at a point and the computational complexity of the evaluation of the initial value is $O(1)$. The following complexity statement is an easy consequence of Theorem 15 in [3], but it is fundamental.

Theorem 2. *The complexity of evaluating the d-dimensional orthant probability by the HGM is*

$$O(2^{2d}) \times \text{(steps of the Runge-Kutta method)}.$$

The function hgm.ncorthant(sigma, mu, ...) in our R package evaluates the orthant probability by the HGM. The first variable sigma is the covariance matrix, and the second variable mu is the mean vector. This function calls a program written by the language C internally, which solves an ordinary differential equation with rank 2^d by a routine in the GNU scientific library (GSL) [2].

For example, when

$$d = 2, \quad \Sigma = \begin{pmatrix} 1 & 1/2 \\ 1/2 & 1 \end{pmatrix}, \quad \mu = \begin{pmatrix} 1 \\ 2 \end{pmatrix},$$

the orthant probability can be computed by the following script of R:

```
sigma <- matrix(c(1, 0.5, 0.5, 1), nrow =2)
mu <- c(1,2)
hgm.ncorthant(sigma,mu)
```

In this example, the rank of the ordinary differential equation equals to $2^2 = 4$. The performance of our implementation for larger d will be illustrated in 3.3.

3   Implementations

3.1   Building Blocks of Our Package

Our algorithms for the holonomic gradient method require efficient and reliable numerical implementations of the Runge-Kutta method and solving numerically systems of linear equations. Our package uses the GSL [2], the deSolve package in R, BLAS, and LAPACK for this purpose.

Most of our algorithms are implemented in the language C. We provide two interfaces for our C-code. One is a command line interface and the other interface is R, which is a software system for statistics [7]. For example, in the problem mh (the largest roots of Wishart matrices), the function mh_cwishart_gen performs the HGM for mh. The both of the main function for the command line interface and the interface module Rmh_cwishart_gen for R call the common function mh_cwishart_gen.

The system R provides an easy and strong mechanism to include C code into the R system [10]. It is recommended by the CRAN repository policy to minimize the size of code and make effort to provide cross-platform code. Then, we have extracted the source code of some of the functions defined in odeiv.h in GSL for solving ordinary differential equations, and include them in our HGM package.

In the current implementation, both of command line interface and R interface are available for the problems mh, orthant, and so3. We provide only a command line interface for the Fisher-Bingham distribution. Because, our implementation relies on linear algebra functions of the GSL and extracting these functions for R or rewriting them in BLAS and LAPACK need some works, which will generate some new bugs without taking relatively long time of careful porting and debugging. We hope that R officially supports the GSL in a future.

3.2 Use of Computer Algebra Systems for a Reliable Implementation

Since some ordinary differential equations for the HGM contain complicated expressions and also evaluation formulas of initial values are complicated, we utilize computer algebra systems to avoid bugs caused by writing programs by hand and to provide correct code. For example, our C implementation for the HGD (holonomic gradient descent) of the Fisher-Bingham distribution is automatically generated by code in Risa/Asir [8], which is a computer algebra system. Our implementation for the Wishart distribution is firstly written in Risa/Asir and contains several debugging and checking code of correctness of each steps (see tk_jack.rr in our package). After the code by Risa/Asir works correctly, we translate it into code in C.

3.3 Performance

We illustrate the performance of our implementation of the orthant probability.

We evaluate the orthant probability by the HGM for $\Sigma = ((1 + \delta_{ij})/2)$ and $\mu = 0$ where δ_{ij} is the Kronecker's delta. In this case, it is known that the orthant probability equals to $1/(d + 1)$. The table 1 shows the result of the HGM, the exact value of the orthant probability, and the CPU time for each d. The HGM is performed by the command hgm_ko_orthant, and it is compiled by the GNU C compiler v.4.7.2. We performed the experiments on an Intel(R) Xeon(R) CPU E5-4650 0 @ 2.70GHz with 252GB RAM, running Linux.

The computational time of hgm_ko_orthant increases rapidly when the dimension increases. This is a consequence of our complexity result (Theorem 2). However, our algorithm and implementation are faster in comparison with the existing software systems which evaluate the orthant probability with high accuracy. For example, the CPU time to compute the same problem by pmvnorm [5], which is in the R package, for the case $d = 10$ is 61.991 seconds. The function pmvnorm dissects an orthant probability into $(d - 1)!$ orthoscheme probabilities, and apply an effective iterative integration whose complexity is $O(d)$. Thus, the theoretical complexity of pmvnorm is proportional to $d!$. No other algorithms and

Table 1. Computational experiments on `hgm_ko_orthant`

dimension	HGM	exact	CPU time
2	0.333333331	0.333333333	0.00
3	0.249999998	0.250000000	0.00
4	0.199999998	0.200000000	0.00
5	0.166666666	0.166666667	0.02
6	0.142857142	0.142857143	0.06
7	0.125000000	0.125000000	0.14
8	0.111111111	0.111111111	0.39
9	0.100000000	0.100000000	0.91
10	0.090909091	0.090909091	2.40
20	0.047619048	0.047619048	22721.30

implementations achieve our timing with more than the 9 digits accuracy as far as we know.

4 Applications

Normalizing constants are fundamental in statistics. In [6], we demonstrate that Fisher's maximal likelihood estimate can be performed by utilizing the HGM to evaluate normalizing constants and its derivatives. The orthant probability of multivariate normal distribution is also used in various area of statistics. In this section, we sketch an application to Bayesian analysis.

Consider the multinomial distribution of size n

$$f(y|\pi) = \frac{n!}{y_1! \cdots y_p!} \pi_1^{y_1} \cdots \pi_p^{y_p}, \quad y = (y_1, \ldots, y_p) \in \mathbb{Z}_{\geq 0}^p, \quad \sum_{i=1}^p y_i = n,$$

where $\pi = (\pi_1, \ldots, \pi_p)$ belongs to the simplex $\Delta^{p-1} = \{\pi \geq 0 \mid \sum_i \pi_i = 1\}$. For the multinomial distribution, the Dirichlet prior density is often used in the Bayesian context (e.g. [1]).

We introduce a different class of prior densities. Put $\Sigma^{-1} = \text{diag}(\theta_1, \ldots, \theta_p)$ and $\pi_i = x_i^2$ for each i in the Bingham distribution defined in Section 2. The random variable $\pi = (\pi_1, \ldots, \pi_p)$ has the density function

$$f(\pi) = \frac{2\pi_1^{-1/2} \cdots \pi_p^{-1/2} e^{\sum_{i=1}^p \theta_i \pi_i}}{c(\theta)}, \quad \pi \in \Delta^{p-1},$$

with respect to $d\pi = d\pi_1 \cdots d\pi_{p-1}$, where $c(\theta) = Z(\Sigma)$ is the Bingham normalizing constant. We call it the Bingham prior density.

One of important quantities in Bayesian analysis is the marginal likelihood $f_{\text{mar}}(y) = \int f(y|\pi)f(\pi)d\pi$ (see e.g. Section 3.4 of [1]). For the Bingham prior, it is shown that

$$f_{\text{mar}}(y) = \frac{n!}{y_1! \cdots y_p!} \frac{\prod_{i=1}^p \Gamma(y_i + \frac{1}{2})}{\pi^{n+p/2}} \frac{c(\theta, 2y + 1_p)}{c(\theta)}, \tag{1}$$

where $1_p = (1, \ldots, 1) \in \mathbb{R}^p$, and $c(\theta, d)$ for $d = (d_1, \ldots, d_p)$ denotes the Bingham normalizing constant on the $(\sum_{i=1}^{p} d_i - 1)$-dimensional sphere with the multiplicity index d, that is, $c(\theta, d) = c((\theta_1, \ldots, \theta_1, \ldots, \theta_p, \ldots, \theta_p))$, where θ_i appears d_i times. The formula (1) is proved in the same way as Proposition 1 of [4]. The other quantities such as posterior density and predictive density are written in terms of $c(\theta, d)$ as well.

For example, if $p = 4$, $\theta = (1, 3, 5, 0)$ and $y = (2, 0, 3, 1)$, then the marginal likelihood is evaluated by the following R script

```
y = c(2,0,3,1); th = c(1,3,5); n = sum(y); p = length(y)
a0 = lfactorial(n) - sum(lfactorial(y))
a1 = sum(lgamma(y+1/2)) - (n+p/2)*log(pi)
a2 = hgm.ncBingham(th, d=2*y+1, withvol=TRUE, logarithm=TRUE)[1]
a3 = hgm.ncBingham(th, withvol=TRUE, logarithm=TRUE)[1]
exp(a0 + a1 + a2 - a3)
```

where the `withvol` option specifies that the total uniform measure on the sphere is its volume (not normalized to 1) and the `logarithm` option specifies the output is in the logarithmic scale. The result of the script is 0.008963549. One can check that the total of $f_{\mathrm{mar}}(y)$ over possible y's given n is 1 (up to numerical error).

For a given data y, the hyper-parameter θ in (1) can be selected by maximizing the marginal likelihood. This maximization problem is analogous to the maximum likelihood estimation and then the HGD [6] can be applied in principle, but details have not been studied and this MLE has not been implemented yet in our package. It is also an interesting project in progress to derive a holonomic system and a Pfaffian system for other prior densities.

References

1. Bishop, C.M.: Pattern Recognition and Machine Learning. Springer (2006)
2. GNU Scientific Library, http://www.gnu.org/s/gsl
3. Koyama, T., Takemura, A.: Calculation of Orthant Probabilities by the Holonomic Gradient Method, arxiv:1211.6822
4. Kume, A., Wood, A.T.A.: On the Derivatives of the Normalizing Constant of the Bingham Distribution. Statistics and Probability Letters 77, 832–837 (2007)
5. Miwa, T., Hayter, H.J., Kuriki, S.: The evaluation of general non-centered orthant probabilities. Journal of the Royal Statistical Society: Series B 65, 223–234 (2003)
6. Nakayama, H., Nishiyama, K., Noro, M., Ohara, K., Sei, T., Takayama, N., Takemura, A.: Holonomic Gradient Descent and its Application to the Fisher-Bingham Integral. Advances in Applied Mathematics 47, 639–658 (2011)
7. The R: Project for Statistical Computing, http://www.r-project.org
8. Risa/Asir, a computer algebra system, http://www.math.kobe-u.ac.jp/Asir
9. Sei, T., Kume, A.: Calculating the Normalizing Constant of the Bingham Distribution on the Sphere using the Holonomic Gradient Method. Statistics and Computing (2013)
10. Writing R Extensions, http://cran.r-project.org/doc/manuals/R-exts.html
11. Zeilberger, D.: A Holonomic Systems Approach to Special Function Identities. Journal of Computational and Applied Mathematics 32, 321–368 (1990)
12. http://www.math.kobe-u.ac.jp/OpenXM/Math/hgm/ref-hgm.html

Metalibm: A Mathematical Functions Code Generator

Olga Kupriianova and Christoph Lauter

Sorbonne Universités, UPMC Univ. Paris 06,
UMR 7606, LIP6, F-75005 Paris, France
{olga.kupriianova,christoph.lauter}@lip6.fr
http://www.kupriianova.info/
http://www.christoph-lauter.org/

Abstract. There are several different libraries with code for mathematical functions such as exp, log, sin, cos, etc. They provide only one implementation for each function. As there is a link between accuracy and performance, that approach is not optimal. Sometimes there is a need to rewrite a function's implementation with the respect to a particular specification.

In this paper we present a code generator for parametrized implementations of mathematical functions. We discuss the benefits of code generation for mathematical libraries and present how to implement mathematical functions. We also explain how the mathematical functions are usually implemented and generalize this idea for the case of arbitrary function with implementation parameters.

Our code generator produces C code for parametrized functions within a known scheme: range reduction (domain splitting), polynomial approximation and reconstruction. This approach can be expanded to generate code for black-box functions, e.g. defined only by differential equations.

Keywords: code generation, elementary functions, mathematical libraries.

1 Introduction

Each time we evaluate mathematical functions in some programming language a corresponding function from a mathematical library (libm) is called. There are several examples of existing libms: glibc libm [1], crlibm by ENS-Lyon [2], libmcr by Sun[1], libultim by IBM[2], etc. They differ not only by the developer company or supported language but also by final accuracy. The common fact for all the versions is that they provide only one manually coded implementation of each supported function and precision. As the codes for elementary functions are

---

[1] http://www.math.utah.edu/cgi-bin/man2html.cgi?/usr/local/man/man3/
 libmcr.3
[2] http://www.math.utah.edu/cgi-bin/man2html.cgi?/usr/local/man/man3/
 libultim.3

H. Hong and C. Yap (Eds.): ICMS 2014, LNCS 8592, pp. 713–717, 2014.

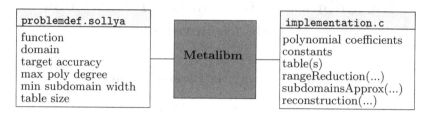

Fig. 1. Metalibm scheme

used in various applications, the implementations are done for the widest possible domain, for maximum possible accuracy, etc. However, such generalization complicates the algorithms and leads to poor performance: for some particular tasks there is no need to compute a precise result. For example, most physical measurements have only small number of digits after the decimal point, so computation of the result with 53 mantissa bits (about 15 decimal digits) is a waste of time. Another example is a small domain known beforehand, which means that that the result is a finite number (no overflows/underflows are possible). In this case, handling the special values like NaNs (Not-A-Number [3]) or infinities in the beginning of the implementation can be skipped, hence the implementation gets faster.

There is also a link between speed and accuracy, so when we process big amounts of data and the needed accuracy is only about several bits, there is no reason to compute a precise result [4].

So, there is a growing need to provide several different versions (flavors) for each libms function. Manual implementation of different function flavors is almost impossible due to the quantity of all the possible parameters and coding time. Thus, we propose to write a code generator that produces parametrized implementations. The name of this prototype is Metalibm[3]. Besides producing different versions of standard libm functions, Metalibm generates implementations for composite functions as well.

The paper is organized as following: in Section 2 we explain in general the generation of "black-box" parametrized functions, Section 3 explains the workflow of the generator, in Section 4 we show how to detect a known type of function, and finally Section 5 shows the importance of the Metalibm, future work on the project and its possible application.

2   A Black-box Function Generator

The Metalibm code is a collection of scripts written in Sollya[4], a software tool for safe floating-point (FP) development with plenty of rigorous numerical algorithms [5].

Among the existing versions of mathematical libraries we are mostly interested in improvement of the current `glibc` `libm`. It runs on all *nix-powered machines

---

[3] http://lipforge.ens-lyon.fr/www/metalibm/
[4] http://sollya.gforge.inria.fr/

(from supercomputers to mobile phones), so tends to be the most used one. Thus, Metalibm generates implementations in C. For each input set of parameters Metalibm generates the C code within the same scheme. On Fig. 1 there is an illustration of Metalibm routine. Metalibm computes and stores the constants with the needed accuracy, and as we use table-driven methods [4], it computes and stores the tables. Then Metalibm generates the code for range reduction (domain splitting), polynomial approximation on each of the small domains, and the final reconstruction procedure. The purpose of these procedures is explained later in the paper.

As all the computations for the code generation are done in Sollya, a function to be implemented can be considered as black-box. On the step of range reduction we need to evaluate the function in some values. Sollya provides elementary functions and theirs combinations. The generation of code for some "exotic functions" like functions purely defined by differential equations (e.g. Dickman's) gets possible as soon as we have a corresponding Sollya implementation, even if it is bound to Sollya only dynamically [6].

Precision of all the constants and accuracy of the interim computations are specially selected in order to obtain the final result of the specified accuracy (target accuracy parameter). Besides the generated code, Metalibm verifies the final results accuracy and generates a Gappa proof [7].

A simple example of the parametrization file is provided in Listing 1.1. The sense of all the parameters will be explained later with the implementation details of Metalibm.

```
f = exp(x); //we want to get the code for exp(x)
dom = [-70, 70]; //on domain [-70, 70]
target = 2^{-42}; //the final error has to be not more that 2^{-42}
maxDegree = 5; //degree of approximating polynomials is not more than 5
minWidth = (sup(dom) - inf(dom)) * 1/4096; //minimal size of the subdomain
tableIndexWidth = 5; //the size of table index is 5 bits, so 32 entries in table
```

Listing 1.1. Example of the parametrization file

3 Function Generation Workflow

In order to implement a mathematical function on a given domain we have to care first about special cases (NaNs and infinities). So, the first or even precomputing step is always filtering the special inputs, handling exceptions, too large or too small inputs unless the domain is so small that these special cases cannot occur.

When designing algorithms for mathematical functions evaluation we usually start with writing the inputs and output in a form of floating-point (FP) numbers $2^E m$ and trying to separate all the factors into two groups: a power of two $n = 2^E$ that represents the results exponential part and one non-integer number with values in a small range to represent significand. This technique of emphasizing a non-integer part with a small range (results significand) is usually called range reduction [8]. The next step is building an approximation. There are different approximation techniques, but we only consider polynomial ones. The larger the range of argument is, the higher polynomial degrees are required. This means that the computation time will be high and the error analysis becomes more difficult with the growing number of FP operations. This is the main reason why we need argument reduction. In Metalibm, we use a parameter `maxDegree` to limit the maximal degree of constructed approximations. These polynomials are computed with Remez algorithm that is implemented in Sollya [6], [9].

All the transformations on the first step fully depend on mathematical properties of the implemented function, i.e. $e^{a+b} = e^a e^b$ [10]. It can happen that it is impossible to reduce the range using only mathematical properties. In this situation the required domain for the implementation is divided into subdomains. On each of the subdomains the argument range is small and the degree of the approximating polynomial gets lower. In this case, one has to build approximations for each of the subdomains and then the reconstruction step gets more complex. In Metalibm we bound minimal size of subdomain by a parameter `minWidth`.

Once the generation process is launched, Metalibm checks whether it is possible to build a polynomial for the specified function, domain and other parameters. If it is not possible, it tries to reduce argument for the set of known functions (exponential, logarithm, periodic). Then there is symmetry detection and expression decomposition for composite functions. When it is still not possible to build a polynomial approximation for the reduced domain, domain splitting is performed. Metalibm adapts computational precision [11] in order to get the needed accuracy, so in some cases it uses double-double or triple-double arithmetic [12], [13].

4   Some Technical Details

As it was already mentioned, Metalibm tries to detect some known properties of the function to perform range reduction. For example, let us have a look on how Metalibm tests hypothesis that the function is exponential and finds its base.

In order to detect the exponential function and to perform the appropriate argument reduction we accept the hypothesis that the function has a form of $f(x) = b^x$ on the implementation domain $\mathcal{I}$ with the unknown base b. It means that we can determine the base from the following:

$$b = \exp\left(\frac{\ln f(x)}{x}\right) = const \ \forall x \in \mathcal{I}.$$

The base b can be computed in Sollya without any information about the function $f(x)$. Sollya will provide the needed value for the function. This is why we were talking about "black-box" functions: we do not know what code are we generating, but we have a mean to evaluate some function values.

As we have accepted the function type as $f(x) = b^x$, it means that for another argument x_1 from the implementation domain the function value is $f(x_1) = b^{x_1}$. Of course the base b stays the same if the hypothesis was correct and we know that we can perform the exponential argument reduction now. If the value $\tilde{\varepsilon} = \left\|\frac{b^x}{f(x)} - 1\right\|_{\infty}^{\mathcal{I}}$ is sufficiently small we accept the initial hypothesis.

The detection of other types of functions is done in the analogous way.

5   Conclusions

The Metalibm code generator is still under development, but it already produces code for basic functions. The next goals are an optimized domain splitting procedure, producing vectorizable implementations and addition of range reduction for other functions.

As it was told, the generated code tends to improve and expand current `glibc` `libm`. So, we will try to provide generated code on the whole code generator to the GNU community for integration with the `glibc` `libm`.

References

1. Loosemore, S., Stallman, R.M., et al.: The GNU C Library Reference Manual for version 2.19, Free Software Foundation, Inc.
2. Daramy-Loirat, C., Defour, D., de Dinechin, F., et al.: CR-LIBM. A library of correctly rounded elementary functions in double-precision, user's manual (2013)
3. IEEE Computer Society, IEEE Standard for Floating-Point Arithmetic, IEEE Standard 754-2008 (August 2008)
4. Muller, J.-M.: Elementary Functions: Algorithms and Implementation. Birkhauser Boston, Inc., Secaucus (1997)
5. Chevillard, S., Joldeş, M., Lauter, C.: Sollya: An Environment for the Development of Numerical Codes. In: Fukuda, K., van der Hoeven, J., Joswig, M., Takayama, N. (eds.) ICMS 2010. LNCS, vol. 6327, pp. 28–31. Springer, Heidelberg (2010)
6. Chevillard, S., Lauter, C., Joldeş, M.: Users manual for the Sollya tool, Release 4.0 (May 2013)
7. de Dinechin, F., Lauter, C., Melquiond, G.: Certifying the Floating-Point Implementation of an Elementary Function Using Gappa. IEEE Transactions on Computers 60, 242–253 (2011)
8. Muller, J.-M., Brisebarre, N., de Dinechin, F., et al.: Handbook of Floating-Point Arithmetic. Birkhäuser (2010)
9. Chevillard, S.: Évaluation efficace de fonctions numériques. Outils et exemples, PhD Thesis, ENS de Lyon (2009)
10. Tang, P.T.P.: Table-driven implementation of the exponential function in IEEE floating-point arithmetic. ACM Transactions on Mathematical Software 18(2), 211–222 (1989)
11. Lauter, C.: Arrondi correct de fonctions mathématiques, PhD Thesis, ENS de Lyon (2008)
12. Schewchuk, J.R.: Adaptive Precision Floating-Point Arithmetic and Fast Robust Geometric Predicates. Discrete & Computational Geometry (1997)
13. de Dinechin, F., Defour, D., Lauter, C.: Fast correct rounding of elementary functions in double precision using double-extended arithmetic, Research report, ENS de Lyon (2004)

From Calculus to Algorithms without Errors

Norbert Müller[1] and Martin Ziegler[2]

[1] Universität Trier, Germany
mailto:mueller@uni-trier.de
http://uni-trier.de/?id=3591
[2] TU Darmstadt, Germany
mailto:ziegler@mathematik.tu-darmstadt.de
http://m.zie.de

Abstract. Using mathematics within computer software almost always includes the necessity to compute with real (or complex) numbers. However, implementations often just use the 64-bit double precision data type. This may lead to serious stability problems even for mathematically correct algorithms. There are many ways to reduce these software-induced stability problems, for example quadruple or multiple-precision data types, interval arithmetic, or even symbolic computation. We propagate *Exact Real Arithmetic* (ERA) as a both convenient and practically efficient framework for rigorous numerical algorithms.

Keywords: Recursive Analysis, Rigorous Numerics.

1 Introduction and Motivation

With numerical methods reaching unprecedented levels of sophistication, there arises the need for a systematic approach to software development over real numbers, smooth functions, Euclidean domains, and operators – accompanied and guided by a sound yet practical theory of computing over continuous universes. Indeed, contemporary algorithms for advanced applications like PDEs regularly involve, and – often only implicitly – build on reliable solutions to, an entire hierarchy of intermediate problems: ODEs (method of characteristics), Riemann integration, Taylor and Fourier expansion, mesh generation and interpolation, computational linear algebra, down to basic operations on single numbers. ('Best'?) Practice in Numerical Engineering generally neglects questions of correctness of such advanced functionality modularly combined and composed from elementary ones, leading to a mix of criticism and fatalism [Linz88, p.412]:

> How do engineers deal with the problem of assigning some measure of reliability to the numbers that the computer produces? Over the years, I have sat on many Ph.D. qualifying examinations or dissertation defenses for engineering students whose work involved a significant amount of numerical computing. In one form or another, I invariably ask two questions "Why did you choose that particular algorithm?" and "How do you

H. Hong and C. Yap (Eds.): ICMS 2014, LNCS 8592, pp. 718–724, 2014.
© Springer-Verlag Berlin Heidelberg 2014

know that your answers are as accurate as you claim?". The first question is usually answered confidently, using such terms as "second-order convergence" or "von Neumann stability criterion". The next question, alas, tends to be embarrassing. After an initial blank or hostile stare, I usually get an answer like "I tested the method with some simple examples and it worked", "I repeated the computation with several values of n and the results agreed to three decimal places", or more lamely, "the answers looked like what I expected". So far, I have not faulted any student for the unsatisfactory nature of such a response. One reason for my reluctance to criticize is that I have really nothing better to offer. Rigorous analysis is out of the question. [...] What I do find disturbing is the pragmatics that are used are often ill-considered. Take for example the common practice of repeating the computations with several values of the discretization parameters. The reasoning behind this is that, if the method converges and we observe that the solution has "settled down" in the first few decimal digits, we can be confident that it is actually exact to this accuracy. Sometimes this makes good sense, but unfortunately it does not always work. For example, it may not catch systematic errors such as a wrong sign somewhere or a dropped factor of two. But such errors are very common, particularly in the computer programs that are eventually written. There are many instances of programs that delivered incorrect results for a considerable period of time before the error was found.

We propagate *Exact Real Arithmetic* (ERA) as a both convenient and practically efficient framework for rigorous numerical algorithms. ERA consists of, and combines, four aspects:

i) *Recursive Analysis* — the Theory of Computing over real numbers, (smooth) functions, and (closed) Euclidean subsets — and its uniform generalization to generic continuous universes [Weih00] called *Type-2 Theory of Effectivity* (TTE).

ii) *Real Complexity Theory* as resource-oriented refinement of (i), including asymptotic runtime analyses and formal proofs of algorithmic optimality [Ko91, Ko98, Weih03].

iii) An imperative programming language with rigorous semantics of computable operations on continuous objects appearing as entities (ɛRA) [BrHe98].

iv) A C++ library [Müll01] implementing, and efficiently realizing, (much of) the semantics according to (iii).

ERA differs from Exact *Geometric* Computation [LPY05] in permitting both output and input being approximated up to prescribable absolute error, thus achieving closure under composition [Yap04, p.325]. The sequel of this work recalls the computability and complexity theoretic foundations (i+ii) to ERA and their implications to a rigorous semantics — particularly of tests/branches, that (necessarily!) differs from the naïve one in the discrete case. We demonstrate the ease and convenience of coding in said language, obtaining both efficient and validated implementations for turnkey calculations and rapid prototyping in transcendental arithmetic. This will lead to a revolution of numerical software development.

2   Algorithmic Foundations of Real Computing

Recursive Analysis was initiated by Alan Turing (1937) in the very same work that introduced 'his' machine — 11 years before he invented matrix condition numbers. It is the theory of computation of real numbers and functions by rational approximations up to prescribable absolute error 2^{-n}. This field provides for a sound algorithmic foundation to numerical calculations — and contains some of the too blatant folklore claims about their performance. For instance [Spec49] constructed an increasing, recursive sequence of fractions converging (with necessarily non-recursive rate of convergence) to a real number that encodes the halting problem. Or [Spec59] constructed a (polynomial-time) computable smooth function $f : [0; 1] \to [0; 1]$ attaining its minimum/zero (frequently, but) at no computable point. Observe how both debunk (allusions behind) the 'specification' of `nag_opt_one_var_deriv` to *normally compute(s) a sequence of x values which <u>tend in the limit</u> to a minimum of Fx subject to the given bounds.* In fact Recursive Analysis has since then developed a rich variety of formal notions of computability over various specific and general continuous universes X. TTE unifies and facilitates comparing and uniformly classifying such notions: for example for continuous functions [Weih00, §6], for closed Euclidean subsets [BrWe99], or for advanced spaces [ZhWe03].

3   Theory and Implementation

By the *Main Theorem* of Recursive Analysis every computable function is necessarily continuous. This renders the sign function, and thus also tests/branches, incomputable. However two concepts from logic permit to avoid this obstacle:

Non-*extensional (a.k.a. multivalued) computation* may return different values $y \in f(x)$ for the same argument x, depending on which (sequence of) approximations to x is given [Weih08]. Consider for instance a computable version of the *Fundamental Theorem of Algebra* [Spec69]:

> Given $(a_0, \ldots, a_{d-1}) \in X := \mathbb{C}^d$, output *some* d-tuple $z_1, \ldots, z_d \in \mathbb{C}$ such that $a_0 + a_1 z + \cdots + a_{d-1}z^{d-1} + z^d = (z - z_1) \cdots (z - z_d)$.

Any tuple $(z_1, \ldots, z_d)$ with this property is an admissible output of such a computation! Or consider for $k \in \mathbb{N}$ the multivalued — a.k.a. *soft* [YSS13] — test "$x >_k$ 0?" may return `true` in case $x \geq 2^{-k}$ and `false` in case $x \leq -2^{-k}$ and *any of* `false`,`true` in case $|x| < 2^{-k}$. Only such a *modified* semantics renders tests (total and) computable. ERA supports multivalued tests in many forms. A simple version in the iRRAM library is the statement `choose(x>0,y>0)`: returning 1 in case

$x > 0$ and $y \leq 0$, 2 in case $x \leq 0$ and $y > 0$, and any 1 or 2 in case $x > 0$ and $y > 0$. It is the programmers responsibility to prevent the case $x \leq 0$ and $y \leq 0$ (similarly to division by zero). See §4.2 for a practical application. . .

Discrete enrichment captures the effect that some practical real (multi-)functions f are incomputable due to discontinuity [PaZi13] — yet become both continuous and computable when providing, in addition to approximations to the continuous argument x, some suitable integer [KrMa82, p.238/239]. For example from the entries of a real symmetric $d \times d$–matrix A one cannot continuously deduce some (!) basis of eigenvectors; whereas restricted to non-degenerate A and, more generally, when *given the number $k := Card\sigma(A) \in \{1, \ldots, d\}$ of distinct eigenvalues*, one can [ZiBr04, §3.5]! Such necessary and sufficient discrete enrichment yield canonical interface declarations in ERA. The following C++ fragment for example can*not* belong to an implementation returning some eigenvector to any given real symmetric 2×2–matrix:

```
void EV(REAL A11, REAL A12, REAL A22, REAL &EVecY, REAL &EVecY);
```

See §4.5 for another practical example. . .

Based on the above multivalued semantics and enrichment, iRRAM implements data types for real numbers and functions: internally based on sequences of intervals with endpoints in unbounded precision, but appearing to the user as single and exact entities.

4   Seven Examples of Practical Programming in ERA

In the sequel we illustrate some core ideas where ERA can be of advantage when compared to algorithms using double precision or even multiple precision. They all have been implemented in the iRRAM library freely available at http://irram.uni-trier.de/.

4.1   Simple Algorithms: Range Reduction and Logistic Map

In the evaluation of transcendental functions, identities like $e^x = (e^{x/2})^2$ or $\sin(x) = 3 \cdot \sin(x/3) - 4 \cdot \sin(x/3)^3$ are frequently used for range reduction. In ERA, such identities can be used without worrying about loss of precision. This is even true for iterated functions systems like the logistic map, where the iteration $x_{i+1} = 3.75 \cdot x_i \cdot (1-x_i)$ (implemented just as x= c*x*(REAL(1)-x)) can be carried out 100000 times for an initial value like $x_0 = 0.5$ within less than three minutes in spite of its chaotic behavior, whereas a double precision computation gives nonsense already after 100 iterations.

4.2   Gaussian Elimination and Inverting Ill-Conditioned Matrices

Gaussian elimination needs some kind of pivoting, i.e. a non-continuous choice of an integer index from two (or more) real parameters. In the iRRAM library such choices can be implemented as a multivalued function in the form `choose(x>0,y>0)`, where a return value of 1 ensures $x > 0$, while 2 ensures $y > 0$. Based on this, matrix inversion turns out to be a simple programming example. A Hilbert matrix of dimension 200×200 can be inverted in about two minutes, where `double precision` arithmetic would already fail for dimension 20×20.

4.3 Root Finding and Wilkinson's Polynomials

Traditional algorithms for root finding have problems when applied to Wilkinson's polynomials, as their coefficients quickly get very large. Of course, ERA allows numbers much bigger that the maximal `double precision` number of approximately 10^{308}. Nevertheless, ERA would also fail if a simple bisection algorithm were used and accidentally hit one of the roots. Using the `choose` method mentioned before, a multi-valued k-section can easily be implemented, using k larger than the degree of the polynomial under consideration.

4.4 Transcendental Calculations: Power Series and Matrix Exponentials

A central component of ERA is the ability to get limits of converging computable sequences, in case an upper bound for the speed of convergence can be computed: If a sequence (x_n) converges to the limit y with $|x_n - y| \le 2^{-n}$, then the statement `y=limit(x)` will give this limit in iRRAM.

If (x_n) is not just a simple sequence of real numbers, but even a sequence of functions with a limit function f (and the same speed of convergence as before), then `fz=limit(x,z)` will compute $f(z)$. An important application are power series $\sum a_j x^j$, where a lower bound R for the radius of convergence and an upper bound M for $\sum |a_j| R^j$ lead to a rigorous bound for the speed of convergence, such that `f=taylor_sum(a,R,M)` in iRRAM really computes the Taylor sum $f(z) = \sum a_j z^j$ for $|z| < R$.

This can also be applied in higher dimensions, leading to a usable implementation of the matrix exponential that is known to be hard to compute otherwise.

4.5 Algorithms with Discrete Enrichment

Evaluating a power series $\sum_j a_j x^j$, say converging on the closed interval $[-1; 1]$, provably requires information in addition to the coefficient sequence $(a_j)_j$ [Müll95]. In fact it suffices to provide an integer k satisfying the following form of the Cauchy–Hadamard Bound: $|a_j| \le (2j)^k$. Moreover this integer governs, in addition to the output precision n, the running time of natural operations on analytic functions: evaluation, addition, multiplication, differentiation, integration, and maximization [Müll87, KMRZ14].

4.6 High-Degree Newton–Cotes Quadrature

The paradigmatic chains to hardware-supported `double` precision induces a blind spot against possibly more efficient algorithms that involve intermediate

of higher accuracy [MüKo10]. Newton–Cotes Quadrature for instance is generally considered 'unstable' in high degrees, but performs rather well on iRRAM: which, transparent to the user, automatically takes care of, and adjusts, said intermediate precision; see §4.7.

4.7 Tuning

The enrichment mentioned before can also be used efficiently for speeding up computations: One such type of enrichment are Lipschitz bounds which can easily be used to reduce error propagation in the underlying multiple precision interval arithmetic to a minimum, which in reduces computation time. For the logistic map for above the use the Lipschitz bound $3.75 - 7.5 \cdot x$ reduces the computation speed by a factor of 3.

References

[BrHe98]   Brattka, V., Hertling, P.: Feasible real random access machines. Journal of Complexity 14(4), 490–526 (1998)

[BrWe99]   Brattka, V., Weihrauch, K.: Computability on Subsets of Euclidean Space I: Closed and Compact Subsets. Theoretical Computer Science 219, 65–93 (1999)

[KMRZ14]   Kawamura, A., Müller, N., Rösnick, C., Ziegler, M.: Computational Benefit of Smoothness: Rigorous Parameterized Complexity Analysis in High-Precision Numerics of Operators on Gevrey's Hierarchy (submitted)

[Ko91]   Ko, K.-I.: Computational Complexity of Real Functions. Birkhäuser (1991)

[Ko98]   Ko, K.-I.: Polynomial-Time Computability in Analysis. In: Ershov, Y.L., et al. (eds.) Handbook of Recursive Mathematics, vol. 2, pp. 1271–1317 (1998)

[KrMa82]   Kreisel, G., Macintyre, A.: Constructive Logic versus Algebraization I. In: Troelstra, van Dalen (eds.) Proc. L.E.J. Brouwer Centenary Symposium, pp. 217–260. North-Holland (1982)

[Linz88]   Linz, P.: A critique of numerical analysis. Bulletin of the American Mathematical Society 19(2), 407–416 (1988)

[LPY05]   Li, C., Pion, S., Yap, C.: Recent progress in exact geometric computation. J. Log. Algebr. Program. 64(1), 85–111 (2005)

[Müll87]   Müller, N.T.: Uniform Computational Complexity of Taylor Series. In: Ottmann, T. (ed.) ICALP 1987. LNCS, vol. 267, pp. 435–444. Springer, Heidelberg (1987)

[Müll95]   Müller, N.T.: Constructive Aspects of Analytic Functions. In: Proc. Workshop on Computability and Complexity in Analysis (CCA), vol. 190, pp. 105–114. InformatikBerichte FernUniversität Hagen (1995)

[Müll01]   Müller, N.T.: The iRRAM: Exact Arithmetic in C++. In: Blank, J., Brattka, V., Hertling, P. (eds.) CCA 2000. LNCS, vol. 2064, pp. 222–252. Springer, Heidelberg (2001)

[MüKo10]   Müller, N.T., Korovina, M.: Making big steps in trajectories. In: Proc. 7th Int. Conf. on Computability and Complexity in Analysis (CCA 2010). Electronic Proceedings in Theoretical Computer Science, vol. 24, pp. 106–119 (2010)

[PaZi13]   Pauly, A., Ziegler, M.: Relative Computability and Uniform Continuity of Relations. Journal of Logic and Analysis 5 (2013)

[Spec49]   Specker, E.: Nicht konstruktiv beweisbare Sätze der Analysis. Journal of Symbolic Logic 14(3), 145–158 (1949)

[Spec59]   Specker, E.: Der Satz vom Maximum in der rekursiven Analysis. In: Heyting, A. (ed.) Constructivity in Mathematics. Studies in Logic and The Foundations of Mathematics, pp. 254–265. North-Holland (1959)

[Spec69]   Specker, E.: The fundamental theorem of algebra in recursive analysis. In: Constructive Aspects of the Fundamental Theorem of Algebra, pp. 321–329. Wiley-Interscience (1969)

[Weih00]   Weihrauch, K.: Computable Analysis. Springer (2000)

[Weih03]   Weihrauch, K.: Computational Complexity on Computable Metric Spaces. Mathematical Logic Quarterly 49(1), 3–21 (2003)

[Weih08]   Weihrauch, K.: The Computable Multi-Functions on Multi-represented Sets are Closed under Programming. Journal of Universal Computer Science 14(6), 801–844 (2008)

[Yap04]   Yap, C.-K.: On Guaranteed Accuracy Computation. In: Geometric Computation, pp. 322–373. World Scientific Publishing (2004)

[YSS13]   Yap, C., Sagraloff, M., Sharma, V.: Analytic Root Clustering: A Complete Algorithm Using Soft Zero Tests. In: Bonizzoni, P., Brattka, V., Löwe, B. (eds.) CiE 2013. LNCS, vol. 7921, pp. 434–444. Springer, Heidelberg (2013)

[ZiBr04]   Ziegler, M., Brattka, V.: Computability in Linear Algebra. Theoretical Computer Science 326, 187–211 (2004)

[ZhWe03]   Weihrauch, K., Zhong, N.: Computability theory of generalized functions. Journal of the ACM 50(4), 469–505 (2003)

Dense Arithmetic over Finite Fields
with the CUMODP Library

Sardar Anisul Haque[1], Xin Li[2], Farnam Mansouri[1], Marc Moreno Maza[1],
Wei Pan[3], and Ning Xie[1]

[1] University of Western Ontario, Canada
{shaque4,fmansou3,moreno,nxie6}@csd.uwo.ca
[2] Universidad Carlos III, Spain
xli@inf.uc3m.es
[3] Intel Corporation, Canada
wei.pan@intel.com

Abstract. CUMODP is a CUDA library for exact computations with
dense polynomials over finite fields. A variety of operations like multiplication, division, computation of subresultants, multi-point evaluation,
interpolation and many others are provided. These routines are primarily
designed to offer GPU support to polynomial system solvers and a bivariate system solver is part of the library. Algorithms combine FFT-based
and plain arithmetic, while the implementation strategy emphasizes reducing parallelism overheads and optimizing hardware usage.

Keywords: Polynomial arithmetic, parallel processing, many-core GPUs.

1 Overview

Polynomial multiplication and matrix multiplication are at the core of many algorithms in symbolic computation. Expressing, in terms of multiplication time,
the algebraic complexity of an operation like univariate polynomial division or
the computation of a characteristic polynomial is a standard practice, see for instance the landmark book [4]. At the software level, the motto "reducing everything to multiplication"[1] is also common, see for instance the computer algebra
systems Magma[2] [1], NTL[3] or FLINT[4].

---

[1] Quoting a talk title by Allan Steel, from the Magma Project.
[2] Magma: http://magma.maths.usyd.edu.au/magma/
[3] NTL: http://www.shoup.net/ntl/
[4] FLINT: http://www.flintlib.org/

H. Hong and C. Yap (Eds.): ICMS 2014, LNCS 8592, pp. 725–732, 2014.
© Springer-Verlag Berlin Heidelberg 2014

With the advent of hardware accelerator technologies, multi-core processors and Graphics Processing Units (GPUs), this reduction to multiplication is, of course, still desirable, but becomes more complex since both algebraic complexity and parallelism need to be considered when selecting and implementing a multiplication algorithm. In fact, other performance factors, such as cache usage or CPU pipeline optimization, should be taken into account on modern computers, even on single-core processors. These observations guide the developers of projects like SPIRAL[5] [16] or FFTW[6] [3].

The CUMODP library provides arithmetic operations for dense matrices and dense polynomials primarily with modular integer coefficients, targeting many-core GPUs. Some operations are available for integer or floating point coefficients as well. A large portion of the CUMODP library code is devoted to polynomial multiplication and the integration of that operation into higher-level algorithms.

Typical CUMODP operations are matrix determinant computation, polynomial multiplication (both plain and FFT-based), univariate polynomial division, the Euclidean algorithm for univariate polynomial GCDs, subproduct tree techniques for multi-point evaluation and interpolation, subresultant chain computation for multivariate polynomials, bivariate system solving. The CUMODP library is written in CUDA [15] and its source code is publicly available at www.cumodp.org.

In this note, we give an overview of the implementation techniques of the CUMODP library. In Section 2, we discuss a model of multithreaded computation, combining fork-join and single-instruction-multiple-data parallelisms, with an emphasis on estimating parallelism overheads of programs written for modern many-core architectures. For each key routine of the CUMODP library this model is used to minimize parallelism overheads by determining an appropriate value range for a given program parameter, e.g. number of threads per block. Experimentation confirms the effectiveness of this model.

Secondly, the design of the CUMODP library emphasizes the importance of adaptive algorithms in the context of many-core GPUs, see Section 3. that is, algorithms which adapt their behavior according to the available computing resources. Based on these techniques, we have obtained the first GPU implementation of subproduct tree techniques for multi-point evaluation and interpolation of univariate polynomials. Hence we compare our code against probably the best serial C code, namely the FLINT library, for the same operations. For sufficiently large input data and on NVIDIA Tesla C2050, our code outperforms its serial counterpart by a factor ranging between 20 to 30.

We conclude in Section 4 by presenting an application of the CUMODP library to bivariate system solving.

---

[5] http://www.spiral.net/
[6] http://www.fftw.org/

2   A Many-Core Machine Model for Designing Algorithms with Minimum Parallelism Overheads

Our model of multithreaded computation [9] extends the following previous works for which we summarize key features and limitations. The PRAM (parallel random access machine) model [5] supports data parallelism but not task parallelism. Moreover, this model cannot support memory traffic issues like cache complexity and memory contention. The Queue Read Queue Write PRAM [6] considers memory contention, however, it unifies in a single quantity time spent in arithmetic operations and time spent in read/write accesses. We believe that this unification is not appropriate for recent many-core processors, such as GPUs, for which the ratio between one global memory read/write access and one floating point operation can be in the 100's. The TMM (Threaded Many-core Memory) model [12] retains many important characteristics of GPU-type architectures, however, the running time estimate on P cores is not given by a Graham-Brent theorem [7]. We believe that, for the purpose of code optimization, this latter theorem is an essential tool.

Our proposed *many-core machine model* (MMM) aims at optimizing algorithms targeting implementation on GPUs. Our abstract machine possesses an unbounded number of streaming multiprocessors (SMs). However, each SM has a finite number of processing cores and a fixed-size local memory. An MMM machine has a two-level memory hierarchy, comprising an unbounded global memory with high latency and low throughput while SMs local memories have low latency and high throughput. Similarly to a CUDA program, an MMM program specifies for each kernel the number of thread-blocks and the number of threads per thread-block. An MMM machine has two parameters:

U: time (expressed in clock cycles) to transfer one machine word between the global memory and the local memory of any SM,

Z: size (expressed in machine words) of the local memory of any SM.

An MMM program $\mathcal{P}$ is a directed acyclic graph (DAG), called the *kernel DAG*, whose vertices are kernels and edges indicate serial dependencies. Since each kernel of the program $\mathcal{P}$ decomposes into a finite number of thread-blocks, we map $\mathcal{P}$ to a second graph, called the *thread-block DAG* of $\mathcal{P}$, whose vertex set consists of all thread-blocks of $\mathcal{P}$. We consider three complexity measures:

- the *work* $W(\mathcal{P})$, which is the total number of local operations (arithmetic operation, read/write requests in the local memory) performed by all threads,
- the *span* $S(\mathcal{P})$, which is the longest path, counting the weight (span) of each vertex (kernel), in the kernel DAG,
- the *parallelism overhead* $O(\mathcal{P})$, which is the total data transfer time (between global and local memories) of all its kernels.

Using these complexity measures, we derive a Graham-Brent theorem with parallelism overhead.

Theorem 1. *Let* K *be the maximum number of thread blocks along an anti-chain of the thread-block DAG of* $\mathcal{P}$. *Then the running time* $T_{\mathcal{P}}$ *of the program* $\mathcal{P}$ *satisfies:*

$$T_{\mathcal{P}} \leq (N(\mathcal{P})/K + L(\mathcal{P}))\, C(\mathcal{P}). \tag{1}$$

where $N(\mathcal{P})$, $L(\mathcal{P})$ *and* $C(\mathcal{P})$ *are respectively: the number of vertices in the thread-block DAG, the critical path length (where length of a path is the number of edges in that path) in the thread-block DAG and the maximum running time of local operations by a thread among all the thread-blocks.*

We have applied the MMM model for optimizing the CUDA implementation of operations like plain univariate polynomial division, plain univariate polynomial multiplication and the Euclidean algorithm. for dense polynomials over small prime fields. In each case, a program $\mathcal{P}(s)$ depends on a parameter s which varies in a range $\mathcal{S}$ around an initial value s_0, such that the work ratio W_{s_0}/W_s remains essentially constant meanwhile the parallelism overhead O_s varies more substantially, say $O_{s_0}/O_s \in \Theta(s - s_0)$. Then, we determine a value $s_{\min} \in \mathcal{S}$ maximizing the ratio O_{s_0}/O_s. Next, we use our version of Graham-Brent theorem to check whether the upper bound for the running time of $\mathcal{P}(s_{\min})$ is less than that of $\mathcal{P}(s_o)$. If this holds, we view $\mathcal{P}(s_{\min})$ as a solution of our problem of algorithm optimization (in terms of parallelism overheads).

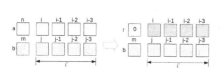

Fig. 1. Naive division algorithm of a thread-block with $s = 1$: each kernel performs 1 division step

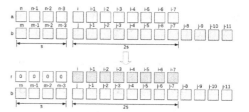

Fig. 2. Optimized division algorithm a thread-block with $s > 1$: each kernel performs s division steps

For each operation, the program parameter s controls the amount of work and parallelism overheads of a thread-block. Figures 1 and 2 illustrate the role of this parameter in our implementation of plain division. See [9] for details.

Applying the optimization strategy described above lead us to determine an optimum value of s among those implied by constraints like the size of the local memory Z or the data transfer time U. For plain polynomial multiplication, this analysis suggested to minimize s which was verified experimentally, as illustrated by Figure 3. For the Euclidean algorithm, our analysis suggested to maximize the program parameter s, which was again verified experimentally, as illustrated by Figure 4. Our experimental results were obtained on a GPU card NVIDIA Tesla C2050.

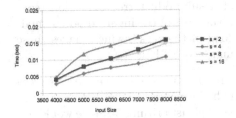

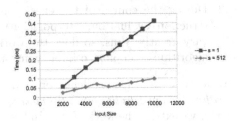

Fig. 3. Plain polynomial multiplication: varying the program parameter *s*

Fig. 4. Euclidean algorithm: varying the program parameter *s*

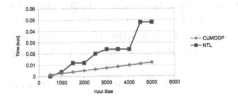

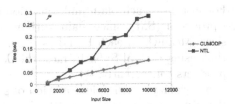

Fig. 5. CUMODP plain polynomial division vs NTL FFT-based (asymptotically fast) polynomial division.

Fig. 6. CUMODP plain Euclidean algorithm vs NTL FFT-based polynomial GCD

Figures 5 and 6 show that the optimized CUMODP implementation of the plain division and the Euclidean algorithm outperforms the NTL implementation of the FFT-based plain division and polynomial GCD computation. Of course, CUMODP code is multithreaded while NTL code is serial. On the other hand, NTL uses asymptotically fast algorithms. The key observation is that optimized implementation of multithreaded plain algorithms provide useful alternative to any serial code. In fact, as we will see in the next section, multithreaded plain algorithms play an essential in higher-level applications targeting many-core GPUs.

3   Adaptive Algorithms

Up to our knowledge, the CUMODP library offers the first GPU implementation of subproduct tree techniques [4][Chapter 10] for multi-point evaluation and interpolation of univariate polynomials. The parallelization of those techniques raises the following challenges on hardware accelerators.

1. The divide-and-conquer formulation of operations on subproduct-trees is not sufficient to provide enough parallelism and one must also parallelize the underlying polynomial arithmetic operations, in particular polynomial multiplication.

2. During the course of the execution of a subproduct tree operation (construction, evaluation, interpolation), the degrees of the involved polynomials vary greatly; thus, so does the work load of the tasks, which makes those algorithms complex to implement on many-core GPUs.

To address the first challenge on many-core GPUs, we combine *parallel plain arithmetic* and *parallel fast arithmetic*. For the former we rely on [8] and, for the latter we extend the work of [13]. Indeed, parallel fast arithmetic alone would not suffice to provide good speedup factors since subproduct tree operations require lots of calculations with low-degree polynomials.

To address the second challenge, we employ *adaptive algorithms*. That is, algorithms that adapt their behavior according to the available computing resources. For instance, each plain multiplication is performed by a single streaming multiprocessor (SM), since plain arithmetic is used for input polynomials of small sizes. Meanwhile, each FFT-based multiplication is computed by a kernel call, thus using several SMs. In fact, this kernel computes a number of FFT-based products concurrently.

To evaluate our implementation of subproduct tree techniques, we measured the effective memory bandwidth of our GPU code for parallel multi-point evaluation and interpolation on a card with a theoretical maximum memory bandwidth of 148 GB/S, our code reaches peaks at 64 GB/S. Since the arithmetic intensity of our algorithms is high, we believe that this is a promising result.

All implementation of subproduct tree techniques that we are aware of are pure serial code. This includes [2] for $GF(2)[x]$, the FLINT library [10] and the Modpn library [11]. Hence we compare our code against probably the best serial C code (namely the FLINT library). For sufficiently large input data, running on NVIDIA Tesla C2050, our code outperforms its serial counterpart by a factor ranging between 20 to 30. Experimental data can be found in Table 1.

Table 1. Multi-point evaluation and interpolation: FLINT vs CUMODP

Deg.	Evaluation			Interpolation		
	CUMODP	FLINT	SpeedUp	CUMODP	FLINT	SpeedUp
2^{12}	0.1361	0.02	0.1468	0.1671	0.03	0.1794
2^{13}	0.1580	0.07	0.4429	0.1963	0.09	0.4584
2^{14}	0.2034	0.17	0.8354	0:2548	0.22	0.8631
2^{15}	0.2415	0.41	1.6971	0.3073	0.53	1.7242
2^{16}	0.3126	0.99	3.1666	0.4026	1.26	3.1294
2^{17}	0.4285	2.33	5.4375	0.5677	2.94	5.1780
2^{18}	0.7106	5.43	7.6404	0.9034	6.81	7.5379
2^{19}	1.0936	12.63	11.5484	1.3931	15.85	11.3768
2^{20}	1.9412	29.2	15.0420	2.4363	36.61	15.0268
2^{21}	3.6927	67.18	18.1923	4.5965	83.98	18.2702
2^{22}	7.4855	153.07	20.4486	9.2940	191.32	20.5851
2^{23}	15.796	346.44	21.9321	19.6923	432.13	21.9441

4   Application

In [14], two of the co-authors of this note reported on the implementation of a bivariate polynomial system solver (based on the theory of *regular chains* and working with coefficients in small prime fields) partially written in CUDA and partially written in C. In that implementation, polynomial subresultant chains were calculated in CUDA while univariate polynomial GCDs were computed in C either by means of the plain Euclidean algorithm or an asymptotically fast algorithm).

The authors observed that about 90% of the overall running time of their solver was spent in univariate GCD computations. They also noted that most of these GCD calculations were using the plain algorithm since the degrees of the input polynomials were not large enough for using the FFT-based algorithm.

Table 2. Bivariate system solving over a small prime field: timings in sec

System	Pure C	Mostly CUDA code	SpeedUp
dense-70	5.22	0.50	10.26
dense-80	6.63	0.77	8.59
dense-90	8.39	1.16	7.19
dense-100	19.53	1.80	10.79
dense-110	21.41	2.57	8.33
dense-120	25.71	3.48	7.39
sparse-70	0.89	0.31	2.81
sparse-80	3.64	1.18	3.09
sparse-90	3.13	0.92	3.40
sparse-100	8.86	1.20	7.38

These observations have lead to a CUDA implementation of the plain Euclidean algorithm which is reported in [8]. More recently, the same authors have put together in a single CUDA application the work reported in [14] and [8], leading to a bivariate polynomial system solver which is mostly written in CUDA. Table 2 compares this latter with an implementation of our bivariate system solver (presented in [14]) entirely written in C. Some of the input systems are random dense and the others are sparse. The number attached to each system name is the total degree of each input polynomial. For each system, the total number of solutions is essentially the square of that degree.

One can see that for a complex application like a polynomial system solver, a CUDA implementation can provide substantial benefit w.r.t. a pure C implementation. We should also point out that our CUDA implementation can be further improved. In particular, the top-level algorithm is still implemented in C and lots of data transfers are still taking place between the host (CPU) and the device (GPU). This performance bottleneck can be removed by using the latest programming model of CUDA.

References

1. Bosma, W., Cannon, J., Playoust, C.: The Magma algebra system. I. The user language. J. Symbolic Comput. 24(3-4), 235–265 (1997); Computational algebra and number theory (London, 1993)
2. Brent, R.P., Gaudry, P., Thomé, E., Zimmermann, P.: Faster multiplication in GF(2)[x]. In: van der Poorten, A.J., Stein, A. (eds.) ANTS-VIII 2008. LNCS, vol. 5011, pp. 153–166. Springer, Heidelberg (2008)
3. Frigo, M., Johnson, S.G.: The design and implementation of FFTW3 93(2), 216–231 (2005)
4. von zur Gathen, J., Gerhard, J.: Modern Computer Algebra, 2nd edn. Cambridge University Press, New York (2003)
5. Gibbons, P.B.: A more practical PRAM model. In: Proc. of SPAA, pp. 158–168 (1989)
6. Gibbons, P.B., Matias, Y., Ramachandran, V.: The Queue-Read Queue-Write PRAM model: Accounting for contention in parallel algorithms. SIAM J. on Comput. 28(2), 733–769 (1998)
7. Graham, R.L.: Bounds on multiprocessing timing anomalies. SIAM J. on Applied Mathematics 17(2), 416–429 (1969)
8. Haque, S.A., Moreno Maza, M.: Plain polynomial arithmetic on GPU. J. of Physics: Conf. Series 385, 12014 (2012)
9. Haque, S.A., Moreno Maza, M., Xie, N.: A Many-core Machine Model for Designing Algorithms with Minimum Parallelism Overheads. Computing Research Repository, abs/1402.0264 (2014), http://arxiv.org/abs/1402.0264
10. Hart, W.B.: Fast library for number theory: An introduction. In: Fukuda, K., van der Hoeven, J., Joswig, M., Takayama, N. (eds.) ICMS 2010. LNCS, vol. 6327, pp. 88–91. Springer, Heidelberg (2010)
11. Li, X., Moreno Maza, M., Rasheed, R., Schost, É.: The modpn library: Bringing fast polynomial arithmetic into maple. J. Symb. Comput. 46(7), 841–858 (2011)
12. Ma, L., Agrawal, K., Chamberlain, R.D.: A memory access model for highly-threaded many-core architectures. In: Proc. of ICPADS, pp. 339–347 (2012)
13. Moreno Maza, M., Pan, W.: Fast polynomial arithmetic on a gpu. J. of Physics: Conference Series 256 (2010)
14. Moreno Maza, M., Pan, W.: Solving bivariate polynomial systems on a gpu. J. of Physics: Conference Series 341 (2011)
15. Nickolls, J., Buck, I., Garland, M., Skadron, K.: Scalable parallel programming with CUDA. Queue 6(2), 40–53 (2008)
16. Püschel, M., Moura, J.M.F., Johnson, J., Padua, D., Veloso, M., Singer, B., Xiong, J., Franchetti, F., Gacic, A., Voronenko, Y., Chen, K., Johnson, R.W., Rizzolo, N.: SPIRAL: Code generation for DSP transforms. Proceedings of the IEEE, Special issue on "Program Generation, Optimization, and Adaptation" 93(2), 232–275 (2005)

Author Index